ITEM LOCATION - U.C.I. FORMAT - PART A

IMPORTANT

PRICES ARE LISTED TWO WAYS, ONE WITH BARE COSTS, THE OTHER INCLUDES THE SUBCONTRACTORS OVERHEAD & PROFIT

GENERAL RECOMMENDATION

When using total costs including Subcontractors O & P, add 10%
[see further discussion in Foreword]

EXAMPLES OF HOW FIGURES ARE DEVELOPED
on p. 134 under VALVES, IRON BODY, 8″ Gate, 125 lb., N.R.S., flanged is shown as follows

15.1	Pipe & Fittings	CREW	DAILY OUTPUT	UNIT	BARE COSTS MAT.	BARE COSTS INST.	BARE COSTS TOTAL	TOTAL INCL O&P
82	VALVES, IRON BODY							
232	C8.1 -081 Gate, 125 lb., N.R.S., flanged, 8″ size	Q-2	2.50	Ea.	425	200	625	755

Reference code (location of additional related data in book sections B or C.)

General Meaning:
Crew Q-2 will install 2.5 8″ Iron body gate valves per 8 hour day

Each

Crew Q-2 is shown on p. x as follows:

Crew No.	Bare Costs		Incl. Subs O & P		Cost Per Man-hour	
Crew Q-2	Hr.	Daily	Hr.	Daily	Bare Costs	Incl. O&P
2 Plumbers	$22.20	$355.20	$32.10	$513.60	$20.72	$29.95
1 Plumber Apprentice	17.76	142.08	25.65	205.20		
24 M.H., Daily Totals		$497.28		$718.80	$20.72	$29.95

These figures are printed above:

BARE COSTS are developed as follows:

MAT. is $425 (rounded to nearest $5)

INST. $\frac{\text{Crew Q-2 Daily Bare Costs}}{\text{Daily Output}} = \frac{\$497.28}{2.5} = \$198.91$ say $200

TOTAL = **MAT.** + **INST.** =$425 + $200 = $625 per each valve

Only the total of these figures is printed above

TOTAL COSTS INCL. O&P are developed as follows:

MAT. is **BARE MAT.** + 10% = $425 + $42.50 = $467.50

INST. $\frac{\text{Crew Q-2 Daily Incl. Subs O\&P}}{\text{Daily Output}} = \frac{\$718.80}{2.5} = \$287.52$

TOTAL = **MAT.** + **INST.** = $467.50 + $287.52 = $755.02 rounded to $755 per each valve.

Line number determination:

Major UCI Subdivision = **15.1**
(2 digits plus decimal point plus last digit)

Major classification within UCI division = **82**
(2 digits)

Item line number = **232**
(3 digits)

Complete line number
= **15.1-82-232**

FOREWORD

OBJECT OF THE BOOK

The object of this book is to provide average unit prices and "systems" prices for Mechanical construction items to make construction cost estimates. Presently there are over 18,000 unit prices plus an equal number of "systems" prices included in the book.

PRICE VARIATIONS

The prices quoted are usually averaged for the major cities in the United States, although in some cases they are obtained from a definite locality. Data accumulated from actual job costs in 1984 and material dealers 1985 quotations have been combined with negotiated 1985 labor rates to revise previous figures. Over 300 companies furnished cost information on their products.

Variations in wage rates, labor efficiency, union restrictions and material prices will result in local fluctuations. The relative material, labor and total construction cost index listed by major division for 162 largest U.S. and Canadian cities is shown in Division C13. Indexes for Canadian cities reflect Canadian dollars. Also shown in Division C13 is the historic trend of construction costs from Jan. 1, 1985 back to 1941.

Local, regional or national shortages of construction materials can severely influence material costs as well as cause job delays with a corresponding increase in indirect job costs.

The specific cost figures when totaled and multiplied by the appropriate city cost index should give a total cost for an entire building that will be close to the low bid and inside the bid range. That is, total cost derived from prices in this book will seldom be below the low bidder or above the average of all bidders. **Sales taxes are not included.**

The square foot and cubic foot cost ranges of 57 building types are shown in Division C14. A more detailed description will be found below in this Foreword.

No guaranty or warranty is made by Robert Snow Means Co., Inc. or the editors as to the correctness or sufficiency of the information contained in this book. Robert Snow Means Co., Inc. or the editors assume no responsibility nor liability in connection with use of this book.

FACTORS AFFECTING COSTS

QUALITY: The prices given in this book are in line with U.S. Government specifications and represent good sound construction.

OVERTIME: No allowance has been made for overtime.

SIZE OF JOB: The book is aimed primarily at industrial and commercial buildings costing $400,000 and up or large housing projects. The costs are also for new construction of complete buildings rather than repairs and minor alterations. Material prices given are usually trade quantity purchases. With reasonable exercise of judgment the figures can be used for any building work, but do not apply to civil engineering structures such as bridges, dams, highways or the like.

REPAIR AND REMODELING: Cost modification information is in Division C18.

LOCATION: Material prices are for metropolitan areas. Beyond a 20 mile radius of large cities, extra trucking or other transportation charges will increase the material costs slightly. This may be offset by lower wage rates, but these two factors should be given consideration in preparing an estimate especially if the job site is remote. Highly specialized subcontract items may require high travel and per diem expenses for mechanics.

OTHER FACTORS which affect costs are weather, season of year, contractor management, local union restrictions, building code requirements, floods, availability of adequate energy, manpower and building materials as well as general business conditions which influence the "in place" costs of all building items. Substitute materials and construction methods may have to be employed that may increase installed cost and/or life cycle costs above those anticipated. These factors, difficult to evaluate, are not dependent on the section of the country where the job is located, and leave a margin of error that is unavoidable and increasingly important.

LABOR COSTS

Labor rates used are averaged from 30 major U.S. cities and rounded to the nearest 5¢. Figures used are the latest available from Trade Union Agreements negotiated by the unions and the employers in the various major cities. The labor rates used are those in effect on January 1, 1985. Labor costs should be factored as the new wage agreements for 1985 are negotiated. There are several reliable national indexes available which will give precise indications of the percentage change of labor rates as the year advances. ("Means Construction Cost Indexes" is one service available.)

Labor rates are listed two ways: **Base rate** which includes fringe benefits but does not include insurance, taxes or contractors overhead and profit. **Billing rate** which includes fringe benefits, insurance, taxes and contractors overhead and profit. Both rates for the various trades are listed on the inside rear cover. Detailed breakdown of how the billing rate was figured is shown at C10.3-200. Labor costs can be factored to correspond with local rates (see Division C13 for labor factors for 162 largest cities listed by major construction division) but relative worker productivity must also be considered. On large jobs built outside the metropolitan area or in the smaller cities, skilled labor is generally imported and wages tend toward larger city averages regardless of local rates.

It may be noticed that in some cases labor unit costs over the last several years have sometimes held steady or even decreased in spite of the increase in hourly wages. This is a result of new mechanical equipment being used in the field or of new methods of handling materials. Factory prefabrication of components and many other innovations have reduced both helpers and skilled mechanics time. Severe competition between subcontractors may tend to keep bid prices from rising as rapidly as the labor rates.

SQUARE FOOT AND CUBIC FOOT BUILDING COSTS

The figures in Division A17 are included to facilitate rapid preliminary budget estimates. See Division C14 for details of how the figures were derived.

There are two times when square foot costs are useful. The first is in the schematic stage when no details are available, when square foot costs make a useful starting point. The second is after the bids are in and the costs can be worked back into their appropriate units for information purposes. As soon as details become available in the project design, the square foot approach should be discontinued and the project priced by its particular components.

In using the figures in Division A17, it is recommended that the median column be used for preliminary figures if no additional information is available. The median figures, when multiplied by the total city construction cost index figures in Division C13 and then multiplied by the project size modifier in Division C14, should present a fairly accurate base figure which would then have to be adjusted in view of the estimator's experience, local economic conditions, code requirements and the owner's particular requirements. There is no need to factor the percentage figures as these remain constant from city to city.

All tabulations mentioning air conditioning had at least partial air conditioning.

HOW THE BOOK IS ARRANGED

NUMBERING SYSTEM The basic numbering format used for unit prices is 16 major divisions of building construction plus a S.F. & C.F. Cost Division as well as a Repair and Remodeling Division. These divisions are patterned after the Uniform System adopted by the American Institute of Architects, Associated General Contractors of America, Inc., and the Construction Specifications Institute, Inc., and the system is widely used by most segments of the building construction industry. These major divisions are easily located in this book by using the flip tabs on each page. The major divisions are then divided into their major component parts shown in bold face type at the top of each page along with the major component subdivision number. The individual items making up the major component part are then listed alphabetically. Subdivisions and other sizes are indented below each descriptive item. An outline of the major components and their numbering system is shown on the first page of the book.

The MEANS Mechanical Cost Data book is divided into three major sections A, B, and C. The A Section is the unit price portion, and is numbered in accordance with the UCI Format divisions. The B section contains the "systems" pricing information. It is numbered by divisions as shown under the Systems Format heading. Section C is an appendix of code extracts and useful technical/engineering data for sizing and quantifying construction components. It is also numbered by divisions as shown under the Systems Format heading.

The table below lists the UCI format for unit prices and the corresponding "systems" format used for this book.

UNIT PRICE ARRANGEMENT The cost and descriptive information for each item in the unit price portions of the book is arranged as follows:

LINE NUMBER Each item has its own 3 part line number made up as follows: (see page 134) The subdivision number at the top of the page is the first part of the number (15.1 for page 134). **VALVES, IRON BODY** has a bold type face number **82** associated with it. The appropriate line number for an 8" gate valve, 125 lb., N.R.S., flanged is 232. The entire 3 part line number for the item is thus 15.1-82-232 (see also second page of the book for a pictorial analysis).

DESCRIPTION Major items of each subdivision are listed in bold type face more or less alphabetically within the subdivisions. Components of the major items are listed after the major items with individual sizes listed after each particular description.

CREW To the right of the item described is the typical crew needed to install the item. Typical crews are tabulated on p. vii to xi of the Foreword. When installations are made by one trade and require no power equipment, the appropriate trade is listed directly in the crew column. For instance "2 Plum" would indicate that the installation is made with 2 Plumbers. If the installing crew was Q-2 (shown on p. x of the Foreword) this would be made up of 2 Plumbers and 1 Apprentice.

The crew costs are listed two ways, one with bare labor rates, the other with billing rates as shown in Div. C10 and the inside rear cover. Local labor rates may be substituted to arrive at precise local daily crew costs. Abbreviations for the individual trades are shown on the inside rear cover.

EQUIPMENT The power equipment required for each crew is included in the particular crew cost. The daily equipment costs are usually the sum of the weekly rental rate divided by 5 plus the hourly operating cost times 8. (See Div. C10 for further discussion.) For daily costs incl. Subs O & P, 10% is added to the bare costs to allow for the contractors mark-up on the equipment costs.

DAILY OUTPUT To the right of the crew column is the Daily Output columns. This is the number of units that a given crew will install in an 8 hour day. In other words, the total daily crew cost divided by the Daily Output will produce the unit installation costs. To calculate the unit equipment costs, divide the daily equipment costs by the Daily Output.

UNIT To the right of the Daily Output column is the Unit column which describes the unit upon which the price, production and crew are based, as S.F. (square foot), etc. See list of abbreviations in back of book.

BARE COSTS The next 3 columns contain costs that do not include the Subcontractors Overhead and Profit. The first column under the **Bare Cost** heading is MAT. which is the unit material cost for the item (without Subs O & P). The next column is the INST. figures. This column is the installation cost which includes labor and equipment. As mentioned above this cost is arrived at by dividing the daily crew or labor cost (using bare costs in both cases) by the Daily Output. The last column under the BARE COST heading is TOTAL which is the arithmetical sum of the two previous columns. The last column is the TOTAL INCL O&P. This column is the sum of the bare material cost plus 10%, which is then added to the installation costs incl. Subs O & P. In this case the crew costs and the labor costs used would include the Subs O & P. When these costs are divided by the Daily Output, a unit installation cost incl. Subs O & P is developed. A sample calculation is shown opposite p. i of the Foreword. In a few cases, an M, L or E will be found next to the prices in this column. This indicates that the particular cost figures includes only Material (M), Labor (L) or Equipment (E).

SPECIAL On a few pages the column headings are EQUIP. and LABOR. The EQUIP. column indicates equipment, rental and operating cost while the LABOR column represents the labor to operate the equipment. These costs are bare costs and do not include the Subs O & P.

GENERAL ARRANGEMENT In the margin and preceding the item there are frequently found numbers in squares which are reference numbers. These refer to "systems" information and supplementary engineering design and code data in the rear portion of the book.

The breakdowns and "systems" costs show how the editors arrived at the tabulated figure and also indicate the cost of typical complete systems commonly encountered in Building Construction. The breakdowns and "systems" information in the rear is generally arranged in the same sequence as the items in the unit price portion of each section.

Repair and Remodeling costs and information are shown in Division A18.

CROSS REFERENCES Numerous cross references are included to aid in the rapid location of items that could logically be listed under 2 or more subdivisions. A cross reference referring to, say, Division 15.2 would mean that the entire division (Plumbing Fixtures) was the intended cross reference. A cross reference of 15.2-56 would indicate showers (see p. 152) and a cross reference of 15.2-56-300 would indicate line 300 Division 15.2-56 or, in this case, Fiberglass showers.

ROUNDING OF COSTS In general, all unit prices in excess of $5.00 have been rounded to make them easier to use and still maintain adequate precision of the results. The rounding rules generally followed in this book are as follows:

Price from $.01 to $5.00 rounded to nearest 1¢
from $5.01 to $20.00 rounded to nearest 5¢
from $20.01 to $100.00 rounded to nearest $1

from $100.01 to $1,000.00 rounded to nearest $5
from $1,000.01 to $10,000.00 rounded to nearest $25
from $10,000.01 to $50,000.00 rounded to nearest $100
over $50,000.01 rounded to nearest $500

MAJOR SECTIONS OF THIS BOOK	UCI FORMAT		SYSTEMS FORMAT*	
	Division Number	Description	Division Number	Description
MECHANICAL RELATED ITEMS	1	General Requirements	1	Foundation/Substructure
	2	Site Work	2	Substructure
	3	Concrete	3	Superstructure
	4	Masonry	4	Exterior Closure
	5	Metals	5	Roofing
	6	Wood & Plastics	6	Interior Construction
	7	Moisture Protection	7	Conveying
	8	Doors, Windows & Glass	10	General Conditions
	9	Finishes	11	Specialties
	10	Specialties	12	Site Work
	11	Equipment		
	12	Furnishings		
	13	Special Construction		
	14	Conveying Systems		
MECHANICAL	15	Mechanical	8	Mechanical
ELECTRICAL	16	Electrical	9	Electrical
COST MODIFICATIONS	17	Square Foot	14	Square Foot Costs
	18	Repair & Remodeling	18	Repair and Remodeling
	19	City Cost Index	13	City Adjustments

*Items have been re-arranged to be consistent with major sections of this book.

OVERHEAD, PROFIT, CONTINGENCIES

GENERAL Prices given in this book are of two kinds: (1) BARE COSTS to which an allowance for Contractor Overhead and Profit must be added, and (2) COSTS INCL. SUBS O & P, to which an allowance covering only the general contractors mark-up must be applied. The general contractor includes this price in his bid with a normal mark-up ranging from 0 to 15%. This mark-up depends on economic conditions plus the supervision and trouble-shooting expected by the general contractor. For purposes of this book it is best to add an allowance of 10% to the figures including O & P. In using the bare cost figures, experience shows that an addition of 35% for general contractors O & P gives best results.

The figures in the column on the right, TOTAL INCL. O & P, contain the subcontractor's overhead and profit, thus it is necessary to add a percentage for all subcontracted items unless the owner is receiving multiple prime contracts.

OVERHEAD AND PROFIT Subcontractors overhead and profit allowances are detailed in Division C10. These are general figures and local deviations are certainly to be expected. A subcontractor should be able to tell from his records very closely what his own overhead will be. He will then add whatever profit and contingency percentage circumstances dictate.

SUBCONTRACTORS Usually a considerable portion of all large jobs is subcontracted. In fact the percentage done by subs is constantly increasing and may run over 90%. Since the workmen employed by these companies do nothing else but install their particular product they soon become experts in that line. The result is, installation by these firms is accomplished so efficiently that the total in-place cost, even with the subcontractor's overhead and profit, is no more and often less than if the principal contractor had handled the installation himself. There is, moreover, the advantage of having the work done right. Companies who sell specialties are anxious that their product perform well and consequently the installation will be the best possible.

CONTINGENCIES The allowance for contingencies generally is to provide for unforeseen construction difficulties. On alterations or repair jobs 20% is none too much. If drawings are final and only field contingencies are being considered 2% or 3% is probably sufficient and often nothing need be added. As far as the contract is concerned, future changes in plans will be covered by extras. The contractor should consider inflationary price trends and possible material shortages during the course of the job. If drawings are not complete or approved or a budget cost wanted, it is wise to add 5% to 10%. Contingencies, then, are a matter of judgment. Additional allowances are shown on page 4 for contingencies and job conditions and Division A18 for factors to convert prices for Repair and Remodeling jobs.

ESTIMATING GUIDELINES

The following suggestions are made to enable the estimator to perform unit price estimating in a logical, easy to check and thorough manner.

1. Use pre-printed or columnar forms for orderly sequence of dimensions and locations and for recording telephone quotations.
2. Use only the front side of each paper or form except for certain pre-printed summary forms.
3. Be consistent in listing dimensions: for example, length x width x height. This helps in re-checking to insure that, say, the total length of partitions is appropriate for the building area.
4. Use printed (rather than measured) dimensions where given.
5. Add up multiple printed dimensions for a single entry where possible.
6. Measure all other dimensions carefully.
7. Use each set of dimensions to calculate multiple related quantities.

8. Convert foot and inch measurements to decimal feet when listing. Memorize decimal equivalents to .01 parts of a foot (1/8" equals approximately .01').
9. Do not "round off" quantities until the final summary.
10. Mark drawings with different colors as items are taken off.
11. Keep similar items together; different items separate.
12. Identify location and drawing numbers to aid in future checking for completeness.
13. Measure or list everything on the drawings or mentioned in the specifications.
14. It may be necessary to list items not called for to make the job complete.
15. Be alert for: notes on plans such as N.T.S. (not to scale); changes in scale throughout the drawings; reduced size drawings; discrepancies between the specifications and the drawings.
16. Develop a consistent pattern of performing an estimate, for example:
    a. Start the quantity take-off at the lower floor and move to the next higher floor.
    b. Proceed from the main section of the building to the wings.
    c. Proceed from south to north or vice versa, clockwise or counterclockwise.
    d. Take off floor plan quantities first, elevations next, then detail drawings.
17. List all gross dimensions that can be either used again for different quantities, or used as a rough check of other quantities for verification (exterior perimeter, gross floor area, individual floor areas, etc.).
18. Utilize design symmetry or repetition (repetitive floors, repetitive wings, symmetrical design around a center line, similar room layouts, etc.).
    Note: extreme caution is needed here so as not to omit or duplicate an area.
19. Do not convert units until the final total is obtained. For instance, when estimating concrete work, keep all units to the nearest cubic foot, then summarize and convert to cubic yards.
20. When figuring alternates, it is best to total all items involved in the basic system, then total all items involved in the alternates. Thus you work with positive numbers in all cases. When adds and deducts are used, it often is confusing whether to add or subtract a portion of an item, especially on a complicated or involved alternate.

DOING THE WORK

In making an engineering estimate ignore the cents column. Just give total per unit to the nearest dollar. The cents will average up in a column of figures. An engineering estimate of $457,323.37 is ridiculous:$457,325 is certainly more sensible and $457,000 is better and just as likely to be right.

If you follow this simple instruction the time saved is tremendous with an added important advantage: using round figures your mind is left free to exercise judgment and common sense rather than being overcome and befuddled by a mass of computations.

When you have finished, roughly check the big items for location of the decimal point. That is important. A large error can creep in if you write down $300 when it should be $3,000. Also check the list to be sure you have not omitted any large item. A common error is to overlook, let us say, heating or to forget finished flooring or painting or otherwise commit a gross omission. No amount of accuracy in prices can compensate for such an oversight.

It is important to keep the bare costs and costs that already include the Subs O & P separate since different mark-ups will have to be applied to each category. Organize your estimating procedures to minimize confusion and simplify checking to insure against omissions and/or duplications.

By using printed forms listing the usual building items, the chance of omissions and duplications is lessened, but never completely eliminated.

The Editors

Mechanical

Melville J. Mossman, a registered professional engineer and chief mechanical engineer for the company, seminar lecturer, member of ASHRAE, ASPE, AACE and AEE.

John J. Moylan, a former commercial/industrial mechanical contractor with over 39 years of experience in estimating, project managing, purchasing and ownership. A member of ASHRAE, ASPE, AEE, and IDHA.

THE COMPANY AND THE EDITORS

Robert S. Means Co., Inc., is a 43-year old firm actively engaged in construction cost publishing and consulting throughout North America. The company helps you cut down on the time and energy you use when organizing and utilizing cost data. A thoroughly experienced and highly qualified staff of over ninety professionals at R. S. Means works daily at collecting, analyzing and disseminating reliable cost information for your needs.

Each editor contributing to this publication is a full-time Means employee. All these staff members have years of practical contractor experience and thorough engineering training prior to joining the firm. Each Means construction expert contributes to the maintenance of a complete, continually-updated construction cost data system.

Constant flow of new products means that even the most sophisticated construction professional finds little time to examine and evaluate all the construction cost possibilities when they are so many and diverse. R. S. Means fills the gap between theoretical and practical knowledge in the field of construction engineering. The Means organization is always prepared to assist you and help in the solution of construction problems through the services of its four major divisions: Construction and Cost Data Publishing, Computer Software Services, Consulting, and Educational Seminars.

CONSTRUCTION AND COST DATA PUBLISHING

BUILDING CONSTRUCTION COST DATA, published annually for 43 years. Contains over 20,000 unit prices for building construction items broken down into material, labor, and total costs as well as total costs including subcontractors' overhead and profit.

MEANS ELECTRICAL COST DATA, published annually. Highly-detailed treatment of electrical unit and systems costs, including design reference tables and demolition figures, illustrated estimating procedures adaptable to customized projects and an efficient man-hour, materials and labor cost format for fast scheduling and materials procurement.

REPAIR & REMODELING COST DATA, published annually. An illustrated book detailing construction costs for conversion, reuse, fire-damage repair and renovation for residential, commercial and industrial construction.

MEANS SQUARE FOOT COSTS, published annually. A budgeting cost resource and replacement cost guide with over 6,000 costs for typical buildings and attachments applicable to residential, commercial, industrial and institutional structures. Examples, illustrations, photographs.

MEANS SYSTEMS COSTS, published annually. Offers thousands of detailed systems costs with hundreds of drawings, explanations, component breakdowns and tables. Uses the expanded and improved "UNIFORMAT" numbering system, assisting in the conceptual phase of estimating any job.

MEANS SITE WORK COST DATA, published annually. Provides answers to the most demanding site work estimating questions. Complete unit and systems costs enable you to confidently handle every aspect of site work estimating, from excavation to planting the grass. Illustrated.

RESIDENTIAL/LIGHT COMMERCIAL COST DATA, published annually. For construction professionals needing the utmost in cost estimating assistance. Includes the most commonly-used residential and light commercial building systems with a wide variety of popular alternatives. Contains unit costs listed in the 16 major UCI divisions. Illustrated.

CONCRETE & MASONRY COST DATA, published annually. In-depth price data for estimating concrete and masonry construction. Covers every concrete/masonry function. Includes a unit cost section plus an illustrated systems cost section that actually pictures the concrete/masonry system.

LABOR RATES FOR THE CONSTRUCTION INDUSTRY, published annually. Contains detailed union labor rates and factors for 46 construction trades in more than 300 U.S. and Canadian cities. Ideally suited for making comparisons between cities, combinations of cities and for rapidly determining wage scales above or below national averages. Includes historical wage rates and a five-year national average chart.

INTERIOR COST DATA, published annually. Provides up-to-date cost facts for solving the complex estimating problems of modern interior construction. Included are detailed material, equipment and labor costs for hardware, custom interior work and furnishings for both new and remodeling commercial and industrial construction.

BOATING COST GUIDE, published annually. Provides average installed cost for tasks involved in the repair, maintenance, outfitting and improvement of power and sail boats up to 65 feet in length. Specifications and cost data are derived from the actual experience of marine service firms and suppliers, coast-to-coast.

MEANS HISTORICAL COST INDEXES, published annually. Indexes historical construction costs for each of 209 United States and Canadian cities from 1940 through 1985. Derived from years of research, this manual gives functional and reliable assistance in all historical cost estimating activities.

MEANS CONSTRUCTION COST INDEXES, published quarterly, every January, April, July and October. This handy report provides cost adjustment factors for the preparation of more precise estimates no matter how late in the year. It is also the ideal method for making continuous cost revisions on on-going projects as the year progresses.

MEANS MAN-HOUR STANDARDS, a working encyclopedia of labor productivity information. Offers detailed description of crews, man-hours with units, daily output, convenient U.C.I. format, many alternate man-hour selections, introductory guidance and information.

SQUARE FOOT ESTIMATING, provides user with the expertise to perform conceptual and design stage estimates. Demonstrates how to minimize time and effort to achieve accurate estimates. Offers methodology for reliability in square foot costing. Illustrated.

MEANS SCHEDULING MANUAL, discusses today's most advanced methods for scheduling and managing building construction projects in the mid-sized range. PERT, CPM, precedence scheduling, computer-based control, preparing and using the network are all included. Illustrated with charts, tables and graphs.

ESTIMATING FOR THE GENERAL CONTRACTOR, a comprehensive book designed to give estimating a refreshing breath of new ideas, insights and better techniques. Presents logical explanations, detailed examples, tables and graphs conveying all components of estimating activities - from analyzing a project's desirability through the final bid for the job.

MEANSCO FORMS are preprinted forms designed for architects, contractors and owners. The forms facilitate construction estimating, cost control and appraisals.

COMPUTER SOFTWARE SERVICES

GALAXY (an automated quantity takeoff and pricing system) — simplifies the process of doing a takeoff using the technology of electronic drafting systems. You can reduce the time it takes to do an estimate as much as 50%, and more for many projects, thanks to GALAXY'S electronic digitizer table. Designed for use with the STAR General Estimating Program, GALAXY provides you with access to your own files of known costs as well as any of the R.S. Means standard cost information files of up-to-date unit price information.

STAR (Means General Estimating Program) — helps develop fast and accurate estimates and manage vital cost data. Easy to use, STAR reduces errors, provides more accurate estimates and enables you to increase productivity and improve profit margins. STAR allows you to customize estimates to specific requirements, and is available for both daily output and man-hours analysis. STAR is designed to access Means Unit Price Cost files listed below.

UNIT PRICE COST FILES — complete cost files on floppy diskettes. Contains item description, crew makeup, daily productivity, unit of measure and unit prices for material, labor, equipment, and total for Divisions 1-16, and separate in-depth files for Building Construction, Repair/Remodeling, Site Work, Mechanical, Electrical, Residential/Light Commercial and Interiors.

COMET — designed for the construction and facilities professional with a growing need for customized reports, COMET Report Writer Sorting program will help you create winning proposals in a convenient form to meet the needs of any required format. This program enables you to calculate, accumulate and print individual items for estimating data in the STAR and GALAXY programs.

VEGA — The VEGA Scheduling Program is designed to create a realistic schedule early in the conceptual stage, or later on during the bidding and negotiating phase. VEGA can provide you with a checklist of the different items involved in the project, and can even estimate resource excesses or predict the target date. You can determine resource allocation schedules based upon availability while estimating duration times and cost data for each selected activity. VEGA will also provide a printout of the schedule, which includes early start and early finish, late start and late finish, critical path and resource leveling.

The VEGA Scheduling Program communicates with both STAR and GALAXY programs - it's an absolute must for the professional who has to stay on top of his scheduling situation.

CONSULTING SERVICES

Through direct consultation with clients associated with the construction industry, R. S. Means performs a wide range of assignments. As a client, you may retain segments of our staff for detailed estimates, appraisals, and assistance for large or unique projects in related areas. Typical assignments include: Feasibility Studies, Estimating Services, Appraising Services, Cost Engineering Research, Independent Cost Evaluations, Energy Conservation Programs, Insurance Claim Settlement Programs, Quantity Surveys, and Value Engineering Studies.

EDUCATIONAL SEMINARS

The Means Continuing Education Department conducts professional seminars throughout the year in selected North American cities, and also provides on-site programs for corporations and others. All Means seminars focus on the latest estimating and management approaches, including computerized estimating and scheduling systems as applied specifically to the construction industry. The program covers the following topics:

UNIT PRICE ESTIMATING — shows how today's advanced estimating techniques and cost information sources can be used to efficiently develop comprehensive, detailed estimates for projects of any size. Course material includes *Building Construction Cost Data* and a workbook.

REPAIR & REMODELING ESTIMATING —demonstrates how to produce detailed and accurate cost estimates for building reuse, conversion, repair, and remodeling. Course participants use *Repair & Remodeling Cost Data* and a seminar workbook for definitive estimating.

SQUARE FOOT COST ESTIMATING — provides expert guidance for those needing extremely rapid estimates based on logical cost relationships of major systems. Course participants can perfect their skills for conceptual stage estimating, budget planning, building planning, building valuation, and replacement cost studies. Seminar presentation is based on *Means Square Foot Costs, Means Systems Costs,* and a workbook.

COMPUTERS IN CONSTRUCTION — prepared for those interested in using computers for construction activities, this seminar describes and demonstrates the latest available hardware and software products for construction cost estimating, project scheduling, and accounting applications. Actual demonstrations of the Means line of construction-oriented software — as well as other construction and business software — highlight this Seminar. Optimum use is made of *Building Construction Cost Data*

SCHEDULING & PROJECT MANAGEMENT — explains how to make efficient use of manpower, equipment, and money to ensure timely completion of projects. *Means Scheduling Manual, Building Construction Cost Data,* and a comprehensive workbook assist attendees in applying the most successful management techniques.

ROBERT SNOW MEANS COMPANY, INC.
CONSTRUCTION CONSULTANTS & PUBLISHERS
P.O. BOX 800
100 CONSTRUCTION PLAZA
KINGSTON, MA 02364-0800
617-747-1270

CREWS

Crew No.	Bare Costs		Incl. Subs O & P		Cost Per Man-hour	
Crew A-1	Hr.	Daily	Hr.	Daily	Bare Costs	Incl. O&P
1 Building Laborer	$15.50	$124.00	$22.40	$179.20	$15.50	$22.40
1 Gas Eng. Power Tool		35.20		38.70	4.40	4.83
8 M.H., Daily Totals		$159.20		$217.90	$19.90	$27.23
Crew A-4	Hr.	Daily	Hr.	Daily	Bare Costs	Incl. O&P
2 Carpenters	$19.60	$313.60	$28.30	$452.80	$19.35	$27.78
1 Painter, Ordinary	18.85	150.80	26.75	214.00		
24 M.H., Daily Totals		$464.40		$666.80	$19.35	$27.78
Crew A-5	Hr.	Daily	Hr.	Daily	Bare Costs	Incl. O&P
2 Building Laborers	$15.50	$248.00	$22.40	$358.40	$15.55	$22.45
.25 Truck Driver (light)	16.00	32.00	22.90	45.80		
.25 Light Truck, 1.5 Ton		14.30		15.70	.79	.87
18 M.H., Daily Totals		$294.30		$419.90	$16.34	$23.32
Crew B-6	Hr.	Daily	Hr.	Daily	Bare Costs	Incl. O&P
2 Building Laborers	$15.50	$248.00	$22.40	$358.40	$16.68	$24.11
1 Equip. Oper. (light)	19.05	152.40	27.55	220.40		
1 Backhoe Loader, 48 H.P.		152.20		167.40	6.34	6.97
24 M.H., Daily Totals		$552.60		$746.20	$23.02	$31.08
Crew B-9	Hr.	Daily	Hr.	Daily	Bare Costs	Incl. O&P
1 Labor Foreman (outside)	$17.50	$140.00	$25.25	$202.00	$15.90	$22.97
4 Building Laborers	15.50	496.00	22.40	716.80		
1 Air Compr., 250 C.F.M.		106.20		116.80		
Air Tools & Accessories		23.70		26.05		
2-50 Ft. Air Hoses, 1.5" Dia.		11.60		12.75	3.53	3.89
40 M.H., Daily Totals		$777.50		$1074.40	$19.43	$26.86
Crew B-10A	Hr.	Daily	Hr.	Daily	Bare Costs	Incl. O&P
1 Equip. Oper. (med.)	$20.20	$161.60	$29.25	$234.00	$18.63	$26.96
.5 Building Laborer	15.50	62.00	22.40	89.60		
1 Roll. Compact., 2K Lbs.		72.10		79.30	6.00	6.60
12 M.H., Daily Totals		$295.70		$402.90	$24.63	$33.56
Crew B-10B	Hr.	Daily	Hr.	Daily	Bare Costs	Incl. O&P
1 Equip. Oper. (med.)	$20.20	$161.60	$29.25	$234.00	$18.63	$26.96
.5 Building Laborer	15.50	62.00	22.40	89.60		
1 Dozer, 200 H.P.		614.40		675.85	51.20	56.32
12 M.H., Daily Totals		$838.00		$999.45	$69.83	$83.28
Crew B-10C	Hr.	Daily	Hr.	Daily	Bare Costs	Incl. O&P
1 Equip. Oper. (med.)	$20.20	$161.60	$29.25	$234.00	$18.63	$26.96
.5 Building Laborer	15.50	62.00	22.40	89.60		
1 Dozer, 200 H.P.		614.40		675.85		
1 Vibratory Roller, Towed		71.50		78.65	57.15	62.87
12 M.H., Daily Totals		$909.50		$1078.10	$75.78	$89.83
Crew B-10D	Hr.	Daily	Hr.	Daily	Bare Costs	Incl. O&P
1 Equip. Oper. (med.)	$20.20	$161.60	$29.25	$234.00	$18.63	$26.96
.5 Building Laborer	15.50	62.00	22.40	89.60		
1 Dozer, 200 H.P		614.40		675.85		
1 Sheepsft. Roller, Towed		78.70		86.55	57.75	63.53
12 M.H., Daily Totals		$916.70		$1086.00	$76.38	$90.49

Crew No.	Bare Costs		Incl. Subs O & P		Cost Per Man-hour	
Crew B-10 I	Hr.	Daily	Hr.	Daily	Bare Costs	Incl. O&P
1 Equip. Oper. (med.)	$20.20	$161.60	$29.25	$234.00	$18.63	$26.96
.5 Building Laborer	15.50	62.00	22.40	89.60		
1 Diaphr. Water Pump, 4"		39.05		42.95		
1-20 Ft. Suction Hose, 4"		6.60		7.25		
2-50 Ft. Disch. Hoses, 4"		14.00		15.40	4.97	5.46
12 M.H., Daily Totals		$283.25		$389.20	$23.60	$32.42
Crew B-10J	Hr.	Daily	Hr.	Daily	Bare Costs	Incl. O&P
1 Equip. Oper. (med.)	$20.20	$161.60	$29.25	$234.00	$18.63	$26.96
.5 Building Laborer	15.50	62.00	22.40	89.60		
1 Centr. Water Pump, 3"		24.10		26.50		
1-20 Ft. Suction Hose, 3"		4.60		5.05		
2-50 Ft. Disch. Hoses, 3"		9.20		10.10	3.15	3.47
12 M.H., Daily Totals		$261.50		$365.25	$21.78	$30.43
Crew B-11B	Hr.	Daily	Hr.	Daily	Bare Costs	Incl. O&P
1 Equipment Oper. (med.)	$20.20	$161.60	$29.25	$234.00	$17.85	$25.82
1 Building Laborer	15.50	124.00	22.40	179.20		
1 Dozer, 200 H.P.		614.40		675.85		
1 Air Powered Tamper		12.45		13.70		
1 Air Compr. 365 C.F.M.		148.80		163.70		
2-50 Ft. Air Hoses, 1.5" Dia.		11.60		12.75	49.20	54.12
16 M.H., Daily Totals		$1072.85		$1279.20	$67.05	$79.94
Crew B-11C	Hr.	Daily	Hr.	Daily	Bare Costs	Incl. O&P
1 Equipment Oper. (med.)	$20.20	$161.60	$29.25	$234.00	$17.85	$25.82
1 Building Laborer	15.50	124.00	22.40	179.20		
1 Backhoe Loader, 48 H.P.		152.20		167.40	9.51	10.46
16 M.H., Daily Totals		$437.80		$580.60	$27.36	$36.28
Crew B-11M	Hr.	Daily	Hr.	Daily	Bare Costs	Incl. O&P
1 Equipment Oper. (med.)	$20.20	$161.60	$29.25	$234.00	$17.85	$25.82
1 Building Laborer	15.50	124.00	22.40	179.20		
1 Backhoe Loader, 80 H.P.		228.00		250.80	14.25	15.67
16 M.H., Daily Totals		$513.60		$664.00	$32.10	$41.49
Crew B-12A	Hr.	Daily	Hr.	Daily	Bare Costs	Incl. O&P
1 Equip. Oper. (crane)	$20.65	$165.20	$29.90	$239.20	$18.87	$27.32
1 Equip. Oper. Oiler	17.10	136.80	24.75	198.00		
1 Hyd. Excavator, 1 C.Y.		402.20		442.40	25.13	27.65
16 M.H., Daily Totals		$704.20		$879.60	$44.00	$54.97
Crew B-12C	Hr.	Daily	Hr.	Daily	Bare Costs	Incl. O&P
1 Equip. Oper. (crane)	$20.65	$165.20	$29.90	$239.20	$18.87	$27.32
1 Equip. Oper. Oiler	17.10	136.80	24.75	198.00		
1 Hyd. Excavator, 2 C.Y.		773.00		850.30	48.31	53.14
16 M.H., Daily Totals		$1075.00		$1287.50	$67.18	$80.46
Crew B-12F	Hr.	Daily	Hr.	Daily	Bare Costs	Incl. O&P
1 Equip. Oper. (crane)	$20.65	$165.20	$29.90	$239.20	$18.87	$27.32
1 Equip. Oper. Oiler	17.10	136.80	24.75	198.00		
1 Hyd. Excavator, .75 C.Y.		323.20		355.50	20.20	22.21
16 M.H., Daily Totals		$625.20		$792.70	$39.07	$49.53
Crew B-12J	Hr.	Daily	Hr.	Daily	Bare Costs	Incl. O&P
1 Equip. Oper. (crane)	$20.65	$165.20	$29.90	$239.20	$18.87	$27.32
1 Equip. Oper. Oiler	17.10	136.80	24.75	198.00		
1 Gradall, 3 Ton, .5 C.Y.		432.20		475.40	27.01	29.71
16 M.H., Daily Totals		$734.20		$912.60	$45.88	$57.03

CREWS

Crew B-12K	Bare Costs Hr.	Bare Costs Daily	Incl. Subs O & P Hr.	Incl. Subs O & P Daily	Cost Per Man-hour Bare Costs	Cost Per Man-hour Incl. O&P
1 Equip. Oper. (crane)	$20.65	$165.20	$29.90	$239.20	$18.87	$27.32
1 Equip. Oper. Oiler	17.10	136.80	24.75	198.00		
1 Gradall, 3 Ton, 1 C.Y.		619.00		680.90	38.68	42.55
16 M.H., Daily Totals		$921.00		$1118.10	$57.55	$69.87

Crew B-12Q	Hr.	Daily	Hr.	Daily	Bare Costs	Incl. O&P
1 Equip. Oper. (crane)	$20.65	$165.20	$29.90	$239.20	$18.87	$27.32
1 Equip. Oper. Oiler	17.10	136.80	24.75	198.00		
1 Hyd. Excavator, 5/8 C.Y.		268.00		294.80	16.75	18.42
16 M.H., Daily Totals		$570.00		$732.00	$35.62	$45.74

Crew B-12R	Hr.	Daily	Hr.	Daily	Bare Costs	Incl. O&P
1 Equip. Oper. (crane)	$20.65	$165.20	$29.90	$239.20	$18.87	$27.32
1 Equip. Oper. Oiler	17.10	136.80	24.75	198.00		
1 Hyd. Excavator, 1.5 C.Y.		560.20		616.20	35.01	38.51
16 M.H., Daily Totals		$862.20		$1053.40	$53.88	$65.83

Crew B-12S	Hr.	Daily	Hr.	Daily	Bare Costs	Incl. O&P
1 Equip. Oper. (crane)	$20.65	$165.20	$29.90	$239.20	$18.87	$27.32
1 Equip. Oper. Oiler	17.10	136.80	24.75	198.00		
1 Hyd. Excavator, 2.5 C.Y.		1143.00		1257.30	71.43	78.58
16 M.H., Daily Totals		$1445.00		$1694.50	$90.30	$105.90

Crew B-13	Hr.	Daily	Hr.	Daily	Bare Costs	Incl. O&P
1 Labor Foreman (outside)	$17.50	$140.00	$25.25	$202.00	$16.75	$24.21
4 Building Laborers	15.50	496.00	22.40	716.80		
1 Equip. Oper. (crane)	20.65	165.20	29.90	239.20		
1 Equip. Oper. Oiler	17.10	136.80	24.75	198.00		
1 Hyd. Crane, 25 Ton		384.00		422.40	6.85	7.54
56 M.H., Daily Totals		$1322.00		$1778.40	$23.60	$31.75

Crew B-14	Hr.	Daily	Hr.	Daily	Bare Costs	Incl. O&P
1 Labor Foreman (outside)	$17.50	$140.00	$25.25	$202.00	$16.42	$23.73
4 Building Laborers	15.50	496.00	22.40	716.80		
1 Equip. Oper. (light)	19.05	152.40	27.55	220.40		
1 Backhoe Loader, 48 H.P.		152.20		167.40	3.17	3.48
48 M.H., Daily Totals		$940.60		$1306.60	$19.59	$27.21

Crew B-20	Hr.	Daily	Hr.	Daily	Bare Costs	Incl. O&P
1 Plumber Foreman (out)	$24.20	$193.60	$34.95	$279.60	$20.63	$29.81
1 Plumber	22.20	177.60	32.10	256.80		
1 Building Laborer	15.50	124.00	22.40	179.20		
24 M.H., Daily Totals		$495.20		$715.60	$20.63	$29.81

Crew B-21	Hr.	Daily	Hr.	Daily	Bare Costs	Incl. O&P
1 Plumber Foreman (out)	$24.20	$193.60	$34.95	$279.60	$20.63	$29.82
1 Plumber	22.20	177.60	32.10	256.80		
1 Building Laborer	15.50	124.00	22.40	179.20		
.5 Equip. Oper. (crane)	20.65	82.60	29.90	119.60		
.5 S.P. Crane, 5 Ton		87.90		96.70	3.13	3.45
28 M.H., Daily Totals		$665.70		$931.90	$23.76	$33.27

Crew B-22	Hr.	Daily	Hr.	Daily	Bare Costs	Incl. O&P
1 Plumber Foreman (out)	$24.20	$193.60	$34.95	$279.60	$20.63	$29.83
1 Plumber	22.20	177.60	32.10	256.80		
1 Building Laborer	15.50	124.00	22.40	179.20		
.75 Equip. Oper. (crane)	20.65	123.90	29.90	179.40		
.75 S.P. Crane, 5 Ton		131.85		145.05	4.39	4.83
30 M.H., Daily Totals		$750.95		$1040.05	$25.02	$34.66

Crew B-23	Hr.	Daily	Hr.	Daily	Bare Costs	Incl. O&P
1 Labor Foreman (outside)	$17.50	$140.00	$25.25	$202.00	$15.90	$22.97
4 Building Laborers	15.50	496.00	22.40	716.80		
Truck & Drill Rig		210.60		231.65	5.26	5.79
40 M.H., Daily Totals		$846.60		$1150.45	$21.16	$28.76

Crew B-25	Hr.	Daily	Hr.	Daily	Bare Costs	Incl. O&P
1 Labor Foreman (outside)	$17.50	$140.00	$25.25	$202.00	$16.64	$24.05
7 Building Laborers	15.50	868.00	22.40	1254.40		
2 Equip. Oper. (med.)	20.20	323.20	29.25	468.00		
1 Paving Machine		534.00		587.40		
1 Tandem Roller, 10 Ton		172.80		190.10	8.83	9.71
80 M.H., Daily Totals		$2038.00		$2701.90	$25.47	$33.76

Crew B-31	Hr.	Daily	Hr.	Daily	Bare Costs	Incl. O&P
1 Labor Foreman (outside)	$17.50	$140.00	$25.25	$202.00	$16.72	$24.15
3 Building Laborers	15.50	372.00	22.40	537.60		
1 Carpenter	19.60	156.80	28.30	226.40		
1 Air Compr., 250 C.F.M.		106.20		116.80		
1 Sheeting Driver		4.85		5.35		
2-50 Ft Air Hoses, 1.5" Dia.		11.60		12.75	3.06	3.37
40 M.H., Daily Totals		$791.45		$1100.90	$19.78	$27.52

Crew B-34A	Hr.	Daily	Hr.	Daily	Bare Costs	Incl. O&P
1 Truck Driver (heavy)	$16.20	$129.60	$23.20	$185.60	$16.20	$23.20
1 Dump Truck, 12 Ton		219.80		241.80	27.47	30.22
8 M.H., Daily Totals		$349.40		$427.40	$43.67	$53.42

Crew B-34B	Hr.	Daily	Hr.	Daily	Bare Costs	Incl. O&P
1 Truck Driver (heavy)	$16.20	$129.60	$23.20	$185.60	$16.20	$23.20
1 Dump Truck, 16 Ton		279.20		307.10	34.90	38.38
8 M.H., Daily Totals		$408.80		$492.70	$51.10	$61.58

Crew B-35	Hr.	Daily	Hr.	Daily	Bare Costs	Incl. O&P
1 Plumber Foreman (ins)	$22.70	$181.60	$32.80	$262.40	$20.43	$29.55
1 Plumber	22.20	177.60	32.10	256.80		
1 Welder (plumber)	22.20	177.60	32.10	256.80		
1 Plumber Apprentice	17.76	142.08	25.65	205.20		
1 Equip. Oper. (crane)	20.65	165.20	29.90	239.20		
1 Equip. Oper. Oiler	17.10	136.80	24.75	198.00		
1 Electric Welding Mach.		18.15		19.95		
1 Hyd. Excavator, .75 C.Y.		323.20		355.50	7.11	7.82
48 M.H., Daily Totals		$1322.23		$1793.85	$27.54	$37.37

Crew B-37	Hr.	Daily	Hr.	Daily	Bare Costs	Incl. O&P
1 Labor Foreman (outside)	$17.50	$140.00	$25.25	$202.00	$16.42	$23.73
4 Building Laborers	15.50	496.00	22.40	716.80		
1 Equip. Oper. (light)	19.05	152.40	27.55	220.40		
1 Tandem Roller, 5 Ton		83.85		92.25	1.74	1.92
48 M.H., Daily Totals		$872.25		$1231.45	$18.16	$25.65

Crew B-38	Hr.	Daily	Hr.	Daily	Bare Costs	Incl. O&P
2 Building Laborers	$15.50	$248.00	$22.40	$358.40	$16.68	$24.11
1 Equip. Oper. (light)	19.05	152.40	27.55	220.40		
1 Backhoe Loader, 48 H.P.		152.20		167.40		
1 Demol. Hammer, Hyd.		209.20		230.10	15.05	16.56
24 M.H., Daily Totals		$761.80		$976.30	$31.73	$40.67

CREWS

Crew No.	Bare Costs		Incl. Subs O & P		Cost Per Man-hour	
Crew B-39	Hr.	Daily	Hr.	Daily	Bare Costs	Incl. O&P
1 Labor Foreman (outside)	$17.50	$140.00	$25.25	$202.00	$16.42	$23.73
4 Building Laborers	15.50	496.00	22.40	716.80		
1 Equipment Oper. (light)	19.05	152.40	27.55	220.40		
1 Air Compr., 250 C.F.M.		106.20		116.80		
Air Tools & Accessories		23.70		26.05		
2-50 Ft. Air Hoses, 1.5" Dia		11.60		12.75	2.94	3.24
48 M.H., Daily Totals		$929.90		$1294.80	$19.36	$26.97
Crew B-40	Hr.	Daily	Hr.	Daily	Bare Costs	Incl. O&P
1 Pile Driver Foreman	$21.70	$173.60	$33.95	$271.60	$19.86	$30.21
4 Pile Drivers	19.70	630.40	30.80	985.60		
2 Equip. Oper. (crane)	20.65	330.40	29.90	478.40		
1 Equip. Oper. Oiler	17.10	136.80	24.75	198.00		
1 Crane, 40 Ton		482.60		530.85		
Vibratory Hammer & Gen.		983.00		1081.30	22.90	25.18
64 M.H., Daily Totals		$2736.80		$3545.75	$42.76	$55.39
Crew B-42	Hr.	Daily	Hr.	Daily	Bare Costs	Incl. O&P
1 Labor Foreman (outside)	$17.50	$140.00	$25.25	$202.00	$17.31	$25.37
4 Building Laborers	15.50	496.00	22.40	716.80		
1 Equip. Oper. (crane)	20.65	165.20	29.90	239.20		
1 Equip. Oper. Oiler	17.10	136.80	24.75	198.00		
1 Welder	21.30	170.40	33.50	268.00		
1 Hyd. Crane, 25 Ton		384.00		422.40		
1 Gas Welding Machine		56.40		62.05		
1 Horz. Boring Csg. Mch.		477.20		524.90	14.33	15.77
64 M.H., Daily Totals		$2026.00		$2633.35	$31.64	$41.14
Crew B-46	Hr.	Daily	Hr.	Daily	Bare Costs	Incl. O&P
1 Pile Driver Foreman	$21.70	$173.60	$33.95	$271.60	$17.93	$27.12
2 Pile Drivers	19.70	315.20	30.80	492.80		
3 Building Laborers	15.50	372.00	22.40	537.60		
1 Chain Saw, 36" Long		28.20		31.00	.58	.64
48 M.H., Daily Totals		$889.00		$1333.00	$18.51	$27.76
Crew B-47	Hr.	Daily	Hr.	Daily	Bare Costs	Incl. O&P
1 Blast Foreman (outside)	$17.50	$140.00	$25.25	$202.00	$17.35	$25.06
1 Driller	15.50	124.00	22.40	179.20		
1 Equip. Oper. (light)	19.05	152.40	27.55	220.40		
1 Crawler Type Drill, 4"		168.80		185.70		
1 Air Compr., 600 C.F.M.		236.20		259.80		
2-50 Ft. Air Hoses, 3" Dia		16.70		18.35	17.57	19.32
24 M.H., Daily Totals		$838.10		$1065.45	$34.92	$44.38
Crew B-50	Hr.	Daily	Hr.	Daily	Bare Costs	Incl. O&P
2 Pile Driver Foremen	$21.70	$347.20	$33.95	$543.20	$19.03	$28.88
6 Pile Drivers	19.70	945.60	30.80	1478.40		
2 Equip. Oper. (crane)	20.65	330.40	29.90	478.40		
1 Equip. Oper. Oiler	17.10	136.80	24.75	198.00		
3 Building Laborers	15.50	372.00	22.40	537.60		
1 Crane, 40 Ton		482.60		530.85		
60 L.F. Leads, 15K Ft. Lbs.		42.00		46.20		
1 Hammer, 15K Ft. Lbs.		197.20		216.90		
1 Air Compr., 600 C.F.M.		236.20		259.80		
2-50 Ft. Air Hoses, 3" Dia		16.70		18.35		
1 Chain Saw, 36" Long		28.20		31.00	8.95	9.84
112 M.H., Daily Totals		$3134.90		$4338.70	$27.98	$38.72

Crew No.	Bare Costs		Incl. Subs O & P		Cost Per Man-hour	
Crew B-53	Hr.	Daily	Hr.	Daily	Bare Costs	Incl. O&P
1 Equip. Oper. (light)	$19.05	$152.40	$27.55	$220.40	$19.05	$27.55
1 Trencher, Chain, 12 H.P.		133.40		146.75	16.67	18.34
8 M.H., Daily Totals		$285.80		$367.15	$35.72	$45.89
Crew B-54	Hr.	Daily	Hr.	Daily	Bare Costs	Incl. O&P
1 Equip. Oper. (light)	$19.05	$152.40	$27.55	$220.40	$19.05	$27.55
1 Trencher, Chain, 40 H.P.		197.80		217.60	24.72	27.20
8 M.H., Daily Totals		$350.20		$438.00	$43.77	$54.75
Crew C-1	Hr.	Daily	Hr.	Daily	Bare Costs	Incl. O&P
3 Carpenters	$19.60	$470.40	$28.30	$679.20	$18.57	$26.82
1 Building Laborer	15.50	124.00	22.40	179.20		
Power Tools		21.00		23.10	.65	.72
32 M.H., Daily Totals		$615.40		$881.50	$19.22	$27.54
Crew C-14	Hr.	Daily	Hr.	Daily	Bare Costs	Incl. O&P
1 Carpenter Foreman (out)	$21.60	$172.80	$31.20	$249.60	$19.00	$27.81
5 Carpenters	19.60	784.00	28.30	1132.00		
4 Building Laborers	15.50	496.00	22.40	716.80		
4 Rodmen (reinf.)	21.30	681.60	32.75	1048.00		
2 Cement Finishers	18.80	300.80	26.40	422.40		
1 Equip. Oper. (crane)	20.65	165.20	29.90	239.20		
1 Equip. Oper. Oiler	17.10	136.80	24.75	198.00		
1 Crane, 80 Ton, & Tools		836.00		919.60		
Power Tools		21.00		23.10		
2 Gas Finishing Mach.		54.20		59.60	6.32	6.96
144 M.H., Daily Totals		$3648.40		$5008.30	$25.32	$34.77
Crew C-17	Hr.	Daily	Hr.	Daily	Bare Costs	Incl. O&P
2 Skilled Worker Foremen	$22.10	$353.60	$32.20	$515.20	$20.50	$29.84
8 Skilled Workers	20.10	1286.40	29.25	1872.00		
80 M.H., Daily Totals		$1640.00		$2387.20	$20.50	$29.84
Crew C-17D	Hr.	Daily	Hr.	Daily	Bare Costs	Incl. O&P
2 Skilled Worker Foremen	$22.10	$353.60	$32.20	$515.20	$20.50	$29.84
8 Skilled Workers	20.10	1286.40	29.25	1872.00		
.5 Crane, 80 Ton & Tools		435.05		478.55	5.43	5.98
80 M.H., Daily Totals		$2075.05		$2865.75	$25.93	$35.82
Crew C-17E	Hr.	Daily	Hr.	Daily	Bare Costs	Incl. O&P
2 Skilled Worker Foremen	$22.10	$353.60	$32.20	$515.20	$20.50	$29.84
8 Skilled Workers	20.10	1286.40	29.25	1872.00		
1 Hyd. Jack with Rods		46.65		51.30	.58	.64
80 M.H., Daily Totals		$1686.65		$2438.50	$21.08	$30.48
Crew D-1	Hr.	Daily	Hr.	Daily	Bare Costs	Incl. O&P
1 Bricklayer	$20.05	$160.40	$28.45	$227.60	$17.85	$25.32
1 Bricklayer Helper	15.65	125.20	22.20	177.60		
16 M.H., Daily Totals		$285.60		$405.20	$17.85	$25.32
Crew E-1	Hr.	Daily	Hr.	Daily	Bare Costs	Incl. O&P
1 Welder Foreman	$23.30	$186.40	$36.65	$293.20	$21.21	$32.56
1 Welder	21.30	170.40	33.50	268.00		
1 Equip. Oper. (light)	19.05	152.40	27.55	220.40		
1 Gas Welding Machine		56.40		62.05	2.35	2.58
24 M.H., Daily Totals		$565.60		$843.65	$23.56	$35.14

CREWS

Crew No.	Bare Costs		Incl. Subs O & P		Cost Per Man-hour	
Crew E-4	**Hr.**	**Daily**	**Hr.**	**Daily**	**Bare Costs**	**Incl. O&P**
1 Struc. Steel Foreman	$23.30	$186.40	$36.65	$293.20	$21.80	$34.28
3 Struc. Steel Workers	21.30	511.20	33.50	804.00		
1 Gas Welding Machine		56.40		62.05	1.76	1.93
32 M.H., Daily Totals		$754.00		$1159.25	$23.56	$36.21

Crew F-1	Hr.	Daily	Hr.	Daily	Bare Costs	Incl. O&P
1 Carpenter	$19.60	$156.80	$28.30	$226.40	$19.60	$28.30
Power Tools		7.00		7.70	.87	.96
8 M.H., Daily Totals		$163.80		$234.10	$20.47	$29.26

Crew F-2	Hr.	Daily	Hr.	Daily	Bare Costs	Incl. O&P
2 Carpenters	$19.60	$313.60	$28.30	$452.80	$19.60	$28.30
Power Tools		14.00		15.40	.87	.96
16 M.H., Daily Totals		$327.60		$468.20	$20.47	$29.26

Crew F-5	Hr.	Daily	Hr.	Daily	Bare Costs	Incl. O&P
1 Carpenter Foreman	$21.60	$172.80	$31.20	$249.60	$20.10	$29.02
3 Carpenters	19.60	470.40	28.30	679.20		
Power Tools		14.00		15.40	.43	.48
32 M.H., Daily Totals		$657.20		$944.20	$20.53	$29.50

Crew G-2	Hr.	Daily	Hr.	Daily	Bare Costs	Incl. O&P
1 Plasterer	$19.60	$156.80	$27.85	$222.80	$17.08	$24.40
1 Plasterer Helper	16.15	129.20	22.95	183.60		
1 Building Laborer	15.50	124.00	22.40	179.20		
Grouting Equipment		103.75		114.15	4.32	4.75
24 M.H., Daily Totals		$513.75		$699.75	$21.40	$29.15

Crew L-1	Hr.	Daily	Hr.	Daily	Bare Costs	Incl. O&P
1 Electrician	$22.10	$176.80	$31.75	$254.00	$22.15	$31.92
1 Plumber	22.20	177.60	32.10	256.80		
16 M.H., Daily Totals		$354.40		$510.80	$22.15	$31.92

Crew L-2	Hr.	Daily	Hr.	Daily	Bare Costs	Incl. O&P
1 Carpenter	$19.60	$156.80	$28.30	$226.40	$17.40	$25.25
1 Helper	15.20	121.60	22.20	177.60		
16 M.H., Daily Totals		$278.40		$404.00	$17.40	$25.25

Crew L-3	Hr.	Daily	Hr.	Daily	Bare Costs	Incl. O&P
1 Carpenter	$19.60	$156.80	$28.30	$226.40	$20.87	$30.17
.5 Electrician	22.10	88.40	31.75	127.00		
.5 Sheet Metal Worker	22.20	88.80	32.35	129.40		
16 M.H., Daily Totals		$334.00		$482.80	$20.87	$30.17

Crew L-4	Hr.	Daily	Hr.	Daily	Bare Costs	Incl. O&P
2 Skilled Workers	$20.10	$321.60	$29.25	$468.00	$18.46	$26.90
1 Helper	15.20	121.60	22.20	177.60		
24 M.H., Daily Totals		$443.20		$645.60	$18.46	$26.90

Crew L-6	Hr.	Daily	Hr.	Daily	Bare Costs	Incl. O&P
1 Plumber	$22.20	$177.60	$32.10	$256.80	$22.16	$31.98
.5 Electrician	22.10	88.40	31.75	127.00		
12 M.H., Daily Totals		$266.00		$383.80	$22.16	$31.98

Crew L-7	Hr.	Daily	Hr.	Daily	Bare Costs	Incl. O&P
2 Carpenters	$19.60	$313.60	$28.30	$452.80	$18.78	$27.10
1 Building Laborer	15.50	124.00	22.40	179.20		
.5 Electrician	22.10	88.40	31.75	127.00		
28 M.H., Daily Totals		$526.00		$759.00	$18.78	$27.10

Crew No.	Bare Costs		Incl. Subs O & P		Cost Per Man-hour	
Crew L-9	**Hr.**	**Daily**	**Hr.**	**Daily**	**Bare Costs**	**Incl. O&P**
1 Labor Foreman (inside)	$16.00	$128.00	$23.10	$184.80	$17.63	$26.06
2 Building Laborers	15.50	248.00	22.40	358.40		
1 Struc. Steel Worker	21.30	170.40	33.50	268.00		
.5 Electrician	22.10	88.40	31.75	127.00		
36 M.H., Daily Totals		$634.80		$938.20	$17.63	$26.06

Crew Q-1	Hr.	Daily	Hr.	Daily	Bare Costs	Incl. O&P
1 Plumber	$22.20	$177.60	$32.10	$256.80	$19.98	$28.87
1 Plumber Apprentice	17.76	142.08	25.65	205.20		
16 M.H., Daily Totals		$319.68		$462.00	$19.98	$28.87

Crew Q-2	Hr.	Daily	Hr.	Daily	Bare Costs	Incl. O&P
2 Plumbers	$22.20	$355.20	$32.10	$513.60	$20.72	$29.95
1 Plumber Apprentice	17.76	142.08	25.65	205.20		
24 M.H., Daily Totals		$497.28		$718.80	$20.72	$29.95

Crew Q-3	Hr.	Daily	Hr.	Daily	Bare Costs	Incl. O&P
1 Plumber Foreman (ins)	$22.70	$181.60	$32.80	$262.40	$21.21	$30.66
2 Plumbers	22.20	355.20	32.10	513.60		
1 Plumber Apprentice	17.76	142.08	25.65	205.20		
32 M.H., Daily Totals		$678.88		$981.20	$21.21	$30.66

Crew Q-4	Hr.	Daily	Hr.	Daily	Bare Costs	Incl. O&P
1 Plumber Foreman (ins)	$22.70	$181.60	$32.80	$262.40	$21.21	$30.66
1 Plumber	22.20	177.60	32.10	256.80		
1 Welder (plumber)	22.20	177.60	32.10	256.80		
1 Plumber Apprentice	17.76	142.08	25.65	205.20		
1 Electric Welding Mach.		18.15		19.95	.56	.62
32 M.H., Daily Totals		$697.03		$1001.15	$21.77	$31.28

Crew Q-5	Hr.	Daily	Hr.	Daily	Bare Costs	Incl. O&P
1 Steamfitter	$22.30	$178.40	$32.20	$257.60	$20.07	$29.00
1 Steamfitter Apprentice	17.84	142.72	25.80	206.40		
16 M.H., Daily Totals		$321.12		$464.00	$20.07	$29.00

Crew Q-6	Hr.	Daily	Hr.	Daily	Bare Costs	Incl. O&P
2 Steamfitters	$22.30	$356.80	$32.20	$515.20	$20.81	$30.06
1 Steamfitter Apprentice	17.84	142.72	25.80	206.40		
24 M.H., Daily Totals		$499.52		$721.60	$20.81	$30.06

Crew Q-7	Hr.	Daily	Hr.	Daily	Bare Costs	Incl. O&P
1 Steamfitter Foreman (ins)	$22.80	$182.40	$32.95	$263.60	$21.31	$30.78
2 Steamfitters	22.30	356.80	32.20	515.20		
1 Steamfitter Apprentice	17.84	142.72	25.80	206.40		
32 M.H., Daily Totals		$681.92		$985.20	$21.31	$30.78

Crew Q-8	Hr.	Daily	Hr.	Daily	Bare Costs	Incl. O&P
1 Steamfitter Foreman (ins)	$22.80	$182.40	$32.95	$263.60	$21.31	$30.78
1 Steamfitter	22.30	178.40	32.20	257.60		
1 Welder (steamfitter)	22.30	178.40	32.20	257.60		
1 Steamfitter Apprentice	17.84	142.72	25.80	206.40		
1 Electric Welding Mach.		18.15		19.95	.56	.62
32 M.H., Daily Totals		$700.07		$1005.15	$21.87	$31.40

Crew Q-9	Hr.	Daily	Hr.	Daily	Bare Costs	Incl. O&P
1 Sheet Metal Worker	$22.20	$177.60	$32.35	$258.80	$19.98	$29.12
1 Sheet Metal Apprentice	17.76	142.08	25.90	207.20		
16 M.H., Daily Totals		$319.68		$466.00	$19.98	$29.12

CREWS

Crew No.	Bare Costs Hr.	Bare Costs Daily	Incl. Subs O & P Hr.	Incl. Subs O & P Daily	Cost Per Man-hour Bare Costs	Cost Per Man-hour Incl. O&P
Crew Q-10						
2 Sheet Metal Workers	$22.20	$355.20	$32.35	$517.60	$20.72	$30.20
1 Sheet Metal Apprentice	17.76	142.08	25.90	207.20		
24 M.H., Daily Totals		$497.28		$724.80	$20.72	$30.20
Crew Q-11						
1 Sheet Metal Foreman (ins)	$22.70	$181.60	$33.10	$264.80	$21.21	$30.92
2 Sheet Metal Workers	22.20	355.20	32.35	517.60		
1 Sheet Metal Apprentice	17.76	142.08	25.90	207.20		
32 M.H., Daily Totals		$678.88		$989.60	$21.21	$30.92
Crew Q-12						
1 Sprinkler Installer	$22.60	$180.80	$32.85	$262.80	$20.34	$29.57
1 Sprinkler Apprentice	18.08	144.64	26.30	210.40		
16 M.H., Daily Totals		$325.44		$473.20	$20.34	$29.57
Crew Q-13						
1 Sprinkler Foreman (ins)	$23.10	$184.80	$33.60	$268.80	$21.59	$31.40
2 Sprinkler Installers	22.60	361.60	32.85	525.60		
1 Sprinkler Apprentice	18.08	144.64	26.30	210.40		
32 M.H., Daily Totals		$691.04		$1004.80	$21.59	$31.40
Crew Q-14						
1 Asbestos Worker	$22.20	$177.60	$32.70	$261.60	$19.98	$29.42
1 Asbestos Apprentice	17.76	142.08	26.15	209.20		
16 M.H., Daily Totals		$319.68		$470.80	$19.98	$29.42
Crew Q-15						
1 Plumber	$22.20	$177.60	$32.10	$256.80	$19.98	$28.87
1 Plumber Apprentice	17.76	142.08	25.65	205.20		
1 Electric Welding Mach.		18.15		19.95	1.13	1.24
16 M.H., Daily Totals		$337.83		$481.95	$21.11	$30.11

Crew No.	Bare Costs Hr.	Bare Costs Daily	Incl. Subs O & P Hr.	Incl. Subs O & P Daily	Cost Per Man-hour Bare Costs	Cost Per Man-hour Incl. O&P
Crew Q-16						
2 Plumbers	$22.20	$355.20	$32.10	$513.60	$20.72	$29.95
1 Plumber Apprentice	17.76	142.08	25.65	205.20		
1 Electric Welding Mach.		18.15		19.95	.75	.83
24 M.H., Daily Totals		$515.43		$738.75	$21.47	$30.78
Crew Q-17						
1 Steamfitter	$22.30	$178.40	$32.20	$257.60	$20.07	$29.00
1 Steamfitter Apprentice	17.84	142.72	25.80	206.40		
1 Electric Welding Mach.		18.15		19.95	1.13	1.24
16 M.H., Daily Totals		$339.27		$483.95	$21.20	$30.24
Crew Q-18						
2 Steamfitters	$22.30	$356.80	$32.20	$515.20	$20.81	$30.06
1 Steamfitter Apprentice	17.84	142.72	25.80	206.40		
1 Electric Welding Mach.		18.15		19.95	.75	.83
24 M.H., Daily Totals		$517.67		$741.55	$21.56	$30.89
Crew Q-19						
1 Steamfitter	$22.30	$178.40	$32.20	$257.60	$20.74	$29.91
1 Steamfitter Apprentice	17.84	142.72	25.80	206.40		
1 Electrician	22.10	176.80	31.75	254.00		
24 M.H., Daily Totals		$497.92		$718.00	$20.74	$29.91
Crew Q-20						
1 Sheet Metal Worker	$22.20	$177.60	$32.35	$258.80	$20.40	$29.65
1 Sheet Metal Apprentice	17.76	142.08	25.90	207.20		
.5 Electrician	22.10	88.40	31.75	127.00		
20 M.H., Daily Totals		$408.08		$593.00	$20.40	$29.65
Crew Q-21						
2 Steamfitters	$22.30	$356.80	$32.20	$515.20	$21.13	$30.48
1 Steamfitter Apprentice	17.84	142.72	25.80	206.40		
1 Electrician	22.10	176.80	31.75	254.00		
32 M.H., Daily Totals		$676.32		$975.60	$21.13	$30.48

TRADES

Trade	Bare Costs		Incl. Subs O & P		Cost Per Man-hour	
Skwk	Hr.	Daily	Hr.	Daily	Bare Costs	Incl. O&P
1 Skilled Worker	$20.10	$160.80	$29.25	$234.00	$20.10	$29.25
8 M.H., Daily Total		$160.80		$234.00	$20.10	$29.25
Clab	Hr.	Daily	Hr.	Daily	Bare Costs	Incl. O&P
1 Building Laborer	$15.50	$124.00	$22.40	$179.20	$15.50	$22.40
8 M.H., Daily Total		$124.00		$179.20	$15.50	$22.40
Asbe	Hr.	Daily	Hr.	Daily	Bare Costs	Incl. O&P
1 Asbestos Worker	$22.20	$177.60	$32.70	$261.60	$22.20	$32.70
8 M.H., Daily Total		$177.60		$261.60	$22.20	$32.70
Boil	Hr.	Daily	Hr.	Daily	Bare Costs	Incl. O&P
1 Boilermaker	$22.30	$178.40	$32.65	$261.20	$22.30	$32.65
8 M.H., Daily Total		$178.40		$261.20	$22.30	$32.65
Bric	Hr.	Daily	Hr.	Daily	Bare Costs	Incl. O&P
1 Bricklayer	$20.05	$160.40	$28.45	$227.60	$20.05	$28.45
8 M.H., Daily Total		$160.40		$227.60	$20.05	$28.45
Brhe	Hr.	Daily	Hr.	Daily	Bare Costs	Incl. O&P
1 Bricklayer Helper	$15.65	$125.20	$22.20	$177.60	$15.65	$22.20
8 M.H., Daily Total		$125.20		$177.60	$15.65	$22.20
Carp	Hr.	Daily	Hr.	Daily	Bare Costs	Incl. O&P
1 Carpenter	$19.60	$156.80	$28.30	$226.40	$19.60	$28.30
8 M.H., Daily Total		$156.80		$226.40	$19.60	$28.30
Cefi	Hr.	Daily	Hr.	Daily	Bare Costs	Incl. O&P
1 Cement Finisher	$18.80	$150.40	$26.40	$211.20	$18.80	$26.40
8 M.H., Daily Total		$150.40		$211.20	$18.80	$26.40
Elec	Hr.	Daily	Hr.	Daily	Bare Costs	Incl. O&P
1 Electrician	$22.10	$176.80	$31.75	$254.00	$22.10	$31.75
8 M.H., Daily Total		$176.80		$254.00	$22.10	$31.75
Elev	Hr.	Daily	Hr.	Daily	Bare Costs	Incl. O&P
1 Elevator Constructor	$22.15	$177.20	$32.15	$257.20	$22.15	$32.15
8 M.H., Daily Total		$177.20		$257.20	$22.15	$32.15
Eqhv	Hr.	Daily	Hr.	Daily	Bare Costs	Incl. O&P
1 Equipment Oper. (crane)	$20.65	$165.20	$29.90	$239.20	$20.65	$29.90
8 M.H., Daily Total		$165.20		$239.20	$20.65	$29.90
Eqmd	Hr.	Daily	Hr.	Daily	Bare Costs	Incl. O&P
1 Equipment Oper. (med.)	$20.20	$161.60	$29.25	$234.00	$20.20	$29.25
8 M.H., Daily Total		$161.60		$234.00	$20.20	$29.25
Eqlt	Hr.	Daily	Hr.	Daily	Bare Costs	Incl. O&P
1 Equipment Oper. (light)	$19.05	$152.40	$27.55	$220.40	$19.05	$27.55
8 M.H., Daily Total		$152.40		$220.40	$19.05	$27.55
Eqol	Hr.	Daily	Hr.	Daily	Bare Costs	Incl. O&P
1 Equipment Oper. Oiler	$17.10	$136.80	$24.75	$198.00	$17.10	$24.75
8 M.H., Daily Total		$136.80		$198.00	$17.10	$24.75
Eqmm	Hr.	Daily	Hr.	Daily	Bare Costs	Incl. O&P
1 Master Mechanic	$21.45	$171.60	$31.05	$248.40	$21.45	$31.05
8 M.H., Daily Total		$171.60		$248.40	$21.45	$31.05
Glaz	Hr.	Daily	Hr.	Daily	Bare Costs	Incl. O&P
1 Glazier	$19.85	$158.80	$28.25	$226.00	$19.85	$28.25
8 M.H., Daily Total		$158.80		$226.00	$19.85	$28.25
Lath	Hr.	Daily	Hr.	Daily	Bare Costs	Incl. O&P
1 Lather	$19.75	$158.00	$27.75	$222.00	$19.75	$27.75
8 M.H., Daily Total		$158.00		$222.00	$19.75	$27.75
Marb	Hr.	Daily	Hr.	Daily	Bare Costs	Incl. O&P
1 Marble Setter	$19.55	$156.40	$27.75	$222.00	$19.55	$27.75
8 M.H., Daily Total		$156.40		$222.00	$19.55	$27.75
Mill	Hr.	Daily	Hr.	Daily	Bare Costs	Incl. O&P
1 Millwright	$20.35	$162.80	$28.75	$230.00	$20.35	$28.75
8 M.H., Daily Total		$162.80		$230.00	$20.35	$28.75
Mstz	Hr.	Daily	Hr.	Daily	Bare Costs	Incl. O&P
1 Mosaic/Terrazzo Worker	$19.35	$154.80	$27.10	$216.80	$19.35	$27.10
8 M.H., Daily Total		$154.80		$216.80	$19.35	$27.10
Pord	Hr.	Daily	Hr.	Daily	Bare Costs	Incl. O&P
1 Painter (Ordinary)	$18.85	$150.80	$26.75	$214.00	$18.85	$26.75
8 M.H., Daily Total		$150.80		$214.00	$18.85	$26.75

Trade	Bare Costs		Incl. Subs O & P		Cost Per Man-hour	
Psst	Hr.	Daily	Hr.	Daily	Bare Costs	Incl. O&P
1 Painter (Struc. Steel)	$19.60	$156.80	$31.50	$252.00	$19.60	$31.50
8 M.H., Daily Total		$156.80		$252.00	$19.60	$31.50
Pape	Hr.	Daily	Hr.	Daily	Bare Costs	Incl. O&P
1 Paper Hanger	$19.15	$153.20	$27.20	$217.60	$19.15	$27.20
8 M.H., Daily Total		$153.20		$217.60	$19.15	$27.20
Pile	Hr.	Daily	Hr.	Daily	Bare Costs	Incl. O&P
1 Pile Driver	$19.70	$157.60	$30.80	$246.40	$19.70	$30.80
8 M.H., Daily Total		$157.60		$246.40	$19.70	$30.80
Plas	Hr.	Daily	Hr.	Daily	Bare Costs	Incl. O&P
1 Plasterer	$19.60	$156.80	$27.85	$222.80	$19.60	$27.85
8 M.H., Daily Total		$156.80		$222.80	$19.60	$27.85
Plah	Hr.	Daily	Hr.	Daily	Bare Costs	Incl. O&P
1 Plasterer Helper	$16.15	$129.20	$22.95	$183.60	$16.15	$22.95
8 M.H., Daily Total		$129.20		$183.60	$16.15	$22.95
Plum	Hr.	Daily	Hr.	Daily	Bare Costs	Incl. O&P
1 Plumber	$22.20	$177.60	$32.10	$256.80	$22.20	$32.10
8 M.H., Daily Total		$177.60		$256.80	$22.20	$32.10
Rodm	Hr.	Daily	Hr.	Daily	Bare Costs	Incl. O&P
1 Rodman, (Reinf.)	$21.30	$170.40	$32.75	$262.00	$21.30	$32.75
8 M.H., Daily Total		$170.40		$262.00	$21.30	$32.75
Rofc	Hr.	Daily	Hr.	Daily	Bare Costs	Incl. O&P
1 Roofer (Composition)	$18.35	$146.80	$27.85	$222.80	$18.35	$27.85
8 M.H., Daily Total		$146.80		$222.80	$18.35	$27.85
Rots	Hr.	Daily	Hr.	Daily	Bare Costs	Incl. O&P
1 Roofer (Tile & Slate)	$18.50	$148.00	$28.10	$224.80	$18.50	$28.10
8 M.H., Daily Total		$148.00		$224.80	$18.50	$28.10
Rohe	Hr.	Daily	Hr.	Daily	Bare Costs	Incl. O&P
1 Roofer (Helper)	$13.45	$107.60	$20.45	$163.60	$13.45	$20.45
8 M.H., Daily Total		$107.60		$163.60	$13.45	$20.45
Shee	Hr.	Daily	Hr.	Daily	Bare Costs	Incl. O&P
1 Sheet Metal Worker	$22.20	$177.60	$32.35	$258.80	$22.20	$32.35
8 M.H., Daily Total		$177.60		$258.80	$22.20	$32.35
Spri	Hr.	Daily	Hr.	Daily	Bare Costs	Incl. O&P
1 Sprinkler Installer	$22.60	$180.80	$32.85	$262.80	$22.60	$32.85
8 M.H., Daily Total		$180.80		$262.80	$22.60	$32.85
Stpi	Hr.	Daily	Hr.	Daily	Bare Costs	Incl. O&P
1 Steamfitter/Pipefitter	$22.30	$178.40	$32.20	$257.60	$22.30	$32.20
8 M.H., Daily Total		$178.40		$257.60	$22.30	$32.20
Ston	Hr.	Daily	Hr.	Daily	Bare Costs	Incl. O&P
1 Stone Mason	$19.85	$158.80	$28.20	$225.60	$19.85	$28.20
8 M.H., Daily Total		$158.80		$225.60	$19.85	$28.20
Sswk	Hr.	Daily	Hr.	Daily	Bare Costs	Incl. O&P
1 Struc. Steel Worker	$21.30	$170.40	$33.50	$268.00	$21.30	$33.50
8 M.H., Daily Total		$170.40		$268.00	$21.30	$33.50
Tilf	Hr.	Daily	Hr.	Daily	Bare Costs	Incl. O&P
1 Tile Layer (Floor)	$19.25	$154.00	$26.95	$215.60	$19.25	$26.95
8 M.H., Daily Total		$154.00		$215.60	$19.25	$26.95
Tilh	Hr.	Daily	Hr.	Daily	Bare Costs	Incl. O&P
1 Tile Layer Helper	$15.25	$122.00	$21.35	$170.80	$15.25	$21.35
8 M.H., Daily Total		$122.00		$170.80	$15.25	$21.35
Trlt	Hr.	Daily	Hr.	Daily	Bare Costs	Incl. O&P
1 Truck Driver (light)	$16.00	$128.00	$22.90	$183.20	$16.00	$22.90
8 M.H., Daily Total		$128.00		$183.20	$16.00	$22.90
Trhv	Hr.	Daily	Hr.	Daily	Bare Costs	Incl. O&P
1 Truck Driver (heavy)	$16.20	$129.60	$23.20	$185.60	$16.20	$23.20
8 M.H., Daily Total		$129.60		$185.60	$16.20	$23.20
Sswl	Hr.	Daily	Hr.	Daily	Bare Costs	Incl. O&P
1 Welder (Struc. Steel)	$21.30	$170.40	$33.50	$268.00	$21.30	$33.50
8 M.H., Daily Total		$170.40		$268.00	$21.30	$33.50

1.0 Crews

		CREW	DAILY OUTPUT	UNIT	BARE COSTS EQUIP.	BARE COSTS LABOR	BARE COSTS TOTAL	TOTAL INCL O&P
05-001	**GENERAL CONDITIONS** see Foreword p. vii							
025	Crew A-1	A-1	1	Day	35.20	124	159.20	217.90
100	Crew A-4	A-4	1			464.40	464.40	666.80
125	Crew A-5	A-5	1		14.30	280	294.30	419.90
10-001	**SITE WORK** see Foreword p. vii to ix							
010	Crew B-1	B-1	1	Day		388	388	560.40
025	Crew B-2	B-2	1			636	636	918.80
085	Crew B-6	B-6	1		152.20	400.40	552.60	746.20
130	Crew B-9	B-9	1		141.50	636	777.50	1,074.45
145	Crew B-10	B-10	1			223.60	223.60	323.60
148	Crew B-10A	B-10A	1		72.10	223.60	295.70	402.90
150	Crew B-10B	B-10B	1		614.40	223.60	838	999.45
152	Crew B-10C	B-10C	1		685.90	223.60	909.50	1,078.10
154	Crew B-10D	B-10D	1		693.10	223.60	916.70	1,086
156	Crew B-10E	B-10E	1		83.85	223.60	307.45	415.80
158	Crew B-10F	B-10F	1		172.80	223.60	396.40	513.70
160	Crew B-10G	B-10G	1		283.20	223.60	506.80	635.10
162	Crew B-10H	B-10H	1		29.05	223.60	252.65	355.55
164	Crew B-10I	B-10I	1		59.65	223.60	283.25	389.20
166	Crew B-10J	B-10J	1		37.90	223.60	261.50	365.30
168	Crew B-10K	B-10K	1		98.85	223.60	322.45	432.30
200	Crew B-11	B-11	1			285.60	285.60	413.20
202	Crew B-11A	B-11A	1		614.40	285.60	900	1,089.05
204	Crew B-11B	B-11B	1		787.25	285.60	1,072.85	1,279.15
206	Crew B-11C	B-11C	1		152.20	285.60	437.80	580.60
224	Crew B-11M	B-11M	1		228	285.60	513.60	664
250	Crew B-12	B-12	1			302	302	437.20
252	Crew B-12A	B-12A	1		402.20	302	704.20	879.60
254	Crew B-12B	B-12B	1		560.20	302	862.20	1,053.40
256	Crew B-12C	B-12C	1		773	302	1,075	1,287.50
258	Crew B-12D	B-12D	1		1,389	302	1,691	1,965.10
260	Crew B-12E	B-12E	1		255	302	557	717.70
262	Crew B-12F	B-12F	1		323.20	302	625.20	792.70
270	Crew B-12J	B-12J	1		432.20	302	734.20	912.60
272	Crew B-12K	B-12K	1		619	302	921	1,118.10
276	Crew B-12M	B-12M	1		422.30	302	724.30	901.70
284	Crew B-12Q	B-12Q	1		268	302	570	732
286	Crew B-12R	B-12R	1		560.20	302	862.20	1,053.40
288	Crew B-12S	B-12S	1		1,143	302	1,445	1,694.50
300	Crew B-13	B-13	1		384	938	1,322	1,778.40
315	Crew B-14	B-14	1		152.20	788.40	940.60	1,306.60
405	Crew B-20	B-20	1			495.20	495.20	715.60
420	Crew B-21	B-21	1		87.90	577.80	665.70	931.90
435	Crew B-22	B-22	1		131.85	619.10	750.95	1,040
450	Crew B-23	B-23	1		210.60	636	846.60	1,150.45
480	Crew B-25	B-25	1		706.80	1,331.20	2,038	2,701.90
570	Crew B-31	B-31	1		122.65	668.80	791.45	1,100.90
630	Crew B-34	B-34	1			129.60	129.60	185.60
632	Crew B-34A	B-34A	1		219.80	129.60	349.40	427.40
634	Crew B-34B	B-34B	1		279.20	129.60	408.80	492.70
650	Crew B-35	B-35	1		341.35	980.88	1,322.23	1,793.90
680	Crew B-37	B-37	1		83.85	788.40	872.25	1,231.40
695	Crew B-38	B-38	1		361.40	400.40	761.80	976.35
710	Crew B-39	B-39	1		141.50	788.40	929.90	1,294.85
725	Crew B-40	B-40	1		1,465.60	1,271.20	2,736.80	3,545.75
755	Crew B-42	B-42	1		917.60	1,108.40	2,026	2,633.35
815	Crew B-46	B-46	1		28.20	860.80	889	1,333
830	Crew B-47	B-47	1		421.70	416.40	838.10	1,065.45
870	Crew B-50	B-50	1		1,002.90	2,132	3,134.90	4,338.80
900	Crew B-53	B-53	1		133.40	152.40	285.80	367.15

For expanded coverage of these items see *Building Construction Cost Data 1985*

1.0 Crews

		Crew	Daily Output	Unit	Bare Costs Equip.	Bare Costs Labor	Bare Costs Total	Total Incl O&P
10 902	Crew B-54	B-54	1	Day	197.80	152.40	350.20	438
15-001	**CONCRETE** see Foreword p. ix to x							
010	Crew C-1	C-1	1	Day	21	594.40	615.40	881.50
335	Crew C-14	C-14	1		911.20	2,737.20	3,648.40	5,008.30
410	Crew C-17	C-17	1			1,640	1,640	2,387.20
430	Crew C-17D	C-17D	1		435.05	1,640	2,075.05	2,865.75
435	Crew C-17E	C-17E	1		46.65	1,640	1,686.65	2,438.50
20-001	**MASONRY** see Foreword p. x							
010	Crew D-1	D-1	1	Day		285.60	285.60	405.20
25-001	**METALS** see Foreword p. x							
010	Crew E-1	E-1	1	Day	56.40	509.20	565.60	843.65
085	Crew E-4	E-4	1	"	56.40	697.60	754	1,159.25
30-001	**WOOD & PLASTICS** see Foreword p. x							
010	Crew F-1	F-1	1	Day	7	156.80	163.80	234.10
035	Crew F-2	F-2	1		14	313.60	327.60	468.20
110	Crew F-5	F-5	1		14	643.20	657.20	944.20
35-001	**MOISTURE PROTECTION** see Foreword p. x							
035	Crew G-2	G-2	1	Day	103.75	410	513.75	699.70
60-001	**EQUIPMENT** see Foreword p. x							
010	Crew L-1	L-1	1	Day		354.40	354.40	510.80
035	Crew L-2	L-2	1			278.40	278.40	404
060	Crew L-3	L-3	1			334	334	482.80
085	Crew L-4	L-4	1			443.20	443.20	645.60
110	Crew L-5	L-5	1		384	1,203.60	1,587.60	2,294.80
135	Crew L-6	L-6	1			266	266	383.80
160	Crew L-7	L-7	1			526	526	759
210	Crew L-9	L-9	1			634.80	634.80	938.20
75-001	**MECHANICAL** see Foreword p. x & xi							
010	Crew Q-1	Q-1	1	Day		319.68	319.68	462
035	Crew Q-2	Q-2	1			497.28	497.28	718.80
060	Crew Q-3	Q-3	1			678.88	678.88	981.20
085	Crew Q-4	Q-4	1		18.15	678.88	697.03	1,001.15
110	Crew Q-5	Q-5	1			321.12	321.12	464
135	Crew Q-6	Q-6	1			499.52	499.52	721.60
160	Crew Q-7	Q-7	1			681.92	681.92	985.20
185	Crew Q-8	Q-8	1		18.15	681.92	700.07	1,005.15
210	Crew Q-9	Q-9	1			319.68	319.68	466
235	Crew Q-10	Q-10	1			497.28	497.28	724.80
260	Crew Q-11	Q-11	1			678.88	678.88	989.60
285	Crew Q-12	Q-12	1			325.44	325.44	473.20
310	Crew Q-13	Q-13	1			691.04	691.04	1,004.80
335	Crew Q-14	Q-14	1			319.68	319.68	470.80
360	Crew Q-15	Q-15	1		18.15	319.68	337.83	481.95
385	Crew Q-16	Q-16	1		18.15	497.28	515.43	738.75
410	Crew Q-17	Q-17	1		18.15	321.12	339.27	483.95
435	Crew Q-18	Q-18	1		18.15	499.52	517.67	741.55
460	Crew Q-19	Q-19	1			497.92	497.92	718
485	Crew Q-20	Q-20	1			408.08	408.08	593
510	Crew Q-21	Q-21	1			676.32	676.32	975.60
80-001	**ELECTRICAL** see Foreword p. xi							
060	Crew R-3	R-3	1	Day	87.90	440.20	528.10	730.30
90-001	**TRADE RATES** see inside back cover							
010	Skilled average	1 Skwk	1	Day		160.80	160.80	234.10
015	Helpers average	1 Help	1			121.60	121.60	177.60
020	Foremen, inside	1 Fori	1			164.80	164.80	239.90
025	Foremen, outside	1 Foro	1			176.80	176.80	257.35
030	Common building laborers	1 Clab	1			124	124	179.05

For expanded coverage of these items see *Building Construction Cost Data 1985*

1.0 Crews

		Crew	Daily Output	Unit	Bare Costs Equip.	Bare Costs Labor	Bare Costs Total	Total Incl O&P
90 035	Air tool laborer	1 Cair	1	Day		125.60	125.60	181.35
040	Asbestos workers	1 Asbe	1			177.60	177.60	261.35
045	Boilermakers	1 Boil	1			178.40	178.40	261.30
050	Bricklayers	1 Bric	1			160.40	160.40	227.75
055	Bricklayer helpers	1 Brhe	1			125.20	125.20	177.75
060	Carpenters	1 Carp	1			156.80	156.80	226.40
065	Carpet & linoleum layer	1 Crpt	1			153.20	153.20	217.50
070	Cement finishers	1 Cefi	1			150.40	150.40	210.95
075	Electricians	1 Elec	1			176.80	176.80	254
080	Elevator constructors	1 Elev	1			177.20	177.20	257.05
085	Equip. oper. crane or shovel	1 Eqhv	1			165.20	165.20	239.05
090	Equip. operators med. equip	1 Eqmd	1			161.60	161.60	233.75
095	Equip. operators lt. equip	1 Eqlt	1			152.40	152.40	220.50
100	Equip. operators, oilers	1 Eqol	1			136.80	136.80	197.90
105	Equip. oper., master mechanics	1 Eqmm	1			171.60	171.60	248.25
110	Glaziers	1 Glaz	1			158.80	158.80	226.10
115	Lathers	1 Lath	1			158	158	222.10
120	Marble setters	1 Marb	1			156.40	156.40	222.10
125	Millwrights	1 Mill	1			162.80	162.80	229.85
130	Mosaic & terrazzo workers	1 Mstz	1			154.80	154.80	216.70
135	Mosaic & terrazzo helpers	1 Mthe	1			124	124	173.60
140	Painters, ordinary	1 Pord	1			150.80	150.80	214.10
145	Painters, spray	1 Pspr	1			154.40	154.40	248.25
150	Painters, structural steel	1 Psst	1			156.80	156.80	252.10
155	Paperhangers	1 Pape	1			153.20	153.20	217.50
160	Pile drivers	1 Pile	1			157.60	157.60	246.50
165	Plasterers	1 Plas	1			156.80	156.80	222.65
170	Plasterer helpers	1 Plah	1			129.20	129.20	183.45
175	Plumbers	1 Plum	1			177.60	177.60	256.55
180	Plumbers helpers	1 Pluh	1			142	142	205.10
185	Rodmen (reinforcing)	1 Rodm	1			170.40	170.40	261.85
190	Roofers composition	1 Rofc	1			146.80	146.80	222.95
195	Roofers, precast	1 Rofp	1			149.20	149.20	226.55
200	Roofers, tile & slate	1 Rots	1			148	148	224.80
205	Roofer helpers (composition)	1 Rohe	1			107.60	107.60	163.45
210	Sheet metal workers	1 Shee	1			177.60	177.60	258.90
215	Sprinkler installers	1 Spri	1			180.80	180.80	262.90
220	Steamfitters or pipefitters	1 Stpi	1			178.40	178.40	257.75
225	Stone masons	1 Ston	1			158.80	158.80	225.45
230	Structural steel workers	1 Sswk	1			170.40	170.40	268
235	Tile layers (floor)	1 Tilf	1			154	154	215.45
240	Tile layer helpers	1 Tilh	1			122	122	170.65
245	Truck drivers, light	1 Trlt	1			128	128	183.30
250	Truck drivers, heavy	1 Trhv	1			129.60	129.60	185.50
255	Welders, structural steel	1 Sswl	1			170.40	170.40	268
260	Wreckers	1 Wrck	1			124	124	192.80

1.1 Overhead

		Crew	Daily Output	Unit	Bare Costs Mat.	Bare Costs Inst.	Bare Costs Total	Total Incl O&P
04-001	**CLEANING UP** After job completion, minimum			Job Cost				.30%
004	Maximum			Project				1%
005	Cleanup of floor area, continuous, per day	A-5	12	M.S.F.	1.44	25	26.44	37
010	Final	"	11.50	"	1.52	26	27.52	38
100	Mechanical demolition, see division 2.2							

For expanded coverage of these items see *Building Construction Cost Data 1985*

1.1 Overhead

		CREW	DAILY OUTPUT	UNIT	BARE COSTS MAT.	BARE COSTS INST.	BARE COSTS TOTAL	TOTAL INCL O&P
06-001	**CONSTRUCTION COST INDEX** For 162 major U.S. and							
002	Canadian cities, total cost, min. (Greensboro N.C.)			%				81.70
005	C13.1 -100 Average							100%
010	Maximum (Anchorage, AK.)							133.90
11-001	**CONTINGENCIES** Allowance to add at conceptual stage			Project				15%
005	Schematic stage							10%
010	Preliminary working drawing stage							7%
015	Final working drawing stage							2%
12-001	C10.3 -300 **CONTRACTOR EQUIPMENT** See division 1.5							
14-001	**CREWS** For building construction, see foreword p. vii to xi							
010	See also division 1.0 for daily costs							
15-001	**ENGINEERING FEES** Educational planning consultant, minimum			Project				.50%
010	Maximum			"				2.50%
020	C10.1 -103 Electrical, minimum			Contract				4.10%
030	Maximum							10.10%
040	Elevator & conveying systems, minimum							2.50%
050	Maximum							5%
060	Food service & kitchen equipment, minimum							8%
070	Maximum							12%
100	Mechanical (plumbing & HVAC), minimum							4.10%
110	Maximum							10.10%
16-001	C13.1 -100 **HISTORICAL COST INDEXES** Back to 1941							
18-001	**INSURANCE** Builders risk, standard, minimum			Job Cost				.10%
005	Maximum							.50%
020	C10.1 -301 All-risk type, minimum							.12%
025	Maximum							.68%
040	Contractor's equipment floater, minimum			Value				.50%
045	Maximum			"				2.50%
060	Public liability, average			Job Cost				.82%
061								
080	C10.2 -200 Workers' compensation & employer's liability, average							
085	by trade, carpentry, general			Payroll		9%		
100	Electrical					3.87%		
115	Insulation					7.35%		
145	Plumbing					4.71%		
155	Sheet metal work (HVAC)					6.01%		
19-001	**JOB CONDITIONS** Modifications to total							
002	project cost summaries							
010	Economic conditions, favorable, deduct			Project				2%
020	Unfavorable, add							5%
030	Hoisting conditions, favorable, deduct							1%
040	Unfavorable, add							5%
070	Labor availability, surplus, deduct							1%
080	Shortage, add							10%
090	Material storage area, available, deduct							1%
100	Not available, add							2%
110	Subcontractor availability, surplus, deduct							5%
120	Shortage, add							12%
130	Work space, available, deduct							1%
140	Not available, add							4%
20-001	**LABOR INDEX** For 162 major U.S. and Canadian cities							
002	Minimum (Charleston, S.C.)			%		70.30%	70.30%	
005	C13.1 -100 Average					100%		
010	Maximum (Anchorage, AK.)					140.90%	140.90%	
24-001	**MARK-UP** For General Contractors for change							
010	of scope of job as bid							
020	Extra work, by subcontractors, add			%				10%
025	By General Contractor, add			"				15%

For expanded coverage of these items see *Building Construction Cost Data 1985*

1.1 Overhead		CREW	DAILY OUTPUT	UNIT	BARE COSTS MAT.	BARE COSTS INST.	BARE COSTS TOTAL	TOTAL INCL O&P
24 040	Omitted work, by subcontractors, deduct			%				5%
045	By General Contractor, deduct							7.50%
060	Overtime work, by subcontractor, add							15%
065	By General Contractor, add							10%
100	C10.3 -200 Subcontractors, on their own labor, minimum					40%		
110	Maximum					60.80%		
26-001	**MATERIAL INDEX** For 162 major U.S. and Canadian cities							
002	Minimum (Greensboro, NC.)			%	94%			
004	C13.1 -100 Average				100%			
006	Maximum (Anchorage, AK.)				125.40%			
28-001	C10.3 -200 **OVERHEAD** As percent of direct costs, minimum						5%	
005	Average						12%	
010	Maximum						22%	
30-001	**OVERHEAD & PROFIT** Allowance to add to items in this							
002	book that do not include Subs O&P, average			%			25%	
010	C10.3 -200 Allowance to add to items in this book that							
011	do include Subs O&P, minimum			%				5%
015	Average							10%
020	Maximum							15%
030	Typical, by size of project, under $100,000						30%	
035	$500,000 project						25%	
040	$2,000,000 project						20%	
045	Over $10,000,000 project						15%	
32-001	C10.2 -400 **OVERTIME** For early completion of projects or where							
002	labor shortages exist, add to usual labor, up to			Costs		118%		
34-001	C10.1 -302 **PERFORMANCE BOND** For buildings, minimum			Job Cost				.39%
010	Maximum							1.20%
36-001	**PERMITS** Rule of thumb, most cities, minimum							.50%
010	Maximum							2%
38-001	**PHOTOGRAPHS** 8" x 10", 4 shots, 2 prints ea., std. mounting			Set	74		74	81M
010	Hinged linen mounts				84		84	92M
020	8" x 10", 4 shots, 2 prints each, in color				150		150	165M
030	For I.D. slugs, add to all above				2.30		2.30	2.53M
150	Time lapse equipment, camera and projector, buy				3,100		3,100	3,400M
155	Rent per month				400		400	440M
170	Cameraman and film, including processing, B.&W.			Day	440		440	485M
172	Color			"	500		500	550M
43-001	**SCAFFOLD** Steel tubular, regular, buy							
009	Building exterior, 1 to 5 stories	3 Carp	16.80	C.S.F.	11.70	28	39.70	53
020	C10.3 -300 6 to 12 stories	4 Carp	15		10.65	42	52.65	72
031	13 to 20 stories	5 Carp	16.75		9.15	47	56.15	78
046	Building interior walls, (area) up to 16' high	3 Carp	22.70		11.05	21	32.05	42
056	16' to 40' high		18.70		11.35	25	36.35	49
080	Building interior floor area, up to 30' high		85		3.70	5.55	9.25	12.05
090	Over 30' high	4 Carp	96.50		4.20	6.50	10.70	14
600	Scaffold steel tubular, suspended slab form supports to 8'-2" high							
610	1 use per month	4 Carp	31	C.S.F.	10.60	20	30.60	41
615	2 uses per month	"	43	"	5.30	14.60	19.90	27
650	Steel tubular, suspended slab form supports to 14'-8" high							
660	1 use per month	4 Carp	16	C.S.F.	16.60	39	55.60	75
665	2 uses per month	"	22	"	8.30	29	37.30	50
44-001	**SCAFFOLDING SPECIALTIES**							
002								
120	Sidewalk bridge, heavy duty steel posts & beams, including							
121	parapet protection & waterproofing							
122	8' to 10' wide, 2 posts	3 Carp	15	L.F.	10	31	41	56
123	3 posts	"	10	"	15	47	62	84
150	Sidewalk bridge using tubular steel							
151	scaffold frames, including planking	3 Carp	45	L.F.	3.35	10.45	13.80	18.80

For expanded coverage of these items see *Building Construction Cost Data 1985*

1.1 Overhead		CREW	DAILY OUTPUT	UNIT	BARE COSTS MAT.	BARE COSTS INST.	BARE COSTS TOTAL	TOTAL INCL O&P
44 160	For 2 uses per month, deduct from all above				50%			
170	For 1 use every 2 months, add to all above				100%			
190	Catwalks, 32" wide, no guardrails, 6' span, buy			Ea.	100		100	110M
200	10' span, buy			"	165		165	180M
280	Hand winch-operated masons							
281	scaffolding, no plank moving required							
290	98' long, 10'-6" high, buy			Ea.	4,500		4,500	4,950M
300	Rent per month				270		270	295M
310	28'-6" high, buy				7,875		7,875	8,675M
320	Rent per month				415		415	455M
340	196' long, 28'-6" high, buy				15,300		15,300	16,800M
350	Rent per month				800		800	880M
360	64'-6" high, buy				26,000		26,000	28,600M
370	Rent per month				1,600		1,600	1,750M
372	Putlog, standard, 8' span, with hangers, buy				70		70	77M
373	Rent per month				10		10	11M
375	12' span, buy			Ea.	100		100	110M
376	Trussed type, 14' span, buy			"	160		160	175M
377	Rent per month				15		15	16.50M
379	20' span, buy			Ea.	185		185	205M
380	Rolling ladders with handrails, 30" wide, buy 2 step				140		140	155M
400	7 step				315		315	345M
405	10 step				410		410	450M
406								
410	Rolling towers, buy, 5' wide, 7' long, 9' high			Ea.	820		820	900M
420	For 5' high added sections, add			"	160		160	175M
430	Complete incl. wheels, railings, etc.							
440	up to 20' high, rent per month			Ea.	90		90	99M
650	Stair unit, interior, for scaffolding, buy				255		255	280M
655	Rent per month				23		23	25M
48-001	**SMALL TOOLS** As % of contractor's work, minimum			Total				.50%
010	Maximum			"				1%
54-001	**TARPAULINS** Cotton duck, 10 oz. to 13.13 oz. per S.Y., minimum			S.F.	.25		.25	.27M
005	Maximum				.42		.42	.46M
010	Polyvinyl coated nylon, 14 oz. to 18 oz., minimum				.34		.34	.37M
015	Maximum				.42		.42	.46M
020	Reinforced polyethylene 3 mils thick, white				.04		.04	.04M
030	4 mils thick, white, clear or black				.05		.05	.05M
040	5.5 mils thick, clear				.06		.06	.06M
050	White, fire retardant				.11		.11	.12M
060	7.5 mils, oil resistant, fire retardant				.12		.12	.13M
070	8.5 mils, black				.15		.15	.16M
071	Woven polyethylene, 6 mils thick				.26		.26	.28M
074	Mylar polyester, non-reinforced, 7 mils thick				.82		.82	.90M
56-001	C10.2 -300 **TAXES** Sales tax, State, County & City, average			%	4.11%			
005	Maximum				8%			
020	Social Security, on first $40,000 of wages					7.05%		
030	Unemployment, Ma., combined Federal and State, minimum			%		2.30%		
035	Average					5.50%		
040	Maximum					6.50%		
58-001	**TEMPORARY CONSTRUCTION** See also Division 1.5-15							
002								
030	Barricades, 5' high, 3 rail @ 2" x 8", fixed	2 Carp	30	L.F.	8.50	10.45	18.95	24
035	Movable		20		9.10	15.70	24.80	33
210	Catwalks, no handrails, 3 joists 2" x 4"		55		2.80	5.70	8.50	11.30
215	3 joists, 3" x 6"		40		5.25	7.85	13.10	17.10
230	Fencing, chain link, 5' high	2 Sswk	100		4.50	3.41	7.91	10.30
235	6' high	"	75		7.15	4.54	11.69	15

For expanded coverage of these items see *Building Construction Cost Data 1985*

1.1 Overhead		CREW	DAILY OUTPUT	UNIT	BARE COSTS MAT.	BARE COSTS INST.	BARE COSTS TOTAL	TOTAL INCL O&P
58 250	Rented chain link, 6' high, to 500'	2 Sswk	100	L.F.	1.75	3.41	5.16	7.30
255	Over 1000' (up to 12 mo.)	"	110		1.47	3.10	4.57	6.50
270	Plywood, painted, 2" x 4" frame, 4' high	A-4	135		3.18	3.44	6.62	8.45
275	4" x 4" frame, 8' high	"	110		5.50	4.22	9.72	12.10
290	Wire mesh on 4" x 4" posts, 4' high	2 Carp	100		3.60	3.14	6.74	8.50
295	8' high		80		5.40	3.92	9.32	11.60
310	Guardrail, wooden, 3' high, 1" x 6" on 2" x 4" posts		200		1.24	1.57	2.81	3.63
315	2" x 6" on 4" x 4" posts	↓	165		2.08	1.90	3.98	5.05
320	Portable metal with base pads, buy				10.50		10.50	11.55M
321	Typical installation, assume 10 reuses	2 Carp	600	↓	1.22	.52	1.74	2.10
330	Heat, incl. fuel and operation, per week, 12 hrs. per day	1 Skwk	8.75	CSF Flr	13.75	18.40	32.15	42
335	24 hrs. per day	"	4.50		18.25	36	54.25	72
350	Lighting, incl. service lamps, wiring & outlets, minimum	1 Elec	34		1.69	5.20	6.89	9.35
355	Maximum	"	17		3.90	10.40	14.30	19.25
370	Power for temporary lighting only, per month, minimum/month						.82	.88
375	Maximum/month						2.15	2.55
390	Power for job duration incl. elevator, etc., minimum						41	49
395	Maximum			↓			84	100
420	Office trailer, furnished, no hookups, 20' x 8', buy	2 Skwk	1	Ea.	3,300	320	3,620	4,100
425	Rent per month				115		115	125M
427	32' x 8', buy	2 Skwk	.70		5,500	460	5,960	6,725
428	Rent per month				165		165	180M
430	50' x 10', buy	2 Skwk	.60		9,100	535	9,635	10,800
440	Rent per month				300		300	330M
443	50' x 12', buy	2 Skwk	.50		10,000	645	10,645	11,900
444	Rent per month				330		330	365M
446	For air conditioning, rent per month, add			↓	30		30	33M
447	For delivery, add per			Mile	1.30		1.30	1.43M
450	Portable buildings, prefab, on skids, economy, 8' x 8'	2 Carp	265	S.F.	50	1.18	51.18	57
460	Deluxe, 8' x 12'		150		65	2.09	67.09	75
470	Ramp, 3/4" plywood on 2" x 6" joists, 16" O.C.		300		1.25	1.05	2.30	2.88
475	On 2" x 10" joists, 16" O.C.	↓	275	↓	1.85	1.14	2.99	3.68
480	Roads, gravel fill, no surfacing, 4" gravel thickness	B-14	715	S.Y.	.57	1.32	1.89	2.45
490	8" gravel thickness	"	615	"	1.10	1.53	2.63	3.33
520	Sidewalks, 2" x 12" planks, 2 uses	1 Carp	350	S.F.	.35	.45	.80	1.03
522	Exterior plywood, 2 uses, 1/2" thick		750		.18	.21	.39	.50
523	5/8" thick		650		.24	.24	.48	.61
525	3/4" thick	↓	600		.32	.26	.58	.73
526	Signs, hi-intensity reflectorized, no posts, buy			↓	8.40		8.40	9.25M
527	Stair tread protection, 2" x 12" planks, 1 use	1 Carp	75	Tread	.91	2.09	3	4.02
529	Exterior plywood, 1/2" thick, 1 use		65		.42	2.41	2.83	3.95
530	3/4" thick, 1 use	↓	60	↓	.76	2.61	3.37	4.61
550	Storage vans, trailer mounted, 16' x 8', buy	2 Skwk	1.80	Ea.	2,200	180	2,380	2,675
552	Rent per month				75		75	83M
555	28' x 10', buy	2 Skwk	1.40		2,600	230	2,830	3,200
557	Rent per month			↓	80		80	88M
580	Toilet, portable, see Division 1.5-15-641							
68-001	**WINTER PROTECTION** Reinforced plastic on wood							
010	framing to close openings	2 Clab	750	S.F.	.22	.33	.55	.72
020	Tarpaulins hung over scaffolding, 8 uses, not incl. scaffolding		1,500		.10	.17	.27	.35
030	Prefab, fiberglass panels, steel frame, 8 uses	↓	1,200	↓	.43	.21	.64	.77

For expanded coverage of these items see *Building Construction Cost Data 1985*

1.5 Contractor Equipment

1

		UNIT	HOURLY OPER. COST	RENT PER DAY	RENT PER WEEK	RENT PER MONTH	CREWS EQUIPMENT COST
05-001	**EARTHWORK EQUIPMENT RENTAL** Without operators unless						
002	noted by*						
005	Augers for truck or trailer mounting, vertical drilling						
006	4" to 36" diam., 54 H.P., gas, 10' spindle travel	Ea.	6.65	150	455	1,450	144.20
007	14' spindle travel	"	6.95	180	540	1,625	163.60
010	Backhoe, diesel hydraulic, crawler mounted, 1/2 C.Y. capacity	Ea.	6.25	480*	1,025	3,325	255
012	5/8 C.Y. capacity		7.25	505*	1,050	3,400	268
014	3/4 C.Y. capacity		10.40	560*	1,200	3,875	323.20
015	1 C.Y. capacity 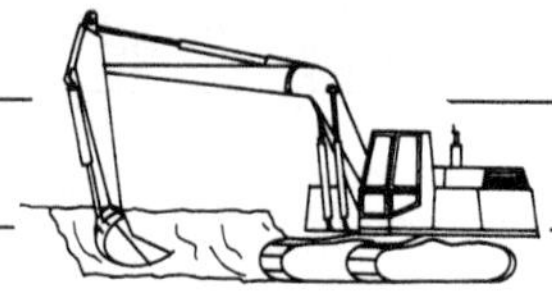		13.40	810*	1,475	4,300	402.20
020	1-1/2 C.Y. capacity		16.90	960*	2,125	6,175	560.20
030	2 C.Y. capacity		26	1,150*	2,825	8,550	773
032	2-1/2 C.Y. capacity		41	1,450*	4,075	11,500	1,143
034	3-1/2 C.Y. capacity		48	1,750*	5,025	15,000	1,389
035	Gradall type, truck mounted, 3 ton @ 15' radius, 5/8 C.Y.		17.15	850*	1,475	4,450	432.20
037	1 C.Y. capacity		23	1,025*	2,175	6,500	619
040	Backhoe-loader, wheel type, 40 to 45 H.P., 5/8 C.Y. capacity		4.66	325*	390	1,200	115.30
045	45 H.P. to 60 H.P., 3/4 C.Y. capacity 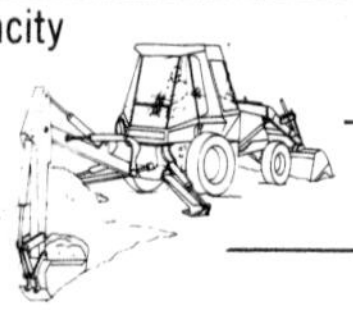		5.65	395*	535	1,650	152.20
046	80 H.P., 1-1/4 C.Y. capacity		8.50	450*	800	2,250	228
075	Bucket, clamshell, general purpose, 3/8 C.Y.		.40	33	100	290	23.20
080	1/2 C.Y.		.55	38	120	360	28.40
085	3/4 C.Y.		.62	50	145	440	33.95
090	1 C.Y.		.73	52	160	480	37.85
095	1-1/2 C.Y.		1.05	65	200	600	48.40
100	2 C.Y.		1.18	82	270	820	63.45
120	Compactor, roller, 2 drum, 2000 lb., operator walking		2.39	75	265	800	72.10
125	Rammer compactor, gas, 1000 lb. blow		.97	50	145	445	36.75
130	Vibratory plate, gas, 13" plate, 1000 lb. blow		.52	40	115	345	27.15
135	24" plate, 5000 lb. blow		.69	49	165	480	38.50
490	Trencher, chain, boom type, gas, operator walking, 12 H.P.		8.05	98	345	1,050	133.40
491	Operator riding, 40 H.P.		12.85	165	475	1,375	197.80
500	Wheel type, gas, 4' deep, 12" wide		11.25	440*	730	2,200	236
510	Diesel, 6' deep, 20" wide		15	610*	1,450	4,400	410
515	Ladder type, gas, 5' deep, 8" wide		8.55	485*	950	2,775	258.40
520	Diesel, 8' deep, 16" wide		10.70	650*	1,650	5,150	415.60
525	Truck, dump, tandem, 12 ton payload		13.10	350*	575	1,725	219.80
530	Three axle dump, 16 ton payload		16.15	435*	750	2,275	279.20
535	Dump trailer only, rear dump, 16-1/2 C.Y.		3.91	80	240	715	79.30
540	20 C.Y.		4.76	92	280	845	94.10
545	Flatbed, single axle, 1-1/2 ton rating 		3.77	46	135	415	57.15
550	3 ton rating		4.30	58	160	505	66.40
555	Off highway rear dump, 25 ton capacity		25	840*	2,200	6,750	640
560	35 ton capacity		34	1,150*	3,150	9,475	902
15-001	**GENERAL EQUIPMENT RENTAL**						
010	Without operators unless noted by*						
015	Aerial lift, scissor type, to 15' high, 1000 lb. capacity	Ea.	.15	77	248	745	50.80
016	To 25' high, 2000 lb. capacity		.18	115	365	1,100	74.45
017	Telescoping boom to 40' high, 750 lb. capacity		.21	240	700	2,250	141.70
018	2000 lb. capacity		.24	300	1,000	3,000	201.90
019	To 60' high, 750 lb. capacity		.27	540	1,250	3,300	252.15
019							
020	Air compressor, portable, gas engine, 60 C.F.M.	Ea.	3.54	34	95	290	47.30
030	160 C.F.M.		4.28	51	155	460	65.25
040	Diesel engine, rotary screw, 250 C.F.M.		6.90	83	255	745	106.20
050	365 C.F.M.		10.35	105	330	935	148.80
060	600 C.F.M.		17.15	160	495	1,400	236.20
070	750 C.F.M.		21	205	605	1,800	289
080	For silenced models, small sizes, add		3%	5%	5%	5%	25
090	Large sizes, add		4%	7%	7%	7%	33.40

For expanded coverage of these items see *Building Construction Cost Data 1985*

1.5 Contractor Equipment

15	Item	Ref.	Description	Unit	Hourly Oper. Cost	Rent per Day	Rent per Week	Rent per Month	Crews Equipment Cost
	092		Air tools and accessories						
	093		Breaker, pavement, 60 lb.	Ea.	.43	15	42	125	11.85
	094		80 lb.		.56	17	50	145	14.50
	098		Dust control per drill		.12	10	25	75	5.95
	100		Hose, air with couplings, 50' long, 3/4" diameter		.10	4	12	25	3.20
	110		1" diameter		.10	5.50	14	35	3.60
	120		1-1/2" diameter		.10	9.35	25	72	5.80
	130		2" diameter		.11	12	35	100	7.90
	140		2-1/2" diameter		.12	12	37	105	8.35
	141		3" diameter		.12	12.30	37	110	8.35
	156	C12.1 -100	Tamper, single, 35 lb.		.43	15	45	135	12.45
	157		Triple, 140 lb.		1.41	30	95	285	30.30
	158		Wrenches, impact, air powered up to 3/4" bolt		.08	19	58	170	12.25
	159		Up to 1-1/4" bolt		.09	26	76	215	15.90
	210		Generator, electric, gas engine, 1.5 KW to 3 KW		.50	22	67	195	17.40
	220		5 KW		.82	30	125	315	31.55
	230		10 KW		1.93	55	180	540	51.45
	240		25 KW		3.64	84	300	780	89.10
	250		Diesel engine, 20 KW		3.25	59	190	575	64
	260		50 KW		3.50	130	330	985	94
	270		100 KW		8.10	205	520	1,500	168.80
	280		250 KW		14.40	285	860	2,575	287.20
	290		Heaters, space, oil or electric, 50 MBH		.99	14	40	125	15.90
	300		100 MBH		1.24	20	56	160	21.10
	310		300 MBH		1.70	32	95	280	32.60
	315		500 MBH		2.15	46	135	405	44.20
	320		Hose, water, suction with coupling, 20' long, 2" diameter			7	17	45	3.40
	321		3" diameter			9	23	65	4.60
	322		4" diameter			12	33	100	6.60
	323		6" diameter			25	65	170	13
	324		8" diameter			34	83	255	16.60
	325		Discharge hose with coupling, 50' long, 2" diameter			7	17	45	3.40
	326		3" diameter			9	23	60	4.60
	327		4" diameter			12	35	90	7
	328		6" diameter			24	62	160	12.40
	329		8" diameter			40	87	180	17.40
	330		Ladders, extension type, 16' to 36' long			9	27	65	5.40
	340		40' to 60' long			18	56	165	11.20
	350		Light towers, towable, with generator, 1000 watt		1.28	78	200	610	50.25
	360		2000 watt		1.93	110	335	1,000	82.45
	410		Pump, centrifugal gas pump, 1-1/2"-4 MGPH		.42	16	48	135	12.95
	420		2"-8 MGPH		.46	22	64	190	16.50
	430		3"-15 MGPH		1.01	27	80	225	24.10
	440		6"-90 MGPH		1.63	80	240	700	61.05
	450		Submersible electric pump, 1-1/4"-55 GPM		.14	18	50	150	11.10
	460		1-1/2"-83 GPM		.22	22	67	185	15.15
	470		2"-120 GPM		.34	31	85	240	19.70
	480		3"-300 GPM		.60	41	115	350	27.80
	490		4"-560 GPM		1.60	51	150	425	42.80
	500		6"-1590 GPM		2.05	135	400	1,200	96.40
	510		Diaphragm pump, gas, single, 1-1/2" diameter		.44	14	40	121	11.50
	520		2" diameter		.58	23	71	200	18.85
	530		3" diameter		.91	29	99	245	27.10
	540		Double, 4" diameter		1.13	49	150	440	39.05
	550		Trash pump, self-priming, gas, 2" diameter		.43	23	68	210	17.05
	560		Diesel, 4" diameter		1.23	58	180	550	45.85
	565		Diesel, 6" diameter		2.60	89	280	815	76.80
	570		Salamanders, L.P. gas fired, 100,000 B.T.U.		.50	14	35	110	11
	580		Saw, chain, gas engine, 18" long		.50	26	93	260	22.60
	590		36" long		.65	31	115	325	28.20

For expanded coverage of these items see *Building Construction Cost Data 1985*

1.5 Contractor Equipment

1

15

		UNIT	HOURLY OPER. COST	RENT PER DAY	RENT PER WEEK	RENT PER MONTH	CREWS EQUIPMENT COST
595	60" long	Ea.	1.10	38	145	400	37.80
600	Masonry, table mounted, 14" diameter, 5 HP		.79	31	110	220	28.30
610	Circular, hand held, electric, 7" diameter		.15	11	29	85	7
620	12" diameter		.16	20	72	185	15.70
641	Toilet, portable chemical				22	70	4.40
642	Recycle flush type				27	85	5.40
643	Toilet, fresh water flush, garden hose,					130	8.65
644	Hoisted, non-flush, for high rise					70	4.65
645	Toilet trailers, minimum					60	4
646	Maximum					95	6.30
650	Trailers, platform, flush deck, 2 axle, 25 ton capacity		1	82	285	1,000	65
660	40 ton capacity		1.20	120	420	1,475	93.60
670	3 axle, 50 ton capacity		2.30	150	525	1,825	123.40
680	75 ton capacity		2.90	220	770	2,700	177.20
710	Truck, pickup, 3/4 ton, 2 wheel drive		4.90	37	120	375	63.20
720	4 wheel drive		5.05	48	175	520	75.40
730	Tractor, 4 x 2, 30 ton capacity, 195 HP		9.10	210	740	2,525	220.80
741	250 H.P.		13.50	235	820	2,900	272
770	Welder, electric, 200 amp		.52	16	50	145	14.15
780	300 amp		.67	20	64	175	18.15
790	Gas engine, 200 amp		3.08	35	100	300	44.65
800	300 amp		4.30	37	110	320	56.40
810	Wheelbarrow, any size			8	22	52	4.40
20-001	**LIFTING & HOISTING EQUIPMENT RENTAL**						
010	without operators unless noted by*						
020	Crane, climbing, 106 ft. jib, 6000 lb. capacity, 410 FPM	Ea.	12.60			7,750	617.45
030	101 ft. jib, 10,250 lb. capacity, 270 FPM	"	16.60			10,000	799.45
040	Tower, static, 130' high, 106 ft. jib,						
050	6200 lb. capacity at 400 FPM	Ea.	12.60			8,300	654.10
065	Crawler, cable, 1/2 C.Y., 15 tons at 12' radius		12.15	660*	900	2,700	277.20
070	3/4 C.Y., 20 tons at 12' radius		12.60	810*	1,225	3,800	345.80
080	1 C.Y., 25 tons at 12' radius		13.40	895*	1,400	4,375	387.20
090	1-1/2 C.Y., 40 tons at 12' radius		15.95	935*	1,775	5,625	482.60
100	2 C.Y., 50 tons at 12' radius		19.95	1,225*	2,400	6,275	639.60
110	3 C.Y., 75 tons at 12' radius		25	1,375*	2,700	7,100	740
120	100 ton capacity, standard boom		32	1,075	2,925	7,950	841
130	165 ton capacity, standard boom		46	1,250	4,100	12,200	1,188
140	200 ton capacity, 150' boom		52	1,750	5,425	16,200	1,501
150	450' boom		53	1,950	6,100	18,400	1,644
160	Truck mounted, cable operated, 6 x 4, 20 tons at 10' radius		9.80	780*	1,125	3,375	303.40
170	25 tons at 10' radius		11.90	940*	1,475	4,375	390.20
180	8 x 4, 30 tons at 10' radius		14.60	1,000*	1,600	4,850	436.80
190	40 tons at 12' radius		14.75	1,050*	1,825	5,500	483
200	8 x 4, 60 tons at 15' radius		19.40	1,100*	2,000	6,075	555.20
205	82 tons at 15' radius		21	1,150*	2,150	6,500	598
210	90 tons at 15' radius		25	1,425*	2,950	8,875	790
220	115 tons at 15' radius		28	1,575*	3,350	10,100	894
230	150 tons at 18' radius		34	1,700*	3,750	11,200	1,022
235	165 tons at 18' radius		35	1,750*	3,925	11,700	1,065
240	Truck mounted, hydraulic, 12 ton capacity		8.70	695*	825	2,500	234.60
250	25 ton capacity		13	880*	1,400	4,375	384
255	33 ton capacity		13.50	925*	1,475	4,625	403
260	55 ton capacity		18.35	1,075*	1,900	5,900	526.80
270	80 ton capacity		22	1,425*	3,300	10,700	836
280	Self-propelled, 4 x 4, with telescoping boom, 5 ton		5.35	535*	665	1,900	175.80
290	12-1/2 ton capacity		9.35	440*	725	2,075	219.80
300	15 ton capacity		9.50	480*	875	2,725	251
310	25 ton capacity		11.75	770*	1,425	4,250	379

For expanded coverage of these items see *Building Construction Cost Data 1985*

1.5 Contractor Equipment

			UNIT	HOURLY OPER. COST	RENT PER DAY	RENT PER WEEK	RENT PER MONTH	CREWS EQUIPMENT COST
20 320		Derricks, guy, 20 ton capacity, 60' boom, 75' mast	Ea.	3.80	135	410	1,225	112.40
330		100' boom, 115' mast		4.32	175	515	1,525	137.55
340		Stiffleg, 20 ton capacity, 70' boom, 37' mast		3.14	180	535	1,625	132.10
350		100' boom, 47' mast		3.67	220	660	2,000	161.35
355		Helicopter, small, lift to 1250 lbs. maximum		180	1,875*			1,440
357								
360		Hoists, chain type, overhead, manual, 3/4 ton	Ea.		9	28	83	5.60
390		10 ton			18	59	160	11.80
400		Hoist and tower, 4000 lb. capacity, portable, 40' high		3			1,050	94
410		For each added 10' section, add					52	3.45
420		Hoist and single tubular tower, 5000 lb., 100' high		3.69			1,650	139.50
430		For each added 6'-6" section, add					44	2.90
440		Hoist and double tubular tower, 5000 lb., 100' high		3.14			1,975	156.80
450		For each added 6'-6" section, add					86	5.70
455		Hoist and tower, mast type, 6000 lb., 100' high		5.05			1,575	145.40
457		For each added 10' section, add					59	3.90
460	C10.3 -300	Hoist and tower, personnel, electric, 2000 lb., 100' @ 125 FPM		3.75			2,325	185
470		3000 lb., 100' @ 200 FPM		4.23			2,550	203.85
480		3000 lb., 150' @ 300 FPM		4.66			2,650	213.95
490		4000 lb., 100' @ 300 FPM		4.93			3,625	281.10
500		6000 lb., 100' @ 275 FPM		5.85			3,725	295.10
510		For added heights up to 500', add	L.F.				6	.40
520		Jacks, hydraulic, 20 ton	Ea.		10	29	70	5.80
550		100 ton	"		15	53	150	10.60
600		Jacks, hydraulic, climbing with 50' jackrods						
601		and control consoles, minimum 3 mo. rental						
610		30 ton capacity	Ea.				700	46.65
615		For each added 10' jackrod section, add					17.30	1.15
630		50 ton capacity					1,300	86.65
635		For each added 10' jackrod section, add					24	1.60
650		125 ton capacity					3,675	245
655		For each added 10' jackrod section, add					180	12
660		Cable jack, 10 ton capacity with 200' cable					635	42.30
665		For each added 50' of cable, add					37	2.45

		UNIT	HOURLY OPER. COST	RENT 1ST MO.	RENT 2ND MO.	RENT 3RD MO.
25-001	**WELLPOINT EQUIPMENT RENTAL** See also division 2.3-55					
002						
010	Combination jetting & wellpoint pump, 60 H.P. diesel	Ea.	2.50	2,125	1,475	1,200
020	High pressure gas jet pump, 200 H.P., 300 psi	"	6.95	1,550	1,400	1,100
030	Discharge pipe, 8" diameter	L.F.		3.51	1.46	1.13
035	12" diameter			4.60	2.27	1.73
040	Header pipe, flows up to 150 G.P.M., 4" diameter			1.89	1.08	.81
050	400 G.P.M., 6" diameter			2.92	1.18	.86
060	800 G.P.M., 8" diameter			4.60	2.27	1.51
070	1500 G.P.M., 10" diameter			4.76	2.43	2.16
080	2500 G.P.M., 12" diameter			5.30	4.21	2.33
090	4500 G.P.M., 16" diameter			9.20	5.72	4
095	For quick coupling aluminum and plastic pipe, add			10%	10%	10%
110	Wellpoint, 25' long, with fittings & riser pipe, 1-1/2" or 2" diameter	Ea.		25	17.85	14.05
120	Wellpoint pump, diesel powered, 4" diameter, 20 H.P.		2.45	1,750	1,200	760
130	6" diameter, 30 H.P.		3.59	2,000	1,300	1,200
140	8" suction, 40 H.P.		3.92	2,350	1,400	1,275
150	10" suction, 75 H.P.		4.72	2,600	1,600	1,450
160	12" suction, 100 H.P.		5.85	3,550	2,350	2,150
170	12" suction, 175 H.P.		6.15	4,200	2,350	2,150

For expanded coverage of these items see *Building Construction Cost Data 1985*

2.1 Exploration & Clearing

		CREW	DAILY OUTPUT	UNIT	BARE COSTS EQUIP.	BARE COSTS LABOR	BARE COSTS TOTAL	TOTAL INCL O&P
20-001	**CORE DRILLING** Reinforced concrete slab, up to 6″ thick slab							
002								
010	1″ diameter core	B-89	50	Ea.		18.90	18.90	22
015	Each added inch thick, add		600			1.57	1.57	1.82
030	3″ diameter core		42			22	22	26
035	Each added inch thick, add		500			1.89	1.89	2.18
050	4″ diameter core		34			28	28	32
055	Each added inch thick, add		400			2.36	2.36	2.73
070	6″ diameter core		27			35	35	40
075	Each added inch thick, add		300			3.15	3.15	3.64
090	8″ diameter core		21			45	45	52
095	Each added inch thick, add		200			4.72	4.72	5.45
110	10″ diameter core		15			63	63	73
115	Each added inch thick, add		150			6.30	6.30	7.30
130	12″ diameter core		10			94	94	110
135	Each added inch thick, add		90			10.50	10.50	12.15
150	14″ diameter core		7.50			125	125	145
155	Each added inch thick, add		70			13.50	13.50	15.60
170	18″ diameter core		6.50			145	145	170
175	Each added inch thick, add		60			15.75	15.75	18.20
176	For horizontal holes, add to above						30%	30%
177	Prestressed hollow core plank, 6″ thick							
178	1″ diameter core	B-89	110	Ea.		8.60	8.60	9.90
179	Each added inch thick, add		2,750			.34	.34	.40
180	3″ diameter core		90			10.50	10.50	12.15
181	Each added inch thick, add		2,500			.38	.38	.44
182	4″ diameter core		70			13.50	13.50	15.60
183	Each added inch thick, add		2,200			.43	.43	.50
184	6″ diameter core		60			15.75	15.75	18.20
185	Each added inch thick, add		2,000			.47	.47	.55
186	8″ diameter core		50			18.90	18.90	22
187	Each added inch thick, add		1,700			.56	.56	.64
188	10″ diameter core		40			24	24	27
189	Each added inch thick, add		1,500			.63	.63	.73
190	12″ diameter core		30			31	31	36
191	Each added inch thick, add		1,000			.94	.94	1.09
195	Minimum charge for above, 3″ diameter core		9			105	105	120
200	4″ diameter core		8.50			110	110	130
205	6″ diameter core		7.75			120	120	140
210	8″ diameter core		7			135	135	155
215	10″ diameter core		6			155	155	180
220	12″ diameter core		5			190	190	220
225	14″ diameter core		4.33			220	220	250
230	18″ diameter core		4			235	235	275
32-001	**DUMP CHARGES** Typical urban city, fees only							
010	Building construction materials			C.Y.				3.49
020	Demolition lumber, trees, brush							6
030	Rubbish only							3.49
100	Reclamation station, usual charge			Ton				16
45-001	**SITE DEMOLITION** Abandon catch basin or manhole	B-6	7	Ea.	22	57	79	105
002	Remove existing catch basin or manhole		4		38	100	138	185
003	Catch basin or manhole frames & covers stored		13		11.70	31	42.70	57
004	Remove and reset		7		22	57	79	105
090	Hydrants, fire, remove only	2 Plum	4.70			76	76	110L
095	Remove and reset	"	1.40			255	255	365L
170	Pavement, with power equipment, bituminous roads	B-38	690	S.Y.	.52	.58	1.10	1.41
180	Bituminous driveways	"	680	"	.53	.59	1.12	1.44

For expanded coverage of these items see *Means' Site Work Cost Data 1985*

2.1 Exploration & Clearing

		CREW	DAILY OUTPUT	UNIT	BARE COSTS EQUIP.	LABOR	TOTAL	TOTAL INCL O&P
45 190	Concrete to 6" thick, mesh reinforced	B-38	255	S.Y.	1.42	1.57	2.99	3.83
200	Rod reinforced		200	"	1.81	2	3.81	4.88
210	Concrete 7" to 24" thick, plain		13.10	C.Y.	28	31	59	75
220	Reinforced	↓	9.50	"	38	42	80	105
230	With hand held air equipment, bituminous	B-39	1,900	S.F.	.07	.41	.48	.68
232	Concrete to 6" thick, no reinforcing		1,200		.12	.66	.78	1.08
234	Mesh reinforced		830		.17	.95	1.12	1.56
236	Rod reinforced	↓	765	↓	.18	1.03	1.21	1.69
400	Sidewalks, bituminous, 2-1/2" thick	B-6	325	S.Y.	.47	1.23	1.70	2.30
405	Brick, set in mortar		185		.82	2.16	2.98	4.03
410	Concrete, plain		160		.95	2.50	3.45	4.66
420	Mesh reinforced	↓	150	↓	1.01	2.67	3.68	4.97
60-001	**TORCH CUTTING** Steel, 1" thick plate	A-1	95	L.F.	.37	1.31	1.68	2.29
010	1" diameter bar	"	210	Ea.	.17	.59	.76	1.04
050	Oxygen lance cutting reinforced concrete walls							
055	12" to 16" thick walls	1 Clab	10	L.F.		12.40	12.40	17.90L
060	24" thick walls	"	6	"		21	21	30L

2

2.2 Building Demolition

		CREW	DAILY OUTPUT	UNIT	BARE COSTS EQUIP.	LABOR	TOTAL	TOTAL INCL O&P
02-001	**ASBESTOS CONTROL METHODS**							
300	Encapsulation, with sealants							
350	Pipes, 12" diameter or less, minimum	B-2	3,076	L.F.		.21	.21	.30
360	Maximum	"	1,538	"		.41	.41	.60
400	Dust protection, polyethylene sheeting, 6 mil thick, 4" duct tape	1 Carp	800	S.F.		.20	.20	.39
48-001	**HVAC DEMOLITION**							
002								
010	Air conditioner, split unit, 3 ton	Q-5	2	Ea.		160	160	230L
015	Package unit, 3 ton	Q-6	3			165	165	240L
030	Boiler, electric	Q-19	2			250	250	360L
034	Gas or oil, steel, under 150 MBH	Q-6	3			165	165	240L
038	Over 150 MBH	"	2	↓		250	250	360L
100	Ductwork, 4" high, 8" wide	1 Clab	200	L.F.		.62	.62	.90L
102	10" wide		190			.65	.65	.94L
104	14" wide		180			.69	.69	.99L
110	6" high, 8" wide		165			.75	.75	1.09L
112	12" wide		150			.83	.83	1.19L
114	18" wide		135			.92	.92	1.33L
120	10" high, 12" wide		125			.99	.99	1.43L
122	18" wide		115			1.08	1.08	1.56L
124	24" wide		110			1.13	1.13	1.63L
130	12" high, 18" wide		85			1.46	1.46	2.11L
132	24" wide		75			1.65	1.65	2.39L
134	48" wide		71			1.75	1.75	2.52L
140	18" high, 24" wide		67			1.85	1.85	2.67L
142	36" wide		63			1.97	1.97	2.84L
144	48" wide		59			2.10	2.10	3.03L
150	30" high, 36" wide		56			2.21	2.21	3.20L
152	48" wide		53			2.34	2.34	3.38L
154	72" wide	↓	50	↓		2.48	2.48	3.58L
155	Duct heater, electric strip	1 Elec	8	Ea.		22	22	32L
220	Furnace, electric	Q-20	2			205	205	295L
230	Gas or oil, under 120 MBH	Q-9	4	↓		80	80	115L

2.2 Building Demolition

	2.2 Building Demolition	CREW	DAILY OUTPUT	UNIT	BARE COSTS EQUIP.	BARE COSTS LABOR	BARE COSTS TOTAL	TOTAL INCL O&P
48 234	Over 120 MBH	Q-9	3	Ea.		105	105	155L
280	Heat pump package unit, 3 ton	Q-5	2.40			135	135	195L
284	Split unit, 3 ton		2			160	160	230L
300	Mechanical equipment, light items		.90	Ton		355	355	515L
360	Heavy items		1.10	"		290	290	420L
370	Deduct for salvage (when applicable), minimum							50
375	Maximum							200
60-001	**PLUMBING DEMOLITION**							
002								
102	Fixtures, including 10' piping							
110	Bath tubs, cast iron	1 Plum	4	Ea.		44	44	64L
112	Fiberglass		6			30	30	43L
114	Steel		5			36	36	51L
120	Lavatory, wall hung		10			17.75	17.75	26L
122	Counter top		8			22	22	32L
130	Sink, steel or cast iron, single		8			22	22	32L
132	Double		7			25	25	37L
140	Water closet, floor mounted		8			22	22	32L
142	Wall mounted		7			25	25	37L
150	Urinal, floor mounted		4			44	44	64L
152	Wall mounted		7			25	25	37L
160	Water fountains, free standing		8			22	22	32L
162	Recessed		6			30	30	43L
200	Piping, metal, to 2" diameter		200	L.F.		.89	.89	1.28L
205	To 4" diameter		150			1.18	1.18	1.71L
210	To 8" diameter	2 Plum	100			3.55	3.55	5.15L
215	To 16" diameter	"	60			5.90	5.90	8.55L
216	Deduct for salvage, aluminum scrap			Ton				700
217	Brass scrap							600
218	Copper scrap							900
220	Lead scrap							200
222	Steel scrap							50
225	Water heater, 40 gal.	1 Plum	6	Ea.		30	30	43L
600	Remove & reset fixtures, minimum		6			30	30	43L
610	Maximum		4			44	44	64L

2.3 Earthwork

	2.3 Earthwork	CREW	DAILY OUTPUT	UNIT	BARE COSTS EQUIP.	BARE COSTS LABOR	BARE COSTS TOTAL	TOTAL INCL O&P
03-001	**BACKFILL** By hand, no compaction, light soil	1 Clab	14	C.Y.		8.85	8.85	12.80L
010	Heavy soil		12			10.35	10.35	14.90L
030	B12.3 -110 Compaction in 6" layers, hand tamp, add		20.60			6	6	8.70L
040	Roller compaction, add	B-10A	100		.72	2.24	2.96	4.03
050	Air tamp, add	B-9	190		.74	3.35	4.09	5.65
060	Vibrating plate, add	A-1	60		.59	2.07	2.66	3.63
080	Compaction in 12" layers, hand tamp, add	1 Clab	34			3.65	3.65	5.25L
090	Roller compaction, add	B-10A	150		.48	1.49	1.97	2.69
100	Air tamp, add	B-9	285		.50	2.23	2.73	3.77
110	Vibrating plate, add	A-1	90		.39	1.38	1.77	2.42
130	Dozer backfilling, bulk, up to 300' haul, no compaction	B-10B	1,200		.51	.19	.70	.83
140	Air tamped	B-11B	240		3.28	1.19	4.47	5.35
160	Compacted backfill, 6" to 12" lifts, vibrating roller	B-10C	800		.86	.28	1.14	1.35
170	Sheepsfoot roller	B-10D	750		.92	.30	1.22	1.45
190	Dozer backfilling, trench, up to 300' haul, no compaction	B-10B	900		.68	.25	.93	1.11
200	Air tamped	B-11B	235		3.35	1.22	4.57	5.45

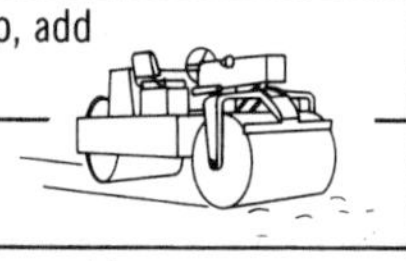

For expanded coverage of these items see *Means' Site Work Cost Data 1985*

2.3 Earthwork

		CREW	DAILY OUTPUT	UNIT	BARE COSTS EQUIP.	BARE COSTS LABOR	BARE COSTS TOTAL	TOTAL INCL O&P
03 220	Compacted backfill, 6" to 12" lifts, vibrating roller	B-10C	700	C.Y.	.98	.32	1.30	1.54
230	Sheepsfoot roller	B-10D	650		1.07	.34	1.41	1.67
10-001	**DEWATERING** Excavate drainage trench, 2' wide, 2' deep	B-11C	90		1.69	3.17	4.86	6.45
010	2' wide, 3' deep, with backhoe loader	"	100		1.52	2.86	4.38	5.80
020	Excavate sump pits by hand, light soil	1 Clab	7.10			17.45	17.45	25L
030	Heavy soil	"	3.50			35	35	51L
050	Pumping 8 hr., attended 2 hrs. per day, including 20 L.F.							
055	of suction hose & 100 L.F. discharge hose							
060	2" diaphragm pump used for 8 hours	B-10H	4	Day	7.25	56	63.25	89
065	4" diaphragm pump used for 8 hours	B-10I	4		14.90	56	70.90	97
080	8 hrs. attended, 2" diaphragm pump	B-10H	1		29	225	254	355
090	3" centrifugal pump	B-10J	1		38	225	263	365
100	4" diaphragm pump	B-10I	1		60	225	285	390
110	6" centrifugal pump	B-10K	1		99	225	324	430
					BARE COSTS MAT.	BARE COSTS INST.	BARE COSTS TOTAL	
130	Re-lay CMP, incl. excavation 3' deep, 12" diameter	B-6	115	L.F.	4.90	4.81	9.71	11.90
140	18" diameter		100	"	7.40	5.55	12.95	15.60
160	Sump hole construction, incl. excavation and gravel, pit		1,250	C.F.	.25	.44	.69	.87
170	With 12" gravel collar, 12" pipe, corrugated, 16 ga.		70	L.F.	8.20	7.90	16.10	19.70
180	15" pipe, corrugated, 16 ga.		55		10.05	10.05	20.10	25
190	18" pipe, corrugated, 16 ga.		50		11.75	11.05	22.80	28
200	24" pipe, corrugated, 14 ga.		40		17.75	13.80	31.55	38
220	Wood lining, up to 4' x 4', add		300	S.F.C.A.	.78	1.84	2.62	3.35
15-001	**DRILLING AND BLASTING** only, rock, under 1500 C.Y.	B-47	75	C.Y.	.95	11.15	12.10	15.25
010	Over 1500 C.Y.		85		.95	9.85	10.80	13.60
220	Trenches, up to 1500 C.Y.		22		.95	38	38.95	49
230	Over 1500 C.Y.		26		.95	32	32.95	42
					BARE COSTS EQUIP.	BARE COSTS LABOR	BARE COSTS TOTAL	
17-001	**EXCAVATING, STRUCTURAL** Hand, pits to 6' deep, sandy soil	1 Clab	8	C.Y.		15.50	15.50	22L
010	Heavy soil or clay		4			31	31	45L
030	Pits 6' to 12' deep, sandy soil		5			25	25	36L
050	Heavy soil or clay		3			41	41	60L
070	Pits 12' to 18' deep, sandy soil		4			31	31	45L
090	Heavy soil or clay		2			62	62	90L
150	For wet or muck hand excavation, add to above				50%	50%		
18-001	**EXCAVATING, TRENCH** or continuous footing, 4' wide							
002	bottom with banks, sloped 1/2 to 1, damp sandy loam							
003	B12.3 -110 3' deep, 3/8 C.Y. tractor backhoe = 245 L.F./day	B-11C	150	C.Y.	1.01	1.90	2.91	3.87
004	1/2 C.Y. tractor backhoe = 325 L.F./day	B-11M	200		1.14	1.43	2.57	3.32
005	4' deep, 3/8 C.Y. tractor backhoe = 170 L.F./day	B-11C	150		1.01	1.90	2.91	3.87
006	1/2 C.Y. tractor backhoe = 225 L.F./day	B-11M	200		1.14	1.43	2.57	3.32
007	5' deep, 3/8 C.Y. tractor backhoe = 125 L.F./day	B-11C	150		1.01	1.90	2.91	3.87
008	1/2 C.Y. tractor backhoe = 165 L.F./day	B-11M	200		1.14	1.43	2.57	3.32
009	6' deep, 1/2 C.Y. tractor backhoe = 130 L.F./day	"	200		1.14	1.43	2.57	3.32
010	5/8 C.Y. hydraulic backhoe = 160 L.F./day	B-12Q	250		1.07	1.21	2.28	2.93
020	8' deep, 1/2 C.Y. tractor backhoe = 85 L.F./day	B-11M	200		1.14	1.43	2.57	3.32
030	1/2 C.Y. Gradall = 85 L.F./day	B-12J	200		2.16	1.51	3.67	4.56
040	3/4 C.Y. hydraulic backhoe = 120 L.F./day	B-12F	300		1.08	1.01	2.09	2.64
050	10' deep, 1 C.Y. hydraulic backhoe = 120 L.F./day	B-12A	400		1.01	.76	1.77	2.20
060	1 C.Y. Gradall = 120 L.F./day	B-12K	400		1.55	.76	2.31	2.80
070	12' deep, 1 C.Y. hydraulic backhoe = 90 L.F./day	B-12A	400		1.01	.76	1.77	2.20
080	1-1/4 C.Y. hydraulic backhoe = 115 L.F./day	B-12R	500		1.12	.60	1.72	2.11
090	14' deep, 1-1/4 C.Y. hydraulic backhoe = 90 L.F./day	"	500		1.12	.60	1.72	2.11
100	1-1/2 C.Y. hydraulic backhoe = 105 L.F./day	B-12B	600		.93	.50	1.43	1.76
110	16' deep, 2 C.Y. hydraulic backhoe = 115 L.F./day	B-12C	800		.97	.38	1.35	1.61
120	18' deep, 2-1/2 C.Y. hydraulic backhoe = 115 L.F./day	B-12S	1,000		1.14	.30	1.44	1.69
130	20' deep, 3-1/2 C.Y. hydraulic backhoe = 135 L.F./day	B-12D	1,400		.99	.22	1.21	1.40

For expanded coverage of these items see *Means' Site Work Cost Data 1985*

2.3 Earthwork

		CREW	DAILY OUTPUT	UNIT	BARE COSTS EQUIP.	BARE COSTS LABOR	BARE COSTS TOTAL	TOTAL INCL O&P
18	140	By hand with pick and shovel to 6' deep, light soil	1 Clab	8	C.Y.		15.50	15.50	22L
	150	Heavy soil	"	4			31	31	45L
	170	For tamping backfilled trenches, air tamp, add	A-1	100		.35	1.24	1.59	2.18
	190	Vibrating plate, add		90	↓	.39	1.38	1.77	2.42
	210	Trim sides and bottom for concrete pours, regular		600	S.F.	.06	.21	.27	.36
	230	Hardpan	↓	180	"	.20	.69	.89	1.21
	19-001	**EXCAVATING, UTILITY TRENCH** Sandy clay soil							
	002								
	005	Trenching with chain trencher, 12 H.P., operator walking							
	010	4" wide trench, 12" deep	B-53	800	L.F.	.17	.19	.36	.46
	015	18" deep		750		.18	.20	.38	.49
	020	24" deep		700		.19	.22	.41	.52
	030	6" wide trench, 12" deep		650		.21	.23	.44	.56
	035	18" deep		600		.22	.25	.47	.61
	040	24" deep		550		.24	.28	.52	.67
	045	36" deep		450		.30	.34	.64	.82
	060	8" wide trench, 12" deep		475		.28	.32	.60	.77
	065	18" deep		400		.33	.38	.71	.92
	070	24" deep		350		.38	.44	.82	1.05
	075	36" deep	↓	300	↓	.44	.51	.95	1.22
	100	Backfill by hand including compaction, add							
	105	4" wide trench, 12" deep	A-1	800	L.F.	.04	.16	.20	.27
	110	18" deep		530		.07	.23	.30	.41
	115	24" deep		400		.09	.31	.40	.54
	130	6" wide trench, 12" deep		540		.07	.23	.30	.40
	135	18" deep		405		.09	.31	.40	.54
	140	24" deep		270		.13	.46	.59	.81
	145	36" deep		180		.20	.69	.89	1.21
	160	8" wide trench, 12" deep		400		.09	.31	.40	.54
	165	18" deep		265		.13	.47	.60	.82
	170	24" deep		200		.18	.62	.80	1.09
	175	36" deep	↓	135	↓	.26	.92	1.18	1.61
	200	Chain trencher, 40 H.P. operator riding							
	205	6" wide trench and backfill, 12" deep	B-54	1,200	L.F.	.16	.13	.29	.36
	210	18" deep		1,000		.20	.15	.35	.44
	215	24" deep		975		.20	.16	.36	.45
	220	36" deep		900		.22	.17	.39	.49
	225	48" deep		750		.26	.20	.46	.58
	230	60" deep		650		.30	.23	.53	.67
	240	8" wide trench and backfill, 12" deep		1,000		.20	.15	.35	.44
	245	18" deep		950		.21	.16	.37	.46
	250	24" deep		900		.22	.17	.39	.49
	255	36" deep		800		.25	.19	.44	.55
	260	48" deep		650		.30	.23	.53	.67
	270	12" wide trench and backfill, 12" deep		975		.20	.16	.36	.45
	275	18" deep		860		.23	.18	.41	.51
	280	24" deep		800		.25	.19	.44	.55
	285	36" deep		725		.27	.21	.48	.60
	300	16" wide trench and backfill, 12" deep		900		.22	.17	.39	.49
	305	18" deep		750		.26	.20	.46	.58
	310	24" deep	↓	700	↓	.28	.22	.50	.63
	320	Compaction with vibratory plate, add				50%	50%		
	20-001	**FILL** Spread dumped material, no compaction, by dozer	B-10B	1,000	C.Y.	.61	.22	.83	1
	010	By hand	1 Clab	12	"		10.35	10.35	14.90L

		CREW	DAILY OUTPUT	UNIT	BARE COSTS MAT.	BARE COSTS INST.	BARE COSTS TOTAL	TOTAL INCL O&P
050	Gravel fill,compacted, under floor slabs, 3" deep	B-14	10,000	S.F.	.06	.09	.15	.20
060	6" deep (gravel at $4.05 ton)		9,000		.12	.10	.22	.28
070	9" deep		7,200		.18	.13	.31	.38
080	12" deep	↓	6,000	↓	.24	.16	.40	.48

For expanded coverage of these items see *Means' Site Work Cost Data 1985*

2.3 Earthwork		CREW	DAILY OUTPUT	UNIT	BARE COSTS MAT.	BARE COSTS INST.	BARE COSTS TOTAL	TOTAL INCL O&P
20 100	Alternate pricing method, 3" deep	B-14	90	C.Y.	6.50	10.45	16.95	22
110	6" deep		160		6.50	5.90	12.40	15.30
120	9" deep		200		6.50	4.70	11.20	13.70
130	12" deep		220		6.50	4.28	10.78	13.10
					BARE COSTS EQUIP.	BARE COSTS LABOR	BARE COSTS TOTAL	
30-001	**HAULING** Earth 6 C.Y. dump truck 1/4 mile round trip, 5.0 loads/hr.	B-34A	240	C.Y.	.92	.54	1.46	1.78
003	1/2 mile round trip, 4.1 loads/hr.		197		1.12	.66	1.78	2.17
004	1 mile round trip, 3.3 loads/hr.		160		1.37	.81	2.18	2.67
010	2 mile round trip, 2.6 loads/hr.		125		1.76	1.04	2.80	3.42
015	3 mile round trip, 2.1 loads/hr.		100		2.20	1.30	3.50	4.27
020	4 mile round trip, 1.8 loads/hr.		85		2.59	1.52	4.11	5.05
030	C10.3-300 C.Y. dump truck, 1 mile round trip, 2.7 loads/hr.	B-34B	260		1.07	.50	1.57	1.90
040	2 mile round trip, 2.2 loads/hr.		210		1.33	.62	1.95	2.35
045	3 mile round trip, 1.9 loads/hr.		180		1.55	.72	2.27	2.74
050	4 mile round trip, 1.6 loads/hr.		150		1.86	.86	2.72	3.28
130	Hauling in medium traffic, add					20%	20%	20%
140	Heavy traffic, add					30%	30%	30%
160	Grading at dump, if required, by dozer	B-10B	1,000		.61	.22	.83	1
180	Spotter at dump or cut, if required	1 Clab	8	Hr.		15.50	15.50	22L
					BARE COSTS MAT.	BARE COSTS INST.	BARE COSTS TOTAL	
32-001	**HORIZONTAL BORING** Casing only, 100' minimum,							
002	not incl. jacking pits or dewatering							
010	Roadwork, 1/2" thick wall, 24" diameter casing	B-42	10	L.F.	35	205	240	300
020	36" diameter		9.50		50	215	265	330
030	48" diameter		9		85	225	310	385
050	Railroad work, 24" diameter		7		35	290	325	415
060	36" diameter		6.50		50	310	360	460
070	48" diameter		6		85	340	425	530
090	For ledge, add						125	155
38-001	**SHEET PILING** Steel, not incl. wales, 15' excav., left in place	B-40	10.80	Ton	580	255	835	965
010	Pull & salvage, 22 psf, 15' excavation		7.20		145	380	525	650
030	C12.1-100 Deep excavation, 27 psf, left in place		12.90		580	210	790	915
040	Pull & salvage		8.60		145	320	465	570
120	15' deep excavation, 22 psf, left in place		985	S.F.	6.40	2.78	9.18	10.65
130	Pull & salvage		655		1.60	4.18	5.78	7.15
150	20' deep excavation, 27 psf, left in place		960		7.85	2.85	10.70	12.35
160	Pull & salvage		640		1.95	4.28	6.23	7.70
210	Rent sheet piling and wales, first month			Ton	150		150	165M
220	Per added month				14		14	15.40M
230	Rental piling left in place, add to rental				460		460	505M
250	Wales, connections & struts, 2/3 salvage				150		150	165M
390	Wood, solid sheeting, incl. wales, braces and spacers,							
391	pull & salvage, 8' deep excavation	B-31	330	S.F.	.80	2.40	3.20	4.22
400	10' deep, 50 S.F./hr. in & 150 S.F./hr. out		300		.90	2.64	3.54	4.66
410	12' deep, 45 S.F./hr. in & 135 S.F./hr. out		270		1	2.93	3.93	5.20
420	14' deep, 42 S.F./hr. in & 126 S.F./hr. out		250		1.25	3.17	4.42	5.80
430	16' deep, 40 S.F./hr. in & 120 S.F./hr. out		240		1.35	3.30	4.65	6.05
440	18' deep, 38 S.F./hr. in & 114 S.F./hr. out		230		1.35	3.44	4.79	6.25
452	Left in place, 8' deep, 55 S.F./hr.		440		1.55	1.80	3.35	4.21
454	10' deep, 50 S.F./hr.		400		1.75	1.98	3.73	4.68
456	12' deep, 45 S.F./hr.		360		1.95	2.20	4.15	5.20
457	16' deep, 40 S.F./hr.		320		2.75	2.47	5.22	6.45
470	Alternate pricing, left in place, 8' deep		1.76	M.B.F.	390	450	840	1,050
480	Pull and salvage, 8' deep		1.32	"	195	600	795	1,050
40-001	**SOLDIER BEAMS & LAGGING** H piles with 3" wood sheeting							
002	horizontal between piles, including removal of wales & braces							

2.3 Earthwork

		CREW	DAILY OUTPUT	UNIT	BARE COSTS MAT.	BARE COSTS INST.	BARE COSTS TOTAL	TOTAL INCL O&P
40 010	No hydrostatic head, 15' deep, 1 line of braces , minimum	B-50	545	S.F.	4.90	5.75	10.65	13.35
020	Maximum		495		5.40	6.35	11.75	14.70
130	No hydrostatic head, left in place, 15' deep, 1 line of braces, min.		635		6.90	4.94	11.84	14.40
140	Maximum		575		7.60	5.45	13.05	15.90
235	Lagging only, 3" thick wood between piles 8' O.C., minimum	B-46	400		1.30	2.22	3.52	4.76
237	Maximum		250		1.50	3.56	5.06	7
240	Open sheeting for trenches to 10' deep, minimum		1,736		.36	.51	.87	1.16
245	Maximum		1,510		.42	.59	1.01	1.34
250	Tie-back method of sheeting, add, minimum						20%	20%
255	Maximum						60%	60%
270	Tie-backs only, based on tie-backs total length, minimum	B-46	86.80	L.F.	7.50	10.25	17.75	24
275	Maximum		38.50	"	16.50	23	39.50	53
350	Tie-backs only, typical average, 25' long		2	Ea.	320	445	765	1,025
360	35' long		1.58	"	410	565	975	1,300
					BARE COSTS EQUIP.	BARE COSTS LABOR	BARE COSTS TOTAL	
55-001	**WELLPOINTS** For wellpoint equipment rental, see division 1.5-25							
010	Installation and removal of single stage system							
011	Labor only, .75 man-hours per L.F., minimum	1 Clab	10.70	L.F.Hdr.		11.60	11.60	16.75L
020	2.0 man-hours per L.F., maximum	"	4	"		31	31	45L
040	Pump operation, 4 @ 6 hr. shifts							
041	Per 24 hour day	4 Eqlt	1.27	Day		480	480	695L
050	Per 168 hour week, 160 hr. straight, 8 hr. double time		.18	Week		3,375	3,375	4,900L
055	Per 4.3 week month		.04	Month		15,200	15,200	22,000L
060	[C12.1 -200] Complete installation, operation, equipment rental, fuel &							
061	removal of system with 2" wellpoints 5' O.C.							
070	100' long header, 6" diameter, first month			L.F.Hdr.	83	185	268	360L
080	Thereafter, per month				59	145	204	275L
100	200' long header, 8" diameter, first month				49	105	154	210L
110	Thereafter, per month				34	72	106	140L
130	500' long header, 8" diameter, first month				26	57	83	115L
140	Thereafter, per month				17	29	46	61L
160	1,000' long header, 10" diameter, first month				19	52	71	96L
170	Thereafter, per month				12	14	26	33L
190	Note: above figures include pumping 168 hrs. per week							
191	and include the pump operator and one stand-by pump.							
					BARE COSTS MAT.	BARE COSTS INST.	BARE COSTS TOTAL	
57-001	**WELLS** Dewatering excavations 10' to 20' deep, 2' diameter							
002	with steel casing, minimum	B-6	165	V.L.F.	1.50	3.35	4.85	6.15
005	Average		98		3	5.65	8.65	10.90
010	Maximum		49		8	11.30	19.30	24
030	For pumps for above, see division 1.5-15-410							
050	For other types of wells, see division 2.5-40							

2.5 Site Drainage & Utilities

		CREW	DAILY OUTPUT	UNIT	BARE COSTS MAT.	BARE COSTS INST.	BARE COSTS TOTAL	TOTAL INCL O&P
02-001	**CATCH BASINS OR MANHOLES** Including footing & excavation,							
002	not including frame and cover							
005	Brick, 4' inside diameter, 4' deep	D-1	1	Ea.	300	285	585	735
010	6' deep		.70		425	410	835	1,050
015	8' deep		.50		540	570	1,110	1,400
020	For depths over 8', add		4	V.L.F.	70	71	141	180
040	Concrete blocks (radial), 4' I.D., 4' deep		1.50	Ea.	195	190	385	485
050	6' deep		1	"	260	285	545	690

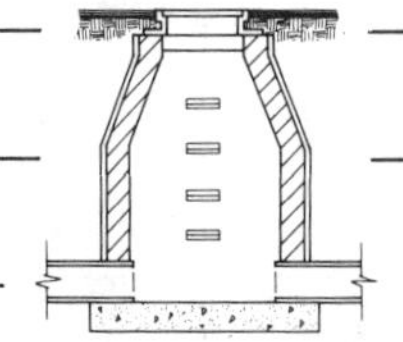

	2.5 Site Drainage & Utilities	CREW	DAILY OUTPUT	UNIT	BARE COSTS MAT.	BARE COSTS INST.	BARE COSTS TOTAL	TOTAL INCL O&P
02 060	8' deep	D-1	.70	Ea.	330	410	740	940
070	For depths over 8', add	"	5.50	V.L.F.	50	52	102	130
080	Concrete, cast in place, 8" thick, 4' deep	B-6	2	Ea.	270	275	545	670
090	6' deep		1.50		355	370	725	890
100	8' deep		1		470	555	1,025	1,275
110	For depths over 8', add		8	V.L.F.	60	69	129	160
111	Precast 4', I.D., riser @ $27.75 per V.L.F., 4' deep		4.10	Ea.	210	135	345	415
112	6' deep		3		290	185	475	570
113	8' deep		2		365	275	640	775
114	For depths over 8', add		16	V.L.F.	50	35	85	100
115	5' I.D. riser @ $46.75 per V.L.F., 4' deep		3	Ea.	315	185	500	595
116	6' deep		2		435	275	710	850
117	8' deep		1.50		550	370	920	1,100
118	For depths over 8', add		12	V.L.F.	70	46	116	140
119	6' I.D. riser @ $69.75 per V.L.F., 4' deep		2	Ea.	520	275	795	945
120	6' deep		1.50		675	370	1,045	1,250
121	8' deep		1		840	555	1,395	1,675
122	For depths over 8', add		8	V.L.F.	110	69	179	215
125	Slab tops, precast, 8" thick							
130	4' diameter manhole	B-6	8	Ea.	75	69	144	175
140	5' diameter manhole		7.50		85	74	159	195
150	6' diameter manhole		7		150	79	229	270
160	Frames and covers, 24" square, 500 lb.		7.80		140	71	211	250
170	26" D shape, 600 lb.		7		155	79	234	275
180	Light traffic, 18" diameter, 100 lb.		10		50	55	105	130
190	24" diameter, 300 lb.		8.70		110	64	174	205
200	36" diameter, 900 lb.		5.80		280	95	375	435
210	Heavy traffic, 24" diameter, 400 lb.		7.80		135	71	206	245
220	36" diameter, 1150 lb.		3		380	185	565	665
230	Mass. State standard, 26" diameter, 475 lb.		7		150	79	229	270
240	30" diameter, 620 lb.		7		190	79	269	315
250	Watertight, 24" diameter, 350 lb.		7.80		225	71	296	345
260	26" diameter, 500 lb.		7		265	79	344	400
270	32" diameter, 575 lb.		6		280	92	372	430
280	3 piece cover & frame, 10" deep,							
290	1200 lbs., for heavy equipment	B-6	3	Ea.	375	185	560	660
300	Raised for paving 1-1/4" to 2" high,							
310	4 piece expansion ring							
320	20" to 26" diameter	1 Clab	3	Ea.	75	41	116	140
330	30" to 36" diameter	"	3		95	41	136	165
332	Frames and covers, existing, raised for paving 2", including							
334	row of brick, concrete collar, up to 12" wide frame	B-9	18		21	43	64	83
336	20" to 26" wide frame		11		26	71	97	125
338	30" to 36" wide frame		9		37	86	123	160
340	Inverts, single channel brick	D-1	3		42	95	137	180
350	Concrete		5		32	57	89	115
360	Triple channel, brick		2		57	145	202	265
370	Concrete		3		36	95	131	175
380	Steps, heavyweight cast iron, 7" x 9"	1 Bric	40		5.50	4.01	9.51	11.75
390	8" x 9"		40		8.25	4.01	12.26	14.75
400	Standard sizes, galvanized steel		40		8.25	4.01	12.26	14.75
410	Aluminum		40		7.50	4.01	11.51	13.95
07-001	**EXCAVATION AND BACKFILL** See division 2.3-18							
010	Hand excavate and trim for bells after trench excavation							
020	B12.3 -110 8" pipe	1 Clab	155	L.F.		.80	.80	1.16L
030	18" pipe	"	130	"		.95	.95	1.38L
08-001	**GAS SERVICE AND DISTRIBUTION** Not including excavation							
005	or backfill.							
010	Polyethylene, 60 psi, coils, 5/8" diameter, SDR 7	B-20	450	L.F.	.12	1.10	1.22	1.72
015	1-1/4" diameter, SDR 10	"	400	"	.49	1.24	1.73	2.33

For expanded coverage of these items see *Means' Site Work Cost Data 1985*

2.5 Site Drainage & Utilities

	2.5 Site Drainage & Utilities	CREW	DAILY OUTPUT	UNIT	BARE COSTS MAT.	BARE COSTS INST.	BARE COSTS TOTAL	TOTAL INCL O&P
08 020	2" diameter, SDR 11	B-20	360	L.F.	.80	1.38	2.18	2.87
025	3" diameter, SDR 11.5	"	300		1.69	1.65	3.34	4.24
030	40' joints with coupling, 3" diameter, SDR 11.5	B-21	300		1.81	2.22	4.03	5.10
035	4" diameter, SDR 11		260		3	2.56	5.56	6.90
040	6" diameter, SDR 21		240		3.35	2.77	6.12	7.55
045	8" diameter, SDR 21		200		6.35	3.33	9.68	11.65
050	Steel, schedule 40, plain end, tar coated & wrapped							
055	1" diameter	Q-4	300	L.F.	2.20	2.32	4.52	5.75
060	2" diameter		280		3.50	2.49	5.99	7.45
065	3" diameter		260		6.30	2.68	8.98	10.80
070	4" diameter	B-35	255		8.90	5.20	14.10	16.80
075	5" diameter		220		22	6	28	32
080	6" diameter		180		23	7.35	30.35	35
085	8" diameter		140		29	9.45	38.45	45
10-001	**FILL** In trench, crushed bank run	B-6	260	C.Y.	5.75	2.13	7.88	9.20
010	Screened 3/4" to 1/2"	"	260	"	7	2.13	9.13	10.55
12-001	**LINING PIPE** With cement, including bypass & cleaning, 6" to 12" pipe	C-17E	160	L.F.	5.20	10.55	15.75	21
020	16" to 20" pipe		145		5.70	11.65	17.35	23
025	24" to 36" pipe		130		6.65	12.95	19.60	26
030	48" to 72" pipe for transmission line		110		10.50	15.35	25.85	34
27-001	**PIPE, DRAINAGE AND SEWAGE** Excavation & backfill							
002	not included							
010	Asbestos Cement, class 2400, 6" diameter	B-20	330	L.F.	3	1.50	4.50	5.45
012	8" diameter	B-21	380		3.90	1.75	5.65	6.75
014	10" diameter		330		5.25	2.02	7.27	8.60
016	12" diameter		280		6.65	2.38	9.03	10.65
018	14" diameter		245		7.65	2.72	10.37	12.20
020	16" diameter		210		9.45	3.17	12.62	14.85
030	Class 4000, truckload lots, 10" diameter		320		4.70	2.08	6.78	8.10
032	12" diameter		265		5.95	2.51	8.46	10.05
034	14" diameter		235		8	2.83	10.83	12.75
036	16" diameter		200		10.20	3.33	13.53	15.90
038	18" diameter		150		11.95	4.44	16.39	19.35
040	24" diameter	B-22	115		19.40	6.55	25.95	30
042	30" diameter		80		27	9.40	36.40	43
044	36" diameter		65		36	11.55	47.55	56
090	Bituminous fiber, plain, 2" diameter	2 Plum	400		.84	.89	1.73	2.21
092	3" diameter		400		1.05	.89	1.94	2.44
094	4" diameter		380		1.25	.93	2.18	2.73
096	5" diameter		360		2.10	.99	3.09	3.74
098	6" diameter		340		2.50	1.04	3.54	4.26
100	8" diameter		300		5.95	1.18	7.13	8.25
130	Concrete, non-reinforced, extra strength, B&S or T&G joints							
132	6" diameter	B-20	265	L.F.	2.25	1.87	4.12	5.20
134	8" diameter	B-21	300		2.45	2.22	4.67	5.80
136	10" diameter		260		2.80	2.56	5.36	6.65
138	12" diameter		198		3.55	3.36	6.91	8.60
140	15" diameter		165		4.20	4.03	8.23	10.25
142	18" diameter		130		5.40	5.10	10.50	13.10
144	21" diameter		105		6.75	6.35	13.10	16.30
146	24" diameter		90		8.35	7.40	15.75	19.55
160	Reinforced Concrete Culvert, class 3, no gaskets							
162	12" diameter, 106#/L.F.	B-21	190	L.F.	4.25	3.50	7.75	9.60
164	15" diameter, 150#/L.F.		155		5.30	4.29	9.59	11.85
166	18" diameter, 200#/L.F.		122		6.70	5.45	12.15	15
168	24" diameter, 325#/L.F.		88		10.50	7.55	18.05	22
170	30" diameter, 385#/L.F.	B-13	80		15	16.55	31.55	39
172	36" diameter, 525#/L.F.		60		22	22	44	54
174	42" diameter, 685#/L.F.		55		28	24	52	63
176	48" diameter, 865#/L.F.		50		36	26	62	75

For expanded coverage of these items see *Means' Site Work Cost Data 1985*

2.5 Site Drainage & Utilities

27

		CREW	DAILY OUTPUT	UNIT	BARE COSTS MAT.	BARE COSTS INST.	BARE COSTS TOTAL	TOTAL INCL O&P
178	60" diameter, 1070#/L.F.	B-13	35	L.F.	58	38	96	115
180	72" diameter, 1880#/L.F.		30		80	44	124	145
182	84" diameter, 2500#/L.F.		24		120	55	175	205
184	96" diameter, 3000#/L.F.		20		145	66	211	250
190	With gaskets, class 3, 12" diameter	B-21	210		5.15	3.17	8.32	10.10
192	15" diameter		175		6.20	3.80	10	12.15
195	18" diameter		150		7.75	4.44	12.19	14.75
197	24" diameter		100		12	6.65	18.65	23
200	30" diameter	B-13	85		16.50	15.55	32.05	39
203	36" diameter		70		24	18.90	42.90	52
205	48" diameter		58		38	23	61	72
210	72" diameter		38		100	35	135	155
211	Flared ends, 6'-1" long, 12" diameter	B-21	190		28	3.50	31.50	36
212	15" diameter		155		32	4.29	36.29	41
213	6'-2" long, 18" diameter		122		38	5.45	43.45	49
214	24" diameter		88		45	7.55	52.55	60
215	30" diameter	B-13	80		56	16.55	72.55	84
216	36" diameter	"	60		80	22	102	120
240	Vitrified plate lined, add to above, 30" to 36" diameter			S.F.C.A.	2.50		2.50	2.75M
245	42" to 54" diameter, add				2.30		2.30	2.53M
250	60" to 72" diameter, add				2.10		2.10	2.31M
255	Over 72" diameter, add				2		2	2.20M
260	Radius pipe, add to pipe prices, 12" to 60" diameter			L.F.	50%	20%		
265	Over 60" diameter, add			"	20%	20%		
270	Reinforced elliptical, 8' lengths, C507 class 3							
275	14" x 23" inside, round equivalent 18" diameter	B-21	82	L.F.	28	8.10	36.10	42
280	24" x 38" inside, round equivalent 30" diameter	B-22	58		38	12.95	50.95	60
285	29" x 45" inside, round equivalent 36" diameter		52		47	14.45	61.45	72
290	38" x 60" inside, round equivalent 48" diameter		38		68	19.75	87.75	100
295	48" x 76" inside, round equivalent 60" diameter		26		100	29	129	150
300	58" x 91" inside, round equivalent 72" diameter		22		145	34	179	205
325								
350	Corrugated Metal Pipe, galvanized or aluminum, bituminous							
351	coated with paved invert, 20' to 30' lengths							
355	8" diameter 16 ga.	B-20	330	L.F.	4.35	1.50	5.85	6.95
360	10" diameter 16 ga.		260		5.50	1.90	7.40	8.80
365	12" diameter 16 ga.		210		6.45	2.36	8.81	10.50
370	15" diameter 16 ga.	B-21	210		8.10	3.17	11.27	13.35
375	18" diameter 16 ga.		190		9.75	3.50	13.25	15.65
380	24" diameter 14 ga.		160		14.90	4.16	19.06	22
385	30" diameter 14 ga.		120		18.70	5.55	24.25	28
390	36" diameter 12 ga.	B-13	120		28	11	39	46
395	48" diameter 12 ga.		100		37	13.20	50.20	58
400	60" diameter 10 ga.		75		59	17.65	76.65	89
405	72" diameter 8 ga.		45		83	29	112	130
410	Corrugated Metal Pipe, galvanized or aluminum, plain							
413	8" diameter 16 ga.	B-20	355	L.F.	3.55	1.39	4.94	5.90
415	10" diameter 16 ga.		280		4.50	1.77	6.27	7.50
417	12" diameter 16 ga.		220		5.20	2.25	7.45	8.95
420	15" diameter 16 ga.	B-21	220		6.45	3.03	9.48	11.35
423	18" diameter 16 ga.		205		7.75	3.25	11	13.05
425	24" diameter 14 ga.		175		12.55	3.80	16.35	19.15
427	30" diameter 14 ga.		130		15.65	5.10	20.75	24
430	36" diameter 12 ga.	B-13	130		24	10.15	34.15	40
433	48" diameter 12 ga.		110		33	12	45	52
435	60" diameter 10 ga.		78		53	16.95	69.95	81
450	Corrugated Metal Pipe, end sections, 18" diameter	B-21	16	Ea.	75	42	117	140
455	30" diameter	"	12	"	150	55	205	245
465	Corrugated steel or alum. oval arch culverts, coated & paved							
470	17" x 11" 16 ga., 15" equivalent	B-22	200	L.F.	8.50	3.75	12.25	14.55

For expanded coverage of these items see *Means' Site Work Cost Data 1985*

2.5 Site Drainage & Utilities

		CREW	DAILY OUTPUT	UNIT	BARE COSTS MAT.	BARE COSTS INST.	BARE COSTS TOTAL	TOTAL INCL O&P
27 475	21" x 15" 16 ga., 18" equivalent	B-22	150	L.F.	9.95	5	14.95	17.90
480	28" x 20" 14 ga., 24" equivalent		125		15.35	6	21.35	25
485	35" x 24" 14 ga., 30" equivalent		100		19.25	7.50	26.75	32
490	42" x 29" 12 ga., 36" equivalent	B-13	100		28	13.20	41.20	49
495	49" x 33" 12 ga., 42" equivalent		90		34	14.70	48.70	57
500	57" x 38" 12 ga., 48" equivalent		70		38	18.90	56.90	67
510	Corrugated steel or alum. oval arch culverts, plain							
513	17" x 13" 16 ga., 15" equivalent	B-22	225	L.F.	6.70	3.34	10.04	12
515	21" x 15" 16 ga., 18" equivalent		175		8.10	4.29	12.39	14.85
517	28" x 20" 14 ga., 24" equivalent		150		13	5	18	21
520	35" x 24" 14 ga., 30" equivalent		108		16.25	6.95	23.20	28
523	42" x 29" 12 ga., 36" equivalent	B-13	108		25	12.25	37.25	44
525	49" x 33" 12 ga., 42" equivalent		92		30	14.35	44.35	52
527	57" x 38" 12 ga., 48" equivalent		75		33	17.65	50.65	60
580	End sections, 17" x 13"		22	Ea.	50	60	110	135
585	42" x 29"		17	"	150	78	228	270
595	Multi-plate arch, steel	B-20	1,690	Lb.	.52	.29	.81	1
605	Polyvinyl chloride pipe, 13' lengths, S.D.R. 35, 4" diameter		375	L.F.	.65	1.32	1.97	2.62
615	6" diameter		350		1.25	1.41	2.66	3.42
625	8" diameter		335		2.20	1.48	3.68	4.56
630	10" diameter	B-21	330		3.60	2.02	5.62	6.80
635	12" diameter		320		5	2.08	7.08	8.40
640	15" diameter		190		7.85	3.50	11.35	13.55
650	Vitrified clay sewer pipe, premium joint, C-200, 4" diameter	B-20	265		1.60	1.87	3.47	4.46
655	4' and 5' lengths, 6" diameter	"	200		2.50	2.48	4.98	6.35
660	8" diameter	B-21	200		3.50	3.33	6.83	8.50
665	10" diameter		190		5	3.50	8.50	10.40
670	12" diameter		150		7	4.44	11.44	13.90
675	15" diameter		110		13	6.05	19.05	23
680	18" diameter		88		19	7.55	26.55	31
685	24" diameter		45		40	14.80	54.80	65
690	30" diameter	B-22	31		44	24	68	82
695	36" diameter	"	20		60	38	98	120
705	3' lengths, add to above				30%	30%		
710	2' lengths, add to above				40%	60%		
715	For die cast slip joints compared to premium joints				85%	105%		
720	For plain joints compared to premium joints				75%	105%		
725	For standard strength compared to extra strength				80%	90%		
29-001	**SUBDRAINAGE** Foundation underdrains							
005	Asbestos Cement, class 4000 underdrain, perforated							
010	4" diameter	B-20	390	L.F.	1.80	1.27	3.07	3.81
015	6" diameter	"	380		2.85	1.30	4.15	5
020	8" diameter	B-21	370		4.50	1.80	6.30	7.45
025	10" diameter		365		5.55	1.82	7.37	8.65
030	12" diameter		360		7.65	1.85	9.50	11
035	Unperforated, 4" diameter	B-20	390		1.25	1.27	2.52	3.21
040	6" diameter	"	380		2.40	1.30	3.70	4.52
045	8" diameter	B-21	370		3.25	1.80	5.05	6.10
050	10" diameter		365		4.40	1.82	6.22	7.40
055	12" diameter		360		5.75	1.85	7.60	8.90
060	Bituminous fiber, perforated underdrain, 3" diameter	2 Plum	400		1.05	.89	1.94	2.44
065	4" diameter		380		1.20	.93	2.13	2.67
070	5" diameter		360		2.10	.99	3.09	3.74
075	6" diameter		340		2.60	1.04	3.64	4.37
080	Corrugated steel or aluminum perforated, asphalt coated							
085	6" diameter, 18 ga.	B-20	380	L.F.	2.50	1.30	3.80	4.63
090	8" diameter, 16 ga.	"	370		3.90	1.34	5.24	6.20
095	10" diameter, 16 ga.	B-21	360		4.90	1.85	6.75	8
100	12" diameter, 16 ga.		285		5.90	2.34	8.24	9.75
105	18" diameter, 16 ga.		205		8.90	3.25	12.15	14.35

For expanded coverage of these items see *Means' Site Work Cost Data 1985*

2.5 Site Drainage & Utilities

	2.5 Site Drainage & Utilities	CREW	DAILY OUTPUT	UNIT	BARE COSTS MAT.	BARE COSTS INST.	BARE COSTS TOTAL	TOTAL INCL O&P
29 110	Plain steel or aluminum, corrugated							
115	6" diameter, 18 ga.	B-20	380	L.F.	2.15	1.30	3.45	4.25
120	8" diameter, 16 ga.	"	370		3.30	1.34	4.64	5.55
125	10" diameter, 16 ga.	B-21	360		4.25	1.85	6.10	7.25
130	12" diameter, 16 ga.		285		5.25	2.34	7.59	9.05
135	18" diameter, 16 ga.		205		7.25	3.25	10.50	12.50
140	Porous wall concrete underdrain, std. strength, 4" diameter	B-20	335		1.30	1.48	2.78	3.57
145	6" diameter	"	315		1.40	1.57	2.97	3.81
150	8" diameter	B-21	310		2.25	2.15	4.40	5.50
155	12" diameter		285		4.80	2.34	7.14	8.55
160	15" diameter		230		5.50	2.89	8.39	10.10
165	18" diameter		165		7.20	4.03	11.23	13.55
170	Extra strength, 6" diameter	B-20	315		1.70	1.57	3.27	4.14
175	8" diameter	B-21	310		2.55	2.15	4.70	5.80
180	10" diameter		285		5	2.34	7.34	8.75
185	12" diameter		230		5.40	2.89	8.29	10
190	15" diameter		200		6.10	3.33	9.43	11.35
195	18" diameter		165		8.90	4.03	12.93	15.45
200	Vitrified clay, perforated, 2' lengths (C-211), 4" diameter	B-20	400		1.80	1.24	3.04	3.77
205	6" diameter		315		3.05	1.57	4.62	5.65
210	8" diameter		290		5.25	1.71	6.96	8.25
215	12" diameter	B-21	275		7.35	2.42	9.77	11.45
220	Channel pipe, 4" diameter	B-20	430		1.30	1.15	2.45	3.09
225	6" diameter		335		1.75	1.48	3.23	4.06
230	8" diameter		295		3.20	1.68	4.88	5.95
235	12" diameter	B-21	280		6.20	2.38	8.58	10.15
320	PVC, perforated, 4" diameter	B-20	430		.55	1.15	1.70	2.27
330	6" diameter	"	400		1.20	1.24	2.44	3.11
30-001	**PIPING, WATER DISTRIBUTION SYSTEMS** Pipe laid in trench,							
002	excavation and backfill not included							
010	Asbestos cement pipe, 150 psi, C.L. lots, 4" diameter	B-20	300	L.F.	2.90	1.65	4.55	5.60
015	6" diameter	"	290		4.15	1.71	5.86	7.05
020	8" diameter	B-21	290		5.65	2.30	7.95	9.45
025	10" diameter		270		8.65	2.47	11.12	12.95
030	12" diameter		250		11	2.66	13.66	15.85
035	14" diameter		200		14.15	3.33	17.48	20
040	16" diameter		180		17.55	3.70	21.25	24
045	18" diameter		150		19.40	4.44	23.84	28
060	For 100 psi pipe, 8" diameter, deduct				10%			
070	16" diameter, deduct				15%			
090	For 200 psi pipe, 8" diameter, add				15%			
100	16" diameter, add				20%			
140	Ductile Iron pipe, class 250 water piping, 18' lengths							
141	Mechanical joint, 4" diameter	B-20	144	L.F.	5	3.44	8.44	10.45
142	6" diameter	"	126		5.50	3.93	9.43	11.75
143	8" diameter	B-21	108		7.50	6.15	13.65	16.90
144	10" diameter		90		10	7.40	17.40	21
145	12" diameter		72		12.75	9.25	22	27
146	14" diameter		54		16.50	12.35	28.85	35
147	16" diameter		46		19.20	14.45	33.65	41
148	18" diameter	B-22	38		22	19.75	41.75	52
149	24" diameter	"	36		33	21	54	65
155	Tyton joint, 4" diameter	B-20	158		4.50	3.13	7.63	9.50
156	6" diameter	"	138		5.10	3.59	8.69	10.80
157	8" diameter	B-21	118		7.05	5.65	12.70	15.65
158	10" diameter		100		9.50	6.65	16.15	19.75
159	12" diameter		80		12	8.30	20.30	25
160	14" diameter		60		15	11.10	26.10	32
161	16" diameter		54		17.75	12.35	30.10	37
162	18" diameter	B-22	44		21	17.05	38.05	47

For expanded coverage of these items see *Means' Site Work Cost Data 1985*

2

2.5 Site Drainage & Utilities

		CREW	DAILY OUTPUT	UNIT	BARE COSTS MAT.	BARE COSTS INST.	BARE COSTS TOTAL	TOTAL INCL O&P
30 163	20" diameter	B-22	42	L.F.	23	17.90	40.90	50
164	24" diameter	"	40		29	18.75	47.75	58
170	Copper tubing, type K, 20' joints, 1-1/2" diameter	B-20	360		2.55	1.38	3.93	4.79
171	2" diameter		315		3.98	1.57	5.55	6.65
172	2-1/2" diameter		220		5.70	2.25	7.95	9.50
173	3" diameter		200		7.65	2.48	10.13	12
175	Gate valves with boxes, cast iron							
178	4" diameter	B-20	6	Ea.	250	83	333	395
180	6" diameter	"	5		320	99	419	495
182	8" diameter	B-21	4		485	165	650	765
184	10" diameter		3.50		750	190	940	1,100
186	12" diameter		3		900	220	1,120	1,300
188	14" diameter		2		2,125	335	2,460	2,800
190	16" diameter		2		2,850	335	3,185	3,600
195	Butterfly valves with boxes, cast iron							
197	4" diameter	B-20	6	Ea.	195	83	278	335
199	6" diameter	"	5		200	99	299	365
201	8" diameter	B-21	4		400	165	565	675
203	10" diameter		3.50		575	190	765	900
205	12" diameter		3		600	220	820	970
207	14" diameter		2		1,200	335	1,535	1,775
209	16" diameter		2		1,400	335	1,735	2,000
215	Hydrant, fire, no excavation or backfill included	2 Plum	1.70		780	210	990	1,150
220	Wall, bronze, non-freeze type, 3/4" diam. for 12" wall		11		120	32	152	180
225	For 18" wall		11		145	32	177	205
230	For 24" wall		11		160	32	192	225
235	Box or post type, 3/4" diameter		4		180	89	269	325
240	Polyvinyl chloride pipe, class 150, S.D.R.-18, 4" diameter	B-20	200	L.F.	2.93	2.48	5.41	6.80
245	6" diameter	"	180		5.15	2.75	7.90	9.65
250	8" diameter	B-21	160		7.90	4.16	12.06	14.50
255	10" diameter		140		11.70	4.76	16.46	19.55
260	12" diameter		100		15.50	6.65	22.15	26
265	Class 160, S.D.R.-26, 1-1/2" diameter	B-20	300		.40	1.65	2.05	2.83
270	2" diameter		250		.52	1.98	2.50	3.43
275	2-1/2" diameter		250		.75	1.98	2.73	3.69
280	3" diameter		200		1.08	2.48	3.56	4.77
285	4" diameter		200		1.76	2.48	4.24	5.50
290	6" diameter		180		2.50	2.75	5.25	6.75
295	8" diameter	B-21	160		5.50	4.16	9.66	11.85
800	Fittings, ductile iron, mechanical joint							
801	90° bend 4" diameter	B-20	37	Ea.	65	13.40	78.40	91
802	6" diameter		25		80	19.80	99.80	115
804	8" diameter		21		115	24	139	160
806	10" diameter	B-21	21		165	32	197	225
808	12" diameter		18		220	37	257	295
810	14" diameter		16		375	42	417	470
812	16" diameter		14		425	48	473	535
814	18" diameter		10		720	67	787	885
816	20" diameter		8		850	83	933	1,050
818	24" diameter		6		1,200	110	1,310	1,475
820	Wye or tee, 4" diameter	B-20	25		90	19.80	109.80	130
822	6" diameter		17		120	29	149	175
824	8" diameter		14		170	35	205	240
826	10" diameter	B-21	14		265	48	313	360
828	12" diameter		12		350	55	405	465
830	14" diameter		10		560	67	627	710
832	16" diameter		8		665	83	748	850
834	18" diameter		6		1,000	110	1,110	1,250
836	20" diameter		4		1,200	165	1,365	1,550
838	24" diameter		3		1,800	220	2,020	2,300

For expanded coverage of these items see *Means' Site Work Cost Data 1985*

2.5 Site Drainage & Utilities

		CREW	DAILY OUTPUT	UNIT	BARE COSTS MAT.	BARE COSTS INST.	BARE COSTS TOTAL	TOTAL INCL O&P
32-001	**SEPTIC TANKS** Not incl. excav. or piping, precast, 1,000 gallon	B-21	8	Ea.	275	83	358	420
010	2,000 gallon		5		530	135	665	770
020	5,000 gallon		1.70		2,200	390	2,590	2,975
030	15,000 gallon	B-13	1.30		7,500	1,025	8,525	9,625
040	25,000 gallon		.80		11,000	1,650	12,650	14,300
050	40,000 gallon		.60		16,500	2,200	18,700	21,100
060	Fiberglass, 1,000 gallon	B-21	6		385	110	495	580
070	1,500 gallon	"	4		490	165	655	770
100	Distribution boxes, concrete, 5 outlets	2 Plum	16		25	22	47	60
110	12 outlets	"	8		175	44	219	255
115	Leaching field chambers, 13' x 3'-7" x 1'-4", standard	B-13	16		170	83	253	300
120	Heavy duty, 8' x 4' x 1'-6"		14		130	94	224	270
130	13' x 3'-9" x 1'-6"		12		200	110	310	370
135	20' x 4' x 1'-6"		5		450	265	715	850
140	Leaching pit, precast concrete, 3' pit	B-21	8		170	83	253	305
150	6' pit		4.70		220	140	360	440
200	Velocity reducing pit, precast conc., 6' diameter, 3' deep		4.70		165	140	305	380
210								
220	Excavation for septic tank, 3/4 C.Y. backhoe	B-12F	145	C.Y.		4.31	4.31	5.45
240	4' trench for disposal field, 3/4 C.Y. backhoe	"	335	L.F.		1.87	1.87	2.37
260	Gravel fill, run of the bank	B-6	150	C.Y.	3	3.68	6.68	8.25
280	Crushed stone, 3/4"	"	150	"	7	3.68	10.68	12.65
35-001	**SEWAGE PUMPING STATIONS** Prefabricated steel, concrete							
002	or fiberglass, 200 GPM	C-17D	.17	Total	21,000	12,200	33,200	40,000
020	1,000 GPM		.07		31,500	29,600	61,100	75,500
050	Add for generator unit, 200 GPM, steel		.34		16,000	6,100	22,100	26,000
060	Concrete		.51		10,000	4,075	14,075	16,600
100	Add for generator unit, 1,000 GPM, steel		.30		17,000	6,925	23,925	28,300
120	Concrete		.38		14,000	5,450	19,450	22,900
150	For wet well, if required, add	B-23	.50		4,200	1,700	5,900	6,925
37-001	**SEWAGE TREATMENT** Not incl. fencing or external piping							
002	Steel packaged, blown air aeration plants							
010	1,000 GPD	C-17D	.69	Total	6,300	3,000	9,300	11,100
020	5,000 GPD	"	.19		19,000	10,900	29,900	36,000
030	15,000 GPD	C-14	.20		32,000	18,200	50,200	60,000
040	30,000 GPD		.10		60,000	36,500	96,500	116,000
050	50,000 GPD		.08		70,000	45,600	115,600	139,500
060	100,000 GPD		.04		120,000	91,000	211,000	257,000
070	200,000 GPD		.03		175,000	121,500	296,500	359,500
080	500,000 GPD		.01		450,000	365,000	815,000	996,000
100	Concrete, extended aeration, primary and secondary treatment							
101	10,000 GPD	C-14	.10	Total	31,500	36,500	68,000	84,500
110	30,000 GPD		.07		40,000	52,000	92,000	115,500
120	50,000 GPD		.04		70,000	91,000	161,000	202,000
140	100,000 GPD		.03		8,500	121,500	206,500	260,500
150	500,000 GPD		.01		400,000	365,000	765,000	941,000
170	Municipal wastewater treatment facility							
172	1.0 MGD (million gallons per day)			Total	1,100,000	1,260,000	2,360,000	3,150,000
174	1.5 MGD (million gallons per day)				1,600,000	1,900,000	3,500,000	4,700,000
176	2.0 MGD (million gallons per day)				1,900,000	2,100,000	4,000,000	5,400,000
178	3.0 MGD (million gallons per day)				2,200,000	2,500,000	4,700,000	6,300,000
180	5.0 MGD (million gallons per day)				3,400,000	3,700,000	7,100,000	9,450,000
200	Holding tank system, not incl. excavation or backfill							
201	Recirculating chemical water closet	2 Plum	4	Ea.	495	89	584	675
210	For voltage converter, add	"	16		130	22	152	175
220	For high level alarm, add	1 Plum	7.80		65	23	88	105
39-001	**WASTEWATER TREATMENT SYSTEM** Fiberglass, 1,000 gallon	B-21	1.29		2,100	515	2,615	3,025
010	1,500 gallon	"	1.03		4,600	645	5,245	5,975
40-001	**WELLS** Domestic water, drilled and cased, including casing							
010	4" to 6" diameter	B-23	160	V.L.F.	7	5.30	12.30	14.90

For expanded coverage of these items see *Means' Site Work Cost Data 1985*

2

2.5 Site Drainage & Utilities

		CREW	DAILY OUTPUT	UNIT	BARE COSTS MAT.	INST.	TOTAL	TOTAL INCL O&P
40 020	8" diameter	B-23	127	V.L.F.	8.75	6.65	15.40	18.70
040	Gravel pack well, 40' deep, incl. gravel & casing, complete							
050	24" diameter casing x 18" diameter screen	B-23	.13	Total	16,000	6,500	22,500	26,400
060	36" diameter casing x 18" diameter screen	↓	.12	"	18,000	7,050	25,050	29,400
080	Observation wells, 1-1/4" riser pipe	↓	163	V.L.F.	8.75	5.20	13.95	16.70
090	For flush Buffalo roadway box, add	1 Skwk	16.60	Ea.	23	9.70	32.70	39
120	Test well, 2-1/2" diameter, up to 50' deep (15 to 50 GPM)	B-23	1.51	"	400	560	960	1,200
130	Over 50' deep, add	"	121.80	L.F.	11.50	6.95	18.45	22
150	Pumps, installed in wells to 100' deep, 4" submersible							
151	1/2 H.P., 216 to 1020 GPH	B-21	1.80	Ea.	350	370	720	905
152	3/4 H.P., 144 to 1110 GPH		1.65		400	405	805	1,000
160	To 180' deep, 1 H.P., 102 to 1356 GPH		1.10		450	605	1,055	1,350
170	1-1/2 H.P., 330 to 1410 GPH		1		500	665	1,165	1,475
180	2 H.P., 840 to 1440 GPH		1		650	665	1,315	1,650
190	3 H.P., 1098 to 1464 GPH		.90		800	740	1,540	1,925
200	5 H.P., 1302 to 1494 GPH		.90		1,000	740	1,740	2,125
300	6" submersible, 25' to 150' deep, 25 H.P., 249 to 297 GPM		.70		2,000	950	2,950	3,525
310	25' to 500' deep, 30 H.P., 100 to 300 GPM	↓	.50	↓	2,500	1,325	3,825	4,625

2.6 Roads & Walks

		CREW	DAILY OUTPUT	UNIT	BARE COSTS MAT.	INST.	TOTAL	TOTAL INCL O&P
10-001	**BITUMINOUS** Paving, wearing course, @ $26.50 per ton, 1-1/2" thick	B-25	3,300	S.Y.	2.16	.62	2.78	3.19
010	3" thick	"	1,650		4.32	1.24	5.56	6.40
140	Patching utility trenches, 2" thick asphalt	B-37	390		3	2.24	5.24	6.45
150	3" thick asphalt	"	260	↓	4.50	3.35	7.85	9.70
12-001	**BITUMINOUS CONCRETE** At the plant (145 lb. per C.F.)			Ton	26.50		26.50	29M
020	All weather patching mix				27.50		27.50	30M
030	Berm mix			↓	27.50		27.50	30M

2.7 Site Improvements

		CREW	DAILY OUTPUT	UNIT	BARE COSTS MAT.	INST.	TOTAL	TOTAL INCL O&P
15-001	**FOUNTAINS** Incl. fiberglass pools, pumps, piping and lights							
020	4' diameter pool, 18" diameter spray ring	Q-1	2	Ea.	525	160	685	810
030	6' diameter pool, 24" diameter spray ring		1.50		860	215	1,075	1,250
040	7.5' diameter pool, 48" diameter spray ring		1		1,300	320	1,620	1,900
050	Rain curtains, 3' rain bar, 2' x 4' x 1' pool		2		495	160	655	775
060	7' rain bar, 2' x 8' x 1' pool	↓	1	↓	1,075	320	1,395	1,650
37-001	**SPRINKLER SYSTEM** For lawns							
010	Golf courses with fully automatic system	C-17	.05	9 Holes	65,000	32,800	97,800	119,000
020	24' diam. head at 15' O.C incl. piping, minimum	B-20	70	Head	13.75	7.05	20.80	25
030	Maximum		40		30.50	12.40	42.90	51
050	60' diameter head, automatic operation, minimum		28		44	17.70	61.70	74
060	Maximum		23	↓	114	22	136	155
080	Residential system, custom, 1" supply		2,619	S.F.Grnd	.19	.19	.38	.48
090	1-1/2" supply	↓	2,311	"	.17	.21	.38	.50

For expanded coverage of these items see *Means' Site Work Cost Data 1985*

3.3 Cast in Place Concrete

	3.3 Cast in Place Concrete	CREW	DAILY OUTPUT	UNIT	BARE COSTS MAT.	BARE COSTS INST.	BARE COSTS TOTAL	TOTAL INCL O&P
12-001	**CONCRETE, READY MIX** Regular weight, 2000 psi			C.Y.	43		43	47M
010	2500 psi				44.60		44.60	49M
015	3000 psi				46.15		46.15	51M
020	3500 psi				47.70		47.70	52M
025	3750 psi				48.60		48.60	53M
030	4000 psi				49.25		49.25	54M
035	4500 psi				51.10		51.10	56M
040	5000 psi				52.05		52.05	57M
100	For high early strength cement, add				10%			
200	For all lightweight aggregate, add			↓	50%			
37-001	**PATCHING CONCRETE**							
002								
010	Floors, 1/4" thick, small areas, regular grout	1 Cefi	170	S.F.	.70	.88	1.58	2.01
015	Epoxy grout	"	100	"	2.60	1.50	4.10	4.97
200	Walls, including chipping, cleaning and epoxy grout							
210	Minimum	1 Cefi	65	S.F.	.11	2.31	2.42	3.37
215	Average		50		.17	3.01	3.18	4.41
220	Maximum	↓	40	↓	.28	3.76	4.04	5.60

3
4
5

4.1 Mortar & Masonry Accessories

	4.1 Mortar & Masonry Accessories	CREW	DAILY OUTPUT	UNIT	BARE COSTS MAT.	BARE COSTS INST.	BARE COSTS TOTAL	TOTAL INCL O&P
80-001	**VENT BOX** Extruded aluminum, 4" deep, 8" x 2-1/2"	1 Bric	30	Ea.	18	5.35	23.35	27
005	8" x 5"		25		27	6.40	33.40	39
010	16" x 2-1/2"		25		32	6.40	38.40	44
015	16" x 8"		22		45	7.30	52.30	60
020	24" x 5"		22		50	7.30	57.30	65
025	24" x 8"	↓	20		60	8	68	77
040	For baked enamel finish, add				35%			
050	For cast aluminum, painted, add				60%			
100	Stainless steel ventilators, 6" x 6"	1 Bric	25		42	6.40	48.40	55
105	8" x 8"		24		45	6.70	51.70	59
110	12" x 12"		23		46	6.95	52.95	61
115	12" x 6"		24		48	6.70	54.70	62
120	Foundation vent, galv., 1-1/4" thick, 8"H, 16"L, no damper	↓	30		5.95	5.35	11.30	14.15
125	For damper, add			↓	.60		.60	.66M

For expanded coverage of these items see Means' *Concrete & Masonry Cost Data 1985*

5.4 Misc. & Ornamental Metals

	5.4 Misc. & Ornamental Metals	CREW	DAILY OUTPUT	UNIT	BARE COSTS MAT.	BARE COSTS INST.	BARE COSTS TOTAL	TOTAL INCL O&P
56-001	**PIPE SUPPORT** Framing, under 10#/L.F.	E-4	3,900	Lb.	.55	.19	.74	.90
020	10.1 to 15#/L.F.		4,300		.48	.18	.66	.80
040	15.1 to 20#/L.F.		4,800		.42	.16	.58	.70
060	Over 20#/L.F.	↓	5,400	↓	.38	.14	.52	.63
68-001	**TOILET PARTITION SUPPORTS** Overhead framing for ceiling							
002	hung partitions, 3' wide	E-4	16	Stall	60	47	107	140

For expanded coverage of these items see *Building Construction Cost Data 1985*

5.8 Expansion Control & Fasteners

		CREW	DAILY OUTPUT	UNIT	BARE COSTS MAT.	BARE COSTS INST.	BARE COSTS TOTAL	TOTAL INCL O&P
03-001	**DRILLING** And layout for anchors, per							
005	inch of depth, concrete or brick walls.							
010	1/4" diameter	1 Carp	32	Ea.	.05	4.90	4.95	7.15
020	3/8" diameter		29		.05	5.40	5.45	7.85
030	1/2" diameter		26		.05	6.05	6.10	8.75
040	5/8" diameter		25		.05	6.25	6.30	9.10
050	3/4" diameter		23		.06	6.80	6.86	9.90
060	7/8" diameter		21		.07	7.45	7.52	10.85
070	1" diameter		20		.08	7.85	7.93	11.40
080	1-1/4" diameter		18		.09	8.70	8.79	12.70
090	1-1/2" diameter		16		.10	9.80	9.90	14.25
100	For ceiling installations add					40%		
110	Drilling & layout for drywall or plaster walls							
120	Holes, 1/4" diameter	1 Carp	150	Ea.	.05	1.05	1.10	1.56
130	3/8" diameter		140		.05	1.12	1.17	1.67
140	1/2" diameter		130		.05	1.21	1.26	1.80
150	3/4" diameter		120		.05	1.31	1.36	1.94
160	1" diameter		110		.06	1.43	1.49	2.12
170	1-1/4" diameter		100		.07	1.57	1.64	2.34
180	1-1/2" diameter		90		.08	1.74	1.82	2.60
190	For ceiling installations add					40%		
200	Minimum labor/equipment charge	1 Carp	4	Job		39	39	57L
12-001	**EXPANSION ANCHORS** Bolts & shields							
010	Bolt anchors for concrete, brick or stone, no layout and drilling							
020	Expansion shields, zinc, 1/4" diameter, 1" long, single	1 Carp	90	Ea.	.40	1.74	2.14	2.96
030	1-3/8" long, double		85		.45	1.84	2.29	3.16
040	3/8" diameter, 1-1/2" long, single		85		.63	1.84	2.47	3.36
050	2" long, double		80		.82	1.96	2.78	3.73
060	1/2" diameter, 2-1/4" long, single		80		1.05	1.96	3.01	3.99
070	2-1/2" long, double		75		1.06	2.09	3.15	4.18
080	5/8" diameter, 2-5/8" long, single		75		1.50	2.09	3.59	4.67
090	3" long, double		70		1.65	2.24	3.89	5.05
100	3/4" diameter, 2-3/4" long, single		70		2.20	2.24	4.44	5.65
110	4" long, double		65		3.05	2.41	5.46	6.85
120	7/8" diameter, 5-1/2" long, double		65		8.10	2.41	10.51	12.40
130	1" diameter, 6" long, double		60		8.80	2.61	11.41	13.45
150	Self drilling, steel, 1/4" diameter bolt		26		.45	6.05	6.50	9.20
160	3/8" diameter bolt		23		.65	6.80	7.45	10.55
170	1/2" diameter bolt		20		.95	7.85	8.80	12.35
180	5/8" diameter bolt		18		1.73	8.70	10.43	14.50
190	3/4" diameter bolt		16		2.88	9.80	12.68	17.30
200	7/8" diameter bolt		14		4.30	11.20	15.50	21
210	Hollow wall anchors for gypsum board,							
220	plaster, tile or wall board							
230	1/8" diameter, short			Ea.	.12		.12	.13M
240	Long				.13		.13	.14M
250	3/16" diameter, short				.28		.28	.30M
260	Long				.30		.30	.33M
270	1/4" diameter, short				.34		.34	.37M
280	Long				.39		.39	.42M
300	Toggle bolts, bright steel, 1/8" diameter, 2" long	1 Carp	85		.16	1.84	2	2.84
310	4" long		80		.21	1.96	2.17	3.06
320	3/16" diameter, 3" long		80		.19	1.96	2.15	3.04
330	6" long		75		.30	2.09	2.39	3.35
340	1/4" diameter, 3" long		75		.22	2.09	2.31	3.26
350	6" long		70		.33	2.24	2.57	3.60
360	3/8" diameter, 3" long		70		.52	2.24	2.76	3.81
370	6" long		60		.72	2.61	3.33	4.57
380	1/2" diameter, 4" long		60		1.50	2.61	4.11	5.40
390	6" long		50		1.96	3.14	5.10	6.70

For expanded coverage of these items see *Building Construction Cost Data 1985*

5.8 Expansion Control & Fasteners		CREW	DAILY OUTPUT	UNIT	BARE COSTS MAT.	BARE COSTS INST.	BARE COSTS TOTAL	TOTAL INCL O&P
12 500	Screw anchors for concrete, masonry,							
510	stone & tile, no layout or drilling included							
520	Jute fiber, #6, #8, & #10, 1" long			Ea.	.06		.06	.06M
530	#12, 2" long				.11		.11	.12M
540	#14, 2" long				.13		.13	.14M
550	#16, 2" long				.16		.16	.17M
560	#20, 2" long				.22		.22	.24M
570	Lag screw, 1/4" diameter, short				.21		.21	.23M
580	Long				.24		.24	.26M
590	3/8" diameter, short				.37		.37	.40M
600	Long				.42		.42	.46M
610	1/2" diameter, short				.73		.73	.80M
620	Long				.86		.86	.94M
630	3/4" diameter, short				1.17		1.17	1.28M
640	Long				1.50		1.50	1.65M
660	Lead, #6 & #8, 3/4" long				.08		.08	.08M
670	#10 & #12, 1-1/2" long				.12		.12	.13M
680	#16 & #18, 1-1/2" long				.16		.16	.17M
690	Plastic, #6 & #8, 3/4" long				.02		.02	.02M
700	#8 & #10, 7/8" long				.03		.03	.03M
710	#10 & #12, 1" long				.03		.03	.03M
720	#14 & #16, 1-1/2" long				.04		.04	.04M
750								
800	Wedge bolts, not including layout or drilling							
805	Carbon steel, 1/4" diameter, 1-3/4" long	1 Carp	150	Ea.	.29	1.05	1.34	1.83
810	3" long		145		.45	1.08	1.53	2.06
815	3/8" diameter, 2-1/4" long		150		.45	1.05	1.50	2
820	5" long		145		.71	1.08	1.79	2.34
825	1/2" diameter, 2-3/4" long		140		.72	1.12	1.84	2.41
830	7" long		130		1.35	1.21	2.56	3.23
835	5/8" diameter, 3-1/2" long		130		1.35	1.21	2.56	3.23
840	8-1/2" long		115		2.15	1.36	3.51	4.33
845	3/4" diameter, 4-1/4" long		115		1.90	1.36	3.26	4.06
850	10" long		100		3.85	1.57	5.42	6.50
855	1" diameter, 6" long		100		5.80	1.57	7.37	8.65
860	12" long		80		8.45	1.96	10.41	12.15
865	1-1/4" diameter, 9" long		70		11	2.24	13.24	15.35
870	12" long		60		12.60	2.61	15.21	17.65
875	For type 303 stainless steel, add				400%			
880	For type 316 stainless steel, add				450%			
17-001	**LAG SCREWS** 1/4" steel, diameter, 2" long	1 Carp	140	Ea.	.12	1.12	1.24	1.75
010	3/8" diameter, 3" long 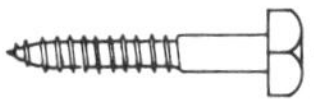		105		.24	1.49	1.73	2.42
020	1/2" diameter, 3" long		95		.38	1.65	2.03	2.80
030	5/8" diameter, 3" long		85		.75	1.84	2.59	3.49
20-001	**MACHINE SCREWS** Steel, #8, 1" long			C	2	2.84	4.84	6.45
011	#8, 2" long				3.85	5.70	9.55	12.10
020	#10 x 1" long				2.35	2.84	5.19	6.55
030	#10 x 2" long				4.20	5.70	9.90	12.25
					HEAD	ANCHOR BOLT	STUD & NUT	
22-001	**MACHINERY ANCHORS** Flush mounted heads to receive studs							
002	Material only (add 3 components for total price)							
020	For 1/2" stud, minimum			Ea.	15	2.50	2.20	22
030	Maximum				15	2.80	2.40	22
050	For 3/4" stud, minimum				19.60	3.80	3.50	30
060	Maximum				19.60	4.20	3.85	30
080	For 1" stud, minimum				34	7	5.25	51
090	Maximum				34	7.55	5.65	52
110	For 1-1/4" stud, minimum				52	10.25	6.50	77
120	Maximum				52	11	6.85	77

5

5.8 **Expansion Control & Fasteners**		CREW	DAILY OUTPUT	UNIT	HEAD	ANCHOR BOLT	STUD & NUT	TOTAL INCL O&P
22 140	For 1-1/2" stud, minimum			Ea.			13	14.30
150	Maximum			"			13.50	14.85
					BARE MAT. ONLY			
					PLAIN	GALV.	ALUM.	
25-001	**NAILS** Prices of material only, copper			Lb.	3.25			
040	Stainless steel				4.40			
060	Common 3d to 20d				.39	.48	1.85	
070	30d to 60d				.38	.47	1.85	
080	Annular or spiral thread, 4d to 60d				.62	.74		
100	Finish nails, 4d to 10d				.41	.49	1.95	
120	Drywall nails				.80	.96		
140	Flooring nails, hardened steel, 2d to 10d				.70	.84		
160	Masonry nails, hardened steel, 3/4" to 3" long				.70	.84		
180	Concrete nails, hardened steel				.70	.84		
290	Add to prices above for cement coating				.04			
310	Zinc or tin plating			↓	.10			
27-001	**RIVETS** 1/2" grip length							
010	Aluminum rivet & mandrel, 1/8" diameter			C			3.02	
020	3/8" diameter						4.65	
030	Aluminum rivet, steel mandrel, 1/8" diameter						3.20	
040	3/8" diameter						5.12	
050	Copper rivet, steel mandrel, 1/8" diameter				4			
120	Steel rivet and mandrel, 1/8" diameter				3.05			
130	3/8" diameter			↓	5.05			
140	Hand riveting tool, minimum			Ea.	28			
150	Maximum				275			
160	Power riveting tool, minimum				200			
170	Maximum			↓	775			
30-001	**SHEET METAL SCREWS** Steel, standard, #8 x 3/4"			C	2.60	3		
010	#10 x 1"				3.60	4.15		
030	With washers, #14 x 1"				6.65	7.65		
040	#14 x 3"				14.65	16.85		
060	Self-drilling, with washers, (pinch point), #8, 3/4" long				7.35	8.45		
070	#10 x 3/4" long				9.70	11		
090	Stainless steel with aluminum or neoprene washers, #14 x 1"				13.30			
100	#14 x 3" long			↓	26			
					BARE COSTS			
					MAT.	INST.	TOTAL	
32-001	**STUDS** .22 caliber stud driver, buy, minimum			Ea.	280		280	310M
010	Maximum			"	330		330	365M
030	Powder charges for above, low velocity			C	12		12	13.20M
040	Standard velocity				16		16	17.60M
060	Drive pins & studs, 1/4" & 3/8" diam., to 3" long, minimum				29		29	32M
070	Maximum			↓	48		48	53M
080	Pneumatic stud driver for 1/8" diameter studs			Ea.	800		800	880M
090	Drive pins for above, 1/2" to 3/4" long			M	34		34	37M
					BARE MAT. ONLY			
					STEEL	BRASS		
50-001	**WOOD SCREWS** #9, 1" long			C	2	9.75		
010	#9, 2" long				3.20	15.60		
020	#12, 1" long				2.60	13.70		
030	#12, 2" long				4.70	24		
040	#12, 3" long				6.30			
050	#14, 2" long				7.45	38		
060	#14, 3" long				8.70			
070	#14, 4" long			↓	14.50			

For expanded coverage of these items see *Building Construction Cost Data 1985*

6.1 Rough Carpentry

		CREW	DAILY OUTPUT	UNIT	BARE COSTS MAT.	BARE COSTS INST.	BARE COSTS TOTAL	TOTAL INCL O&P
70-001	**SHEATHING** Plywood on roof, CDX							
003	5/16" thick	F-2	1,600	S.F.	.26	.20	.46	.58
005	3/8" thick		1,525		.29	.21	.50	.63
010	1/2" thick		1,400		.37	.23	.60	.74
020	5/8" thick		1,300		.41	.25	.66	.81
030	3/4" thick		1,200		.46	.27	.73	.90
050	Plywood on walls with exterior standard, 3/8" thick		1,200		.29	.27	.56	.71
060	1/2" thick		1,125		.37	.29	.66	.82
070	5/8" thick		1,050		.41	.31	.72	.90
080	3/4" thick		975		.46	.34	.80	.99

6.2 Finish Carpentry

		CREW	DAILY OUTPUT	UNIT	BARE COSTS MAT.	BARE COSTS INST.	BARE COSTS TOTAL	TOTAL INCL O&P
06-001	**CABINETS** Corner china cabinets, stock pine,							
002	80" high, unfinished, minimum	2 Carp	6.60	Ea.	225	48	273	315
010	Maximum		4.40		365	71	436	505
150	Range or sink base, two doors below, 30" wide		21.40		120	14.65	134.65	155
154	36" wide		20.30		135	15.45	150.45	170
158	48" wide		18.90		150	16.60	166.60	190
180	For sink front units, deduct				37		37	41M
900	For deluxe models of all cabinets, add to above				40%			
950	For custom built in place, add to above				25%	10%		
10-001	**CONVECTOR COVERS** Laminated plastic on 3/4"							
002	thick particle board, 12" wide, minimum	1 Carp	16	L.F.	16.85	9.80	26.65	33
005	Average		14		21	11.20	32.20	39
010	Maximum		12		27	13.05	40.05	49
030	Add to above for grille, minimum		150	S.F.	1.35	1.05	2.40	2.99
040	Maximum		75	"	3.87	2.09	5.96	7.30
32-001	**LOUVERS** Redwood, 2'-0" opening, full circle		16	Ea.	70	9.80	79.80	91
010	Half circle, 2'-0" diameter		16		60	9.80	69.80	80
020	Octagonal		16		55	9.80	64.80	75
030	Triangular, 5/12 pitch, 5'-0" at base		16		73	9.80	82.80	94

For expanded coverage of these items see Means' *Interior Cost Data 1985*

6

7

7.6 Sheet Metal Work

		CREW	DAILY OUTPUT	UNIT	BARE COSTS MAT.	BARE COSTS INST.	BARE COSTS TOTAL	TOTAL INCL O&P
10-001	**DOWNSPOUTS** Aluminum 2" x 3", .020" thick, embossed	1 Shee	190	L.F.	.50	.93	1.43	1.91
010	Enameled		190		.55	.93	1.48	1.97
480	Steel, galvanized, round, corrugated, 2" or 3" diam., 28 gauge		190		.45	.93	1.38	1.86
490	4" diameter, 28 gauge		145		.57	1.22	1.79	2.41
510	5" diameter, 28 gauge		130		.71	1.37	2.08	2.77
520	26 gauge		130		.90	1.37	2.27	2.98
540	6" diameter, 28 gauge		105		1.03	1.69	2.72	3.60
550	26 gauge		105		1.17	1.69	2.86	3.75
750	Steel pipe, black, extra heavy, 4" diameter		20		11.25	8.90	20.15	25
760	6" diameter		18		24	9.85	33.85	41
25-001	**FLASHING** Aluminum, mill finish, .013" thick		145	S.F.	.27	1.22	1.49	2.08
003	.016" thick		145		.31	1.22	1.53	2.13
006	.019" thick		145		.63	1.22	1.85	2.48
010	.032" thick		145		.77	1.22	1.99	2.63

For expanded coverage of these items see *Building Construction Cost Data 1985*

7.6 Sheet Metal Work		CREW	DAILY OUTPUT	UNIT	BARE COSTS MAT.	BARE COSTS INST.	BARE COSTS TOTAL	TOTAL INCL O&P
25 020	.040" thick	1 Shee	145	S.F.	1.30	1.22	2.52	3.22
030	.050" thick	"	145		1.60	1.22	2.82	3.55
040	Painted finish, add				.16		.16	.17M
130	Asphalt flashing cement, 5 gallon			Gal.	13.40		13.40	14.75M
160	Copper, 16 oz., sheets, under 6000 lbs.	1 Shee	115	S.F.	1.11	1.54	2.65	3.47
170	Over 6000 lbs.		155		1.11	1.15	2.26	2.89
190	20 oz. sheets, under 6000 lbs.		110		1.40	1.61	3.01	3.89
200	Over 6000 lbs.		145		1.40	1.22	2.62	3.33
220	24 oz. sheets, under 6000 lbs.		105		1.67	1.69	3.36	4.30
230	Over 6000 lbs.		135		1.67	1.32	2.99	3.75
250	32 oz. sheets, under 6000 lbs.		100		2.22	1.78	4	5.05
260	Over 6000 lbs.		130		2.22	1.37	3.59	4.43
430	Copper-clad stainless steel, .015" thick, under 500 lbs.		115		2.10	1.54	3.64	4.56
440	Over 2000 lbs.		155		2.05	1.15	3.20	3.93
460	.018" thick, under 500 lbs.		100		2.70	1.78	4.48	5.55
470	Over 2000 lbs.		145		2.40	1.22	3.62	4.43
580	Lead, 2.5 lb. per S.F., up to 12" wide	1 Rofc	135		2.20	1.09	3.29	4.07
590	Over 12" wide	"	135		2.95	1.09	4.04	4.90
850	Shower pan, bituminous membrane, 7 oz.	1 Shee	155		1.40	1.15	2.55	3.21
855	3 ply copper and fabric, 3 oz.		155		1.70	1.15	2.85	3.54
860	7 oz.		155		3.65	1.15	4.80	5.70
865	Copper, 16 oz.		100		2.75	1.78	4.53	5.60
870	Lead on copper and fabric, 5 oz.		155		1.40	1.15	2.55	3.21
880	7 oz.		155		2.75	1.15	3.90	4.70
885	Polyvinyl chloride, .030" thick		160		.45	1.11	1.56	2.11
890	Stainless steel sheets, 32 ga., .010" thick		155		1.70	1.15	2.85	3.54
900	28 ga., .015" thick		155		2.10	1.15	3.25	3.98
910	26 ga., .018" thick		155		2.60	1.15	3.75	4.53
920	24 ga., .025" thick		155		3.20	1.15	4.35	5.20
930								
940	Terne coated stainless steel, .015" thick, 28 ga.	1 Shee	155	S.F.	2.10	1.15	3.25	3.98
950	.018" thick, 26 ga.		155		2.15	1.15	3.30	4.04
960	Zinc and copper alloy, .020" thick		155		1.05	1.15	2.20	2.83
970	.027" thick		155		1.35	1.15	2.50	3.16
980	.032" thick		155		1.60	1.15	2.75	3.43
990	.040" thick		155		2	1.15	3.15	3.87
42-001	**LOUVERS** Aluminum with screen, residential, 8" x 8"	1 Carp	38	Ea.	3.50	4.13	7.63	9.80
010	12" x 12"		38		4.90	4.13	9.03	11.35
020	12" x 18"		35		7.55	4.48	12.03	14.75
025	14" x 24"		30		9.65	5.25	14.90	18.15
030	18" x 24"		27		11.50	5.80	17.30	21
050	30" x 24"		24		24	6.55	30.55	36
070	Triangle, adjustable, small		20		12.25	7.85	20.10	25
080	Large		15		12.90	10.45	23.35	29
120	Extruded aluminum, see division 15.7-79							
210	Midget, aluminum, 3/4" deep, 1" diameter	1 Carp	85	Ea.	.43	1.84	2.27	3.14
215	3" diameter		60		1.14	2.61	3.75	5.05
220	4" diameter		50		1.43	3.14	4.57	6.10
225	6" diameter		30		2.01	5.25	7.26	9.75
230	Ridge vent strip, mill finish	1 Shee	155	L.F.	1.51	1.15	2.66	3.33
240	Under eaves vent, aluminum, mill finish, 16" x 4"	1 Carp	75	Ea.	.85	2.09	2.94	3.95
250	16" x 8"	"	75	"	1.05	2.09	3.14	4.17

For expanded coverage of these items see *Building Construction Cost Data 1985*

7

10.1 Specialties

		CREW	DAILY OUTPUT	UNIT	BARE COSTS MAT.	BARE COSTS INST.	BARE COSTS TOTAL	TOTAL INCL O&P
02-001	**BATHROOM ACCESSORIES**							
020	Curtain rod, stainless steel, 5' long, 1" diameter	1 Carp	13	Ea.	12.85	12.05	24.90	32
030	1-1/4" diameter	"	13	"	9.65	12.05	21.70	28
050	Dispenser units, combined soap & towel dispensers,							
051	mirror and shelf, flush mounted	1 Carp	10	Ea.	195	15.70	210.70	235
060	Towel dispenser and waste receptacle,							
061	flush mounted	1 Carp	10	Ea.	280	15.70	295.70	330
080	Grab bar, straight, 1" diameter, stainless steel, 12" long		24		16.05	6.55	22.60	27
090	18" long		23		18	6.80	24.80	30
100	24" long		22		18.20	7.15	25.35	30
110	36" long		20		21	7.85	28.85	34
120	1-1/2" diameter, 18" long		23		23	6.80	29.80	35
130	36" long		20		30	7.85	37.85	44
150	Tub bar, 1" diameter, horizontal		14		51	11.20	62.20	72
160	Plus vertical arm		12		73	13.05	86.05	99
190	End tub bar, 1" diameter, 90° angle		12		46	13.05	59.05	69
200	For 1-1/4" diameter bars, add				20%			
210	For 1-1/2" diameter bars, add				30%			
230	Hand dryer, surface mounted, electric, 110 volt	1 Carp	4		270	39	309	355
240	220 volt		4		270	39	309	355
260	Hat and coat strip, stainless steel, 4 hook, 36" long		24		36	6.55	42.55	49
270	6 hook, 60" long		20		60	7.85	67.85	77
300	Mirror with stainless steel, 3/4" square frame, 18" x 24"		20		56	7.85	63.85	73
310	36" x 24"		15		98	10.45	108.45	125
320	48" x 24"		10		125	15.70	140.70	160
330	72" x 24"		6		180	26	206	235
350	Mirror with 5" stainless steel shelf, 3/4" sq. frame, 18" x 24"		20		73	7.85	80.85	92
360	36" x 24"		15		135	10.45	145.45	165
370	48" x 24"		10		180	15.70	195.70	220
380	72" x 24"		6		265	26	291	330
410	Mop holder strip, stainless steel, 6 holders, 60" long		20		64	7.85	71.85	82
420	Napkin/tampon dispenser, surface mounted		15		245	10.45	255.45	285
430	Robe hook, single, regular		36		10.70	4.36	15.06	18.05
440	Heavy duty, concealed mounting		36		13.90	4.36	18.26	22
460	Soap dispenser, chrome, surface mounted, liquid		20		47	7.85	54.85	63
470	Powder		20		39	7.85	46.85	54
500	Recessed stainless steel, liquid		10		54	15.70	69.70	82
510	Powder		10		83	15.70	98.70	115
530	Soap tank, stainless steel, 1 gallon		10		68	15.70	83.70	97
540	5 gallon		5		120	31	151	175
560	Shelf, stainless steel, 5" wide, 18 ga., 24" long		24		20	6.55	26.55	31
570	72" long		16		58	9.80	67.80	78
580	8" wide shelf, 18 ga., 24" long		22		30	7.15	37.15	43
590	72" long		14		79	11.20	90.20	105
600	Toilet seat cover dispenser, S.S., recessed		20		60	7.85	67.85	77
605	Surface mounted		15		26	10.45	36.45	44
610	Toilet tissue dispenser, surface mounted, S.S., single roll		30		17.10	5.25	22.35	26
620	Double roll		24		28	6.55	34.55	40
640	Towel bar, stainless steel, 18" long		23		17.10	6.80	23.90	29
650	30" long		21		19	7.45	26.45	32
670	Towel dispenser, stainless steel, surface mounted		16		41	9.80	50.80	59
680	Flush mounted, recessed		10		75	15.70	90.70	105
700	Towel holder, hotel type, 2 guest size		20		8.55	7.85	16.40	21
720	Towel shelf, stainless steel, 24" long, 8" wide		20		26	7.85	33.85	40
740	Tumbler holder, tumbler only		30		9.65	5.25	14.90	18.15
750	Soap, tumbler & toothbrush		30		16.05	5.25	21.30	25
770	Wall urn ash receiver, recessed, 14" long		12		75	13.05	88.05	100
780	Surface, 8" long		18		47	8.70	55.70	64
800	Waste receptacles, stainless steel, with top, 13 gallon		10		130	15.70	145.70	165
810	36 gallon		8		220	19.60	239.60	270

For expanded coverage of these items see *Building Construction Cost Data 1985*

10.1 Specialties

		CREW	DAILY OUTPUT	UNIT	BARE COSTS MAT.	BARE COSTS INST.	BARE COSTS TOTAL	TOTAL INCL O&P
30-001	**FIREPLACE, PREFABRICATED** Free standing or wall hung							
.010	with hood & screen, minimum	F-1	1.30	Ea.	500	125	625	730
015	Average		1		900	165	1,065	1,225
020	Maximum		.90		3,000	180	3,180	3,550
050	For chimney heights over 8'-6", 7" diameter, add		33	V.L.F.	19	4.96	23.96	28
060	10" diameter, add		32		31	5.10	36.10	41
070	12" diameter, add		31		46	5.30	51.30	58
080	14" diameter, add		30		65	5.45	70.45	79
100	Simulated brick chimney top, 4' high, 16" x 16"		10	Ea.	125	16.40	141.40	160
110	24" x 24"		7		200	23	223	255
130	Woodburning stoves, cast iron, minimum	F-2	1.30		400	250	650	800
131	Average		1		550	330	880	1,075
132	Maximum		.80		1,300	410	1,710	2,025
135	For gas log lighter, add				28		28	31M
150	Simulated logs, gas fired, 40,000 BTU, 2' long, minimum	F-1	7	Set	125	23	148	170
160	Maximum		6		300	27	327	370
170	Electric, 1,500 BTU, 1'-6" long, minimum		7		75	23	98	115
180	11,500 BTU, maximum		6		120	27	147	170
57-001	**PARTITIONS, SHOWER** Economy, painted steel, steel							
010	base, no door or plumbing included	2 Shee	5	Ea.	115	71	186	230
030	Square, 32" x 32", stock, with receptor & door, fiberglass		4.50		255	79	334	395
060	Galvanized and painted steel		5		175	71	246	295
070	Shower stall, double wall, incl. receptor but not including							
080	door or plumbing, enameled steel	2 Shee	5	Ea.	570	71	641	730
110	Porcelain enameled steel		5		1,000	71	1,071	1,200
120	Stainless steel		5		1,025	71	1,096	1,225
140	For double entry type, no doors, deduct				10%			
150	Circular fiberglass, 36" diameter, no plumbing included	2 Shee	4		225	89	314	375
170	One piece, 36" diameter, less door		4		255	89	344	410
180	With door		3.50		410	100	510	600
200	Curved shell shower, no door needed		3		420	120	540	635
230	For fiberglass seat, add to both above				53		53	58M
240	Glass stalls, with doors, no receptors, chrome on brass	2 Shee	3		630	120	750	865
270	Anodized aluminum	"	4		390	89	479	560
290	Marble shower stall, stock design, with shower door	2 Marb	1.20		830	260	1,090	1,275
300	With curtain		1.30		685	240	925	1,100
320	Receptors, precast terrazzo, 32" x 32"		14		135	22	157	180
330	48" x 34"		12		165	26	191	220
350	Plastic, simulated terrazzo receptor, 32" x 32"		14		60	22	82	98
360	32" x 48"		12		84	26	110	130
380	Precast concrete, colors, 32" x 32"		14		110	22	132	155
390	48" x 48"		12		130	26	156	180
410	Doors, economy plastic, 24" wide	1 Shee	9		63	19.75	82.75	98
420	Tempered glass door, economy		8		94	22	116	135
440	Folding, tempered glass, aluminum frame		6		140	30	170	195
450	Sliding, tempered glass, 48" opening		6		100	30	130	155
470	Deluxe, tempered glass, chrome on brass frame, minimum		5		265	36	301	345
500	Maximum		5		490	36	526	590
530	On anodized aluminum frame, minimum		5		74	36	110	135
540	Maximum		5		265	36	301	345
560	Tub enclosure, plastic panels, economy, sliding panel		4		65	44	109	135
570	Folding panel		4		90	44	134	165
590	Deluxe, tempered glass, anodized alum. frame, minimum		2		155	89	244	300
620	Maximum		1.50		335	120	455	540
650	On chrome-plated brass frame, minimum		2		245	89	334	400
660	Maximum		1.50		515	120	635	740
60-001	**PARTITIONS, TOILET** For ceiling framing, see division 5.4-68							
010	Cubicles, ceiling hung, marble	2 Marb	2	Ea.	450	155	605	715
020	Painted metal	2 Carp	4		200	78	278	335
030	Plastic laminate on particle board	"	4		295	78	373	440

For expanded coverage of these items see *Building Construction Cost Data 1985*

10.1 Specialties		CREW	DAILY OUTPUT	UNIT	BARE COSTS MAT.	BARE COSTS INST.	BARE COSTS TOTAL	TOTAL INCL O&P
60 040	Porcelain enamel	2 Carp	4	Ea.	605	78	683	780
050	Stainless steel	″	4		555	78	633	725
060	For handicap units, add				66		66	73M
070								
080	Floor & ceiling anchored, marble	2 Marb	2.50	Ea.	465	125	590	690
100	Painted metal	2 Carp	5		200	63	263	310
110	Plastic laminate on particle board		5		350	63	413	475
120	Porcelain enamel		5		615	63	678	765
130	Stainless steel		5		555	63	618	700
140	For handicap units, add				66		66	73M
160	Floor mounted, marble	2 Marb	3		400	105	505	590
170	Painted metal	2 Carp	7		180	45	225	265
180	Plastic laminate on particle board		7		170	45	215	250
190	Porcelain enamel		7		605	45	650	730
200	Stainless steel		7		555	45	600	675
210	For handicap units, add				62		62	68M
220	For juvenile units, deduct				20		20	22M
230								
240	Floor mounted, headrail braced, marble	2 Marb	3	Ea.	425	105	530	615
250	Painted metal	2 Carp	6		190	52	242	285
260	Plastic laminate on particle board		6		180	52	232	275
270	Porcelain enamel		6		600	52	652	735
280	Stainless steel		6		545	52	597	675
290	For handicap units, add				66		66	73M
300	Wall hung partitions, painted metal	2 Carp	7		370	45	415	470
320	Porcelain enamel		7		600	45	645	725
330	Stainless steel		7		650	45	695	780
340	For handicap units, add				66		66	73M
400	Screens, entrance, floor mounted, 54" high							
410	Marble	D-1	35	L.F.	72	8.15	80.15	91
420	Painted metal	2 Carp	60		46	5.25	51.25	58
430	Plastic laminate on particle board		60		56	5.25	61.25	69
440	Porcelain enamel		60		110	5.25	115.25	130
450	Stainless steel		60		97	5.25	102.25	115
460	Urinal screen, 18" wide, ceiling braced, marble	D-1	6	Ea.	215	48	263	305
470	Painted metal	2 Carp	8		110	39	149	180
480	Plastic laminate on particle board		8		145	39	184	215
490	Porcelain enamel		8		210	39	249	290
500	Stainless steel		8		210	39	249	290
505								
510	Floor mounted, head rail braced							
520	Marble	D-1	6	Ea.	190	48	238	275
530	Painted metal	2 Carp	8		115	39	154	185
540	Plastic laminate on particle board		8		135	39	174	205
550	Porcelain enamel		8		210	39	249	290
560	Stainless steel		8		210	39	249	290
570	Pilaster, flush, marble	D-1	9		235	32	267	305
580	Painted metal	2 Carp	10		60	31	91	110
590	Plastic laminate on particle board		10		75	31	106	130
600	Porcelain enamel		10		150	31	181	210
610	Stainless steel		10		150	31	181	210
615								
620	Post braced, marble	D-1	9	Ea.	230	32	262	300
630	Painted metal	2 Carp	10		60	31	91	110
640	Plastic laminate on particle board		10		75	31	106	130
650	Porcelain enamel		10		150	31	181	210
660	Stainless steel		10		150	31	181	210
670	Wall hung, bracket supported							
680	Painted metal	2 Carp	10	Ea.	92	31	123	145
690	Plastic laminate on particle board	″	10	″	87	31	118	140

For expanded coverage of these items see *Building Construction Cost Data 1985*

10.1 Specialties

		Crew	Daily Output	Unit	Bare Costs Mat.	Bare Costs Inst.	Bare Costs Total	Total Incl O&P
60	700	Porcelain enamel	2 Carp	10	Ea.	180	31	211	245

No.	Description	Crew	Daily Output	Unit	Bare Costs Mat.	Bare Costs Inst.	Bare Costs Total	Total Incl O&P
60 700	Porcelain enamel	2 Carp	10	Ea.	180	31	211	245
710	Stainless steel		10		170	31	201	230
740	Flange supported, painted metal		10		185	31	216	250
750	Plastic laminate on particle board		10		105	31	136	160
760	Porcelain enamel		10		220	31	251	285
770	Stainless steel		10		210	31	241	275
780	Wedge type, painted metal		10		125	31	156	185
800	Porcelain enamel		10		195	31	226	260
810	Stainless steel		10		195	31	226	260

11.1 Architectural Equipment

No.	Description	Crew	Daily Output	Unit	Bare Costs Mat.	Bare Costs Inst.	Bare Costs Total	Total Incl O&P
03-001	**APPLIANCES** Cooking range, 30" free standing, 1 oven, minimum	2 Clab	10	Ea.	320	25	345	390
005	Maximum		4		1,200	62	1,262	1,400
015	2 oven, minimum		10		780	25	805	895
020	Maximum		4		1,500	62	1,562	1,750
150	Combination range, refrigerator and sink, 30" wide, minimum	L-1	2		640	175	815	960
155	Maximum		1		1,180	355	1,535	1,800
157	60" wide, average		1.40		1,650	255	1,905	2,175
159	72" wide, average		1.20		1,870	295	2,165	2,475
160	Office model, 48" wide		2		1,470	175	1,645	1,875
162	Refrigerator and sink only		2.40		1,300	150	1,450	1,650
164	Combination range, refrigerator, sink, microwave							
166	oven and ice maker	L-1	.80	Ea.	3,175	445	3,620	4,125
245	Dehumidifier, portable, automatic, 14 pint				175		175	195M
255	30 pint				245		245	270M
275	Dishwasher, built-in, 2 cycles, minimum	L-1	4		270	89	359	425
280	Maximum		2		360	175	535	650
295	4 or more cycles, minimum		4		410	89	499	580
300	Maximum		2		700	175	875	1,025
320	Dryer, automatic, minimum	L-2	3		260	93	353	420
325	Maximum	"	2		530	140	670	785
330	Garbage disposer, sink type, minimum	L-1	5		50	71	121	155
335	Maximum	"	3		180	120	300	370
355	Heater, electric, built-in, 1250 watt, ceiling type, minimum	1 Elec	4		40	44	84	110
360	Maximum		3		100	59	159	195
370	Wall type, minimum		4		38	44	82	105
375	Maximum		3		75	59	134	165
390	1500 watt wall type, with blower		4		75	44	119	145
395	3000 watt		3		130	59	189	230
415	Hood for range, 2 speed, vented, 30" wide, minimum	L-3	5		40	67	107	140
420	Maximum		3		230	110	340	415
430	42" wide, minimum		5		155	67	222	265
435	Maximum		3		270	110	380	460
450	For ventless hood, 2 speed, add				12		12	13.20M
465	For vented 1 speed, deduct from maximum				25		25	28M
485	Humidifier, portable, 7 gallons per day				80		80	88M
500	15 gallons per day				160		160	175M
520	Icemaker, automatic, 3 lb. per day	1 Plum	7		300	25	325	365
535	35 lb. per day		2		725	89	814	925
640	Sump pump cellar drainer, 1/3 HP, minimum		3		75	59	134	170
645	Maximum		2		200	89	289	350
665	Washing machine, automatic, minimum		3		360	59	419	480
670	Maximum		1		690	180	870	1,025

For expanded coverage of these items see *Building Construction Cost Data 1985*

11.1 Architectural Equipment

		CREW	DAILY OUTPUT	UNIT	BARE COSTS MAT.	BARE COSTS INST.	BARE COSTS TOTAL	TOTAL INCL O&P
03 690	Water heater, electric, glass lined, 30 gallon, minimum	L-1	5	Ea.	120	71	191	235
695	Maximum		3		210	120	330	400
710	82 gallon, minimum		2		225	175	400	505
715	Maximum		1		380	355	735	930
718	Water heater, gas, glass lined, 30 gallon, minimum	2 Plum	5		115	71	186	230
722	Maximum		3		240	120	360	435
726	50 gallon, minimum		2.50		185	140	325	410
730	Maximum		1.50		335	235	570	710
735	Water softener, automatic, to 30 grains per gallon		5		330	71	401	465
740	To 75 grains per gallon		4		650	89	739	845
745	Vent kits for dryers	1 Carp	10		12	15.70	27.70	36
10-001	**BARBER EQUIPMENT** Chair, hydraulic, minimum	1 Carp	1.30	Ea.	400	120	520	615
005	Maximum	"	1		2,000	155	2,155	2,425
050	Sink, hair washing basin	1 Plum	1		315	180	495	605
100	Sterilizer, liquid solution for tools				80		80	88M
15-001	**DARKROOM EQUIPMENT** Developing tanks, 5" deep, 24"x 48"	Q-1	2		860	160	1,020	1,175
005	48" x 52"		1.70		1,545	190	1,735	1,975
020	10" deep, 24" x 48"		1.70		970	190	1,160	1,350
025	24" x 108"		1.50		1,750	215	1,965	2,225
350	Washers, round, maximum sheet 11" x 14"		2		1,080	160	1,240	1,425
355	Maximum sheet 20" x 24"		1		1,280	320	1,600	1,875
380	Square, maximum sheet 20" x 24"		1		1,170	320	1,490	1,750
390	Maximum sheet 50" x 56"		.80		1,850	400	2,250	2,625
450	Combination tank sink, tray sink, washers, with							
451	dry side tables, average	Q-1	1	Ea.	3,800	320	4,120	4,650
16-001	**DENTAL EQUIPMENT** Central suction system, minimum	1 Plum	1.20		1,475	150	1,625	1,825
010	Maximum	"	.90		3,000	195	3,195	3,575
060	Chair, electric or hydraulic, minimum	1 Skwk	.50		2,500	320	2,820	3,225
070	Maximum		.25		4,800	645	5,445	6,225
200	Light, floor or ceiling mounted, minimum		3.60		800	45	845	945
210	Maximum		1.20		1,200	135	1,335	1,525
230	Sterilizers, steam portable, minimum				1,300		1,300	1,425M
235	Maximum				2,000		2,000	2,200M
260	Steam, institutional, walk-in				8,000		8,000	8,800M
265	Dry heat, electric, portable, 3 trays				450		450	495M
28-001	**KITCHEN EQUIPMENT** Bake oven, single deck	Q-1	8		1,630	40	1,670	1,850
030	Double deck		7		3,190	46	3,236	3,575
060	Triple deck		6		4,760	53	4,813	5,325
090	Electric convection, 40" x 45" x 57"	L-7	4		2,430	130	2,560	2,875
235	Cooler, reach-in, beverage, 6 ft. long	Q-1	6		2,200	53	2,253	2,500
250								
270	Dishwasher, commercial, rack type							
272	10 to 12 racks per hour	Q-1	3.20	Ea.	2,000	100	2,100	2,350
275	Semi-automatic 38 to 50 racks per hour	"	1.30		4,200	245	4,445	4,975
280	Automatic 190 to 230 racks per hour	Q-2	.70		9,000	710	9,710	10,900
282	235 to 275 racks per hour		.50		13,000	995	13,995	15,700
284	8750 to 12,500 dishes per hour		.20		33,900	2,475	36,375	40,900
430	Freezers, reach-in, 40 C.F.	Q-1	4		4,200	80	4,280	4,725
450	70 C.F.		3		5,100	105	5,205	5,775
580	Ice cube maker, 50 pounds per day		6		1,375	53	1,428	1,600
605	500 pounds per day		4		3,500	80	3,580	3,975
635	Kettles, steam-jacketed, 20 gallons	L-7	7		3,900	75	3,975	4,400
660	60 gallons	"	6		4,800	88	4,888	5,400
795	Hood fire protection system, minimum	Q-1	3		2,300	105	2,405	2,675
805	Maximum		1		17,000	320	17,320	19,200
830	Refrigerators, reach-in type, 44 C.F.		5		2,760	64	2,824	3,125
855	With glass doors, 68 C.F.		4		4,100	80	4,180	4,625
31-001	**LABORATORY EQUIPMENT** Cabinets, base, door units, metal	2 Carp	18	L.F.	90	17.40	107.40	125
030	Drawer units	"	18	"	175	17.40	192.40	220

For expanded coverage of these items see *Building Construction Cost Data 1985*

11.1 Architectural Equipment

		CREW	DAILY OUTPUT	UNIT	BARE COSTS MAT.	BARE COSTS INST.	BARE COSTS TOTAL	TOTAL INCL O&P
31 200	Fume hood, with countertop & base, not including HVAC							
202	Simple, minimum	2 Carp	5.40	L.F.	450	58	508	580
205	Complex, including fixtures		2.40		700	130	830	960
210	Special, maximum	↓	1.70	↓	910	185	1,095	1,275
220	Ductwork, minimum	2 Shee	.1	Hood	580	355	935	1,150
225	Maximum	"	.50	"	3,100	710	3,810	4,450
230	Service fixtures, average			Ea.	30		30	33M
250	For sink assembly with hot and cold water, add	1 Plum	1.40		325	125	450	540
255	Glassware washer, distilled water rinse, minimum	L-1	1.80		2,275	195	2,470	2,775
260	Maximum	"	1		13,500	355	13,855	15,400
280	Sink, one piece plastic, flask wash, hose, free standing	1 Plum	1.60	↓	875	110	985	1,125
285	Epoxy resin sink, 25" x 16" x 10"	"	2		195	89	284	345
420	Alternate pricing method: as percent of lab furniture							
440	Installation, not incl. plumbing & duct work			% Furn.				22%
480	Plumbing, final connections, simple system							10%
500	Moderately complex system							15%
520	Complex system							20%
540	Electrical, simple system						10%	10%
560	Moderately complex system						20%	20%
580	Complex system			↓			35%	35%
600	Safety equipment, eye wash, hand held			Ea.	180		180	200M
620	Deluge shower			"	105		105	115M
630	Rule of thumb: lab furniture including installation & connection							
632	High school			S.F.				20
634	College							30
636	Clinical, health care							25
638	Industrial			↓				40
32-001	**LAUNDRY EQUIPMENT** Not incl. rough-in. Dryers, gas fired							
050	Residential, 16 lb. capacity, average	1 Plum	3	Ea.	420	59	479	550
100	Commercial, 30 lb. capacity, coin operated, single		3		1,600	59	1,659	1,850
110	Double stacked		2		1,300	89	1,389	1,550
150	Industrial, 30 lb. capacity		3		1,600	59	1,659	1,850
160	50 lb. capacity		3		1,950	59	2,009	2,225
500	Washers, residential, 4 cycle, average		3		560	59	619	700
530	Commercial, coin operated, average	↓	3		650	59	709	800
600	Combination washer/extractor, 20 lb. capacity	L-6	2		2,200	135	2,335	2,600
610	30 lb. capacity		.80		5,500	335	5,835	6,525
620	50 lb. capacity		.80		6,500	335	6,835	7,625
630	75 lb. capacity		.40		12,000	665	12,665	14,200
635	125 lb. capacity	↓	.40	↓	19,000	665	19,665	21,900
37-001	**MEDICAL EQUIPMENT** Autopsy table, standard	1 Plum	1	Ea.	5,500	180	5,680	6,300
020	Deluxe		.60		7,500	295	7,795	8,675
070	Distiller, water, steam heated, 50 gal. capacity	↓	1.40		12,500	125	12,625	13,900
180	Heat therapy unit, humidified, 26" x 78" x 28"				1,600		1,600	1,750M
560	Sterilizers, floor loading, 28" x 67" x 52", single door, steam				85,000		85,000	93,500M
565	Double door, steam				130,000		130,000	143,000
580	General purpose, 20" x 20" x 28", single door				6,000		6,000	6,600M
600	Portable, counter top, steam, minimum			↓	1,000		1,000	1,100M
602	Maximum				3,000		3,000	3,300M
605	Portable, counter top, gas			Ea.	10,000		10,000	11,000M
615	Automatic washer/sterilizer	1 Plum	2	"	21,500	89	21,589	23,800
620	Steam generators, electric 10 KW to 180 KW							
625	Minimum	1 Elec	3	Ea.	2,100	59	2,159	2,400
630	Maximum	"	.70		18,000	255	18,255	20,200
810	Utensil washer-sanitizer	1 Plum	2		4,800	89	4,889	5,400
840	Whirlpool bath, mobile, 18" x 24" x 60"				2,400		2,400	2,650M
845	Fixed, incl. mixing valves	1 Plum	2	↓	2,800	89	2,889	3,200

11.1 Architectural Equipment

		CREW	DAILY OUTPUT	UNIT	BARE COSTS MAT.	BARE COSTS INST.	BARE COSTS TOTAL	TOTAL INCL O&P
42-001	**REFRIGERATED FOOD CASES** Dairy, multi-deck, 12' long	Q-5	3	Ea.	4,690	105	4,795	5,325
010	For rear sliding doors, add				700		700	770M
020	Delicatessen case, service deli, 12' long, single deck	Q-5	3.90		5,100	82	5,182	5,725
030	Multi-deck, 18 S.F. shelf display		3		5,675	105	5,780	6,400
040	Freezer, self-contained, chest-type, 30 C.F.		3.90		2,180	82	2,262	2,525
050	Glass door, upright, 78 C.F.		3.30		4,250	97	4,347	4,825
060	Frozen food, chest type, 12' long		3.30		3,800	97	3,897	4,325
070	Glass door, reach-in, 5 door		3		6,800	105	6,905	7,625
080	Island case, 12' long, single deck		3.30		4,100	97	4,197	4,650
090	Multi-deck		3		8,600	105	8,705	9,625
100	Meat case, 12' long, single deck		3.30		2,900	97	2,997	3,325
105	Multi-deck		3.10		3,200	105	3,305	3,675
110	Produce, 12' long, single deck		3.30		3,150	97	3,247	3,600
120	Multi-deck		3.10		3,600	105	3,705	4,100
46-001	**SAUNA** Prefabricated, incl. heater & controls, 7' high, 6' x 4'	L-7	2.20		2,300	240	2,540	2,875
040	6' x 5'		2		2,600	265	2,865	3,250
060	6' x 6'		1.80		2,850	290	3,140	3,550
080	6' x 9'		1.60		3,500	330	3,830	4,325
100	8' x 12'		1.10		5,100	480	5,580	6,300
120	8' x 8'		1.40		4,000	375	4,375	4,950
140	8' x 10'		1.20		4,450	440	4,890	5,525
160	10' x 12'		1		5,500	525	6,025	6,800
250	Heaters only (incl. above), wall mounted, to 200 C.F.				325		325	360M
275	To 300 C.F.				475		475	525M
300	Floor standing, to 720 C.F., 10,000 watts	1 Elec	3		600	59	659	745
325	To 1,000 C.F., 12,500 watts	"	3		675	59	734	825
58-001	**STEAM BATH** Heater, timer & head, single, to 140 C.F.	1 Plum	1.20		725	150	875	1,000
050	To 300 C.F.		1.10		800	160	960	1,125
100	Commercial size, to 750 C.F.		.90		1,800	195	1,995	2,275
150	To 2500 C.F.		.80		3,900	220	4,120	4,600
200	Multiple baths, motels, apartment, 2 baths	Q-1	1.30		1,125	245	1,370	1,600
250	4 baths	"	.70		1,300	455	1,755	2,100
270	Conversion unit for residential tub				700		700	770M
61-001	**VACUUM CLEANING** Central, 3 valve, residential	1 Skwk	.90	Total	540	180	720	855
020	Commercial		.70		680	230	910	1,075
040	5 valve system		.50		575	320	895	1,100
060	7 valve system		.40		750	400	1,150	1,400
080	9 valve system		.30		860	535	1,395	1,725
64-001	**WASTE HANDLING** Compactors, 115 volt, 250#/hr., chute fed	L-4	1	Ea.	6,200	445	6,645	7,475
010	Hand fed		2.40		4,150	185	4,335	4,825
030	Multi-bag, hand or chute fed, 230 volt, 600#/hr.		1		5,700	445	6,145	6,925
050	Containerized, hand fed, 2 to 6 C.Y. containers, 250#/hr.		1		6,000	445	6,445	7,250
055	For chute fed, add per floor		1		850	445	1,295	1,575
056								
100	Heavy duty industrial compactor, 0.5 C.Y. capacity	L-4	1	Ea.	4,500	445	4,945	5,600
105	1.0 C.Y. capacity		1		7,250	445	7,695	8,625
110	2.5 C.Y. capacity		.50		11,500	885	12,385	13,900
115	5.0 C.Y. capacity		.50		20,000	885	20,885	23,300
120	Combination shredder/compactor (5,000 lbs./hr.)		.50		25,000	885	25,885	28,800
150	Crematory, not including building, 1 place	Q-3	.20		42,000	3,400	45,400	51,000
175	2 place	"	.10		60,000	6,800	66,800	76,000
375	Incinerator, electric, 100 lb. per hr., minimum	L-9	.75		13,000	845	13,845	15,600
385	Maximum		.70		26,000	905	26,905	29,900
400	400 lb. per hr., minimum		.60		25,000	1,050	26,050	29,100
410	Maximum		.50		60,000	1,275	61,275	68,000
425	1000 lb. per hr., minimum		.25		68,000	2,550	70,550	78,500
435	Maximum		.20		140,000	3,175	143,175	158,500
440	Gas, not incl. chimney, elec. or pipe, 50#/hr., minimum	Q-3	.80		14,500	850	15,350	17,200

For expanded coverage of these items see *Building Construction Cost Data 1985*

11.1 Architectural Equipment

		CREW	DAILY OUTPUT	UNIT	BARE COSTS MAT.	BARE COSTS INST.	BARE COSTS TOTAL	TOTAL INCL O&P
64 442	Maximum	Q-3	.70	Ea.	19,000	970	19,970	22,300
444	200 lb. per hr., minimum (batch type)		.60		19,000	1,125	20,125	22,500
446	Maximum (with feeder)		.50		37,000	1,350	38,350	42,700
448	400 lb. per hr., minimum (batch type)		.30		22,500	2,275	24,775	28,000
450	Maximum (with feeder)		.25		42,500	2,725	45,225	50,500
452	800 lb. per hr., with feeder, minimum		.20		55,000	3,400	58,400	65,500
454	Maximum		.17		75,000	4,000	79,000	88,500
456	1200 lb. per hr., with feeder, minimum		.15		80,000	4,525	84,525	94,500
458	Maximum		.11		96,000	6,175	102,175	114,500
460	2000 lb. per hr., with feeder, minimum		.10		140,000	6,800	146,800	164,000
462	Maximum		.05		235,000	13,600	248,600	278,000
470	For heat recovery system, add, minimum		.25		50,000	2,725	52,725	59,000
471	Add, maximum		.11		160,000	6,175	166,175	185,000
472	For automatic ash conveyer, add	↓	.50	↓	20,000	1,350	21,350	24,000
473								
475	Large municipal incinerators, incl. stack, minimum	Q-3	.25	Ton/Day	12,800	2,725	15,525	18,000
485	Maximum	"	.10	"	34,000	6,800	40,800	47,200
580	Shredder, industrial, minimum			Ea.	15,000		15,000	16,500M
585	Maximum				80,000		80,000	88,000M
590	Baler, industrial, minimum				6,000		6,000	6,600M
595	Maximum			↓	350,000		350,000	385,000

13.1 Special Construction

		CREW	DAILY OUTPUT	UNIT	BARE COSTS MAT.	BARE COSTS INST.	BARE COSTS TOTAL	TOTAL INCL O&P
02-001	**AIR CURTAINS** Not incl. motor starters, transformers,							
005	door switches or temperature controls							
010	Shipping and receiving doors, unheated, minimal wind stoppage							
015	8' high, multiples of 3' wide	2 Shee	3.60	L.F.	170	99	269	330
016	Multiples of 5' wide		3.60		125	99	224	280
021	10' high, multiples of 4' wide		3.60		173	99	272	335
025	12' high, 3'-6" wide		3.30		170	110	280	345
026	12' wide		3.30		125	110	235	295
035	16' high, 3'-6" wide		3.30		177	110	287	350
036	12' wide	↓	3.30	↓	137	110	247	310
150	Customer entrance doors, unheated, minimal wind stoppage							
155	10' high, multiples of 3' wide	2 Shee	3.60	L.F.	165	99	264	325
156	Multiples of 5' wide		3.60		110	99	209	265
165	Maximum wind stoppage, 12' high, multiples of 4' wide	↓	3.30	↓	270	110	380	455
170	Heated, minimal wind stoppage, electric heat							
175	8' high, multiples of 3' wide	2 Shee	3.10	L.F.	342	115	457	545
185	10' high, multiples of 3' wide		3.10		388	115	503	595
186	Multiples of 5' wide	↓	3.10	↓	272	115	387	465
195	Maximum wind stoppage, steam heat							
196	12' high, multiples of 4' wide	2 Shee	2.80	L.F.	370	125	495	590
200	Walk-in coolers and freezers, ambient air, minimal wind stoppage							
205	8' high, multiples of 3' wide	2 Shee	3.60	L.F.	175	99	274	335
206	Multiples of 5' wide		3.60		144	99	243	300
225	Maximum wind stoppage, 12' high, multiples of 3' wide		3.30		278	110	388	465
245	Conveyor openings or service windows, unheated, 5' high		4		115	89	204	255
246	Heated, electric, 5' high, 2'-6" wide	↓	3.30	↓	190	110	300	365
17-001	**CHIMNEY** Foundations, add to all prices below							
180	Metal, steel, guyed, 2' diameter, 1/4" thick shell	E-2	65	V.L.F.	67.50	30	97.50	115
190	5' diameter, 1/2" thick shell	"	30	"	500	65	565	640

For expanded coverage of these items see *Building Construction Cost Data 1985*

13.1	**Special Construction**	CREW	DAILY OUTPUT	UNIT	BARE COSTS MAT.	BARE COSTS INST.	BARE COSTS TOTAL	TOTAL INCL O&P
17 210	Poured concrete, brick lining, no foundation, 200' x 10' diameter			Ea.				700,000
220	200' x 18' diameter							960,000
240	250' x 9' diameter							885,000
250	300' x 14' diameter							1,200,000
270	400' x 18' diameter							2,115,000
280	500' x 20' diameter							3,435,000
300	Radial brick, foundations not incl., 75' x 3'-6" I.D.							125,000
310	85' x 3'-6" I.D.							150,000
330	85' x 5'-0" I.D.							175,000
340	100' x 5'-0" I.D.							205,000
360	125' x 5'-6" I.D.							225,000
370	150' x 6'-6" I.D.							260,000
390	160' x 6'-6" I.D.							342,000
400	175' x 7'-0" I.D.			↓				375,000
20-001	**COMFORT STATIONS** Prefab., stock, with doors, windows and fixtures							
010	Not incl. interior finish or electrical							
030	Mobile, on steel frame, minimum			S.F.	24		24	26M
035	Maximum				35		35	39M
040	Permanent, including concrete slab, minimum	B-12J	50		120	14.70	134.70	150
050	Maximum	"	43	↓	160	17.05	177.05	195
060	Alternate pricing method, mobile, minimum			Fixture	1,500		1,500	1,650M
065	Maximum				2,250		2,250	2,475M
070	Permanent, minimum	B-12J	.70		8,550	1,050	9,600	10,700
075	Maximum	"	.50	↓	14,500	1,475	15,975	17,800
42-001	**ICE SKATING** Equipment incl. refrigeration and plumbing, not							
002	including building or slab, 85' x 200' rink							
030	55° system, 5 mos., 100 ton			Total				195,000
070	90° system, 12 mos., 135 ton			"				205,000
100	Dasher boards, polyethylene coated plywood, 3' acrylic							
102	screen at sides, 5' acrylic ends, 85' x 200'	F-5	.09	Ea.	62,500	7,300	69,800	79,000
110	Fiberglass & aluminum construction, same sides and ends	"	.08		80,000	8,225	88,225	100,000
120	Subsoil heating system (recycled from compressor), 85' x 200'	Q-7	.30		13,100	2,275	15,375	17,700
130	Subsoil insulation, 2 lb. polystyrene with vapor barrier, 85' x 200'	F-2	.16	↓	18,250	2,050	20,300	23,000
45-001	**INTEGRATED CEILINGS** Lighting, ventilating & acoustical							
010	Luminaire, 5' x 5' modules, suspended, 50% lighted	L-3	90	S.F.	2.90	3.71	6.61	8.55
020	100% lighted	"	50		4.35	6.70	11.05	14.45
040	For ventilating capacity with perforations, add			↓				.25
060	For supply air diffuser, add			Ea.				55
070	Dimensionaire, 2' x 4' board system	L-3	50	L.F.	6.75	6.70	13.45	17.10
090	Tile system	1 Carp	250	S.F.	1.75	.63	2.38	2.83
100	For air bar suspension, add, minimum							.50
110	Maximum			↓				.85
120	For vaulted coffer, including fixture, stock, add			Ea.	66		66	73M
130	Custom, add, minimum							115
140	Average							300
150	Maximum			↓				500
180	Radiant hot water system with finished acoustic ceiling,							
181	not including supply piping. Heating only (gross S.F.)							
210	Elementary schools, minimum			S.F.				4.25
220	Maximum							5.35
240	High schools and colleges, minimum							3.75
250	Maximum							5.35
270	Libraries, minimum							3.80
280	Maximum							5.05
300	Hospitals, minimum							5.20
310	Maximum							6.50
330	Office buildings, minimum							3.70
340	Maximum							4.90
360	For combined heating and cooling, add, minimum			↓	30%	30%		

13.1 Special Construction		CREW	DAILY OUTPUT	UNIT	BARE COSTS MAT.	BARE COSTS INST.	BARE COSTS TOTAL	TOTAL INCL O&P
45 370	Maximum			S.F.	40%	40%		
400	Radiant electric ceiling board, strapped between joists	1 Elec	250	"	1.40	.71	2.11	2.56
410	2' x 4' heating panel for grid system, manila finish		25	Ea.	32.35	7.05	39.40	46
420	Textured epoxy finish		22		33.50	8.05	41.55	48
430	Vinyl finish		19		34.80	9.30	44.10	52
440	Hair cell, ABS plastic finish	↓	13	↓	57.25	13.60	70.85	83
450	2' x 4' alternate blank panel, for use with above							
460	Manila finish	1 Elec	50	Ea.	7.70	3.54	11.24	13.55
470	Textured epoxy finish		45		17.05	3.93	20.98	24
480	Vinyl finish		40		18.92	4.42	23.34	27
490	Hair cell, ABS plastic finish	↓	25	↓	51	7.05	58.05	66
50-001	**PEDESTAL ACCESS FLOORS** Computer room application, metal							
002	Particle bd. or steel panels, no covering, under 6000 S. F.	2 Carp	1,000	S.F.	6	.31	6.31	7.05
030	Metal covered, over 6000 S.F.		1,100		4.90	.29	5.19	5.80
040	Aluminum, 24" panels	↓	500		15	.63	15.63	17.40
060	For carpet covering, add				2.25		2.25	2.47M
070	For vinyl floor covering, add				.75		.75	.82M
090	For high pressure laminate covering, add				1.25		1.25	1.37M
091	For snap on stringer system, add	2 Carp	1,000	↓	.35	.31	.66	.84
095	Office applications, to 8" high, steel panels,							
096	no covering, over 6000 S.F.	2 Carp	400	S.F.	4.75	.78	5.53	6.35
100	Machine cutouts after initial installation	1 Carp	10	Ea.	5	15.70	20.70	28
110	Air conditioning grilles, 4" x 12"		17		22	9.20	31.20	38
115	6" x 18"	↓	14	↓	35	11.20	46.20	55
120	Approach ramps, minimum	2 Carp	85	S.F.	9.60	3.69	13.29	15.90
130	Maximum	"	60	"	18.20	5.25	23.45	28
150	Handrail, 2 rail aluminum	1 Carp	15	L.F.	25	10.45	35.45	43
57-001	**REFRIGERATORS** Curbs, 12" high, 4" thick, concrete	2 Carp	58	"	2.65	5.40	8.05	10.70
100								
240	Finishes, 2 coat portland cement plaster, 1/2" thick	1 Plas	48	S.F.	.75	3.27	4.02	5.45
250	For galvanized reinforcing mesh, add	1 Lath	335		.50	.47	.97	1.21
270	3/16" thick latex cement	1 Plas	88		1.25	1.78	3.03	3.91
290	For glass cloth reinforced ceilings, add	"	450		.30	.35	.65	.82
310	Fiberglass panels, 1/8" thick	1 Carp	135		1.52	1.16	2.68	3.35
320	Polystyrene, plastic finish ceiling, 1" thick		230		1.40	.68	2.08	2.52
340	2" thick		230		1.63	.68	2.31	2.78
350	4" thick	↓	205		1.75	.76	2.51	3.03
380	Floors, concrete, 4" thick	1 Cefi	93		.73	1.62	2.35	3.07
390	6" thick	"	85	↓	1.10	1.77	2.87	3.69
400	Insulation, 1" to 6" thick, cork			B.F.	.57		.57	.62M
410	Urethane				.50		.50	.55M
430	Polystyrene, regular				.37		.37	.40M
440	Bead board			↓	.27		.27	.29M
460	Installation of above, add per layer	2 Carp	680	S.F.	.17	.46	.63	.85
470	Wall and ceiling juncture		270	L.F.	.96	1.16	2.12	2.73
490	Partitions, galvanized sandwich panels, 4" thick, stock		220	S.F.	4.40	1.43	5.83	6.90
500	Aluminum or fiberglass	↓	220	"	4.70	1.43	6.13	7.25
520	Prefab walk-in, 7'-6" high, aluminum, incl. door & floors,							
521	not incl. partitions or refrigeration, 6' x 6' O.D. nominal	2 Carp	27	S.F.Flr.	47.35	11.60	58.95	69
550	10' x 10' O.D. nominal		29		33	10.80	43.80	52
570	12' x 14' O.D. nominal		31		27	10.10	37.10	44
580	12' x 20' O.D. nominal	↓	33		25	9.50	34.50	41
600	For aluminum exterior and interior, add				22%			
610	For 8'-6" high, add				5%			
630	Rule of thumb for complete units, not incl. doors, cooler	2 Carp	31		68	10.10	78.10	89
640	Freezer		29	↓	82	10.80	92.80	105
660	Shelving, plated or galvanized, steel wire type		360	S.F.Hor.	5	.87	5.87	6.75
670	Slat shelf type	↓	375		7.25	.84	8.09	9.20
690	For stainless steel shelving, add			↓	290%			

For expanded coverage of these items see *Building Construction Cost Data 1985*

13.1 Special Construction		CREW	DAILY OUTPUT	UNIT	BARE COSTS MAT.	BARE COSTS INST.	BARE COSTS TOTAL	TOTAL INCL O&P
57 700	Vapor barrier, on wood walls	2 Carp	1,815	S.F.	.07	.17	.24	.33
720	On masonry walls	"	1,360	"	.22	.23	.45	.57
750	For air curtain doors, see division 13.1-02							
72-001	**SWIMMING POOL EQUIPMENT** Diving stand, stainless steel, 3 meter	2 Carp	.40	Ea.	2,525	785	3,310	3,900
030	1 meter	"	2.70	"	1,700	115	1,815	2,050
090	Filter system, sand or diatomite type, incl. pump, 6000 gal./hr.	2 Plum	1.80	Total	740	195	935	1,100
102	Add for chlorination system, 800 S.F. pool	↓	3	Ea.	255	120	375	450
104	5000 S.F. pool	↓	3	"	535	120	655	760
110	Gutter system, stainless steel, with grating, stock,							
111	contains supply and drainage system	E-1	20	L.F.	120	28	148	175
112	Integral gutter and 5' high wall system, stainless steel	"	10	"	175	57	232	275
210	Lights, underwater, 12 volt, with transformer, 300 watt	1 Elec	.40	Ea.	175	440	615	830
220	110 volt, 500 watt, standard	↓	.40	↓	60	440	500	700
240	Low water cutoff type	↓	.40	↓	69	440	509	710
280	Heaters, see division 15.5-86							
75-001	**SWIMMING POOLS** Residential in-ground, vinyl lined, concrete sides							
002	Sides including equipment, sand bottom			S.F.Surf				16
010	Metal or polystyrene sides			↓				12
020	Add for vermiculite bottom			↓	.65		.65	.71M
050	Gunite bottom and sides, white plaster finish							
060	350 S.F.			S.F.Surf				25
075	800 S.F.			"				16
110	Motel, gunite with plaster finish, incl. medium							
115	capacity filtration & chlorination			S.F.Surf				22
120	Municipal, gunite with plaster finish, incl. high							
125	capacity filtration & chlorination			S.F.Surf				35
135	Add for formed gutters			L.F.	46		46	51M
136	Add for stainless steel gutters			"	125		125	140M
170	Filtration and deck equipment only, as % of			Total	20%	20%		
180	Deck equipment, rule of thumb, 20' x 40' pool			S.F.Pool				1.30
190	5000 S.F. pool			"				1.90
300	Painting pools, preparation + 3 coats, 20' x 40' pool, epoxy	2 Pord	.33	Total	525	915	1,440	1,875
310	Rubber base paint, 18 gallons	"	.33	↓	395	915	1,310	1,725
350	42' x 82' pool, 75 gallons, epoxy paint	3 Pord	.14	↓	2,425	3,225	5,650	7,250
360	Rubber base paint	"	.14	↓	1,700	3,225	4,925	6,450
77-001	**TANKS** Not incl. pipe or pumps, prestressed concrete, 250,000 gallons			Ea.				175,000
010	500,000 gallons			↓				280,000
030	1,000,000 gallons			↓				380,000
040	2,000,000 gallons			↓				595,000
060	4,000,000 gallons			↓				935,000
070	6,000,000 gallons			↓				1,250,000
090	Steel, ground level, not incl. foundations, 100,000 gallons			↓				80,000
100	250,000 gallons			↓				100,000
120	500,000 gallons			↓				175,000
125	750,000 gallons			↓				205,000
130	1,000,000 gallons			↓				265,000
150	2,000,000 gallons			↓				495,000
160	4,000,000 gallons			↓				795,000
180	6,000,000 gallons			↓				1,075,000
185	8,000,000 gallons			↓				1,475,000
190	10,000,000 gallons			↓				1,975,000
210	Steel standpipes, 100 ft. to overflow, no foundations							
220	500,000 gallons			Ea.				235,000
240	750,000 gallons			↓				295,000
250	1,000,000 gallons			↓				350,000
270	1,500,000 gallons			↓				480,000
280	2,000,000 gallons			↓				585,000
300	Elevated water tanks, 100 ft. to bottom capacity line							
301	50,000 gallons			Ea.				170,000

13.1 Special Construction

		CREW	DAILY OUTPUT	UNIT	BARE COSTS MAT.	BARE COSTS INST.	BARE COSTS TOTAL	TOTAL INCL O&P
77 330	100,000 gallons			Ea.				215,000
340	250,000 gallons							295,000
360	500,000 gallons							475,000
370	750,000 gallons							657,000
390	1,000,000 gallons							835,000
391								
400	Fixed roof oil storage tanks, steel, (1 barrel=42 gallons)							
420	5,000 barrels			Ea.				66,000
430	25,000 barrels							190,000
450	55,000 barrels							320,000
460	100,000 barrels							515,000
480	150,000 barrels							725,000
490	225,000 barrels							1,090,000
491								
510	Floating roof gasoline tanks, steel, 5,000 barrels			Ea.				115,000
520	25,000 barrels							255,000
540	55,000 barrels							385,000
550	100,000 barrels							585,000
570	150,000 barrels							790,000
580	225,000 barrels							1,100,000
600	Wood tanks, ground level, 2" cypress, 3,000 gallons	C-1	.19		3,900	3,250	7,150	8,925
610	2-1/2" cypress, 10,000 gallons		.12		10,700	5,125	15,825	19,100
630	3" redwood or 3" fir, 20,000 gallons		.10		13,600	6,150	19,750	23,800
640	30,000 gallons		.08		18,800	7,700	26,500	31,700
660	45,000 gallons		.07		23,800	8,800	32,600	38,800
661								
670	Larger sizes, minimum			Gal.				.63
690	Maximum			"				.81
700	Vinyl coated fabric pillow tanks, freestanding, 5000 gallons	4 Clab	4	Ea.	1,350	125	1,475	1,675
710	Supporting embankment not included, 25,000 gallons	6 Clab	2		4,750	370	5,120	5,750
720	50,000 gallons	8 Clab	1.50		6,900	660	7,560	8,550
730	100,000 gallons	9 Clab	.90		10,800	1,250	12,050	13,700
740	150,000 gallons		.50		13,500	2,225	15,725	18,100
750	200,000 gallons		.40		16,200	2,800	19,000	21,800
760	250,000 gallons		.30		18,900	3,725	22,625	26,200

For expanded coverage of these items see *Building Construction Cost Data 1985*

13

15

15.1 Pipe & Fittings

		CREW	DAILY OUTPUT	UNIT	BARE COSTS MAT.	BARE COSTS INST.	BARE COSTS TOTAL	TOTAL INCL O&P
01-001	**AVERAGE** Square foot and percent of total							
010	job cost for plumbing, see division 17.1							
04-001	**BACKFLOW PREVENTER** Includes gate valves,							
002	and four test cocks, corrosion resistant, automatic operation							
100	Double check principle							
108	Threaded							
110	 3/4" pipe size	1 Plum	16	Ea.	110	11.10	121.10	135
112	1" pipe size		14		125	12.70	137.70	155
114	1-1/2" pipe size		10		235	17.75	252.75	285
116	2" pipe size		7		310	25	335	380
130	Flanged							
138	3" pipe size	Q-1	4.50	Ea.	835	71	906	1,025
140	4" pipe size	"	3		1,250	105	1,355	1,525
142	6" pipe size	Q-2	3		1,900	165	2,065	2,325
400	Reduced pressure principle							
410	Threaded							

	15.1 Pipe & Fittings	CREW	DAILY OUTPUT	UNIT	BARE COSTS MAT.	BARE COSTS INST.	BARE COSTS TOTAL	TOTAL INCL O&P
04 412	3/4" pipe size	1 Plum	16	Ea.	150	11.10	161.10	180
414	1" pipe size		14		180	12.70	192.70	215
415	1-1/4" pipe size		12		295	14.80	309.80	345
416	1-1/2" pipe size		10		325	17.75	342.75	385
418	2" pipe size		7		400	25	425	475
500	Flanged, bronze							
506	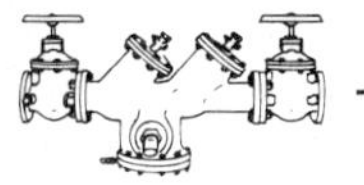 2-1/2" pipe size	Q-1	5	Ea.	1,200	64	1,264	1,400
508	3" pipe size		4.50		1,525	71	1,596	1,775
510	4" pipe size		3		2,125	105	2,230	2,500
512	6" pipe size	Q-2	3		4,475	165	4,640	5,150
560	Flanged, iron							
566	2-1/2" pipe size	Q-1	5	Ea.	1,100	64	1,164	1,300
568	3" pipe size		4.50		1,250	71	1,321	1,475
570	4" pipe size		3		1,750	105	1,855	2,075
572	6" pipe size	Q-2	3		2,675	165	2,840	3,175
574	8" pipe size		2		5,850	250	6,100	6,800
576	10" pipe size		1		7,750	495	8,245	9,250
07-001	**CLEANOUTS**							
006	Floor type							
008	Round or square, scoriated nickel bronze top							
010	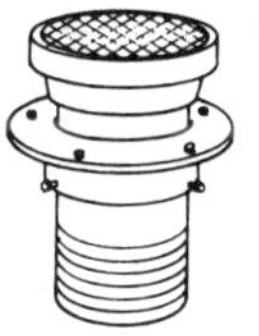 2" pipe size	1 Plum	10	Ea.	34	17.75	51.75	63
012	3" pipe size		8		36	22	58	72
014	4" pipe size		6		48	30	78	96
016	5" pipe size		4		79	44	123	150
018	6" pipe size	Q-1	6		79	53	132	165
020	8" pipe size	"	4		96	80	176	220
034	Recessed for tile, same price							
098	Round top, recessed for terrazzo							
100	2" pipe size	1 Plum	9	Ea.	54	19.75	73.75	88
108	3" pipe size		6		57	30	87	105
110	4" pipe size		4		70	44	114	140
112	5" pipe size	Q-1	6		110	53	163	200
114	6" pipe size		5		115	64	179	220
116	8" pipe size		4		140	80	220	270
200	Round scoriated nickel bronze top, extra heavy duty							
206	2" pipe size	1 Plum	9	Ea.	43	19.75	62.75	76
208	3" pipe size		6		47	30	77	94
210	4" pipe size		4		59	44	103	130
212	5" pipe size	Q-1	6		94	53	147	180
214	6" pipe size		5		100	64	164	200
216	8" pipe size		4		155	80	235	285
400	Wall type, square smooth cover, over wall frame							
406	 2" pipe size	1 Plum	14	Ea.	39	12.70	51.70	61
408	3" pipe size		12		42	14.80	56.80	68
410	4" pipe size		10		52	17.75	69.75	83
412	5" pipe size		9		57	19.75	76.75	91
414	6" pipe size		8		64	22	86	100
416	8" pipe size	Q-1	11		95	29	124	145
500	Extension, C.I., bronze countersunk plug, 8" long							
504	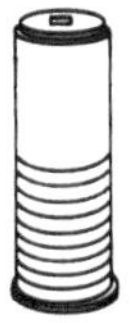 2" pipe size	1 Plum	16	Ea.	9.45	11.10	20.55	26
506	3" pipe size		14		12.55	12.70	25.25	32
508	4" pipe size		13		13.20	13.65	26.85	34
510	5" pipe size		12		27	14.80	41.80	51
512	6" pipe size		11		35	16.15	51.15	62
10-001	**CLEANOUT TEE** Cast iron with countersunk plug							
020	 2" pipe size	1 Plum	4	Ea.	12.35	44	56.35	78
022	3" pipe size		3.60		15.45	49	64.45	88
024	4" pipe size		3.30		24	54	78	105

15

15.1 Pipe & Fittings

		CREW	DAILY OUTPUT	UNIT	BARE COSTS MAT.	BARE COSTS INST.	BARE COSTS TOTAL	TOTAL INCL O&P
10 026	5" pipe size	Q-1	5.50	Ea.	45	58	103	135
028	6" pipe size	"	5		69	64	133	170
030	8" pipe size	Q-3	5		86	135	221	290
050	For round smooth access cover, add				20%			
060	For round scoriated access cover, add				75%			
070	For square smooth access cover, add				200%			
13-001	**CONNECTORS** Flexible, corrugated, 7/8" O.D., 1/2" I.D.							
005	Gas, seamless brass, steel fittings							
020	12" long	1 Plum	36	Ea.	4.60	4.93	9.53	12.20
022	18" long		36		5.70	4.93	10.63	13.40
024	24" long		34		6.80	5.20	12	15.05
026	30" long		34		7.30	5.20	12.50	15.60
028	36" long		32		8.05	5.55	13.60	16.85
032	48" long		30		10.25	5.90	16.15	19.85
034	60" long		30		12.20	5.90	18.10	22
036	72" long		30		14.10	5.90	20	24
200	Water, copper tubing, dielectric separators							
210	12" long	1 Plum	36	Ea.	4.34	4.93	9.27	11.90
222	15" long		36		4.81	4.93	9.74	12.40
224	18" long		36		5.25	4.93	10.18	12.90
226	24" long		34		6.45	5.20	11.65	14.65
16-001	**DRAINS**							
014	Cornice, C.I., 45° or 90° outlet							
018	B8.1 -310 1-1/2" & 2" pipe size	Q-1	14	Ea.	29	23	52	65
020	3" and 4" pipe size	"	12		40	27	67	83
026	For galvanized body, add				7.70		7.70	8.45M
028	For polished bronze dome, add				6.60		6.60	7.25M
040	Deck, auto park, C.I., 13" top							
044	 3", 4", 5", and 6" pipe size	Q-1	8	Ea.	140	40	180	210
048	For galvanized body, add			"	63		63	69M
080	Promenade, heelproof grate, C.I., 14" top							
084	2", 3", and 4" pipe size	Q-1	10	Ea.	66	32	98	120
086	5" and 6" pipe size		9		82	36	118	140
088	8" pipe size		8		97	40	137	165
094	For galvanized body, add				42		42	46M
096	For polished bronze top, add				44		44	48M
120	Promenade, heelproof grate, C.I., lateral, 14" top							
124	2", 3" and 4" pipe size	Q-1	10	Ea.	89	32	121	145
126	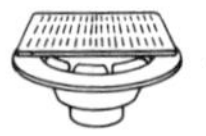 5" and 6" pipe size		9		110	36	146	170
128	8" pipe size		8		130	40	170	200
134	For galvanized body, add				43		43	47M
136	For polished bronze top, add				44		44	48M
150	Promenade, slotted grate, C.I., 11" top							
154	2", 3", 4", 5", and 6" pipe size	Q-1	12	Ea.	59	27	86	105
160	For galvanized body, add				24		24	26M
164	For polished bronze top, add				35		35	39M
200	Floor, medium duty, C.I., deep flange, 7" top							
204	2" and 3" pipe size	Q-1	12	Ea.	24	27	51	65
208	For galvanized body, add				10.30		10.30	11.35M
212	For polished bronze top, add				13.40		13.40	14.75M
216	Heavy duty, C.I., 12" antitilt grate							
218	2", 3", 4", 5" and 6" pipe size	Q-1	10	Ea.	56	32	88	110
222	 For galvanized body, add				30		30	33M
224	For polished bronze top, add				35		35	39M
230	X-Heavy duty, C.I., 15" antitilt grate							
232	4", 5", 6", and 8" pipe size	Q-1	8	Ea.	150	40	190	225
236	For galvanized body, add				51		51	56M
238	For polished bronze top, add				59		59	65M
240	Heavy duty, with sediment bucket, C.I., 12" loose grate							

15.1 Pipe & Fittings

		CREW	DAILY OUTPUT	UNIT	BARE COSTS MAT.	BARE COSTS INST.	BARE COSTS TOTAL	TOTAL INCL O&P
16 242	3", 4", 5", and 6" pipe size	Q-1	9	Ea.	82	36	118	140
244	For galvanized body, add				42		42	46M
246	For polished bronze top, add			↓	35		35	39M
250	Heavy duty, cleanout & trap w/bucket, C.I., 15" top							
254	2", 3", and 4" pipe size	Q-1	6	Ea.	1,075	53	1,128	1,250
256	For galvanized body, add				175		175	195M
258	For polished bronze top, add			↓	110		110	120M
260	Medium duty, with perforated SS basket, C.I., body,							
261	18" top for refuse container washing area							
262	2" thru 6" pipe size	Q-1	4	Ea.	995	80	1,075	1,200
263	Polyethylene, corrosion-resistant							
264	3" and 4" pipe size	Q-1	16	Ea.	200	20	220	250
265	PVC or ABS thermoplastic							
266	3" and 4" pipe size	Q-1	16	Ea.	35	20	55	67
268	Extra heavy duty, oil intercepting, gas seal cone,							
269	with cleanout, loose grate, C.I. body 16" top							
270	3" and 4" diameter outlet, 4" slab depth	Q-1	4	Ea.	2,075	80	2,155	2,400
272	3" and 4" diameter outlet, 8" slab depth		3		2,075	105	2,180	2,425
274	3" and 4" diam. outlet, 10"-12" slab depth	↓	2	↓	2,075	160	2,235	2,525
278	Shower, with strainer, uniform diam. trap, bronze top							
280	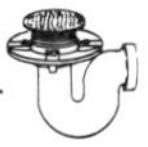 1-1/2", 2" and 3" pipe size	Q-1	8	Ea.	61	40	101	125
282	4" pipe size	"	7		90	46	136	165
284	For galvanized body, add			↓	21		21	23M
286	With strainer, backwater valve, drum trap							
288	1-1/2", 2", and 3" pipe size	Q-1	8	Ea.	72	40	112	135
289	4" pipe size	"	7		120	46	166	200
290	For galvanized body, add				23		23	25M
291	Prison cell, vandal-proof, 1-1/2", and 2" diam. pipe	Q-1	12		15.15	27	42.15	55
292	3" pipe size	"	10	↓	42	32	74	92
293	Trap drain, light duty, backwater valve C.I. top							
295	8" diameter top, 2" pipe size	Q-1	12	Ea.	46	27	73	89
296	10" diameter top, 3" pipe size		10		63	32	95	115
297	12" diameter top, 4" pipe size	↓	8	↓	90	40	130	155
299	Pool main discharge, polished bronze top, C.I. body							
300	8" diam. grate, 2", 3" & 4" pipe size	Q-1	8	Ea.	60	40	100	125
310	12" diam. grate, 4", 5" & 6" pipe size	"	8		135	40	175	205
316	For galvanized body, add			↓	30		30	33M
320	Gutter, bronze body, polished bronze top							
322	9" grate, 1-1/2" & 2" pipe size	Q-1	8	Ea.	24	40	64	84
324	12" grate, 2", 2-1/2" & 3" pipe size	"	7	"	56	46	102	130
338	Overflow, 5" dome & 6" high pipe polished bronze,							
340	cast iron body, 2", 3", & 4" pipe size	Q-1	10	Ea.	91	32	123	145
358	Recirculating inlet, 6" sq face polished bronze,							
360	C.I. body, 1-1/2", 2", 3" & 4" pipe size	Q-1	10	Ea.	36	32	68	86
362	For galvanized body, add			"	9.35		9.35	10.30M
386	Roof, flat metal deck, C.I. body, 10" aluminum dome							
388	2" pipe size	Q-1	15	Ea.	83	21	104	120
389	3" pipe size		14		86	23	109	130
390	4" pipe size		13		91	25	116	135
391	5" pipe size		12		110	27	137	160
392	6" pipe size	↓	10	↓	155	32	187	215
398	Precast plank deck, C.I. body, aluminum dome							
400	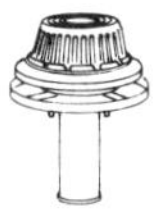 10" top, 2" pipe size	Q-1	14	Ea.	40	23	63	77
410	10" top, 3" pipe size		13		47	25	72	87
412	13" top, 4" pipe size		12		62	27	89	105
414	13" top, 5" pipe size		10		73	32	105	125
416	16" top, 6" pipe size		8		87	40	127	155
418	16" top, 8" pipe size	↓	7		125	46	171	205
422	For galvanized body, add			↓	34		34	37M
428	Integral expansion joint, C.I. body, 12" polypropylene dome							

15

		15.1 Pipe & Fittings	CREW	DAILY OUTPUT	UNIT	BARE COSTS MAT.	INST.	TOTAL	TOTAL INCL O&P
16	430	2" pipe size	Q-1	8	Ea.	72	40	112	135
	432	3" pipe size		7		75	46	121	150
	434	4" pipe size		6		80	53	133	165
	436	5" pipe size		4		115	80	195	240
	438	6" pipe size		3		125	105	230	290
	440	8" pipe size		3		195	105	300	370
	444	For galvanized body, add				51		51	56M
	448	Flexible neoprene bellows, no-hub connection							
	450	3" and 4" pipe size	Q-1	5	Ea.	98	64	162	200
	452	5" and 6" pipe size	"	4	"	105	80	185	230
	462	Main, all aluminum, 12" low profile dome							
	464	2", 3" and 4" pipe size	Q-1	14	Ea.	120	23	143	165
	466	5" and 6" pipe size		13		135	25	160	185
	468	8" pipe size		10		170	32	202	235
	469	Main, C.I. body, 12" polyprop. dome, 2", 3", & 4" pipe size		8		36	40	76	97
	471	5" and 6" pipe size		6		51	53	104	135
	472	8" pipe size		4		67	80	147	190
	473	For underdeck clamp, add		22		12.35	14.55	26.90	35
	474	For vandalproof dome, add				6.20		6.20	6.80M
	475	For galvanized body, add				53		53	58M
	476	Main, PVC body and dome, 2" pipe size	Q-1	14	Ea.	32	23	55	68
	478	3" pipe size		14		32	23	55	68
	480	4" pipe size		14		37	23	60	74
	482	For underdeck clamp, add		24		8.45	13.30	21.75	29
	490	Terrace planting area, with perforated overflow, C.I.							
	492	2", 3" and 4" pipe size	Q-1	8	Ea.	97	40	137	165
	498	Scupper floor, oblique strainer, C.I.							
	500	6" x 7" top, 2", 3" and 4" pipe size	Q-1	16	Ea.	33	20	53	65
	510	 8" x 12" top, 5" and 6" pipe size	"	14		64	23	87	105
	516	For galvanized body, add				40%			
	520	For polished bronze strainer, add				100%			
	598	Trench, floor, heavy duty, modular, C.I., 12" x 12" top							
	600	2", 3", 4", 5", & 6" pipe size	Q-1	8	Ea.	135	40	175	205
	610	For polished bronze top, add				80		80	88M
	620	For 12" extension section, C.I. top, add				105		105	115M
	624	For polished bronze top, add				170		170	185M
	696	Backwater valve, in soil pipe, C.I. body,							
	698	bronze gate and automatic flapper valves							
	700	3" and 4" pipe size	Q-1	13	Ea.	215	25	240	270
	710	5" and 6" pipe size	"	13	"	345	25	370	415
	724	Bronze flapper valve, bolted cover							
	726	2" pipe size	Q-1	16	Ea.	58	20	78	93
	728	3" pipe size		14.50		90	22	112	130
	730	4" pipe size		13		110	25	135	155
	732	5" pipe size	Q-2	18		135	28	163	190
	734	6" pipe size	"	17		160	29	189	220
	736	8" pipe size	Q-3	10		210	68	278	330
	738	10" pipe size	"	9		425	75	500	575
	750	For threaded cover, same cost							
	754	Revolving disk type, same cost as flapper type							
	796	Drainguard, floating plug, brass							
	800	3" pipe size	1 Plum	14	Ea.	16.90	12.70	29.60	37
	804	4" pipe size	"	14	"	25	12.70	37.70	46
	900	Sewer control system, valve unit, guardrail, leak detector							
	901	With fittings, monitor, 40' cable. No pit fabrication							
	902	4" pipe size	Q-2	.25	Ea.	4,350	2,000	6,350	7,650
	912	5" pipe size		.22		4,750	2,250	7,000	8,500
	914	6" pipe size		.20		5,650	2,475	8,125	9,800
	916	8" pipe size		.17		6,450	2,925	9,375	11,300

15.1 Pipe & Fittings

		CREW	DAILY OUTPUT	UNIT	BARE COSTS MAT.	BARE COSTS INST.	BARE COSTS TOTAL	TOTAL INCL O&P
19-001	**DIELECTRIC UNIONS** Standard gaskets for water and air							
002	250 psi maximum pressure							
028	Female IPT to sweat, straight							
030	1/2" pipe size	1 Plum	24	Ea.	2.34	7.40	9.74	13.25
034	3/4" pipe size		20		2.34	8.90	11.24	15.40
036	1" pipe size		19		4.17	9.35	13.52	18.10
038	1-1/4" pipe size		15		6.85	11.85	18.70	25
040	1-1/2" pipe size		13		10.25	13.65	23.90	31
042	2" pipe size		11		14.35	16.15	30.50	39
058	Female IPT to brass pipe thread, straight							
060	1/2" pipe size	1 Plum	24	Ea.	3.95	7.40	11.35	15.05
064	3/4" pipe size		20		5.05	8.90	13.95	18.40
066	1" pipe size		19		7.60	9.35	16.95	22
068	1-1/4" pipe size		15		11.35	11.85	23.20	30
070	1-1/2" pipe size		13		15.55	13.65	29.20	37
072	2" pipe size		11		21	16.15	37.15	46
078	Female IPT to female IPT, straight							
080	1/2" pipe size	1 Plum	24	Ea.	3.16	7.40	10.56	14.15
084	3/4" pipe size		20		3.52	8.90	12.42	16.70
086	1" pipe size		19		4.68	9.35	14.03	18.65
088	1-1/4" pipe size		15		6.40	11.85	18.25	24
090	1-1/2" pipe size		13		9.85	13.65	23.50	31
092	2" pipe size		11		14.30	16.15	30.45	39
200	175 psi maximum pressure, flanged							
218	Female IPT to sweat							
220	1-1/2" pipe size	1 Plum	11	Ea.	25	16.15	41.15	51
224	2" pipe size	″	9		29	19.75	48.75	60
226	2-1/2" pipe size	Q-1	15		32	21	53	66
228	3" pipe size		14		44	23	67	81
230	4" pipe size		11		115	29	144	170
232	5" pipe size	Q-2	14		190	36	226	260
234	6" pipe size		12		220	41	261	300
236	8" pipe size		10		670	50	720	810
248	Female IPT to brass pipe							
250	1-1/2" pipe size	1 Plum	11	Ea.	38	16.15	54.15	65
254	2" pipe size	″	9		46	19.75	65.75	79
256	2-1/2" pipe size	Q-1	15		66	21	87	105
258	3" pipe size		14		77	23	100	120
260	4" pipe size		11		130	29	159	185
262	5" pipe size	Q-2	14		150	36	186	215
264	6" pipe size		12		170	41	211	245
268	8" pipe size		10		280	50	330	380
288	Female IPT to female IPT							
290	1-1/2" pipe size	1 Plum	11	Ea.	23	16.15	39.15	49
294	2" pipe size	″	9		25	19.75	44.75	56
296	2-1/2" pipe size	Q-1	15		31	21	52	65
298	3" pipe size		14		39	23	62	76
300	4" pipe size		11		58	29	87	105
302	5" pipe size	Q-2	14		67	36	103	125
304	6" pipe size		12		86	41	127	155
306	8" pipe size		10		96	50	146	175
338	Copper to copper							
340	1-1/2" pipe size	1 Plum	11	Ea.	30	16.15	46.15	56
344	2" pipe size	″	9		35	19.75	54.75	67
346	2-1/2" pipe size	Q-1	15		39	21	60	74
348	3" pipe size		14		48	23	71	86
350	4" pipe size		11		145	29	174	200
352	5" pipe size	Q-2	14		235	36	271	310
354	6" pipe size		12		295	41	336	385
356	8" pipe size		10		800	50	850	950

15.1 Pipe & Fittings

	Pipe & Fittings	CREW	DAILY OUTPUT	UNIT	BARE COSTS MAT.	BARE COSTS INST.	BARE COSTS TOTAL	TOTAL INCL O&P
22-001	FAUCETS/FITTINGS							
015	Bath, faucets, diverter spout combination, sweat	1 Plum	8	Ea.	37	22	59	73
020	For integral stops, IPS unions, add				18		18	19.80M
030	Three valve combinations, spout, head, arm, flange, sweat	1 Plum	6		37	30	67	83
040	For integral stops, IPS unions, add				24		24	26M
050	Drain, central lift, 1-1/2" IPS male	1 Plum	20		23	8.90	31.90	38
060	Trip lever, 1-1/2" IPS male		20		22	8.90	30.90	37
070	Pop up, 1-1/2" IPS male		18		27	9.85	36.85	44
080	Chain and stopper, 1-1/2" IPS male		24		18.55	7.40	25.95	31
084	Flush valves, with vacuum breaker							
085	Water closet							
086	Exposed, rear spud	1 Plum	8	Ea.	74	22	96	115
087	Top spud		8		67	22	89	105
088	Concealed, rear spud		8		76	22	98	115
089	Top spud		8		87	22	109	130
090	Wall hung		8		73	22	95	110
092	Urinal							
093	Exposed, stall	1 Plum	8	Ea.	67	22	89	105
094	Wall, (washout)		8		67	22	89	105
095	Pedestal, top spud		8		69	22	91	110
096	Concealed, stall		8		69	22	91	110
097	Wall (washout)		8		69	22	91	110
100	Kitchen sink faucets, top mount, cast spout		10		27	17.75	44.75	55
110	For spray, add				8	10%		
120	Wall type, swing tube spout	1 Plum	10		33	17.75	50.75	62
124	For soap dish, add				2		2	2.20M
200	Laundry faucets, shelf type, I.P.S. or copper unions	1 Plum	12		25	14.80	39.80	49
210	Lavatory faucet, centerset, without drain	″	10		19	17.75	36.75	47
220	For pop-up drain, add				8.95	15%		
225	For acrylic handles, add				3.17		3.17	3.48M
240	Concealed, 12" centers	1 Plum	10		45	17.75	62.75	75
245	For pop-up drain, add				7.90	15%		
260	Shelfback, 4" to 6" centers, 17 Ga. tailpiece	1 Plum	10		33	17.75	50.75	62
265	For pop-up drain, add				8.95	15%		
270	For shampoo faucet with supply tube, add				22	10%		
280	Self-closing, center set	1 Plum	10		53	17.75	70.75	84
285	Medical, bedpan cleanser, with pedal valve,		12		295	14.80	309.80	345
286	With screwdriver stop valve		12		150	14.80	164.80	185
287	With self-closing spray valve		12		115	14.80	129.80	150
290	Faucet, gooseneck spout, wrist handles, grid drain		10		175	17.75	192.75	220
294	Mixing valve, knee action, screwdriver stops		4		220	44	264	305
300	Service sink faucet, cast spout, pail hook, hose end		14		38	12.70	50.70	60
400	Shower by-pass valve with union		18		29	9.85	38.85	46
410	Shower arm with flange and head		22		10.80	8.05	18.85	24
420	Shower thermostatic mixing valve, concealed		8		140	22	162	185
430	For inlet strainer, check, and stops, add				40	5%		
500	Sillcock, compact, brass, I.P.S. or copper to hose	1 Plum	24		4.07	7.40	11.47	15.15
600	Stop and waste valves, bronze							
610	Angle, solder end 1/2"	1 Plum	24	Ea.	4.85	7.40	12.25	16.05
611	3/4"		20		5.25	8.90	14.15	18.60
630	Straightway, solder end 3/8"		24		3.46	7.40	10.86	14.50
631	1/2"		24		3.09	7.40	10.49	14.10
632	3/4"		20		3.50	8.90	12.40	16.70
633	1"		19		12.10	9.35	21.45	27
640	Straightway, threaded 3/8"		24		3.70	7.40	11.10	14.75
641	1/2"		24		3.70	7.40	11.10	14.75
642	3/4"		20		4.01	8.90	12.91	17.25
643	1"		19		12.40	9.35	21.75	27
800	Water supply stops, polished chrome plate							
820	Angle, 3/8"	1 Plum	24	Ea.	9.75	7.40	17.15	21

15.1	Pipe & Fittings	CREW	DAILY OUTPUT	UNIT	BARE COSTS MAT.	BARE COSTS INST.	BARE COSTS TOTAL	TOTAL INCL O&P
22 830	1/2″	1 Plum	22	Ea.	10.85	8.05	18.90	24
840	Straight, 3/8″		26		9.75	6.85	16.60	21
850	1/2″		24		10.85	7.40	18.25	23
860	Water closet, angle, w/flex riser, 3/8″		24		12.80	7.40	20.20	25
25-001	**FLOOR RECEPTORS** For connection to 2″, 3″ & 4″ diameter pipe							
020	12-1/2″ square top, 25 sq. in. open area	Q-1	10	Ea.	155	32	187	215
030	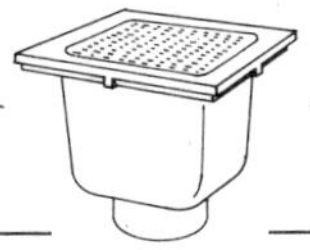 For grate with 4″ diam. x 3-3/4″ high funnel, add				23		23	25M
040	For grate with 6″ diameter x 6″ high funnel, add				34		34	37M
050	For full hinged grate with open center, add				29		29	32M
060	For aluminum bucket, add				17.15		17.15	18.85M
070	 For acid-resisting bucket, add				33		33	36M
080	For galvanized bucket, add				21		21	23M
090	For stainless steel mesh bucket liner, add				26		26	29M
100	For bronze antisplash dome strainer, add				12.35		12.35	13.60M
110	For partial solid cover, add				12.35		12.35	13.60M
120	For trap primer connection, add				13.40		13.40	14.75M
200	12-5/8″ diameter top, 40 sq. in. open area	Q-1	10		135	32	167	195
210	For options, add same prices as square top							
300	8″ x 4″ rectangular top, 7.5 sq. in. open area	Q-1	14		89	23	112	130
310	For trap primer connections, add				10.30		10.30	11.35M
400	24″ x 16″ rectangular top, 70 sq. in. open area	Q-1	4		335	80	415	485
410	For trap primer connection, add				21		21	23M
28-001	**HYDRANTS**							
005	Wall type, moderate climate, bronze, encased							
020	3/4″ IPS connection	1 Plum	16	Ea.	91	11.10	102.10	115
030	1″ IPS connection	″	14		105	12.70	117.70	135
040	For 3/4″ adapter type vacuum breaker, add				41		41	45M
050	For anti-siphon type, add				50		50	55M
100	Non-freeze, bronze, exposed							
110	3/4″ IPS connection, 4″ to 9″ thick wall	1 Plum	14	Ea.	64	12.70	76.70	89
112	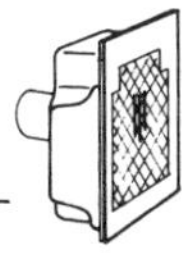 10″ to 14″ thick wall		12		69	14.80	83.80	97
114	15″ to 19″ thick wall		12		77	14.80	91.80	105
116	20″ to 24″ thick wall		10		83	17.75	100.75	115
120	For 1″ IPS connection, add				12	10%		
124	For 3/4″ adapter type vacuum breaker, add				7.70		7.70	8.45M
128	For anti-siphon type, add				16.30		16.30	17.95M
200	Non-freeze bronze, encased							
210	3/4″ IPS connection, 5″ to 9″ thick wall	1 Plum	14	Ea.	110	12.70	122.70	140
212	10″ to 14″ thick wall		12		115	14.80	129.80	150
214	15″ to 19″ thick wall		12		120	14.80	134.80	155
216	20″ to 24″ thick wall		10		130	17.75	147.75	170
220	For 1″ IPS connection, add				12.05	10%		
224	For 3/4″ adapter type vacuum breaker, add				41		41	45M
228	For anti-siphon type, add				50		50	55M
300	Ground box type, bronze frame, 3/4″ IPS connection							
308	Non-freeze, all bronze, polished face, set flush							
310	 2 feet depth of bury	1 Plum	8	Ea.	130	22	152	175
312	3 feet depth of bury		8		135	22	157	180
314	4 feet depth of bury		8		145	22	167	190
316	5 feet depth of bury		7		155	25	180	205
318	6 feet depth of bury		7		165	25	190	220
320	7 feet depth of bury		6		175	30	205	235
322	8 feet depth of bury		5		185	36	221	255
324	9 feet depth of bury		4		195	44	239	280
326	10 feet depth of bury		4		210	44	254	295
340	For 1″ IPS connection, add				18	10%		
345	For 1-1/4″ IPS connection, add				300	14%		
350	For 1-1/2″ connection, add				350	18%		
355	For 2″ connection, add				425	24%		
360	For tapped drain port in box, add				10.30		10.30	11.35M

15.1 Pipe & Fittings

			DAILY		BARE COSTS			TOTAL
		CREW	OUTPUT	UNIT	MAT.	INST.	TOTAL	INCL O&P
28 400	Non-freeze, CI body, bronze frame & scoriated cover							
401	with hose storage							
410	2 feet depth of bury	1 Plum	7	Ea.	385	25	410	460
412	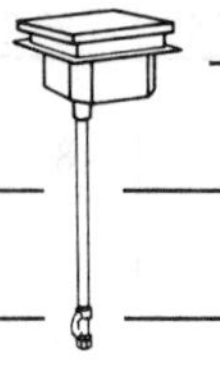 3 feet depth of bury		7		395	25	420	470
414	4 feet depth of bury		7		400	25	425	475
416	5 feet depth of bury		6.50		405	27	432	485
418	6 feet depth of bury		6		410	30	440	495
420	7 feet depth of bury		5.50		420	32	452	510
422	8 feet depth of bury		5		430	36	466	525
424	9 feet depth of bury		4.50		435	39	474	535
426	10 feet depth of bury		4		445	44	489	555
428	For 1" IPS connection, add				47		47	52M
430	For tapped drain port in box, add				10.30		10.30	11.35M
500	Moderate climate, all bronze, polished face							
502	and scoriated cover, set flush							
510	3/4" IPS connection	1 Plum	16	Ea.	100	11.10	111.10	125
512	1" IPS connection	"	14		115	12.70	127.70	145
520	For tapped drain port in box, add				10.30		10.30	11.35M
600	Ground post type,all non-freeze, all bronze, aluminum casing							
601	guard, exposed head, 3/4" IPS connection							
610	2 feet depth of bury	1 Plum	8	Ea.	93	22	115	135
612	3 feet depth of bury		8		100	22	122	140
614	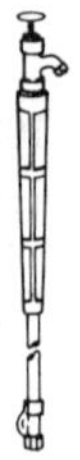 4 feet depth of bury		8		110	22	132	155
616	5 feet depth of bury		7		115	25	140	165
618	6 feet depth of bury		7		125	25	150	175
620	7 foot depth of bury		6		130	30	160	185
622	8 feet depth of bury		5		140	36	176	205
624	9 feet depth of bury		4		150	44	194	230
626	10 feet depth of bury		4		160	44	204	240
630	For 1" IPS connection, add				28	10%		
635	For 1-1/4" IPS connection, add				205	14%		
640	For 1-1/2" IPS connection, add				300	18%		
645	For 2" IPS connection, add				390	24%		
30-001	**PIPING** See also division 2.5 for sitework							
100	Add to labor for elevated installation							
108	10' to 15' high					10%		
110	15' to 20' high					20%		
112	20' to 25' high					25%		
114	25' to 30' high					35%		
116	30' to 35' high					40%		
118	35' to 40' high					50%		
120	Over 40' high					55%		
31-001	**PIPE, ALUMINUM** Prices for 1 to 2-1/2 ton quantity							
085	For 5 or more tons, deduct			L.F.	10%			
090	Does not include hangers or fittings							
098	Schedule 5, type 6063-T6							
100	1-1/4" diameter			L.F.	.71		.71	.78M
110	1-1/2" diameter				.82		.82	.90M
112	2" diameter				1.03		1.03	1.13M
114	2-1/2" diameter				1.49		1.49	1.63M
116	3" diameter				1.79		1.79	1.96M
118	4" diameter				2.41		2.41	2.65M
120	5" diameter				3.80		3.80	4.18M
122	6" diameter				4.79		4.79	5.25M
200	Schedule 10, type 6063-T6, 3/4" diameter				.84		.84	.92M
210	1" diameter				.82		.82	.90M
212	1-1/4" diameter				1.04		1.04	1.14M
214	1-1/2" diameter				1.20		1.20	1.32M

15

15.1 Pipe & Fittings

		CREW	DAILY OUTPUT	UNIT	BARE COSTS MAT.	BARE COSTS INST.	BARE COSTS TOTAL	TOTAL INCL O&P
31 216	2" diameter			L.F.	1.53		1.53	1.68M
218	2-1/2" diameter				2.04		2.04	2.24M
220	3" diameter				2.46		2.46	2.70M
222	4" diameter				3.19		3.19	3.50M
224	5" diameter				4.45		4.45	4.89M
226	6" diameter				5.75		5.75	6.30M
300	Schedule 40, type 6063-T6, 1" diameter				.97		.97	1.06M
310	1-1/4" diameter				1.29		1.29	1.41M
312	1-1/2" diameter				1.54		1.54	1.69M
314	2" diameter				2.07		2.07	2.27M
316	2-1/2" diameter				3.29		3.29	3.61M
318	3" diameter				4.24		4.24	4.66M
320	3-1/2" diameter				5.10		5.10	5.60M
322	4" diameter				6.05		6.05	6.65M
324	5" diameter				8.40		8.40	9.25M
326	6" diameter				11.80		11.80	13M
328	8" diameter				17.75		17.75	19.50M
330	10" diameter				27		27	30M
332	12" diameter				55		55	61M
398	Type 6061-T6							
400	1/4" diameter			L.F.	.54		.54	.59M
410	3/8" diameter				.71		.71	.78M
412	1/2" diameter				.92		.92	1.01M
414	3/4" diameter				1.06		1.06	1.16M
416	1" diameter				1.06		1.06	1.16M
418	1-1/4" diameter				1.38		1.38	1.51M
420	1-1/2" diameter				1.66		1.66	1.82M
422	2" diameter				2.23		2.23	2.45M
424	2-1/2" diameter				3.54		3.54	3.89M
426	3" diameter				4.63		4.63	5.10M
428	3-1/2" diameter				5.60		5.60	6.15M
430	4" diameter				6.60		6.60	7.25M
432	5" diameter				9		9	9.90M
434	6" diameter				12.65		12.65	13.90M
436	8" diameter				19.05		19.05	21M
438	10" diameter				29		29	32M
440	12" diameter				58		58	64M
442	14" diameter				80		80	88M
444	16" diameter				93		93	100M
600	Schedule 80, type 6063-T6							
602	1" diameter			L.F.	1.25		1.25	1.37M
610	1-1/4" diameter				1.70		1.70	1.87M
612	1-1/2" diameter				2.06		2.06	2.26M
614	2" diameter				2.85		2.85	3.13M
616	2-1/2" diameter				4.35		4.35	4.78M
618	3" diameter				5.75		5.75	6.30M
620	3-1/2" diameter				7		7	7.70M
622	4" diameter				8.40		8.40	9.25M
624	5" diameter				11.90		11.90	13.10M
626	6" diameter				17.75		17.75	19.50M
628	8" diameter				27		27	30M
630	10" diameter				42		42	46M
632	12" diameter				88		88	97M
698	Type 6061-T6							
700	1" diameter			L.F.	1.37		1.37	1.50M
710	1-1/4" diameter				1.83		1.83	2.01M
712	1-1/2" diameter				2.22		2.22	2.44M
714	2" diameter				3.07		3.07	3.37M
716	2-1/2" diameter				4.68		4.68	5.15M
718	3" diameter				6.30		6.30	6.95M

15.1 Pipe & Fittings

	15.1 Pipe & Fittings	CREW	DAILY OUTPUT	UNIT	BARE COSTS MAT.	BARE COSTS INST.	BARE COSTS TOTAL	TOTAL INCL O&P
31 720	3-1/2" diameter			L.F.	7.65		7.65	8.40M
722	4" diameter				9.15		9.15	10.05M
724	5" diameter				12.80		12.80	14.10M
726	6" diameter				19.05		19.05	21M
728	8" diameter				29		29	32M
730	10" diameter				44		44	48M
732	12" diameter				91		91	100M
34-001	**PIPE, BRASS** Plain end,							
090	Field threaded, coupling & clevis hanger 10' OC							
092	Regular weight							
098	1/8" diameter	1 Plum	62	L.F.	1.77	2.86	4.63	6.10
100	1/4" diameter		57		2.48	3.12	5.60	7.25
110	3/8" diameter		52		3.17	3.42	6.59	8.40
112	1/2" diameter		48		4.25	3.70	7.95	10
114	3/4" diameter		46		5.72	3.86	9.58	11.85
116	1" diameter		43		8.05	4.13	12.18	14.80
118	1-1/4" diameter	Q-1	72		11.97	4.44	16.41	19.60
120	1-1/2" diameter		65		13.98	4.92	18.90	22
122	2" diameter		53		18.17	6.05	24.22	29
124	2-1/2" diameter		41		27.41	7.80	35.21	41
126	3" diameter		31		38.09	10.30	48.39	57
128	3-1/2" diameter	Q-2	39		56.80	12.75	69.55	81
130	4" diameter		37		56.88	13.45	70.33	82
132	5" diameter		31		84.39	16.05	100.44	115
134	6" diameter		24		105.60	21	126.60	145
136	8" diameter		20		195.67	25	220.67	250
193	To delete coupling & hanger, subtract							
194	1/8" diam. to 1/2" diam.				14%	46%		
195	3/4" diam. to 1-1/2" diam.				8%	47%		
196	2" diam. to 5" diam.				10%	37%		
197	6" diam. to 8" diam.				13%	29%		
200	Extra heavy weight							
210	1/8" diameter	1 Plum	58	L.F.	2.35	3.06	5.41	7
212	1/4" diameter		55		3.46	3.23	6.69	8.45
214	3/8" diameter		50		4.34	3.55	7.89	9.90
216	1/2" diameter		46		6.05	3.86	9.91	12.25
218	3/4" diameter		44		8.25	4.04	12.29	14.90
220	1" diameter		41		11.90	4.33	16.23	19.35
222	1-1/4" diameter	Q-1	69		16.42	4.63	21.05	25
224	1-1/2" diameter		62		19.81	5.15	24.96	29
226	2" diameter		51		26.64	6.25	32.89	38
228	2-1/2" diameter		39		39.51	8.20	47.71	55
230	3" diameter		30		54.59	10.65	65.24	75
232	3-1/2" diameter	Q-2	39		79.90	12.75	92.65	105
234	4" diameter	"	35		79.98	14.20	94.18	110
295	To delete coupling & hanger, subtract							
296	1/8" diam. to 1/2" diam.				16%	46%		
297	3/4" diam. to 1-1/2" diam.				10%	47%		
298	2" diam. to 4" diam.				12%	37%		
35-001	**PIPE, BRASS, FITTINGS,** Rough bronze, threaded							
100	Standard wt., 90° Elbow							
104	1/8"	1 Plum	13	Ea.	1.19	13.65	14.84	21
106	1/4"		13		1.19	13.65	14.84	21
108	3/8"		13		1.19	13.65	14.84	21
110	1/2"		12		1.76	14.80	16.56	23
112	3/4"		11		2.40	16.15	18.55	26
114	1"		10		3.72	17.75	21.47	30
116	1-1/4"	Q-1	17		6	18.80	24.80	34
118	1-1/2"	"	16		7.50	20	27.50	37

15.1 Pipe & Fittings

		CREW	DAILY OUTPUT	UNIT	BARE COSTS MAT.	INST.	TOTAL	TOTAL INCL O&P
35 120	2″	Q-1	14	Ea.	12.20	23	35.20	46
122	2-1/2″		11		24	29	53	68
124	3″		8		36	40	76	97
126	4″	Q-2	11		84	45	129	160
128	5″		8		160	62	222	265
130	6″		7		235	71	306	360
132	8″		5		475	99	574	665
150	45° Elbow, 1/8″	1 Plum	13		1.30	13.65	14.95	21
154	1/4″		13		1.30	13.65	14.95	21
156	3/8″		13		1.30	13.65	14.95	21
158	1/2″		12		1.63	14.80	16.43	23
160	3/4″		11		2.40	16.15	18.55	26
162	1″		10		4.10	17.75	21.85	30
164	1-1/4″	Q-1	17		6.55	18.80	25.35	34
166	1-1/2″		16		8.20	20	28.20	38
168	2″		14		13.30	23	36.30	48
170	2-1/2″		11		26	29	55	71
172	3″		8		39	40	79	100
174	4″	Q-2	11		91	45	136	165
176	5″		8		170	62	232	275
178	6″		7		250	71	321	380
200	Tee, 1/8″	1 Plum	9		1.65	19.75	21.40	30
204	1/4″		9		1.65	19.75	21.40	30
206	3/8″		9		1.65	19.75	21.40	30
208	1/2″		8		2.06	22	24.06	34
210	3/4″		7		2.94	25	27.94	40
212	1″		6		5.30	30	35.30	49
214	1-1/4″	Q-1	10		7.50	32	39.50	54
216	1-1/2″		9		10.30	36	46.30	63
218	2″		8		16.90	40	56.90	76
220	2-1/2″		7		32	46	78	100
222	3″		5		49	64	113	145
224	4″	Q-2	7		115	71	186	230
226	5″		5		200	99	299	365
228	6″		4		330	125	455	545
250	Coupling, 1/8″	1 Plum	26		1.19	6.85	8.04	11.20
254	1/4″		22		1.19	8.05	9.24	12.95
256	3/8″		18		1.19	9.85	11.04	15.55
258	1/2″		15		1.48	11.85	13.33	18.75
260	3/4″		14		1.98	12.70	14.68	21
262	1″		13		3	13.65	16.65	23
264	1-1/4″	Q-1	22		4.69	14.55	19.24	26
266	1-1/2″		20		6.40	16	22.40	30
268	2″		18		10.50	17.75	28.25	37
270	2-1/2″		14		17.05	23	40.05	52
272	3″		10		24	32	56	73
273	3-1/2″		8		44	40	84	105
274	4″	Q-2	12		44	41	85	110
276	5″		10		73	50	123	150
278	6″		9		105	55	160	195
279	8″		8		170	62	232	275
300	Union, 150 lb.							
302	1/8″	1 Plum	12	Ea.	3.84	14.80	18.64	26
304	1/4″		12		3.84	14.80	18.64	26
306	3/8″		12		3.84	14.80	18.64	26
308	1/2″		11		4.53	16.15	20.68	28
310	3/4″		10		6.25	17.75	24	33
312	1″		9		8.40	19.75	28.15	38
314	1-1/4″	Q-1	16		12.40	20	32.40	43
316	1-1/2″	″	15		14.65	21	35.65	47

15

15.1 Pipe & Fittings

35		CREW	DAILY OUTPUT	UNIT	BARE COSTS MAT.	BARE COSTS INST.	BARE COSTS TOTAL	TOTAL INCL O&P
318	2"	Q-1	13	Ea.	23	25	48	61
320	2-1/2"		10		42	32	74	92
322	3"		7		64	46	110	135
324	4"	Q-2	10		145	50	195	230
332	For 250 lb. (navy pattern), add				32%			
500	Extra heavy, 90° Elbow							
504	1/8"	1 Plum	13	Ea.	2.08	13.65	15.73	22
506	1/4"		13		2.08	13.65	15.73	22
508	3/8"		13		2.08	13.65	15.73	22
510	1/2"		12		3.13	14.80	17.93	25
512	3/4"		11		4.23	16.15	20.38	28
514	1"		10		7.35	17.75	25.10	34
516	1-1/4"	Q-1	17		11.25	18.80	30.05	40
518	1-1/2"		16		16.75	20	36.75	47
520	2"		14		28	23	51	64
522	2-1/2"		11		45	29	74	92
524	3"		8		76	40	116	140
526	4"	Q-2	11		140	45	185	220
550	45° Elbow, 1/8"	1 Plum	13		2.34	13.65	15.99	22
554	1/4"		13		2.34	13.65	15.99	22
556	3/8"		13		2.34	13.65	15.99	22
558	1/2"		12		3.24	14.80	18.04	25
560	3/4"		11		4.46	16.15	20.61	28
562	1"		10		7.80	17.75	25.55	34
564	1-1/4"	Q-1	17		11.80	18.80	30.60	40
566	1-1/2"		16		17.80	20	37.80	48
568	2"		14		29	23	52	65
570	2-1/2"		11		47	29	76	94
572	3"		8		80	40	120	145
574	4"	Q-2	11		150	45	195	230
600	Tee, 1/8"	1 Plum	9		3.35	19.75	23.10	32
604	1/4"		9		3.35	19.75	23.10	32
606	3/8"		9		3.35	19.75	23.10	32
608	1/2"		8		4.42	22	26.42	37
610	3/4"		7		6.15	25	31.15	43
612	1"		6		9.60	30	39.60	53
614	1-1/4"	Q-1	10		15.75	32	47.75	64
616	1-1/2"		9		22	36	58	76
618	2"		8		38	40	78	100
620	2-1/2"		7		57	46	103	130
622	3"		5		91	64	155	195
624	4"	Q-2	7		155	71	226	275
650	Coupling, 1/8"	1 Plum	26		2.64	6.85	9.49	12.75
654	1/4"		22		2.64	8.05	10.69	14.55
656	3/8"		18		2.64	9.85	12.49	17.15
658	1/2"		15		3.69	11.85	15.54	21
660	3/4"		14		4.65	12.70	17.35	23
662	1"		13		7.05	13.65	20.70	27
664	1-1/4"	Q-1	22		10.20	14.55	24.75	32
666	1-1/2"		20		15	16	31	40
668	2"		18		21	17.75	38.75	49
670	2-1/2"		14		34	23	57	70
672	3"		10		50	32	82	100
673	3-1/2"		8		76	40	116	140
674	4"	Q-2	12		76	41	117	145
700	Union, 300 lb.							
702	1/8"	1 Plum	12	Ea.	10.30	14.80	25.10	33
704	1/4"		12		10.30	14.80	25.10	33
706	3/8"		12		10.30	14.80	25.10	33
708	1/2"		11		11.60	16.15	27.75	36

15

15.1 Pipe & Fittings

		CREW	DAILY OUTPUT	UNIT	BARE COSTS MAT.	BARE COSTS INST.	BARE COSTS TOTAL	TOTAL INCL O&P
35 710	3/4"	1 Plum	10	Ea.	16.40	17.75	34.15	44
712	1"	"	9		22	19.75	41.75	53
714	1-1/4"	Q-1	16		32	20	52	64
716	1-1/2"		15		39	21	60	74
718	2"	↓	13	↓	58	25	83	99
37-001	**PIPE, CAST IRON** Soil, on hangers 5' O.C.							
002	Single hub, service wt., lead & oakum joints 10' O.C.							
212	2" diameter	Q-1	50	L.F.	2.82	6.40	9.22	12.35
214	3" diameter		48		4.01	6.65	10.66	14.05
216	4" diameter	↓	44		5.34	7.25	12.59	16.35
218	5" diameter	Q-2	62		6.99	8	14.99	19.30
220	6" diameter	"	59		10.20	8.45	18.65	23
222	8" diameter	Q-3	49		15.55	13.85	29.40	37
224	10" diameter		44		23.27	15.45	38.72	48
226	12" diameter	↓	40		29.45	16.95	46.40	57
232	For service weight, double hub, add				44%			
234	For extra heavy, single hub, add				47%	4%		
236	For extra heavy, double hub, add			↓	95%	4%		
240	Lead for caulking			Lb.	.90		.90	.99M
242	Oakum for caulking			"	1.17		1.17	1.28M
296	To delete hangers, subtract							
297	2" diam. to 4" diam.				16%	19%		
298	5" diam. to 8" diam.				14%	14%		
299	10" diam. to 12" diam.				13%	19%		
400	No hub, couplings 10' O.C.							
410	1-1/2" diameter	Q-1	65	L.F.	2.93	4.92	7.85	10.35
412	2" diameter		61		3.01	5.25	8.26	10.90
414	3" diameter		58		4	5.50	9.50	12.35
416	4" diameter	↓	53		5.17	6.05	11.22	14.40
418	5" diameter	Q-2	74		7.70	6.70	14.40	18.20
420	6" diameter	"	70		9.42	7.10	16.52	21
422	8" diameter	Q-3	60		15.99	11.30	27.29	34
424	10" diameter	"	53	↓	25.96	12.80	38.76	47
428	To delete hangers, subtract							
429	1-1/2" diam. to 6" diam.				22%	47%		
430	8" diam. to 10" diam.				21%	44%		
38-001	**PIPE, CAST IRON, FITTINGS,** Soil							
004	Hub and spigot, service weight							
008	1/4 Bend, 2"	Q-1	16	Ea.	3.01	20	23.01	32
012	3"		14		5.60	23	28.60	39
014	4"	↓	13		8.20	25	33.20	45
016	5"	Q-2	18		10.75	28	38.75	52
018	6"	"	17		13.90	29	42.90	58
020	8"	Q-3	11		44	62	106	140
022	10"		10		66	68	134	170
024	12"	↓	9		83	75	158	200
025	Closet bend, 3" diameter with ring 10" x 16"	Q-1	14		24	23	47	59
026	16"x16"		12		27	27	54	68
027	Closet bend, 4" diameter, 1" x 4" ring, 6" x 16"		13		27	25	52	65
028	8" x 16"		13		24	25	49	62
029	10" x 12"		12		23	27	50	64
030	10" x 18"		11		31	29	60	76
031	12" x 16"		11		27	29	56	72
033	16" x 16"		10		32	32	64	81
034	1/8 Bend, 2"		16		2.27	20	22.27	31
035	3"		14		4.48	23	27.48	38
036	4"	↓	13		6.65	25	31.65	43
038	5"	Q-2	18	↓	9	28	37	50

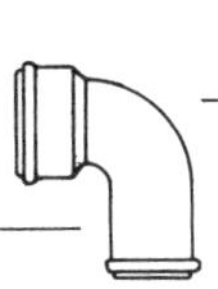

15

15.1 Pipe & Fittings

		CREW	DAILY OUTPUT	UNIT	BARE COSTS MAT.	BARE COSTS INST.	BARE COSTS TOTAL	TOTAL INCL O&P
38 040	 6"	Q-2	17	Ea.	11.05	29	40.05	54
042	8"	Q-3	11		32	62	94	125
044	10"		10		51	68	119	155
046	12"	↓	9		81	75	156	200
050	Sanitary Tee, 2" 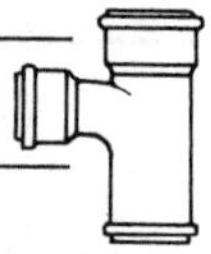	Q-1	10		5.20	32	37.20	52
054	3"		9		9.75	36	45.75	62
062	4"	↓	8		11.65	40	51.65	71
070	5"	Q-2	12		21	41	62	83
080	6"	"	11		25	45	70	93
088	8"	Q-3	7		66	97	163	215
100	Tee, 2" 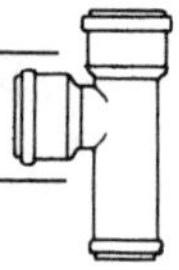	Q-1	10		6.40	32	38.40	53
106	3"		9		10	36	46	62
112	4"	↓	8		13.55	40	53.55	73
120	5"	Q-2	12		21	41	62	83
130	6"	"	11		27	45	72	95
138	8"	Q-3	7	↓	62	97	159	210
140	Combination Y and 1/8 Bend 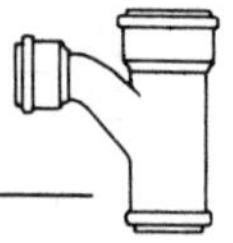							
142	2"	Q-1	10	Ea.	7.40	32	39.40	54
146	3"		9		11	36	47	63
152	4"	↓	8		15.10	40	55.10	74
154	5"	Q-2	12		28	41	69	91
156	6"		11		35	45	80	105
158	8"	↓	7		86	71	157	195
160	Double Y, 2" 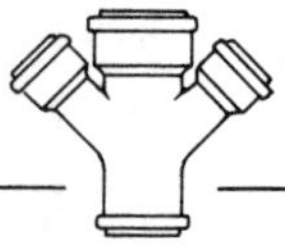	Q-1	8		7.95	40	47.95	67
161	3"		7		13.45	46	59.45	81
162	4"	↓	6.50		19.30	49	68.30	92
163	5"	Q-2	9		28	55	83	110
164	6"	"	8		41	62	103	135
165	8"	Q-3	5.50		105	125	230	295
166	10"		5		145	135	280	355
167	12"	↓	4.50		210	150	360	450
174	Reducer, 3" x 2" 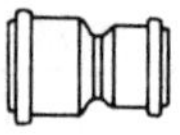	Q-1	15		3.85	21	24.85	35
175	4" x 2"		14.50		3.96	22	25.96	36
176	4" x 3"		14		5.15	23	28.15	39
177	5" x 2"		14		6.05	23	29.05	40
178	5" x 3"		13.50		6.05	24	30.05	41
179	5" x 4"		13		6.25	25	31.25	42
180	6" x 2"		13.50		7.30	24	31.30	42
181	6" x 3"		13		8.15	25	33.15	45
184	6" x 5"	↓	11		9.25	29	38.25	52
186	8" x 2"	Q-2	14		16.45	36	52.45	69
188	8" x 3"		13		16.70	38	54.70	74
15 190	8" x 4"		13		17.10	38	55.10	74
192	8" x 5"		12		17.30	41	58.30	79
194	8" x 6"	↓	12		17.45	41	58.45	79
196	Increaser, 2" x 3" 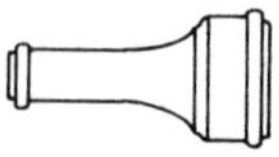	Q-1	15		5.15	21	26.15	36
198	2" x 4"		14		6.95	23	29.95	41
200	2" x 5"		13		8	25	33	44
202	3" x 4"		13		7.55	25	32.55	44
204	3" x 5"		13		8.10	25	33.10	44
206	3" x 6"		12		9.60	27	36.60	49
207	4" x 5"		13		9.25	25	34.25	46
208	4" x 6"	↓	12		12.05	27	39.05	52
209	4" x 8"	Q-2	13		24	38	62	82
210	5" x 6"	Q-1	11		17.40	29	46.40	61
211	5" x 8"	Q-2	12		27	41	68	90
212	6" x 8"		12		28	41	69	91
213	6" x 10"		8		40	62	102	135
214	8" x 10"		6.50		48	77	125	165
215	10" x 12"	↓	5.50	↓	76	90	166	215

15.1 Pipe & Fittings

		CREW	DAILY OUTPUT	UNIT	BARE COSTS MAT.	BARE COSTS INST.	BARE COSTS TOTAL	TOTAL INCL O&P
38 250	Y, 2"	Q-1	10	Ea.	5.20	32	37.20	52
251	3"		9		9.75	36	45.75	62
252	4"		8		12.60	40	52.60	72
253	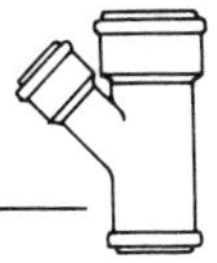 5"	Q-2	12		22	41	63	84
254	6"	"	11		29	45	74	97
255	8"	Q-3	7		73	97	170	220
256	10"		6		110	115	225	285
257	12"		5		175	135	310	390
300	For extra heavy, add				33%	4%		
599	No hub							
600	Cplg. & labor required at joints not incl. in fitting							
601	price. Add 1 per joint for installed price							
602	1/4 Bend, 1-1/2"			Ea.	2.58		2.58	2.83M
606	2"				2.64		2.64	2.90M
608	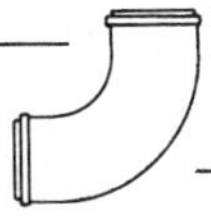 3"				3.43		3.43	3.77M
612	4"				5.10		5.10	5.60M
614	5"				11.30		11.30	12.45M
616	6"				12.30		12.30	13.55M
618	8"				29		29	32M
620	1/8 Bend, 1-1/2"				1.95		1.95	2.14M
624	2"				2		2	2.20M
626	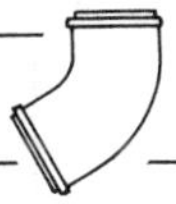 3"				2.85		2.85	3.13M
628	4"				3.75		3.75	4.12M
630	5"				7.15		7.15	7.85M
632	6"				8.35		8.35	9.20M
634	8"				19.20		19.20	21M
636	10"				37		37	41M
640	Sanitary Tee, 1-1/2"				3.22		3.22	3.54M
646	2"				3.59		3.59	3.94M
652	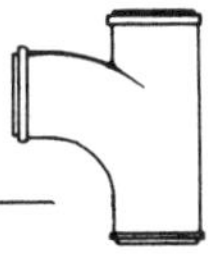 3"				4.48		4.48	4.92M
660	4"				6.90		6.90	7.60M
668	5"				16.30		16.30	17.95M
670	6"				19.60		19.60	22M
672	8"				50		50	55M
673	Y, 1-1/2"				3.32		3.32	3.65M
674	2"				3.59		3.59	3.94M
675	 3"				4.43		4.43	4.87M
676	4"				7.55		7.55	8.30M
677	5"				16.30		16.30	17.95M
678	6"				20		20	22M
679	8"				39		39	43M
680	Double Y, 2"				4.91		4.91	5.40M
692	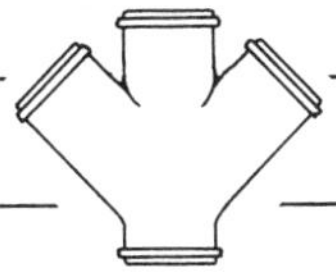 3"				8.75		8.75	9.60M
700	4"				18.15		18.15	19.95M
710	6"				31		31	34M
712	8"				59		59	65M
720	Combination Y and 1/8 Bend							
722	1-1/2"			Ea.	3.53		3.53	3.88M
726	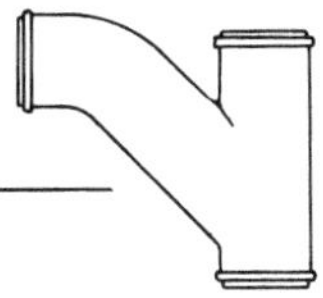 2"				3.75		3.75	4.12M
732	3"				5.70		5.70	6.25M
740	4"				10.80		10.80	11.90M
748	5"				21		21	23M
750	6"				27		27	30M
752	8"				62		62	68M
760	Reducer, 3" x 2"				3.01		3.01	3.31M
782	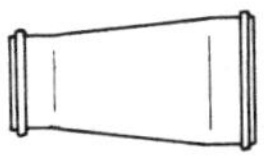 4" x 2"				4.06		4.06	4.46M
784	4" x 3"				4.59		4.59	5.05M
800	Coupling, standard (by CISPI Mfrs.)							
802	1-1/2"	Q-1	28	Ea.	2.20	11.40	13.60	18.90
804	2"	"	26	"	2.20	12.30	14.50	20

15

15.1 Pipe & Fittings

		CREW	DAILY OUTPUT	UNIT	BARE COSTS MAT.	BARE COSTS INST.	BARE COSTS TOTAL	TOTAL INCL O&P
38 808	3″	Q-1	24	Ea.	2.60	13.30	15.90	22
812	4″	″	22		3	14.55	17.55	24
816	5″	Q-2	28		7.30	17.75	25.05	34
818	6″	″	24		7.70	21	28.70	38
820	8″	Q-3	19		14.50	36	50.50	68
822	10″	″	16		21	42	63	84
830	Coupling, cast iron clamp & neoprene gasket (by MG)							
831	1-1/2″	Q-1	28	Ea.	5.90	11.40	17.30	23
832	2″		26		6	12.30	18.30	24
833	3″		24		6.90	13.30	20.20	27
834	4″		22		8.50	14.55	23.05	30
835	5″	Q-2	28		17.20	17.75	34.95	45
836	6″	″	24		19.45	21	40.45	51
838	8″	Q-3	19		39	36	75	95
840	10″	″	16		64	42	106	130
860	Coupling, Stainless steel, by Clamp-All Corp.							
862	1-1/2″	Q-1	28	Ea.	6.15	11.40	17.55	23
863	2″		26		6.35	12.30	18.65	25
864	2″ x 1-1/2″		26		7.05	12.30	19.35	26
865	3″		24		7.35	13.30	20.65	27
866	4″		22		8.50	14.55	23.05	30
867	4″ x 3″		22		10.25	14.55	24.80	32
868	5″	Q-2	28		17.65	17.75	35.40	45
869	6″	″	24		20	21	41	52
870	8″	Q-3	19		42	36	78	98
871	10″	″	16		55	42	97	120
40-001	**PIPE, COPPER** 50/50 solder joints,							
002	Type K tubing, couplings & clevis hangers 10′ O.C.							
110	1/4″ diameter	1 Plum	55	L.F.	.53	3.23	3.76	5.25
112	3/8″ diameter		54		.77	3.29	4.06	5.60
114	C8.1 -032 1/2″ diameter		50		.88	3.55	4.43	6.10
116	5/8″ diameter		47		1.15	3.78	4.93	6.70
118	3/4″ diameter		42		1.47	4.23	5.70	7.75
120	1″ diameter		34		1.98	5.20	7.18	9.70
122	1-1/4″ diameter		28		2.47	6.35	8.82	11.90
124	1-1/2″ diameter		26		3.14	6.85	9.99	13.30
126	2″ diameter		22		4.78	8.05	12.83	16.90
128	2-1/2″ diameter	Q-1	31		6.88	10.30	17.18	22
130	3″ diameter		28		9.18	11.40	20.58	27
132	3-1/2″ diameter		24		12.33	13.30	25.63	33
133	4″ diameter		22		15.13	14.55	29.68	38
134	5″ diameter		18		36.98	17.75	54.73	66
136	6″ diameter	Q-2	23		54.23	22	76.23	91
138	8″ diameter	″	20		100	25	125	145
139	For other than full hard temper, add				25%			
140	For 95/5 solder, add					6%		
144	For silver solder, add					15%		
180	For medical clean, (oxygen class), add				3.50%			
195	To delete cplgs. & hngrs., 1/4″-5/8″ pipe, subtract				27%	40%		
196	3/4″-1-1/2″ pipe, subtract				14%	29%		
197	2″-6″ pipe, subtract				10%	31%		
198	8″ pipe, subtract			L.F.	19%	31%		
200	Type L tubing, couplings & hangers 10′ O.C.							
210	1/4″ diameter	1 Plum	59	L.F.	.47	3.01	3.48	4.87
212	3/8″ diameter		56		.64	3.17	3.81	5.30
214	1/2″ diameter		53		.76	3.35	4.11	5.70
216	5/8″ diameter		49		1.05	3.62	4.67	6.40
218	3/4″ diameter		44		1.11	4.04	5.15	7.05
220	1″ diameter		36		1.61	4.93	6.54	8.90
222	1-1/4″ diameter		29		2.08	6.10	8.18	11.15

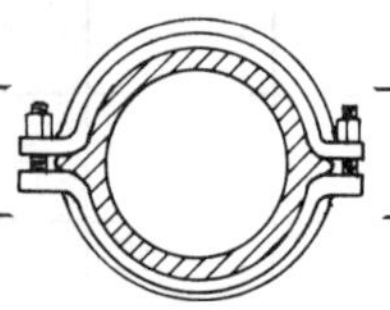

15.1 Pipe & Fittings

	15.1 Pipe & Fittings	CREW	DAILY OUTPUT	UNIT	BARE COSTS MAT.	BARE COSTS INST.	BARE COSTS TOTAL	TOTAL INCL O&P
40 224	1-1/2" diameter	1 Plum	27	L.F.	2.64	6.60	9.24	12.40
226	2" diameter	"	23		3.98	7.70	11.68	15.55
228	2-1/2" diameter	Q-1	33		5.74	9.70	15.44	20
230	3" diameter		30		7.58	10.65	18.23	24
232	3-1/2" diameter		25		10.51	12.80	23.31	30
234	4" diameter		23		12.71	13.90	26.61	34
236	5" diameter		19		32.58	16.85	49.43	60
238	6" diameter	Q-2	24		42.13	21	63.13	76
240	8" diameter	"	21		78	24	102	120
241	For other than full hard temper, add				25%			
258	For 95/5 solder, add					6%		
259	For silver solder, add					15%		
290	For medical clean, (oxygen class), add				3.50%			
294	To delete cplgs. & hngrs., 1/4"-3/8" pipe, subtract				37%	43%		
295	1/2"-1" pipe, subtract				22%	38%		
296	1-1/4"-3" pipe, subtract				12%	27%		
297	3-1/2"-6" pipe, subtract				12%	37%		
298	8" pipe, subtract				24%	33%		
300	Type M tubing, couplings & hangers 10' O.C.							
310	1/4" diameter	1 Plum	61	L.F.	.50	2.91	3.41	4.76
312	3/8" diameter		59		.53	3.01	3.54	4.93
314	1/2" diameter		56		.61	3.17	3.78	5.25
316	5/8" diameter		51		.82	3.48	4.30	5.95
318	3/4" diameter		46		.87	3.86	4.73	6.55
320	1" diameter		37		1.26	4.80	6.06	8.30
322	1-1/4" diameter		30		1.71	5.90	7.61	10.45
324	1-1/2" diameter		28		2.30	6.35	8.65	11.70
326	2" diameter		24		3.46	7.40	10.86	14.50
328	2-1/2" diameter	Q-1	35		4.89	9.15	14.04	18.60
330	3" diameter		32		6.36	10	16.36	21
332	3-1/2" diameter		27		8.91	11.85	20.76	27
334	4" diameter		24		11.28	13.30	24.58	32
336	5" diameter		20		30.38	16	46.38	57
337	6" diameter	Q-2	25		38.83	19.90	58.73	71
338	8" diameter	"	22		68	23	91	110
339	For other than full hard temper, add				25%			
340	For 95/5 solder, add					6%		
344	For silver solder, add					15%		
395	For medical clean, (oxygen class), add				3.50%			
396	To delete cplgs. & hngrs., 1/4"-5/8" pipe, subtract				35%	44%		
397	3/4"-2" pipe, subtract				19%	31%		
398	2-1/2"-6" pipe, subtract				13%	35%		
399	8" pipe, subtract				28%	35%		
400	Type DWV tubing, couplings & hangers 10' O.C.							
410	1-1/4" diameter	1 Plum	30	L.F.	1.68	5.90	7.58	10.40
412	1-1/2" diameter		28		2.08	6.35	8.43	11.45
414	2" diameter		24		2.69	7.40	10.09	13.65
416	3" diameter	Q-1	32		4.41	10	14.41	19.30
418	4" diameter		24		7.60	13.30	20.90	28
420	5" diameter		20		22.57	16	38.57	48
422	6" diameter	Q-2	25		32.23	19.90	52.13	64
424	8" diameter	"	20		82.37	25	107.37	125
426	For other than full hard temper, add				25%			
473	To delete cplgs. & hngrs., 1-1/4"-1-1/2" pipe, subt.				16%	31%		
474	2"-4" pipe, subtract				13%	31%		
475	5"-8" pipe, subtract				23%	23%		
520	ACR tubing, type L, hard temper, cleaned and							
522	capped, no couplings or hangers							
524	3/8" OD			L.F.	.30		.30	.33M
525	1/2" OD			"	.43		.43	.47M

15.1 Pipe & Fittings

		CREW	DAILY OUTPUT	UNIT	BARE COSTS MAT.	BARE COSTS INST.	BARE COSTS TOTAL	TOTAL INCL O&P
40 526	5/8" OD			L.F.	.56		.56	.61M
527	3/4" OD				.77		.77	.84M
528	7/8" OD				.89		.89	.97M
529	1-1/8" OD				1.29		1.29	1.41M
530	1-3/8" OD				1.71		1.71	1.88M
531	1-5/8" OD				2.16		2.16	2.37M
532	2-1/8" OD				3.35		3.35	3.68M
533	2-5/8" OD				4.82		4.82	5.30M
534	3-1/8" OD				6.45		6.45	7.10M
535	3-5/8" OD				8.65		8.65	9.50M
536	4-1/8" OD				11.05		11.05	12.15M
544	For soft temper, add				25%			
580	Refrigeration tubing, dryseal, 50' coils							
584	1/8" OD			Coil	8.75		8.75	9.60M
585	3/16" OD				9.40		9.40	10.35M
586	1/4" OD				11.50		11.50	12.65M
587	5/16" OD				14.70		14.70	16.15M
588	3/8" OD				16.95		16.95	18.65M
589	1/2" OD				22		22	24M
590	5/8" OD				29		29	32M
591	3/4" OD				34		34	37M
592	7/8" OD				50		50	55M
593	1-1/8" OD				74		74	81M
594	1-3/8" OD				100		100	110M
595	1-5/8" OD				130		130	145M
41-001	**PIPE, COPPER, FITTINGS,** Wrought unless otherwise noted							
004	Solder joints, copper x copper							
007	90° Elbow, 1/4"	1 Plum	22	Ea.	.66	8.05	8.71	12.40
009	3/8"		22		.69	8.05	8.74	12.40
010	1/2"		20		.13	8.90	9.03	12.95
011	5/8"		19		1.07	9.35	10.42	14.70
012	3/4"		19		.30	9.35	9.65	13.85
013	1"		16		.78	11.10	11.88	16.90
014	1-1/4"		15		1.38	11.85	13.23	18.60
015	1-1/2"		13		1.71	13.65	15.36	22
016	2"		11		3.28	16.15	19.43	27
017	2-1/2"	Q-1	13		7	25	32	43
018	3"		11		10.05	29	39.05	53
019	3-1/2"		10		30	32	62	79
020	4"		9		21	36	57	74
021	5" cast brass		6		71	53	124	155
022	6" cast brass	Q-2	9		110	55	165	200
023	8" cast brass	"	8		600	62	662	750
025	45° Elbow, 1/4"	1 Plum	22		.94	8.05	8.99	12.70
027	3/8"		22		.85	8.05	8.90	12.60
028	1/2"		20		.30	8.90	9.20	13.15
029	5/8"		19		1.39	9.35	10.74	15.05
030	3/4"		19		.52	9.35	9.87	14.10
031	1"		16		1.19	11.10	12.29	17.35
032	1-1/4"		15		1.77	11.85	13.62	19.05
033	1-1/2"		13		2.13	13.65	15.78	22
034	2"		11		3.56	16.15	19.71	27
035	2-1/2"	Q-1	13		7.55	25	32.55	44
036	3"		13		9.65	25	34.65	46
037	3-1/2"		10		21	32	53	69
038	4"		9		20	36	56	73
039	5" cast brass		6		71	53	124	155
040	6" cast brass	Q-2	9		110	55	165	200
041	8" cast brass	"	8		560	62	622	705

15

15.1 Pipe & Fittings

			CREW	DAILY OUTPUT	UNIT	BARE COSTS MAT.	BARE COSTS INST.	BARE COSTS TOTAL	TOTAL INCL O&P
41	045	Tee, 1/4"	1 Plum	14	Ea.	1.19	12.70	13.89	19.65
	047	3/8"		14		1.07	12.70	13.77	19.50
	048	1/2"		13		.22	13.65	13.87	20
	049	5/8"		12		1.63	14.80	16.43	23
	050	3/4"		12		.55	14.80	15.35	22
	051	1"		10		2.15	17.75	19.90	28
	052	1-1/4"		9		3.12	19.75	22.87	32
	053	1-1/2"		8		4.07	22	26.07	37
	054	2"		7		6.60	25	31.60	44
	055	2-1/2"	Q-1	8		13.70	40	53.70	73
	056	3"		7		18.85	46	64.85	87
	057	3-1/2"		6		55	53	108	140
	058	4"		5		45	64	109	140
	059	5" cast brass		4		120	80	200	250
	060	6" cast brass	Q-2	6		155	83	238	290
	061	8" cast brass	"	5		880	99	979	1,100
	065	Coupling, 1/4"	1 Plum	24		.07	7.40	7.47	10.75
	067	3/8"		24		.35	7.40	7.75	11.10
	068	1/2"		22		.12	8.05	8.17	11.80
	069	5/8"		21		.44	8.45	8.89	12.70
	070	3/4"		21		.21	8.45	8.66	12.45
	071	1"		18		.84	9.85	10.69	15.20
	072	1-1/4"		17		.95	10.45	11.40	16.15
	073	1-1/2"		15		1.30	11.85	13.15	18.55
	074	2"		13		1.94	13.65	15.59	22
	075	2-1/2"	Q-1	15		3.40	21	24.40	35
	076	3"		13		4.70	25	29.70	41
	077	3-1/2"		8		8.25	40	48.25	67
	078	4"		7		9.50	46	55.50	76
	079	5"		6		32	53	85	110
	080	6"	Q-2	8		58	62	120	155
	081	8"	"	7		160	71	231	280
	085	Unions, 1/4"	1 Plum	21		2.98	8.45	11.43	15.50
	087	3/8"		21		2.98	8.45	11.43	15.50
	088	1/2"		19		1.27	9.35	10.62	14.90
	089	5/8"		18		6.70	9.85	16.55	22
	090	3/4"		18		1.81	9.85	11.66	16.25
	091	1"		15		2.88	11.85	14.73	20
	092	1-1/4"		14		4.32	12.70	17.02	23
	093	1-1/2"		12		6	14.80	20.80	28
	094	2"		10		10.25	17.75	28	37
	095	2-1/2" cast brass	Q-1	12		54	27	81	98
	096	3" cast brass	"	10		69	32	101	120
	098	Adapter, copper x male, 1/4" IPS	1 Plum	24		1.68	7.40	9.08	12.55
	099	3/8" IPS		24		1.08	7.40	8.48	11.90
	100	1/2" IPS		22		.32	8.05	8.37	12
	101	3/4" IPS		21		.50	8.45	8.95	12.75
	102	1" IPS		18		1.50	9.85	11.35	15.90
	103	1-1/4" IPS		17		2.26	10.45	12.71	17.60
	104	1-1/2" IPS		15		2.40	11.85	14.25	19.75
	105	2" IPS		13		3.94	13.65	17.59	24
	106	2-1/2" IPS	Q-1	15		10.60	21	31.60	42
	107	3" IPS		13		12.65	25	37.65	49
	108	3-1/2" IPS		8		43	40	83	105
	109	4" IPS		7		43	46	89	115
	120	5" IPS		6		125	53	178	215
	121	6" IPS	Q-2	8		215	62	277	325
	125	Cross, 1/2"	1 Plum	10		2.20	17.75	19.95	28
	126	3/4"		9.50		4.27	18.70	22.97	32
	127	1"		8		7.20	22	29.20	40

15

15.1 Pipe & Fittings

		Description	Crew	Daily Output	Unit	Bare Costs Mat.	Bare Costs Inst.	Bare Costs Total	Total Incl O&P
41	128	1-1/4"	1 Plum	7.50	Ea.	9.75	24	33.75	45
	129	1-1/2"		6.50		14.75	27	41.75	56
	130	2"		5.50		28	32	60	77
	131	2-1/2"	Q-1	6.50		54	49	103	130
	132	3"		5.50		130	58	188	225
	134	4"		4.70		160	68	228	275
	135	5"		3.20		445	100	545	635
	136	6"	Q-2	4.70		720	105	825	945
	150	Tee, mechanically formed, 3/8"	1 Plum	80			2.22	2.22	3.21L
	152	1/2"		80			2.22	2.22	3.21L
	153	3/4"		60			2.96	2.96	4.28L
	154	1"		48			3.70	3.70	5.35L
	155	1-1/4"		48			3.70	3.70	5.35L
	156	1-1/2"		40			4.44	4.44	6.40L
	157	2"		24			7.40	7.40	10.70L
	200	DWV, solder joints, copper x copper							
	203	90° Elbow, 1-1/4"	1 Plum	13	Ea.	1.58	13.65	15.23	21
	205	1-1/2"		12		1.58	14.80	16.38	23
	207	2"		10		3.05	17.75	20.80	29
	209	3"	Q-1	10		7.35	32	39.35	54
	210	4"	"	9		24	36	60	78
	215	45° Elbow, 1-1/4"	1 Plum	13		1.41	13.65	15.06	21
	217	1-1/2"		12		1.18	14.80	15.98	23
	218	2"		10		2.78	17.75	20.53	29
	219	3"	Q-1	10		5.70	32	37.70	52
	220	4"	"	9		14.95	36	50.95	68
	225	Tee, Sanitary, 1-1/4"	1 Plum	9		2.63	19.75	22.38	31
	227	1-1/2"		8		2.42	22	24.42	35
	229	2"		7		4.49	25	29.49	42
	231	3"	Q-1	7		10.20	46	56.20	77
	233	4"	"	6		31	53	84	110
	240	Coupling, 1-1/4"	1 Plum	14		.60	12.70	13.30	19
	242	1-1/2"		13		.91	13.65	14.56	21
	244	2"		11		1.24	16.15	17.39	25
	246	3"	Q-1	11		2.08	29	31.08	44
	248	4"	"	10		5.30	32	37.30	52
	260	Traps, see division 15.1-73-480 thru 546							
	300	Refrigeration fittings 90° Elbow							
	301	3/8" outside diameter	1 Plum	22	Ea.	.66	8.05	8.71	12.40
	302	1/2" outside diameter		20		.69	8.90	9.59	13.60
	303	5/8" outside diameter		19		.13	9.35	9.48	13.65
	304	3/4" outside diameter		18		1.07	9.85	10.92	15.45
	305	7/8" outside diameter		17		.30	10.45	10.75	15.40
	306	1-1/8" outside diameter		16		.78	11.10	11.88	16.90
	307	1-3/8" outside diameter		15		1.38	11.85	13.23	18.60
	308	1-5/8" outside diameter		13		1.71	13.65	15.36	22
	309	2-1/8" outside diameter		11		3.28	16.15	19.43	27
	310	2-5/8" outside diameter	Q-1	13		7	25	32	43
	312	3-1/8" outside diameter		11		10.05	29	39.05	53
	313	3-5/8" outside diameter		10		30	32	62	79
	314	4-1/8" outside diameter		9		21	36	57	74
	699	Polybutylene/polyethylene pipe, See 15.1-50-799 for plastic fittings							
	700	Insert type Brass/copper, 100psi @ 180°F, CTS							
	701	Adapter MPT 3/8" x 3/8" CTS	1 Plum	29	Ea.	.54	6.10	6.64	9.45
	702	1/2" x 1/2"		26		.66	6.85	7.51	10.60
	703	3/4" x 1/2"		26		.89	6.85	7.74	10.85
	704	3/4" x 3/4"		25		.96	7.10	8.06	11.30
	705	Adapter CTS 1/2" x 1/2" sweat		24		.40	7.40	7.80	11.15
	706	3/4" x 3/4" sweat		22		.72	8.05	8.77	12.45
	707	Coupler center set 3/8" CTS		25		.38	7.10	7.48	10.70

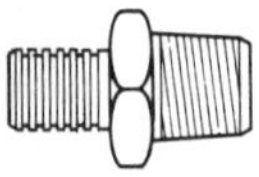

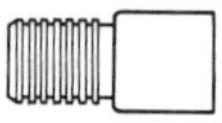

15

15.1 Pipe & Fittings

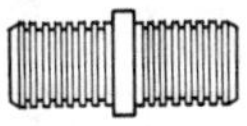

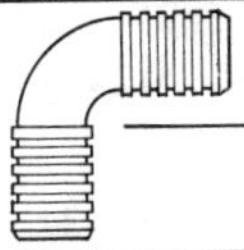

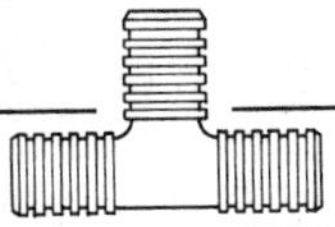

		CREW	DAILY OUTPUT	UNIT	BARE COSTS MAT.	BARE COSTS INST.	BARE COSTS TOTAL	TOTAL INCL O&P
41 708	1/2" CTS	1 Plum	23	Ea.	.43	7.70	8.13	11.65
709	3/4" CTS		22		.61	8.05	8.66	12.35
710	Elbow 90°, copper 3/8"		25		.59	7.10	7.69	10.90
711	1/2" CTS		23		.64	7.70	8.34	11.85
712	3/4" CTS		22		.83	8.05	8.88	12.55
713	Tee copper 3/8" CTS		17		.75	10.45	11.20	15.90
714	1/2" CTS		15		.83	11.85	12.68	18
715	3/4" CTS		14		.96	12.70	13.66	19.40
716	3/8" x 3/8" x 1/2"		16		1.18	11.10	12.28	17.35
717	1/2" x 3/8" x 1/2"		15		.94	11.85	12.79	18.15
718	3/4" x 1/2" x 3/4"		14		1.16	12.70	13.86	19.60
43-001	**PIPE, CORROSION RESISTANT** No couplings or hangers							
002	Iron alloy, drain, mechanical joint							
100	1-1/2" diameter	Q-1	70	L.F.	12.45	4.57	17.02	20
110	2" diameter		66		14.30	4.84	19.14	23
112	3" diameter		60		19.90	5.35	25.25	30
114	4" diameter		52		25	6.15	31.15	36
198	Iron alloy, drain, B&S joint							
200	2" diameter	Q-1	54	L.F.	18.70	5.90	24.60	29
210	3" diameter		52		27	6.15	33.15	39
212	4" diameter		48		37	6.65	43.65	50
214	6" diameter	Q-2	59		62	8.45	70.45	80
216	8" diameter	"	54		110	9.20	119.20	135
298	Plastic, Epoxy, fiberglass filament wound							
300	2" diameter	Q-1	62	L.F.	4.23	5.15	9.38	12.10
310	3" diameter		51		5.80	6.25	12.05	15.45
312	4" diameter		45		7.40	7.10	14.50	18.40
314	6" diameter		32		11.15	10	21.15	27
316	8" diameter	Q-2	38		17.45	13.10	30.55	38
318	10" diameter		32		26	15.55	41.55	51
320	12" diameter		28		34	17.75	51.75	63
398	Polyester, fiberglass filament wound							
400	2" diameter	Q-1	62	L.F.	9.15	5.15	14.30	17.50
410	3" diameter		51		11.75	6.25	18	22
412	4" diameter		45		14.45	7.10	21.55	26
414	6" diameter		32		20	10	30	36
416	8" diameter	Q-2	38		28	13.10	41.10	50
418	10" diameter		32		33	15.55	48.55	59
420	12" diameter		28		40	17.75	57.75	70
498	Polypropylene, acid resistant, Schedule 40							
500	1-1/2" diameter	Q-1	68	L.F.	1.46	4.70	6.16	8.40
510	2" diameter		62		1.88	5.15	7.03	9.50
512	3" diameter		51		3.34	6.25	9.59	12.75
514	4" diameter		45		5.05	7.10	12.15	15.80
516	6" diameter		32		9.60	10	19.60	25
598	Proxylene, fire retardent, Schedule 40							
600	1-1/2" diameter	Q-1	68	L.F.	1.94	4.70	6.64	8.95
610	2" diameter		62		2.60	5.15	7.75	10.30
612	3" diameter		51		4.72	6.25	10.97	14.25
614	4" diameter		45		6.70	7.10	13.80	17.65
616	6" diameter		32		12.35	10	22.35	28
660	Polyvinylidene fluoride, Schedule 40							
664	1/2" diameter	1 Plum	44	L.F.	3.83	4.04	7.87	10.05
666	3/4" diameter		34		5.05	5.20	10.25	13.10
668	1" diameter		31		6.80	5.75	12.55	15.75
670	1-1/2" diameter		25		10.80	7.10	17.90	22
672	2" diameter		22		14.25	8.05	22.30	27
682	For schedule 80, add				30%	2%		
710	Steel pipe, plastic lined for temperature and chemical							

15

15.1 Pipe & Fittings

	15.1 Pipe & Fittings	CREW	DAILY OUTPUT	UNIT	BARE COSTS MAT.	BARE COSTS INST.	BARE COSTS TOTAL	TOTAL INCL O&P
43 712	resistance, flanged joints							
714	Polypropylene lining for temperatures to 225°F.							
716	1" diameter	Q-1	76	L.F.	7.30	4.21	11.51	14.10
718	1-1/2" diameter		64		8.40	5	13.40	16.45
720	2" diameter		51		9.50	6.25	15.75	19.50
722	3" diameter		34		16.65	9.40	26.05	32
724	4" diameter		29		23	11	34	41
726	6" diameter	Q-2	26		42	19.15	61.15	74
728	8" diameter		23		66	22	88	105
730	10" diameter		20		92	25	117	135
760	Fluorinated ethylene propylene lining							
762	for temperatures to 300°F.							
764	1" diameter	Q-1	76	L.F.	16.70	4.21	20.91	24
766	1-1/2" diameter		64		21	5	26	30
768	2" diameter		51		24	6.25	30.25	35
770	3" diameter		34		37	9.40	46.40	54
772	4" diameter		29		58	11	69	80
774	6" diameter	Q-2	26		110	19.15	129.15	150
776	8" diameter		23		200	22	222	250
778	10" diameter		20		285	25	310	350
800	Polytetrafluoroethylene lining							
802	for temperatures to 500°F.							
804	1" diameter	Q-1	76	L.F.	23	4.21	27.21	31
806	1-1/2" diameter		64		28	5	33	38
808	2" diameter		51		32	6.25	38.25	44
810	3" diameter		34		48	9.40	57.40	66
812	4" diameter		29		71	11	82	94
814	6" diameter	Q-2	26		135	19.15	154.15	175
816	8" diameter		23		235	22	257	290
818	10" diameter		20		335	25	360	405
830	For heavy duty series, add				5%			
900	External pipe protection							
904	Bituminous coating, per diameter inch, 1 coat, add			L.F.	.19		.19	.20M
905	3 coat, add				.29		.29	.31M
908	Coal tar epoxy coating, per diameter inch, 1 coat, add				.16		.16	.17M
910	3 coat, add				.24		.24	.26M
920	Coating and wrapping, 1" diameter, add				2.81		2.81	3.09M
921	2" diameter, add				1.06		1.06	1.16M
922	2-1/2" diameter, add				1.40		1.40	1.54M
923	3" diameter, add				1.40		1.40	1.54M
924	4" diameter, add				1.80		1.80	1.98M
925	5" diameter, add				1.94		1.94	2.13M
926	6" diameter, add				1.94		1.94	2.13M
927	8" diameter, add				1.94		1.94	2.13M
928	10" diameter, add				2.70		2.70	2.97M
929	12" diameter, add				2.70		2.70	2.97M
930	14" diameter, add				3		3	3.30M
931	16" diameter, add				3.36		3.36	3.69M
932	18" diameter, add				3.98		3.98	4.37M
933	20" diameter, add				4.48		4.48	4.92M
934	24" diameter, add				5.75		5.75	6.30M
938	Fittings are covered in field			S.F.	5.25		5.25	5.75M
939	Minimum cost per fitting			Ea.	35		35	39M
954	Polyethylene high density coating,							
955	extruded and bonded, yellow							
956	1/2" diam. pipe, .025" plastic thkns, add			L.F.	.19		.19	.20M
957	3/4" diam. pipe, .025" plastic thkns, add				.20		.20	.22M
958	1" diam. pipe, .025" plastic thkns, add				.22		.22	.24M
959	1-1/4" diam. pipe, .025" plastic thkns, add				.25		.25	.27M
960	1-1/2" diam. pipe, .025" plastic thkns, add				.28		.28	.30M

15

15.1 Pipe & Fittings		CREW	DAILY OUTPUT	UNIT	BARE COSTS MAT.	BARE COSTS INST.	BARE COSTS TOTAL	TOTAL INCL O&P
43 961	2" diam. pipe, .030" plastic thkns, add			L.F.	.31		.31	.34M
962	2-1/2" diam. pipe, .030" plastic thkns, add				.38		.38	.41M
963	3" diam. pipe, .035" plastic thkns, add				.46		.46	.50M
964	3-1/2" diam. pipe, .035" plastic thkns, add				.53		.53	.58M
965	4" diam. pipe, .035" plastic thkns, add				.60		.60	.66M
966	5" diam. pipe, .040" plastic thkns, add				.80		.80	.88M
967	6" diam. pipe, .040" plastic thkns, add				.84		.84	.92M
968	8" diam. pipe, .040" plastic thkns, add				1.10		1.10	1.21M
969	10" diam. pipe, .040" plastic thkns, add				1.35		1.35	1.48M
970	12" diam. pipe, .040" plastic thkns, add				1.60		1.60	1.76M
971	14" diam. pipe, .060" plastic thkns, add				2.22		2.22	2.44M
972	16" diam. pipe, .060" plastic thkns, add				2.60		2.60	2.86M
973	18" diam. pipe, .060" plastic thkns, add				3.07		3.07	3.37M
974	20" diam. pipe, .060" plastic thkns, add				3.46		3.46	3.80M
977	Fittings are wrapped in field			S.F.	10		10	11M
44-001	**PIPE, CORROSION RESISTANT, FITTINGS,**							
003	Iron alloy							
005	Mechanical joint							
006	1/4 Bend, 1-1/2"	Q-1	12	Ea.	15.20	27	42.20	55
008	2"		10		19.20	32	51.20	67
009	3"		9		28	36	64	82
010	4"		8		38	40	78	100
011	1/8 Bend, 1-1/2"		12		15.20	27	42.20	55
013	2"		10		19.20	32	51.20	67
014	3"		9		28	36	64	82
015	4"		8		38	40	78	100
016	Tee and Y, sanitary, straight							
017	1-1/2"	Q-1	8	Ea.	20	40	60	80
018	2"		7		27	46	73	96
019	3"		6		41	53	94	120
020	4"		5		70	64	134	170
036	Coupling, 1-1/2"		14		14.40	23	37.40	49
038	2"		12		16.80	27	43.80	57
039	3"		11		17.60	29	46.60	61
040	4"		10		19.20	32	51.20	67
050	Bell & Spigot							
051	1/4 and 1/16 Bend, 2"	Q-1	16	Ea.	26	20	46	57
052	3"		14		38	23	61	75
053	4"		13		49	25	74	89
054	6"	Q-2	17		180	29	209	240
055	8"	"	12		430	41	471	535
062	1/8 Bend, 2"	Q-1	16		26	20	46	57
064	3"		14		38	23	61	75
065	4"		13		49	25	74	89
066	6"	Q-2	17		120	29	149	175
068	8"	"	12		355	41	396	450
070	Tee, sanitary, 2"	Q-1	10		34	32	66	84
071	3"		9		48	36	84	105
072	4"		8		75	40	115	140
073	6"	Q-2	11		150	45	195	230
074	8"	"	8		540	62	602	685
180	Y, sanitary, 2"	Q-1	10		33	32	65	83
182	3"		9		56	36	92	115
183	4"		8		75	40	115	140
184	6"	Q-2	11		150	45	195	230
185	8"	"	8		700	62	762	860
300	Epoxy, filament wound							
303	Quick-lock joint							
304	90° Elbow, 2"	Q-1	28	Ea.	28	11.40	39.40	47

15.1 Pipe & Fittings

Line	Item	Crew	Daily Output	Unit	Bare Costs Mat.	Bare Costs Inst.	Bare Costs Total	Total Incl O&P
44 306	3"	Q-1	16	Ea.	32	20	52	64
307	4"		13		39	25	64	78
308	6"		8		71	40	111	135
309	8"	Q-2	9		140	55	195	235
310	10"		7		205	71	276	330
311	12"		6		275	83	358	420
312	45° Elbow, 2"	Q-1	28		20	11.40	31.40	39
313	3"		16		27	20	47	59
314	4"		13		35	25	60	74
315	6"		8		71	40	111	135
316	8"	Q-2	9		135	55	190	230
317	10"		7		170	71	241	290
318	12"		6		210	83	293	350
319	Tee, 2"	Q-1	19		30	16.85	46.85	57
320	3"		11		32	29	61	77
321	4"		9		42	36	78	98
322	6"		5		120	64	184	225
323	8"	Q-2	6		195	83	278	335
324	10"		5		315	99	414	490
325	12"		4		385	125	510	605
400	Polypropylene, acid resistant							
402	Non-pressure							
405	1/4 bend, 1-1/2"	1 Plum	16	Ea.	8.55	11.10	19.65	25
406	2"	Q-1	28		10.15	11.40	21.55	28
408	3"		17		19.45	18.80	38.25	49
409	4"		14		29	23	52	65
411	6"		8		72	40	112	135
415	1/4 bend, long sweep							
417	1-1/2"	1 Plum	16	Ea.	9	11.10	20.10	26
418	2"	Q-1	28		12.30	11.40	23.70	30
420	3"		17		21	18.80	39.80	50
421	4"		14		30	23	53	66
425	1/8 bend, 1-1/2"	1 Plum	16		8.55	11.10	19.65	25
426	2"	Q-1	28		10.45	11.40	21.85	28
428	3"		17		19.15	18.80	37.95	48
429	4"		14		22	23	45	57
431	6"		8		60	40	100	125
440	Tee, sanitary							
442	1-1/2"	1 Plum	10	Ea.	10.85	17.75	28.60	38
443	2"	Q-1	17		13	18.80	31.80	41
445	3"		11		26	29	55	71
446	4"		9		39	36	75	94
448	6"		5		66	64	130	165
450	Tee/wye, long turn							
452	1-1/2"	1 Plum	10	Ea.	14.40	17.75	32.15	42
453	2"	Q-1	17		19.20	18.80	38	48
455	3"		11		31	29	60	76
457	4"		9		44	36	80	100
465	Wye 45°, 1-1/2"	1 Plum	10		11.05	17.75	28.80	38
467	2"	Q-1	17		15.75	18.80	34.55	45
468	3"		11		28	29	57	73
470	4"		9		40	36	76	95
472	6"		5		100	64	164	200
480	Wye, double 45°							
482	1-1/2"	1 Plum	8	Ea.	18.85	22	40.85	53
483	2"	Q-1	14		32	23	55	68
485	3"		9		45	36	81	100
487	4"		4		61	80	141	185
600	Plastic lined, ductile iron, flanged (for bolt sets see 15.1-57)							
601	Polypropylene (PP) lined							

15.1 Pipe & Fittings

		CREW	DAILY OUTPUT	UNIT	BARE COSTS MAT.	BARE COSTS INST.	BARE COSTS TOTAL	TOTAL INCL O&P
44	602	90° Elbow, 1"	Q-1	16	Ea.	62	20	82	97
	603	1-1/2"		14		72	23	95	110
	604	2"		13		87	25	112	130
	606	3"		11		120	29	149	175
	607	4"		8		155	40	195	230
	609	6"	Q-2	9		305	55	360	415
	610	8"		8		495	62	557	635
	611	10"		7		730	71	801	905
	619	45° Elbow, 1"	Q-1	16		61	20	81	96
	621	1-1/2"		14		69	23	92	110
	622	2"		13		82	25	107	125
	624	3"		11		120	29	149	175
	625	4"		8		150	40	190	225
	627	6"		9		300	36	336	380
	628	8"		8		465	40	505	570
	629	10"	Q-2	7		670	71	741	840
	637	Tee, 1"	Q-1	11		90	29	119	140
	639	1-1/2"		10		110	32	142	165
	640	2"		9		135	36	171	200
	642	3"		7		150	46	196	230
	643	4"		5		220	64	284	335
	645	6"	Q-2	6		395	83	478	555
	646	8"		5		640	99	739	850
	647	10"		4		1,075	125	1,200	1,350
	656	Reducer, 1-1/2"	Q-1	16		110	20	130	150
	657	2"		15		115	21	136	155
	659	3"		13		135	25	160	185
	666	4"		10		160	32	192	220
	668	6"	Q-2	11		260	45	305	350
	669	8"		10		480	50	530	600
	670	10"		9		845	55	900	1,000
	680	Fluorinated ethylene propylene (FEP)lined							
	681	90° Elbow, 1"	Q-1	16	Ea.	97	20	117	135
	683	1-1/2"		14		100	23	123	145
	684	2"		13		120	25	145	170
	686	3"		11		190	29	219	250
	687	4"		8		280	40	320	365
	689	6"	Q-2	9		595	55	650	735
	690	8"		8		910	62	972	1,100
	691	10"		7		1,500	71	1,571	1,750
	700	45° Elbow, 1"	Q-1	16		89	20	109	125
	702	1-1/2"		14		100	23	123	145
	703	2"		13		110	25	135	155
	705	3"		11		185	29	214	245
	706	4"		8		250	40	290	335
	708	6"	Q-2	9		500	55	555	630
	709	8"		8		795	62	857	965
	710	10"		7		1,325	71	1,396	1,550
	718	Tee, 1"	Q-1	11		175	29	204	235
	720	1-1/2"		10		200	32	232	265
	721	2"		9		225	36	261	300
	723	3"		7		310	46	356	405
	724	4"		5		400	64	464	530
	726	6"	Q-2	6		745	83	828	940
	727	8"		5		1,025	99	1,124	1,275
	728	10"		4		1,775	125	1,900	2,125
	737	Reducer, 1-1/2"	Q-1	16		120	20	140	160
	738	2"		15		130	21	151	175
	740	3"		13		180	25	205	235
	741	4"		10		245	32	277	315

15.1 Pipe & Fittings

		CREW	DAILY OUTPUT	UNIT	BARE COSTS MAT.	BARE COSTS INST.	BARE COSTS TOTAL	TOTAL INCL O&P
44 743	6"	Q-2	11	Ea.	390	45	435	495
744	8"		10		665	50	715	805
745	10"		9		1,200	55	1,255	1,400
755	Polytetrafluoroethylene (TFE) lined							
756	90° Elbow, 1"	Q-1	16	Ea.	110	20	130	150
758	1-1/2"		14		115	23	138	160
759	2"		13		140	25	165	190
761	3"		11		215	29	244	280
762	4"		8		320	40	360	410
764	6"	Q-2	9		675	55	730	820
765	8"		8		1,025	62	1,087	1,225
766	10"		7		1,500	71	1,571	1,750
774	45° Elbow, 1"	Q-1	16		100	20	120	140
776	1-1/2"		14		115	23	138	160
777	2"		13		125	25	150	175
779	3"		11		210	29	239	275
780	4"		8		285	40	325	370
782	6"	Q-2	9		570	55	625	705
783	8"		8		905	62	967	1,075
784	10"		7		1,325	71	1,396	1,550
792	Tee, 1"	Q-1	11		200	29	229	260
794	1-1/2"		10		225	32	257	295
795	2"		9		255	36	291	330
797	3"		7		350	46	396	450
798	4"		5		455	64	519	595
800	6"	Q-2	6		840	83	923	1,050
801	8"		5		1,175	99	1,274	1,425
802	10"		4		1,975	125	2,100	2,350
811	Reducer, 1-1/2"	Q-1	16		135	20	155	175
812	2"		15		145	21	166	190
814	3"		13		205	25	230	260
815	4"		10		275	32	307	350
817	6"	Q-2	11		445	45	490	555
818	8"		10		755	50	805	900
819	10"		9		1,350	55	1,405	1,575
46-001	**PIPE, GLASS** Borosilicate, couplings & hangers 10' O.C. (15.1-65-141)							
002	Drainage							
110	1-1/2" diameter	Q-1	47	L.F.	4.91	6.80	11.71	15.25
112	2" diameter		39		6.57	8.20	14.77	19.05
114	3" diameter		34		8.77	9.40	18.17	23
116	4" diameter		25		15.07	12.80	27.87	35
118	6" diameter		21		27.12	15.20	42.32	52
187	To delete coupling & hanger, subtract							
188	1-1/2" diam. to 2" diam.				19%	22%		
189	3" diam. to 6" diam.				20%	17%		
200	Process supply (pressure), beaded joints							
204	1/2" diameter	1 Plum	36	L.F.	3.83	4.93	8.76	11.35
206	3/4" diameter		31		4.71	5.75	10.46	13.45
208	1" diameter		27		6.17	6.60	12.77	16.30
210	1-1/2" diameter	Q-1	47		7.63	6.80	14.43	18.20
212	2" diameter		39		10.27	8.20	18.47	23
214	3" diameter		34		13.90	9.40	23.30	29
216	4" diameter		25		20.73	12.80	33.53	41
218	6" diameter		21		34.65	15.20	49.85	60
286	To delete coupling & hanger, subtract							
287	1/2" diam. to 1" diam.				25%	33%		
288	1-1/2" diam. to 3" diam.				22%	21%		
289	4" diam. to 6" diam.				23%	15%		
300	Beaded joint, armored, translucent							

Line	15.1 Pipe & Fittings	CREW	DAILY OUTPUT	UNIT	BARE COSTS MAT.	BARE COSTS INST.	BARE COSTS TOTAL	TOTAL INCL O&P
46 304	1/2" diameter	1 Plum	36	L.F.	9.41	4.93	14.34	17.50
306	3/4" diameter		31		10.77	5.75	16.52	20
308	1" diameter		27		11.64	6.60	18.24	22
310	1-1/2" diameter	Q-1	47		14.11	6.80	20.91	25
312	2" diameter		39		17.98	8.20	26.18	32
314	3" diameter		34		28.15	9.40	37.55	45
316	4" diameter		25		45.37	12.80	58.17	68
318	6" diameter		21		77.55	15.20	92.75	105
371	To delete coupling & hanger, subtract							
372	1/2" diam. to 3/4" diam.				30%	27%		
373	1" diam. to 3" diam.				28%	22%		
374	4" diam. to 6" diam.				24%	15%		
380	Conical joint, transparent							
388	1" diameter	1 Plum	27	L.F.	6.70	6.60	13.30	16.85
390	1-1/2" diameter	Q-1	47		8.87	6.80	15.67	19.60
392	2" diameter		39		11.25	8.20	19.45	24
394	3" diameter		34		18.47	9.40	27.87	34
396	4" diameter		25		27	12.80	39.80	48
398	6" diameter		21		46.53	15.20	61.73	73
450	To delete couplings & hangers, subtract							
451	1" diam.				23%	27%		
452	1-1/2" diam. to 4" diam.				17%	23%		
453	6" diam.				22%	26%		
47-001	**PIPE, GLASS, FITTINGS**							
002	Drainage, beaded ends							
004	Cplg. & labor required at joints not incl. in fitting							
005	price. Add 1 per joint for installed price							
007	90° Bend or sweep, 1-1/2"			Ea.	10.45		10.45	11.50M
009	2"				13.20		13.20	14.50M
010	3"				22		22	24M
011	4"				34		34	37M
012	6" (sweep only)				105		105	115M
020	45° Bend or sweep same as 90°							
035	Tee, single sanitary, 1-1/2"			Ea.	16.90		16.90	18.60M
037	2"				16.90		16.90	18.60M
038	3"				25		25	28M
039	4"				44		44	48M
040	6"				120		120	130M
041	Tee, straight, 1-1/2"				21		21	23M
043	2"				21		21	23M
044	3"				30		30	33M
045	4"				52		52	57M
046	6"				130		130	145M
050	Coupling, stainless steel, TFE seal ring							
052	1-1/2"	Q-1	32	Ea.	6.90	10	16.90	22
053	2"		30		8.85	10.65	19.50	25
054	3"		25		11.85	12.80	24.65	32
055	4"		23		20	13.90	33.90	42
056	6"		20		46	16	62	74
200	Process supply (pressure), beaded ends							
205	90° Sweep elbow, 1/2"			Ea.	10.40		10.40	11.45M
207	3/4"				12.30		12.30	13.55M
208	1"				16.30		16.30	17.95M
209	1-1/2"				17.15		17.15	18.85M
210	2"				23		23	25M
212	3"				36		36	40M
213	4"				51		51	56M
214	6"				135		135	150M
215	45° Sweep elbow, same as 90°							

15

	15.1 Pipe & Fittings	CREW	DAILY OUTPUT	UNIT	BARE COSTS MAT.	BARE COSTS INST.	BARE COSTS TOTAL	TOTAL INCL O&P
47 225	Tee, straight, 1/2"			Ea.	18		18	19.80M
227	3/4"				18		18	19.80M
228	1"				20		20	22M
229	1-1/2"				29		29	32M
230	2"				34		34	37M
231	3"				42		42	46M
232	4"				71		71	78M
233	6"				125		125	140M
235	Coupling, Hydrin liner, for temperatures to 300°F							
237	1/2"	Q-1	40	Ea.	7.95	8	15.95	20
238	3/4"		37		9.40	8.65	18.05	23
239	1"		35		9.70	9.15	18.85	24
240	1-1/2"		32		11.95	10	21.95	28
241	2"		30		14.30	10.65	24.95	31
242	3"		25		22	12.80	34.80	43
243	4"		23		33	13.90	46.90	56
244	6"		20		50	16	66	78
250	Coupling, for systems to 400°F, add				100%			
255	For beaded joint armored fittings, add				200%			
260	Conical ends. Flange set, gasket & labor not incl. in fitting							
262	price. Add 1 per joint for installed price.							
265	90° sweep elbow, 1"			Ea.	26		26	29M
267	1-1/2"				40		40	44M
268	2"				43		43	47M
269	3"				63		63	69M
270	4"				100		100	110M
275	Cross (straight), add				50%			
285	Tee, add				20%			
295	Lateral "Y", 1"				48		48	53M
297	1-1/2"				55		55	61M
298	2"				67		67	74M
299	3"				96		96	105M
300	4"				150		150	165M
301	6"				315		315	345M
305	Flange set and Gasket, 1"	Q-1	28		11.40	11.40	22.80	29
307	1-1/2"		26		11.75	12.30	24.05	31
308	2"		24		14.25	13.30	27.55	35
309	3"		22		25	14.55	39.55	49
310	4"		16		44	20	64	77
311	6"		12		98	27	125	145
49-001	**PIPE, PLASTIC** (See also division 15.1-43)							
002	Fiberglass reinforced, couplings 10' O.C., hangers 3 per 10'							
008	General service							
012	2" diameter	Q-1	59	L.F.	5.21	5.40	10.61	13.55
014	3" diameter		52		6.43	6.15	12.58	15.95
015	4" diameter		48		8.68	6.65	15.33	19.15
016	6" diameter		39		12.18	8.20	20.38	25
017	8" diameter	Q-2	49		19.67	10.15	29.82	36
018	10" diameter		41		29.93	12.15	42.08	50
019	12" diameter		36		38.47	13.80	52.27	62
020	High strength							
024	2" diameter	Q-1	58	L.F.	6.55	5.50	12.05	15.15
026	3" diameter		51		8.65	6.25	14.90	18.55
028	4" diameter		47		10.88	6.80	17.68	22
030	6" diameter		38		16.59	8.40	24.99	30
032	8" diameter	Q-2	48		26.45	10.35	36.80	44
034	10" diameter		40		39.43	12.45	51.88	61
036	12" diameter		36		50.59	13.80	64.39	76
055	To delete coupling & hangers, subtract							
056	2" diam. to 6" diam.				33%	56%		

15

15.1 Pipe & Fittings		CREW	DAILY OUTPUT	UNIT	BARE COSTS MAT.	BARE COSTS INST.	BARE COSTS TOTAL	TOTAL INCL O&P
49 057	8" diam. to 12" diam.				31%	52%		
060	PVC, high impact/pressure, cplgs. 10' O.C., hangers 3 per 10'							
062	Schedule 40							
064	1/4" diameter	1 Plum	59	L.F.	.97	3.01	3.98	5.40
066	3/8" diameter		56		1.11	3.17	4.28	5.80
067	1/2" diameter		54		.82	3.29	4.11	5.65
068	3/4" diameter		51		.97	3.48	4.45	6.10
069	1" diameter		46		1.21	3.86	5.07	6.90
070	1-1/4" diameter		42		1.41	4.23	5.64	7.65
071	1-1/2" diameter		36		1.54	4.93	6.47	8.80
072	2" diameter	Q-1	59		1.84	5.40	7.24	9.85
073	2-1/2" diameter		56		2.84	5.70	8.54	11.35
074	3" diameter		53		3.52	6.05	9.57	12.60
075	4" diameter		48		4.94	6.65	11.59	15.05
076	5" diameter		43		8.22	7.45	15.67	19.80
077	6" diameter		39		9.57	8.20	17.77	22
078	8" diameter	Q-2	48		15.34	10.35	25.69	32
079	10" diameter		43		25.47	11.55	37.02	45
080	12" diameter		42		32.26	11.85	44.11	53
093	To delete coupling & hangers, subtract							
094	1/4" diam. to 1/2" diam.				64%	81%		
095	3/4" diam. to 1-1/4" diam.				61%	72%		
096	1-1/2" diam. to 6" diam.				44%	57%		
097	8" diam. to 12" diam.				46%	53%		
102	Schedule 80							
104	1/4" diameter	1 Plum	58	L.F.	1.21	3.06	4.27	5.75
106	3/8" diameter		55		1.42	3.23	4.65	6.25
107	1/2" diameter		50		1.01	3.55	4.56	6.25
108	3/4" diameter		47		1.18	3.78	4.96	6.75
109	1" diameter		43		1.36	4.13	5.49	7.45
110	1-1/4" diameter		39		1.76	4.55	6.31	8.50
111	1-1/2" diameter		34		1.90	5.20	7.10	9.65
112	2" diameter	Q-1	55		2.49	5.80	8.29	11.15
113	2-1/2" diameter		52		3.40	6.15	9.55	12.60
114	3" diameter		50		4.65	6.40	11.05	14.35
115	4" diameter		46		7.01	6.95	13.96	17.75
116	5" diameter		42		11.30	7.60	18.90	23
117	6" diameter		38		13.39	8.40	21.79	27
118	8" diameter	Q-2	47		21.27	10.60	31.87	39
119	10" diameter		42		30.53	11.85	42.38	51
120	12" diameter		38		41.22	13.10	54.32	64
173	To delete coupling & hangers, subtract							
174	1/4" diam. to 1/2" diam.				62%	80%		
175	3/4" diam. to 1-1/4" diam.				58%	73%		
176	1-1/2" diam. to 6" diam.				40%	57%		
177	8" diam. to 12" diam.				34%	50%		
180	PVC, couplings 10' O.C., hangers 3 per 10'							
182	Schedule 40							
184	1/4" diameter	1 Plum	59	L.F.	.90	3.01	3.91	5.35
185	3/8" diameter		56		1.01	3.17	4.18	5.70
186	1/2" diameter		54		.71	3.29	4	5.55
187	3/4" diameter		51		.79	3.48	4.27	5.90
188	1" diameter		46		.95	3.86	4.81	6.60
189	1-1/4" diameter		42		1.07	4.23	5.30	7.30
190	1-1/2" diameter		36		1.10	4.93	6.03	8.35
191	2" diameter	Q-1	59		1.31	5.40	6.71	9.25
192	2-1/2" diameter		56		2.22	5.70	7.92	10.70
193	3" diameter		53		2.77	6.05	8.82	11.75
194	4" diameter		48		3.26	6.65	9.91	13.20
195	5" diameter		43		5.56	7.45	13.01	16.85

15

15.1 Pipe & Fittings

Line	Description	CREW	DAILY OUTPUT	UNIT	BARE COSTS MAT.	BARE COSTS INST.	BARE COSTS TOTAL	TOTAL INCL O&P
49 196	6" diameter	Q-1	39	L.F.	6.91	8.20	15.11	19.45
197	8" diameter	Q-2	48		12.36	10.35	22.71	29
198	10" diameter		43		24.48	11.55	36.03	44
199	12" diameter		42		30.94	11.85	42.79	51
200	14" diameter		31		54.42	16.05	70.47	83
201	16" diameter		23		66.84	22	88.84	105
234	To delete coupling & hangers, subtract							
235	1/4" diam. to 3/8" diam.				63%	81%		
236	1/2" diam. to 1-1/4" diam.				65%	74%		
237	1-1/2" diam. to 6" diam.				44%	57%		
238	8" diam. to 12" diam.				41%	53%		
239	14" diam. to 16" diam.				48%	45%		
242	Schedule 80							
244	1/4" diameter	1 Plum	58	L.F.	1.05	3.06	4.11	5.60
245	3/8" diameter		55		1.23	3.23	4.46	6
246	1/2" diameter		50		.85	3.55	4.40	6.05
247	3/4" diameter		47		.96	3.78	4.74	6.50
248	1" diameter		43		1.14	4.13	5.27	7.20
249	1-1/4" diameter		39		1.37	4.55	5.92	8.10
250	1-1/2" diameter		34		1.50	5.20	6.70	9.20
251	2" diameter	Q-1	55		1.87	5.80	7.67	10.45
252	2-1/2" diameter		52		2.71	6.15	8.86	11.85
253	3" diameter		50		3.44	6.40	9.84	13
254	4" diameter		46		5.26	6.95	12.21	15.85
255	5" diameter		42		6.90	7.60	14.50	18.60
256	6" diameter		33		9.98	9.70	19.68	25
257	8" diameter	Q-2	47		15.93	10.60	26.53	33
258	10" diameter		42		29.43	11.85	41.28	49
259	12" diameter		38		39.02	13.10	52.12	62
283	To delete coupling & hangers, subtract							
284	1/4" diam. to 1/2" diam.				66%	80%		
285	3/4" diam. to 1-1/4" diam.				61%	73%		
286	1-1/2" diam. to 6" diam.				41%	57%		
287	8" diam. to 12" diam.				31%	50%		
290	Schedule 120							
291	1/2" diameter	1 Plum	50	L.F.	1.14	3.55	4.69	6.40
295	3/4" diameter		47		1.33	3.78	5.11	6.90
296	1" diameter		43		1.68	4.13	5.81	7.80
297	1-1/4" diameter		39		2.12	4.55	6.67	8.90
298	1-1/2" diameter		33		2.43	5.40	7.83	10.45
299	2" diameter	Q-1	54		3.21	5.90	9.11	12.10
300	2-1/2" diameter		52		4.70	6.15	10.85	14.05
301	3" diameter		49		6.16	6.50	12.66	16.20
302	4" diameter		45		9.89	7.10	16.99	21
303	6" diameter		37		18.78	8.65	27.43	33
324	To delete coupling & hangers, subtract							
325	1/2" diam. to 1-1/4" diam.				52%	74%		
326	1-1/2" diam. to 4" diam.				30%	57%		
327	6" diam.				17%	50%		
330	PVC, pressure, couplings 10' O.C., hangers 3 per 10'							
331	SDR 26, 160 psi							
335	1-1/4" diameter	1 Plum	42	L.F.	.81	4.23	5.04	7
336	1-1/2" diameter	"	36		.88	4.93	5.81	8.10
337	2" diameter	Q-1	59		1	5.40	6.40	8.95
338	2-1/2" diameter		56		1.52	5.70	7.22	9.90
339	3" diameter		53		1.85	6.05	7.90	10.75
340	4" diameter		48		2.75	6.65	9.40	12.65
342	6" diameter		39		6.27	8.20	14.47	18.75
343	8" diameter	Q-2	48		11.15	10.35	21.50	27
366	To delete coupling & hangers, subtract							

15.1 Pipe & Fittings

			DAILY		BARE COSTS			TOTAL
		CREW	OUTPUT	UNIT	MAT.	INST.	TOTAL	INCL O&P
49 367	1-1/4" diam.				63%	68%		
368	1-1/2" diam. to 5" diam.				48%	57%		
369	6" diam.				60%	54%		
372	SDR 21, 200 psi, 1/2" diameter	1 Plum	54	L.F.	.71	3.29	4	5.55
374	3/4" diameter		51		.76	3.48	4.24	5.85
375	1" diameter		46		.83	3.86	4.69	6.50
376	1-1/4" diameter		42		.96	4.23	5.19	7.15
377	1-1/2" diameter		36		1.06	4.93	5.99	8.30
378	2" diameter	Q-1	59		1.30	5.40	6.70	9.25
379	2-1/2" diameter		56		2.06	5.70	7.76	10.50
380	3" diameter		53		2.34	6.05	8.39	11.30
381	4" diameter		48		3.81	6.65	10.46	13.80
383	6" diameter		39		7.70	8.20	15.90	20
384	8" diameter	Q-2	48		12.47	10.35	22.82	29
400	To delete coupling & hangers, subtract							
401	1/2" diam. to 3/4" diam.				71%	77%		
402	1" diam. to 1-1/4" diam.				63%	70%		
403	1-1/2" diam. to 6" diam.				44%	57%		
404	8" diam.				46%	54%		
410	DWV type, schedule 40, couplings 10' O.C., hangers 3 per 10'							
412	ABS							
414	1-1/4" diameter	1 Plum	42	L.F.	.97	4.23	5.20	7.20
415	1-1/2" diameter	"	36		.98	4.93	5.91	8.20
416	2" diameter	Q-1	59		1.12	5.40	6.52	9.05
417	3" diameter		53		1.98	6.05	8.03	10.90
418	4" diameter		48		2.78	6.65	9.43	12.70
419	6" diameter		39		6.33	8.20	14.53	18.80
436	To delete coupling & hangers, subtract							
437	1-1/4" diam.				64%	68%		
438	1-1/2" diam. to 6" diam.				54%	57%		
440	PVC							
441	1-1/4" diameter	1 Plum	42	L.F.	1.14	4.23	5.37	7.35
442	1-1/2" diameter	"	36		1.17	4.93	6.10	8.40
446	2" diameter	Q-1	59		1.38	5.40	6.78	9.35
447	3" diameter		53		2.48	6.05	8.53	11.45
448	4" diameter		48		3.45	6.65	10.10	13.40
449	6" diameter		39		7.49	8.20	15.69	20
475	To delete coupling & hangers, subtract							
476	1-1/4" diam. to 1-1/2" diam.				71%	64%		
477	2" diam. to 6" diam.				60%	57%		
480	PVC, clear pipe, couplings 10' O.C., hangers 3 per 10', schedule 40							
484	1/4" diameter	1 Plum	59	L.F.	.93	3.01	3.94	5.35
485	3/8" diameter		56		1.07	3.17	4.24	5.75
486	1/2" diameter		54		1.34	3.29	4.63	6.25
487	3/4" diameter		51		1.63	3.48	5.11	6.80
488	1" diameter		46		2.20	3.86	6.06	8
489	1-1/4" diameter		42		2.78	4.23	7.01	9.15
490	1-1/2" diameter		36		3.24	4.93	8.17	10.70
491	2" diameter	Q-1	59		4.16	5.40	9.56	12.40
492	2-1/2" diameter		56		6.49	5.70	12.19	15.40
493	3" diameter		53		8.33	6.05	14.38	17.90
494	3-1/2" diameter		50		10.48	6.40	16.88	21
495	4" diameter		48		12.35	6.65	19	23
525	To delete coupling & hangers, subtract							
526	1/4" diam. to 3/8" diam.				60%	81%		
527	1/2" diam. to 3/4" diam.				41%	77%		
528	1" diam. to 1-1/2" diam.				26%	67%		
529	2" diam. to 4" diam.				16%	58%		
536	CPVC, couplings 10' O.C., hangers 3 per 10'							
538	Schedule 40							

15

15.1 Pipe & Fittings

		CREW	DAILY OUTPUT	UNIT	BARE COSTS MAT.	BARE COSTS INST.	BARE COSTS TOTAL	TOTAL INCL O&P
49 546	1/2" diameter	1 Plum	54	L.F.	1.41	3.29	4.70	6.30
547	3/4" diameter		51		1.71	3.48	5.19	6.90
548	1" diameter		46		2.26	3.86	6.12	8.05
549	1-1/4" diameter		42		2.87	4.23	7.10	9.25
550	1-1/2" diameter		36		3.32	4.93	8.25	10.80
551	2" diameter	Q-1	59		4.11	5.40	9.51	12.35
552	2-1/2" diameter		56		7.19	5.70	12.89	16.15
553	3" diameter		53		8.45	6.05	14.50	18
554	4" diameter		48		11.66	6.65	18.31	22
555	6" diameter		43		22.08	7.45	29.53	35
573	To delete coupling & hangers, subtract							
574	1/2" diam. to 3/4" diam.				37%	77%		
575	1" diam. to 1-1/4" diam.				27%	70%		
576	1-1/2" diam. to 3" diam.				21%	57%		
577	4" diam. to 6" diam.				16%	57%		
580	Schedule 80							
586	1/2" diameter	1 Plum	50	L.F.	1.46	3.55	5.01	6.75
587	3/4" diameter		47		1.80	3.78	5.58	7.45
588	1" diameter		43		2.37	4.13	6.50	8.55
589	1-1/4" diameter		39		3.19	4.55	7.74	10.10
590	1-1/2" diameter		34		3.73	5.20	8.93	11.65
591	2" diameter	Q-1	55		4.74	5.80	10.54	13.60
592	2-1/2" diameter		52		8.22	6.15	14.37	17.95
593	3" diameter		50		9.76	6.40	16.16	20
594	4" diameter		46		13.75	6.95	20.70	25
595	6" diameter		38		26.09	8.40	34.49	41
596	8" diameter	Q-2	47		43.10	10.60	53.70	63
606	To delete couplings & hangers, subtract							
607	1/2" diam. to 3/4" diam.				44%	77%		
608	1" diam. to 1-1/4" diam.				32%	71%		
609	1-1/2" diam. to 4" diam.				25%	58%		
610	6" diam. to 8" diam.				20%	53%		
624	CTS, 1/2" diameter	1 Plum	54	L.F.	.95	3.29	4.24	5.80
625	3/4" diameter	"	51	"	1.19	3.48	4.67	6.35
637	To delete coupling & hangers, subtract							
638	1/2" diam.				51%	79%		
639	3/4" diam.				40%	76%		
641	Polybutylene, flexible, no couplings or hangers, 100' coils							
642	For plastic hangers see 15.1-65-800							
643	Cold water							
644	SDR 13.5, 160 psi @ 73°F, 3/4" CTS	1 Plum	250	L.F.	.18	.71	.89	1.22
645	1" CTS		204		.29	.87	1.16	1.58
646	1-1/4" CTS		151		.46	1.18	1.64	2.21
647	1-1/2" CTS		107		.62	1.66	2.28	3.08
648	2" CTS		87		1.05	2.04	3.09	4.10
650	SDR 9, 250 psi @ 73°F, 3/4" CTS		250		.27	.71	.98	1.32
651	1" CTS		204		.44	.87	1.31	1.74
652	1-1/4" CTS		151		.64	1.18	1.82	2.40
653	1-1/2" CTS		107		.92	1.66	2.58	3.41
654	2" CTS		87		1.58	2.04	3.62	4.69
656	SDR 11.5, 160 psi @ 73°F, 3/4" IPS		239		.24	.74	.98	1.34
657	1" IPS		195		.35	.91	1.26	1.70
658	1-1/4" IPS		143		.64	1.24	1.88	2.50
659	1-1/2" IPS		102		.88	1.74	2.62	3.48
660	2" IPS		83		1.46	2.14	3.60	4.70
663	SDR 11, 200 psi @ 73°F, 1" IPS		195		.43	.91	1.34	1.79
665	1-1/4" IPS		143		.67	1.24	1.91	2.53
666	1-1/2" IPS		102		.88	1.74	2.62	3.48
667	2" IPS		83		1.37	2.14	3.51	4.60
670	SDR 7, 250 psi @ 73°F, 3/4" IPS		239		.33	.74	1.07	1.44

15.1 Pipe & Fittings		CREW	DAILY OUTPUT	UNIT	BARE COSTS MAT.	BARE COSTS INST.	BARE COSTS TOTAL	TOTAL INCL O&P
49 671	1" IPS	1 Plum	195	L.F.	.53	.91	1.44	1.90
672	1-1/4" IPS		143		.91	1.24	2.15	2.80
673	1-1/2" IPS		102		1.24	1.74	2.98	3.88
674	2" IPS		83		2.03	2.14	4.17	5.30
682	Hot and cold water							
683	SDR 11, 100 psi @ 180°F, 1/8" CTS	1 Plum	395	L.F.	.10	.45	.55	.76
684	1/4" CTS		390		.11	.46	.57	.78
685	3/8" CTS		342		.12	.52	.64	.88
686	1/2" CTS		294		.14	.60	.74	1.03
687	3/4" CTS		212		.25	.84	1.09	1.49
690	SDR 11, 100 psi @ 180°F, fusion joints, 1" CTS		164		.42	1.08	1.50	2.03
691	1" IPS		157		.58	1.13	1.71	2.27
692	1-1/4" IPS		117		.89	1.52	2.41	3.17
693	1-1/2" IPS		83		1.19	2.14	3.33	4.40
694	2" IPS		67	Ea.	1.87	2.65	4.52	5.90
695	3" IPS		61	L.F.	5.20	2.91	8.11	9.95
700	SDR 11.5, 80 psi @ 180°F, 3/4" IPS		193		.32	.92	1.24	1.68
701	1" IPS		157		.50	1.13	1.63	2.18
702	1-1/4" IPS		117		.86	1.52	2.38	3.14
703	1-1/2" IPS		83		1.19	2.14	3.33	4.40
704	2" IPS		67		1.95	2.65	4.60	5.95
710	Straight lengths same price as coil							
728	Polyethylene, flexible, no couplings or hangers							
730	SDR 15, 125 psi							
731	3/4" diameter			L.F.	.10		.10	.11M
735	1" diameter				.14		.14	.15M
736	1-1/4" diameter				.25		.25	.27M
737	1-1/2" diameter				.33		.33	.36M
738	2" diameter				.55		.55	.60M
770	SDR 11.5, 160 psi							
771	1/2" diameter			L.F.	.08		.08	.08M
775	3/4" diameter				.12		.12	.13M
776	1" diameter				.18		.18	.19M
777	1-1/4" diameter				.32		.32	.35M
778	1-1/2" diameter				.44		.44	.48M
779	2" diameter				.73		.73	.80M
812	SDR 9, 200 psi							
814	1/2" diameter			L.F.	.09		.09	.09M
815	3/4" diameter				.16		.16	.17M
816	1" diameter				.25		.25	.27M
817	1-1/4" diameter				.43		.43	.47M
818	1-1/2" diameter				.56		.56	.61M
842	SDR 7							
844	3/4" diameter			L.F.	.21		.21	.23M
845	1" diameter				.33		.33	.36M
846	1-1/4" diameter				.58		.58	.63M
880	PVC, type PSP, drain & sewer, belled end gasket joint, no hangers							
884	3" diameter			L.F.	.36		.36	.39M
885	4" diameter				.44		.44	.48M
886	6" diameter				1		1	1.10M
900	Perforated							
904	4" diameter			L.F.	.44		.44	.48M
50-001	PIPE, PLASTIC, FITTINGS							
003	Epoxy resin, fiberglass reinforced, general service							
009	Elbow, 90°, 2"	Q-1	23	Ea.	26	13.90	39.90	49
010	3"		16		28	20	48	60
011	4"		13		35	25	60	74
012	6"		8		64	40	104	130
013	8"	Q-2	9		130	55	185	225
014	10"	"	7		190	71	261	310

15

15.1 Pipe & Fittings

			CREW	DAILY OUTPUT	UNIT	BARE COSTS MAT.	INST.	TOTAL	TOTAL INCL O&P
50	015	12"	Q-2	5	Ea.	255	99	354	425
	016	45° Elbow, same as 90°							
	029	Tee, 2"	Q-1	17	Ea.	30	18.80	48.80	60
	030	3"		10		31	32	63	80
	031	4"		8		40	40	80	100
	032	6"		5		115	64	179	220
	033	8"	Q-2	6		180	83	263	320
	034	10"		5		300	99	399	475
	035	12"		4		360	125	485	575
	038	Couplings							
	041	2"	Q-1	28	Ea.	10.50	11.40	21.90	28
	042	3"		20		11.70	16	27.70	36
	043	4"		17		13.45	18.80	32.25	42
	044	6"		12		21	27	48	62
	045	8"	Q-2	15		34	33	67	85
	046	10"		11		53	45	98	125
	047	12"		10		68	50	118	145
	049	High strength couplings, add				15%			
	050	PVC, high impact/pressure, schedule 40							
	053	90° Elbow, 1/2"	1 Plum	20	Ea.	.42	8.90	9.32	13.30
	055	3/4"		19		.52	9.35	9.87	14.10
	056	1"		16		.92	11.10	12.02	17.05
	057	1-1/4"		15		1.30	11.85	13.15	18.55
	058	1-1/2"		14		1.61	12.70	14.31	20
	059	2"	Q-1	23		2.46	13.90	16.36	23
	060	3"		16		5.85	20	25.85	35
	061	4"		13		8.70	25	33.70	45
	062	6"		8		38	40	78	100
	063	8"	Q-2	9		87	55	142	175
	066	45° Elbow, 1/2"	1 Plum	20		.54	8.90	9.44	13.40
	068	3/4"		19		.90	9.35	10.25	14.50
	069	1"		16		1.02	11.10	12.12	17.15
	070	1-1/4"		15		1.47	11.85	13.32	18.70
	071	1-1/2"		14		1.81	12.70	14.51	20
	072	2"	Q-1	23		2.44	13.90	16.34	23
	073	3"		16		5.85	20	25.85	35
	074	4"		13		10.50	25	35.50	47
	075	6"		8		38	40	78	100
	076	8"	Q-2	9		87	55	142	175
	080	Tee, 1/2"	1 Plum	13		.43	13.65	14.08	20
	082	3/4"		12		.54	14.80	15.34	22
	083	1"		11		1.01	16.15	17.16	24
	084	1-1/4"		10		1.25	17.75	19	27
	085	1-1/2"		10		1.60	17.75	19.35	27
	086	2"	Q-1	17		2.26	18.80	21.06	30
	087	3"		10		6.85	32	38.85	54
	088	4"		8		11.40	40	51.40	70
	089	6"		5		54	64	118	150
	090	8"	Q-2	6		125	83	208	255
	091	10"		4		160	125	285	355
	092	12"		3		235	165	400	500
	105	Coupling, 1/4"	1 Plum	26		.41	6.85	7.26	10.30
	106	3/8"		24		.41	7.40	7.81	11.15
	107	1/2"		22		.21	8.05	8.26	11.90
	108	3/4"		21		.28	8.45	8.73	12.55
	109	1"		18		.49	9.85	10.34	14.80
	110	1-1/4"		17		.56	10.45	11.01	15.70
	111	1-1/2"		16		.68	11.10	11.78	16.80
	112	2"	Q-1	28		.60	11.40	12	17.15
	113	2-1/2"	"	25		1.59	12.80	14.39	20

15.1 Pipe & Fittings

		Description	Crew	Daily Output	Unit	Bare Costs Mat.	Bare Costs Inst.	Bare Costs Total	Total Incl O&P
50	114	3"	Q-1	22	Ea.	2.69	14.55	17.24	24
	115	3-1/2"		19		3.85	16.85	20.70	29
	116	4"		17		3.85	18.80	22.65	31
	117	5"		14		6.45	23	29.45	40
	118	6"		12		16.30	27	43.30	56
	119	8"	Q-2	14		34	36	70	89
	120	10"		13		68	38	106	130
	121	12"		12		85	41	126	155
	122	14"		10		240	50	290	335
	123	16"		7		265	71	336	395
	210	Schedule 80							
	211	90° Elbow, 1/2"	1 Plum	18	Ea.	1.16	9.85	11.01	15.55
	213	3/4"		17		1.53	10.45	11.98	16.75
	214	1"		15		1.84	11.85	13.69	19.15
	215	1-1/4"		14		1.97	12.70	14.67	20
	216	1-1/2"		13		2.27	13.65	15.92	22
	217	2"	Q-1	22		3.10	14.55	17.65	24
	218	3"		14		11.30	23	34.30	45
	219	4"		12		16.60	27	43.60	57
	220	6"		7		53	46	99	125
	221	8"	Q-2	8		45	62	107	140
	225	45° Elbow, 1/2"	1 Plum	18		1.06	9.85	10.91	15.40
	227	3/4"		17		1.37	10.45	11.82	16.60
	228	1"		15		1.68	11.85	13.53	18.95
	229	1-1/4"		14		1.97	12.70	14.67	20
	230	1-1/2"		13		2.17	13.65	15.82	22
	231	2"	Q-1	22		3.10	14.55	17.65	24
	232	3"		14		9.45	23	32.45	43
	233	4"		12		14.05	27	41.05	54
	234	6"		7		53	46	99	125
	235	8"	Q-2	8		45	62	107	140
	240	Tee, 1/2"	1 Plum	12		1.29	14.80	16.09	23
	242	3/4"		11		1.45	16.15	17.60	25
	243	1"		10		1.78	17.75	19.53	28
	244	1-1/4"		9		2.24	19.75	21.99	31
	245	1-1/2"		8		2.68	22	24.68	35
	246	2"	Q-1	14		3.59	23	26.59	37
	247	3"		9		17.45	36	53.45	71
	248	4"		8		26	40	66	86
	249	6"		5		76	64	140	175
	250	8"	Q-2	6		64	83	147	190
	255	Coupling, 1/2"	1 Plum	18		1.07	9.85	10.92	15.45
	257	3/4"		17		1.25	10.45	11.70	16.45
	258	1"		15		1.35	11.85	13.20	18.60
	259	1-1/4"		14		1.58	12.70	14.28	20
	260	1-1/2"		13		1.81	13.65	15.46	22
	261	2"	Q-1	22		2.53	14.55	17.08	24
	262	3"		19		5.20	16.85	22.05	30
	263	4"		16		10.45	20	30.45	40
	264	6"		12		23	27	50	64
	265	8"	Q-2	14		37	36	73	92
	266	10"		13		40	38	78	99
	267	12"		12		53	41	94	120
	270	PVC (white), schedule 40, socket joints							
	273	Elbow 90°, 1/4" diameter	1 Plum	26	Ea.	.11	6.85	6.96	10
	275	3/8"		24		.11	7.40	7.51	10.80
	276	1/2"		22		.14	8.05	8.19	11.80
	277	3/4"		21		.16	8.45	8.61	12.40
	278	1"		18		.28	9.85	10.13	14.55
	279	1-1/4"		17		.47	10.45	10.92	15.60

		CREW	DAILY OUTPUT	UNIT	BARE COSTS MAT.	BARE COSTS INST.	BARE COSTS TOTAL	TOTAL INCL O&P
50 280	1-1/2"	1 Plum	16	Ea.	.49	11.10	11.59	16.55
281	2"	Q-1	28		.77	11.40	12.17	17.35
282	2-1/2"		22		2.39	14.55	16.94	24
283	3"		17		2.37	18.80	21.17	30
284	4"		14		5.10	23	28.10	39
285	5"		12		16.30	27	43.30	56
286	6"		8		16.30	40	56.30	76
287	8"	Q-2	10		42	50	92	120
295	Elbow 45°, 1/4"	1 Plum	26		.16	6.85	7.01	10.05
297	3/8"		24		.16	7.40	7.56	10.85
298	1/2"		22		.21	8.05	8.26	11.90
299	3/4"		21		.34	8.45	8.79	12.60
300	1"		18		.40	9.85	10.25	14.70
301	1-1/4"		17		.57	10.45	11.02	15.70
302	1-1/2"		16		.70	11.10	11.80	16.80
303	2"	Q-1	28		.93	11.40	12.33	17.50
304	2-1/2"		22		2.39	14.55	16.94	24
305	3"		17		3.70	18.80	22.50	31
306	4"		14		6.70	23	29.70	40
307	5"		12		16.50	27	43.50	57
308	6"		8		16.50	40	56.50	76
309	8"	Q-2	10		40	50	90	115
315	Tee, 1/4"	1 Plum	17		.14	10.45	10.59	15.25
317	3/8"		16		.14	11.10	11.24	16.20
318	1/2"		14		.17	12.70	12.87	18.50
319	3/4"		13		.19	13.65	13.84	19.95
320	1"		12		.35	14.80	15.15	22
321	1-1/4"		11		.54	16.15	16.69	24
322	1-1/2"		10		.68	17.75	18.43	26
323	2"	Q-1	17		.93	18.80	19.73	28
324	2-1/2"		14		3.38	23	26.38	37
325	3"		11		4.57	29	33.57	47
326	4"		9		7.60	36	43.60	60
327	5"		8		26	40	66	86
328	6"		5		26	64	90	120
329	8"	Q-2	6		60	83	143	185
335	Coupling, 1/4"	1 Plum	26		.06	6.85	6.91	9.95
337	3/8"		24		.06	7.40	7.46	10.75
338	1/2"		22		.08	8.05	8.13	11.75
339	3/4"		21		.12	8.45	8.57	12.35
340	1"		18		.20	9.85	10.05	14.45
341	1-1/4"		17		.26	10.45	10.71	15.40
342	1-1/2"		16		.28	11.10	11.38	16.35
343	2"	Q-1	28		.46	11.40	11.86	17
344	2-1/2"		20		1.04	16	17.04	24
345	3"		19		1.63	16.85	18.48	26
346	4"		16		2.31	20	22.31	31
347	5"		14		7.45	23	30.45	41
348	6"		12		7.45	27	34.45	47
349	8"	Q-2	14		13.85	36	49.85	67
450	DWV, ABS, non pressure, socket joints							
454	1/4 Bend, 1-1/4"	1 Plum	17	Ea.	.62	10.45	11.07	15.75
456	1-1/2"	"	16		.46	11.10	11.56	16.55
457	2"	Q-1	28		.62	11.40	12.02	17.20
458	3"		17		1.98	18.80	20.78	29
459	4"		14		3.62	23	26.62	37
460	6"		8		25	40	65	85
465	1/8 Bend, same as 1/4 bend							
480	Tee, sanitary							
482	1-1/4"	1 Plum	11	Ea.	.72	16.15	16.87	24

15.1 Pipe & Fittings

		CREW	DAILY OUTPUT	UNIT	BARE COSTS MAT.	INST.	TOTAL	TOTAL INCL O&P
50 483	1-1/2"	1 Plum	10	Ea.	.72	17.75	18.47	26
484	2"	Q-1	17		1.06	18.80	19.86	28
485	3"		11		2.58	29	31.58	45
486	4"		9		5	36	41	57
500	PVC, schedule 40, socket joints							
504	1/4 Bend, 1-1/4" diameter	1 Plum	17	Ea.	1.40	10.45	11.85	16.65
506	1-1/2"	"	16		.86	11.10	11.96	17
507	2"	Q-1	28		1.16	11.40	12.56	17.80
508	3"		17		3.71	18.80	22.51	31
509	4"		14		6.80	23	29.80	40
510	6"		8		46	40	86	110
515	1/8 Bend, 1-1/4"	1 Plum	17		1.36	10.45	11.81	16.60
517	1-1/2"	"	16		.83	11.10	11.93	16.95
518	2"	Q-1	28		1.08	11.40	12.48	17.70
519	3"		17		3.13	18.80	21.93	31
520	4"		14		5.80	23	28.80	39
521	6"		8		42	40	82	105
525	Tee, sanitary 1-1/4"	1 Plum	11		1.86	16.15	18.01	25
527	1-1/2"	"	10		1.34	17.75	19.09	27
528	2"	Q-1	17		2	18.80	20.80	29
529	3"		11		4.84	29	33.84	47
530	4"		9		9.45	36	45.45	62
535	Coupling, 1-1/4"	1 Plum	17		.73	10.45	11.18	15.90
536	1-1/2"	"	16		.50	11.10	11.60	16.60
537	2"	Q-1	28		.69	11.40	12.09	17.25
538	3"		22		1.70	14.55	16.25	23
539	4"		17		2.84	18.80	21.64	30
540	6"		12		18.55	27	45.55	59
545	Solvent cement for PVC, industrial grade, per quart				8.50		8.50	9.35M
550	CPVC, schedule 80, socket or threaded joints							
554	90° Elbow, 1/4"	1 Plum	20	Ea.	4.69	8.90	13.59	18
556	1/2"		18		1.54	9.85	11.39	15.95
557	3/4"		17		2.40	10.45	12.85	17.75
558	1"		15		3.83	11.85	15.68	21
559	1-1/4"		14		8.85	12.70	21.55	28
560	1-1/2"		13		7.30	13.65	20.95	28
561	2"	Q-1	22		11.15	14.55	25.70	33
562	2-1/2"		18		27	17.75	44.75	55
563	3"		14		29	23	52	65
564	4"		12		53	27	80	97
565	6"		7		105	46	151	180
570	45° Elbow same as 90° Elbow							
585	Tee, 1/4"	1 Plum	14	Ea.	5.65	12.70	18.35	25
587	1/2"		12		2.78	14.80	17.58	24
588	3/4"		11		4.50	16.15	20.65	28
589	1"		10		5.50	17.75	23.25	32
590	1-1/4"		9		12.55	19.75	32.30	42
591	1-1/2"		8		13.35	22	35.35	47
592	2"	Q-1	14		14.85	23	37.85	49
593	2-1/2"		12		37	27	64	79
594	3"		9		41	36	77	96
595	4"		8		42	40	82	105
596	6"		5		130	64	194	235
600	Coupling, 1/4"	1 Plum	20		5.45	8.90	14.35	18.80
602	1/2"		18		2.03	9.85	11.88	16.50
603	3/4"		17		2.78	10.45	13.23	18.15
604	1"		15		3.71	11.85	15.56	21
605	1-1/4"		14		6.05	12.70	18.75	25
606	1-1/2"		13		7.10	13.65	20.75	28
607	2"	Q-1	22		8.20	14.55	22.75	30

15.1 Pipe & Fittings

	15.1 Pipe & Fittings	CREW	DAILY OUTPUT	UNIT	BARE COSTS MAT.	BARE COSTS INST.	BARE COSTS TOTAL	TOTAL INCL O&P
50 608	2-1/2"	Q-1	20	Ea.	19.80	16	35.80	45
609	3"		19		20	16.85	36.85	46
610	4"		16		26	20	46	57
611	6"		12		51	27	78	95
612	8"	Q-2	14		65	36	101	125
620	CTS, 100 psi at 180°F, hot and cold water							
623	90° Elbow, 1/2"	1 Plum	20	Ea.	.09	8.90	8.99	12.95
625	3/4"		19		.20	9.35	9.55	13.70
626	45° Elbow, 1/2"		20		.12	8.90	9.02	12.95
628	3/4"		19		.20	9.35	9.55	13.70
629	Tee, 1/2"		13		.14	13.65	13.79	19.90
631	3/4"		12		.28	14.80	15.08	22
632	Coupling, 1/2"		22		.10	8.05	8.15	11.75
634	3/4"		21		.13	8.45	8.58	12.35
636	Solvent cement for CPVC, commercial grade, per quart				6.65		6.65	7.30M
799	Polybutylene/polyethylene pipe, for copper fittings see 15.1-41-70							
800	Compression type, PVC, 160 psi cold water							
801	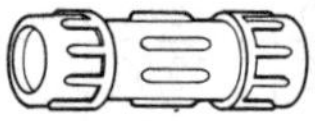 Coupling, 3/4" CTS	1 Plum	21	Ea.	2.88	8.45	11.33	15.40
802	1" CTS		18		3.60	9.85	13.45	18.20
803	1-1/4" CTS		17		5.60	10.45	16.05	21
804	1-1/2" CTS		16		6.45	11.10	17.55	23
805	2" CTS		15		8.60	11.85	20.45	27
806	Female adapter, 3/4" FPT x 3/4" CTS		23		2.12	7.70	9.82	13.50
807	3/4" FPT x 1" CTS		21		3.16	8.45	11.61	15.70
808	1" FPT x 1" CTS		20		3.16	8.90	12.06	16.30
809	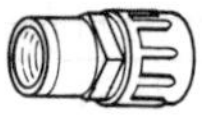 1-1/4" FPT x 1-1/4" CTS		18		4.17	9.85	14.02	18.85
810	1-1/2" FPT x 1-1/2" CTS		16		4.65	11.10	15.75	21
811	2" FPT x 2" CTS		13		7.25	13.65	20.90	28
813	Male adapter, 3/4" MPT x 3/4" CTS		23		1.90	7.70	9.60	13.25
814	 3/4" MPT x 1" CTS		21		2.37	8.45	10.82	14.80
815	1" MPT x 1" CTS		20		2.37	8.90	11.27	15.45
816	1-1/4" MPT x 1-1/4" CTS		18		3.70	9.85	13.55	18.30
817	1-1/2" MPT x 1-1/2" CTS		16		4.27	11.10	15.37	21
818	2" MPT x 2" CTS		13		6.80	13.65	20.45	27
820	Spigot adapter, 3/4" IPS x 3/4" CTS		23		1.90	7.70	9.60	13.25
821	 3/4" IPS x 1" CTS		21		1.90	8.45	10.35	14.30
822	1" IPS x 1" CTS		20		2.05	8.90	10.95	15.10
823	1-1/4" IPS x 1-1/4" CTS		18		3.70	9.85	13.55	18.30
824	1-1/2" IPS x 1-1/2" CTS		16		4.27	11.10	15.37	21
825	2" IPS x 2" CTS		13		6.80	13.65	20.45	27
827	Price includes insert stiffeners							
828	250 psi is same price as 160 psi							
830	Insert type, nylon, 160 & 250 psi, cold water							
831	Clamp ring stainless steel, 3/4" IPS	1 Plum	115	Ea.	.33	1.54	1.87	2.59
832	1" IPS		107		.35	1.66	2.01	2.78
833	 1-1/4" IPS		101		.38	1.76	2.14	2.96
834	1-1/2" IPS		95		.38	1.87	2.25	3.12
835	2" IPS		85		.39	2.09	2.48	3.45
837	Coupling, 3/4" IPS		22		.25	8.05	8.30	11.95
838	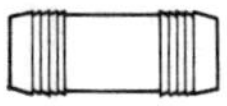 1" IPS		19		.34	9.35	9.69	13.90
839	1-1/4" IPS		18		.62	9.85	10.47	14.95
840	1-1/2" IPS		17		.80	10.45	11.25	15.95
841	2" IPS		16		.98	11.10	12.08	17.10
843	Elbow, 90°, 3/4" IPS		22		.43	8.05	8.48	12.15
844	1" IPS		19		.50	9.35	9.85	14.05
845	 1-1/4" IPS		18		.84	9.85	10.69	15.20
846	1-1/2" IPS		17		1.22	10.45	11.67	16.45
847	2" IPS		16		2.05	11.10	13.15	18.30
849	Male adapter, 3/4" IPS x 3/4" MPT		25		.34	7.10	7.44	10.65
850	1" IPS x 1" MPT		21		.44	8.45	8.89	12.70

15.1 Pipe & Fittings

		CREW	DAILY OUTPUT	UNIT	BARE COSTS MAT.	BARE COSTS INST.	BARE COSTS TOTAL	TOTAL INCL O&P
50 851	1-1/4" IPS x 1-1/4" MPT	1 Plum	20	Ea.	.74	8.90	9.64	13.65
852	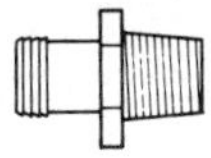 1-1/2" IPS x 1-1/2" MPT		18		.98	9.85	10.83	15.35
853	2" IPS x 2" MPT		15		1.26	11.85	13.11	18.50
855	Tee, 3/4" IPS		14		.64	12.70	13.34	19.05
856	1" IPS		13		1.05	13.65	14.70	21
857	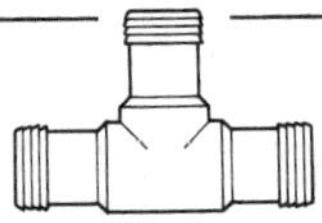 1-1/4" IPS		12		1.87	14.80	16.67	23
858	1-1/2" IPS		11		2.37	16.15	18.52	26
859	2" IPS		10		3.18	17.75	20.93	29
861	Insert type, acetal, 100 psi @ 180°F, hot & cold water							
862	Coupler, male, 3/8" CTS x 3/8" MPT	1 Plum	29	Ea.	.27	6.10	6.37	9.15
863	3/8" CTS x 1/2" MPT		28		.27	6.35	6.62	9.45
864	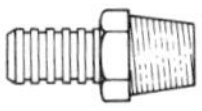 1/2" CTS x 1/2" MPT		27		.29	6.60	6.89	9.80
865	1/2" CTS x 3/4" MPT		26		.34	6.85	7.19	10.25
866	3/4" CTS x 1/2" MPT		25		.34	7.10	7.44	10.65
867	3/4" CTS x 3/4" MPT		25		.34	7.10	7.44	10.65
870	Coupling, 3/8" CTS x 1/2" CTS		25		.15	7.10	7.25	10.45
871	1/2" CTS		23		.15	7.70	7.85	11.30
872	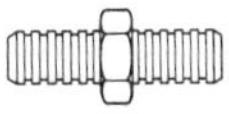 1/2" CTS x stub		23		.22	7.70	7.92	11.40
873	3/4" CTS		22		.25	8.05	8.30	11.95
875	Elbow 90°, 3/8" CTS		25		.26	7.10	7.36	10.55
876	1/2" CTS		23		.26	7.70	7.96	11.45
877	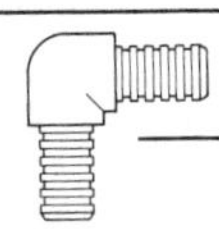 3/4" CTS		22		.37	8.05	8.42	12.05
880	Rings, crimp, copper, 3/8" CTS		120		.07	1.48	1.55	2.22
881	1/2" CTS		117		.08	1.52	1.60	2.28
882	3/4" CTS		115		.10	1.54	1.64	2.34
885	Reducer tee, 3/8"x3/8"x1/2" CTS		17		.32	10.45	10.77	15.45
886	1/2"x3/8"x1/2" CTS		15		.32	11.85	12.17	17.45
887	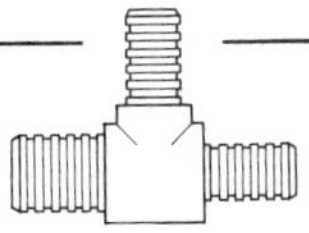 3/4"x1/2"x1/2" CTS		14		.46	12.70	13.16	18.85
889	3/4"x3/4"x1/2" CTS		14		.46	12.70	13.16	18.85
890	3/4"x1/2"x3/8" CTS		14		.46	12.70	13.16	18.85
893	Tee, 3/8" CTS		17		.46	10.45	10.91	15.60
894	1/2" CTS		15		.46	11.85	12.31	17.60
895	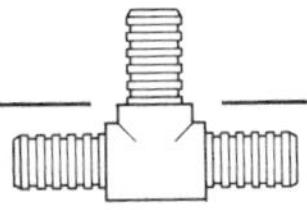 3/4" CTS		14		.46	12.70	13.16	18.85
896	Copper rings included in fitting price							
900	Flare type, assembled, acetal, hot & cold water							
901	Coupling, 1/4" & 3/8" CTS	1 Plum	24	Ea.	.99	7.40	8.39	11.80
902	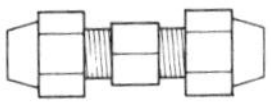 1/2" CTS		22		1.14	8.05	9.19	12.90
903	3/4" CTS		21		1.68	8.45	10.13	14.05
904	1" CTS		18		2.22	9.85	12.07	16.70
905	Elbow 90°, 1/4" CTS		26		1.10	6.85	7.95	11.10
906	3/8" CTS		24		1.18	7.40	8.58	12
907	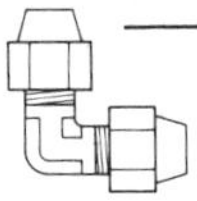 1/2" CTS		22		1.40	8.05	9.45	13.20
908	3/4" CTS		21		2.15	8.45	10.60	14.60
909	1" CTS		18		2.71	9.85	12.56	17.25
911	Tee ,1/4" & 3/8" CTS		15		1.22	11.85	13.07	18.45
912	1/2" CTS		14		1.60	12.70	14.30	20
913	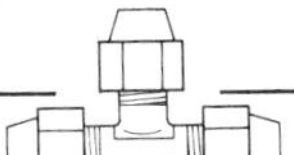 3/4" CTS		13		2.44	13.65	16.09	22
914	1" CTS		12		3.26	14.80	18.06	25
916	Fusion weld fittings, "hot and cold water"							
917	Bushing, reducing, 3/4" CTS x smaller size	1 Plum	22	Ea.	.29	8.05	8.34	12
918	1" CTS x smaller size		19		.68	9.35	10.03	14.25
919	1" IPS x smaller size		19		.80	9.35	10.15	14.40
920	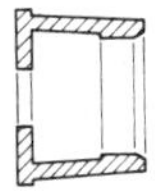 1-1/4" IPS x smaller size		18		.88	9.85	10.73	15.20
921	1-1/2" IPS x smaller size		17		1.10	10.45	11.55	16.30
922	2" IPS x smaller size		16		3.27	11.10	14.37	19.65
923	3" IPS x smaller size		14		7.70	12.70	20.40	27
925	Coupling, 1" CTS		19		.81	9.35	10.16	14.40
926	1" IPS		19		.83	9.35	10.18	14.40
927	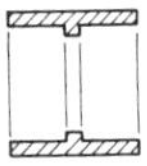 1-1/4" IPS		18		.87	9.85	10.72	15.20
928	1-1/2" IPS		17		1.01	10.45	11.46	16.20

15.1 Pipe & Fittings

		CREW	DAILY OUTPUT	UNIT	BARE COSTS MAT.	BARE COSTS INST.	BARE COSTS TOTAL	TOTAL INCL O&P
50 929	2" IPS	1 Plum	16	Ea.	1.74	11.10	12.84	17.95
930	3" IPS		10		6.80	17.75	24.55	33
931	Elbow, 90°, 1" CTS		19		.91	9.35	10.26	14.50
932	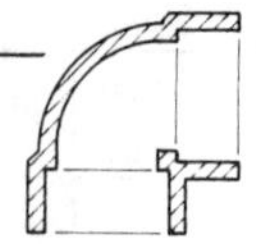 1" IPS		19		1.16	9.35	10.51	14.80
933	1-1/4" IPS		18		1.31	9.85	11.16	15.70
934	1-1/2" IPS		17		1.74	10.45	12.19	17
935	2" IPS		16		3.09	11.10	14.19	19.45
936	3" IPS		10		25	17.75	42.75	53
937	Elbow, 45°, 1" CTS		19		.74	9.35	10.09	14.30
938	1" IPS		19		.94	9.35	10.29	14.55
939	1-1/4" IPS		18		1.20	9.85	11.05	15.55
940	1-1/2" IPS		17		1.48	10.45	11.93	16.70
941	2" IPS		16		2.34	11.10	13.44	18.60
942	3" IPS		10		18	17.75	35.75	45
943	Flange, 1" CTS		40		4.61	4.44	9.05	11.50
944	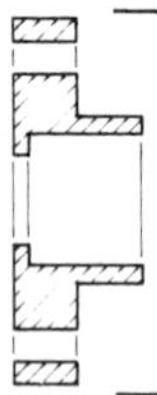 1" IPS		38		4.61	4.67	9.28	11.80
945	1-1/4" IPS		36		5.90	4.93	10.83	13.60
946	1-1/2" IPS		34		6.25	5.20	11.45	14.40
947	2" IPS		32		6.65	5.55	12.20	15.35
948	Tee, 1" CTS		13		1.67	13.65	15.32	22
949	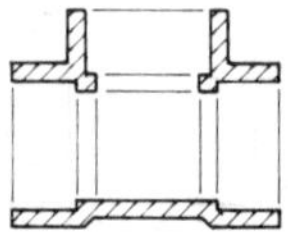 1" IPS		13		2.62	13.65	16.27	23
950	1-1/4" IPS		12		2.87	14.80	17.67	25
951	1-1/2" IPS		11		3.63	16.15	19.78	27
952	2" IPS		10		5.85	17.75	23.60	32
953	3" IPS	↓	7	↓	27	25	52	66
955	For plastic hangers see 15.1-65-800							
956	For copper/brass fittings see 15.1-41-700							
52-001	**PIPE, STAINLESS STEEL**							
002	Welded, with clevis type hangers 10' O.C.							
050	Schedule 5, type 304							
054	1/2" diameter	Q-15	128	L.F.	1.69	2.64	4.33	5.60
055	3/4" diameter		116		1.91	2.91	4.82	6.25
056	1" diameter		103		2.12	3.28	5.40	7
057	1-1/4" diameter		93		2.47	3.63	6.10	7.90
058	1-1/2" diameter		85		2.71	3.97	6.68	8.65
059	2" diameter		69		3.41	4.90	8.31	10.75
060	2-1/2" diameter		53		4.82	6.35	11.17	14.40
061	3" diameter		48		5.52	7.05	12.57	16.10
062	4" diameter		44		9.45	7.70	17.15	21
063	5" diameter	↓	36		14.55	9.40	23.95	29
064	6" diameter	Q-16	42		13.53	12.25	25.78	32
065	8" diameter		34		22.47	15.15	37.62	46
066	10" diameter		26		33.27	19.80	53.07	65
067	12" diameter	↓	21	↓	43.32	25	68.32	83
070	To delete hangers, subtract							
071	1/2" diam. to 1-1/2" diam.				8%	19%		
072	2" diam. to 5" diam.				4%	9%		
073	6" diam. to 12" diam.				3%	4%		
075	For small quantities, add			L.F.	10%			
125	Schedule 5, type 316							
129	1/2" diameter	Q-15	128	L.F.	2.13	2.64	4.77	6.10
130	3/4" diameter		116		2.57	2.91	5.48	7
131	1" diameter		103		2.90	3.28	6.18	7.85
132	1-1/4" diameter		93		3.40	3.63	7.03	8.90
133	1-1/2" diameter		85		3.85	3.97	7.82	9.90
134	2" diameter		69		4.80	4.90	9.70	12.25
135	2-1/2" diameter		53		6.95	6.35	13.30	16.75
136	3" diameter		48		8.12	7.05	15.17	18.95
137	4" diameter	↓	44	↓	10.19	7.70	17.89	22

15.1 Pipe & Fittings

		CREW	DAILY OUTPUT	UNIT	BARE COSTS MAT.	BARE COSTS INST.	BARE COSTS TOTAL	TOTAL INCL O&P
52 138	5" diameter	Q-15	36	L.F.	16.80	9.40	26.20	32
139	6" diameter	Q-16	42		19.14	12.25	31.39	39
140	8" diameter		34		29.34	15.15	44.49	54
141	10" diameter		26		44.27	19.80	64.07	77
142	12" diameter		21		57.62	25	82.62	99
149	For small quantities, add				10%			
194	To delete hanger, subtract							
195	1/2" diam. to 1-1/2" diam.				5%	19%		
196	2" diam. to 5" diam.				3%	9%		
197	6" diam. to 12" diam.				2%	4%		
200	Schedule 10, type 304							
204	1/4" diameter	Q-15	131	L.F.	1.61	2.58	4.19	5.45
205	3/8" diameter		128		1.68	2.64	4.32	5.60
206	1/2" diameter		125		1.95	2.70	4.65	6
207	3/4" diameter		113		2.22	2.99	5.21	6.70
208	1" diameter		100		2.96	3.38	6.34	8.10
209	1-1/4" diameter		91		3.81	3.71	7.52	9.50
210	1-1/2" diameter		83		3.85	4.07	7.92	10.05
211	2" diameter		67		4.92	5.05	9.97	12.60
212	2-1/2" diameter		51		6.29	6.60	12.89	16.35
213	3" diameter		46		7.40	7.35	14.75	18.60
214	4" diameter		42		9.69	8.05	17.74	22
215	5" diameter		35		13.28	9.65	22.93	28
216	6" diameter	Q-16	40		15.79	12.90	28.69	36
217	8" diameter		33		28.24	15.60	43.84	53
218	10" diameter		25		38.77	21	59.77	72
219	12" diameter		21		48.82	25	73.82	89
225	For small quantities, add				10%			
265	To delete hanger, subtract							
266	1/4" diam. to 3/4" diam.				9%	22%		
267	1" diam. to 2" diam.				4%	15%		
268	2-1/2" diam. to 5" diam.				3%	8%		
269	6" diam. to 12" diam.				3%	4%		
275	Schedule 10, type 316							
279	1/4" diameter	Q-15	131	L.F.	1.79	2.58	4.37	5.65
280	3/8" diameter		128		2	2.64	4.64	5.95
281	1/2" diameter		125		2.55	2.70	5.25	6.65
282	3/4" diameter		113		2.85	2.99	5.84	7.40
283	1" diameter		100		4.02	3.38	7.40	9.25
284	1-1/4" diameter		91		5.17	3.71	8.88	11
285	1-1/2" diameter		83		5.50	4.07	9.57	11.85
286	2" diameter		67		7.01	5.05	12.06	14.90
287	2-1/2" diameter		51		9.26	6.60	15.86	19.65
288	3" diameter		46		10.81	7.35	18.16	22
289	4" diameter		42		13.60	8.05	21.65	26
290	5" diameter		35		19.27	9.65	28.92	35
291	6" diameter	Q-16	40		22.22	12.90	35.12	43
292	8" diameter		33		38.14	15.60	53.74	64
293	10" diameter		25		51.97	21	72.97	87
294	12" diameter		21		64.22	25	89.22	105
299	For small quantities, add				10%			
343	To delete hanger, subtract							
344	1/4" diam. to 3/4" diam.				6%	22%		
345	1" diam. to 2" diam.				3%	15%		
346	2-1/2" diam. to 5" diam.				2%	8%		
347	6" diam. to 12" diam.				2%	4%		
350	Threaded, couplings and hangers 10' O.C.							
352	Schedule 40, type 304							
354	1/4" diameter	1 Plum	54	L.F.	1.82	3.29	5.11	6.75
355	3/8" diameter	"	53	"	2.22	3.35	5.57	7.30

15.1 Pipe & Fittings

		Crew	Daily Output	Unit	Bare Costs Mat.	Bare Costs Inst.	Bare Costs Total	Total Incl O&P
52 356	1/2" diameter	1 Plum	52	L.F.	2.73	3.42	6.15	7.95
357	3/4" diameter		51		3.08	3.48	6.56	8.40
358	1" diameter		45		3.89	3.95	7.84	10
359	1-1/4" diameter	Q-1	76		5.13	4.21	9.34	11.70
360	1-1/2" diameter		69		5.77	4.63	10.40	13.05
361	2" diameter		57		7.47	5.60	13.07	16.30
362	2-1/2" diameter		44		13.22	7.25	20.47	25
363	3" diameter		38		17.08	8.40	25.48	31
364	4" diameter	Q-2	51		25.09	9.75	34.84	42
374	For small quantities, add				10%			
420	To delete couplings & hangers, subtract							
421	1/4" diam. to 3/4" diam.				15%	56%		
422	1" diam. to 2" diam.				18%	49%		
423	2-1/2" diam. to 4" diam.				34%	40%		
425	Schedule 40, type 316							
429	1/4" diameter	1 Plum	54	L.F.	2.27	3.29	5.56	7.25
430	3/8" diameter		53		2.73	3.35	6.08	7.85
431	1/2" diameter		52		3.46	3.42	6.88	8.75
432	3/4" diameter		51		3.96	3.48	7.44	9.40
433	1" diameter		45		5.28	3.95	9.23	11.50
434	1-1/4" diameter	Q-1	76		7.04	4.21	11.25	13.80
435	1-1/2" diameter		69		7.94	4.63	12.57	15.45
436	2" diameter		57		10.45	5.60	16.05	19.60
437	2-1/2" diameter		44		18.67	7.25	25.92	31
438	3" diameter		38		23.98	8.40	32.38	39
439	4" diameter	Q-2	51		34.30	9.75	44.05	52
449	For small quantities, add				10%			
490	To delete couplings & hangers, subtract							
491	1/4" diam. to 3/4" diam.				12%	56%		
492	1" diam. to 2" diam.				14%	49%		
493	2-1/2" diam. to 4" diam.				27%	40%		
500	Schedule 80, type 304							
504	1/4" diameter	1 Plum	53	L.F.	2.82	3.35	6.17	7.95
505	3/8" diameter		52		3.32	3.42	6.74	8.60
506	1/2" diameter		51		4.33	3.48	7.81	9.80
507	3/4" diameter		48		5.19	3.70	8.89	11.05
508	1" diameter		43		7.19	4.13	11.32	13.90
509	1-1/4" diameter	Q-1	73		9.56	4.38	13.94	16.85
510	1-1/2" diameter		67		10.71	4.77	15.48	18.70
511	2" diameter		54		14.11	5.90	20.01	24
519	For small quantities, add				10%			
570	To delete couplings & hangers, subtract							
571	1/4" diam. to 3/4" diam.				10%	53%		
572	1" diam. to 2" diameter				14%	47%		
575	Schedule 80, type 316							
579	1/4" diameter	1 Plum	53	L.F.	3.78	3.35	7.13	9
580	3/8" diameter		52		4.67	3.42	8.09	10.05
581	1/2" diameter		51		6.05	3.48	9.53	11.70
582	3/4" diameter		48		7.57	3.70	11.27	13.65
583	1" diameter		43		10.47	4.13	14.60	17.50
584	1-1/4" diameter	Q-1	73		14.27	4.38	18.65	22
585	1-1/2" diameter		67		17.68	4.77	22.45	26
586	2" diameter		54		22.19	5.90	28.09	33
595	For small quantities, add				10%			
700	To delete couplings & hangers, subtract							
701	1/4" diam. to 3/4" diam.				9%	53%		
702	1" diam. to 2" diam.				14%	47%		
800	Weld joints with clevis type hangers 10' O.C.							
801	Schedule 40, type 304							
805	1/8" pipe size	Q-15	126	L.F.	1.47	2.68	4.15	5.45

15.1 Pipe & Fittings

		CREW	DAILY OUTPUT	UNIT	BARE COSTS MAT.	BARE COSTS INST.	BARE COSTS TOTAL	TOTAL INCL O&P
52 806	1/4" pipe size	Q-15	125	L.F.	1.73	2.70	4.43	5.75
807	3/8" pipe size		122		2.10	2.77	4.87	6.25
808	1/2" pipe size		118		2.55	2.86	5.41	6.90
809	3/4" pipe size		109		2.81	3.10	5.91	7.50
810	1" pipe size		95		3.51	3.56	7.07	8.95
811	1-1/4" pipe size		86		4.51	3.93	8.44	10.55
812	1-1/2" pipe size		78		4.91	4.33	9.24	11.60
813	2" pipe size		62		6.35	5.45	11.80	14.75
814	2-1/2" pipe size		49		9.59	6.90	16.49	20
815	3" pipe size		44		12.24	7.70	19.94	24
816	3-1/2" pipe size		44		15.08	7.70	22.78	28
817	4" pipe size		39		16.95	8.65	25.60	31
818	5" pipe size		32		23.56	10.55	34.11	41
819	6" pipe size	Q-16	37		29.15	13.95	43.10	52
820	8" pipe size		29		55.74	17.75	73.49	87
821	10" pipe size		24		93.77	21	114.77	135
822	12" pipe size		20		107.12	26	133.12	155
830	Schedule 40, type 316							
831	1/8" pipe size	Q-15	126	L.F.	1.83	2.68	4.51	5.85
832	1/4" pipe size		125		2.16	2.70	4.86	6.25
833	3/8" pipe size		122		2.59	2.77	5.36	6.80
834	1/2" pipe size		118		3.25	2.86	6.11	7.65
835	3/4" pipe size		109		3.63	3.10	6.73	8.40
836	1" pipe size		95		4.82	3.56	8.38	10.40
837	1-1/4" pipe size		86		6.29	3.93	10.22	12.50
838	1-1/2" pipe size		78		6.90	4.33	11.23	13.75
839	2" pipe size		62		9.10	5.45	14.55	17.80
840	2-1/2" pipe size		49		14.32	6.90	21.22	26
841	3" pipe size		44		18.18	7.70	25.88	31
842	3-1/2" pipe size		44		22.29	7.70	29.99	35
843	4" pipe size		39		24.54	8.65	33.19	39
844	5" pipe size		32		34.56	10.55	45.11	53
845	6" pipe size	Q-16	37		43.45	13.95	57.40	68
846	8" pipe size		29		83.24	17.75	100.99	115
847	10" pipe size		24		122.37	21	143.37	165
848	12" pipe size		20		155.52	26	181.52	210
850	Schedule 80, type 304							
851	1/4" pipe size	Q-15	110	L.F.	2.63	3.07	5.70	7.25
852	3/8" pipe size		109		3.12	3.10	6.22	7.85
853	1/2" pipe size		106		4.03	3.19	7.22	9
854	3/4" pipe size		96		4.76	3.52	8.28	10.25
855	1" pipe size		87		6.51	3.88	10.39	12.70
856	1-1/4" pipe size		81		8.21	4.17	12.38	15
857	1-1/2" pipe size		74		8.99	4.57	13.56	16.40
858	2" pipe size		58		11.58	5.80	17.38	21
859	2-1/2" pipe size		46		18.94	7.35	26.29	31
860	3" pipe size		41		26.65	8.25	34.90	41
861	4" pipe size		33		41.04	10.25	51.29	60
863	6" pipe size	Q-16	30		85.25	17.20	102.45	120
864	Schedule 80, type 316							
865	1/4" pipe size	Q-15	110	L.F.	3.54	3.07	6.61	8.30
866	3/8" pipe size		109		4.41	3.10	7.51	9.25
867	1/2" pipe size		106		5.63	3.19	8.82	10.75
868	3/4" pipe size		96		6.95	3.52	10.47	12.65
869	1" pipe size		87		9.42	3.88	13.30	15.90
870	1-1/4" pipe size		81		12.28	4.17	16.45	19.45
871	1-1/2" pipe size		74		15.04	4.57	19.61	23
872	2" pipe size		58		18.34	5.80	24.14	28
873	2-1/2" pipe size		46		29.94	7.35	37.29	43
874	3" pipe size		41		37.65	8.25	45.90	53

15.1 Pipe & Fittings

		CREW	DAILY OUTPUT	UNIT	BARE COSTS MAT.	BARE COSTS INST.	BARE COSTS TOTAL	TOTAL INCL O&P
52 876	4" pipe size	Q-15	33	L.F.	49.84	10.25	60.09	69
877	6" pipe size	Q-16	30	"	132.55	17.20	149.75	170
879	Schedule 160, type 304							
880	1/2" pipe size	Q-15	96	L.F.	5.10	3.52	8.62	10.65
881	3/4" pipe size		88		6.78	3.84	10.62	12.95
882	1" pipe size		80		8.43	4.22	12.65	15.30
884	1-1/2" pipe size		67		13.12	5.05	18.17	22
885	2" pipe size		53		19.99	6.35	26.34	31
887	3" pipe size		37		35.45	9.15	44.60	52
890	Schedule 160, type 316							
891	1/2" pipe size	Q-15	96	L.F.	7.55	3.52	11.07	13.35
892	3/4" pipe size		88		10.58	3.84	14.42	17.10
893	1" pipe size		80		12.45	4.22	16.67	19.70
894	1-1/4" pipe size		73		14.26	4.63	18.89	22
895	1-1/2" pipe size		67		20.10	5.05	25.15	29
896	2" pipe size		53		29.89	6.35	36.24	42
899	4" pipe size		30		79.54	11.25	90.79	105
925	Welding labor per joint for stainless steel							
926	Schedule 5 and 10							
927	1/4" pipe size	Q-15	36	Ea.		9.40	9.40	13.40
928	3/8" pipe size		35			9.65	9.65	13.75
929	1/2" pipe size		35			9.65	9.65	13.75
930	3/4" pipe size		28			12.05	12.05	17.20
931	1" pipe size		25			13.50	13.50	19.30
932	1-1/4" pipe size		22			15.35	15.35	22
933	1-1/2" pipe size		21			16.10	16.10	23
934	2" pipe size		18			18.75	18.75	27
935	2-1/2" pipe size		14			24	24	34
936	3" pipe size		13			26	26	37
937	4" pipe size		12			28	28	40
938	5" pipe size		10			34	34	48
939	6" pipe size		9			38	38	54
940	8" pipe size		6			56	56	80
941	10" pipe size		4			84	84	120
942	12" pipe size		3			115	115	160
950	Schedule 40							
951	1/4" pipe size	Q-15	35	Ea.		9.65	9.65	13.75
952	3/8" pipe size		35			9.65	9.65	13.75
953	1/2" pipe size		34			9.95	9.95	14.20
954	3/4" pipe size		28			12.05	12.05	17.20
955	1" pipe size		24			14.10	14.10	20
956	1-1/4" pipe size		21			16.10	16.10	23
957	1-1/2" pipe size		20			16.90	16.90	24
958	2" pipe size		17			19.85	19.85	28
959	2-1/2" pipe size		14			24	24	34
960	3" pipe size		13			26	26	37
961	4" pipe size		11			31	31	44
962	5" pipe size		9			38	38	54
963	6" pipe size		8			42	42	60
964	8" pipe size		5			68	68	96
965	10" pipe size		4			84	84	120
966	12" pipe size		3			115	115	160
975	Schedule 80							
976	1/4" pipe size	Q-15	28	Ea.		12.05	12.05	17.20
977	3/8" pipe size		28			12.05	12.05	17.20
978	1/2" pipe size		28			12.05	12.05	17.20
979	3/4" pipe size		24			14.10	14.10	20
980	1" pipe size		21			16.10	16.10	23
981	1-1/4" pipe size		20			16.90	16.90	24
982	1-1/2" pipe size		19			17.80	17.80	25

		CREW	DAILY OUTPUT	UNIT	BARE COSTS MAT.	BARE COSTS INST.	BARE COSTS TOTAL	TOTAL INCL O&P
52 983	2" pipe size	Q-15	16	Ea.		21	21	30
984	2-1/2" pipe size		13			26	26	37
985	3" pipe size		12			28	28	40
986	4" pipe size		8			42	42	60
987	5" pipe size		6			56	56	80
988	6" pipe size		5			68	68	96
989	8" pipe size		3			115	115	160
990	10" pipe size		2			170	170	240
991	12" pipe size		1			340	340	480
992	Schedule 160, 1/2" pipe size		25			13.50	13.50	19.30
993	3/4" pipe size		22			15.35	15.35	22
994	1" pipe size		19			17.80	17.80	25
995	1-1/4" pipe size		18			18.75	18.75	27
996	1-1/2" pipe size		17			19.85	19.85	28
997	2" pipe size		14			24	24	34
998	3" pipe size		11			31	31	44
999	4" pipe size		7			48	48	69
53-001	**PIPE, STAINLESS STEEL, FITTINGS**							
010	Butt weld joint, schedule 5, type 304							
012	90° Elbow, long							
014	1/2"	Q-15	16	Ea.	5.30	21	26.30	36
015	3/4"		16		5.30	21	26.30	36
016	1"		15		6.10	23	29.10	39
017	1-1/4"		14		10	24	34	45
018	1-1/2"		13		6.50	26	32.50	44
019	2"		11		9.10	31	40.10	54
020	2-1/2"		8		19.20	42	61.20	81
021	3"		6		19	56	75	100
022	3-1/2"		5		80	68	148	185
023	4"		4		34	84	118	160
024	5"		3		145	115	260	320
025	6"	Q-16	4		92	130	222	285
026	8"		3		145	170	315	405
027	10"		2		480	260	740	895
028	12"		1.57		545	330	875	1,075
032	For schedule 5, type 316, add				25%			
060	45° Elbow, long							
062	1/2"	Q-15	16	Ea.	12	21	33	43
063	3/4"		16		12	21	33	43
064	1"		15		9.15	23	32.15	42
065	1-1/4"		14		11	24	35	47
066	1-1/2"		13		7.15	26	33.15	45
067	2"		11		9.15	31	40.15	54
068	2-1/2"		8		19.20	42	61.20	81
069	3"		6		15.20	56	71.20	97
070	3-1/2"		5		72	68	140	175
071	4"		4		28	84	112	150
072	5"		3		135	115	250	310
073	6"	Q-16	4		60	130	190	250
074	8"		3		94	170	264	350
075	10"		2		210	260	470	600
076	12"		1.57		325	330	655	830
080	For schedule 5, type 316, add				25%			
110	Tee, straight							
113	1/2"	Q-15	10	Ea.	16.20	34	50.20	66
114	3/4"		10		16.20	34	50.20	66
115	1"		10		16.20	34	50.20	66
116	1-1/4"		9		36	38	74	93
117	1-1/2"		8		19.40	42	61.40	82

15.1 Pipe & Fittings

		CREW	DAILY OUTPUT	UNIT	BARE COSTS MAT.	BARE COSTS INST.	BARE COSTS TOTAL	TOTAL INCL O&P
53 118	2"	Q-15	7	Ea.	24	48	72	95
119	2-1/2"		5		79	68	147	185
120	3"		4		43	84	127	170
121	3-1/2"		3		135	115	250	310
122	4"		2		70	170	240	320
123	5"		2		360	170	530	635
124	6"	Q-16	2.60		145	200	345	445
125	8"		2		320	260	580	720
126	10"		1.80		540	285	825	1,000
127	12"		1.40		705	370	1,075	1,300
132	For schedule 5, type 316, add				30%			
200	Butt weld joint, schedule 10, type 304							
202	90° Elbow, long							
204	1/2"	Q-15	16	Ea.	5.60	21	26.60	36
205	3/4"		16		5.60	21	26.60	36
206	1"		15		6.40	23	29.40	39
207	1-1/4"		14		10	24	34	45
208	1-1/2"		13		6.80	26	32.80	45
209	2"		11		9.60	31	40.60	54
210	2-1/2"		8		19.20	42	61.20	81
211	3"		6		20	56	76	100
212	3-1/2"		5		80	68	148	185
213	4"		4		36	84	120	160
214	5"		3		145	115	260	320
215	6"	Q-16	4		92	130	222	285
216	8"		3		145	170	315	405
217	10"		2		320	260	580	720
218	12"		1.57		545	330	875	1,075
222	For schedule 10, type 316, add				25%			
250	45° Elbow, long							
252	1/2"	Q-15	16	Ea.	8.40	21	29.40	39
253	3/4"		16		8.40	21	29.40	39
254	1"		15		9.60	23	32.60	43
255	1-1/4"		14		11	24	35	47
256	1-1/2"		13		7.50	26	33.50	45
257	2"		11		9.60	31	40.60	54
258	2-1/2"		8		19.20	42	61.20	81
259	3"		6		16	56	72	98
260	3-1/2"		5		72	68	140	175
261	4"		4		29	84	113	150
262	5"		3		135	115	250	310
263	6"	Q-16	4		60	130	190	250
264	8"		3		94	170	264	350
265	10"		2		210	260	470	600
266	12"		1.57		325	330	655	830
270	For schedule 10, type 316, add				25%			
300	Tee, straight							
303	1/2"	Q-15	10	Ea.	16.20	34	50.20	66
304	3/4"		10		16.20	34	50.20	66
305	1"		10		16.20	34	50.20	66
306	1-1/4"		9		36	38	74	93
307	1-1/2"		8		19.40	42	61.40	82
308	2"		7		24	48	72	95
309	2-1/2"		5		79	68	147	185
310	3"		4		43	84	127	170
311	3-1/2"		3		135	115	250	310
312	4"		2		70	170	240	320
313	5"		2		360	170	530	635
314	6"	Q-16	2.60		145	200	345	445
315	8"	"	2		320	260	580	720

15

15.1 Pipe & Fittings

Line	Description	CREW	DAILY OUTPUT	UNIT	BARE COSTS MAT.	BARE COSTS INST.	BARE COSTS TOTAL	TOTAL INCL O&P
53 316	10"	Q-16	1.80	Ea.	540	285	825	1,000
317	12"	"	1.40		705	370	1,075	1,300
322	For schedule 10, type 316, add				30%			
325	Butt weld joint, schedule 40, type 304							
326	90° Elbow, long, 1/2"	Q-15	17	Ea.	6	19.85	25.85	35
327	3/4"		14		6	24	30	41
328	1"		12		6.80	28	34.80	48
329	1-1/4"		11		10.40	31	41.40	55
330	1-1/2"		10		8.40	34	42.40	57
331	2"		9		13.20	38	51.20	68
332	2-1/2"		7		28	48	76	100
333	3"		6		32	56	88	115
334	3-1/2"		5.50		100	61	161	200
335	4"		5		56	68	124	160
336	5"		4		165	84	249	300
337	6"	Q-16	6		170	86	256	310
338	8"		4		380	130	510	605
339	10"		3		680	170	850	995
340	12"		2		960	260	1,220	1,425
341	For schedule 40, type 316, add				29%			
346	45° Elbow, long, 1/2"	Q-15	17	Ea.	9	19.85	28.85	38
347	3/4"		14		9	24	33	44
348	1"		12		10.20	28	38.20	51
349	1-1/4"		11		12.50	31	43.50	58
350	1-1/2"		10		8.40	34	42.40	57
351	2"		9		11.90	38	49.90	67
352	2-1/2"		7		28	48	76	100
353	3"		6		24	56	80	105
354	3-1/2"		5.50		90	61	151	185
355	4"		5		42	68	110	145
356	5"		4		135	84	219	270
357	6"	Q-16	6		105	86	191	240
358	8"		4		230	130	360	440
359	10"		3		410	170	580	695
360	12"		2		575	260	835	1,000
361	For schedule 40, type 316, add				28%			
366	Tee, straight 1/2"	Q-15	11	Ea.	18	31	49	64
367	3/4"		9		18	38	56	73
368	1"		8		18	42	60	80
369	1-1/4"		7		48	48	96	120
370	1-1/2"		6.50		32	52	84	110
371	2"		6		33	56	89	115
372	2-1/2"		5		88	68	156	195
373	3"		4		70	84	154	195
374	3-1/2"		3.70		205	91	296	355
375	4"		3.50		96	97	193	245
376	5"		3		405	115	520	605
377	6"	Q-16	4		290	130	420	505
378	8"		3		505	170	675	800
379	10"		2		920	260	1,180	1,375
380	12"		1.50		1,300	345	1,645	1,925
381	For schedule 40, type 316, add				25%			
400	Socket weld joint, 3000 lb., type 304							
410	90° Elbow							
414	1/4"	Q-15	20	Ea.	3.81	16.90	20.71	28
415	3/8"		18		4.47	18.75	23.22	32
416	1/2"		17		5.70	19.85	25.55	35
417	3/4"		16		8	21	29	39
418	1"		15		11.90	23	34.90	45
419	1-1/4"		14		19.80	24	43.80	56

15

15.1 Pipe & Fittings

		CREW	DAILY OUTPUT	UNIT	BARE COSTS MAT.	BARE COSTS INST.	BARE COSTS TOTAL	TOTAL INCL O&P
53 420	1-1/2"	Q-15	13	Ea.	31	26	57	71
421	2"	"	11	"	51	31	82	100
430	45° Elbow							
434	1/4"	Q-15	20	Ea.	4.13	16.90	21.03	29
435	3/8"		18		5.05	18.75	23.80	32
436	1/2"		17		6.75	19.85	26.60	36
437	3/4"		16		10.80	21	31.80	42
438	1"		15		15.60	23	38.60	49
439	1-1/4"		14		24	24	48	61
440	1-1/2"		13		30	26	56	70
441	2"		11		50	31	81	99
450	Tee							
454	1/4"	Q-15	13	Ea.	4.33	26	30.33	42
455	3/8"		12		5.50	28	33.50	46
456	1/2"		12		7.20	28	35.20	48
457	3/4"		11		10.25	31	41.25	55
458	1"		10		15.05	34	49.05	65
459	1-1/4"		9		28	38	66	84
460	1-1/2"		8		49	42	91	115
461	2"		7		68	48	116	145
500	Socket weld joint, 3000 lb., type 316							
510	90° Elbow							
514	1/4"	Q-15	20	Ea.	4.69	16.90	21.59	29
515	3/8"		18		5.45	18.75	24.20	33
516	1/2"		17		7.60	19.85	27.45	37
517	3/4"		16		10	21	31	41
518	1"		15		15.35	23	38.35	49
519	1-1/4"		14		28	24	52	65
520	1-1/2"		13		39	26	65	80
521	2"		11		67	31	98	120
530	45° Elbow							
534	1/4"	Q-15	20	Ea.	4.95	16.90	21.85	30
535	3/8"		18		6.55	18.75	25.30	34
536	1/2"		17		8.25	19.85	28.10	37
537	3/4"		16		13.05	21	34.05	44
538	1"		15		19.65	23	42.65	54
539	1-1/4"		14		30	24	54	67
540	1-1/2"		13		39	26	65	80
541	2"		11		58	31	89	110
550	Tee							
554	1/4"	Q-15	13	Ea.	5.20	26	31.20	43
555	3/8"		12		6.95	28	34.95	48
556	1/2"		12		9.70	28	37.70	51
557	3/4"		11		13.30	31	44.30	58
558	1"		10		21	34	55	71
559	1-1/4"		9		35	38	73	92
560	1-1/2"		8		58	42	100	125
561	2"		7		85	48	133	160
570	For socket weld joint, 6000 lb., type 304 and 316, add				100%			
600	Threaded companion flange							
601	Stainless steel, 150 lb., type 304							
602	1/2" diam.	1 Plum	30	Ea.	14.05	5.90	19.95	24
603	3/4" diam.		28		15.50	6.35	21.85	26
604	1" diam.		27		19.90	6.60	26.50	31
605	1-1/4" diam.	Q-1	44		22	7.25	29.25	35
606	1-1/2" diam.		40		28	8	36	42
607	2" diam.		36		32	8.90	40.90	48
608	2-1/2" diam.		28		45	11.40	56.40	66
609	3" diam.		20		48	16	64	76
611	4" diam.		12		79	27	106	125

15

15.1 Pipe & Fittings

Line	Description	Crew	Daily Output	Unit	Bare Costs Mat.	Bare Costs Inst.	Bare Costs Total	Total Incl O&P
53 613	6" diam.	Q-2	14	Ea.	165	36	201	235
614	8" diam.	"	12	"	295	41	336	385
615	For type 316 add				27%			
626	Weld flanges, stainless steel, type 304							
627	Slip on, 150 lb. (welded, front and back)							
628	1/2" diam.	Q-15	16	Ea.	14.05	21	35.05	46
629	3/4" diam.		16		15.50	21	36.50	47
630	1" diam.		13		19.90	26	45.90	59
631	1-1/4" diam.		12		22	28	50	64
632	1-1/2" diam.		11		28	31	59	75
633	2" diam.		10		32	34	66	83
634	2-1/2" diam.		9		45	38	83	105
635	3" diam.		8		48	42	90	115
637	4" diam.		6		79	56	135	165
639	6" diam.	Q-16	6		165	86	251	305
640	8" diam.	"	4		295	130	425	510
641	For type 316 add				27%			
653	Weld neck 150 lb.							
654	1/2" diam.	Q-15	35	Ea.	18.15	9.65	27.80	34
655	3/4" diam.		32		20	10.55	30.55	37
656	1" diam.		24		26	14.10	40.10	49
657	1-1/4"		23		29	14.70	43.70	53
658	1-1/2" diam.		20		34	16.90	50.90	62
659	2" diam.		18		44	18.75	62.75	75
660	2-1/2" diam.		14		54	24	78	94
661	3" diam.		12		63	28	91	110
663	4" diam.		9		98	38	136	160
665	6" diam.	Q-16	9		200	57	257	300
666	8" diam.	"	6		360	86	446	520
667	For type 316 add				23%			
700	Threaded joint, 150 lb., type 304							
703	90° Elbow							
704	1/8"	1 Plum	13	Ea.	2.45	13.65	16.10	22
705	1/4"		13		2.45	13.65	16.10	22
707	3/8"		13		2.65	13.65	16.30	23
708	1/2"		12		3.54	14.80	18.34	25
709	3/4"		11		5.25	16.15	21.40	29
710	1"		10		6.30	17.75	24.05	33
711	1-1/4"	Q-1	17		8.85	18.80	27.65	37
712	1-1/2"		16		10.50	20	30.50	40
713	2"		14		15.95	23	38.95	51
714	2-1/2"		11		39	29	68	85
715	3"		8		52	40	92	115
716	4"	Q-2	11		100	45	145	175
718	45° Elbow							
719	1/8"	1 Plum	13	Ea.	2.85	13.65	16.50	23
720	1/4"		13		2.85	13.65	16.50	23
721	3/8"		13		3.33	13.65	16.98	23
722	1/2"		12		4.02	14.80	18.82	26
723	3/4"		11		5.70	16.15	21.85	30
724	1"		10		6.35	17.75	24.10	33
725	1-1/4"	Q-1	17		9.45	18.80	28.25	38
726	1-1/2"		16		11.80	20	31.80	42
727	2"		14		16.70	23	39.70	51
728	2-1/2"		11		44	29	73	90
729	3"		8		64	40	104	130
730	4"	Q-2	11		120	45	165	195
732	Tee, straight							
733	1/8"	1 Plum	9	Ea.	3.04	19.75	22.79	32
734	1/4"	"	9	"	3.04	19.75	22.79	32

15

15.1 Pipe & Fittings

		CREW	DAILY OUTPUT	UNIT	BARE COSTS MAT.	BARE COSTS INST.	BARE COSTS TOTAL	TOTAL INCL O&P
53 735	3/8"	1 Plum	9	Ea.	3.54	19.75	23.29	32
736	1/2"		8		5.25	22	27.25	38
737	3/4"		7		7	25	32	44
738	1"		6.50		8.15	27	35.15	48
739	1-1/4"	Q-1	11		12.15	29	41.15	55
740	1-1/2"		10		15.30	32	47.30	63
741	2"		9		22	36	58	76
742	2-1/2"		7		58	46	104	130
743	3"		5		83	64	147	185
744	4"	Q-2	7		135	71	206	250
746	Coupling, straight							
747	1/8"	1 Plum	19	Ea.	.85	9.35	10.20	14.45
748	1/4"		19		.85	9.35	10.20	14.45
749	3/8"		19		1.13	9.35	10.48	14.75
750	1/2"		19		1.65	9.35	11	15.30
751	3/4"		18		2.52	9.85	12.37	17.05
752	1"		15		3.50	11.85	15.35	21
753	1-1/4"	Q-1	26		5.70	12.30	18	24
754	1-1/2"		24		7.90	13.30	21.20	28
755	2"		21		10.25	15.20	25.45	33
756	2-1/2"		18		33	17.75	50.75	62
757	3"		14		44	23	67	81
758	4"	Q-2	16		74	31	105	125
771	Union							
772	1/8"	1 Plum	12	Ea.	4.56	14.80	19.36	26
773	1/4"		12		4.56	14.80	19.36	26
774	3/8"		12		5.20	14.80	20	27
775	1/2"		11		6.20	16.15	22.35	30
776	3/4"		10		7.60	17.75	25.35	34
777	1"		9		11.85	19.75	31.60	42
778	1-1/4"	Q-1	16		24	20	44	55
779	1-1/2"		15		30	21	51	64
780	2"		13		36	25	61	75
781	2-1/2"		10		99	32	131	155
782	3"		7		145	46	191	225
783	4"	Q-2	10		225	50	275	320
785	For 150 lb., type 316, add				25%			
875	Threaded joint, 2000 lb., type 304							
877	90° Elbow							
878	1/8"	1 Plum	13	Ea.	3.02	13.65	16.67	23
879	1/4"		13		3.02	13.65	16.67	23
880	3/8"		13		3.64	13.65	17.29	24
881	1/2"		12		5.40	14.80	20.20	27
882	3/4"		11		7.35	16.15	23.50	31
883	1"		10		11.35	17.75	29.10	38
884	1-1/4"	Q-1	17		19.75	18.80	38.55	49
885	1-1/2"		16		28	20	48	60
886	2"		14		50	23	73	88
888	45° Elbow							
889	1/8"	1 Plum	13	Ea.	4.57	13.65	18.22	25
890	1/4"		13		4.57	13.65	18.22	25
891	3/8"		13		4.83	13.65	18.48	25
892	1/2"		12		6.50	14.80	21.30	29
893	3/4"		11		10.30	16.15	26.45	35
894	1"		10		14.30	17.75	32.05	41
895	1-1/4"	Q-1	17		22	18.80	40.80	51
896	1-1/2"		16		29	20	49	61
897	2"		14		48	23	71	86
899	Tee, straight							
900	1/8"	1 Plum	9	Ea.	3.81	19.75	23.56	33

15.1 Pipe & Fittings

		CREW	DAILY OUTPUT	UNIT	BARE COSTS MAT.	BARE COSTS INST.	BARE COSTS TOTAL	TOTAL INCL O&P
53 901	1/4"	1 Plum	9	Ea.	3.81	19.75	23.56	33
902	3/8"		9		4.66	19.75	24.41	34
903	1/2"		8		6.50	22	28.50	39
904	3/4"		7		10	25	35	48
905	1"		6.50		14.30	27	41.30	55
906	1-1/4"	Q-1	11		27	29	56	72
907	1-1/2"		10		44	32	76	95
908	2"		9		48	36	84	105
910	For couplings and unions use 3000 lb., type 304							
912	2000 lb., type 316							
913	90° Elbow							
914	1/8"	1 Plum	13	Ea.	4.34	13.65	17.99	25
915	1/4"		13		4.34	13.65	17.99	25
916	3/8"		13		4.54	13.65	18.19	25
917	1/2"		12		6.80	14.80	21.60	29
918	3/4"		11		9.85	16.15	26	34
919	1"		10		14.50	17.75	32.25	42
920	1-1/4"	Q-1	17		25	18.80	43.80	55
921	1-1/2"		16		35	20	55	67
922	2"		14		65	23	88	105
924	45° Elbow							
925	1/8"	1 Plum	13	Ea.	5.05	13.65	18.70	25
926	1/4"		13		5.05	13.65	18.70	25
927	3/8"		13		6	13.65	19.65	26
928	1/2"		12		7.85	14.80	22.65	30
930	3/4"		11		12.40	16.15	28.55	37
931	1"		10		18.20	17.75	35.95	46
932	1-1/4"	Q-1	17		28	18.80	46.80	58
933	1-1/2"		16		38	20	58	71
934	2"		14		59	23	82	98
936	Tee, straight							
937	1/8"	1 Plum	9	Ea.	4.60	19.75	24.35	34
938	1/4"		9		4.60	19.75	24.35	34
939	3/8"		9		5.50	19.75	25.25	35
940	1/2"		8		8.25	22	30.25	41
941	3/4"		7		12.60	25	37.60	51
942	1"		6.50		19.25	27	46.25	61
943	1-1/4"	Q-1	11		33	29	62	78
944	1-1/2"		10		56	32	88	110
945	2"		9		84	36	120	145
947	For couplings and unions use 3000 lb., type 316							
949	3000 lb., type 304							
951	Coupling							
952	1/8"	1 Plum	19	Ea.	1.63	9.35	10.98	15.30
953	1/4"		19		1.79	9.35	11.14	15.45
954	3/8"		19		1.89	9.35	11.24	15.60
955	1/2"		19		2.77	9.35	12.12	16.55
956	3/4"		18		3.96	9.85	13.81	18.60
957	1"		15		6.20	11.85	18.05	24
958	1-1/4"	Q-1	26		12.30	12.30	24.60	31
959	1-1/2"		24		15.70	13.30	29	37
960	2"		21		23	15.20	38.20	47
962	Union							
963	1/8"	1 Plum	12	Ea.	10.05	14.80	24.85	32
964	1/4"		12		10.05	14.80	24.85	32
965	3/8"		12		10.80	14.80	25.60	33
966	1/2"		11		12.55	16.15	28.70	37
967	3/4"		10		19.15	17.75	36.90	47
968	1"		9		23	19.75	42.75	54
969	1-1/4"	Q-1	16		40	20	60	73

	15.1 Pipe & Fittings	CREW	DAILY OUTPUT	UNIT	BARE COSTS MAT.	BARE COSTS INST.	BARE COSTS TOTAL	TOTAL INCL O&P
53 970	1-1/2"	Q-1	15	Ea.	54	21	75	90
971	2"	"	13	"	69	25	94	110
973	3000 lb., type 316							
975	Coupling							
977	1/8"	1 Plum	19	Ea.	2.06	9.35	11.41	15.75
978	1/4"		19		2.26	9.35	11.61	16
979	3/8"		19		2.40	9.35	11.75	16.15
980	1/2"		19		3.89	9.35	13.24	17.80
981	3/4"		18		5.70	9.85	15.55	21
982	1"		15		9.55	11.85	21.40	28
983	1-1/4"	Q-1	26		18.10	12.30	30.40	38
984	1-1/2"		24		24	13.30	37.30	46
985	2"		21		35	15.20	50.20	61
987	Union							
988	1/8"	1 Plum	12	Ea.	12.30	14.80	27.10	35
989	1/4"		12		12.30	14.80	27.10	35
990	3/8"		12		13.20	14.80	28	36
991	1/2"		11		15.60	16.15	31.75	40
992	3/4"		10		25	17.75	42.75	53
993	1"		9		32	19.75	51.75	64
994	1-1/4"	Q-1	16		51	20	71	85
995	1-1/2"		15		75	21	96	115
996	2"		13		95	25	120	140
55-001	**PIPE, STEEL**							
002	Diameters thru 4" are Spec. A-120, 5" up are Spec. A-53							
005	Schedule 40, threaded, with couplings, and clevis type							
006	hangers sized for covering, 10' O.C.							
054	C8.1-032 Black, 1/4" diameter	1 Plum	66	L.F.	1.13	2.69	3.82	5.15
055	3/8" diameter		65		1.26	2.73	3.99	5.35
056	1/2" diameter		63		1.08	2.82	3.90	5.25
057	3/4" diameter		61		1.24	2.91	4.15	5.55
058	1" diameter		53		1.78	3.35	5.13	6.80
059	1-1/4" diameter	Q-1	89		2.23	3.59	5.82	7.65
060	1-1/2" diameter		80		2.61	4	6.61	8.65
061	2" diameter		64		3.58	5	8.58	11.15
062	2-1/2" diameter		50		5.83	6.40	12.23	15.65
063	3" diameter		43		7.53	7.45	14.98	19.05
064	3-1/2" diameter		40		10.03	8	18.03	23
065	4" diameter		36		11.45	8.90	20.35	25
066	5" diameter		26		27.80	12.30	40.10	48
067	6" diameter	Q-2	31		31.50	16.05	47.55	58
068	8" diameter		27		44.64	18.40	63.04	76
069	10" diameter		23		67.45	22	89.45	105
070	12" diameter		18		80.20	28	108.20	130
122	To delete coupling & hanger, subtract							
123	1/4" diam. to 3/4" diam.				31%	56%		
124	1" diam. to 1-1/2" diam.				23%	51%		
125	2" diam. to 4" diam.				23%	41%		
126	5" diam. to 12" diam.				21%	45%		
129	Galvanized, 1/4" diameter	1 Plum	66	L.F.	1.46	2.69	4.15	5.50
130	3/8" diameter		65		1.60	2.73	4.33	5.70
131	1/2" diameter		63		1.26	2.82	4.08	5.45
132	3/4" diameter		61		1.43	2.91	4.34	5.80
133	1" diameter		53		2.04	3.35	5.39	7.10
134	1-1/4" diameter	Q-1	89		2.57	3.59	6.16	8
135	1-1/2" diameter		80		3.01	4	7.01	9.10
136	2" diameter		64		4.13	5	9.13	11.75
137	2-1/2" diameter		50		6.63	6.40	13.03	16.55
138	3" diameter		43		8.57	7.45	16.02	20

15.1 Pipe & Fittings

		Description	Crew	Daily Output	Unit	Bare Costs Mat.	Bare Costs Inst.	Bare Costs Total	Total Incl O&P
55	139	3-1/2" diameter	Q-1	40	L.F.	11.29	8	19.29	24
	140	4" diameter		36		12.99	8.90	21.89	27
	141	5" diameter		26		31.43	12.30	43.73	52
	142	6" diameter	Q-2	31		36.45	16.05	52.50	63
	143	8" diameter		27		51.13	18.40	69.53	83
	144	10" diameter		23		77.90	22	99.90	115
	145	12" diameter		18		92.85	28	120.85	140
	175	To delete coupling & hanger, subtract							
	176	1/4" diam. to 3/4" diam.				31%	56%		
	177	1" diam. to 1-1/2" diam.				23%	51%		
	178	2" diam. to 4" diam.				23%	41%		
	179	5" diam. to 12" diam.				21%	45%		
	200	Welded, on yoke & roll hangers							
	201	sized for covering, 10' O.C.							
	204	Black, 1" diameter	Q-15	93	L.F.	1.92	3.63	5.55	7.30
	205	1-1/4" diameter		84		2.50	4.02	6.52	8.50
	206	1-1/2" diameter		76		2.81	4.45	7.26	9.45
	207	2" diameter		61		3.58	5.55	9.13	11.85
	208	2-1/2" diameter		47		5.05	7.20	12.25	15.80
	209	3" diameter		43		6.31	7.85	14.16	18.15
	210	3-1/2" diameter		39		7.77	8.65	16.42	21
	211	4" diameter		37		9.08	9.15	18.23	23
	212	5" diameter		32		23.81	10.55	34.36	41
	213	6" diameter	Q-16	36		25.27	14.30	39.57	48
	214	8" diameter		29		32.83	17.75	50.58	62
	215	10" diameter		24		51.97	21	72.97	88
	216	12" diameter		19		57.47	27	84.47	100
	256	To delete hanger, subtract							
	257	1" diam. to 1-1/2" diam.				15%	34%		
	258	2" diam. to 3-1/2" diam.				9%	21%		
	259	4" diam. to 12" diam.				5%	12%		
	264	Galvanized, 1" diameter	Q-15	93	L.F.	2.14	3.63	5.77	7.55
	265	1-1/4" diameter		84		2.78	4.02	6.80	8.80
	266	1-1/2" diameter		76		3.13	4.45	7.58	9.80
	267	2" diameter		61		4.02	5.55	9.57	12.30
	268	2-1/2" diameter		47		5.68	7.20	12.88	16.50
	269	3" diameter		43		7.13	7.85	14.98	19.05
	270	3-1/2" diameter		39		8.76	8.65	17.41	22
	271	4" diameter		37		10.24	9.15	19.39	24
	272	5" diameter		32		26.01	10.55	36.56	44
	273	6" diameter	Q-16	36		29.67	14.30	43.97	53
	274	8" diameter		29		39.43	17.75	57.18	69
	275	10" diameter		24		60.77	21	81.77	98
	276	12" diameter		19		67.37	27	94.37	115
	316	To delete hanger, subtract							
	317	1" diam. to 1-1/2" diam.				30%	34%		
	318	2" diam. to 3-1/2" diam.				19%	21%		
	319	4" diam. to 12" diam.				10%	12%		
	325	Flanged, 150 lb. weld neck, on yoke & roll hangers							
	326	sized for covering, 10' O.C.							
	329	Black, 1" diameter	Q-15	70	L.F.	5.56	4.83	10.39	13
	330	1-1/4" diameter		64		6.28	5.30	11.58	14.45
	331	1-1/2" diameter		58		6.59	5.80	12.39	15.55
	332	2" diameter		45		7.90	7.50	15.40	19.40
	333	2-1/2" diameter		36		9.92	9.40	19.32	24
	334	3" diameter		32		11.85	10.55	22.40	28
	335	3-1/2" diameter		29		16.90	11.65	28.55	35
	336	4" diameter		26		16.50	13	29.50	37
	337	5" diameter		21		33.42	16.10	49.52	60
	338	6" diameter	Q-16	25		36.49	21	57.49	70

15

15.1 Pipe & Fittings		CREW	DAILY OUTPUT	UNIT	BARE COSTS MAT.	INST.	TOTAL	TOTAL INCL O&P
55 339	8" diameter	Q-16	19	L.F.	50.07	27	77.07	94
340	10" diameter		16		87.17	32	119.17	140
341	12" diameter		14		108.73	37	145.73	170
347	For 300 lb. flanges, add				63%			
348	For 600 lb. flanges, add				310%			
396	To delete flanges & hanger, subtract							
397	1" diam. to 2" diam.				76%	65%		
398	2-1/2" diam. to 4" diam.				62%	59%		
399	5" diam. to 12" diam.				60%	46%		
404	Galvanized, 1" diameter	Q-15	70	L.F.	5.78	4.83	10.61	13.25
405	1-1/4" diameter		64		6.42	5.30	11.72	14.60
406	1-1/2" diameter		58		6.91	5.80	12.71	15.90
407	2" diameter		45		7.91	7.50	15.41	19.40
408	2-1/2" diameter		36		10.15	9.40	19.55	25
409	3" diameter		32		12.01	10.55	22.56	28
410	3-1/2" diameter		29		14.59	11.65	26.24	33
411	4" diameter		26		19.42	13	32.42	40
412	5" diameter		21		33.86	16.10	49.96	60
413	6" diameter	Q-16	25		39.35	21	60.35	73
414	8" diameter		19		50.51	27	77.51	94
415	10" diameter		16		80.13	32	112.13	135
416	12" diameter		14		88.49	37	125.49	150
422	For 300 lb. flanges, add				63%			
424	For 600 lb. flanges, add				310%			
466	To delete flanges & hanger, subtract							
467	1" diam. to 2" diam.				73%	65%		
468	2-1/2" diam. to 4" diam.				58%	59%		
469	5" diam. to 12" diam.				43%	46%		
475	Schedule 80, threaded, with couplings, and clevis type hangers							
476	sized for covering, 10' O.C.							
479	Black, 1/4" diameter	1 Plum	54	L.F.	1.30	3.29	4.59	6.20
480	3/8" diameter		53		1.46	3.35	4.81	6.45
481	1/2" diameter		52		1.30	3.42	4.72	6.35
482	3/4" diameter		50		1.56	3.55	5.11	6.85
483	1" diameter		45		2.20	3.95	6.15	8.10
484	1-1/4" diameter	Q-1	75		2.98	4.26	7.24	9.45
485	1-1/2" diameter		69		3.40	4.63	8.03	10.45
486	2" diameter		56		4.75	5.70	10.45	13.50
487	2-1/2" diameter		44		7.33	7.25	14.58	18.55
488	3" diameter		38		9.38	8.40	17.78	22
489	3-1/2" diameter		35		11.60	9.15	20.75	26
490	4" diameter		32		14.03	10	24.03	30
491	5" diameter		23		30.50	13.90	44.40	54
492	6" diameter	Q-2	28		33.70	17.75	51.45	63
493	8" diameter		23		52.88	22	74.88	89
494	10" diameter		20		69.10	25	94.10	110
495	12" diameter		15		84.60	33	117.60	140
543	To delete coupling & hanger, subtract							
544	1/4" diam. to 1/2" diam.				31%	54%		
545	3/4" diam. to 1-1/2" diam.				28%	49%		
546	2" diam. to 4" diam.				21%	40%		
547	5" diam. to 12" diam.				17%	43%		
551	Galvanized, 1/4" diameter	1 Plum	54	L.F.	1.69	3.29	4.98	6.60
552	3/8" diameter		53		1.85	3.35	5.20	6.90
553	1/2" diameter		52		1.50	3.42	4.92	6.60
554	3/4" diameter		50		1.79	3.55	5.34	7.10
555	1" diameter		45		2.55	3.95	6.50	8.50
556	1-1/4" diameter	Q-1	75		3.42	4.26	7.68	9.90
557	1-1/2" diameter		69		3.89	4.63	8.52	10.95
558	2" diameter		56		5.59	5.70	11.29	14.40

15.1 Pipe & Fittings

		CREW	DAILY OUTPUT	UNIT	BARE COSTS MAT.	BARE COSTS INST.	BARE COSTS TOTAL	TOTAL INCL O&P
55 559	2-1/2" diameter	Q-1	44	L.F.	8.34	7.25	15.59	19.65
560	3" diameter		38		10.67	8.40	19.07	24
561	3-1/2" diameter		35		13.25	9.15	22.40	28
562	4" diameter		32		16.12	10	26.12	32
563	5" diameter		23		36.77	13.90	50.67	61
564	6" diameter	Q-2	28		44.48	17.75	62.23	75
565	8" diameter		23		67.29	22	89.29	105
566	10" diameter		20		87.80	25	112.80	135
567	12" diameter		15		107.15	33	140.15	165
593	To delete coupling & hanger, subtract							
594	1/4" diam. to 1/2" diam.				31%	54%		
595	3/4" diam. to 1-1/2" diam.				28%	49%		
596	2" diam. to 4" diam.				21%	40%		
597	5" diam. to 12" diam.				17%	43%		
600	Welded, on yoke & roller hangers							
601	sized for covering, 10' O.C.							
604	Black, 1" diameter	Q-15	85	L.F.	2.25	3.97	6.22	8.15
605	1-1/4" diameter		79		3.02	4.28	7.30	9.40
606	1-1/2" diameter		72		3.44	4.69	8.13	10.50
607	2" diameter		57		4.46	5.95	10.41	13.35
608	2-1/2" diameter		44		6.64	7.70	14.34	18.25
609	3" diameter		40		8.51	8.45	16.96	21
610	3-1/2" diameter		34		10.40	9.95	20.35	26
611	4" diameter		33		12.33	10.25	22.58	28
612	5" diameter		26		28.16	13	41.16	50
613	6" diameter	Q-16	30		31.23	17.20	48.43	59
614	8" diameter		25		47.36	21	68.36	82
615	10" diameter		20		60.77	26	86.77	105
616	12" diameter		15		73.42	34	107.42	130
654	To delete hanger, subtract							
655	1" diam. to 1-1/2" diam.				30%	14%		
656	2" diam. to 3" diam.				23%	9%		
657	3-1/2" diam. to 5" diam.				12%	6%		
658	6" diam. to 12" diam.				10%	4%		
661	Galvanized, 1" diameter	Q-15	85	L.F.	2.51	3.97	6.48	8.45
665	1-1/4" diameter		79		3.35	4.28	7.63	9.80
666	1-1/2" diameter		72		3.83	4.69	8.52	10.90
667	2" diameter		57		5	5.95	10.95	13.95
668	2-1/2" diameter		44		7.46	7.70	15.16	19.15
669	3" diameter		40		9.61	8.45	18.06	23
670	3-1/2" diameter		34		11.78	9.95	21.73	27
671	4" diameter		33		13.98	10.25	24.23	30
672	5" diameter		26		33.66	13	46.66	56
673	6" diameter	Q-16	30		41.13	17.20	58.33	70
674	8" diameter		25		60.56	21	81.56	96
675	10" diameter		20		78.37	26	104.37	125
676	12" diameter		15		94.32	34	128.32	155
715	To delete hanger, subtract							
716	1' diam. to 1-1/2" diam.				26%	14%		
717	2" diam. to 3" diam.				16%	9%		
718	3-1/2" diam. to 5" diam.				10%	6%		
719	6" diam. to 12" diam.				7%	4%		
725	Flanged, 300 lb. weld neck, on yoke & roll hangers							
726	sized for covering, 10' O.C.							
729	Black, 1" diameter	Q-15	66	L.F.	6.18	5.10	11.28	14.10
730	1-1/4" diameter		61		7.24	5.55	12.79	15.85
731	1-1/2" diameter		54		7.95	6.25	14.20	17.65
732	2" diameter		42		9.69	8.05	17.74	22
733	2-1/2" diameter		33		12.99	10.25	23.24	29
734	3" diameter		29		15.53	11.65	27.18	34

15

15.1 Pipe & Fittings

		CREW	DAILY OUTPUT	UNIT	BARE COSTS MAT.	BARE COSTS INST.	BARE COSTS TOTAL	TOTAL INCL O&P
55 735	3-1/2" diameter	Q-15	24	L.F.	20.69	14.10	34.79	43
736	4" diameter		23		22.66	14.70	37.36	46
737	5" diameter		19		44.86	17.80	62.66	75
738	6" diameter	Q-16	23		49.25	22	71.25	86
739	8" diameter		17		78.01	30	108.01	130
740	10" diameter		14		126.77	37	163.77	190
741	12" diameter		12		154.93	43	197.93	230
747	For 600 lb. flanges, add				35%			
794	To delete flanges & hanger, subtract							
795	1" diam. to 1-1/2" diam.				75%	66%		
796	2" diam. to 3" diam.				62%	60%		
797	3-1/2" diam. to 5" diam.				54%	66%		
798	6" diam. to 12" diam.				55%	62%		
804	Galvanized, 1" diameter	Q-15	66	L.F.	6.44	5.10	11.54	14.40
805	1-1/4" diameter		61		7.57	5.55	13.12	16.25
806	1-1/2" diameter		54		8.34	6.25	14.59	18.10
807	2" diameter		42		10.23	8.05	18.28	23
808	2-1/2" diameter		33		13.81	10.25	24.06	30
809	3" diameter		29		16.63	11.65	28.28	35
810	3-1/2" diameter		24		22.07	14.10	36.17	44
811	4" diameter		23		24.31	14.70	39.01	48
812	5" diameter		19		50.36	17.80	68.16	81
813	6" diameter	Q-16	23		59.15	22	81.15	97
814	8" diameter		17		91.21	30	121.21	145
815	10" diameter		14		144.37	37	181.37	210
816	12" diameter		12		175.83	43	218.83	255
824	For 600 lb. flanges, add				35%			
890	To delete flanges & hangers, subtract							
891	1" diam. to 1-1/2" diam.				72%	66%		
892	2" diam. to 3" diam.				59%	60%		
893	3-1/2" diam. to 5" diam.				51%	66%		
894	6" diam. to 12" diam.				49%	62%		
900	Threading pipe labor, one end, all schedules through 80							
901	1/4" through 3/4" pipe size	1 Plum	80	Ea.		2.22	2.22	3.21L
902	1" through 2" pipe size		73			2.43	2.43	3.51L
903	2-1/2" pipe size		53			3.35	3.35	4.84L
904	3" pipe size		50			3.55	3.55	5.15L
905	3-1/2" pipe size	Q-1	89			3.59	3.59	5.20L
906	4" pipe size		73			4.38	4.38	6.35L
907	5" pipe size		53			6.05	6.05	8.70L
908	6" pipe size		46			6.95	6.95	10.05L
909	8" pipe size		29			11	11	15.95L
910	10" pipe size		21			15.20	15.20	22L
911	12" pipe size		13			25	25	36L
920	Welding labor per joint							
921	Schedule 40,							
923	1/2" pipe size	Q-15	32	Ea.		10.55	10.55	15.05
924	3/4" pipe size		27			12.50	12.50	17.85
925	1" pipe size		23			14.70	14.70	21
926	1-1/4" pipe size		20			16.90	16.90	24
927	1-1/2" pipe size		19			17.80	17.80	25
928	2" pipe size		16			21	21	30
929	2-1/2" pipe size		13			26	26	37
930	3" pipe size		12			28	28	40
931	4" pipe size		10			34	34	48
932	5" pipe size		9			38	38	54
933	6" pipe size		8			42	42	60
934	8" pipe size		5			68	68	96
935	10" pipe size		4			84	84	120
936	12" pipe size		3			115	115	160

15.1 Pipe & Fittings

		CREW	DAILY OUTPUT	UNIT	BARE COSTS MAT.	BARE COSTS INST.	BARE COSTS TOTAL	TOTAL INCL O&P
55 937	14″ pipe size	Q-15	2.60	Ea.		130	130	185
938	16″ pipe size		2.20			155	155	220
939	18″ pipe size		2			170	170	240
940	20″ pipe size		1.80			190	190	270
941	22″ pipe size		1.70			200	200	285
942	24″ pipe size		1.50			225	225	320
945	Schedule 80,							
946	1/2″ pipe size	Q-15	27	Ea.		12.50	12.50	17.85
947	3/4″ pipe size		23			14.70	14.70	21
948	1″ pipe size		20			16.90	16.90	24
949	1-1/4″ pipe size		19			17.80	17.80	25
950	1-1/2″ pipe size		18			18.75	18.75	27
951	2″ pipe size		15			23	23	32
952	2-1/2″ pipe size		12			28	28	40
953	3″ pipe size		11			31	31	44
954	4″ pipe size		8			42	42	60
955	5″ pipe size		6			56	56	80
956	6″ pipe size		5			68	68	96
957	8″ pipe size		4			84	84	120
958	10″ pipe size		3			115	115	160
959	12″ pipe size		2			170	170	240
960	14″ pipe size	Q-16	2.60			200	200	285
961	16″ pipe size		2.30			225	225	320
962	18″ pipe size		2			260	260	370
963	20″ pipe size		1.80			285	285	410
964	22″ pipe size		1.60			320	320	460
965	24″ pipe size		1.50			345	345	495
56-001	**PIPE, STEEL, FITTINGS,** Threaded							
002	Cast Iron,							
004	Standard weight, black							
006	90° Elbow, straight							
007	1/4″	1 Plum	16	Ea.	.89	11.10	11.99	17
008	3/8″		16		.89	11.10	11.99	17
009	1/2″		15		.89	11.85	12.74	18.10
010	3/4″		14		1.13	12.70	13.83	19.55
011	1″		13		1.39	13.65	15.04	21
012	1-1/4″	Q-1	22		1.94	14.55	16.49	23
013	1-1/2″		20		2.71	16	18.71	26
014	2″		18		4.19	17.75	21.94	30
015	2-1/2″		14		7.95	23	30.95	42
016	3″		10		12	32	44	59
017	3-1/2″		8		18.75	40	58.75	78
018	4″		6		22	53	75	100
019	5″		5		37	64	101	135
020	6″	Q-2	7		48	71	119	155
021	8″	″	6		96	83	179	225
025	45° Elbow, straight							
026	1/4″	1 Plum	16	Ea.	1.04	11.10	12.14	17.20
027	3/8″		16		1.13	11.10	12.23	17.30
028	1/2″		15		1.04	11.85	12.89	18.25
030	3/4″		14		1.27	12.70	13.97	19.70
032	1″		13		1.61	13.65	15.26	22
033	1-1/4″	Q-1	22		2.18	14.55	16.73	23
034	1-1/2″		20		3.15	16	19.15	27
035	2″		18		4.35	17.75	22.10	30
036	2-1/2″		14		8.45	23	31.45	42
037	3″		10		13.05	32	45.05	61
038	3-1/2″		8		19.80	40	59.80	80
040	4″		6		24	53	77	105

15

		CREW	DAILY OUTPUT	UNIT	BARE COSTS MAT.	INST.	TOTAL	TOTAL INCL O&P
56 042	5"	Q-1	5	Ea.	41	64	105	140
044	6"	Q-2	7		53	71	124	160
046	8"	"	6		105	83	188	235
050	Tee, straight							
051	1/4"	1 Plum	10	Ea.	1.11	17.75	18.86	27
052	3/8"		10		1.11	17.75	18.86	27
053	1/2"		9		1.27	19.75	21.02	30
054	3/4"		9		1.64	19.75	21.39	30
055	1"		8		1.80	22	23.80	34
056	1-1/4"	Q-1	14		3.05	23	26.05	36
057	1-1/2"		13		3.75	25	28.75	40
058	2"		11		5.65	29	34.65	48
059	2-1/2"		9		10.10	36	46.10	62
060	3"		6		15.25	53	68.25	94
061	3-1/2"		5		26	64	90	120
062	4"		4		29	80	109	145
063	5"		3		52	105	157	210
064	6"	Q-2	4		65	125	190	250
065	8"	"	3		135	165	300	390
070	Standard weight, galvanized							
072	90° Elbow, straight							
073	1/4"	1 Plum	16	Ea.	1.39	11.10	12.49	17.55
074	3/8"		16		1.39	11.10	12.49	17.55
075	1/2"		15		1.39	11.85	13.24	18.65
076	3/4"		14		1.74	12.70	14.44	20
077	1"		13		2.04	13.65	15.69	22
078	1-1/4"	Q-1	22		3.12	14.55	17.67	24
079	1-1/2"		20		4.31	16	20.31	28
080	2"		18		6.40	17.75	24.15	33
081	2-1/2"		14		12.30	23	35.30	47
082	3"		10		18.75	32	50.75	67
083	3-1/2"		8		29	40	69	90
084	4"		6		34	53	87	115
085	5"		5		59	64	123	155
086	6"	Q-2	7		75	71	146	185
087	8"	"	6		155	83	238	290
090	45° Elbow, straight							
091	1/4"	1 Plum	16	Ea.	1.61	11.10	12.71	17.80
092	3/8"		16		1.80	11.10	12.90	18
093	1/2"		15		1.61	11.85	13.46	18.90
094	3/4"		14		1.86	12.70	14.56	20
095	1"		13		2.30	13.65	15.95	22
096	1-1/4"	Q-1	22		3.47	14.55	18.02	25
097	1-1/2"		20		4.91	16	20.91	29
098	2"		18		6.85	17.75	24.60	33
099	2-1/2"		14		13.25	23	36.25	48
100	3"		10		20	32	52	68
101	3-1/2"		8		31	40	71	92
102	4"		6		39	53	92	120
103	5"		5		65	64	129	165
104	6"	Q-2	7		83	71	154	195
105	8"	"	6		165	83	248	300
110	Tee, straight							
111	1/4"	1 Plum	10	Ea.	1.80	17.75	19.55	28
112	3/8"		10		1.80	17.75	19.55	28
113	1/2"		9		1.86	19.75	21.61	31
114	3/4"		9		2.38	19.75	22.13	31
115	1"		8		2.67	22	24.67	35
116	1-1/4"	Q-1	14		4.67	23	27.67	38
117	1-1/2"	"	13		5.85	25	30.85	42

15

15.1 Pipe & Fittings

	15.1 Pipe & Fittings	CREW	DAILY OUTPUT	UNIT	BARE COSTS MAT.	BARE COSTS INST.	BARE COSTS TOTAL	TOTAL INCL O&P
56 118	2″	Q-1	11	Ea.	7.70	29	36.70	50
119	2-1/2″		9		15.70	36	51.70	69
120	3″		6		24	53	77	105
121	3-1/2″		5		41	64	105	140
122	4″		4		46	80	126	165
123	5″		3		82	105	187	245
124	6″	Q-2	4		100	125	225	290
125	8″	″	3		210	165	375	470
130	Extra heavy weight, black							
131	Couplings, steel straight							
132	1/4″	1 Plum	19	Ea.	1.07	9.35	10.42	14.70
133	3/8″		19		1.17	9.35	10.52	14.80
134	1/2″		19		2.13	9.35	11.48	15.85
135	3/4″		18		2.24	9.85	12.09	16.70
136	1″		15		3.15	11.85	15	21
137	1-1/4″	Q-1	26		5.05	12.30	17.35	23
138	1-1/2″		24		5	13.30	18.30	25
139	2″		21		7.85	15.20	23.05	31
140	2-1/2″		18		11.45	17.75	29.20	38
141	3″		14		13.15	23	36.15	47
142	3-1/2″		12		17.45	27	44.45	58
143	4″		10		22	32	54	70
144	5″		6		31	53	84	110
145	6″	Q-2	8		36	62	98	130
146	8″		7		69	71	140	180
147	10″		6		110	83	193	240
148	12″		4		145	125	270	340
151	90° Elbow, straight							
152	1/2″	1 Plum	15	Ea.	2.78	11.85	14.63	20
153	3/4″		14		3.33	12.70	16.03	22
154	1″		13		4.03	13.65	17.68	24
155	1-1/4″	Q-1	22		5.75	14.55	20.30	27
156	1-1/2″		20		7.45	16	23.45	31
158	2″		18		10.15	17.75	27.90	37
159	2-1/2″		14		15.75	23	38.75	50
160	3″		10		23	32	55	72
161	4″		6		48	53	101	130
162	6″	Q-2	7		90	71	161	200
165	45° Elbow, straight							
166	1/2″	1 Plum	15	Ea.	3.30	11.85	15.15	21
167	3/4″		14		3.74	12.70	16.44	22
168	1″		13		4.49	13.65	18.14	25
169	1-1/4″	Q-1	22		6.30	14.55	20.85	28
170	1-1/2″		20		8.15	16	24.15	32
171	2″		18		11.55	17.75	29.30	38
172	2-1/2″		14		18.90	23	41.90	54
173	3″		10		26	32	58	75
174	4″		6		54	53	107	135
180	Tee, straight							
181	1/2″	1 Plum	9	Ea.	4.28	19.75	24.03	33
182	3/4″		9		4.76	19.75	24.51	34
183	1″		8		5.75	22	27.75	38
184	1-1/4″	Q-1	14		7.90	23	30.90	42
185	1-1/2″		13		10.15	25	35.15	47
186	2″		11		13.70	29	42.70	57
187	2-1/2″		9		22	36	58	76
188	3″		6		33	53	86	115
189	4″		4		65	80	145	185
190	6″	Q-2	4		130	125	255	325
195	Extra heavy weight, galvanized							

15.1 Pipe & Fittings

		CREW	DAILY OUTPUT	UNIT	BARE COSTS MAT.	BARE COSTS INST.	BARE COSTS TOTAL	TOTAL INCL O&P
56 197	90° Elbow, straight							
198	1/2"	1 Plum	15	Ea.	4.56	11.85	16.41	22
199	3/4"		14		5.40	12.70	18.10	24
200	1"		13		6.60	13.65	20.25	27
201	1-1/4"	Q-1	22		9.30	14.55	23.85	31
202	1-1/2"		20		12	16	28	36
203	2"		18		16.60	17.75	34.35	44
204	2-1/2"		14		26	23	49	62
205	3"		10		37	32	69	87
206	4"		6		76	53	129	160
207	6"	Q-2	7		145	71	216	260
210	45° Elbow, straight							
211	1/2"	1 Plum	15	Ea.	5.35	11.85	17.20	23
212	3/4"		14		6.30	12.70	19	25
213	1"		13		7.30	13.65	20.95	28
214	1-1/4"	Q-1	22		10.30	14.55	24.85	32
215	1-1/2"		20		13.25	16	29.25	38
216	2"		18		18.85	17.75	36.60	46
217	2-1/2"		14		33	23	56	69
218	3"		10		47	32	79	98
219	4"		6		99	53	152	185
225	Tee, straight							
226	1/2"	1 Plum	9	Ea.	6.95	19.75	26.70	36
227	3/4"		9		7.85	19.75	27.60	37
228	1"		8		9.30	22	31.30	42
229	1-1/4"	Q-1	14		12.90	23	35.90	47
230	1-1/2"		13		16.60	25	41.60	54
231	2"		11		22	29	51	66
232	2-1/2"		9		36	36	72	91
233	3"		6		55	53	108	140
234	4"		4		105	80	185	230
236	6"	Q-2	4		210	125	335	410
250	Couplings, steel straight							
251	1/4"	1 Plum	19	Ea.	1.23	9.35	10.58	14.85
252	3/8"		19		1.34	9.35	10.69	15
253	1/2"		19		2.65	9.35	12	16.40
254	3/4"		18		2.80	9.85	12.65	17.35
255	1"		15		3.93	11.85	15.78	21
256	1-1/4"	Q-1	26		6.05	12.30	18.35	24
257	1-1/2"		24		5.95	13.30	19.25	26
258	2"		21		10.55	15.20	25.75	34
259	2-1/2"		18		13.20	17.75	30.95	40
260	3"		14		14.90	23	37.90	49
261	3-1/2"		12		19.95	27	46.95	60
262	4"		10		26	32	58	75
263	5"		6		38	53	91	120
264	6"	Q-2	8		44	62	106	140
265	8"		7		80	71	151	190
266	10"		6		120	83	203	250
267	12"		4		160	125	285	355
500	Malleable iron, 150 lb							
502	Black							
504	90° Elbow, straight							
506	1/4"	1 Plum	16	Ea.	.76	11.10	11.86	16.85
507	3/8"		16		.76	11.10	11.86	16.85
508	1/2"		15		.56	11.85	12.41	17.70
509	3/4"		14		.65	12.70	13.35	19.05
510	1"		13		1.21	13.65	14.86	21
511	1-1/4"	Q-1	22		1.98	14.55	16.53	23
512	1-1/2"	″	20		2.61	16	18.61	26

15

15.1 Pipe & Fittings

56		CREW	DAILY OUTPUT	UNIT	BARE COSTS MAT.	BARE COSTS INST.	BARE COSTS TOTAL	TOTAL INCL O&P
513	2"	Q-1	18	Ea.	3.80	17.75	21.55	30
514	2-1/2"		14		9.25	23	32.25	43
515	3"		10		14.40	32	46.40	62
516	3-1/2"		8		24	40	64	84
517	4"		6		24	53	77	105
518	5"		5		58	64	122	155
519	6"	Q-2	7		71	71	142	180
525	45° Elbow, straight							
527	1/4"	1 Plum	16	Ea.	.98	11.10	12.08	17.10
528	3/8"		16		.98	11.10	12.08	17.10
529	1/2"		15		.88	11.85	12.73	18.05
530	3/4"		14		1.09	12.70	13.79	19.50
531	1"		13		1.40	13.65	15.05	21
532	1-1/4"	Q-1	22		2.34	14.55	16.89	24
533	1-1/2"		20		2.93	16	18.93	26
534	2"		18		4.03	17.75	21.78	30
535	2-1/2"		14		10.25	23	33.25	44
536	3"		10		16.05	32	48.05	64
537	3-1/2"		8		26	40	66	86
538	4"		6		26	53	79	105
539	5"		5		65	64	129	165
540	6"	Q-2	7		78	71	149	190
545	Tee, straight							
547	1/4"	1 Plum	10	Ea.	1.15	17.75	18.90	27
548	3/8"		10		1.15	17.75	18.90	27
549	1/2"		9		.69	19.75	20.44	29
550	3/4"		9		1.01	19.75	20.76	30
551	1"		8		1.87	22	23.87	34
552	1-1/4"	Q-1	14		3.04	23	26.04	36
553	1-1/2"		13		3.74	25	28.74	40
554	2"		11		5.50	29	34.50	48
555	2-1/2"		9		12.25	36	48.25	65
556	3"		6		16.05	53	69.05	95
557	3-1/2"		5		29	64	93	125
558	4"		4		32	80	112	150
559	5"		3		67	105	172	230
560	6"	Q-2	4		92	125	217	280
565	Coupling, straight							
567	1/4"	1 Plum	19	Ea.	.83	9.35	10.18	14.40
568	3/8"		19		.83	9.35	10.18	14.40
569	1/2"		19		.77	9.35	10.12	14.35
570	3/4"		18		.92	9.85	10.77	15.25
571	1"		15		1.39	11.85	13.24	18.65
572	1-1/4"	Q-1	26		1.75	12.30	14.05	19.70
573	1-1/2"		24		2.21	13.30	15.51	22
574	2"		21		3.20	15.20	18.40	26
575	2-1/2"		18		7.10	17.75	24.85	33
576	3"		14		9.75	23	32.75	44
577	3-1/2"		12		18.10	27	45.10	58
578	4"		10		18.10	32	50.10	66
579	5"		6		36	53	89	115
580	6"	Q-2	8		46	62	108	140
581	8"		7		72	71	143	180
582	10"		6		115	83	198	245
583	12"		4		155	125	280	350
600	For galvanized elbows, tees, and couplings add				23%			
700	Union, with brass seat							
701	1/4"	1 Plum	15	Ea.	2.24	11.85	14.09	19.55
702	3/8"		15		2.24	11.85	14.09	19.55
703	1/2"		14		2.24	12.70	14.94	21

15

15.1	Pipe & Fittings	CREW	DAILY OUTPUT	UNIT	BARE COSTS MAT.	BARE COSTS INST.	BARE COSTS TOTAL	TOTAL INCL O&P
56 704	3/4"	1 Plum	13	Ea.	2.29	13.65	15.94	22
705	1"	"	12		3.36	14.80	18.16	25
706	1-1/4"	Q-1	21		4.63	15.20	19.83	27
707	1-1/2"		19		5.65	16.85	22.50	31
708	2"		17		7.15	18.80	25.95	35
709	2-1/2"		13		14.80	25	39.80	52
710	3"		9		24	36	60	78
711	Union, all iron, 1/4"	1 Plum	15		2.73	11.85	14.58	20
713	3/8"		15		2.92	11.85	14.77	20
714	1/2"		14		3.12	12.70	15.82	22
715	3/4"		13		3.56	13.65	17.21	24
716	1"		12		4.66	14.80	19.46	27
717	1-1/4"	Q-1	21		6	15.20	21.20	29
718	1-1/2"		19		7.20	16.85	24.05	32
719	2"		17		9.95	18.80	28.75	38
720	2-1/2"		13		28	25	53	66
721	3"		9		34	36	70	89
725	For galvanized unions, add				5%			
750	Malleable iron, 300 lb							
752	Black							
754	90° Elbow, straight, 1/4"	1 Plum	16	Ea.	2.14	11.10	13.24	18.40
756	3/8"		16		2.14	11.10	13.24	18.40
757	1/2"		15		3.28	11.85	15.13	21
758	3/4"		14		3.78	12.70	16.48	22
759	1"		13		4.85	13.65	18.50	25
760	1-1/4"	Q-1	22		6.40	14.55	20.95	28
761	1-1/2"		20		8.15	16	24.15	32
762	2"		18		11.30	17.75	29.05	38
763	2-1/2"		14		18.55	23	41.55	53
764	3"		10		25	32	57	74
765	4"		6		51	53	104	135
766	5"		5		125	64	189	230
767	6"	Q-2	7		125	71	196	240
768	8"	"	6		250	83	333	395
770	45° Elbow, straight, 1/4"	1 Plum	16		3.02	11.10	14.12	19.35
772	3/8"		16		3.02	11.10	14.12	19.35
773	1/2"		15		4.08	11.85	15.93	22
774	3/4"		14		4.52	12.70	17.22	23
775	1"		13		5	13.65	18.65	25
776	1-1/4"	Q-1	22		7	14.55	21.55	29
777	1-1/2"		20		9.15	16	25.15	33
778	2"		18		13.55	17.75	31.30	41
779	2-1/2"		14		22	23	45	57
780	3"		10		29	32	61	78
781	4"		6		55	53	108	140
785	Tee, straight, 1/4"	1 Plum	10		3.08	17.75	20.83	29
787	3/8"		10		3.08	17.75	20.83	29
788	1/2"		9		4.39	19.75	24.14	33
789	3/4"		9		4.82	19.75	24.57	34
790	1"		8		5.80	22	27.80	38
791	1-1/4"	Q-1	14		7.70	23	30.70	41
792	1-1/2"		13		9.50	25	34.50	46
793	2"		11		14.05	29	43.05	57
794	2-1/2"		9		24	36	60	78
795	[illegible]"		6		40	53	93	120
796	[illegible]"		4		69	80	149	190
797	5"		3		185	105	290	360
798	6"	Q-2	4		185	125	310	385
799	8"	"	3		360	165	525	635
805	Couplings, straight, 1/4"	1 Plum	19		1.84	9.35	11.19	15.55

15

15.1 Pipe & Fittings		CREW	DAILY OUTPUT	UNIT	BARE COSTS MAT.	BARE COSTS INST.	BARE COSTS TOTAL	TOTAL INCL O&P
56 807	3/8"	1 Plum	19	Ea.	1.84	9.35	11.19	15.55
808	1/2"		19		2.12	9.35	11.47	15.85
809	3/4"		18		2.48	9.85	12.33	17
810	1"		15		3.24	11.85	15.09	21
811	1-1/4"	Q-1	26		3.83	12.30	16.13	22
812	1-1/2"		24		5.25	13.30	18.55	25
813	2"		21		7.25	15.20	22.45	30
814	2-1/2"		18		10.95	17.75	28.70	38
815	3"		14		15.15	23	38.15	50
850	For galvanized elbows, tees and couplings add				24%			
950	Union with brass seat, 1/4"	1 Plum	15		3.75	11.85	15.60	21
953	3/8"		15		3.75	11.85	15.60	21
954	1/2"		14		3.86	12.70	16.56	23
955	3/4"		13		3.34	13.65	16.99	23
956	1"		12		5.65	14.80	20.45	28
957	1-1/4"	Q-1	21		7.95	15.20	23.15	31
958	1-1/2"		19		9.50	16.85	26.35	35
959	2"		17		12	18.80	30.80	40
960	2-1/2"		13		25	25	50	63
961	3"		9		41	36	77	96
962	4"		5		80	64	144	180
963	Union, all iron, 1/4"	1 Plum	15		2.73	11.85	14.58	20
965	3/8"		15		2.92	11.85	14.77	20
966	1/2"		14		3.12	12.70	15.82	22
967	3/4"		13		3.56	13.65	17.21	24
968	1"		12		4.66	14.80	19.46	27
969	1-1/4"	Q-1	21		6	15.20	21.20	29
970	1-1/2"		19		7.20	16.85	24.05	32
971	2"		17		9.95	18.80	28.75	38
972	2-1/2"		13		28	25	53	66
973	3"		9		34	36	70	89
975	For galvanized unions, add				5%			
57-001	**PIPE, STEEL, FITTINGS** Flanged, welded and special type							
002	Flanged joints, cast iron, standard weight, black. One gasket & bolt							
004	set required at each joint, not included (see line 062)							
006	90° Elbow, straight, 1-1/2" pipe size	Q-1	14	Ea.	42	23	65	79
008	2" pipe size		13		39	25	64	78
009	2-1/2" pipe size		12		42	27	69	85
010	3" pipe size		11		47	29	76	94
011	4" pipe size		8		62	40	102	125
012	5" pipe size		7		84	46	130	160
013	6" pipe size	Q-2	9		97	55	152	185
014	8" pipe size		8		150	62	212	255
015	10" pipe size		7		275	71	346	405
016	12" pipe size		6		400	83	483	560
020	45° Elbow, straight, 1-1/2" pipe size	Q-1	14		51	23	74	89
022	2" pipe size		13		51	25	76	92
023	2-1/2" pipe size		12		52	27	79	96
024	3" pipe size		11		59	29	88	105
025	4" pipe size		8		77	40	117	140
026	5" pipe size		7		105	46	151	180
027	6" pipe size	Q-2	9		110	55	165	200
028	8" pipe size		8		165	62	227	270
029	10" pipe size		7		305	71	376	440
030	12" pipe size		6		450	83	533	615
035	Tee, straight, 1-1/2" pipe size	Q-1	10		49	32	81	100
037	2" pipe size		9		45	36	81	100
038	2-1/2" pipe size		8		55	40	95	120
039	3" pipe size		7		57	46	103	130
040	4" pipe size		5		94	64	158	195

15.1 Pipe & Fittings

		CREW	DAILY OUTPUT	UNIT	BARE COSTS MAT.	INST.	TOTAL	TOTAL INCL O&P
57 041	5" pipe size	Q-1	4	Ea.	125	80	205	255
042	6" pipe size	Q-2	6		135	83	218	270
043	8" pipe size		5		200	99	299	365
044	10" pipe size		4		420	125	545	640
045	12" pipe size		3		550	165	715	845
050	For galvanized elbows and tees, add				40%			
052	For extra heavy weight elbows and tees, add				8%			
062	Gasket and 4 to 20 bolt set, 1-1/2" pipe size	1 Plum	30		2.03	5.90	7.93	10.80
063	2" pipe size		30		3.58	5.90	9.48	12.50
064	2-1/2" pipe size		30		3.74	5.90	9.64	12.65
065	3" pipe size		30		3.86	5.90	9.76	12.80
066	3-1/2" pipe size		28		7.60	6.35	13.95	17.50
067	4" pipe size		27		7.95	6.60	14.55	18.25
068	5" pipe size		26		11.90	6.85	18.75	23
069	6" pipe size		24		12.05	7.40	19.45	24
070	8" pipe size		20		12.90	8.90	21.80	27
071	10" pipe size		18		30	9.85	39.85	47
072	12" pipe size		16		31	11.10	42.10	50
080	For 300 Lb. gasket set, add				3%			
200	Unions, 125 lb., black, 1/2" pipe size	1 Plum	17	Ea.	8.30	10.45	18.75	24
204	3/4" pipe size		17		10	10.45	20.45	26
205	1" pipe size		16		10.35	11.10	21.45	27
206	1-1/4" pipe size	Q-1	28		12.45	11.40	23.85	30
207	1-1/2" pipe size		27		14.50	11.85	26.35	33
208	2" pipe size		26		18.70	12.30	31	38
209	2-1/2" pipe size		24		23	13.30	36.30	45
210	3" pipe size		22		31	14.55	45.55	55
211	3-1/2" pipe size		18		41	17.75	58.75	71
212	4" pipe size		16		41	20	61	74
213	5" pipe size		14		52	23	75	90
214	6" pipe size	Q-2	19		67	26	93	110
215	8" pipe size	"	16		105	31	136	160
220	For galvanized unions, add				100%			
229	Threaded companion flange,							
230	Cast iron,							
231	Black 125 lb., per flange,							
232	1" pipe size	1 Plum	27	Ea.	7.05	6.60	13.65	17.25
233	1-1/4" pipe size	Q-1	44		7.65	7.25	14.90	18.90
234	1-1/2" pipe size		40		7.90	8	15.90	20
235	2" pipe size		36		8.80	8.90	17.70	23
236	2-1/2" pipe size		28		9.70	11.40	21.10	27
237	3" pipe size		20		13.20	16	29.20	38
238	3-1/2" pipe size		16		17.55	20	37.55	48
239	4" pipe size		12		17.80	27	44.80	58
240	5" pipe size		10		23	32	55	72
241	6" pipe size	Q-2	14		28	36	64	82
242	8" pipe size		12		42	41	83	105
243	10" pipe size		10		61	50	111	140
244	12" pipe size		8		83	62	145	180
246	For galvanized flanges, add				40%			
258	Forged steel,							
259	Black 150 lb., per flange							
260	1/2" pipe size	1 Plum	30	Ea.	12.90	5.90	18.80	23
261	3/4" pipe size		28		13.60	6.35	19.95	24
262	1" pipe size		27		14.30	6.60	20.90	25
263	1-1/4" pipe size	Q-1	44		14.30	7.25	21.55	26
264	1-1/2" pipe size		40		19.50	8	27.50	33
265	2" pipe size		36		23	8.90	31.90	38
266	2-1/2" pipe size		28		24	11.40	35.40	43
267	3" pipe size		20		27	16	43	53

15.1 Pipe & Fittings

		Pipe & Fittings	CREW	DAILY OUTPUT	UNIT	BARE COSTS MAT.	BARE COSTS INST.	BARE COSTS TOTAL	TOTAL INCL O&P
57	269	4" pipe size	Q-1	12	Ea.	32	27	59	74
	270	5" pipe size	"	10		43	32	75	94
	271	6" pipe size	Q-2	14		51	36	87	105
	272	8" pipe size		12		88	41	129	155
	273	10" pipe size		10		150	50	200	235
	286	Black 300 lb., per flange							
	287	1/2" pipe size	1 Plum	30	Ea.	15.60	5.90	21.50	26
	288	3/4" pipe size		28		17.55	6.35	23.90	28
	289	1" pipe size		27		18.85	6.60	25.45	30
	290	1-1/4" pipe size	Q-1	44		22	7.25	29.25	35
	291	1-1/2" pipe size		40		22	8	30	36
	292	2" pipe size		36		25	8.90	33.90	40
	293	2-1/2" pipe size		28		32	11.40	43.40	52
	294	3" pipe size		20		38	16	54	65
	296	4" pipe size		12		55	27	82	99
	297	6" pipe size	Q-2	14		89	36	125	150
	300	Weld joint, butt, carbon steel, standard weight							
	304	90° Elbow, long							
	305	1/2" pipe size	Q-15	16	Ea.	5.15	21	26.15	36
	306	3/4" pipe size		16		5.15	21	26.15	36
	307	1" pipe size		16		5.15	21	26.15	36
	308	1-1/4" pipe size		14		6.20	24	30.20	41
	309	1-1/2" pipe size		13		6.20	26	32.20	44
	310	2" pipe size		10		6.75	34	40.75	56
	311	2-1/2" pipe size		8		8.70	42	50.70	70
	312	3" pipe size		7		11.60	48	59.60	82
	313	4" pipe size		5		18.90	68	86.90	115
	314	6" pipe size	Q-16	5		53	105	158	205
	315	8" pipe size		4		97	130	227	290
	316	10" pipe size		3		175	170	345	440
	317	12" pipe size		2.50		250	205	455	570
	320	45° Elbow, long							
	321	1/2" pipe size	Q-15	16	Ea.	5.15	21	26.15	36
	322	3/4" pipe size		16		5.15	21	26.15	36
	323	1" pipe size		16		5.15	21	26.15	36
	324	1-1/4" pipe size		14		6.20	24	30.20	41
	325	1-1/2" pipe size		13		6.20	26	32.20	44
	326	2" pipe size		10		6.35	34	40.35	55
	327	2-1/2" pipe size		8		7.40	42	49.40	68
	328	3" pipe size		7		9	48	57	79
	329	4" pipe size		5		13.25	68	81.25	110
	330	6" pipe size	Q-16	5		34	105	139	185
	331	8" pipe size		4		61	130	191	250
	332	10" pipe size		3		99	170	269	355
	333	12" pipe size		2.50		145	205	350	455
	335	Tee, straight							
	336	1/2" pipe size	Q-15	10	Ea.	18.30	34	52.30	68
	337	3/4" pipe size		10		18.30	34	52.30	68
	338	1" pipe size		10		18.30	34	52.30	68
	339	1-1/4" pipe size		9		22	38	60	78
	340	1-1/2" pipe size		8		28	42	70	91
	341	2" pipe size		6		22	56	78	105
	342	2-1/2" pipe size		5		27	68	95	125
	343	3" pipe size		4		30	84	114	155
	344	4" pipe size		3		41	115	156	205
	345	6" pipe size	Q-16	3		75	170	245	330
	346	8" pipe size		2.50		135	205	340	445
	347	10" pipe size		2		230	260	490	620
	348	12" pipe size		1.60		350	320	670	845
	350	Weld joint, butt, carbon steel, extra strong							

57

		CREW	DAILY OUTPUT	UNIT	BARE COSTS MAT.	BARE COSTS INST.	BARE COSTS TOTAL	TOTAL INCL O&P
351	90° Elbow, long							
352	1/2" pipe size	Q-15	16	Ea.	7.15	21	28.15	38
353	3/4" pipe size		16		7.15	21	28.15	38
354	1" pipe size		16		7.15	21	28.15	38
355	1-1/4" pipe size		14		8.20	24	32.20	43
356	1-1/2" pipe size		13		8.85	26	34.85	47
357	2" pipe size		10		9.20	34	43.20	58
358	2-1/2" pipe size		8		13.40	42	55.40	75
359	3" pipe size		7		17.25	48	65.25	88
360	4" pipe size		5		29	68	97	130
361	6" pipe size	Q-16	5		79	105	184	235
362	8" pipe size		4		145	130	275	345
363	10" pipe size		3		260	170	430	530
364	12" pipe size		2.50		370	205	575	705
365	45° Elbow, long							
366	1/2" pipe size	Q-15	16	Ea.	7.15	21	28.15	38
367	3/4" pipe size		16		7.15	21	28.15	38
368	1" pipe size		16		7.15	21	28.15	38
369	1-1/4" pipe size		14		7.80	24	31.80	43
370	1-1/2" pipe size		13		8.10	26	34.10	46
371	2" pipe size		10		7.60	34	41.60	57
372	2-1/2" pipe size		8		10.40	42	52.40	72
373	3" pipe size		7		13.25	48	61.25	83
374	4" pipe size		5		20	68	88	120
375	6" pipe size	Q-16	5		55	105	160	210
376	8" pipe size		4		90	130	220	285
377	10" pipe size		3		145	170	315	405
378	12" pipe size		2.50		220	205	425	540
380	Tee, straight							
381	1/2" pipe size	Q-15	10	Ea.	26	34	60	77
382	3/4" pipe size		10		25	34	59	76
383	1" pipe size		10		25	34	59	76
384	1-1/4" pipe size		9		25	38	63	81
385	1-1/2" pipe size		8		29	42	71	92
386	2" pipe size		6		25	56	81	110
387	2-1/2" pipe size		5		33	68	101	135
388	3" pipe size		4		40	84	124	165
389	4" pipe size		3		65	115	180	230
390	6" pipe size	Q-16	3		105	170	275	360
391	8" pipe size		2.50		210	205	415	525
392	10" pipe size		2		340	260	600	745
393	12" pipe size		1.60		495	320	815	1,000
500	Weld joint, socket, forged steel, 3000 lb., Schedule 40							
501	90° Elbow, straight							
505	3/4" pipe size	Q-15	20	Ea.	4.80	16.90	21.70	29
506	1" pipe size		20		6.15	16.90	23.05	31
507	1-1/4" pipe size		18		9.65	18.75	28.40	37
508	1-1/2" pipe size		16		11.70	21	32.70	43
509	2" pipe size		12		17.70	28	45.70	60
510	2-1/2" pipe size		10		43	34	77	96
511	3" pipe size		8		69	42	111	135
512	4" pipe size		6		180	56	236	280
513	45° Elbow, straight							
514	1" pipe size	Q-15	20	Ea.	7.45	16.90	24.35	32
515	1-1/4" pipe size		18		9.45	18.75	28.20	37
516	1-1/2" pipe size		16		11.50	21	32.50	43
517	2" pipe size		12		21	28	49	63
518	2-1/2" pipe size		10		56	34	90	110
519	3" pipe size		8		96	42	138	165
520	4" pipe size		6		185	56	241	285

15.1 Pipe & Fittings		CREW	DAILY OUTPUT	UNIT	BARE COSTS MAT.	BARE COSTS INST.	BARE COSTS TOTAL	TOTAL INCL O&P
57 525	Tee, straight							
526	1" pipe size	Q-15	13	Ea.	8	26	34	46
527	1-1/4" pipe size		12		11.40	28	39.40	53
528	1-1/2" pipe size		11		18	31	49	64
529	2" pipe size		8		26	42	68	89
530	2-1/2" pipe size		6		59	56	115	145
531	3" pipe size		5		110	68	178	215
532	4" pipe size		4		240	84	324	385
535	For reducing sizes, add				100%			
545	Couplings							
546	1" pipe size	Q-15	22	Ea.	3.09	15.35	18.44	25
547	1-1/4" pipe size		20		5.25	16.90	22.15	30
548	1-1/2" pipe size		18		6.70	18.75	25.45	34
549	2" pipe size		14		10.50	24	34.50	46
550	2-1/2" pipe size		12		18.95	28	46.95	61
551	3" pipe size		9		47	38	85	105
552	4" pipe size		7		67	48	115	145
600	Weld-on flanges, forged steel							
602	Slip-on, 150 lb. flange, (welded front and back)							
605	1/2" pipe size	Q-15	18	Ea.	11.05	18.75	29.80	39
606	3/4" pipe size		18		11.70	18.75	30.45	40
607	1" pipe size		17		11.70	19.85	31.55	41
608	1-1/4" pipe size		16		11.70	21	32.70	43
609	1-1/2" pipe size		15		12.35	23	35.35	46
610	2" pipe size		12		13	28	41	54
611	2-1/2" pipe size		10		16.90	34	50.90	67
612	3" pipe size		9		18.85	38	56.85	74
613	3-1/2" pipe size		7		24	48	72	95
614	4" pipe size		6		24	56	80	105
615	5" pipe size		5		33	68	101	135
616	6" pipe size	Q-16	6		38	86	124	165
617	8" pipe size		5		58	105	163	210
618	10" pipe size		4		110	130	240	305
619	12" pipe size		3		165	170	335	430
620	300 lb. flange							
621	1/2" pipe size	Q-15	17	Ea.	12.35	19.85	32.20	42
622	3/4" pipe size		17		13	19.85	32.85	43
623	1" pipe size		16		13.65	21	34.65	45
624	1-1/4" pipe size		13		14.95	26	40.95	54
625	1-1/2" pipe size		12		16.25	28	44.25	58
626	2" pipe size		11		18.20	31	49.20	64
627	2-1/2" pipe size		9		22	38	60	78
628	3" pipe size		7		26	48	74	97
629	4" pipe size		6		40	56	96	125
630	5" pipe size		4		64	84	148	190
631	6" pipe size	Q-16	5		65	105	170	220
632	8" pipe size		4		110	130	240	305
633	10" pipe size		3.40		215	150	365	455
634	12" pipe size		2.80		260	185	445	550
640	Welding neck, 150 lb. flange							
641	1/2" pipe size	Q-15	40	Ea.	14.95	8.45	23.40	28
642	3/4" pipe size		36		15.60	9.40	25	31
643	1" pipe size		32		15.60	10.55	26.15	32
644	1-1/4" pipe size		29		16.25	11.65	27.90	34
645	1-1/2" pipe size		26		16.25	13	29.25	36
646	2" pipe size		20		18.20	16.90	35.10	44
647	2-1/2" pipe size		16		20	21	41	52
648	3" pipe size		14		23	24	47	60
649	3-1/2" pipe size		12		38	28	66	82
650	4" pipe size		10		30	34	64	81

15.1 Pipe & Fittings		CREW	DAILY OUTPUT	UNIT	BARE COSTS MAT.	BARE COSTS INST.	BARE COSTS TOTAL	TOTAL INCL O&P
57 651	5" pipe size	Q-15	8	Ea.	38	42	80	100
652	6" pipe size	Q-16	10		45	52	97	125
653	8" pipe size		7		73	74	147	185
654	10" pipe size		6		145	86	231	285
655	12" pipe size		5		210	105	315	380
656	300 lb. flange							
657	1/2" pipe size	Q-15	36	Ea.	15.60	9.40	25	31
658	3/4" pipe size		34		16.25	9.95	26.20	32
659	1" pipe size		30		16.90	11.25	28.15	35
660	1-1/4" pipe size		28		18.20	12.05	30.25	37
661	1-1/2" pipe size		24		19.50	14.10	33.60	42
662	2" pipe size		18		22	18.75	40.75	51
663	2-1/2" pipe size		14		27	24	51	64
664	3" pipe size		12		30	28	58	73
665	4" pipe size		8		43	42	85	110
666	5" pipe size		7		70	48	118	145
667	6" pipe size	Q-16	9		78	57	135	170
668	8" pipe size		6		135	86	221	270
669	10" pipe size		5		285	105	390	460
670	12" pipe size		4		355	130	485	575
774	Plain ends for plain end pipe, mechanically coupled							
775	Cplg & labor required at joints not included add 1 per							
776	joint for installed price, see 15.1-57-918							
777	Malleable iron, black, unless noted otherwise							
780	90° Elbow 1"			Ea.	5.40		5.40	5.95M
781	1-1/2"				9.15		9.15	10.05M
782	2"				8.25		8.25	9.05M
783	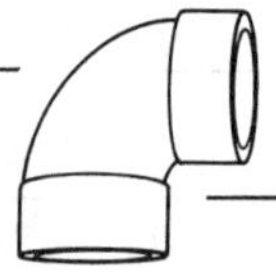 2-1/2"				11.40		11.40	12.55M
784	3"				13.70		13.70	15.05M
786	4"				20		20	22M
787	5" welded steel				70		70	77M
788	6"				55		55	61M
789	8" welded steel				115		115	125M
790	10" welded steel				145		145	160M
791	12" welded steel				185		185	205M
797	45° Elbow 1"				5.40		5.40	5.95M
798	1-1/2"				9.15		9.15	10.05M
799	2"				8.25		8.25	9.05M
800	 2-1/2"				11.40		11.40	12.55M
801	3"				13.70		13.70	15.05M
803	4"				20		20	22M
804	5" welded steel				70		70	77M
805	6"				55		55	61M
806	8"				64		64	70M
807	10" welded steel				90		90	99M
808	12" welded steel				115		115	125M
814	Tee, straight 1"				6.80		6.80	7.50M
815	1-1/2"				13.35		13.35	14.70M
816	2"				12.10		12.10	13.30M
817	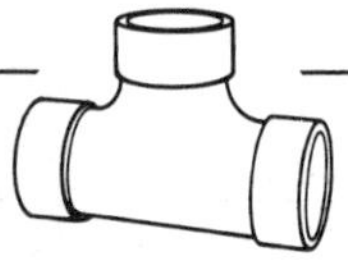 2-1/2"				15.35		15.35	16.90M
818	3"				19.25		19.25	21M
820	4"				31		31	34M
821	5" welded steel				68		68	75M
822	6"				84		84	92M
823	8" welded steel				115		115	125M
824	10" welded steel				180		180	200M
825	12" welded steel 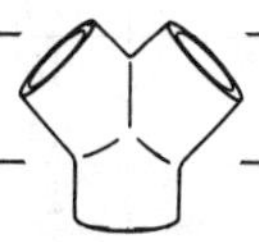				240		240	265M
834	Segmentally welded steel							
839	Wye 2"			Ea.	35		35	39M
840	2-1/2"			"	45		45	50M

15.1 Pipe & Fittings

	15.1 Pipe & Fittings	CREW	DAILY OUTPUT	UNIT	BARE COSTS MAT.	BARE COSTS INST.	BARE COSTS TOTAL	TOTAL INCL O&P
57 841	3"			Ea.	51		51	56M
843	4"				76		76	84M
844	5"				92		92	100M
845	6"				125		125	140M
846	8"				165		165	180M
847	10"				240		240	265M
848	12"				370		370	405M
854	Wye, lateral 2"				43		43	47M
855	2-1/2"				48		48	53M
856	 3"				54		54	59M
858	4"				82		82	90M
859	5"				105		105	115M
860	6"				135		135	150M
861	8"				200		200	220M
862	10"				280		280	310M
863	12"				380		380	420M
869	Cross, 2"				46		46	51M
870	2-1/2"				53		53	58M
871	3"				61		61	67M
873	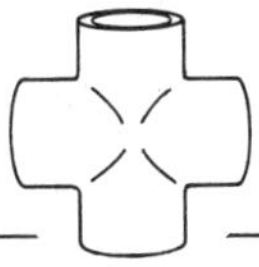 4"				84		84	92M
874	5"				120		120	130M
875	6"				155		155	170M
876	8"				200		200	220M
877	10"				295		295	325M
878	12"				425		425	470M
880	Tees, reducing 2" x 2" x 1"				24		24	26M
881	2" x 2" x 1 1/2"				24		24	26M
882	3" x 3" x 1"				30		30	33M
883	3" x 3" x 1 1/2"				30		30	33M
884	3" x 3" x 2"				30		30	33M
885	4" x 4" x 1"				44		44	48M
886	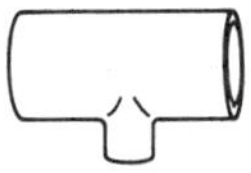 4" x 4" x 1-1/2"				44		44	48M
887	4" x 4" x 2"				44		44	48M
888	4" x 4" x 2-1/2"				46		46	51M
889	4" x 4" x 3"				46		46	51M
890	6" x 6" x 2"				74		74	81M
891	6" x 6" x 3"				78		78	86M
892	6" x 6" x 4"				78		78	86M
893	8" x 8" x 2"				95		95	105M
894	8" x 8" x 3"				100		100	110M
895	8" x 8" x 4"				100		100	110M
896	8" x 8" x 5"				105		105	115M
897	8" x 8" x 6"				105		105	115M
898	10" x 10" x 4"				150		150	165M
899	10" x 10" x 6"				160		160	175M
900	10" x 10" x 8"				160		160	175M
901	12" x 12" x 6"				230		230	255M
902	12" x 12" x 8"				230		230	255M
903	12" x 12" x 10"				240		240	265M
908	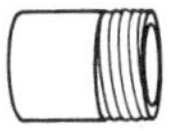 Adapter nipples 3" long							
909	1"			Ea.	2.65		2.65	2.91M
910	1-1/2"				3.20		3.20	3.52M
911	 2"				4		4	4.40M
912	2-1/2"				4.80		4.80	5.30M
913	3"				6		6	6.60M
914	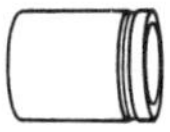 4"				10		10	11M
915	6"				27		27	30M
918	Coupling, mechanical, plain end pipe to plain end pipe or fitting							
919	1"	Q-1	29	Ea.	7.20	11	18.20	24
920	1-1/2"	"	28	"	15.35	11.40	26.75	33

15.1 Pipe & Fittings

		CREW	DAILY OUTPUT	UNIT	BARE COSTS MAT.	BARE COSTS INST.	BARE COSTS TOTAL	TOTAL INCL O&P
57 921	2"	Q-1	27	Ea.	15.35	11.85	27.20	34
922	2-1/2"		26		17	12.30	29.30	36
923	3"		25		25	12.80	37.80	46
924	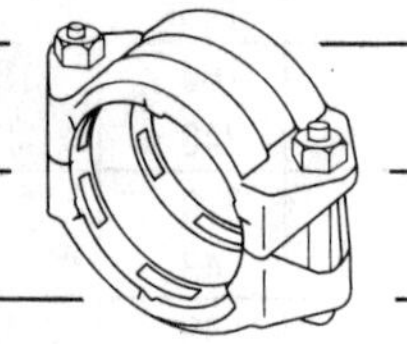 3-1/2"		24		29	13.30	42.30	51
925	4"		22		29	14.55	43.55	53
926	5"	Q-2	28		42	17.75	59.75	72
927	6"		24		51	21	72	86
928	8"		19		91	26	117	140
929	10"		16		120	31	151	175
930	12"		12		150	41	191	225
940	Mechanical joint ends for plain end pipe							
941	Malleable iron, black 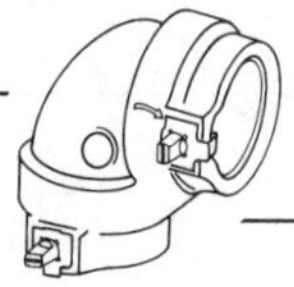							
942	90° Elbow, 1-1/4"	Q-1	29	Ea.	8.05	11	19.05	25
943	1-1/2"		27		8.85	11.85	20.70	27
944	2"		24		10.50	13.30	23.80	31
949	Tee, reducing outlet							
951	1-1/4" x 1-1/4" x 1/2"	Q-1	18	Ea.	6.40	17.75	24.15	33
952	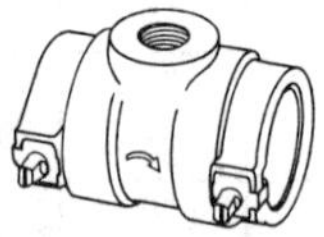 1-1/4" x 1-1/4" x 3/4"		18		6.40	17.75	24.15	33
953	1-1/4" x 1-1/4" x 1"		18		6.40	17.75	24.15	33
954	1-1/2" x 1-1/2" x 1/2"		17		6.80	18.80	25.60	35
955	1-1/2" x 1-1/2" x 3/4"		17		6.80	18.80	25.60	35
956	1-1/2" x 1-1/2" x 1"		17		6.80	18.80	25.60	35
957	2" x 2" x 1/2"		15		8.35	21	29.35	40
958	2" x 2" x 3/4"		15		8.35	21	29.35	40
959	2" x 2" x 1"		15		8.35	21	29.35	40
964	Tee, reducing run and outlet							
966	1-1/4" x 1" x 1/2"	Q-1	18	Ea.	6.90	17.75	24.65	33
967	1-1/4" x 1" x 3/4"		18		6.90	17.75	24.65	33
968	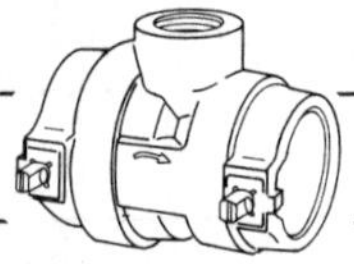 1-1/4" x 1" x 1"		18		6.90	17.75	24.65	33
969	1-1/2" x 1-1/4" x 1/2"		17		7.30	18.80	26.10	35
970	1-1/2" x 1-1/4" x 3/4"		17		7.30	18.80	26.10	35
971	1-1/2" x 1-1/4" x 1"		17		7.30	18.80	26.10	35
972	2" x 1-1/2" x 1/2"		15		8.85	21	29.85	41
973	2" x 1-1/2" x 3/4"		15		8.85	21	29.85	41
974	2" x 1-1/2" x 1"		15		8.85	21	29.85	41
979	Tee, outlet							
981	 3" x 1-1/4"	Q-1	29	Ea.	15.10	11	26.10	33
982	3" x 1-1/2"		28		15.10	11.40	26.50	33
983	3" x 2"		26		17.05	12.30	29.35	37
984	4" x 1-1/4"		28		19	11.40	30.40	37
985	4" x 1-1/2"		26		20	12.30	32.30	40
986	4" x 2"		24		20	13.30	33.30	41
994	For galvanized fittings for plain end pipe, add				14%			
61-001	**PIPE, GROOVED-JOINT STEEL FITTINGS & VALVES**							
002	Pipe includes coupling & clevis type hanger 10' O.C.							
050	Schedule 10, black							
055	2" diameter	1 Plum	43	L.F.	3.68	4.13	7.81	10
056	2-1/2" diameter	Q-1	61		4.50	5.25	9.75	12.50
057	3" diameter		55		5.16	5.80	10.96	14.10
058	3-1/2" diameter		53		6.68	6.05	12.73	16.05
059	4" diameter		49		7.16	6.50	13.66	17.30
060	5" diameter		40		10.55	8	18.55	23
061	6" diameter	Q-2	46		12.38	10.80	23.18	29
062	8" diameter		41		19.99	12.15	32.14	40
063	10" diameter		34		27.28	14.65	41.93	51
064	12" diameter		30		31.22	16.60	47.82	58
070	To delete couplings & hangers, subtract							
071	2" diam. to 5" diam.				25%	20%		
072	6" diam. to 12" diam.				27%	15%		

15

15.1 Pipe & Fittings

	15.1 Pipe & Fittings	CREW	DAILY OUTPUT	UNIT	BARE COSTS MAT.	BARE COSTS INST.	BARE COSTS TOTAL	TOTAL INCL O&P
61 100	Schedule 40, black							
104	3/4" diameter	1 Plum	71	L.F.	1.58	2.50	4.08	5.35
105	1" diameter		63		1.91	2.82	4.73	6.15
106	1-1/4" diameter		58		2.44	3.06	5.50	7.10
107	1-1/2" diameter		51		2.82	3.48	6.30	8.15
108	2" diameter		40		3.56	4.44	8	10.35
109	2-1/2" diameter	Q-1	57		5.23	5.60	10.83	13.85
110	3" diameter		50		6.59	6.40	12.99	16.50
111	4" diameter		45		9.52	7.10	16.62	21
112	5" diameter		37		24.57	8.65	33.22	40
113	6" diameter	Q-2	42		26.18	11.85	38.03	46
114	8" diameter		37		34.51	13.45	47.96	57
115	10" diameter		31		53.40	16.05	69.45	82
116	12" diameter		27		59.82	18.40	78.22	92
174	To delete coupling & hanger, subtract							
175	3/4" diam. to 2" diam.				35%	27%		
176	2-1/2" diam. to 5" diam.				18%	18%		
177	6" diam. to 12" diam.				14%	13%		
180	Galvanized							
184	3/4" diameter	1 Plum	71	L.F.	1.86	2.50	4.36	5.65
185	1" diameter		63		2.22	2.82	5.04	6.50
186	1-1/4" diameter		58		2.84	3.06	5.90	7.55
187	1-1/2" diameter		51		3.26	3.48	6.74	8.60
188	2" diameter		40		4.14	4.44	8.58	10.95
189	2-1/2" diameter	Q-1	57		6.06	5.60	11.66	14.75
190	3" diameter		50		7.63	6.40	14.03	17.65
191	4" diameter		45		11.03	7.10	18.13	22
192	5" diameter		37		27.41	8.65	36.06	43
193	6" diameter	Q-2	42		31.24	11.85	43.09	51
194	8" diameter		37		42.32	13.45	55.77	66
195	10" diameter		31		63.74	16.05	79.79	93
196	12" diameter		27		71.48	18.40	89.88	105
254	To delete coupling & hanger, subtract							
255	3/4" diam. to 2" diam.				36%	27%		
256	2-1/2" diam. to 5" diam.				19%	18%		
257	6" diam. to 12" diam.				14%	13%		
260	Schedule 80, black							
261	3/4" diameter	1 Plum	65	L.F.	1.76	2.73	4.49	5.90
265	1" diameter		61		2.25	2.91	5.16	6.70
266	1-1/4" diameter		55		2.97	3.23	6.20	7.95
267	1-1/2" diameter		49		3.46	3.62	7.08	9.05
268	2" diameter		38		4.51	4.67	9.18	11.70
269	2-1/2" diameter	Q-1	54		6.76	5.90	12.66	16
270	3" diameter		48		8.73	6.65	15.38	19.25
271	4" diameter		44		12.82	7.25	20.07	25
272	5" diameter		35		28.97	9.15	38.12	45
273	6" diameter	Q-2	40		31.68	12.45	44.13	53
274	8" diameter		35		48.81	14.20	63.01	74
275	10" diameter		29		62.20	17.15	79.35	93
276	12" diameter		24		74.12	21	95.12	110
324	To delete coupling & hanger, subtract							
325	3/4" diam. to 2" diam.				30%	25%		
326	2-1/2" diam. to 5" diam.				14%	17%		
327	6" diam. to 12" diam.				12%	12%		
330	Galvanized							
331	3/4" diameter	1 Plum	65	L.F.	2.02	2.73	4.75	6.15
335	1" diameter		61		2.60	2.91	5.51	7.05
336	1-1/4" diameter		55		3.42	3.23	6.65	8.45
337	1-1/2" diameter		46		3.97	3.86	7.83	9.95
338	2" diameter		38		5.19	4.67	9.86	12.45

	15.1 Pipe & Fittings	CREW	DAILY OUTPUT	UNIT	BARE COSTS MAT.	BARE COSTS INST.	BARE COSTS TOTAL	TOTAL INCL O&P
61 339	2-1/2" diameter	Q-1	54	L.F.	7.78	5.90	13.68	17.10
340	3" diameter		48		10.05	6.65	16.70	21
341	4" diameter		44		14.82	7.25	22.07	27
342	5" diameter		35		35.11	9.15	44.26	52
343	6" diameter	Q-2	40		42.24	12.45	54.69	64
344	8" diameter		35		63.22	14.20	77.42	90
345	10" diameter		29		81.34	17.15	98.49	115
346	12" diameter		24		96.78	21	117.78	135
392	To delete coupling & hanger, subtract							
393	3/4" diam. to 2" diam.				30%	25%		
394	2-1/2" diam. to 5" diam.				15%	17%		
395	6" diam. to 12" diam.				11%	12%		
400	Elbow, 90° or 45°, black steel							
403	3/4" diameter	1 Plum	50	Ea.	4.17	3.55	7.72	9.70
404	1" diameter		50		4.17	3.55	7.72	9.70
405	1-1/4" diameter		40		5.55	4.44	9.99	12.50
406	1-1/2" diameter		33		5.95	5.40	11.35	14.30
407	2" diameter		25		5.95	7.10	13.05	16.80
408	2-1/2" diameter	Q-1	40		8.05	8	16.05	20
409	3" diameter		33		10.70	9.70	20.40	26
410	4" diameter		25		15.75	12.80	28.55	36
411	5" diameter		20		38	16	54	65
412	6" diameter	Q-2	25		45	19.90	64.90	78
413	8" diameter		21		94	24	118	140
414	10" diameter		18		145	28	173	200
415	12" diameter		15		235	33	268	305
416	For galvanized elbows, add				15%			
469	Tee, black steel							
470	3/4" diameter	1 Plum	38	Ea.	5.90	4.67	10.57	13.25
474	1" diameter		33		5.90	5.40	11.30	14.25
475	1-1/4" diameter		27		6.90	6.60	13.50	17.10
476	1-1/2" diameter		22		7.40	8.05	15.45	19.80
477	2" diameter		17		9.20	10.45	19.65	25
478	2-1/2" diameter	Q-1	27		12.50	11.85	24.35	31
479	3" diameter		22		17.55	14.55	32.10	40
480	4" diameter		17		27	18.80	45.80	57
481	5" diameter		13		63	25	88	105
482	6" diameter	Q-2	17		73	29	102	125
483	8" diameter		14		160	36	196	225
484	10" diameter		12		285	41	326	375
485	12" diameter		10		400	50	450	510
490	For galvanized tees, add				15%			
494	Couplings, standard, black steel							
495	3/4" diameter	1 Plum	100	Ea.	4.14	1.78	5.92	7.10
496	1" diameter		100		4.14	1.78	5.92	7.10
497	1-1/4" diameter		80		5.55	2.22	7.77	9.30
498	1-1/2" diameter		67		6.10	2.65	8.75	10.55
499	2" diameter		50		6.40	3.55	9.95	12.15
500	2-1/2" diameter	Q-1	80		7.75	4	11.75	14.30
501	3" diameter		67		8.60	4.77	13.37	16.35
502	3-1/2" diameter		57		12.50	5.60	18.10	22
503	4" diameter		50		12.50	6.40	18.90	23
504	5" diameter		40		19.25	8	27.25	33
505	6" diameter	Q-2	50		23	9.95	32.95	40
507	8" diameter		42		37	11.85	48.85	58
509	10" diameter		35		53	14.20	67.20	79
511	12" diameter		32		60	15.55	75.55	88
520	For galvanized couplings, add				26%			
575	Flange, w/groove gasket, black steel(see 15.1-57-062 for bolt sets)							
578	2" pipe size	1 Plum	23	Ea.	24	7.70	31.70	38

15

15.1 Pipe & Fittings

61		CREW	DAILY OUTPUT	UNIT	BARE COSTS MAT.	BARE COSTS INST.	BARE COSTS TOTAL	TOTAL INCL O&P
579	2-1/2" pipe size	Q-1	37	Ea.	30	8.65	38.65	45
580	3" pipe size		31		33	10.30	43.30	51
582	4" pipe size		23		44	13.90	57.90	68
583	5" pipe size		19		51	16.85	67.85	80
584	6" pipe size	Q-2	23		56	22	78	93
585	8" pipe size		17		63	29	92	110
586	10" pipe size		14		100	36	136	160
587	12" pipe size		12		125	41	166	195
800	Butterfly valve, with standard trim							
801	1-1/2" pipe size	Q-1	35	Ea.	44	9.15	53.15	62
802	2" pipe size		34		44	9.40	53.40	62
803	3" pipe size		33		72	9.70	81.70	93
805	4" pipe size		25		97	12.80	109.80	125
807	6" pipe size	Q-2	25		215	19.90	234.90	265
808	8" pipe size		21		365	24	389	435
809	10" pipe size		18		640	28	668	745
820	With stainless steel trim							
824	1-1/2" pipe size	Q-1	35	Ea.	67	9.15	76.15	87
825	2" pipe size		34		67	9.40	76.40	87
827	3" pipe size		33		105	9.70	114.70	130
828	4" pipe size		25		130	12.80	142.80	160
830	6" pipe size	Q-2	25		255	19.90	274.90	310
831	8" pipe size		21		535	24	559	625
832	10" pipe size		18		730	28	758	845
840	Gate valve, iron body, bronze trim, NRS, class 125							
850	3" pipe size	Q-1	33	Ea.	325	9.70	334.70	370
851	4" pipe size	"	25		410	12.80	422.80	470
853	6" pipe size	Q-2	25		730	19.90	749.90	830
854	8" pipe size		21		1,175	24	1,199	1,325
855	10" pipe size		18		1,975	28	2,003	2,200
900	Cut one groove, labor							
901	3/4" pipe size	Q-1	152	Ea.		2.10	2.10	3.04L
902	1" pipe size		140			2.28	2.28	3.30L
903	1-1/4" pipe size		124			2.58	2.58	3.73L
904	1-1/2" pipe size		114			2.80	2.80	4.05L
905	2" pipe size		104			3.07	3.07	4.44L
906	2-1/2" pipe size		96			3.33	3.33	4.81L
907	3" pipe size		88			3.63	3.63	5.25L
908	3-1/2" pipe size		83			3.85	3.85	5.55L
909	4" pipe size		78			4.10	4.10	5.90L
910	5" pipe size		72			4.44	4.44	6.40L
911	6" pipe size		70			4.57	4.57	6.60L
912	8" pipe size		54			5.90	5.90	8.55L
913	10" pipe size		38			8.40	8.40	12.15L
914	12" pipe size		30			10.65	10.65	15.40L
921	Roll one groove, labor							
922	3/4" pipe size	Q-1	266	Ea.		1.20	1.20	1.74L
923	1" pipe size		228			1.40	1.40	2.03L
924	1-1/4" pipe size		200			1.60	1.60	2.31L
925	1-1/2" pipe size		178			1.80	1.80	2.60L
926	2" pipe size		116			2.76	2.76	3.98L
927	2-1/2" pipe size		110			2.91	2.91	4.20L
928	3" pipe size		100			3.20	3.20	4.62L
929	3-1/2" pipe size		94			3.40	3.40	4.91L
930	4" pipe size		86			3.72	3.72	5.35L
931	5" pipe size		84			3.81	3.81	5.50L
932	6" pipe size		80			4	4	5.80L
933	8" pipe size		66			4.84	4.84	7L
934	10" pipe size		58			5.50	5.50	7.95L
935	12" pipe size		46			6.95	6.95	10.05L

15.1 Pipe & Fittings

		CREW	DAILY OUTPUT	UNIT	BARE COSTS MAT.	BARE COSTS INST.	BARE COSTS TOTAL	TOTAL INCL O&P
64-001	PIPE CONDUIT, PREFABRICATED							
002	Does not include trenching, hangers, fittings or crane.							
030	For cathodic protection, add 12 to 14%							
031	of total built-up price (casing plus service pipe)							
058	Polyurethane insulated system, 250°F. max. temp.							
062	Black steel service pipe, standard wt., 1/2" insulation							
066	3/4" diam. pipe size	Q-17	54	L.F.	8.20	6.30	14.50	18
067	1" diam. pipe size		50		9	6.80	15.80	19.60
068	1-1/4" diam. pipe size		47		10	7.20	17.20	21
069	1-1/2" diam. pipe size		45		10.85	7.55	18.40	23
070	2" diam. pipe size		42		10.35	8.10	18.45	23
071	2-1/2" diam. pipe size		34		11.50	10	21.50	27
072	3" diam. pipe size		28		13.30	12.10	25.40	32
073	4" diam. pipe size		22		16.55	15.40	31.95	40
074	5" diam. pipe size		18		21	18.85	39.85	50
075	6" diam. pipe size	Q-18	23		25	23	48	60
076	8" diam. pipe size		19		36	27	63	79
077	10" diam. pipe size		16		48	32	80	99
078	12" diam. pipe size		13		59	40	99	120
079	14" diam. pipe size		11		67	47	114	140
080	16" diam. pipe size		10		76	52	128	160
081	18" diam. pipe size		8		88	65	153	190
082	20" diam. pipe size		7		96	74	170	210
083	24" diam. pipe size		6		120	86	206	255
090	For 1" thick insulation, add				10%			
094	For 1-1/2" thick insulation, add				13%			
098	For 2" thick insulation, add				20%			
150	Gland seal for system, 3/4" diam. pipe size	Q-17	32	Ea.	135	10.60	145.60	165
151	1" diam. pipe size		32		135	10.60	145.60	165
154	1-1/4" diam. pipe size		30		145	11.30	156.30	175
155	1-1/2" diam. pipe size		30		145	11.30	156.30	175
156	2" diam. pipe size		28		165	12.10	177.10	200
157	2-1/2" diam. pipe size		26		180	13.05	193.05	215
158	3" diam. pipe size		24		195	14.15	209.15	235
159	4" diam. pipe size		22		230	15.40	245.40	275
160	5" diam. pipe size		19		280	17.85	297.85	335
161	6" diam. pipe size	Q-18	26		300	19.90	319.90	360
162	8" diam. pipe size		25		355	21	376	420
163	10" diam. pipe size		23		425	23	448	500
164	12" diam. pipe size		21		475	25	500	560
165	14" diam. pipe size		19		525	27	552	615
166	16" diam. pipe size		18		615	29	644	720
167	18" diam. pipe size		16		660	32	692	770
168	20" diam. pipe size		14		750	37	787	880
169	24" diam. pipe size		12		835	43	878	980
200	Elbow, 45° for system							
202	3/4" diam. pipe size	Q-17	14	Ea.	83	24	107	125
204	1" diam. pipe size		13		86	26	112	130
205	1-1/4" diam. pipe size		11		99	31	130	155
206	1-1/2" diam. pipe size		9		110	38	148	175
207	2" diam. pipe size		6		105	57	162	195
208	2-1/2" diam. pipe size		4		115	85	200	245
209	3" diam. pipe size		3.50		135	97	232	285
210	4" diam. pipe size		3		155	115	270	330
211	5" diam. pipe size		2.80		200	120	320	395
212	6" diam. pipe size	Q-18	4		225	130	355	435
213	8" diam. pipe size		3		330	175	505	610
214	10" diam. pipe size		2.40		415	215	630	765
215	12" diam. pipe size		2		550	260	810	975
216	14" diam. pipe size		1.80		685	290	975	1,175

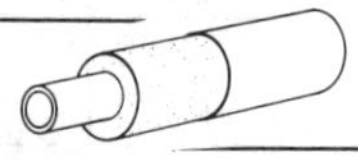

15.1 Pipe & Fittings

64

	Pipe & Fittings	CREW	DAILY OUTPUT	UNIT	BARE COSTS MAT.	BARE COSTS INST.	BARE COSTS TOTAL	TOTAL INCL O&P
217	16" diam. pipe size	Q-18	1.60	Ea.	815	325	1,140	1,350
218	18" diam. pipe size		1.30		995	400	1,395	1,675
219	20" diam. pipe size		1		1,300	520	1,820	2,175
220	24" diam. pipe size		.70		1,750	740	2,490	2,975
226	For elbow, 90°, add				25%			
230	For tee, straight, add				85%	30%		
234	For tee, reducing, add				170%	30%		
238	For weldolet, straight, add				50%			
280	Calcium silicate insulated system, high temp. (1200°F)							
284	Steel casing with protective exterior coating							
285	6-5/8" diameter	Q-18	52	L.F.	15.25	9.95	25.20	31
286	8-5/8" diameter		50		16.80	10.35	27.15	33
287	10-3/4" diameter		47		19.25	11	30.25	37
288	12-3/4" diameter		44		21	11.75	32.75	40
289	14" diameter		41		24	12.65	36.65	44
290	16" diameter		39		26	13.25	39.25	48
291	18" diameter		36		29	14.40	43.40	53
292	20" diameter		34		31	15.25	46.25	56
293	22" diameter		32		45	16.20	61.20	73
294	24" diameter		29		51	17.85	68.85	82
295	26" diameter		26		58	19.90	77.90	92
296	28" diameter		23		72	23	95	110
297	30" diameter		21		77	25	102	120
298	32" diameter		19		86	27	113	135
299	34" diameter		18		87	29	116	135
300	36" diameter		16		93	32	125	150
304	For multi-pipe casings, add				10%			
306	For oversize casings, add				2%			
340	Steel casing gland seal, single pipe							
342	6-5/8" diameter	Q-18	25	Ea.	215	21	236	265
344	8-5/8" diameter		23		250	23	273	305
345	10-3/4" diameter		21		280	25	305	345
346	12-3/4" diameter		19		335	27	362	410
347	14" diameter		17		365	30	395	445
348	16" diameter		16		425	32	457	515
349	18" diameter		15		480	35	515	575
350	20" diameter		13		525	40	565	635
351	22" diameter		12		595	43	638	715
352	24" diameter		11		660	47	707	795
353	26" diameter		10		750	52	802	900
354	28" diameter		9.50		975	54	1,029	1,150
355	30" diameter		9		870	58	928	1,050
356	32" diameter		8.50		980	61	1,041	1,175
357	34" diameter		8		1,075	65	1,140	1,275
358	36" diameter		7		1,125	74	1,199	1,350
362	For multi-pipe casings, add				5%			
400	Steel casing anchors, single pipe							
402	6-5/8" diameter	Q-18	8	Ea.	175	65	240	285
404	8-5/8" diameter		7.50		200	69	269	320
405	10-3/4" diameter		7		265	74	339	395
406	12-3/4" diameter		6.50		280	80	360	420
407	14" diameter		6		330	86	416	485
408	16" diameter		5.50		385	94	479	560
409	18" diameter		5		430	105	535	620
410	20" diameter		4.50		480	115	595	695
411	22" diameter		4		525	130	655	765
412	24" diameter		3.50		580	150	730	850
413	26" diameter		3		650	175	825	960
414	28" diameter		2.50		720	205	925	1,100
415	30" diameter		2		760	260	1,020	1,200

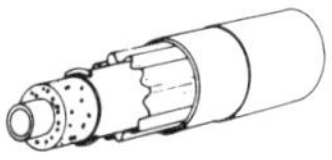

15

15.1 Pipe & Fittings

64

		CREW	DAILY OUTPUT	UNIT	BARE COSTS MAT.	BARE COSTS INST.	BARE COSTS TOTAL	TOTAL INCL O&P
416	32" diameter	Q-18	1.50	Ea.	910	345	1,255	1,500
417	34" diameter		1		1,025	520	1,545	1,875
418	36" diameter		1		1,100	520	1,620	1,950
422	For multi-pipe, add				5%	20%		
480	Steel casing elbow							
482	6-5/8" diameter	Q-18	15	Ea.	280	35	315	355
483	8-5/8" diameter		15		280	35	315	355
485	10-3/4" diameter		14		340	37	377	425
486	12-3/4" diameter		13		395	40	435	490
487	14" diameter		12		425	43	468	530
488	16" diameter		11		470	47	517	585
489	18" diameter		10		530	52	582	655
490	20" diameter		9		570	58	628	710
491	22" diameter		8		615	65	680	770
492	24" diameter		7		675	74	749	850
493	26" diameter		6		740	86	826	940
494	28" diameter		5		795	105	900	1,025
495	30" diameter		4		800	130	930	1,075
496	32" diameter		3		920	175	1,095	1,250
497	34" diameter		2		995	260	1,255	1,475
498	36" diameter		2		1,075	260	1,335	1,550
550	Black steel service pipe, std. wt., 1" thick insulation							
551	3/4" diameter pipe size	Q-17	54	L.F.	6.70	6.30	13	16.35
554	1" diameter pipe size		50		7	6.80	13.80	17.40
555	1-1/4" diameter pipe size		47		7.80	7.20	15	18.90
556	1-1/2" diameter pipe size		45		8.55	7.55	16.10	20
557	2" diameter pipe size		42		8.05	8.10	16.15	20
558	2-1/2" diameter pipe size		34		9.95	10	19.95	25
559	3" diameter pipe size		28		11.35	12.10	23.45	30
560	4" diameter pipe size		22		14.50	15.40	29.90	38
561	5" diameter pipe size		18		19.40	18.85	38.25	48
562	6" diameter pipe size	Q-18	23		21	23	44	55
600	Black steel service pipe, std. wt., 1-1/2" thick insul.							
601	3/4" diameter pipe size	Q-17	54	L.F.	6.80	6.30	13.10	16.45
604	1" diameter pipe size		50		7.60	6.80	14.40	18.05
605	1-1/4" diameter pipe size		47		8.55	7.20	15.75	19.70
606	1-1/2" diameter pipe size		45		9.30	7.55	16.85	21
607	2" diameter pipe size		42		8.75	8.10	16.85	21
608	2-1/2" diameter pipe size		34		10.80	10	20.80	26
609	3" diameter pipe size		28		12.10	12.10	24.20	31
610	4" diameter pipe size		22		15.40	15.40	30.80	39
611	5" diameter pipe size		18		20	18.85	38.85	49
612	6" diameter pipe size	Q-18	23		22	23	45	56
613	8" diameter pipe size		19		31	27	58	73
614	10" diameter pipe size		16		42	32	74	93
615	12" diameter pipe size		13		50	40	90	110
619	For 2" thick insulation, add				15%			
622	For 2-1/2" thick insulation, add				25%			
626	For 3" thick insulation, add				30%			
680	Black steel service pipe, ex. hvy. wt., 1" thick insul.							
682	3/4" diameter pipe size	Q-17	50	L.F.	7.10	6.80	13.90	17.50
684	1" diameter pipe size		47		7.45	7.20	14.65	18.50
685	1-1/4" diameter pipe size		44		8.60	7.70	16.30	20
686	1-1/2" diameter pipe size		42		9.25	8.10	17.35	22
687	2" diameter pipe size		40		8.75	8.50	17.25	22
688	2-1/2" diameter pipe size		31		12.05	10.95	23	29
689	3" diameter pipe size		27		13.60	12.55	26.15	33
690	4" diameter pipe size		21		17.80	16.15	33.95	43
691	5" diameter pipe size		17		25	19.95	44.95	56
692	6" diameter pipe size	Q-18	22		28	24	52	65

15

15.1 Pipe & Fittings

		CREW	DAILY OUTPUT	UNIT	BARE COSTS MAT.	BARE COSTS INST.	BARE COSTS TOTAL	TOTAL INCL O&P
64 740	Black steel service pipe, ex. hvy. wt., 1-1/2" thick insul.							
742	3/4" diameter pipe size	Q-17	50	L.F.	7.10	6.80	13.90	17.50
744	1" diameter pipe size		47		8.05	7.20	15.25	19.15
745	1-1/4" diameter pipe size		44		9.35	7.70	17.05	21
746	1-1/2" diameter pipe size		42		10.25	8.10	18.35	23
747	2" diameter pipe size		40		9.50	8.50	18	23
748	2-1/2" diameter pipe size		31		12.20	10.95	23.15	29
749	3" diameter pipe size		27		14.35	12.55	26.90	34
750	4" diameter pipe size		21		18.80	16.15	34.95	44
751	5" diameter pipe size		17		26	19.95	45.95	57
752	6" diameter pipe size	Q-18	22		29	24	53	66
753	8" diameter pipe size		18		43	29	72	89
754	10" diameter pipe size		15		51	35	86	105
755	12" diameter pipe size		13		62	40	102	125
759	For 2" thick insulation, add				13%			
764	For 2-1/2" thick insulation, add				18%			
768	For 3" thick insulation, add				24%			
65-001	**PIPE HANGERS AND SUPPORTS**							
005	Brackets							
006	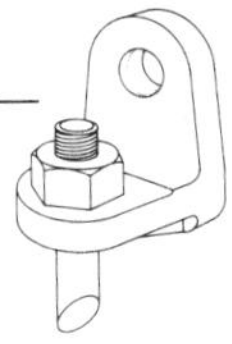 Beam side or wall, malleable iron							
007	3/8" threaded rod size	1 Plum	48	Ea.	.75	3.70	4.45	6.15
008	1/2" threaded rod size		48		1.45	3.70	5.15	6.95
009	5/8" threaded rod size		48		1.94	3.70	5.64	7.50
010	3/4" threaded rod size		48		2.99	3.70	6.69	8.65
011	7/8" threaded rod size		48		4.19	3.70	7.89	9.95
012	For concrete installation, add					30%		
015	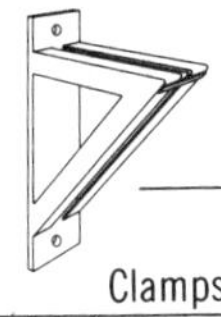 Wall, welded steel							
016	0 size, 12" wide, 18" deep	1 Plum	34	Ea.	47	5.20	52.20	59
017	1 size, 18" wide 24" deep		34		56	5.20	61.20	69
018	2 size, 24" wide, 30" deep		34		74	5.20	79.20	89
030	Clamps							
031	 C-clamp, for mounting on steel beam flange, w/locknut							
032	3/8" threaded rod size	1 Plum	160	Ea.	.92	1.11	2.03	2.62
033	1/2" threaded rod size		160		1.16	1.11	2.27	2.88
034	5/8" threaded rod size		160		1.79	1.11	2.90	3.57
035	3/4" threaded rod size		160		2.43	1.11	3.54	4.28
040	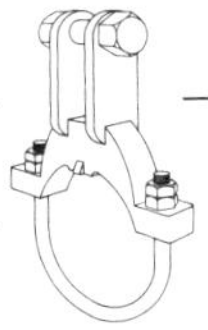 High temperature to 1050°F, alloy steel							
041	4" pipe size	Q-1	106	Ea.	38	3.02	41.02	46
042	6" pipe size		106		50	3.02	53.02	59
043	8" pipe size		97		61	3.30	64.30	72
044	10" pipe size		84		82	3.81	85.81	96
045	12" pipe size		72		135	4.44	139.44	155
046	14" pipe size		64		160	5	165	185
047	16" pipe size		56		210	5.70	215.70	240
050	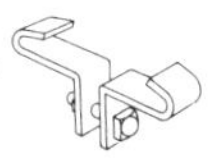 I-beam, for mounting on bottom flange, strap iron							
051	2" flange size	1 Plum	96	Ea.	1.87	1.85	3.72	4.73
052	3" flange size		95		1.94	1.87	3.81	4.83
053	4" flange size		93		1.99	1.91	3.90	4.95
054	5" flange size		92		2.16	1.93	4.09	5.15
055	6" flange size		90		2.26	1.97	4.23	5.35
056	7" flange size		88		2.35	2.02	4.37	5.50
057	8" flange size		86		2.47	2.07	4.54	5.70
060	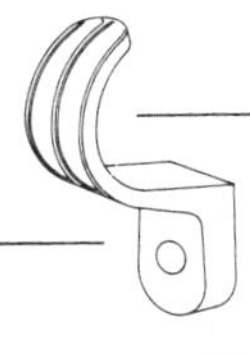 One hole, vertical mounting, malleable iron							
061	1/2" pipe size	1 Plum	160	Ea.	.28	1.11	1.39	1.91
062	3/4" pipe size		145		.35	1.22	1.57	2.15
063	1" pipe size		136		.52	1.31	1.83	2.46
064	1-1/4" pipe size		128		.69	1.39	2.08	2.76
065	1-1/2" pipe size		120		.79	1.48	2.27	3.01
066	2" pipe size		112		1.15	1.59	2.74	3.56

15.1 Pipe & Fittings

65

		CREW	DAILY OUTPUT	UNIT	BARE COSTS MAT.	BARE COSTS INST.	BARE COSTS TOTAL	TOTAL INCL O&P
067	2-1/2" pipe size	1 Plum	104	Ea.	2.16	1.71	3.87	4.84
068	3" pipe size		96		3.46	1.85	5.31	6.50
069	3-1/2" pipe size		90		4.75	1.97	6.72	8.10
070	4" pipe size		84		6.60	2.11	8.71	10.30
075	Riser or extension pipe, carbon steel							
076	3/4" pipe size	1 Plum	48	Ea.	1.32	3.70	5.02	6.80
077	1" pipe size		47		1.33	3.78	5.11	6.90
078	1-1/4" pipe size		46		1.67	3.86	5.53	7.40
079	1-1/2" pipe size		45		1.80	3.95	5.75	7.70
080	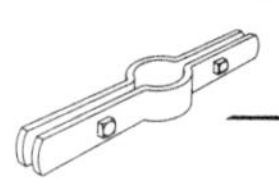 2" pipe size		43		1.83	4.13	5.96	8
081	2-1/2" pipe size		41		1.92	4.33	6.25	8.35
082	3" pipe size		40		2.04	4.44	6.48	8.65
083	3-1/2" pipe size		39		2.57	4.55	7.12	9.40
084	4" pipe size		38		2.59	4.67	7.26	9.60
085	5" pipe size		37		3.77	4.80	8.57	11.10
086	6" pipe size		36		4.37	4.93	9.30	11.95
087	8" pipe size		34		7.05	5.20	12.25	15.30
088	10" pipe size		32		9.75	5.55	15.30	18.75
089	12" pipe size		28		11.75	6.35	18.10	22
090	For plastic coating 3/4" to 4", add				190%			
091	For copper plating 3/4" to 4", add				58%			
095	Two piece, complete, carbon steel, medium weight							
096	1/2" pipe size	Q-1	137	Ea.	.85	2.33	3.18	4.31
097	3/4" pipe size		134		.90	2.39	3.29	4.44
098	1" pipe size		132		.97	2.42	3.39	4.57
099	1-1/4" pipe size		130		1.08	2.46	3.54	4.74
100	1-1/2" pipe size		126		1.35	2.54	3.89	5.15
101	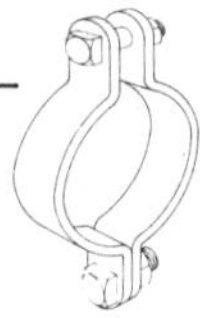 2" pipe size		124		1.43	2.58	4.01	5.30
102	2-1/2" pipe size		120		1.52	2.66	4.18	5.50
103	3" pipe size		117		1.72	2.73	4.45	5.85
104	3-1/2" pipe size		114		1.99	2.80	4.79	6.25
105	4" pipe size		110		2.37	2.91	5.28	6.80
106	5" pipe size		106		3.87	3.02	6.89	8.60
107	6" pipe size		104		5.95	3.07	9.02	11
108	8" pipe size		100		7.10	3.20	10.30	12.45
109	10" pipe size		96		12.80	3.33	16.13	18.90
110	12" pipe size		89		14.40	3.59	17.99	21
111	14" pipe size		82		23	3.90	26.90	31
112	16" pipe size		68		25	4.70	29.70	34
113	For galvanized, add				45%			
115	Insert, concrete							
116	Wedge type, carbon steel body, malleable iron nut							
117	1/4" threaded rod size	1 Plum	96	Ea.	1.01	1.85	2.86	3.78
118	3/8" threaded rod size		96		1.03	1.85	2.88	3.81
119	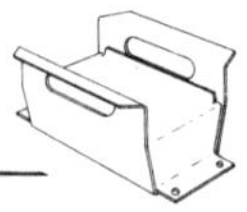 1/2" threaded rod size		96		1.08	1.85	2.93	3.86
120	5/8" threaded rod size		96		1.14	1.85	2.99	3.93
121	3/4" threaded rod size		96		1.20	1.85	3.05	3.99
122	7/8" threaded rod size		96		2.57	1.85	4.42	5.50
123	For galvanized, add				.80		.80	.88M
125	Pipe guide sized for insulation							
126	No. 1, 1" pipe size, 1" thick insulation	1 Stpi	26	Ea.	50	6.85	56.85	65
127	No. 2, 1-1/4"-2" pipe size, 1" thick insulation		23		58	7.75	65.75	75
128	No. 3, 1-1/4"-2" pipe size, 1-1/2" thick insulation		21		66	8.50	74.50	85
129	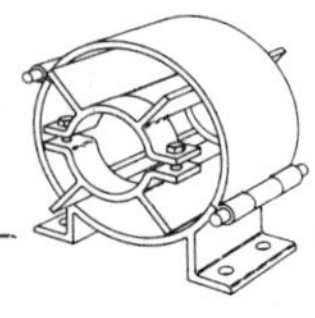 No. 4, 2-1/2"-3-1/2" pipe size, 1-1/2" thick insul.		18		93	9.90	102.90	115
130	No. 5, 4"-5" pipe size, 1-1/2" thick insulation		16		120	11.15	131.15	150
131	No. 6, 5"-6" pipe size, 2" thick insulation	Q-5	21		160	15.30	175.30	200
132	No. 7, 8" pipe size, 2" thick insulation		16		205	20	225	255
133	No. 8, 10" pipe size, 2" thick insulation		12		255	27	282	320
134	No. 9, 12" pipe size, 2" thick insulation	Q-6	17		300	29	329	370
135	No. 10, 12"-14" pipe size, 2-1/2" thick insulation	"	16		375	31	406	460

15

15.1 Pipe & Fittings

		CREW	DAILY OUTPUT	UNIT	BARE COSTS MAT.	BARE COSTS INST.	BARE COSTS TOTAL	TOTAL INCL O&P
65 136	No. 11, 16" pipe size, 2-1/2" thick insulation	Q-6	10.50	Ea.	430	48	478	540
137	No. 12, 16"-18" pipe size, 3" thick insulation		9		470	56	526	595
138	No. 13, 20" pipe size, 3" thick insulation		7.50		495	67	562	640
139	No. 14, 24" pipe size, 3" thick insulation	↓	7	↓	635	71	706	800
140	Rings							
141	Adjustable band, carbon steel, for non-insulated pipe							
142	1/2" pipe size	Q-1	142	Ea.	.38	2.25	2.63	3.67
143	3/4" pipe size		140		.38	2.28	2.66	3.72
144	1" pipe size		137		.42	2.33	2.75	3.83
145	 1-1/4" pipe size		134		.45	2.39	2.84	3.94
146	1-1/2" pipe size		131		.50	2.44	2.94	4.08
147	2" pipe size		129		.67	2.48	3.15	4.32
148	2-1/2" pipe size		125		.73	2.56	3.29	4.50
149	3" pipe size		122		.80	2.62	3.42	4.67
150	3-1/2" pipe size		119		.90	2.69	3.59	4.87
151	4" pipe size		114		1.25	2.80	4.05	5.45
152	5" pipe size		110		1.33	2.91	4.24	5.65
153	6" pipe size		108		2.06	2.96	5.02	6.55
154	8" pipe size	↓	104		2.64	3.07	5.71	7.35
155	For copper plated, add				50%			
156	For galvanized, add				30%			
157	For plastic coating, add			↓	30%			
160	Adjusting nut malleable iron, steel band							
161	1/2" pipe size, galvanized band	Q-1	137	Ea.	.58	2.33	2.91	4.01
162	3/4" pipe size, galvanized band		135		.60	2.37	2.97	4.08
163	1" pipe size, gavanized band		132		.61	2.42	3.03	4.17
164	1-1/4" pipe size, galvanized band		129		.67	2.48	3.15	4.32
165	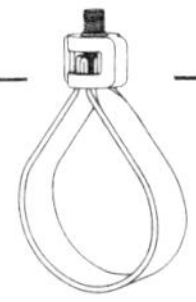 1-1/2" pipe size, galvanized band		126		.70	2.54	3.24	4.44
166	2" pipe size, galvanized band		124		.77	2.58	3.35	4.57
167	2-1/2" pipe size, galvanized band		120		1.30	2.66	3.96	5.30
168	3" pipe size, galvanized band		117		1.41	2.73	4.14	5.50
169	3-1/2" pipe size, galvanized band		114		1.56	2.80	4.36	5.75
170	4" pipe size, cadmium plated band		110		2.02	2.91	4.93	6.40
171	5" pipe size, cadmium plated band		106		2.66	3.02	5.68	7.30
172	6" pipe size, cadmium plated band		104		4.06	3.07	7.13	8.90
173	8" pipe size, cadmium plated band	↓	100		6.35	3.20	9.55	11.60
174	For plastic coated band, add				35%			
175	For completely copper coated, add			↓	45%			
180	Clevis, adjustable, carbon steel, for non-insulated pipe							
181	1/2" pipe size	Q-1	137	Ea.	.38	2.33	2.71	3.79
182	3/4" pipe size		135		.40	2.37	2.77	3.86
183	1" pipe size		132		.41	2.42	2.83	3.95
184	1-1/4" pipe size		129		.43	2.48	2.91	4.05
185	1-1/2" pipe size		126		.45	2.54	2.99	4.16
186	2" pipe size		124		.47	2.58	3.05	4.24
187	2-1/2" pipe size		120		.86	2.66	3.52	4.80
188	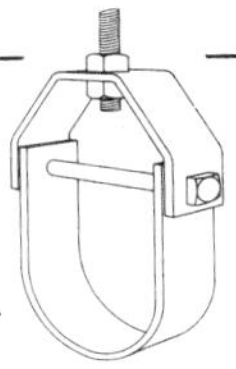 3" pipe size		117		.94	2.73	3.67	4.98
189	3-1/2" pipe size		114		1.01	2.80	3.81	5.15
190	4" pipe size		110		1.24	2.91	4.15	5.55
191	5" pipe size		106		2.33	3.02	5.35	6.90
192	6" pipe size		104		2.69	3.07	5.76	7.40
193	8" pipe size		100		4.44	3.20	7.64	9.50
194	10" pipe size		96		8	3.33	11.33	13.60
195	12" pipe size		89		9.40	3.59	12.99	15.55
196	14" pipe size		82		13.75	3.90	17.65	21
197	16" pipe size	↓	68		19.70	4.70	24.40	28
198	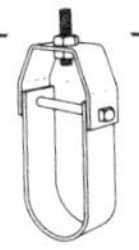 For galvanized, add				40%			
199	For copper plated 1/2" to 4", add				40%			
200	For light weight 1/2" to 4", deduct				15%			
201	Insulated pipe type, 3/4" to 12" pipe, add			↓	180%			

65

	15.1 Pipe & Fittings	CREW	DAILY OUTPUT	UNIT	BARE COSTS MAT.	BARE COSTS INST.	BARE COSTS TOTAL	TOTAL INCL O&P
202	Insulated pipe type, chrome-moly U-strap			Ea.	530%			
225	Split ring, malleable iron, for non-insulated pipe							
226	1/2" pipe size	Q-1	137	Ea.	.79	2.33	3.12	4.24
227	3/4" pipe size		135		.84	2.37	3.21	4.35
228	1" pipe size		132		.89	2.42	3.31	4.48
229	1-1/4" pipe size		129		1.10	2.48	3.58	4.79
230	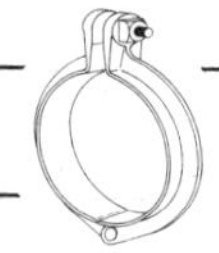 1-1/2" pipe size		126		1.32	2.54	3.86	5.10
231	2" pipe size		124		1.51	2.58	4.09	5.40
232	2-1/2" pipe size		120		2.11	2.66	4.77	6.15
233	3" pipe size		117		2.74	2.73	5.47	6.95
234	3-1/2" pipe size		114		3.79	2.80	6.59	8.20
235	4" pipe size		110		3.79	2.91	6.70	8.35
236	5" pipe size		106		6.15	3.02	9.17	11.10
237	6" pipe size		104		8.50	3.07	11.57	13.80
238	8" pipe size	↓	100		15	3.20	18.20	21
239	For copper plated, add			↓	45%			
265	Rods, carbon steel							
266	Continuous thread							
267	1/4" thread size	1 Plum	144	L.F.	.13	1.23	1.36	1.92
268	3/8" thread size		144		.18	1.23	1.41	1.98
269	1/2" thread size		144		.29	1.23	1.52	2.10
270	5/8" thread size		144		.55	1.23	1.78	2.39
271	3/4" thread size		144		.82	1.23	2.05	2.68
272	7/8" thread size	↓	144		1.32	1.23	2.55	3.23
273	For galvanized add			↓	30%			
275	Both ends machine threaded 18" length							
276	3/8" thread size	1 Plum	240	Ea.	1.06	.74	1.80	2.24
277	1/2" thread size		240		1.70	.74	2.44	2.94
278	5/8" thread size		240		2.45	.74	3.19	3.76
279	3/4" thread size		240		3.60	.74	4.34	5.05
280	7/8" thread size		240		4.41	.74	5.15	5.90
281	1" thread size	↓	240	↓	6.15	.74	6.89	7.85
290	Rolls							
291	Adjustable yoke, carbon steel with CI roll							
292	2-1/2" pipe size	Q-1	137	Ea.	4.39	2.33	6.72	8.20
293	3" pipe size		131		4.54	2.44	6.98	8.50
294	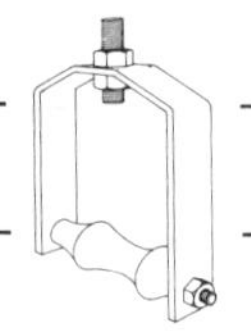 3-1/2" pipe size		124		6.40	2.58	8.98	10.75
295	4" pipe size		117		6.40	2.73	9.13	11
296	5" pipe size		110		7.50	2.91	10.41	12.45
297	6" pipe size		104		9.20	3.07	12.27	14.55
298	8" pipe size		96		14.15	3.33	17.48	20
299	10" pipe size		80		17.40	4	21.40	25
300	12" pipe size		68		24	4.70	28.70	33
301	14" pipe size		56		48	5.70	53.70	61
302	16" pipe size	↓	48	↓	63	6.65	69.65	79
305	Chair, carbon steel with CI roll							
306	2" pipe size	1 Plum	68	Ea.	4.50	2.61	7.11	8.70
307	2-1/2" pipe size		65		4.75	2.73	7.48	9.15
308	3" pipe size		62		4.78	2.86	7.64	9.40
309	3-1/2" pipe size		60		6.90	2.96	9.86	11.85
310	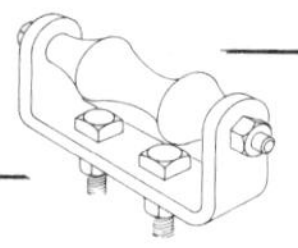 4" pipe size		58		6.90	3.06	9.96	12
311	5" pipe size		56		7.50	3.17	10.67	12.85
312	6" pipe size		53		10.40	3.35	13.75	16.30
313	8" pipe size		50		13.95	3.55	17.50	20
314	10" pipe size		48		17.70	3.70	21.40	25
315	12" pipe size	↓	46		25	3.86	28.86	33
317	Trapeze w/roller,(see line 265 for rods), 1" pipe size	Q-1	137		3.43	2.33	5.76	7.15
318	1-1/4" pipe size		131		3.43	2.44	5.87	7.30
319	1-1/2" pipe size		129		3.43	2.48	5.91	7.35
320	2" pipe size	↓	124	↓	3.58	2.58	6.16	7.65

15

15.1 Pipe & Fittings

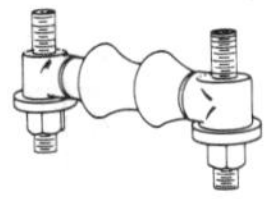

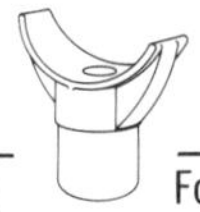

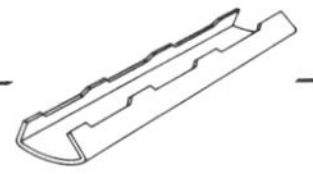
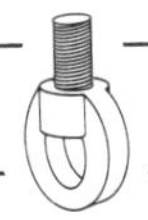

		CREW	DAILY OUTPUT	UNIT	BARE COSTS MAT.	BARE COSTS INST.	BARE COSTS TOTAL	TOTAL INCL O&P
65 321	2-1/2" pipe size	Q-1	118	Ea.	3.86	2.71	6.57	8.15
322	3" pipe size		115		4.06	2.78	6.84	8.50
323	3-1/2" pipe size		113		5.50	2.83	8.33	10.15
324	4" pipe size		112		5.50	2.85	8.35	10.20
325	5" pipe size		110		6.10	2.91	9.01	10.90
326	6" pipe size		101		8.05	3.17	11.22	13.45
327	8" pipe size		90		12.30	3.55	15.85	18.65
328	10" pipe size		80		13.80	4	17.80	21
329	12" pipe size		68		20	4.70	24.70	29
330	Saddles							
331	Pipe support, complete, adjustable, CI saddle							
332	2-1/2" pipe size	1 Plum	96	Ea.	37	1.85	38.85	43
333	3" pipe size		88		38	2.02	40.02	45
334	3-1/2" pipe size		79		38	2.25	40.25	45
335	4" pipe size		68		50	2.61	52.61	59
336	5" pipe size		64		50	2.78	52.78	59
337	6" pipe size		59		51	3.01	54.01	60
338	8" pipe size		53		54	3.35	57.35	64
339	10" pipe size		50		60	3.55	63.55	71
340	12" pipe size		48		62	3.70	65.70	74
345	For standard pipe support, one piece, CI deduct				28%			
346	For stanchion support, CI with steel yoke, deduct				65%			
375	Covering protection							
376	1" covering size							
377	3/4" pipe size	1 Plum	68	Ea.	2.98	2.61	5.59	7.05
378	1" pipe size		68		2.98	2.61	5.59	7.05
379	1-1/4" pipe size		68		2.98	2.61	5.59	7.05
380	1-1/2" pipe size		66		3.14	2.69	5.83	7.35
381	2" pipe size		66		4.15	2.69	6.84	8.45
382	2-1/2" pipe size		64		4.15	2.78	6.93	8.55
383	3" pipe size		64		4.63	2.78	7.41	9.10
384	3-1/2" pipe size		62		4.97	2.86	7.83	9.60
385	4" pipe size		62		4.97	2.86	7.83	9.60
386	5" pipe size		60		4.97	2.96	7.93	9.75
387	6" pipe size		60		5.90	2.96	8.86	10.75
390	1-1/2" covering size							
391	3/4" pipe size	1 Plum	68	Ea.	3.72	2.61	6.33	7.85
392	1" pipe size		68		3.72	2.61	6.33	7.85
393	1-1/4" pipe size		68		3.72	2.61	6.33	7.85
394	1-1/2" pipe size		66		3.72	2.69	6.41	8
395	2" pipe size		66		4.15	2.69	6.84	8.45
396	2-1/2" pipe size		64		4.63	2.78	7.41	9.10
397	3" pipe size		64		4.63	2.78	7.41	9.10
398	3-1/2" pipe size		62		4.97	2.86	7.83	9.60
399	4" pipe size		62		4.97	2.86	7.83	9.60
400	5" pipe size		60		4.97	2.96	7.93	9.75
401	6" pipe size		60		6.90	2.96	9.86	11.85
402	8" pipe size		58		8.70	3.06	11.76	14
403	For 2" covering size, add				7%			
420	Sockets							
421	Rod end, malleable iron							
422	1/4" thread size	1 Plum	240	Ea.	.35	.74	1.09	1.45
423	3/8" thread size		240		.37	.74	1.11	1.48
424	1/2" thread size		230		.45	.77	1.22	1.61
425	5/8" thread size		225		1.15	.79	1.94	2.41
426	3/4" thread size		220		1.87	.81	2.68	3.22
427	7/8" thread size		210		2.40	.85	3.25	3.86
429	Strap, 1/2" pipe size	Q-1	142		.50	2.25	2.75	3.80
430	3/4" pipe size		140		.52	2.28	2.80	3.87
431	1" pipe size		137		.53	2.33	2.86	3.96

15

Line	Description	Crew	Daily Output	Unit	Bare Costs Mat.	Bare Costs Inst.	Bare Costs Total	Total Incl O&P
65 432	1-1/4" pipe size	Q-1	134	Ea.	.58	2.39	2.97	4.09
433	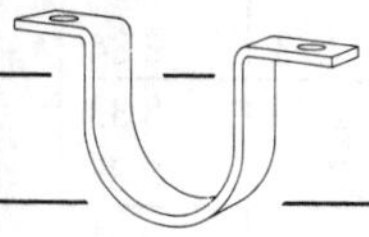 1-1/2" pipe size		131		.72	2.44	3.16	4.32
434	2" pipe size		129		.77	2.48	3.25	4.43
435	2-1/2" pipe size		125		1.32	2.56	3.88	5.15
436	3" pipe size		122		1.46	2.62	4.08	5.40
437	3-1/2" pipe size		119		1.82	2.69	4.51	5.90
438	4" pipe size	↓	114	↓	1.82	2.80	4.62	6.05
440	U-bolt, carbon steel							
441	Standard, with nuts							
442	1/2" pipe size	1 Plum	160	Ea.	.57	1.11	1.68	2.23
443	3/4" pipe size		158		.58	1.12	1.70	2.26
445	1" pipe size		152		.60	1.17	1.77	2.35
446	1-1/4" pipe size		148		.65	1.20	1.85	2.45
447	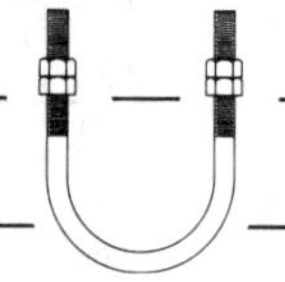 1-1/2" pipe size		143		.67	1.24	1.91	2.53
448	2" pipe size		139		.70	1.28	1.98	2.62
449	2-1/2" pipe size		134		1.10	1.33	2.43	3.12
450	3" pipe size		128		1.15	1.39	2.54	3.27
451	3-1/2" pipe size		122		1.22	1.46	2.68	3.44
452	4" pipe size		117		1.27	1.52	2.79	3.59
453	5" pipe size		114		1.39	1.56	2.95	3.78
454	6" pipe size		111		2.35	1.60	3.95	4.90
455	8" pipe size		109		2.69	1.63	4.32	5.30
456	10" pipe size		107		4.87	1.66	6.53	7.75
457	12" pipe size	↓	104		6.50	1.71	8.21	9.60
458	For plastic coating on 1/2" thru 8" size add			↓	150%			
470	U-hook, carbon steel, requires mounting screws or bolts							
471	3/4" thru 2" pipe size							
472	6" long	1 Plum	96	Ea.	.80	1.85	2.65	3.55
473	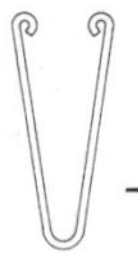 8" long		96		.91	1.85	2.76	3.67
474	10" long		96		1.04	1.85	2.89	3.82
475	12" long	↓	96		1.13	1.85	2.98	3.92
476	For copper plated, add				65%			
800	Pipe clamp, plastic, 1/2 CTS	1 Plum	80		.13	2.22	2.35	3.35
801	3/4" CTS		73		.14	2.43	2.57	3.67
802	1" CTS		68		.22	2.61	2.83	4.01
808	Economy clamp, 1/4" CTS		175		.05	1.01	1.06	1.52
809	3/8" CTS		168		.05	1.06	1.11	1.58
810	1/2" CTS		160		.05	1.11	1.16	1.66
811	3/4" CTS		145		.07	1.22	1.29	1.85
820	Half clamp, 1/2" CTS		80		.08	2.22	2.30	3.30
821	3/4" CTS		73		.09	2.43	2.52	3.61
830	Suspension clamp, 1/2" CTS		80		.17	2.22	2.39	3.39
831	3/4" CTS		73		.20	2.43	2.63	3.73
832	1" CTS		68		.34	2.61	2.95	4.15
840	Insulator, 1/2" CTS		80		.30	2.22	2.52	3.54
841	3/4" CTS		73		.30	2.43	2.73	3.84
842	1" CTS	↓	68	↓	.38	2.61	2.99	4.19
67-001	**SHOCK ABSORBERS**							
050	3/4" male I.P.S. For 1 to 11 fixtures	1 Plum	12	Ea.	25	14.80	39.80	49
060	1" male I.P.S., For 12 to 32 fixtures		8		51	22	73	88
070	For 33 to 60 fixtures		8		77	22	99	115
080	For 61 to 113 fixtures		8		195	22	217	245
090	For 114 to 154 fixtures		8		230	22	252	285
100	For 155 to 330 fixtures	↓	4	↓	265	44	309	355
70-001	**SUPPORTS/CARRIERS** For plumbing fixtures							
050	Drinking fountain, wall mounted							
060	Plate type with studs, top back plate	1 Plum	7	Ea.	10.55	25	35.55	48
070	Top front and back plate	"	7	"	13.20	25	38.20	51

15

15.1	Pipe & Fittings	CREW	DAILY OUTPUT	UNIT	BARE COSTS MAT.	BARE COSTS INST.	BARE COSTS TOTAL	TOTAL INCL O&P
70 080	Top & bottom, front & back plates, w/bearing jacks	1 Plum	7	Ea.	19.15	25	44.15	58
300	Lavatory, concealed arm 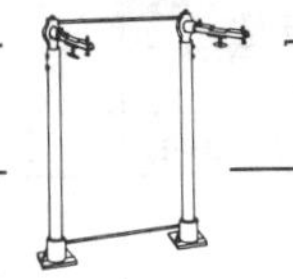							
305	Floor mounted, single							
310	High back fixture	1 Plum	6	Ea.	51	30	81	99
320	Flat slab fixture		6		60	30	90	110
322	Paraplegic		6		73	30	103	125
325	Floor mounted, back to back							
330	High back fixtures	1 Plum	5	Ea.	100	36	136	160
340	Flat slab fixtures		5		90	36	126	150
343	Paraplegic		5		80	36	116	140
350	Wall mounted, in stud or masonry							
360	High back fixture	1 Plum	6	Ea.	30	30	60	76
370	Flat slab fixture	"	6	"	39	30	69	86
400	Exposed arm type, floor mounted							
410	Single high back or flat slab fixture	1 Plum	6	Ea.	100	30	130	155
420	Back to back, high back or flat slab fixtures		5		155	36	191	220
430	Wall mounted, High back or flat slab lavatory		6		69	30	99	120
460	Sink, floor mounted							
465	Exposed arm system							
470	Single heavy fixture	1 Plum	5	Ea.	115	36	151	180
475	Single heavy sink with slab		5		180	36	216	250
480	Back to back, standard fixtures		5		130	36	166	195
485	Back to back, heavy fixtures		5		120	36	156	185
490	Back to back, heavy sink with slab		5		190	36	226	260
495	Exposed offset arm system							
500	Single heavy deep fixture	1 Plum	5	Ea.	88	36	124	150
510	Plate type system							
520	With bearing jacks, single fixture	1 Plum	5	Ea.	66	36	102	125
530	With exposed arms, single heavy fixture		5		99	36	135	160
540	Wall mounted, exposed arms, single heavy fixture		5		48	36	84	105
600	Urinal, floor mounted, 2" or 3" coupling, blowout type		6		55	30	85	105
610	With fixture or hanger bolts, blowout or washout		6		39	30	69	86
620	With bearing plate		6		44	30	74	91
630	Wall mounted, plate type system		6		46	30	76	93
698	Water closet, siphon jet							
700	Horizontal, adjustable, caulk							
704	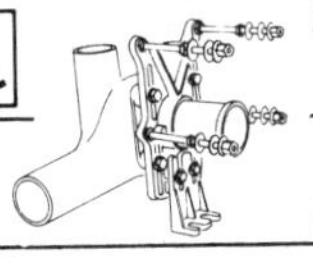 Single, 4" pipe size	1 Plum	6	Ea.	80	30	110	130
705	4" pipe size, paraplegic		6		100	30	130	155
706	5" pipe size		6		105	30	135	160
710	Double, 4" pipe size		5		150	36	186	215
711	4" pipe size, paraplegic		5		190	36	226	260
712	5" pipe size		5		185	36	221	255
716	Horizontal, adjustable, extended, caulk							
718	Single, 4" pipe size	1 Plum	6	Ea.	105	30	135	160
720	5" pipe size		6		135	30	165	190
724	Double, 4" pipe size		5		180	36	216	250
726	5" pipe size		5		230	36	266	305
740	Vertical, adjustable, caulk or thread							
744	Single, 4" pipe size	1 Plum	6	Ea.	94	30	124	145
746	5" pipe size		6		120	30	150	175
748	6" pipe size		5		160	36	196	225
752	Double, 4" pipe size		5		170	36	206	240
754	5" pipe size		5		220	36	256	295
756	6" pipe size		4		250	44	294	340
760	Vertical, adjustable, extended, caulk							
762	Single, 4" pipe size	1 Plum	6	Ea.	175	30	205	235
764	5" pipe size		6		195	30	225	255
768	6" pipe size		5		310	36	346	390
772	Double, 4" pipe size		5		275	36	311	355
774	5" pipe size		5		330	36	366	415

15

15.1 Pipe & Fittings

		CREW	DAILY OUTPUT	UNIT	BARE COSTS MAT.	BARE COSTS INST.	BARE COSTS TOTAL	TOTAL INCL O&P
70 776	6" pipe size	1 Plum	4	Ea.	420	44	464	525
778	Water closet, blow out							
780	Vertical offset, caulk or thread							
782	Single, 4" pipe size	1 Plum	6	Ea.	100	30	130	155
784	Double, 4" pipe size	"	5	"	170	36	206	240
788	Vertical offset, extended, caulk							
790	Single, 4" pipe size	1 Plum	6	Ea.	120	30	150	175
792	Double, 4" pipe size	"	5	"	170	36	206	240
796	Vertical, for floor mounted back-outlet							
798	Single, 4" thread, 2" vent	1 Plum	6	Ea.	70	30	100	120
800	Double, 4" thread, 2" vent	"	6	"	145	30	175	200
804	Vertical, for floor mounted back-outlet, extended							
806	Single, 4" caulk, 2" vent	1 Plum	6	Ea.	95	30	125	145
808	Double, 4" caulk, 2" vent	"	6	"	135	30	165	190
820	Water closet, residential							
822	Vertical centerline, floor mount							
824	Single, 3" caulk, 2" or 3" vent	1 Plum	6	Ea.	56	30	86	105
826	4" caulk, 2" or 4" vent		6		75	30	105	125
828	3" copper sweat, 3" vent		6		56	30	86	105
830	4" copper sweat, 4" vent		6		100	30	130	155
840	Vertical offset, floor mount							
842	Single, 3" or 4" caulk, vent	1 Plum	4	Ea.	100	44	144	175
844	3" or 4" copper sweat, vent		5		145	36	181	210
846	Double, 3" or 4" caulk, vent		4		170	44	214	250
848	3" or 4" copper sweat, vent		5		220	36	256	295
900	Water cooler (electric), floor mounted							
910	Plate type with bearing plate, single	1 Plum	6	Ea.	43	30	73	90
912	Plate type with bearing plate, double		4		63	44	107	135
914	Plate type with bearing plate, back to back		4		69	44	113	140
73-001	**TRAPS**							
003	Cast iron, service weight							
005	Long P trap, 2" pipe size							
110	12" long	Q-1	16	Ea.	8.55	20	28.55	38
114	18" long	"	16	"	8.90	20	28.90	39
118	Running trap, single hub, with vent							
208	3" pipe size, 3" vent	Q-1	14	Ea.	14.75	23	37.75	49
212	4" pipe size, 4" vent	"	13		19.20	25	44.20	57
214	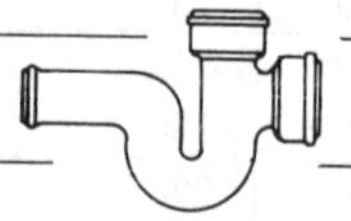 5" pipe size, 4" vent	Q-2	11		40	45	85	110
216	6" pipe size, 4" vent		10		45	50	95	120
218	6" pipe size, 6" vent		8		47	62	109	140
220	8" pipe size, 4" vent	Q-3	10		98	68	166	205
222	8" pipe size, 6" vent	"	8		105	85	190	240
230	For double hub, vent, add				10%	20%		
280	S trap, extra heavy weight, 2" pipe size	Q-1	15		11.50	21	32.50	43
284	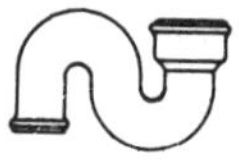 3" pipe size		14		29	23	52	65
285	4" pipe size		13		35	25	60	74
300	P trap, 2" pipe size		16		6.20	20	26.20	36
304	3" pipe size		14		7.75	23	30.75	42
306	4" pipe size		13		12.45	25	37.45	49
308	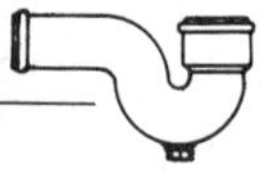 5" pipe size	Q-2	18		27	28	55	70
310	6" pipe size	"	17		36	29	65	82
312	8" pipe size	Q-3	11		99	62	161	200
313	10" pipe size	"	10		170	68	238	285
315	P trap, no hub, 1-1/2" pipe size	Q-1	17		4	18.80	22.80	32
316	2" pipe size		16		4.25	20	24.25	34
317	3" pipe size		14		9.55	23	32.55	44
318	4" pipe size		13		16.50	25	41.50	54
319	6" pipe size	Q-2	17		36	29	65	82
335	Deep seal trap							

15

15.1 Pipe & Fittings

		Description	CREW	DAILY OUTPUT	UNIT	BARE COSTS MAT.	BARE COSTS INST.	BARE COSTS TOTAL	TOTAL INCL O&P
73	340	1-1/4" pipe size	Q-1	14	Ea.	24	23	47	59
	341	1-1/2" pipe size		14		24	23	47	59
	342	2" pipe size		14		24	23	47	59
	344	3" pipe size		12		37	27	64	79
	346	4" pipe size		11		55	29	84	105
	350	For trap primer connection, add				8		8	8.80M
	354	For trap with floor cleanout, add				17	5%		
	358	For trap with adjustable cleanout, add				36		36	40M
	380	Drum trap, 4" x 5", 1-1/2" tapping	Q-2	17	Ea.	9.40	29	38.40	53
	382	2" tapping	"	17		9.40	29	38.40	53
	384	For galvanized, add				100%			
	470	Copper, drainage, drum trap							
	480	3" x 5" solid, 1-1/2" pipe size	1 Plum	16	Ea.	10.45	11.10	21.55	28
	484	3" x 6" swivel, 1-1/2" pipe size		16		14.65	11.10	25.75	32
	490	4" x 8" swivel, 1-1/2" pipe size		13		34	13.65	47.65	57
	492	2" pipe size		13		34	13.65	47.65	57
	510	P trap, standard pattern							
	520	1-1/4" pipe size	1 Plum	18	Ea.	8.70	9.85	18.55	24
	524	1-1/2" pipe size		17		7.55	10.45	18	23
	526	2" pipe size		15		15.25	11.85	27.10	34
	528	3" pipe size		11		38	16.15	54.15	65
	534	With cleanout and slip joint							
	536	1-1/4" pipe size	1 Plum	18	Ea.	8.70	9.85	18.55	24
	540	1-1/2" pipe size		17		7.55	10.45	18	23
	542	2" pipe size		15		15.25	11.85	27.10	34
	546	For swivel, add				1.95		1.95	2.14M
	575	Chromed brass, tubular, P trap, without cleanout, 20 Ga.							
	580	1-1/4" pipe size	1 Plum	18	Ea.	6.35	9.85	16.20	21
	584	1-1/2" pipe size	"	17	"	7.25	10.45	17.70	23
	590	With cleanout, 20 Ga.							
	594	1-1/4" pipe size	1 Plum	18	Ea.	10	9.85	19.85	25
	600	1-1/2" pipe size	"	17	"	10.75	10.45	21.20	27
	635	S trap, without cleanout, 20 Ga.							
	640	1-1/4" pipe size	1 Plum	18	Ea.	11.05	9.85	20.90	26
	644	1-1/2" pipe size	"	17	"	12.25	10.45	22.70	29
	655	With cleanout, 20 Ga.							
	660	1-1/4" pipe size	1 Plum	18	Ea.	16.95	9.85	26.80	33
	664	1-1/2" pipe size	"	17		19.65	10.45	30.10	37
	666	Corrosion resistant, glass, P trap, 1-1/2" pipe size	Q-1	17		24	18.80	42.80	54
	667	2" pipe size		16		31	20	51	63
	668	3" pipe size		14		64	23	87	105
	669	4" pipe size		13		95	25	120	140
	670	6" pipe size	Q-2	17		365	29	394	445
	671	ABS DWV P trap, solvent weld joint							
	672	1-1/2" pipe size	1 Plum	18	Ea.	1.44	9.85	11.29	15.85
	673	2" pipe size		17		2.60	10.45	13.05	17.95
	674	3" pipe size		15		9.75	11.85	21.60	28
	675	4" pipe size		14		19.95	12.70	32.65	40
	676	PP DWV mech. jnt., dilution trap, 1-1/2" pipe size		16		61	11.10	72.10	83
	677	P trap, 1-1/2" pipe size		17		10	10.45	20.45	26
	678	2" pipe size		16		11	11.10	22.10	28
	679	3" pipe size		14		21	12.70	33.70	41
	680	4" pipe size		13		37	13.65	50.65	60
	681	Running trap, 1-1/2" pipe size		16		12	11.10	23.10	29
	682	2" pipe size		15		18	11.85	29.85	37
	683	S trap, 1-1/2" pipe size		16		11.50	11.10	22.60	29
	684	2" pipe size		15		17.80	11.85	29.65	37
	685	Universal trap, 1-1/2" pipe size		14		9.65	12.70	22.35	29
	686	PVC DWV hub x hub, basin trap, 1-1/4" pipe size		18		3	9.85	12.85	17.55
	687	Sink P trap, 1-1/2" pipe size		18		2	9.85	11.85	16.45

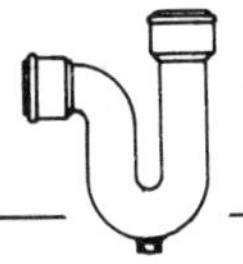

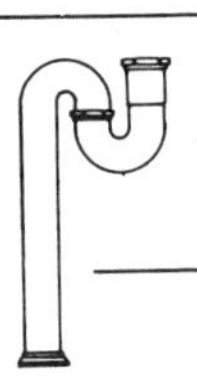

15

		CREW	DAILY OUTPUT	UNIT	BARE COSTS MAT.	BARE COSTS INST.	BARE COSTS TOTAL	TOTAL INCL O&P
73 688	Tubular S trap, 1-1/2" pipe size	1 Plum	17	Ea.	2.66	10.45	13.11	18
689	PVC sch. 40 DWV, drum trap, 3/4" pipe size							
690	1-1/2" pipe size	1 Plum	16	Ea.	3.11	11.10	14.21	19.45
691	P trap, 1-1/2" pipe size		18		.88	9.85	10.73	15.20
692	2" pipe size		17		1.67	10.45	12.12	16.95
693	3" pipe size		15		5.10	11.85	16.95	23
694	4" pipe size		14		8.65	12.70	21.35	28
695	P trap w/clean out, 1-1/2" pipe size		18		1.45	9.85	11.30	15.85
696	2" pipe size		17		2.44	10.45	12.89	17.80
697	P trap adjustable, 1-1/2" pipe size		17		1.19	10.45	11.64	16.40
698	P trap adj. w/union & cleanout, 1-1/2" pipe size		16		1.76	11.10	12.86	17.95
699	P trap fixture, 1-1/2" pipe size		18		1.97	9.85	11.82	16.40
700	Trap primer, flow through type, 1/2" diameter		24		22	7.40	29.40	35
710	With sediment strainer		22		25	8.05	33.05	39
745	Trap primer distribution unit							
750	2 opening	1 Plum	18	Ea.	20	9.85	29.85	36
754	3 opening		17		22	10.45	32.45	39
756	4 opening		16		25	11.10	36.10	44
785	Trap primer manifold							
790	2 outlet	1 Plum	18	Ea.	26	9.85	35.85	43
794	4 outlet		16		41	11.10	52.10	61
796	6 outlet		15		51	11.85	62.85	73
798	8 outlet		13		63	13.65	76.65	89
76-001	**VACUUM BREAKERS** Hot or cold water							
103	Anti-siphon, brass							
104	1/4" size	1 Plum	24	Ea.	9.35	7.40	16.75	21
105	3/8" size		24		9.35	7.40	16.75	21
106	1/2" size		24		10.65	7.40	18.05	22
108	3/4" size		20		12.60	8.90	21.50	27
110	1" size		19		19.70	9.35	29.05	35
112	1-1/4" size		15		33	11.85	44.85	53
114	1-1/2" size		13		38	13.65	51.65	62
116	2" size		11		60	16.15	76.15	89
118	2-1/2" size		9		170	19.75	189.75	215
120	3" size		7		230	25	255	290
130	For polished chrome, (1/4" thru 1"), add				40%			
190	Vacuum relief, water service, bronze							
200	1/2" size	1 Plum	30	Ea.	8.05	5.90	13.95	17.40
204	3/4" size	"	28	"	10.05	6.35	16.40	20
79-001	**VALVES, BRASS**							
003	For motorized valves, see Division 15.7-40							
050	Gas stops, without checks							
051	1/4" size	1 Plum	26	Ea.	5.45	6.85	12.30	15.85
052	3/8" size		24		5.95	7.40	13.35	17.25
053	1/2" size		24		6.65	7.40	14.05	18
054	3/4" size		22		7.65	8.05	15.70	20
055	1" size		19		10.55	9.35	19.90	25
056	1-1/4" size		15		18.50	11.85	30.35	37
057	1-1/2" size		13		25	13.65	38.65	47
058	2" size		11		37	16.15	53.15	64
066	For stop with check, add				5%			
067	For larger sizes use lubricated plug valve, Section 15.1-93							
80-001	**VALVES, BRONZE**							
102	Angle, 150 lb., rising stem, threaded							
103	C8.1 -081 1/8" size	1 Plum	24	Ea.	17.35	7.40	24.75	30
104	1/4" size		24		17.35	7.40	24.75	30
105	3/8" size		24		22	7.40	29.40	35
106	1/2" size		22		22	8.05	30.05	36

15

15.1 Pipe & Fittings

		CREW	DAILY OUTPUT	UNIT	BARE COSTS MAT.	BARE COSTS INST.	BARE COSTS TOTAL	TOTAL INCL O&P
80 107	3/4" size	1 Plum	20	Ea.	26	8.90	34.90	41
108	1" size		19		37	9.35	46.35	54
109	1-1/4" size		15		49	11.85	60.85	71
110	1-1/2" size		13		65	13.65	78.65	91
111	2" size		11		100	16.15	116.15	135
138	Ball, 150 psi, threaded							
140	1/4" size	1 Plum	24	Ea.	3.54	7.40	10.94	14.60
143	3/8" size		24		3.54	7.40	10.94	14.60
145	1/2" size		22		3.54	8.05	11.59	15.55
146	3/4" size		20		5.65	8.90	14.55	19.05
147	1" size		19		7.25	9.35	16.60	21
148	1-1/4" size		15		12.65	11.85	24.50	31
149	1-1/2" size		13		16.10	13.65	29.75	37
150	2" size		11		20	16.15	36.15	45
160	Butterfly, 175 psi, full port, solder or threaded ends							
161	Stainless steel disc and stem							
162	1/4" size	1 Plum	24	Ea.	3.90	7.40	11.30	15
163	3/8" size		24		3.90	7.40	11.30	15
164	1/2" size		22		3.90	8.05	11.95	15.95
165	3/4" size		20		6	8.90	14.90	19.45
166	1" size		19		7.90	9.35	17.25	22
167	1-1/4" size		15		13.10	11.85	24.95	32
168	1-1/2" size		13		16.30	13.65	29.95	38
169	2" size		11		20	16.15	36.15	45
175	Check, swing, class 150, regrinding disc, threaded							
180	1/8" size	1 Plum	24	Ea.	10.25	7.40	17.65	22
183	1/4" size		24		10.25	7.40	17.65	22
184	3/8" size		24		10.25	7.40	17.65	22
185	1/2" size		24		10.70	7.40	18.10	22
186	3/4" size		20		12.45	8.90	21.35	27
187	1" size		19		16.30	9.35	25.65	31
188	1-1/4" size		15		23	11.85	34.85	42
189	1-1/2" size		13		26	13.65	39.65	48
190	2" size		11		40	16.15	56.15	67
191	2-1/2" size	Q-1	15		78	21	99	115
192	3" size	"	13		110	25	135	155
200	For 200 lb., add				20%	10%		
204	For 300 lb., add				60%	15%		
235	Check, lift, class 150, horizontal composition disc, threaded							
240	1/8" size	1 Plum	24	Ea.	15.40	7.40	22.80	28
243	1/4" size		24		15.40	7.40	22.80	28
244	3/8" size		24		15.40	7.40	22.80	28
245	1/2" size		24		15.95	7.40	23.35	28
246	3/4" size		20		21	8.90	29.90	36
247	1" size		19		30	9.35	39.35	47
248	1-1/4" size		15		40	11.85	51.85	61
249	1-1/2" size		13		49	13.65	62.65	74
250	2" size		11		83	16.15	99.15	115
260	For 200 lb. add				15%	5%		
285	Gate, NRS, soldered, 300 psi							
290	3/8" size	1 Plum	24	Ea.	10.45	7.40	17.85	22
292	1/2" size		24		10.45	7.40	17.85	22
294	3/4" size		20		13.55	8.90	22.45	28
295	1" size		19		16.40	9.35	25.75	32
296	1-1/4" size		15		22	11.85	33.85	41
297	1-1/2" size		13		28	13.65	41.65	51
298	2" size		11		39	16.15	55.15	66
299	2-1/2" size	Q-1	15		80	21	101	120
300	3" size	"	13		115	25	140	160
335	Threaded, class 150							

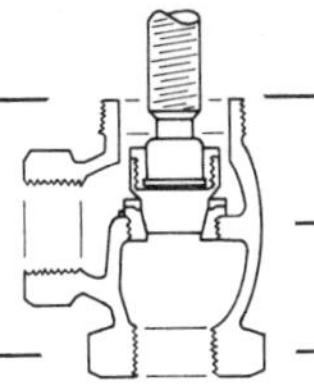

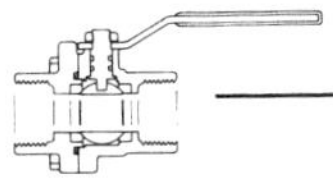

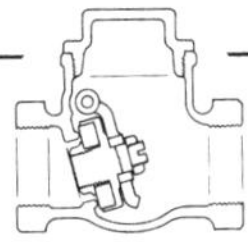

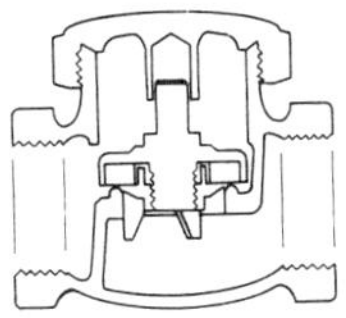

15

15.1 Pipe & Fittings

		CREW	DAILY OUTPUT	UNIT	BARE COSTS MAT.	BARE COSTS INST.	BARE COSTS TOTAL	TOTAL INCL O&P
80 340	1/8" size	1 Plum	24	Ea.	10.55	7.40	17.95	22
341	1/4" size		24		10.55	7.40	17.95	22
342	3/8" size		24		10.55	7.40	17.95	22
343	1/2" size		24		10.55	7.40	17.95	22
344	3/4" size		20		13.10	8.90	22	27
345	1" size		19		16.60	9.35	25.95	32
346	1-1/4" size		15		22	11.85	33.85	41
347	1-1/2" size		13		28	13.65	41.65	51
348	2" size		11		39	16.15	55.15	66
349	2-1/2" size	Q-1	15		85	21	106	125
350	3" size	"	13		120	25	145	170
385	Rising stem, soldered, 300 psi							
390	3/8" size	1 Plum	24	Ea.	10.80	7.40	18.20	23
392	1/2" size		24		10.80	7.40	18.20	23
394	3/4" size		20		13.55	8.90	22.45	28
395	1" size		19		16.30	9.35	25.65	31
396	1-1/4" size		15		22	11.85	33.85	41
397	1-1/2" size		13		28	13.65	41.65	51
398	2" size		11		40	16.15	56.15	67
399	2-1/2" size	Q-1	15		82	21	103	120
400	3" size	"	13		115	25	140	160
425	Threaded, class 150							
430	1/8" size	1 Plum	24	Ea.	10.55	7.40	17.95	22
431	1/4" size		24		10.55	7.40	17.95	22
432	3/8" size		24		10.55	7.40	17.95	22
433	1/2" size		24		10.55	7.40	17.95	22
434	3/4" size		20		13.10	8.90	22	27
435	1" size		19		16.60	9.35	25.95	32
436	1-1/4" size		15		22	11.85	33.85	41
437	1-1/2" size		13		28	13.65	41.65	51
438	2" size		11		39	16.15	55.15	66
439	2-1/2" size	Q-1	15		85	21	106	125
440	3" size	"	13		120	25	145	170
450	For 300 psi, threaded, add				100%	15%		
454	For chain operated type, add				15%			
485	Globe, class 150, rising stem, threaded							
490	1/8" size	1 Plum	24	Ea.	13.85	7.40	21.25	26
492	1/4" size		24		13.85	7.40	21.25	26
494	3/8" size		24		14.30	7.40	21.70	26
495	1/2" size		24		14.65	7.40	22.05	27
496	3/4" size		20		18.40	8.90	27.30	33
497	1" size		19		30	9.35	39.35	47
498	1-1/4" size		15		43	11.85	54.85	64
499	1-1/2" size		13		53	13.65	66.65	78
500	2" size		11		81	16.15	97.15	110
501	2-1/2" size	Q-1	15		155	21	176	200
502	3" size	"	13		215	25	240	270
512	For class 300, threaded, add				150%	15%		
560	Relief, pressure & temperature, self-closing, ASME							
564	3/4" size	1 Plum	28	Ea.	29	6.35	35.35	41
565	1" size		24		42	7.40	49.40	57
566	1-1/4" size		20		89	8.90	97.90	110
567	1-1/2" size		18		160	9.85	169.85	190
568	2" size		16		170	11.10	181.10	205
595	Pressure, poppet type, threaded							
600	1/2" size	1 Plum	30	Ea.	8.70	5.90	14.60	18.10
604	3/4" size	"	28	"	9.35	6.35	15.70	19.45
640	Pressure, water, ASME, threaded							
644	3/4" size	1 Plum	28	Ea.	18.75	6.35	25.10	30
645	1" size	"	24	"	39	7.40	46.40	54

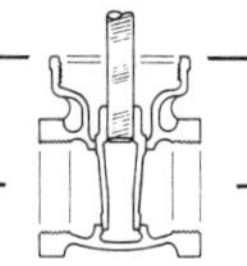

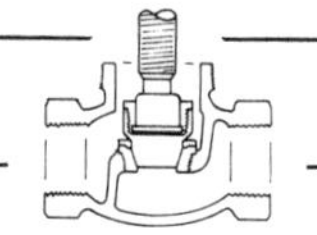

15

15.1 Pipe & Fittings		CREW	DAILY OUTPUT	UNIT	BARE COSTS MAT.	BARE COSTS INST.	BARE COSTS TOTAL	TOTAL INCL O&P
80 646	1-1/4"size	1 Plum	20	Ea.	62	8.90	70.90	81
647	1-1/2" size		18		86	9.85	95.85	110
648	2" size		16		125	11.10	136.10	155
690	Reducing, water pressure							
692	300 psi to 25-75 psi, threaded or sweat							
694	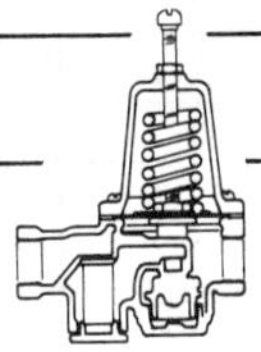 1/2" size	1 Plum	24	Ea.	36	7.40	43.40	50
695	3/4" size		20		43	8.90	51.90	60
696	1" size		19		68	9.35	77.35	88
697	1-1/4" size		15		135	11.85	146.85	165
698	1-1/2" size		13		155	13.65	168.65	190
699	2" size		11		220	16.15	236.15	265
710	For built-in by-pass or 10-35 psi, add				2		2	2.20M
770	High capacity, 250 psi to 25-75 psi, threaded							
774	1/2" size	1 Plum	24	Ea.	50	7.40	57.40	66
778	3/4" size		20		50	8.90	58.90	68
779	1" size		19		77	9.35	86.35	98
780	1-1/4" size		15		130	11.85	141.85	160
781	1-1/2" size		13		205	13.65	218.65	245
782	2" size		11		300	16.15	316.15	355
783	2-1/2" size		9		320	19.75	339.75	380
784	3" size		8		480	22	502	560
785	3" flanged (iron body)	Q-1	10		470	32	502	565
786	4" flanged (iron body)	"	8		690	40	730	815
792	For higher pressure, add				25%			
835	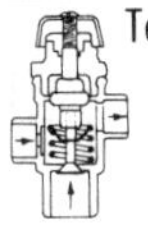 Tempering, water, sweat connections							
840	1/2" size	1 Plum	24	Ea.	17.35	7.40	24.75	30
844	3/4" size	"	20	"	20	8.90	28.90	35
865	Threaded connections							
870	1/2" size	1 Plum	24	Ea.	20	7.40	27.40	33
874	3/4" size		20		25	8.90	33.90	40
875	1" size		19		65	9.35	74.35	85
876	1-1/4" size		15		105	11.85	116.85	135
877	1-1/2" size		13		115	13.65	128.65	145
878	2" size		11		115	16.15	131.15	150
82-001	**VALVES, IRON BODY**							
002	For grooved joint, see Division 15.1-61							
102	C8.1 -081 Butterfly, wafer type, lever actuator							
103	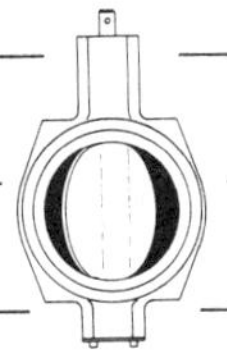 2" size	1 Plum	8	Ea.	64	22	86	100
104	2-1/2" size	Q-1	5		70	64	134	170
105	3" size		4.50		74	71	145	185
106	4" size		3		88	105	193	250
107	5" size	Q-2	3.40		115	145	260	340
108	6" size		3		150	165	315	405
109	8" size		2.50		195	200	395	500
110	10" size		2.20		295	225	520	650
111	12" size		1.70		430	295	725	895
118	For gear actuator, add				65%			
120	Lug type, lever actuator							
122	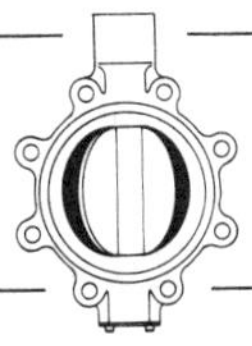 2" size	1 Plum	8	Ea.	59	22	81	97
123	2-1/2" size	Q-1	5		66	64	130	165
124	3" size		4.50		72	71	143	180
125	4" size		3		86	105	191	250
126	5" size	Q-2	3.40		115	145	260	340
127	6" size		3		135	165	300	390
128	8" size		2.50		190	200	390	495
129	10" size		2.20		245	225	470	595
130	12" size		1.70		370	295	665	830
132	For gear actuator, add				60%			
140	Diverter, 150 lb. flanged, bronze or iron plugs							

15

15.1 Pipe & Fittings

82

		CREW	DAILY OUTPUT	UNIT	BARE COSTS MAT.	BARE COSTS INST.	BARE COSTS TOTAL	TOTAL INCL O&P
144	2" pipe size	Q-1	2	Ea.	660	160	820	955
145	3" pipe size	"	1.50	"	915	215	1,130	1,325
165	Gate, 125 lb., NRS, threaded							
170	2" size	1 Plum	11	Ea.	120	16.15	136.15	155
174	2-1/2" size	Q-1	15		130	21	151	175
176	3" size		13		160	25	185	210
178	4" size		10		200	32	232	265
215	Flanged							
220	2" size	1 Plum	5	Ea.	90	36	126	150
224	2-1/2" size	Q-1	5		95	64	159	195
226	3" size		4.50		110	71	181	225
228	4" size		3		150	105	255	320
229	5" size	Q-2	3.40		250	145	395	485
230	6" size		3		250	165	415	515
232	8" size		2.50		425	200	625	755
234	10" size		2.20		890	225	1,115	1,300
236	12" size		1.70		1,200	295	1,495	1,750
242	For 250 lb., flanged, add				200%	10%		
295	OS&Y, threaded							
296	1/2" size	1 Plum	24	Ea.	51	7.40	58.40	67
297	3/4" size		20		55	8.90	63.90	73
298	1" size		19		59	9.35	68.35	78
299	1-1/2" size		13		78	13.65	91.65	105
300	2" size		11		130	16.15	146.15	165
306	2-1/2" size	Q-1	15		150	21	171	195
308	3" size		13		160	25	185	210
310	4" size		10		225	32	257	295
355	OS&Y, flanged							
360	2" size	1 Plum	5	Ea.	93	36	129	155
364	2-1/2" size	Q-1	5		100	64	164	200
366	3" size		4.50		115	71	186	230
367	3-1/2" size		3		175	105	280	345
368	4" size		3		160	105	265	330
369	5" size	Q-2	3.40		255	145	400	490
370	6" size		3		255	165	420	520
372	8" size		2.50		455	200	655	790
374	10" size		2.20		965	225	1,190	1,400
376	12" size		1.70		1,250	295	1,545	1,800
390	For 250 lb flanged, add				120%	10%		
435	Globe, OS&Y, class 125, threaded							
440	2" size	1 Plum	11	Ea.	220	16.15	236.15	265
444	2-1/2" size	Q-1	15		245	21	266	300
445	3" size		13		290	25	315	355
446	4" size		10		400	32	432	485
454	Class 125, flanged							
455	2" size	1 Plum	5	Ea.	150	36	186	215
456	2-1/2" size	Q-1	5		175	64	239	285
457	3" size		4.50		205	71	276	330
458	4" size		3		310	105	415	495
459	5" size	Q-2	3.40		495	145	640	755
460	6" size		3		495	165	660	785
461	8" size		2.50		1,025	200	1,225	1,425
485	Class 250, threaded							
494	2-1/2" size	Q-1	13	Ea.	370	25	395	445
504	Class 250, flanged							
505	2" size	1 Plum	4.50	Ea.	225	39	264	305
506	2-1/2" size	Q-1	4.50		250	71	321	380
507	3" size		4		295	80	375	440
508	4" size		2.70		410	120	530	620
509	5" size	Q-2	3		775	165	940	1,100

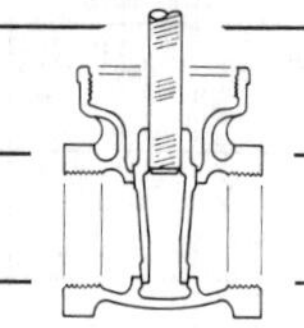

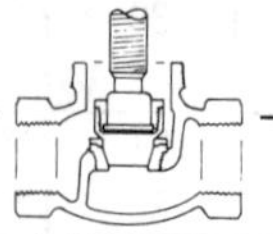

15

15.1 Pipe & Fittings

		CREW	DAILY OUTPUT	UNIT	BARE COSTS MAT.	INST.	TOTAL	TOTAL INCL O&P
82 510	6" size	Q-2	2.70	Ea.	775	185	960	1,125
511	8" size		2.20		2,500	225	2,725	3,075
512	10" size		2		4,300	250	4,550	5,100
513	12" size		1.60		7,000	310	7,310	8,150
545	Swing check, 125 lb., threaded							
550	2" size	1 Plum	11	Ea.	80	16.15	96.15	110
554	2-1/2" size	Q-1	15		96	21	117	135
555	3" size		13		115	25	140	160
556	4" size		10		175	32	207	240
595	Flanged							
600	2" size	1 Plum	5	Ea.	63	36	99	120
604	2-1/2" size	Q-1	5		76	64	140	175
605	3" size		4.50		85	71	156	195
606	4" size		3		130	105	235	295
607	6" size	Q-2	3		220	165	385	480
608	8" size		2.50		475	200	675	810
609	10" size		2.20		835	225	1,060	1,250
610	12" size		1.70		1,175	295	1,470	1,725
616	For 250 lb., flanged, add				200%	20%		
660	Combination pump valve (CPV), 125 psi,							
662	wafer style, silent check							
664	2" size	1 Plum	6	Ea.	77	30	107	125
665	2-1/2" size	Q-1	5		68	64	132	165
666	3" size		4.50		77	71	148	185
667	4" size		3		100	105	205	265
668	5" size	Q-2	3.40		130	145	275	355
669	6" size		3		175	165	340	430
670	8" size		2.50		305	200	505	625
671	10" size		2.20		470	225	695	845
84-001	**VALVES, LINED, CORROSION RESISTANT/HIGH PURITY**							
200	Butterfly 150 lb. ductile iron,							
201	Wafer type							
203	FEP lined 3" lever handle	Q-1	4.50	Ea.	585	71	656	745
205	4" lever handle	"	3		835	105	940	1,075
207	6" gear operated	Q-2	3		1,275	165	1,440	1,650
208	8" gear operated	"	2.50		1,800	200	2,000	2,275
209	10" gear operated	G-2	2.20		2,450	235	2,685	3,025
210	12" gear operated	Q-2	1.70		2,900	295	3,195	3,625
350	Check lift, 125 lb., cast iron flanged							
351	Horizontal PPL or SL lined							
353	1" size	1 Plum	14	Ea.	200	12.70	212.70	240
354	1-1/2" size		11		245	16.15	261.15	295
355	2" size		8		285	22	307	345
356	2-1/2" size	Q-1	5		370	64	434	500
357	3" size		4.50		460	71	531	610
359	4" size		3		585	105	690	800
361	6" size	Q-2	3		1,025	165	1,190	1,375
362	8" size	"	2.50		1,800	200	2,000	2,275
425	Vertical PPL or SL lined							
427	1" size	1 Plum	14	Ea.	180	12.70	192.70	215
429	1-1/2" size		11		225	16.15	241.15	270
430	2" size		8		260	22	282	320
431	2-1/2" size	Q-1	5		350	64	414	475
432	3" size		4.50		390	71	461	530
434	4" size		3		485	105	590	690
436	6" size	Q-2	3		840	165	1,005	1,175
437	8" size	"	2.50		1,500	200	1,700	1,950
500	Clamp type, ductile iron, 150 lb. flanged							
501	TFE lined							

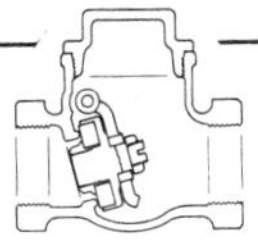

15

15.1	Pipe & Fittings	CREW	DAILY OUTPUT	UNIT	BARE COSTS MAT.	BARE COSTS INST.	BARE COSTS TOTAL	TOTAL INCL O&P
84 503	1" size, lever handle	1 Plum	9	Ea.	385	19.75	404.75	450
505	1-1/2" size, lever handle		6		510	30	540	605
506	2" size, lever handle		5		620	36	656	735
508	3" size, lever handle	Q-1	4.50		775	71	846	955
510	4" size, gear operated	"	3		1,075	105	1,180	1,325
512	6" size, gear operated	Q-2	3		3,100	165	3,265	3,650
513	8" size, gear operated	"	2.50		3,525	200	3,725	4,175
600	Diaphragm type, cast iron, 125 lb. flanged							
601	PPL or SL, lined							
603	1" size, handwheel operated	1 Plum	9	Ea.	76	19.75	95.75	110
605	1-1/2" size, handwheel operated		6		98	30	128	150
606	2" size, handwheel operated		5		110	36	146	170
607	2-1/2" size, handwheel operated	Q-1	5		160	64	224	270
608	3" size, handwheel operated		4.50		180	71	251	300
610	4" size, handwheel operated		3		300	105	405	485
612	6" size, handwheel operated	Q-2	3		570	165	735	865
613	8" size, handwheel operated	"	2.50		1,250	200	1,450	1,675
800	Plug, 150 lb. ductile iron flanged							
802	FEP lined 1" lever handle	1 Plum	14	Ea.	390	12.70	402.70	445
804	1-1/2" lever handle		11		520	16.15	536.15	595
805	2" lever handle		8		645	22	667	740
807	3" lever handle	Q-1	4.50		700	71	771	875
809	4" lever handle	"	3		955	105	1,060	1,200
811	6" gear operated	Q-2	3		2,100	165	2,265	2,550
812	8" gear operated	"	2.50		3,550	200	3,750	4,200
85-001	**VALVES, PLASTIC**							
110	Angle, PVC, threaded							
111	C8.1-081 1/4" size	1 Plum	26	Ea.	14.25	6.85	21.10	26
113	3/4" size		25		29	7.10	36.10	42
114	1" size		23		32	7.70	39.70	46
115	Ball, PVC, socket or threaded, single union							
120	1/4" size	1 Plum	26	Ea.	11.80	6.85	18.65	23
122	3/8" size		26		11.80	6.85	18.65	23
123	1/2" size		26		11.80	6.85	18.65	23
124	3/4" size		25		13.50	7.10	20.60	25
125	1" size		23		16.95	7.70	24.65	30
126	1-1/4" size		21		22	8.45	30.45	36
127	1-1/2" size		20		27	8.90	35.90	43
128	2" size		17		39	10.45	49.45	58
129	2-1/2" size	Q-1	26		65	12.30	77.30	89
130	3" size		24		60	13.30	73.30	85
131	4" size		20		125	16	141	160
136	For PVC, flanged, add				45%	15%		
140	For true union, socket or threaded, add				40%	5%		
165	CPVC, socket or threaded, single union							
170	1/2" size	1 Plum	26	Ea.	14.80	6.85	21.65	26
172	3/4" size		25		18.35	7.10	25.45	30
173	1" size		23		21	7.70	28.70	34
175	1-1/4" size		21		48	8.45	56.45	65
176	1-1/2" size		20		43	8.90	51.90	60
177	2" size		17		52	10.45	62.45	72
178	3" size	Q-1	24		185	13.30	198.30	225
184	For CPVC, flanged, add				100%	15%		
188	For true union, socket or threaded, add				50%	5%		
205	Polypropylene, threaded							
210	1/4" size	1 Plum	26	Ea.	20	6.85	26.85	32
212	3/8" size		26		20	6.85	26.85	32
213	1/2" size		26		20	6.85	26.85	32
214	3/4" size		25		23	7.10	30.10	36

15

15.1 Pipe & Fittings

		CREW	DAILY OUTPUT	UNIT	BARE COSTS MAT.	BARE COSTS INST.	BARE COSTS TOTAL	TOTAL INCL O&P
85 215	1" size	1 Plum	23	Ea.	29	7.70	36.70	43
216	1-1/4" size		21		38	8.45	46.45	54
217	1-1/2" size		20		47	8.90	55.90	65
218	2" size		17		65	10.45	75.45	87
219	3" size	Q-1	24		205	13.30	218.30	245
220	4" size	"	20		545	16	561	625
255	PVC, three way, socket or threaded							
260	1/2" size	1 Plum	26	Ea.	41	6.85	47.85	55
264	3/4" size		25		43	7.10	50.10	58
265	1" size		23		58	7.70	65.70	75
266	1-1/2" size		20		94	8.90	102.90	115
267	2" size		17		110	10.45	120.45	135
268	3" size	Q-1	24		275	13.30	288.30	320
274	For flanged, add				60%	15%		
315	Ball check, PVC, socket or threaded							
320	1/4" size	1 Plum	26	Ea.	15.50	6.85	22.35	27
322	3/8" size		26		15.50	6.85	22.35	27
324	1/2" size		26		15.50	6.85	22.35	27
325	3/4" size		25		18.50	7.10	25.60	31
326	1" size		23		22	7.70	29.70	35
327	1-1/4" size		21		30	8.45	38.45	45
328	1-1/2" size		20		37	8.90	45.90	54
329	2" size		17		50	10.45	60.45	70
331	3" size	Q-1	24		130	13.30	143.30	160
332	4" size	"	20		185	16	201	225
336	For PVC, flanged, add				50%	15%		
375	CPVC, socket or threaded							
380	1/2" size	1 Plum	26	Ea.	25	6.85	31.85	37
384	3/4" size		25		27	7.10	34.10	40
385	1" size		23		32	7.70	39.70	46
386	1-1/2" size		20		56	8.90	64.90	74
387	2" size		17		75	10.45	85.45	98
388	3" size	Q-1	24		155	13.30	168.30	190
392	4" size	"	20		210	16	226	255
393	For CPVC, flanged, add				40%	15%		
434	Polypropylene, threaded							
436	1/2" size	1 Plum	26	Ea.	32	6.85	38.85	45
440	3/4" size		25		44	7.10	51.10	59
444	1" size		23		57	7.70	64.70	74
445	1-1/2" size		20		87	8.90	95.90	110
446	2" size		17		160	10.45	170.45	190
450	For polypropylene flanged, add				55%	15%		
485	Foot valve, PVC, socket or threaded							
490	1/2" size	1 Plum	34	Ea.	25	5.20	30.20	35
493	3/4" size		32		32	5.55	37.55	43
494	1" size		28		41	6.35	47.35	54
495	1-1/4" size		27		80	6.60	86.60	98
496	1-1/2" size		26		80	6.85	86.85	98
497	2" size		24		93	7.40	100.40	115
498	3" size		20		225	8.90	233.90	260
499	4" size		18		300	9.85	309.85	345
500	For flanged, add				25%	10%		
505	CPVC, socket or threaded							
506	1/2" size	1 Plum	34	Ea.	28	5.20	33.20	38
507	3/4" size		32		39	5.55	44.55	51
508	1" size		28		49	6.35	55.35	63
509	1-1/4" size		27		87	6.60	93.60	105
510	1-1/2" size		26		87	6.85	93.85	105
511	2" size		24		105	7.40	112.40	125
512	3" size		20		295	8.90	303.90	335

15.1 Pipe & Fittings

85		CREW	DAILY OUTPUT	UNIT	BARE COSTS MAT.	BARE COSTS INST.	BARE COSTS TOTAL	TOTAL INCL O&P
513	4″ size	1 Plum	18	Ea.	350	9.85	359.85	400
514	For flanged, add			″	25%	10%		
528	Needle valve, PVC, threaded							
530	1/4″ size	1 Plum	26	Ea.	20	6.85	26.85	32
534	3/8″ size		26		23	6.85	29.85	35
536	1/2″ size		26		23	6.85	29.85	35
538	For polypropylene, add				20%			
580	Y check, PVC, socket or threaded							
582	1/2″ size	1 Plum	26	Ea.	22	6.85	28.85	34
584	3/4″ size		25		23	7.10	30.10	36
585	1″ size		23		25	7.70	32.70	39
586	1-1/4″ size		21		41	8.45	49.45	57
587	1-1/2″ size		20		44	8.90	52.90	61
588	2″ size		17		56	10.45	66.45	77
589	2-1/2″ size		15		140	11.85	151.85	170
590	3″ size	Q-1	24		105	13.30	118.30	135
591	4″ size	″	20		190	16	206	230
596	For PVC flanged, add				45%	15%		
635	Y sediment strainer, PVC, socket or threaded							
640	1/2″ size	1 Plum	26	Ea.	17.35	6.85	24.20	29
644	3/4″ size		24		19.35	7.40	26.75	32
645	1″ size		23		21	7.70	28.70	34
646	1-1/4″ size		21		39	8.45	47.45	55
647	1-1/2″ size		20		39	8.90	47.90	56
648	2″ size		17		49	10.45	59.45	69
649	2-1/2″ size		15		94	11.85	105.85	120
650	3″ size	Q-1	24		94	13.30	107.30	125
651	4″ size	″	20		160	16	176	200
656	For PVC, flanged, add				50%	15%		
88-001	VALVES, STEEL							
080	Cast, angle, 150 lb., flanged							
083	C8.1 -081 2″ size	1 Plum	8	Ea.	750	22	772	855
084	2-1/2″ size	Q-1	5		780	64	844	950
085	3″ size		4.50		780	71	851	960
086	4″ size		3		1,150	105	1,255	1,425
087	6″ size	Q-2	3		1,725	165	1,890	2,125
088	8″ size	″	2.50		3,350	200	3,550	3,975
094	For 300 lb., flanged, add				30%	15%		
096	For 600 lb., flanged, add				100%	20%		
135	Check valve, 150 lb., flanged							
140	2″ size	1 Plum	8	Ea.	365	22	387	435
144	2-1/2″ size	Q-1	5		455	64	519	595
145	3″ size		4.50		455	71	526	605
146	4″ size		3		645	105	750	865
147	6″ size	Q-2	3		940	165	1,105	1,275
148	8″ size		2.50		1,650	200	1,850	2,100
149	10″ size		2.20		2,675	225	2,900	3,275
150	12″ size		1.70		3,900	295	4,195	4,725
154	For 300 lb., flanged, add				50%	15%		
156	For 600 lb., flanged, add				110%	20%		
170	Diverter, 150 lb., flanged, bronze or iron plugs							
172	1-1/2″ pipe size	Q-1	3	Ea.	945	105	1,050	1,200
173	2″ pipe size		2		1,400	160	1,560	1,775
174	3″ pipe size		1.50		1,725	215	1,940	2,200
175	4″ pipe size		1		2,650	320	2,970	3,375
176	6″ pipe size	Q-2	1		4,500	495	4,995	5,675
179	For 300 lb., add				20%	5%		
181	Diverter, 150 lb., flanged, stainless steel plugs							
183	1-1/2″ pipe size	Q-1	3	Ea.	1,100	105	1,205	1,375

15

15.1 Pipe & Fittings

88

		CREW	DAILY OUTPUT	UNIT	BARE COSTS MAT.	BARE COSTS INST.	BARE COSTS TOTAL	TOTAL INCL O&P
184	2" pipe size	Q-1	2	Ea.	1,725	160	1,885	2,125
185	3" pipe size		1.50		2,025	215	2,240	2,525
186	4" pipe size	↓	1		3,200	320	3,520	3,975
187	6" pipe size	Q-2	1		5,175	495	5,670	6,400
189	For 300 lb., add			↓	10%	5%		
195	Gate valve, 150 lb., flanged							
200	2" size	1 Plum	8	Ea.	380	22	402	450
204	2-1/2"size	Q-1	5		550	64	614	695
205	3" size		4.50		550	71	621	710
206	4" size	↓	3		645	105	750	865
207	6" size	Q-2	3		1,050	165	1,215	1,400
208	8" size		2.50		1,625	200	1,825	2,075
209	10" size		2.20		2,550	225	2,775	3,125
210	12" size		1.70		3,400	295	3,695	4,175
211	14" size		1.30		5,775	385	6,160	6,900
212	16" size		1		7,900	495	8,395	9,400
213	18" size		.80		9,350	620	9,970	11,200
214	20" size	↓	.60	↓	12,000	830	12,830	14,400
265	300 lb., flanged							
270	2" size	1 Plum	7.40	Ea.	510	24	534	595
274	2-1/2" size	Q-1	4.20		670	76	746	845
275	3" size		4		670	80	750	855
276	4" size	↓	2.80		925	115	1,040	1,175
277	6" size	Q-2	2.90		1,600	170	1,770	2,000
278	8" size		2.40		2,425	205	2,630	2,975
279	10" size		2.10		4,100	235	4,335	4,850
280	12" size		1.60		5,350	310	5,660	6,325
281	14" size		1.20		9,300	415	9,715	10,800
282	16" size		.90		13,100	555	13,655	15,200
283	18" size		.70		16,500	710	17,210	19,200
284	20" size	↓	.50	↓	19,000	995	19,995	22,300
320	600 lb., flanged							
322	2" size	1 Plum	7	Ea.	875	25	900	1,000
324	2-1/2" size	Q-1	4		1,925	80	2,005	2,225
325	3" size		3.60		1,925	89	2,014	2,250
326	4" size	↓	2.50		2,000	130	2,130	2,375
327	6" size	Q-2	2.60		3,725	190	3,915	4,375
328	8" size		2.10		6,300	235	6,535	7,275
329	10" size		1.70		9,325	295	9,620	10,700
330	12" size		1.20		12,000	415	12,415	13,800
331	14" size		.80		15,700	620	16,320	18,200
332	16" size		.60		20,600	830	21,430	23,900
333	18" size		.50		24,500	995	25,495	28,400
334	20" size	↓	.40	↓	34,400	1,250	35,650	39,600
365	Globe valve, 150 lb., flanged							
370	2" size	1 Plum	8	Ea.	505	22	527	590
374	2-1/2" size	Q-1	5		655	64	719	815
375	3" size		4.50		655	71	726	825
376	4" size	↓	3		950	105	1,055	1,200
377	6" size	Q-2	3		1,450	165	1,615	1,825
378	8" size	"	2.50	↓	2,600	200	2,800	3,150
408	300 lb., flanged							
410	2" size	1 Plum	7.40	Ea.	665	24	689	765
414	2-1/2" size	Q-1	4.20		825	76	901	1,025
415	3" size		4		825	80	905	1,025
416	4" size	↓	2.80		1,125	115	1,240	1,400
417	6" size	Q-2	2.90		1,975	170	2,145	2,425
418	8" size	"	2.40	↓	3,050	205	3,255	3,650
468	600 lb., flanged							
470	2" size	1 Plum	7	Ea.	855	25	880	975

15

88

15.1 Pipe & Fittings		CREW	DAILY OUTPUT	UNIT	BARE COSTS MAT.	BARE COSTS INST.	BARE COSTS TOTAL	TOTAL INCL O&P
474	2-1/2" size	Q-1	4	Ea.	1,300	80	1,380	1,550
475	3" size		3.60		1,300	89	1,389	1,550
476	4" size	↓	2.50		2,150	130	2,280	2,550
477	6" size	Q-2	2.60		4,175	190	4,365	4,875
478	8" size	"	2.10	↓	11,600	235	11,835	13,100
515	Forged, angle, class 800, socket or threaded							
520	1/4" size	1 Plum	24	Ea.	77	7.40	84.40	95
522	3/8" size		24		77	7.40	84.40	95
523	1/2" size		24		77	7.40	84.40	95
524	3/4" size		20		91	8.90	99.90	115
525	1" size		19		100	9.35	109.35	125
526	1-1/4" size		15		270	11.85	281.85	315
527	1-1/2" size		13		270	13.65	283.65	315
528	2" size	↓	11	↓	375	16.15	391.15	435
565	Check valve, class 800, horizontal, socket or threaded							
570	1/4" size	1 Plum	24	Ea.	36	7.40	43.40	50
572	3/8" size		24		36	7.40	43.40	50
573	1/2" size		24		36	7.40	43.40	50
574	3/4" size		20		42	8.90	50.90	59
575	1" size		19		49	9.35	58.35	67
576	1-1/4" size		15		95	11.85	106.85	120
577	1-1/2" size		13		95	13.65	108.65	125
578	2" size	↓	11		135	16.15	151.15	170
584	For class 150, flanged, add				100%	15%		
586	For class 300, flanged, add			↓	120%	20%		
610	Gate, class 800, inside screw stem, socket or threaded							
612	3/8" size	1 Plum	24	Ea.	33	7.40	40.40	47
613	1/2" size		24		33	7.40	40.40	47
614	3/4" size		20		37	8.90	45.90	54
615	1" size		19		37	9.35	46.35	54
616	1-1/4" size		15		86	11.85	97.85	110
617	1-1/2" size		13		86	13.65	99.65	115
618	2" size	↓	11		110	16.15	126.15	145
624	For OS&Y, socket or threaded, add				20%			
626	For OS&Y, flanged, add			↓	100%	20%		
670	Globe, OS&Y, class 800, socket or threaded							
671	1/4" size	1 Plum	24	Ea.	41	7.40	48.40	56
672	3/8" size		24		41	7.40	48.40	56
673	1/2" size		24		41	7.40	48.40	56
674	3/4" size		20		47	8.90	55.90	65
675	1" size		19		58	9.35	67.35	77
676	1-1/4" size		15		120	11.85	131.85	150
677	1-1/2" size		13		120	13.65	133.65	150
678	2" size	↓	11		155	16.15	171.15	195
686	For OS&Y, flanged, add			↓	86%	20%		
91-001	**VALVES, STAINLESS STEEL**							
100	Butterfly, 175 lb., full port, threaded							
102	C8.1-081 Type 316 stainless steel							
103	1/4" size	1 Plum	24	Ea.	19.50	7.40	26.90	32
104	3/8" size		24		19.50	7.40	26.90	32
105	1/2" size		22		19.50	8.05	27.55	33
106	3/4" size		20		31	8.90	39.90	47
107	1" size		19		35	9.35	44.35	52
108	1-1/4" size		15		73	11.85	84.85	97
109	1-1/2" size		13		89	13.65	102.65	120
110	2" size	↓	11	↓	95	16.15	111.15	130
170	Check, 200 lb., threaded							
171	1/4" size	1 Plum	24	Ea.	110	7.40	117.40	130
172	1/2" size	"	22	"	110	8.05	118.05	135

15

15.1 Pipe & Fittings

			CREW	DAILY OUTPUT	UNIT	BARE COSTS MAT.	BARE COSTS INST.	BARE COSTS TOTAL	TOTAL INCL O&P
91	173	3/4" size	1 Plum	20	Ea.	125	8.90	133.90	150
	175	1" size		19		160	9.35	169.35	190
	176	1-1/2" size		13		240	13.65	253.65	285
	177	2" size		11		290	16.15	306.15	340
	180	150 lb., flanged							
	181	2-1/2" size	Q-1	5	Ea.	560	64	624	710
	182	3" size		4.50		625	71	696	790
	183	4" size		3		945	105	1,050	1,200
	184	6" size	Q-2	3		1,725	165	1,890	2,125
	185	8" size	"	2.50		3,250	200	3,450	3,875
	190	Diverter, 150 lb. flanged, stainless steel plugs							
	191	1-1/2" pipe size	Q-1	3	Ea.	1,900	105	2,005	2,250
	192	2" pipe size		2		2,550	160	2,710	3,025
	194	3" pipe size		1.50		3,325	215	3,540	3,975
	195	4" pipe size		1		5,575	320	5,895	6,600
	210	Gate, OS&Y, 150 lb., flanged							
	212	1/2" size	1 Plum	18	Ea.	165	9.85	174.85	195
	214	3/4" size		16		175	11.10	186.10	210
	215	1" size		14		215	12.70	227.70	255
	216	1-1/2" size		11		305	16.15	321.15	360
	217	2" size		8		360	22	382	430
	218	2-1/2" size	Q-1	5		485	64	549	625
	219	3" size		4.50		570	71	641	730
	220	4" size		3		815	105	920	1,050
	221	6" size	Q-2	3		1,475	165	1,640	1,850
	222	8" size	"	2.50		2,775	200	2,975	3,350
	226	For 300 lb., flanged, add				120%	15%		
	260	600 lb., flanged							
	262	1/2" size	1 Plum	16	Ea.	310	11.10	321.10	355
	264	3/4" size		14		335	12.70	347.70	385
	265	1" size		12		415	14.80	429.80	480
	266	1-1/2" size		10		395	17.75	412.75	460
	267	2" size		7		755	25	780	865
	268	2-1/2" size	Q-1	4		3,025	80	3,105	3,450
	269	3" size		3.60		3,150	89	3,239	3,600
	270	4" size		2.50		4,800	130	4,930	5,475
	271	6" size	Q-2	2.60		8,050	190	8,240	9,125
	272	8" size		2.10		14,800	235	15,035	16,600
	273	10" size		1.70		22,500	295	22,795	25,200
	274	12" size		1.20		28,500	415	28,915	31,900
	310	Globe, OS&Y, 150 lb., flanged							
	312	1/2" size	1 Plum	18	Ea.	170	9.85	179.85	200
	314	3/4" size		16		185	11.10	196.10	220
	315	1" size		14		225	12.70	237.70	265
	316	1-1/2" size		11		325	16.15	341.15	380
	317	2" size		8		400	22	422	470
	318	2-1/2" size	Q-1	5		645	64	709	800
	319	3" size		4.50		885	71	956	1,075
	320	4" size		3		1,050	105	1,155	1,300
	321	6" size	Q-2	3		1,850	165	2,015	2,275
	93-001	**VALVES, SEMI-STEEL**							
	102	Lubricated plug valve, threaded							
	103	C8.1 -081 1/2" pipe size	1 Plum	18	Ea.	65	9.85	74.85	86
	104	3/4" pipe size		16		65	11.10	76.10	88
	105	1" pipe size		14		77	12.70	89.70	105
	106	1-1/4" pipe size		12		93	14.80	107.80	125
	107	1-1/2" pipe size		11		99	16.15	115.15	130
	108	2" pipe size		8		115	22	137	160
	109	2-1/2" pipe size	Q-1	5		180	64	244	290
	110	3" pipe size	"	4.50		220	71	291	345

15

	15.1 Pipe & Fittings	CREW	DAILY OUTPUT	UNIT	BARE COSTS MAT.	BARE COSTS INST.	BARE COSTS TOTAL	TOTAL INCL O&P
93 699	Flanged							
700	2" pipe size	1 Plum	8	Ea.	100	22	122	140
701	2-1/2" pipe size	Q-1	5		135	64	199	240
702	3" pipe size		4.50		165	71	236	285
703	4" pipe size		3		225	105	330	400
704	6" pipe size	Q-2	3		735	165	900	1,050
705	8" pipe size		2.50		980	200	1,180	1,375
706	10" pipe size		2.20		1,375	225	1,600	1,850
707	12" pipe size		1.70		1,975	295	2,270	2,600
95-001	**VENT FLASHING**							
100	Aluminum with lead ring							
102	1-1/4" pipe	1 Plum	20	Ea.	3.80	8.90	12.70	17
103	1-1/2" pipe		20		3.80	8.90	12.70	17
104	2" pipe		18		3.99	9.85	13.84	18.65
105	3" pipe		17		4.88	10.45	15.33	20
106	4" pipe		16		5.30	11.10	16.40	22
135	Copper with lead ring							
140	1-1/4" pipe	1 Plum	20	Ea.	11.65	8.90	20.55	26
143	1-1/2" pipe		20		11.65	8.90	20.55	26
144	2" pipe		18		12.30	9.85	22.15	28
145	3" pipe		17		14.40	10.45	24.85	31
146	4" pipe		16		15.90	11.10	27	34
200	Galvanized with lead ring							
202	1-1/4" pipe	1 Plum	20	Ea.	3.60	8.90	12.50	16.80
203	1-1/2" pipe		20		3.60	8.90	12.50	16.80
204	2" pipe		18		3.80	9.85	13.65	18.45
205	3" pipe		17		4.56	10.45	15.01	20
206	4" pipe		16		4.92	11.10	16.02	21
238	Lead, cast, flanged							
240	1-1/4" pipe	1 Plum	20	Ea.	8.45	8.90	17.35	22
243	1-1/2" pipe		20		8.45	8.90	17.35	22
244	2" pipe		18		9.40	9.85	19.25	25
245	2-1/2" pipe		18		10.50	9.85	20.35	26
246	3" pipe		17		10.50	10.45	20.95	27
247	4" pipe		16		11.45	11.10	22.55	29
248	5" pipe		14		22	12.70	34.70	43
249	6" pipe		13		31	13.65	44.65	54
298	Neoprene, one piece							
300	1-1/4" pipe	1 Plum	24	Ea.	2.56	7.40	9.96	13.50
303	1-1/2" pipe		24		2.56	7.40	9.96	13.50
304	2" pipe		23		2.69	7.70	10.39	14.10
305	3" pipe		21		3.21	8.45	11.66	15.75
306	4" pipe		20		4.07	8.90	12.97	17.30
97-001	**WATER SUPPLY METERS**							
100	Detector, serves dual systems such as fire and domestic or process							
102	water, wide range capacity, UL and FM approved							
110	3" mainline x 2" by-pass, 400 GPM	Q-1	3.60	Ea.	1,725	89	1,814	2,025
114	4" mainline x 2" by-pass, 700 GPM	"	2.50		1,850	130	1,980	2,225
118	6" mainline x 3" by-pass, 1600 GPM	Q-2	2.60		2,575	190	2,765	3,100
122	8" mainline x 4" by-pass, 2800 GPM		2.10		3,900	235	4,135	4,625
126	10" mainline x 6" by-pass, 4400 GPM		2		6,625	250	6,875	7,650
130	10" x 12" mainlines x 6" by-pass,5400 GPM		1.70		6,900	295	7,195	8,025
200	Domestic/commercial, bronze							
202	Threaded							
206	5/8" diameter, to 20 GPM	1 Plum	16	Ea.	36	11.10	47.10	56
208	3/4" diameter, to 30 GPM		14		45	12.70	57.70	68
210	1" diameter, to 50 GPM		12		75	14.80	89.80	105
230	Threaded/flanged							
234	1-1/2" diameter, to 100 GPM	1 Plum	8	Ea.	150	22	172	195

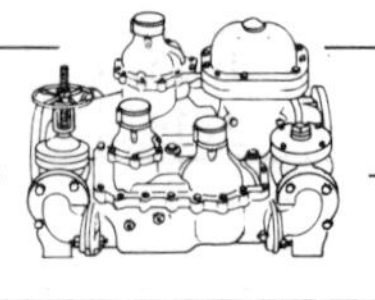

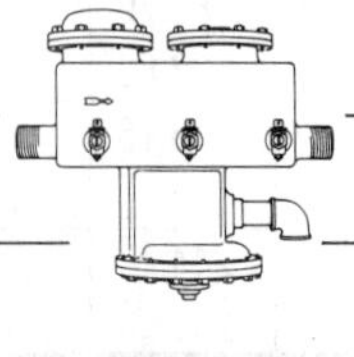

15

15.1 Pipe & Fittings

		CREW	DAILY OUTPUT	UNIT	BARE COSTS MAT.	BARE COSTS INST.	BARE COSTS TOTAL	TOTAL INCL O&P
97 236	2" diameter, to 160 GPM	1 Plum	6	Ea.	220	30	250	285
260	Flanged, compound							
264	3" diameter, 320 GPM	Q-1	3	Ea.	970	105	1,075	1,225
266	4" diameter, to 500 GPM		1.50		1,500	215	1,715	1,950
268	6" diameter, to 1,000 GPM		1		2,850	320	3,170	3,600
270	8" diameter, to 1,800 GPM	↓	.80	↓	5,425	400	5,825	6,550
700	Turbine							
726	Flanged							
730	2" diameter, to 160 GPM	1 Plum	7	Ea.	285	25	310	350
732	3" diameter, to 450 GPM	Q-1	3.60		430	89	519	600
734	4" diameter, to 650 GPM	"	2.50		725	130	855	980
736	6" diameter, to 1800 GPM	Q-2	2.60		1,450	190	1,640	1,875
738	8" diameter, to 2500 GPM		2.10		2,150	235	2,385	2,700
740	10" diameter, to 5500 GPM	↓	1.70	↓	3,150	295	3,445	3,900

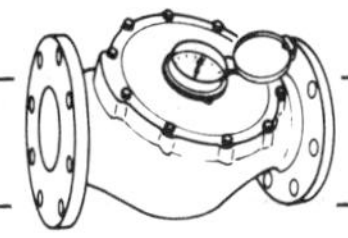

15.2 Plumbing Fixtures

		CREW	DAILY OUTPUT	UNIT	BARE COSTS MAT.	BARE COSTS INST.	BARE COSTS TOTAL	TOTAL INCL O&P
01-001	**FIXTURES** Includes trim fittings unless otherwise noted							
008	For rough-in, supply, waste, and vent, see add for each type							
012	C8.1 -401 For electric water coolers, see Division 15.3-45							
016	For color, unless otherwise noted, add			Ea.	20%			
04-001	**BATHS**							
010	Tubs, recessed porcelain enamel on cast iron, with trim							
014	B8.1 -410 42" x 37"	Q-1	5	Ea.	440	64	504	575
018	48" x 42"		4		680	80	760	865
022	72" x 36"		3		705	105	810	930
030	Mat bottom, 4' long		5.50		585	58	643	730
034	4'-6" long		5		385	64	449	515
038	5' long		4.40		245	73	318	375
042	5'-6" long		4		555	80	635	725
048	Above floor drain, 5' long		4		425	80	505	585
056	Corner 48" x 44"	↓	4.40	↓	785	73	858	970
075	Add for color				25%			
200	Enameled formed steel, 4'-6" long	Q-1	5.80	Ea.	175	55	230	270
220	5' long		5.50		165	58	223	265
230	Above floor drain, 5' long	↓	5.50	↓	180	58	238	280
235	Add for color				15%			
400	Soaking, acrylic with pop-up drain, 40" x 40"	Q-1	5.50	Ea.	630	58	688	775
410	66" x 36" x 18-1/2" deep		5		1,075	64	1,139	1,275
420	72" x 44" x 18" deep		4.80		1,200	67	1,267	1,425
430	72" x 60" x 20" deep	↓	4.40	↓	1,450	73	1,523	1,700
460	Module tub & showerwall surround, molded fiberglass							
461	5' long x 34" wide x 76" high	Q-1	4	Ea.	470	80	550	635
475	Handicap with 1-1/2" OD grab bar, antiskid bottom							
476	60" x 32-3/4" x 72" high	Q-1	4	Ea.	525	80	605	695
477	75" x 40" x 76" high with molded seat	"	3.50	"	730	91	821	935
500	Hospital type without trim (see 15.1-22),							
505	Bathing pool, porcelain enamel on cast iron, grab bars							
506	pop-up drain, 72" x 36"	Q-1	3	Ea.	955	105	1,060	1,200
510	Perineal (sitz), vitreous china		3		500	105	605	705
512	Pedestal, vitreous china		8		90	40	130	155
518	Pier tub, porcelain enamel on cast iron, 66-3/4" x 30"		3		965	105	1,070	1,225
520	Base, porcelain enamel on cast iron		8		620	40	660	740
530	Whirlpool, porcelain enamel on cast iron, 72" x 36"	↓	1	↓	2,250	320	2,570	2,925
600	Whirlpool, bath with vented overflow, molded fiberglass							

15

15.2 Plumbing Fixtures

		CREW	DAILY OUTPUT	UNIT	BARE COSTS MAT.	BARE COSTS INST.	BARE COSTS TOTAL	TOTAL INCL O&P
04 610	56" x 46" x 23"	Q-1	1	Ea.	2,300	320	2,620	3,000
640	72" x 36" x 18"		1		1,700	320	2,020	2,325
650	60" x 40-1/2" x 30"		1		1,775	320	2,095	2,425
660	72" x 66" x 30"		1		3,350	320	3,670	4,150
670	7' x 5-1/2'	↓	.30	↓	2,725	1,075	3,800	4,550
700	Redwood tub system							
705	4' diameter x 4' deep	Q-1	1	Ea.	1,250	320	1,570	1,825
710	5' diameter x 4' deep		1		1,400	320	1,720	2,000
715	6' diameter x 4' deep		.80		1,500	400	1,900	2,225
720	8' diameter x 4' deep		.80		1,750	400	2,150	2,500
960	Rough-in, supply, waste and vent, for all above tubs, add	↓	1.17	↓	87.85	275	362.85	490
08-001	**BIDET**							
018	Vitreous china, with trim on fixture	Q-1	5	Ea.	485	64	549	625
020	With trim for wall mounting		5		465	64	529	605
960	For rough-in, supply, waste and vent, add	↓	1.48	↓	100.98	215	315.98	425
12-001	**DENTAL FOUNTAIN** See also division 15.2-16.							
002	Deck mounted, with cuspidor							
005	Stainless steel receptor	1 Plum	4	Ea.	140	44	184	220
010	Enameled steel receptor		4		105	44	149	180
080	For aspirator, add		8		195	22	217	245
960	For rough-in, supply and waste, add	↓	1.27	↓	41.10	140	181.10	245
16-001	**DRINKING FOUNTAIN** For connection to cold water supply							
080								
100	B8.1-420 Wall mounted, non-recessed							
120	Aluminum,							
128	Dual bubbler type	1 Plum	3.20	Ea.	470	56	526	595
140	Bronze, with no back		4		500	44	544	615
160	C8.1-401 Cast iron, enameled, low back, single bubbler		4		230	44	274	315
164	Dual bubbler type		3.20		385	56	441	505
168	Triple bubbler type		3.20		490	56	546	620
180	Cast iron, enameled, for correctional institutions		4		210	44	254	295
200	 Fiberglass, 12" back, single bubbler unit		4		205	44	249	290
204	Dual bubbler		3.20		295	56	351	405
208	Triple bubbler		3.20		365	56	421	480
220	Polymarble, no back, single bubbler		4		205	44	249	290
224	Dual bubbler		3.20		325	56	381	440
228	Triple bubbler		3.20		480	56	536	610
240	Precast stone, no back		4		200	44	244	285
270	Stainless steel, single bubbler, no back		4		495	44	539	610
274	With back		4		675	44	719	805
278	Dual handle & wheelchair projection type		4		290	44	334	385
282	Dual level for handicapped type		3.20		540	56	596	675
284	Vandal resistant type	↓	4	↓	350	44	394	450
330	Vitreous china							
334	7" back	1 Plum	4	Ea.	215	44	259	300
394	For vandal-resistant bottom plate, add				16		16	17.60M
396	For freeze-proof valve system, add	1 Plum	2		25	89	114	155
398	For rough-in, supply and waste, add	"	1.24	↓	39.93	145	184.93	250
400	Wall mounted, semi-recessed							
420	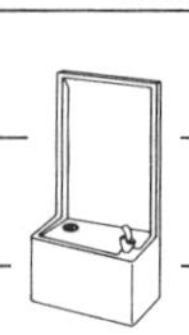 Poly-marble, single bubbler	1 Plum	4	Ea.	265	44	309	355
430	Fountain and cuspidor combination		2		590	89	679	775
460	Stainless steel, satin finish, single bubbler		4		300	44	344	395
468	Fountain and cuspidor combination		2		850	89	939	1,075
490	Vitreous china, single bubbler		4		300	44	344	395
500	Fountain and cuspidor combination		2		590	89	679	775
598	For rough-in, supply and waste, add	↓	1.24	↓	39.93	145	184.93	250
600	Wall mounted, fully recessed							
640	Poly-marble, single bubbler	1 Plum	4	Ea.	395	44	439	500
644	For water glass filler, add			"	60		60	66M

15.2 Plumbing Fixtures

			CREW	DAILY OUTPUT	UNIT	BARE COSTS MAT.	INST.	TOTAL	TOTAL INCL O&P
16	680	Stainless steel, single bubbler	1 Plum	4	Ea.	350	44	394	450
	690	Fountain and cuspidor combination		2		710	89	799	910
	756	For freeze-proof valve system, add		2		140	89	229	280
	758	For rough-in, supply and waste, add		1.24		39.93	145	184.93	250
	760	Floor mounted, pedestal type							
	770	Aluminum, architectural style, C.I. base	1 Plum	2	Ea.	250	89	339	405
	778	Wheelchair handicap unit		2		590	89	679	775
	800	Bronze, architectural style		2		610	89	699	800
	804	Enameled steel cylindrical column style		2		435	89	524	605
	820	Precast stone/concrete, cylindrical column		1		240	180	420	520
	824	Wheelchair handicap unit		1		485	180	665	790
	840	Stainless steel, architectural style		2		625	89	714	815
	860	Enameled iron, heavy duty service, 2 bubblers		2		550	89	639	735
	866	4 bubblers		2		555	89	644	740
	888	For freeze-proof valve system, add		2		150	89	239	295
	890	For rough-in, supply and waste, add		1.24		39.93	145	184.93	250
	910	Deck mounted							
	950	Stainless steel, circular receptor	1 Plum	4	Ea.	175	44	219	255
	954	14" x 9" receptor		4		135	44	179	215
	958	25" x 17" deep receptor, with water glass filler		3		105	59	164	200
	976	White enameled steel, 14" x 9" receptor		4		105	44	149	180
	986	White enameled cast iron, 24" x 16" receptor		3		115	59	174	210
	998	For rough-in, supply and waste, add		1.24		39.93	145	184.93	250
	20-001	**HOT WATER DISPENSERS**							
	016	Commercial, 100 cup, 11.3 amp	1 Plum	14	Ea.	205	12.70	217.70	245
	318	Household, 60 cup	"	14	"	110	12.70	122.70	140
	24-001	**INDUSTRIAL SAFETY FIXTURES** Rough-in not included							
	100	Eye wash fountain, aluminum bowl							
	120	Wall mounted	Q-1	4	Ea.	190	80	270	325
	140	Plastic bowl, pedestal mounted		4		110	80	190	235
	160	Unmounted		4		70	80	150	195
	180	Wall mounted		4		75	80	155	200
	200	Stainless steel, pedestal mounted		4		150	80	230	280
	220	Unmounted		4		96	80	176	220
	240	Wall mounted		4		99	80	179	225
	300	Eye wash, portable, self-contained				225		225	250M
	400	Eye and face wash, combination fountain							
	420	Stainless steel, pedestal mounted	Q-1	4	Ea.	175	80	255	310
	440	Unmounted		4		120	80	200	250
	460	Wall Mounted		4		125	80	205	255
	500	Shower, single head, drench, ball valve, pull, freestanding		4		145	80	225	275
	520	Horizontal or vertical supply		4		125	80	205	255
	600	Multi-nozzle, eye/face wash combination		4		275	80	355	420
	640	Multi-nozzle, 12 spray, shower only		4		785	80	865	980
	660	For freeze-proof, add		6		250	53	303	350
	800	Walk-thru decontamination with eye-face wash		2		1,225	160	1,385	1,575
	820	For freeze proof, add		4		150	80	230	280
	28-001	**INTERCEPTORS**							
	015	Grease, cast iron, 4 GPM, 8 lb. fat capacity	1 Plum	4	Ea.	140	44	184	220
	020	7 GPM, 14 lb. fat capacity		4		195	44	239	280
	100	10 GPM, 20 lb. fat capacity		4		260	44	304	350
	104	15 GPM, 30 lb. fat capacity		4		345	44	389	445
	106	20 GPM, 40 lb. fat capacity		3		420	59	479	550
	108	25 GPM, 50 lb. fat capacity	Q-1	3.50		470	91	561	650
	110	35 GPM, 70 lb. fat capacity		3		585	105	690	800
	112	Fabricated steel, 50 GPM, 100 lb. fat capacity		2		1,250	160	1,410	1,600
	114	75 GPM, 150 lb. fat capacity		2		1,375	160	1,535	1,750

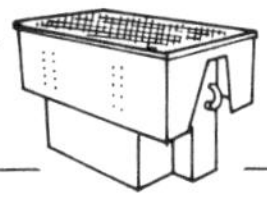

15.2 Plumbing Fixtures

		CREW	DAILY OUTPUT	UNIT	BARE COSTS MAT.	BARE COSTS INST.	BARE COSTS TOTAL	TOTAL INCL O&P
28 116	100 GPM, 200 lb. fat capacity	Q-1	2	Ea.	1,550	160	1,710	1,925
118	150 GPM, 300 lb. fat capacity		2		1,725	160	1,885	2,125
120	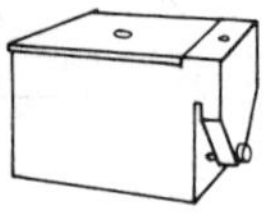 200 GPM, 400 lb. fat capacity		1.50		2,500	215	2,715	3,050
122	250 GPM, 500 lb. fat capacity		1.30		3,025	245	3,270	3,675
124	300 GPM, 600 lb. fat capacity		1		4,000	320	4,320	4,850
126	400 GPM, 800 lb. fat capacity	Q-2	1.20		5,725	415	6,140	6,900
128	500 GPM, 1000 lb. fat capacity	"	1		7,325	495	7,820	8,775
156	For chemical add-port, add				31		31	34M
158	For seepage pan, add				5%			
300	Hair, cast iron, 1-1/4" and 1-1/2" pipe connection	1 Plum	8		46	22	68	83
310	For chrome-plated cast iron, add				37		37	41M
320	For polished bronze, add				51		51	56M
330	For chrome plated bronze, add				64		64	70M
400	Oil, fabricated steel, 10 GPM, 2" pipe size	1 Plum	4		300	44	344	395
410	15 GPM, 2" or 3" pipe size		4		410	44	454	515
412	20 GPM, 2" or 3" pipe size		3		500	59	559	635
414	25 GPM, 2" or 3" pipe size	Q-1	3.50		540	91	631	725
416	35 GPM, 2", 3", or 4" pipe size		3		655	105	760	875
418	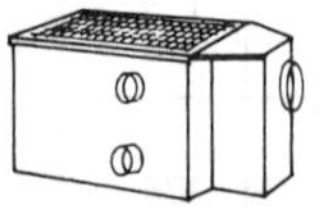 50 GPM, 2", 3", or 4" pipe size		2		885	160	1,045	1,200
420	75 GPM, 3" pipe size		2		1,425	160	1,585	1,800
422	100 GPM, 3" pipe size		2		1,550	160	1,710	1,925
424	150 GPM, 4" pipe size		2		1,925	160	2,085	2,350
426	200 GPM, 4" pipe size		1.50		2,700	215	2,915	3,275
428	250 GPM, 5" pipe size		1.30		3,800	245	4,045	4,525
430	300 GPM, 5" pipe size		1		5,000	320	5,320	5,950
432	400 GPM, 6" pipe size	Q-2	1.20		7,150	415	7,565	8,475
434	500 GPM, 6" pipe size	"	1		9,150	495	9,645	10,800
600	Solids, precious metals recovery, C.I., 1-1/4" to 2" pipe	1 Plum	4		81	44	125	155
610	Dental Lab., large, C.I., 1-1/2" to 2" pipe	"	3		370	59	429	495
30-001	**LABORATORY EQUIPMENT,** Corrosion resistant							
700	Tanks, covers included							
740	Halar E-CTFE, FRP casing, high chemical resistance							
741	Temperature range 110°F to 257°F							
742	7 gallon, 12" x 12" x 12"	Q-1	20	Ea.	150	16	166	190
743	15 gallon, 18" x 12" x 18"		17		280	18.80	298.80	335
744	30 gallon, 24" x 18" x 18"		12		380	27	407	455
750	Polyethylene, neutralization & dilution							
751	Continuous use to 180°F, includes 4" inlet & outlet							
755	55 gallon, upright cylinder	Q-1	5	Ea.	555	64	619	705
757	100 gallon, upright cylinder		4		570	80	650	745
759	250 gallon, upright cylinder		3		680	105	785	900
765	Polyethylene with ultra violet light inhibitor							
766	Continuous use to 180°F, includes saddle & fittings							
768	55 gallon 24" O.D. x 36" long	Q-1	10	Ea.	400	32	432	485
770	200 gallon 33" O.D. x 65" long		6		690	53	743	835
772	300 gallon 38" O.D. x 72" long		5		1,050	64	1,114	1,250
774	500 gallon 48" O.D. x 75" long		3		1,825	105	1,930	2,150
780	Polyethylene liner, fiberglass casing							
781	Continuous service to 220°F							
783	2 gallon, 8" x 8" x 8"	Q-1	20	Ea.	100	16	116	135
785	7 gallon, 12" x 12" x 12"		20		140	16	156	175
787	15 gallon, 24" x 12" x 12"		17		230	18.80	248.80	280
801	30 gallon, 24" x 18" x 18"		12		325	27	352	395
807	45 gallon, 24" x 18 x 24"		10		370	32	402	455
808	90 gallon, 36" x 24" x 24"		8		565	40	605	680
815	Polyethylene, heavy duty walls							
816	Continuous service to 180°F							
818	5 gallon, 11" I.D. x 15" deep	Q-1	20	Ea.	23	16	39	48
821	15 gallon, 14" I.D. x 27" deep	"	17	"	44	18.80	62.80	76

15

15.2 Plumbing Fixtures

		CREW	DAILY OUTPUT	UNIT	BARE COSTS MAT.	BARE COSTS INST.	BARE COSTS TOTAL	TOTAL INCL O&P
30 823	55 gallon, 22" I.D. x 36" deep	Q-1	10	Ea.	94	32	126	150
825	100 gallon 28" I.D. x 42" deep		8		190	40	230	265
827	200 gallon 36" I.D. x 48" deep		6		270	53	323	375
829	360 gallon 48" I.D. x 48" deep		5		445	64	509	580
835	With fiberglass casing							
842	55 gallon tank size	Q-1	20	Ea.	375	16	391	435
844	100 gallon tank size		16		580	20	600	665
846	200 gallon tank size		12		855	27	882	980
848	360 gallon tank size		10		1,275	32	1,307	1,450
32-001	**LAVATORIES** With trim, white unless noted otherwise							
050	Vanity top, porcelain enamel on cast iron							
060	B8.1-433 20" x 18"	Q-1	6.40	Ea.	100	50	150	180
064	26" x 18" oval		6.40		130	50	180	215
068	19" x 16" oval		6.40		94	50	144	175
072	18" round		6.40		92	50	142	175
076	C8.1-401 20" x 12" triangular bowl		6.40		100	50	150	180
086	For color, add				25%			
100	Cultured marble, 19" x 17", single bowl	Q-1	6.40		65	50	115	145
104	25" x 19", single bowl		6.40		77	50	127	155
108	31" x 19", single bowl		6.40		85	50	135	165
112	25" x 22", single bowl		6.40		78	50	128	160
116	37" x 22", single bowl		6.40		95	50	145	175
120	49" x 22", single bowl		6.40		125	50	175	210
124	61" x 22", single bowl		6.20		155	52	207	245
128	61" x 22", double bowl		4.80		185	67	252	300
132	73" x 22", double bowl		4.80		205	67	272	320
158	For color, same price							
190	Stainless steel, self-rimming, 25" x 22", single bowl, ledge	Q-1	6.40	Ea.	125	50	175	210
196	17" x 22", single bowl		6.40		115	50	165	200
204	18-3/4" round		6.40		395	50	445	505
260	Steel, enameled, 20" x 17", single bowl		5.80		60	55	115	145
266	19" round		5.80		58	55	113	145
272	18" round		5.80		53	55	108	140
286	For color, add				10%			
290	Vitreous china, 20" x 16", single bowl	Q-1	5.40		150	59	209	250
296	20" x 17", single bowl		5.40		105	59	164	200
302	19" round, single bowl		5.40		100	59	159	195
308	19" x 16", single bowl		5.40		105	59	164	200
314	17" x 14", single bowl		5.40		94	59	153	190
320	22" x 13", single bowl		5.40		100	59	159	195
356	For color, add				20%			
358	Rough-in, supply, waste and vent for all above lavatories	Q-1	1.56		67.84	205	272.84	370
400	Wall hung							
404	Porcelain enamel on cast iron, 16" x 14", single bowl	Q-1	8	Ea.	165	40	205	240
406	18" x 15" single bowl		8		185	40	225	260
412	19" x 17", single bowl		8		155	40	195	230
418	20" x 18", single bowl		8		115	40	155	185
424	22" x 19", single bowl		8		205	40	245	285
458	For color, add				30%			
600	Vitreous china, 18" x 15", single bowl with backsplash	Q-1	7		120	46	166	200
606	19" x 17", single bowl		7		72	46	118	145
612	24" x 20", single bowl		7		165	46	211	250
618	18" x 18", corner style		8		230	40	270	310
650	For color, add				30%			
670	Hospital type, without trim (see 15.1-22)							
671	20" x 18", contoured splash shield	Q-1	8	Ea.	74	40	114	140
672	26" x 20", patient, corner,side deck & back		7		240	46	286	330
673	28" x 20", surgeon, side decks		8		300	40	340	390
674	28" x 22", surgeon scrub-up, deep bowl		8		315	40	355	405

		CREW	DAILY OUTPUT	UNIT	BARE COSTS MAT.	BARE COSTS INST.	BARE COSTS TOTAL	TOTAL INCL O&P
32 675	20" x 27", patient, wheelchair	Q-1	7	Ea.	210	46	256	295
676	30" x 22", all purpose		7		275	46	321	370
677	30" x 22", plaster work		7		275	46	321	370
682	20" x 24" clinic service, liquid/solid waste		6		460	53	513	585
696	Rough-in, supply, waste and vent for above lavatories		1.36		110.74	235	345.74	460
36-001	**LAUNDRY SINKS** With trim							
002	Porcelain enamel on cast iron, black iron frame							
005	B8.1 -432 24" x 20", single compartment	Q-1	6	Ea.	160	53	213	255
010	24" x 23", single compartment		6		185	53	238	280
020	48" x 20", double compartment		5		285	64	349	405
200	Molded stone, on wall hanger or legs							
202	22" x 21", single compartment	Q-1	6	Ea.	90	53	143	175
210	45" x 21", double compartment	"	5	"	150	64	214	255
300	Plastic, on wall hanger or legs							
302	18" x 23", single compartment	Q-1	6.50	Ea.	51	49	100	125
310	20" x 24", single compartment		6.50		70	49	119	150
320	36" x 23", double compartment		5.50		75	58	133	165
330	40" x 24", double compartment		5.50		120	58	178	215
500	Stainless steel, counter top, 22" x 17" single compartment		6		135	53	188	225
510	19" x 22", single compartment		6		145	53	198	235
520	33" x 22", double compartment		5		220	64	284	335
960	Rough-in, supply, waste and vent, for all laundry sinks		1.34		75.77	240	315.77	430
38-001	**PRISON/INSTITUTION FIXTURES,** Stainless steel							
100	Lavatory, wall hung, push button filler valve							
110	Rectangular bowl	Q-1	8	Ea.	500	40	540	610
120	Oval bowl		8		475	40	515	580
124	Oval bowl, corner mount		8		555	40	595	670
130	For lavatory rough-in, supply, waste and vent		1.09		58.21	295	353.21	490
160	Pressure soap machine, 25 gal. tank		6		1,700	53	1,753	1,950
170	Service sink, with soap dish							
174	24" x 19" size	Q-1	3	Ea.	1,100	105	1,205	1,375
179	For sink rough-in, supply, waste and vent	"	.89	"	123.89	360	483.89	655
180	Shower cabinet, unitized							
184	36" x 36" x 88"	Q-1	2.20	Ea.	3,375	145	3,520	3,925
190	Shower package for built-in							
194	Hot & cold valves, recessed soap dish	Q-1	6	Ea.	415	53	468	535
200	Urinal, back supply and flush							
220	Wall hung	Q-1	4	Ea.	1,525	80	1,605	1,800
224	Stall		2.50		1,650	130	1,780	2,000
230	For urinal rough-in, supply, waste and vent		1.49		80.37	215	295.37	400
300	Water closet, integral seat, back supply and flush							
330	Wall hung, wall outlet	Q-1	5.80	Ea.	1,375	55	1,430	1,600
340	Floor mount, wall outlet		5.80		1,375	55	1,430	1,600
344	Floor mount, floor outlet		5.80		1,425	55	1,480	1,650
348	For recessed tissue holder, add				85		85	94M
350	For water closet rough-in, supply, waste and vent	Q-1	1.19		99.18	270	369.18	495
500	Water closet and lavatory units, push button filler valves,							
501	soap & paper holders, seat							
530	Wall hung	Q-1	5	Ea.	1,350	64	1,414	1,575
540	Floor mount		5		1,350	64	1,414	1,575
630	For unit rough-in, supply, waste and vent		1		110.80	320	430.80	585
40-001	**PUMPS** See also Division 15.4-40							
41-001	**PUMPS, CIRCULATING** Heated or chilled water application							
060	Bronze, sweat connections, 1/40 HP, in line							
064	3/4" size	Q-1	16	Ea.	69	20	89	105
100	Flange connection, 3/4" to 1-1/2" size							

15.2 Plumbing Fixtures

		CREW	DAILY OUTPUT	UNIT	BARE COSTS MAT.	BARE COSTS INST.	BARE COSTS TOTAL	TOTAL INCL O&P
41 104	1/12 HP	Q-1	6	Ea.	180	53	233	275
106	1/8 HP		6		275	53	328	380
110	1/3 HP		6		290	53	343	395
114	2" size, 1/6 HP		5		355	64	419	485
118	2-1/2" size, 1/4 HP		5		535	64	599	680
122	3" size, 1/4 HP		4		560	80	640	730
126	1/3 HP		4		680	80	760	865
130	1/2 HP		4		700	80	780	885
134	3/4 HP		4		760	80	840	950
138	1 HP		4		1,125	80	1,205	1,350
200	Cast iron, flange connection							
204	3/4" to 1-1/2" size, in line, 1/12 HP	Q-1	6	Ea.	115	53	168	205
206	1/8 HP		6		180	53	233	275
210	1/3 HP		6		200	53	253	295
214	2" size, 1/6 HP		5		215	64	279	330
218	2-1/2" size, 1/4 HP		5		295	64	359	415
222	3" size, 1/4 HP		4		300	80	380	445
226	1/3 HP		4		400	80	480	555
230	1/2 HP		4		425	80	505	585
234	3/4 HP		4		505	80	585	670
238	1 HP		4		710	80	790	895
260	For non-ferrous impeller, add				3%			
300	High head, bronze impeller							
303	1-1/2" size 1/2 HP	Q-1	5	Ea.	430	64	494	565
304	1-1/2" size 3/4 HP		5		470	64	534	610
305	2" size 1 HP		4		555	80	635	725
309	2" size 1-1/2 HP		4		630	80	710	810
400	Close coupled, end suction, bronze impeller							
404	1-1/2" size, 1-1/2 HP, to 40 GPM	Q-1	3	Ea.	570	105	675	780
409	2" size, 2 HP, to 50 GPM		3		645	105	750	865
410	2" size, 3 HP, to 90 GPM		2.30		1,025	140	1,165	1,325
419	2-1/2" size, 3 HP, to 150 GPM		2		750	160	910	1,050
430	3" size, 5 HP, to 225 GPM		1.80		985	180	1,165	1,350
441	4" size, 5 HP, to 350 GPM		1.60		955	200	1,155	1,350
442	4" size, 7-1/2 HP, to 350 GPM		1.60		1,150	200	1,350	1,550
452	5" size, 10 HP, to 600 GPM	Q-2	1.70		1,475	295	1,770	2,050
453	5" size, 15 HP, to 1000 GPM		1.70		1,725	295	2,020	2,325
461	6" size, 20 HP, to 1350 GPM		1.50		2,050	330	2,380	2,725
462	6" size, 25 HP, to 1550 GPM		1.50		2,300	330	2,630	3,000
500	Base mounted, bronze impeller, coupling guard							
504	1-1/2" size, 1-1/2 HP, to 40 GPM	Q-1	2.30	Ea.	605	140	745	865
509	2" size, 2 HP, to 50 GPM		2.30		660	140	800	925
510	2" size, 3 HP, to 90 GPM		2		900	160	1,060	1,225
519	2-1/2" size, 3 HP, to 150 GPM		1.80		715	180	895	1,050
530	3" size, 5 HP, to 225 GPM		1.60		805	200	1,005	1,175
541	4" size, 5 HP, to 350 GPM		1.50		865	215	1,080	1,250
542	4" size, 7-1/2 HP, to 350 GPM		1.50		900	215	1,115	1,300
552	5" size, 10 HP, to 600 GPM	Q-2	1.60		1,175	310	1,485	1,750
553	5" size, 15 HP, to 1000 GPM		1.60		1,275	310	1,585	1,850
561	6" size, 20 HP, to 1350 GPM		1.40		1,600	355	1,955	2,275
562	6" size, 25 HP, to 1550 GPM		1.40		1,675	355	2,030	2,350
580	The above pump capacities are based on 1800 RPM,							
581	At a 60 foot head. increasing the RPM							
582	or decreasing the head will increase the GPM.							
43-001	**PUMPS, CONDENSATE RETURN SYSTEM**							
020	Simplex, motor, float switch, controls, cast iron receiver	Q-1	1	Ea.	850	320	1,170	1,400
100	Duplex, 2 pumps, motors, float switch,							
106	alternator asssembly, C.I. receiver	Q-1	.50	Ea.	2,300	640	2,940	3,450
44-001	**PUMP FOOT VALVES** With strainer							
020	4" size	Q-2	2	Ea.	1,300	250	1,550	1,800

15

15.2 Plumbing Fixtures

		Crew	Daily Output	Unit	Bare Costs Mat.	Bare Costs Inst.	Bare Costs Total	Total Incl O&P
44 030	8" size	Q-2	1.30	Ea.	2,000	385	2,385	2,750
040	12" size	Q-3	1		2,900	680	3,580	4,175
050	16" size	"	.60	↓	5,000	1,125	6,125	7,125
45-001	**PUMPS, GENERAL UTILITY** With motor, mounted on base							
020	Multi-stage, horizontal split, for boiler feed applications							
030	Two stage, 3" discharge x 4" suction, 75 HP	Q-7	.30	Ea.	10,000	2,275	12,275	14,300
034	Four stage, 3" discharge x 4" suction, 150 HP	"	.18	"	17,000	3,800	20,800	24,200
200	Single stage							
206	End suction, 1"D. x 2"S., 3 HP	Q-1	.50	Ea.	2,400	640	3,040	3,575
210	1-1/2"D. x 3"S., 10 HP	"	.40		2,600	800	3,400	4,025
214	2"D. x 3"S., 15 HP	Q-2	.60		2,800	830	3,630	4,275
218	3"D. x 4"S., 20 HP		.50		2,900	995	3,895	4,625
222	4"D. x 6"S., 30 HP	↓	.40		3,200	1,250	4,450	5,325
300	Double suction, 2"D. x 2-1/2"S., 10 HP	Q-1	.30		3,300	1,075	4,375	5,175
306	3"D. x 4"S., 15 HP	Q-2	.46		3,400	1,075	4,475	5,300
310	4"D. x 5"S., 30 HP		.40		4,700	1,250	5,950	6,975
314	5"D. x 6"S., 50 HP	↓	.33		5,100	1,500	6,600	7,800
318	6"D. x 8"S., 60 HP	Q-3	.30	↓	5,700	2,275	7,975	9,550
800	Vertical submerged, with non-submerged motor							
806	1"D., 3 HP	Q-1	.50	Ea.	3,400	640	4,040	4,675
810	1-1/2"D., 10 HP	"	.30		3,500	1,075	4,575	5,400
814	2"D., 15 HP	Q-2	.40		3,500	1,250	4,750	5,650
818	3"D., 25 HP		.30		3,500	1,650	5,150	6,250
822	4"D., 30 HP	↓	.20	↓	3,800	2,475	6,275	7,775
47-001	**PUMPS, GRINDER SYSTEM** Complete, includes check valve, tank,							
002	standard controls. Excavation not included							
026	Simplex, 11 GPM at 40 PSIG, 60 gal. tank			Ea.	1,750		1,750	1,925M
030	For manway, 26" I.D., 18" high, add				405		405	445M
034	26" I.D., 36" high, add				560		560	615M
038	43" I.D., 4' high, add				735		735	810M
060	Simplex, 11 GPM at 40 PSIG, 120 gal. tank				2,300		2,300	2,525M
066	For manway 26" I.D., 18" high, add				350		350	385M
070	26" I.D., 36" high, add				415		415	455M
074	43" I.D., 4' high, add				1,250		1,250	1,375M
200	Duplex, 22 GPM at 40 PSIG, 120 gal. tank				3,775		3,775	4,150M
206	For manway 43" I.D., 4' high, add				1,825		1,825	2,000M
240	For core only				1,400		1,400	1,550M
241	For salt water core, add			↓	130		130	145M
260	For addit. manway 5' to 10' high, 60 gal. tank, per ft., add			V.L.F.	200		200	220M
262	For manways 5' to 10', 120 gal tank, add			Ea.	250		250	275M
15 49-001	**PUMPS, PRESSURE BOOSTER SYSTEM** Constant speed							
020	2 pump system with hydrocumulator							
030	100 GPM @ 50 psi	Q-2	.30	Ea.	9,975	1,650	11,625	13,400
040	300 GPM @ 100 psi	"	.26	"	13,900	1,925	15,825	18,100
100	3 pump system without hydrocumulator							
110	300 GPM @ 100 psi	Q-2	.23	Ea.	14,400	2,150	16,550	19,000
120	1000 GPM @ 100 psi		.12		27,300	4,150	31,450	36,000
130	5000 GPM @ 100 psi	↓	.08	↓	46,700	6,225	52,925	60,500
51-001	**PUMPS, PEDESTAL SUMP** With float control							
040	Molded base, 42 GPM at 15' head, 1/3 HP	1 Plum	5	Ea.	100	36	136	160
080	Iron base, 42 GPM at 15' head, 1/3 HP	"	5	"	83	36	119	145
52-001	**PUMPS, SEWAGE EJECTOR** With operating and level controls							
010	Simplex, bitumastic coated steel tank, cover, 10' head, 230 volt							
060	Bronze pump							
064	70 GPM, 1/3 HP	Q-1	3	Ea.	1,125	105	1,230	1,400
068	143 GPM, 1/2 HP	"	2.50	"	1,175	130	1,305	1,475
104	Cast iron pump							

15.2 Plumbing Fixtures

		CREW	DAILY OUTPUT	UNIT	BARE COSTS MAT.	BARE COSTS INST.	BARE COSTS TOTAL	TOTAL INCL O&P
52 106	110 GPM, 1/2 HP	Q-1	2.50	Ea.	1,125	130	1,255	1,425
108	173 GPM, 3/4 HP		2		1,950	160	2,110	2,375
110	218 GPM, 1 HP		1.60		1,975	200	2,175	2,450
112	285 GPM, 2 HP		1.30		2,425	245	2,670	3,025
114	325 GPM, 3 HP	Q-2	1.40		2,575	355	2,930	3,350
116	370 GPM, 5 HP	"	1		2,900	495	3,395	3,900
126	For fiberglass reinforced plastic tank, deduct				10%			
200	Duplex, bitumastic coated steel tank, cover, 10' head, 230 volt							
264	Bronze pump							
266	70 GPM each pump, 1/3 HP	Q-1	2.50	Ea.	2,575	130	2,705	3,025
270	143 GPM each pump, 1/2 HP	"	2	"	2,700	160	2,860	3,200
304	Cast iron pump							
306	110 GPM each pump, 1/2 HP	Q-1	2	Ea.	2,500	160	2,660	2,975
308	173 GPM each pump, 3/4 HP	"	1.60		3,325	200	3,525	3,950
310	218 GPM each pump, 1 HP	Q-2	1.20		3,550	415	3,965	4,500
312	285 GPM each pump, 2 HP		1		3,875	495	4,370	4,975
314	325 GPM each pump, 3 HP		.80		4,125	620	4,745	5,425
316	370 GPM each pump, 5 HP		.50		4,475	995	5,470	6,350
326	For fiberglass reinforced plastic tank, deduct				40%			
53-001	**PUMPS, SPRINKLER** With check valve, steel base, 15' lift							
010	37 GPM, 3/4 HP	1 Plum	4	Ea.	220	44	264	305
014	56 GPM, 1-1/2 HP	"	2		295	89	384	455
018	68 GPM, 2 H.P.	Q-1	2		340	160	500	605
54-001	**PUMPS, SUBMERSIBLE** Dewatering							
002	Sand & sludge, 20' head, starter & level control							
005	Cast iron							
010	2" discharge, 10 GPM	1 Plum	4	Ea.	305	44	349	400
016	60 GPM		3		400	59	459	525
020	120 GPM		2		750	89	839	955
100	160 GPM	Q-1	1.50		1,175	215	1,390	1,600
110	3" discharge, 220 GPM		1.20		1,450	265	1,715	1,975
120	300 GPM		1		1,725	320	2,045	2,350
136	4" discharge, 300 GPM		.90		1,750	355	2,105	2,450
200	Sewage & solids mixture, 8' head, automatic							
202	Bronze, 1/2 HP							
210	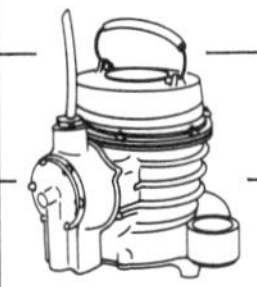 75 GPM, 1-1/4" or 1-1/2" NPT discharge	1 Plum	8	Ea.	325	22	347	390
214	120 GPM, 2" or 3" discharge	"	7	"	400	25	425	475
225	Cast iron							
230	122 GPM, 1/2 HP, 2" or 3" discharge	1 Plum	6	Ea.	295	30	325	365
300	Sewage & Solids mixture, high capacity, 20' head, non-automatic							
310	Cast iron, 3" flanged discharge							
314	40 GPM, 3/4 HP	1 Plum	4	Ea.	1,000	44	1,044	1,175
316	100 GPM, 1 HP	"	3		1,100	59	1,159	1,300
318	175 GPM, 2 HP	Q-1	4.50		1,150	71	1,221	1,375
320	240 GPM, 3 HP		4		1,300	80	1,380	1,550
322	310 GPM, 5 HP		3		1,450	105	1,555	1,750
330	For tandem seal, add				100		100	110M
400	Cast iron, 4" flanged discharge							
410	10 GPM, 3/4 HP	1 Plum	4	Ea.	1,100	44	1,144	1,275
412	40 GPM, 1 HP	"	3		1,150	59	1,209	1,350
414	175 GPM, 2 HP	Q-1	4.50		1,225	71	1,296	1,450
416	275 GPM, 3 HP		4		1,425	80	1,505	1,675
418	450 GPM, 5 HP		3		1,525	105	1,630	1,825
424	For tandem seal, add				100		100	110M
500	Sewage and solids, high head							
510	Bronze, 45 GPM at 25', 1/2 HP	1 Plum	6	Ea.	410	30	440	495
514	28 GPM at 50', 1 HP		5		505	36	541	605
516	25 GPM at 75', 1-1/2 HP		4		825	44	869	970
518	18 GPM at 100', 1-1/2 HP		4		825	44	869	970

15

15.2 Plumbing Fixtures

	15.2 Plumbing Fixtures	CREW	DAILY OUTPUT	UNIT	BARE COSTS MAT.	BARE COSTS INST.	BARE COSTS TOTAL	TOTAL INCL O&P
54 550	Cast iron, 45 GPM at 25', 1/2 HP	1 Plum	6	Ea.	335	30	365	410
554	28 GPM at 50', 1 HP		5		690	36	726	810
556	25 GPM at 75', 1-1/2 HP		4		735	44	779	875
700	Sump pump, 10' head, automatic							
710	Bronze, 22 GPM., 1/4 HP, 1-1/4" discharge	1 Plum	6	Ea.	200	30	230	265
714	68 GPM, 1/2 HP, 1-1/4" or 1-1/2" dis.		6		325	30	355	400
716	94 GPM, 1/2 HP, 1-1/4" or 1-1/2" dis.		5		430	36	466	525
718	105 GPM, 1/2 HP, 2" or 3" discharge		4		400	44	444	505
750	Cast iron, 23 GPM, 1/4 HP, 1-1/4" discharge		6		94	30	124	145
754	35 GPM, 1/3 HP, 1-1/4" discharge		6		105	30	135	160
756	68 GPM, 1/2 HP, 1-1/4" or 1-1/2" dis.		5		220	36	256	295
758	87 GPM, 1/2 HP, 2" or 3" discharge		4		230	44	274	315
55-001	**PUMPS, WELL** Water system, with pressure control							
100	Deep well, multi-stage jet, 42 gal. tank							
104	110' lift, 40 lb. discharge, 5 GPM, 3/4 HP	1 Plum	.80	Ea.	440	220	660	805
200	Shallow well, reciprocating, 25 gal. tank							
204	25' lift, 5 GPM, 1/3 HP	1 Plum	2	Ea.	340	89	429	500
300	Shallow well, single stage jet, 42 gal. tank,							
304	15' lift, 40 lb. discharge, 16 GPM, 3/4 HP	1 Plum	2	Ea.	335	89	424	495
56-001	**SHOWERS**							
150	Stall, with door and trim							
151	B8.1 -440 Baked enamel, molded stone receptor, 30" square	Q-1	2	Ea.	260	160	420	515
152	32" square		2		260	160	420	515
154	Terrazzo receptor, 32" square		2		460	160	620	735
156	36" square		1.80		490	180	670	795
158	36" corner angle		1.80		490	180	670	795
160	For color add				10%			
300	Fiberglass, one piece, with 3 walls, 32" x 32" square	Q-1	2.40		210	135	345	425
310	36" x 36" square	"	2.40		240	135	375	455
320	Handicap, 1-1/2" OD grab bars, nonskid floor							
321	48" x 34-1/2" x 72" corner seat	Q-1	2.40	Ea.	525	135	660	770
322	60" x 34-1/2" x 72" corner seat		2		550	160	710	835
323	36" x 34-1/2" x 72" fold up seat		2.20		835	145	980	1,125
325	64" x 65-3/4" x 81-1/2" fold. seat, whlchr.		1.80		2,225	180	2,405	2,700
400	Polypropylene, with molded-stone floor, 30" x 30"		2		290	160	450	550
410	32" x 32"		2		290	160	450	550
496	Rough-in, supply, waste and vent for above showers		1.24		69.45	260	329.45	450
500	Built-in, head, arm, 4 GPM valve	1 Plum	4		42	44	86	110
520	Head, arm, by-pass, integral stops, handles		3.60		98	49	147	180
550	Head, water economizer, 3.0 GPM		24		32	7.40	39.40	46
580	Mixing valve, Built-in		6		65	30	95	115
590	Exposed		6		230	30	260	295
595	Module, handicap, SS panel, fixed & hand held head, control							
596	valves, grab bar, curtain & rod, folding seat	1 Plum	4	Ea.	1,850	44	1,894	2,100
600	Group, w/valve, rough-in and rigging not included							
680	Column, 6 heads, no receptors, less partitions	Q-1	3	Ea.	1,500	105	1,605	1,800
690	With enameled partitions		1		2,525	320	2,845	3,250
760	5 heads, no receptors, less partitions		3		1,275	105	1,380	1,550
762	4 heads (1 handicap) no receptors, less partitions		3		655	105	760	875
770	With enameled partitions		1		2,075	320	2,395	2,750
800	Corner, 2 heads, no receptors, less partitions		4		825	80	905	1,025
810	With partitions		2		1,025	160	1,185	1,350
60-001	**SINKS** With faucets and drain							
005	Corrosion resistant							
100	B8.1 -434 Polyethylene, single sink, bench mounted, with							
102	plug & waste fitting with 1-1/2" straight threads							
103	2 drainboards, backnut & strainer							
105	18-1/2"x15-1/2"x12-1/2" sink,54"x24"OD	Q-1	3	Ea.	490	105	595	695

15

15.2 Plumbing Fixtures

	15.2 Plumbing Fixtures	CREW	DAILY OUTPUT	UNIT	BARE COSTS MAT.	BARE COSTS INST.	BARE COSTS TOTAL	TOTAL INCL O&P
60 110	Single drainboard, backnut & strainer							
113	18-1/2"x15-1/2"x12-1/2"sink, 47"x24"OD	Q-1	3	Ea.	450	105	555	650
115	18-1/2"x15-1/2"x12-1/2"sink, 70"x24"OD	"	3	"	535	105	640	745
129	Flanged 1-1/4" wide, rectangular with strainer							
130	plug & waste fitting, 1-1/2" straight threads							
132	12"x12"x8" sink, 14-1/2"x14-1/2" O.D.	Q-1	4	Ea.	125	80	205	255
134	16"x16"x8" sink, 18-1/2"x18-1/2" O.D.		4		145	80	225	275
136	21"x18"x10" sink, 23-1/2"x20-1/2" O.D.		4		185	80	265	320
149	For rough-in, supply, waste & vent, add		1.30		81.46	245	326.46	445
160	Polypropylene							
162	Cup sink, oval, integral strainers							
164	6"x3" I.D., 7"x 4" O.D.	Q-1	6	Ea.	32	53	85	110
166	9"x3" I.D., 10"x 4-1/2" O.D.	"	6	"	52	53	105	135
172	Overflow standpipe							
174	1-1/2" diam. x 11" long			Ea.	19.80		19.80	22M
198	For rough-in, supply, waste & vent, add	Q-1	1.30		81.46	245	326.46	445
200	B8.1 -431 Kitchen, counter top, P.E. on C.I., 24" x 21" single bowl		3.20		110	100	210	265
210	30" x 21" single bowl		3.20		130	100	230	285
220	32" x 21" double bowl		2.60		145	125	270	335
230	42" x 21" double bowl		2.60		240	125	365	440
300	Stainless steel, self rimming, 19" x 18" single bowl		3.20		200	100	300	365
310	25" x 22" single bowl		3.20		220	100	320	385
320	33" x 22" double bowl		2.60		235	125	360	435
330	43" x 22" double bowl		2.60		260	125	385	465
340	44" x 22" triple bowl		2.20		360	145	505	605
350	44" x 24" corner double bowl		2.60		275	125	400	480
400	Steel, enameled, with ledge, 24" x 21" single bowl		3.20		59	100	159	210
410	32" x 21" double bowl		2.60		77	125	202	260
496	For color sinks except stainless steel, add				10%			
498	For rough-in, supply, waste and vent, counter top sinks	Q-1	1.18		85.75	270	355.75	485
500	Kitchen, raised deck, P.E. on C.I.							
510	32" x 21", dual level, double bowl	Q-1	1.60	Ea.	220	200	420	530
520	42" x 21", double bowl & disposer well	"	1.20		285	265	550	700
570	For color, add				30%			
579	For rough-in, supply, waste & vent, sinks	Q-1	1.13		89.45	285	374.45	505
665	Service, floor, corner, P.E. on C.I., 28" x 28"	"	4		315	80	395	460
675	B8.1 -434 Vinyl coated rim guard, add				35		35	39M
679	For rough-in, supply, waste & vent, floor service sinks	Q-1	1.05		67.98	305	372.98	515
700	Service, wall, P.E. on C.I., roll rim, 22" x 18"		3		230	105	335	405
710	24" x 20"		3		255	105	360	435
760	For stainless steel rim guard, add				45		45	50M
780	For stainless steel rim guard, front only, add				22		22	24M
860	Vitreous china, 22" x 20"	Q-1	3		250	105	355	430
896	For stainless steel rim guard, front or side, add				15		15	16.50M
898	For rough-in, supply, waste & vent, wall service sinks	Q-1	.89		142.06	360	502.06	675
64-001	**SINK WASTE TREATMENT** System for commercial kitchens							
010	includes clock timer, & fittings							
020	System less chemical, wall mounted cabinet	1 Plum	16	Ea.	200	11.10	211.10	235
200	Chemical, 1 gallon, add				27		27	30M
210	6 gallons, add				110		110	120M
220	15 gallons, add				250		250	275M
230	30 gallons, add				485		485	535M
240	55 gallons, add				825		825	910M
66-001	**TOILET SEATS**							
010	Molded composition, white							
015	Industrial, w/o cover, open front, regular bowl	1 Plum	24	Ea.	13.10	7.40	20.50	25
020	With self-sustaining hinge		24		13.90	7.40	21.30	26
022	With self-sustaining check hinge		24		13.80	7.40	21.20	26
024	Extra heavy, with check hinge		24		18.10	7.40	25.50	31

15.2 Plumbing Fixtures		CREW	DAILY OUTPUT	UNIT	BARE COSTS MAT.	BARE COSTS INST.	BARE COSTS TOTAL	TOTAL INCL O&P
66 026	Elongated bowl, same price							
030	Junior size, w/o cover, open front							
032	Regular primary bowl, open front	1 Plum	24	Ea.	17.95	7.40	25.35	30
034	Regular baby bowl, open front, check hinge		24		17.95	7.40	25.35	30
038	Open back & front, w/o cover, reg. or elongated bowl		24		18.10	7.40	25.50	31
040	Residential							
042	Regular bowl, w/cover, closed front	1 Plum	24	Ea.	17.70	7.40	25.10	30
044	Open front	″	24	″	17.70	7.40	25.10	30
046	Elongated bowl, add				25%			
050	Self-raising hinge, w/o cover, open front							
052	Regular bowl	1 Plum	24	Ea.	43	7.40	50.40	58
054	Elongated bowl	″	24	″	44	7.40	51.40	59
060	Self-sustaining hinge, w/o cover, open front							
062	Regular bowl	1 Plum	24	Ea.	42	7.40	49.40	57
064	Elongated bowl	″	24	″	43	7.40	50.40	58
070	Molded wood, white, with cover							
072	Closed front, regular bowl, square back	1 Plum	24	Ea.	6	7.40	13.40	17.30
074	Extended back		24		8	7.40	15.40	19.50
078	Elongated bowl, square back		24		8.35	7.40	15.75	19.90
080	Open front, same price as closed front							
085	Decorator styles							
087	Cane top	1 Plum	24	Ea.	10.65	7.40	18.05	22
089	Vinyl top, patterned		24		9.85	7.40	17.25	22
090	Vinyl padded, plain colors, regular bowl		24		9.95	7.40	17.35	22
093	Elongated bowl		24		16	7.40	23.40	28
100	Solid plastic, white							
103	Industrial, w/o cover, open front, regular bowl	1 Plum	24	Ea.	12.55	7.40	19.95	25
108	Extra heavy, concealed check hinge		24		17.20	7.40	24.60	30
110	Self-sustaining hinge		24		13.25	7.40	20.65	25
115	Elongated bowl		24		12.60	7.40	20	25
117	Concealed check		24		12.60	7.40	20	25
119	Self-sustaining hinge, concealed check		24		13.40	7.40	20.80	25
122	Residential, with cover, closed front, regular bowl		24		16.85	7.40	24.25	29
124	Elongated bowl		24		21	7.40	28.40	34
126	Open front, regular bowl		24		16.85	7.40	24.25	29
128	Elongated bowl		24		21	7.40	28.40	34
68-001	**URINALS**							
300	Wall hung, vitreous china, with hanger & self-closing valve	Q-1	3	Ea.	275	105	380	455
330	B8.1 -450 Rough-in, supply, waste & vent		1.49		71.21	215	286.21	390
500	Stall type, vitreous china, includes valve		2.50		345	130	475	565
510	C8.1 -401 3″ seam cover, add		12		84	27	111	130
520	6″ seam cover, add		12		115	27	142	165
698	Rough-in, supply, waste and vent		1.36		79.52	235	314.52	425
72-001	**WASH CENTER** prefabricated,stainless steel, semirecessed							
005	Lavatory, storage cabinet, mirror, light & switch, electric							
006	outlet, towel dispenser, waste receptacle & trim							
010	Foot water valve, cup & soap dispenser, 16″ W x 54-3/4″H	Q-1	8	Ea.	960	40	1,000	1,125
020	Handicap, wrist blade handles, 17″ W x 66-1/2″ H		8		675	40	715	800
022	20″ W x 67-3/8″ H		8		900	40	940	1,050
030	Push button metering & thermostatic mixing valves							
032	Handicap 17″ W x 27-1/2″ H	Q-1	8	Ea.	540	40	580	650
76-001	**WASH FOUNTAINS** Rigging not included							
190	Group, foot control							
200	B8.1 -560 Precast terrazzo, circular, 36″ diam., 5 or 6 persons	Q-2	3	Ea.	835	165	1,000	1,150
210	54″ diameter for 8 or 10 persons		2.50		975	200	1,175	1,350
240	Semi-circular, 36″ diam. for 3 persons		3		765	165	930	1,075
242	36″ diam. for 3 persons in wheelchairs		3		1,250	165	1,415	1,625
250	54″ diam. for 4 or 5 persons		2.50		940	200	1,140	1,325
252	54″ diam. for 4 persons in wheelchairs		2.50		1,350	200	1,550	1,775

15.2 Plumbing Fixtures

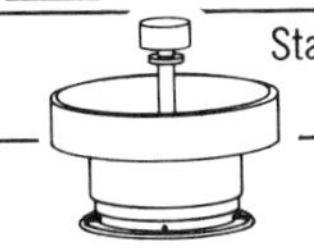

			DAILY		BARE COSTS			TOTAL
	15.2 Plumbing Fixtures	CREW	OUTPUT	UNIT	MAT.	INST.	TOTAL	INCL O&P
76 270	Quarter circle (corner), 54" for 3 persons	Q-2	3.50	Ea.	995	140	1,135	1,300
272	54" diam. for 3 persons in wheelchairs		3.50		1,475	140	1,615	1,825
300	Stainless steel, circular, 36" diameter		3.50		1,000	140	1,140	1,300
310	54" diameter		2.80		1,300	180	1,480	1,675
340	Semi-circular, 36" diameter		3.50		860	140	1,000	1,150
350	54" diameter		2.80		1,125	180	1,305	1,500
500	Thermoplastic, circular, 36" diameter		3.50		965	140	1,105	1,275
510	54" diameter		2.80		1,250	180	1,430	1,625
540	Semi-circular, 36" diameter		3.50		830	140	970	1,125
560	54" diameter		2.80		1,075	180	1,255	1,450
570	Rough-in, supply, waste and vent for above wash fountains	Q-1	1.06		84.21	300	384.21	530
620	Duo for small washrooms, stainless steel		2		465	160	625	745
640	Bowl with backsplash		2		540	160	700	825
650	Rough-in, supply, waste & vent for duo fountains		1.06		84.21	300	384.21	530
80-001	**WATER CLOSETS**							
002	For seats, see 15.2-66							
015	B8.1 -470 Tank type, vitreous china, including seat, supply pipe with stop							
020	C8.1 -401 Wall hung, one piece	Q-1	5.30	Ea.	425	60	485	555
040	Two piece, close coupled		5.30		260	60	320	375
096	For rough-in, supply, waste, vent and carrier		1.68		149.49	190	339.49	440
100	Floor mounted, one piece		5.30		370	60	430	495
102	One piece, low profile		5.30		570	60	630	715
105	One piece combination		5.30		550	60	610	690
110	Two piece, close coupled, water saver		5.30		105	60	165	205
115	With wall outlet		5.30		170	60	230	275
120	With 18" high bowl		5.30		195	60	255	300
196	For color, add				30%			
198	For rough-in, supply, waste and vent	Q-1	1.73		91.19	185	276.19	365
300	Bowl only, with flush valve, seat							
310	Wall hung	Q-1	5.80	Ea.	220	55	275	320
315	Hospital type, slotted rim for bed pan							
316	Elongated bowl	Q-1	5.80	Ea.	240	55	295	345
320	For rough-in, supply, waste and vent, single WC		1.52		158.64	210	368.64	480
330	Floor mounted		5.80		195	55	250	295
335	With wall outlet		5.80		260	55	315	365
336	Hospital type, slotted rim for bed pan							
337	Elongated bowl	Q-1	5	Ea.	225	64	289	340
338	Elongated bowl, 18" high		5		245	64	309	360
340	For rough-in, supply, waste and vent, single WC		1.56		100.34	205	305.34	405
350	B8.1 -510 Gang side by side carrier system, rough-in, supply, waste & vent							
351	For single hook-up	Q-1	.98	Ea.	267.38	325	592.38	765
352	For each additional hook-up, add	"	2.04	"	139.72	155	294.72	380
355	Gang back to back carrier system, rough-in, supply, waste & vent							
356	For pair hook-up	Q-1	.81	Pr.	351.15	395	746.15	955
357	For each additional pair hook-up, add	"	1.42	"	223.49	225	448.49	570
400	Water conserving systems							
410	2-1/2 gallon flush	Q-1	5.40	Ea.	190	59	249	295
490	1 gallon flush		5.40		450	59	509	580
498	For rough-in, supply, waste and vent		1.73		87.89	185	272.89	365
510	2 quart flush		4.60		300	70	370	430
520	For remote valve, add		24		125	13.30	138.30	155
530	For residential air compressor		6		655	53	708	800
540	For light industrial air compressor		4		850	80	930	1,050
550	For heavy duty industrial air compressor		1		1,575	320	1,895	2,200
560	For dual compressor alternator		20		550	16	566	630
570	Waterless, biological action, 110 volt, A.C., requires vent	1 Plum	8		845	22	867	960
90-001	**WATER FILTERS** Purification and treatment							
100	Cartridge style, under sink, dirt and rust type	1 Plum	12	Ea.	31	14.80	45.80	55

15

15.2 Plumbing Fixtures

			CREW	DAILY OUTPUT	UNIT	BARE COSTS MAT.	BARE COSTS INST.	BARE COSTS TOTAL	TOTAL INCL O&P
90	120	Replacement cartridge	1 Plum	32	Ea.	3.51	5.55	9.06	11.90
	160	Taste and odor type		12		37	14.80	51.80	62
	170	Replacement cartridge		32		7.80	5.55	13.35	16.60
	300	Household unit, complete dwelling, 3 GPM		4		280	44	324	370
	310	Replacement cartridge		20		11.45	8.90	20.35	25
	340	4 GPM		4		425	44	469	530
	360	Replacement cartridge		20		16.80	8.90	25.70	31
	430	Oxidizing, removes iron, sulphur, manganese							
	440	5 GPM	1 Plum	1	Ea.	375	180	555	670
	520	Neutralizing acid water							
	540	5 GPM	1 Plum	1	Ea.	310	180	490	600
	620	Sediment, removal of suspended particles							
	640	5 GPM	1 Plum	1	Ea.	300	180	480	585
	720	Carbon filter, to remove taste & odor							
	740	5 GPM	1 Plum	1	Ea.	350	180	530	640
	800	Commercial, fully automatic or push button automatic							
	820	Iron removal, 660 GPH, 1" pipe size	Q-1	1.50	Ea.	870	215	1,085	1,275
	824	1500 GPH, 1-1/4" pipe size		1		1,350	320	1,670	1,950
	828	2340 GPH, 1-1/2" pipe size		.80		1,925	400	2,325	2,700
	832	3420 GPH, 2" pipe size		.60		2,800	535	3,335	3,850
	836	4620 GPH, 2-1/2" pipe size		.50		4,075	640	4,715	5,400
	850	Neutralizer for acid water, 780 GPH, 1" pipe size		1.50		915	215	1,130	1,325
	854	1140 GPH, 1" pipe size		1		1,075	320	1,395	1,650
	858	1740 GPH, 1-1/4" pipe size		.80		1,575	400	1,975	2,300
	862	2520 GPH, 2" pipe size		.60		2,300	535	2,835	3,300
	866	3480 GPH, 2-1/2" pipe size		.50		3,300	640	3,940	4,550
	880	Sediment removal, 780 GPH, 1" pipe size		1.50		885	215	1,100	1,275
	884	1140 GPH, 1-1/4" pipe size		1		1,075	320	1,395	1,650
	888	1740 GPH, 1-1/2" pipe size		.80		1,525	400	1,925	2,250
	892	2520 GPH, 2" pipe size		.60		2,225	535	2,760	3,225
	896	3480 GPH, 2-1/2" pipe size		.50		3,200	640	3,840	4,450
	920	Taste and odor removal, 660 GPH, 1" pipe size		1.50		860	215	1,075	1,250
	924	1500 GPH, 1-1/4" pipe size		1		1,550	320	1,870	2,175
	928	2340 GPH, 1-1/2" pipe size		.80		2,300	400	2,700	3,100
	932	3420 GPH, 2" pipe size		.60		3,225	535	3,760	4,325
	936	4620 GPH, 2-1/2" pipe size		.50		4,475	640	5,115	5,850

15

15.3 Plumbing Appliances

		CREW	DAILY OUTPUT	UNIT	BARE COSTS MAT.	BARE COSTS INST.	BARE COSTS TOTAL	TOTAL INCL O&P
40-001	**WATER CHILLERS REMOTE** ,80°F inlet							
010	Aircooled,50°F outlet, 115V, 4.1 GPH	1 Plum	6	Ea.	260	30	290	330
020	5.7 GPH		5.50		370	32	402	455
030	8.0 GPH		5		285	36	321	365
040	10.3 GPH		4.50		395	39	434	490
050	13.4 GPH		4		470	44	514	580
060	19.5 GPH		3.60		490	49	539	610
070	29 GPH	Q-1	5		905	64	969	1,100
100	230V, 38 GPH	"	5		1,200	64	1,264	1,400
120	For remote grill, add	1 Plum	16		30	11.10	41.10	49
400	Water cooled,50°F outlet, 24 GPH		3		590	59	649	735
410	33 GPH		3		645	59	704	795
490	For remote grill, add		16		105	11.10	116.10	130
45-001	**WATER COOLER**							
003	See line 15.3-45-980 for rough-in, waste & vent							

15.3 Plumbing Appliances

		CREW	DAILY OUTPUT	UNIT	BARE COSTS MAT.	INST.	TOTAL	TOTAL INCL O&P
45 004	for all water coolers							
010	Wall mounted, non-recessed							
014	B8.1 -460  4 GPH	Q-1	4	Ea.	285	80	365	430
018	8.2 GPH		4		335	80	415	485
022	C8.1 -403 14.3 GPH		4		345	80	425	495
026	16.1 GPH		4		370	80	450	525
060	For hot and cold water, add				96		96	105M
064	For stainless steel cabinet, add				35		35	39M
100	Dual height, 8.2 GPH	Q-1	3.80		440	84	524	605
104	14.3 GPH		3.80		465	84	549	635
108	16.1 GPH		3.80		510	84	594	685
124	For stainless steel cabinet, add				60		60	66M
260	Wheelchair type, 8 GPH	Q-1	4		865	80	945	1,075
300	Simulated recessed, 8 GPH		4		390	80	470	545
304	11.5 GPH		4		415	80	495	570
320	For glass filler in addition to bubbler, add				52		52	57M
324	For stainless steel cabinet, add				18		18	19.80M
330	Semi-recessed, 8.1 GPH	Q-1	4		435	80	515	595
332	12 GPH	"	4		460	80	540	620
334	For glass filler, add				31		31	34M
336	For stainless steel cabinet, add				18		18	19.80M
340	Full recessed, stainless steel, 8 GPH	Q-1	3.50		735	91	826	940
342	15 GPH	"	3.50		755	91	846	965
346	For glass filler, add				42		42	46M
360	For mounting can only				80		80	88M
460	Floor mounted, flush-to-wall							
464	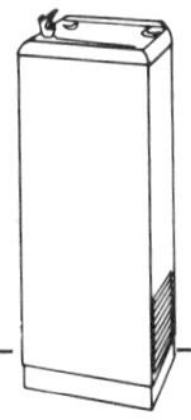 4 GPH	1 Plum	3	Ea.	300	59	359	415
468	8.2 GPH		3		335	59	394	455
472	14.3 GPH		3		355	59	414	475
476	16.0 GPH		3		415	59	474	540
480	19.5 GPH		3		515	59	574	650
496	For hot and cold water, add				110		110	120M
498	For stainless steel cabinet, add				54		54	59M
500	Dual height, 8.2 GPH	1 Plum	2		500	89	589	680
504	14.3 GPH		2		505	89	594	685
508	19.5 GPH		2		515	89	604	695
512	For stainless steel cabinet, add				89		89	98M
560	Explosion Proof, 16 GPH	1 Plum	3		615	59	674	760
564	For stainless steel cabinet, add				54		54	59M
600	Refrigerator Compartment Type, 1.5 GPH	1 Plum	3		670	59	729	825
604	As above but hot and cold		3		725	59	784	885
610	Bottle Supply, cold only, 0.5 GPH		4		475	44	519	585
620	Hot and cold, 0.5 GPH		4		520	44	564	635
660	Bottle Supply Type, 0.5 GPH		4		185	44	229	270
664	2.9 GPH		4		225	44	269	310
770	Cafeteria type, dual glass fillers, 27 GPH	Q-1	2.50		1,250	130	1,380	1,550
774	36 GPH	"	2.50		1,400	130	1,530	1,725
780	For stainless steel cabinet, add				145		145	160M
790	For two stainless steel top glassware shelves, add				225		225	250M
800	For stainless steel side shelf, one side, add				52		52	57M
980	For supply, waste & vent, all coolers	1 Plum	1.24		34.03	145	179.03	245
50-001	**WATER HEATERS**							
002	For solar, see Division 15.5-77							
100	B8.1 -110 Residential, electric, glass lined tank, 10 gal., single element	1 Plum	2.30	Ea.	105	77	182	225
104	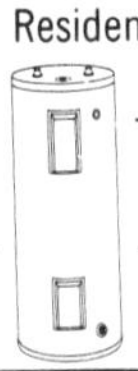 20 gallon, single element		2.20		130	81	211	260
106	30 gallon, double element		2.20		135	81	216	265
108	40 gallon, double element		2		140	89	229	280
110	52 gallon, double element		2		165	89	254	310
112	66 gallon, double element		1.80		225	99	324	390

15

15.3 Plumbing Appliances

50		CREW	DAILY OUTPUT	UNIT	BARE COSTS MAT.	BARE COSTS INST.	BARE COSTS TOTAL	TOTAL INCL O&P
114	80 gallon, double element	1 Plum	1.60	Ea.	265	110	375	450
118	120 gallon, double element		1.40		440	125	565	665
200	B8.1 -120 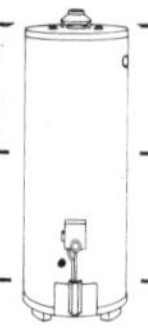 Gas fired, glass lined tank, vent not incl., 20 gallon		2.10		140	85	225	275
204	30 gallon		2		150	89	239	295
206	40 gallon		1.90		160	93	253	310
208	50 gallon		1.80		200	99	299	365
210	75 gallon		1.50		400	120	520	610
212	100 gallon		1.30		635	135	770	895
300	B8.1 -130 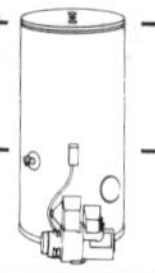 Oil fired, glass lined tank, vent not included, 30 gallon		2		675	89	764	870
304	50 gallon		1.80		890	99	989	1,125
306	70 gallon		1.50		985	120	1,105	1,250
308	85 gallon		1.40		1,375	125	1,500	1,700
400	Commercial, 100° rise. NOTE: for each size tank, a range of heaters							
401	between the ones shown are available							
402	B8.1 -160 Electric							
410	5 gal., 3 KW, 12 GPH	1 Plum	2	Ea.	615	89	704	805
412	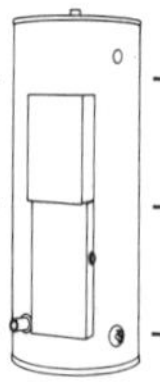 10 gal., 6 KW, 25 GPH		2		680	89	769	875
414	50 gal., 9 KW, 37 GPH		1.80		875	99	974	1,100
416	50 gal., 36 KW, 148 GPH		1.80		1,350	99	1,449	1,625
418	80 gal., 12 KW, 49 GPH		1.50		1,125	120	1,245	1,400
420	80 gal., 36 KW, 148 GPH		1.50		1,550	120	1,670	1,875
422	100 gal., 36 KW, 148 GPH		1.20		1,625	150	1,775	2,000
424	120 gal., 36 KW, 148 GPH		1.20		1,700	150	1,850	2,075
426	150 gal., 15 KW , 61 GPH		1		3,800	180	3,980	4,425
428	150 gal., 120 KW, 490 GPH		1		5,775	180	5,955	6,600
430	200 gal., 15 KW, 61 GPH	Q-1	1.70		4,150	190	4,340	4,825
432	200 gal., 120 KW , 490 GPH		1.70		6,050	190	6,240	6,925
434	250 gal., 15 KW, 61 GPH		1.50		4,275	215	4,490	5,000
436	250 gal., 150 KW, 615 GPH		1.50		6,700	215	6,915	7,675
438	300 gal., 30 KW, 123 GPH		1.30		4,725	245	4,970	5,550
440	300 gal., 180 KW, 738 GPH		1.30		7,325	245	7,570	8,425
442	350 gal., 30 KW, 123 GPH		1.10		5,075	290	5,365	6,000
444	350 gal, 180 KW, 738 GPH		1.10		7,500	290	7,790	8,675
446	400 gal., 30 KW, 123 GPH		1		5,700	320	6,020	6,725
448	400 gal., 210 KW, 860 GPH		1		8,925	320	9,245	10,300
450	500 gal., 30 KW, 123 GPH		.80		6,575	400	6,975	7,800
452	500 gal., 240 KW, 984 GPH		.80		10,500	400	10,900	12,100
454	600 gal., 30 KW, 123 GPH	Q-2	1.20		7,600	415	8,015	8,950
456	600 gal., 300 KW, 1230 GPH		1.20		12,400	415	12,815	14,200
458	700 gal., 30 KW, 123 GPH		1		8,000	495	8,495	9,525
460	700 gal., 300 KW, 1230 GPH		1		12,800	495	13,295	14,800
462	800 gal., 60 KW, 245 GPH		.90		8,775	555	9,330	10,500
464	800 gal., 300 KW, 1230 GPH		.90		13,200	555	13,755	15,300
466	1000 gal., 60 KW, 245 GPH		.70		9,550	710	10,260	11,500
468	1000 gal., 480 KW, 1970 GPH		.70		17,200	710	17,910	19,900
470	1250 gal., 60 KW, 245 GPH		.60		10,800	830	11,630	13,100
472	1250 gal., 480 KW, 1970 GPH		.60		17,700	830	18,530	20,700
474	1500 gal., 60 KW, 245 GPH		.50		14,300	995	15,295	17,200
476	1500 gal., 480 KW, 1970 GPH		.50		20,500	995	21,495	24,000
478	2000 gal., 60 KW, 245 GPH		.30		17,900	1,650	19,550	22,100
480	2000 gal., 480 KW, 1970 GPH		.30		25,000	1,650	26,650	29,900
540	Modulating step control, 2-5 steps	1 Elec	5.30		600	33	633	710
544	6-10 steps		3.20		765	55	820	920
546	11-15 steps		2.70		935	65	1,000	1,125
548	16-20 steps		1.60		1,100	110	1,210	1,375
550	21-25 steps		.30		1,275	590	1,865	2,250
552	26-30 steps		.26		1,450	680	2,130	2,575
600	Gas fired, flush jacket, std. controls, vent not incl.							
604	B8.1 -170 75 MBH input, 63 GPH	1 Plum	1.40	Ea.	640	125	765	885
606	96 MBH input, 81 GPH	"	1.40	"	790	125	915	1,050

15

15.3 Plumbing Appliances

		CREW	DAILY OUTPUT	UNIT	BARE COSTS MAT.	BARE COSTS INST.	BARE COSTS TOTAL	TOTAL INCL O&P
50 608	120 MBH input, 101 GPH	1 Plum	1.20	Ea.	825	150	975	1,125
610	115 MBH input, 110 GPH		1.10		1,000	160	1,160	1,325
612	135 MBH input, 130 GPH		1		1,125	180	1,305	1,500
614	155 MBH input, 150 GPH		.80		1,200	220	1,420	1,650
616	175 MBH input, 168 GPH		.70		1,325	255	1,580	1,825
618	200 MBH input, 192 GPH		.60		1,550	295	1,845	2,125
620	240 MBH input, 230 GPH		.50		1,900	355	2,255	2,600
622	295 MBH input, 278 GPH	Q-1	.80		2,450	400	2,850	3,275
624	365 MBH input, 374 GPH		.80		2,750	400	3,150	3,600
626	500 MBH input, 480 GPH		.70		3,750	455	4,205	4,775
628	600 MBH input, 576 GPH		.60		4,325	535	4,860	5,525
630	800 MBH input, 768 GPH		.50		5,150	640	5,790	6,600
632	1000 MBH input, 960 GPH		.50		6,025	640	6,665	7,550
634	1200 MBH input, 1150 GPH		.40		7,500	800	8,300	9,400
636	1500 MBH input, 1440 GPH	Q-2	.60		9,275	830	10,105	11,400
638	1800 MBH input, 1730 GPH		.50		10,500	995	11,495	13,000
640	2450 MBH input, 2350 GPH		.40		14,000	1,250	15,250	17,200
642	3000 MBH input, 2880 GPH		.30		17,100	1,650	18,750	21,200
644	3750 MBH input, 3600 GPH		.30		21,100	1,650	22,750	25,600
690	For low water cutoff, add	1 Plum	8		120	22	142	165
696	For bronze body hot water circulator, add	"	4		500	44	544	615
800	Oil fired, flush jacket, std. controls, vent not incl.							
806	B8.1 -180 103 MBH gross output, 116 GPH	1 Plum	1.10	Ea.	1,150	160	1,310	1,500
808	122 MBH gross output, 141 GPH		1		1,175	180	1,355	1,550
810	137 MBH gross output, 161 GPH		.80		1,250	220	1,470	1,700
812	168 MBH gross output, 192 GPH		.60		1,525	295	1,820	2,100
814	195 MBH gross output, 224 GPH		.50		1,600	355	1,955	2,275
816	225 MBH gross output, 256 GPH	Q-1	.80		1,850	400	2,250	2,625
818	262 MBH gross output, 315 GPH		.70		2,075	455	2,530	2,950
820	315 MBH gross output, 409 GPH		.70		2,725	455	3,180	3,650
822	420 MBH gross output, 504 GPH		.60		3,125	535	3,660	4,200
824	525 MBH gross output, 630 GPH		.50		4,000	640	4,640	5,325
826	630 MBH gross output, 756 GPH		.50		4,475	640	5,115	5,850
828	735 MBH gross output, 880 GPH		.40		5,500	800	6,300	7,200
830	840 MBH gross output, 1000 GPH		.40		6,300	800	7,100	8,075
832	1050 MBH gross output, 1260 GPH	Q-2	.60		7,675	830	8,505	9,650
834	1365 MBH gross output, 1640 GPH		.50		9,850	995	10,845	12,300
836	1680 MBH gross output, 2000 GPH		.40		11,600	1,250	12,850	14,600
838	2310 MBH gross output, 2780 GPH		.30		15,000	1,650	16,650	18,900
840	2835 MBH gross output, 3400 GPH		.30		18,600	1,650	20,250	22,900
842	3150 MBH gross output, 3780 GPH		.20		20,900	2,475	23,375	26,600
890	For low water cutoff, add	1 Plum	8		120	22	142	165
896	For bronze body hot water circulator, add	"	4		500	44	544	615
60-001	**WATER HEATER PACKAGED SYSTEMS** Commercial, glass lined tank							
100	Car wash package, continuous duty, high recovery, gas fired							
104	100° rise, 180 MBH input, 174 GPH	1 Plum	3	Ea.	2,150	59	2,209	2,450
106	280 MBH input, 270 GPH		2.50		2,400	71	2,471	2,750
108	400 MBH input, 386 GPH		2.50		2,900	71	2,971	3,300
110	480 MBH input, 464 GPH		2		3,300	89	3,389	3,750
112	605 MBH input, 584 GPH		1.50		3,675	120	3,795	4,225
114	700 MBH input, 676 GPH		1		3,875	180	4,055	4,525
116	1000 MBH input, 966 GPH	Q-1	1.60		4,775	200	4,975	5,550
118	1200 MBH input, 1159 GPH		1.40		5,925	230	6,155	6,850
120	1400 MBH input, 1353 GPH		1.20		6,350	265	6,615	7,375
300	Combination dishwasher and general purpose, two temp., gas fired							
304	124 GPH @ 140° rise; 434 GPH @ 40° rise	1 Plum	2	Ea.	2,400	89	2,489	2,775
306	193 GPH @ 140° rise; 677 GPH @ 40° rise		1.60		2,850	110	2,960	3,300
308	276 GPH @ 140° rise; 969 GPH @ 40° rise		1.40		3,350	125	3,475	3,875
310	330 GPH @ 140° rise; 1160 GPH @ 40° rise		1.20		3,800	150	3,950	4,400

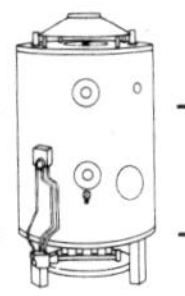

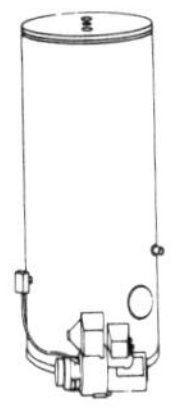

15.3 Plumbing Appliances

60

		CREW	DAILY OUTPUT	UNIT	BARE COSTS MAT.	BARE COSTS INST.	BARE COSTS TOTAL	TOTAL INCL O&P
312	417 GPH @ 140° rise; 1460 GPH @40°rise	1 Plum	1	Ea.	4,150	180	4,330	4,825
314	481 GPH @140° rise; 1685 GPH @40°rise	Q-1	1.40		4,550	230	4,780	5,325
316	688 GPH @140° rise; 2410 GPH @40°rise		1.20		5,450	265	5,715	6,375
318	828 GPH @140° rise; 2790 GPH @40°rise		1		6,650	320	6,970	7,775
320	965 GPH @140° rise; 3370 GPH @40°rise		.80		7,100	400	7,500	8,400
396	For unit base, add				150		150	165M
500	Coin laundry units, gas fired, 100° rise							
502	Single heater,							
504	280 MBH input, 270 GPH	1 Plum	1.60	Ea.	2,775	110	2,885	3,225
506	400 MBH input, 386 GPH		1.30		3,600	135	3,735	4,150
508	480 MBH input, 464 GPH		1		4,050	180	4,230	4,700
510	605 MBH input, 584 GPH	Q-1	1.40		4,950	230	5,180	5,775
512	700 MBH input, 676 GPH		1.20		5,025	265	5,290	5,925
514	1000 MBH input, 966 GPH		1		6,150	320	6,470	7,225
516	1200 MBH input, 1159 GPH		.90		7,400	355	7,755	8,650
518	1400 MBH input, 1353 GPH		.70		8,000	455	8,455	9,450
600	Multiple heater							
604	560 MBH input, 540 GPH	1 Plum	1	Ea.	4,725	180	4,905	5,450
606	800 MBH input, 772 GPH	Q-1	1.30		6,125	245	6,370	7,100
608	960 MBH input, 928 GPH		1.20		7,525	265	7,790	8,675
610	1210 MBH input, 1168 GPH		1		8,075	320	8,395	9,350
612	1400 MBH input, 1352 GPH		1		8,525	320	8,845	9,850
614	1700 MBH input, 1642 GPH		.80		9,850	400	10,250	11,400
616	2000 MBH input, 1932 GPH		.70		10,700	455	11,155	12,400
618	2400 MBH input, 2318 GPH	Q-2	.90		13,200	555	13,755	15,300
620	2600 MBH input, 2512 GPH		.80		13,600	620	14,220	15,900
622	2800 MBH input, 2706 GPH		.70		15,300	710	16,010	17,900
680	Standard system sizing is based on							
682	80% of the washers operating at one time.							
65-001	**WATER HEATER STORAGE TANKS** 125 psi ASME							
100	Copper, lined, 190 gal., 30" diam x 66" LOA	1 Plum	4	Ea.	2,825	44	2,869	3,175
106	225 gal., 30" diam x 78" LOA	"	3		3,275	59	3,334	3,700
108	325 gal., 36" diam x 81" LOA	Q-1	4		4,100	80	4,180	4,625
110	460 gal., 42" diam x 84" LOA		3		5,725	105	5,830	6,450
112	605 gal., 48" diam x 87" LOA		2.50		6,550	130	6,680	7,400
114	740 gal., 54" diam x 91" LOA		2		7,775	160	7,935	8,775
116	940 gal., 60" diam x 93" LOA		1.50		9,000	215	9,215	10,200
118	1505 gal., 66" diam x 119" LOA		1		11,600	320	11,920	13,200
120	1615 gal., 72" diam x 110" LOA	Q-2	1.50		13,100	330	13,430	14,900
122	2275 gal., 84" diam x 116" LOA		1		18,000	495	18,495	20,500
124	3815 gal., 96" diam x 145" LOA		1		23,300	495	23,795	26,300
200	Galvanized steel, 18 gal., 12" diam., 39" LOA	1 Plum	12		215	14.80	229.80	260
206	30 gal., 14" diam. x 40" LOA		11		265	16.15	281.15	315
208	75 gal., 18" diam. x 72" LOA		9		600	19.75	619.75	690
210	140 gal., 24" diam. x 75" LOA		6		890	30	920	1,025
212	225 gal., 30" diam. x 78" LOA		4		1,250	44	1,294	1,450
214	325 gal., 36" diam. x 81" LOA		3		1,775	59	1,834	2,050
216	460 gal., 36" diam. x 117" LOA	Q-1	4		2,475	80	2,555	2,850
218	605 gal., 48" diam., x 87" LOA	"	3		4,000	105	4,105	4,550
300	Glass lined, P.E., 80 gal., 20" diam. x 60" LOA	1 Plum	9		685	19.75	704.75	780
306	140 gal., 24" diam. x 80" LOA		6		945	30	975	1,075
308	225 gal., 30" diam. x 78" LOA		4		1,250	44	1,294	1,450
310	325 gal., 36" diam. x 81" LOA		3		1,850	59	1,909	2,125
312	460 gal., 42" diam. x 84" LOA	Q-1	4		2,350	80	2,430	2,700
314	605 gal., 48" diam. x 87" LOA		3		3,350	105	3,455	3,850
316	740 gal., 54" diam. x 91" LOA		3		3,825	105	3,930	4,350
318	940 gal., 60" diam. x 93" LOA		2.50		4,550	130	4,680	5,200
320	1330 gal., 66" diam. x 107" LOA		2		5,850	160	6,010	6,675
322	1615 gal., 72" diam. x 110" LOA		1.50		6,825	215	7,040	7,825

15

15.3 Plumbing Appliances

		CREW	DAILY OUTPUT	UNIT	BARE COSTS MAT.	BARE COSTS INST.	BARE COSTS TOTAL	TOTAL INCL O&P
65 324	2275 gal., 84" diam. x 116" LOA	Q-1	1	Ea.	9,775	320	10,095	11,200
326	3815 gal., 96" diam. x 145" LOA	Q-2	1.50	"	13,900	330	14,230	15,800
70-001	**WATER TREATMENT, OIL RETRIEVAL**							
160	Disk skimmer style, totally enclosed motor							
164	15 GPH recovery	1 Plum	3.20	Ea.	1,500	56	1,556	1,725
166	25 GPH recovery		3		2,100	59	2,159	2,400
172	45 GPH recovery		2.60		2,400	68	2,468	2,750
178	65 GPH recovery		2.40		2,800	74	2,874	3,175
280	For explosion-proof motor, add				10%			
290	For air motor, add				8%			
296	For pontoons, add				440		440	485M
75-001	**WATER TREATMENT, POTABLE**							
200	Chemical feeders, for chlorine, caustics, mild acids							
260	Air-cooled, automatic, max. output 9 gal. per 24 hrs.	1 Plum	8	Ea.	145	22	167	190
270	30 gal. per 24 hrs		8		150	22	172	195
400	Oil cooled, automatic, max. output 4.3 gal. per 24 hrs.		8		205	22	227	260
410	14 gal. per 24 hrs.		8		220	22	242	275
420	28 gal. per 24 hrs.		8		230	22	252	285
480	For 230V operation, add				8		8	8.80M
540	Chemical solution tank for above feeders, 20 gallon	1 Plum	24		79	7.40	86.40	98
580	Softener systems, automatic, intermediate sizes							
582	available, may be used in multiples.							
600	Hardness capacity between regenerations and flow							
610	150,000 grains, 37 GPM cont., 51 GPM peak	Q-1	1.20	Ea.	1,575	265	1,840	2,125
620	300,000 grains, 81 GPM cont., 113 GPM peak		1		2,500	320	2,820	3,200
630	750,000 grains, 160 GPM cont., 230 GPM peak		.80		5,225	400	5,625	6,325
640	900,000 grains, 185 GPM cont., 270 GPM peak		.70		5,875	455	6,330	7,125

15.4 Fire Extinguishing Systems

		CREW	DAILY OUTPUT	UNIT	BARE COSTS MAT.	BARE COSTS INST.	BARE COSTS TOTAL	TOTAL INCL O&P
02-001	**AUTOMATIC FIRE SUPPRESSION SYSTEMS**							
004	For detectors and control stations, see division 16.8-15							
010	B8.2 -810 Control panel, single zone with batteries	1 Elec	1	Ea.	1,050	175	1,225	1,400
015	Multizone (4) with batteries	"	.50		2,075	355	2,430	2,800
100	Dispersion nozzle, CO2, 3" x 5"	1 Plum	18		70	9.85	79.85	91
110	Halon, 1-1/2"	"	14		63	12.70	75.70	88
200	Extinguisher, CO2 system, high pressure, 75 lb. cylinder	Q-1	6		670	53	723	815
210	100 lb. cylinder	"	5		970	64	1,034	1,150
240	Halon system, filled, with mounting bracket							
246	26 lb. container	Q-1	8	Ea.	1,225	40	1,265	1,400
248	44 lb. container		7		1,325	46	1,371	1,525
250	63 lb. container		6		1,450	53	1,503	1,675
252	101 lb. container		5		1,775	64	1,839	2,050
254	196 lb. container		4		2,600	80	2,680	2,975
300	Electro/mechanical release	L-1	4		160	89	249	305
340	Manual pull station	1 Plum	6		26	30	56	71
400	Pneumatic damper release	"	8		80	22	102	120
600	Average halon system, minimum			C.F.				.65
602	Maximum			"				1.05
10-001	**FIRE EQUIPMENT CABINETS** Not equipped, 20 ga. steel box,							
004	recessed, D.S. glass in door, box size given							
100	B8.2 -310 Portable extinguisher, single, 8" x 12" x 27", alum. door & frame	Q-12	8	Ea.	70	41	111	135
110	Steel door and frame	"	8	"	62	41	103	125

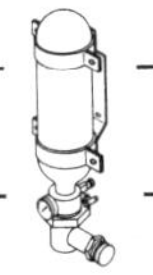

15

15.4 Fire Extinguishing Systems

			CREW	DAILY OUTPUT	UNIT	BARE COSTS MAT.	BARE COSTS INST.	BARE COSTS TOTAL	TOTAL INCL O&P
10	120	Stainless steel door and frame	Q-12	8	Ea.	125	41	166	195
	200	B8.2-320 Portable extinguisher, large, 8" x 12" x 36", alum. door & frame	↓	8	↓	77	41	118	145
	210	B8.2-390 Steel door and frame	↓	8	↓	72	41	113	140
	220	Stainless steel door and frame	↓	8	↓	160	41	201	235
	300	C8.2-301 Hose rack assy., 1-1/2" valve & 100' hose, 24" x 40" x 5-1/2"							
	310	Aluminum door and frame	Q-12	6	Ea.	100	54	154	190
	320	C8.2-302 Steel door and frame	↓	6	↓	97	54	151	185
	330	Stainless steel door and frame	↓	6	↓	215	54	269	315
	400	Hose rack assy., 2-1/2" x 1-1/2" valve, 100' hose, 24" x 40" x 8"							
	410	Aluminum door and frame	Q-12	6	Ea.	105	54	159	195
	420	Steel door and frame	↓	6	↓	100	54	154	190
	430	Stainless steel door and frame	↓	6	↓	215	54	269	315
	500	Hose rack assy., 2-1/2" x 1-1/2" valve, 100' hose							
	501	and extinguisher, 30" x 40" x 8"							
	510	Aluminum door and frame	Q-12	5	Ea.	115	65	180	220
	520	Steel door and frame	↓	5	↓	110	65	175	215
	530	Stainless steel door and frame	↓	5	↓	240	65	305	360
	600	Hose rack assy., 1-1/2" valve, 100' hose							
	601	and 2-1/2" FD valve, 24" x 44" x 8"							
	610	Aluminum door and frame	Q-12	5	Ea.	115	65	180	220
	620	Steel door and frame	↓	5	↓	115	65	180	220
	630	Stainless steel door and frame	↓	5	↓	230	65	295	350
	700	Hose rack assy., 1-1/2" valve & 100' hose, 2-1/2" FD valve							
	701	and extinguisher, 30" x 44" x 8"							
	710	Aluminum door and frame	Q-12	5	Ea.	120	65	185	225
	720	Steel door and frame	↓	5	↓	125	65	190	230
	730	Stainless steel door and frame	↓	5	↓	255	65	320	375
	800	Valve cabinet for 2-1/2" F.D. angle valve, 18" x 18" x 8"							
	810	Aluminum door and frame	Q-12	12	Ea.	70	27	97	115
	820	Steel door and frame	↓	12	↓	61	27	88	105
	830	Stainless steel door and frame	↓	12	↓	125	27	152	175
20-001		FIRE EXTINGUISHERS							
	012	CO2, portable with swivel horn, 5 lb.			Ea.	60		60	66M
	014	With hose and "H" horn, 10 lb.			↓	90		90	99M
	016	15 lb.			↓	100		100	110M
	018	20 lb.			↓	135		135	150M
	036	Wheeled type, cart mounted, 50 lb.			↓	745		745	820M
	038	100 lb. (two 50 lb. cylinders)			↓	1,150		1,150	1,275M
	100	Dry chemical, pressurized							
	104	Standard type, portable, painted, 2-1/2 lb.			Ea.	15.75		15.75	17.30M
	106	5 lb.			↓	28		28	31M
	108	10 lb.			↓	42		42	46M
	110	20 lb.			↓	62		62	68M
	130	Standard type, wheeled, 150 lb.			↓	1,050		1,050	1,150M
	136	350 lb.			↓	1,775		1,775	1,950M
	200	ABC all purpose type, portable, 2-1/2 lb.			↓	17.35		17.35	19.10M
	206	5 lb.			↓	29		29	32M
	208	9-1/2 lb.			↓	41		41	45M
	210	18 lb.			↓	67		67	74M
	212	30 lb.			↓	85		85	94M
	230	Wheeled, 45 lb.			↓	905		905	995M
	236	150 lb.			↓	1,250		1,250	1,375M
	238	315 lb.			↓	1,525		1,525	1,675M
	300	Dry chemical, outside cartridge to -65°F, painted, 9 lb.			↓	130		130	145M
	306	26 lb.			↓	140		140	155M
	400	Halon, painted, wall bracket, 2-1/2 lb.			↓	58		58	64M
	406	5 lb.			↓	79		79	87M
	500	Pressurized water, 2-1/2 gallon, stainless steel			↓	40		40	44M
	506	With anti-freeze			↓	55		55	61M

15

15.4 Fire Extinguishing Systems

		CREW	DAILY OUTPUT	UNIT	BARE COSTS MAT.	BARE COSTS INST.	BARE COSTS TOTAL	TOTAL INCL O&P
20 940	Installation of extinguishers, 12 or more, on wood	1 Carp	30	Ea.		5.25	5.25	7.55L
942	On masonry or concrete	"	15	"		10.45	10.45	15.10L
30-001	**FIRE HOSE AND EQUIPMENT**							
020	Adapters, rough brass, straight hose threads							
022	B8.2-310 One piece, female to male, rocker lugs							
024	1" x 1"			Ea.	8.25		8.25	9.05M
026	B8.2-320 1-1/2" x 1"				9.35		9.35	10.30M
028	1-1/2" x 1-1/2"				10.80		10.80	11.90M
030	B8.2-390 2" x 1-1/2"				13.35		13.35	14.70M
032	2" x 2"				19.20		19.20	21M
034	C8.2-301 2-1/2" x 1-1/2"				14.40		14.40	15.85M
036	2-1/2" x 2"				19.20		19.20	21M
038	C8.2-302 2-1/2" x 2-1/2"				24		24	26M
040	3" x 2-1/2"				27		27	30M
042	3" x 3"				38		38	42M
050	For polished brass, add				25%			
052	For polished chrome, add				65%			
070	One piece, female to male, hexagon							
074	1-1/2" x 3/4"			Ea.	12.40		12.40	13.65M
076	2" x 1"				17.20		17.20	18.90M
078	2-1/2" x 1"				33		33	36M
080	2-1/2" x 1-1/2"				29		29	32M
082	2-1/2" x 2"				29		29	32M
084	3" x 2-1/2"				48		48	53M
090	For chrome, add				60%			
110	Swivel, female to female, pin lugs							
112	1-1/2" x 1-1/2"			Ea.	32		32	35M
114	2" x 2"				38		38	42M
120	2-1/2 x 2-1/2"				41		41	45M
126	For polished brass, add				25%			
128	For polished chrome, add				55%			
140	Couplings, single & double jacket, pin lug or rocker lug, cast brass							
141	1-1/2"			Ea.	12		12	13.20M
142	2-1/2"				25		25	28M
150	For polished brass, add				15%			
152	For polished chrome, add				25%			
158	Reducing,, F x F, interior installation, cast brass							
159	2" x 1-1/2"			Ea.	18		18	19.80M
160	2-1/2" x 1-1/2"				18		18	19.80M
168	For polished brass, add				40%			
172	For polished chrome, add				75%			
190	Escutcheon plate, for angle valves, polished brass, 1-1/2"				10		10	11M
192	2-1/2"				12.50		12.50	13.75M
194	3"				13		13	14.30M
198	For polished chrome, add				15%			
220	Hose, less couplings							
226	Synthetic jacket, lined, 300 lb. test, 1-1/2" diameter			L.F.	1.03		1.03	1.13M
228	2-1/2" diameter				1.69		1.69	1.85M
236	High strength, 500 lb. test, 1-1/2" diameter				1.22		1.22	1.34M
238	2-1/2" diameter				1.96		1.96	2.15M
260	Hose rack, swinging, for 1-1/2" diameter hose,							
262	Enameled steel, 50' & 75' lengths of hose	Q-12	20	Ea.	15.80	16.25	32.05	41
264	100' and 125' lengths of hose		20		15.80	16.25	32.05	41
268	Brass, 50' and 75' lengths of hose		20		58	16.25	74.25	87
270	100' and 125' lengths of hose		20		58	16.25	74.25	87
276	For polished chrome, add				25%			
278	For hose rack nipple, 1-1/2" polished brass, add				7.15		7.15	7.85M
282	2-1/2" polished brass, add				13.10		13.10	14.40M
284	1-1/2" polished chrome, add				7.50		7.50	8.25M
286	2-1/2" polished chrome, add				15		15	16.50M

15.4 Fire Extinguishing Systems

		CREW	DAILY OUTPUT	UNIT	BARE COSTS MAT.	BARE COSTS INST.	BARE COSTS TOTAL	TOTAL INCL O&P
30 299	Hose reel, swinging, for 1-1/2" polyester neoprene lined hose							
300	50' long	Q-12	14	Ea.	42	23	65	80
302	100' long		14		50	23	73	89
306	For 2-1/2" cotton rubber hose, 75' long		14		55	23	78	94
310	150' long		14		64	23	87	105
375	Hydrants, wall, w/caps, single, flush, polished brass							
380	2-1/2" x 2-1/2"	Q-12	5	Ea.	67	65	132	170
384	2-1/2" x 3"		5		96	65	161	200
386	3" x 3"		4.80		93	68	161	200
390	For polished chrome, add				10%			
395	Double, flush, polished brass							
400	2-1/2" x 2-1/2" x 4"	Q-12	5	Ea.	200	65	265	315
404	2-1/2" x 2-1/2" x 6"		4.60		230	71	301	355
408	3" x 3" x 4"		4.90		220	66	286	340
410	3" x 3" x 5"		4.50		285	72	357	420
412	3" x 3" x 6"		4.50		275	72	347	410
420	For polished chrome, add				3%			
435	Double, projecting, polished brass							
440	2-1/2" x 2-1/2" x 4"	Q-12	5	Ea.	185	65	250	300
444	2-1/2" x 2-1/2" x 5"		4.60		265	71	336	395
445	2-1/2" x 2-1/2" x 6"		4.60		210	71	281	335
446	Valve control, dbl. flush/projecting hydrant, cap &							
447	chain, ext. rod & cplg., escutcheon, polished brass	Q-12	8	Ea.	110	41	151	180
448	Four-way square, flush, polished brass							
450	2-1/2"(4) x 5"	Q-12	4	Ea.	510	81	591	680
454	2-1/2"(4) x 6"	"	3.60	"	480	90	570	660
460	Vertical, flush, cast brass							
462	Two-way, 2-1/2" x 2-1/2" x 4"	Q-12	5	Ea.	270	65	335	390
464	Three-way, 2-1/2" x 2-1/2" x 2-1/2" x 5"		4.60		390	71	461	530
468	Four-way, 2-1/2"(4) x 5"		3.80		500	86	586	675
472	Four-way, 2-1/2"(4) x 6"		3.80		445	86	531	615
478	For polished chrome, add				20		20	22M
500	Nipples, straight hose to tapered iron pipe, brass							
504	Female to female, 1" x 1"			Ea.	20		20	22M
506	1-1/2" x 1-1/2"				18.50		18.50	20M
508	2" x 2"				26		26	29M
510	2-1/2" x 2-1/2"				50		50	55M
520	Double male or male to female, 1" x 1"				5		5	5.50M
522	1-1/2" x 1"				15.50		15.50	17.05M
523	1-1/2" x 1-1/2"				7		7	7.70M
524	2" x 1"				22		22	24M
526	2" x 1-1/2"				22		22	24M
527	2" x 2"				13		13	14.30M
528	2-1/2" x 1-1/2"				15		15	16.50M
530	2-1/2" x 2"				20		20	22M
531	2-1/2" x 2-1/2"				15		15	16.50M
534	For polished chrome, add				75%			
560	Nozzles, brass							
562	Adjustable fog, 3/4" booster line			Ea.	58		58	64M
563	1" booster line				58		58	64M
564	1-1/2" leader line				58		58	64M
566	2-1/2" direct connection				115		115	125M
568	2-1/2" playpipe nozzle				94		94	105M
578	For chrome plated, add				6%			
585	Electrical fire, adjustable fog, no shock							
590	1-1/2"			Ea.	205		205	225M
592	2-1/2"				265		265	290M
598	For polished chrome, add				6%			
620	Heavy duty, comb. adj. fog and str. stream, with handle							
621	1" booster line			Ea.	165		165	180M

15

15.4 Fire Extinguishing Systems

	Line	Description	CREW	DAILY OUTPUT	UNIT	BARE COSTS MAT.	BARE COSTS INST.	BARE COSTS TOTAL	TOTAL INCL O&P
30	624	1-1/2"			Ea.	165		165	180M
	626	2-1/2", for playpipe				180		180	200M
	628	2-1/2" direct connection				215		215	235M
	630	2-1/2" playpipe combination				300		300	330M
	648	For polished chrome, add				7%			
	650	Plain fog, polished brass, 1-1/2"				26		26	29M
	654	Chrome plated, 1-1/2"				27		27	30M
	670	Plain stream,polished brass, 1-1/2" x 10"				14.70		14.70	16.15M
	674	1-1/2" x 15" x 5/8"				28		28	31M
	676	2-1/2" x 15" x 13/16" or 1-1/2"				43		43	47M
	686	For polished chrome, add				5%			
	700	Underwriters playpipe, 2-1/2" x 30" with 1-1/8" tip				115		115	125M
	704	Less tip				94		94	105M
	714	Standpipe connections, wall, w/plugs & chains							
	716	Single, flush, brass, 2-1/2" x 2-1/2"	Q-12	5	Ea.	72	65	137	175
	718	2-1/2" x 3"	"	5		82	65	147	185
	724	For polished chrome, add				3%			
	728	Double, flush, polished brass							
	730	2-1/2" x 2-1/2" x 4"	Q-12	5	Ea.	220	65	285	335
	732	2-1/2" x 2-1/2" x 5"		4.60		285	71	356	415
	733	2-1/2" x 2-1/2" x 6"		4.60		255	71	326	385
	734	3" x 3" x 4"		4.90		310	66	376	440
	736	3" x 3" x 5"		4.50		370	72	442	510
	737	3" x 3" x 6"		4.50		325	72	397	465
	740	For polished chrome, add				3%			
	744	For sill cock combination, add				70		70	77M
	758	Double projecting, polished brass							
	760	2-1/2" x 2-1/2" x 4"	Q-12	5	Ea.	185	65	250	300
	762	2-1/2" x 2-1/2" x 5"		4.60		265	71	336	395
	763	2-1/2" x 2-1/2" x 6"		4.60		215	71	286	340
	768	For polished chrome, add				4%			
	770	For sill cock combination, add				60		60	66M
	790	Three way, flush, polished brass							
	792	2-1/2" (3) x 4"	Q-12	4.80	Ea.	405	68	473	545
	793	2-1/2" (3) x 6"		4.80		420	68	488	560
	794	2-1/2" x 3-1/2" x 2-1/2" x 4"		4.80		565	68	633	720
	795	2-1/2" x 3-1/2" x 2-1/2" x 6"		4.80		510	68	578	660
	800	For polished chrome, add				6%			
	802	Three way, projecting, polished brass							
	804	2-1/2"(3) x 4"	Q-12	4.80	Ea.	305	68	373	435
	806	2-1/2"(3) x 5"		4.60		370	71	441	510
	807	2-1/2" (3) x 6"		4.60		370	71	441	510
	810	For polished chrome, add				8%			
	820	Four way, square, flush, polished brass,							
	822	2-1/2"(4) x 5"	Q-12	4	Ea.	755	81	836	950
	824	2-1/2"(4) x 6"	"	3.60		715	90	805	920
	830	For polished chrome, add				3%			
	840	Four way, projecting, siamese, brass							
	844	2-1/2"(4) x 5"	Q-12	3.60	Ea.	385	90	475	555
	845	2-1/2" (4) x 6"	"	3.60		385	90	475	555
	850	For polished chrome, add				15%			
	855	Wall, vertical, flush, cast brass							
	860	Two way, 2-1/2" x 2-1/2" x 4"	Q-12	5	Ea.	380	65	445	515
	862	Three way, 2-1/2" (3) x 5"		4.60		560	71	631	720
	864	Four way, 2-1/2"(4) x 5"		3.80		715	86	801	910
	866	Four way, 2-1/2"(4) x 6"		3.80		670	86	756	860
	868	Six way, 2-1/2"(6) x 6"		3.40		1,675	96	1,771	1,975
	870	For polished chrome, add				15%			
	880	Sidewalk siamese unit, polished brass, two way							
	882	2-1/2" x 2-1/2" x 4"	Q-12	2.50	Ea.	235	130	365	450

15

15.4 Fire Extinguishing Systems

		CREW	DAILY OUTPUT	UNIT	BARE COSTS MAT.	BARE COSTS INST.	BARE COSTS TOTAL	TOTAL INCL O&P
30 884	2-1/2" x 2-1/2" x 5"	Q-12	2	Ea.	325	165	490	595
885	2-1/2" x 2-1/2" x 6"		2		325	165	490	595
886	3" x 3" x 4"		2.50		340	130	470	565
888	3" x 3" x 5"		2		460	165	625	745
889	3" x 3" x 6"		2		345	165	510	615
894	For polished chrome, add				15%			
910	Sidewalk siamese unit, polished brass, three way							
912	2-1/2" x 2-1/2" x 2-1/2" x 6"	Q-12	2	Ea.	590	165	755	885
916	For polished chrome, add				50%			
920	Storage house, hose only, primed steel				310		310	340M
922	Aluminum				445		445	490M
928	Hose and hydrant house, primed steel				345		345	380M
930	Aluminum				480		480	530M
934	Tools, crowbar and brackets	1 Carp	12		33	13.05	46.05	55
936	Combination hydrant wrench and spanner				19.20		19.20	21M
938	Fire axe and brackets							
940	6 lb.	1 Carp	12	Ea.	34	13.05	47.05	56
40-001	**FIRE PUMPS** Including controller, fittings and relief valve							
003	Diesel							
005	500 GPM, 50 psi, 40 HP, 4" pump	Q-13	.64	Ea.	22,300	1,075	23,375	26,100
010	500 GPM, 100 psi, 68 HP, 4" pump		.60		23,400	1,150	24,550	27,400
015	500 GPM, 125 psi, 103 HP, 4" pump		.56		24,600	1,225	25,825	28,900
020	750 GPM, 50 psi, 40 HP, 5" pump		.60		22,800	1,150	23,950	26,800
025	750 GPM, 100 psi, 99 HP, 5" pump		.56		26,900	1,225	28,125	31,400
030	750 GPM, 165 psi, 188 HP, 5" pump		.52		32,800	1,325	34,125	38,000
035	1000 GPM, 50 psi, 53 HP, 5" pump		.58		23,500	1,200	24,700	27,600
040	1000 GPM, 100 psi, 99 HP, 5" pump		.56		26,700	1,225	27,925	31,200
045	1000 GPM, 150 psi, 188 HP, 6" pump		.48		37,400	1,450	38,850	43,200
047	1000 GPM, 200 psi, 238 HP, 6" pump		.40		38,100	1,725	39,825	44,400
050	1500 GPM, 50 psi, 81 HP, 6" pump		.50		25,200	1,375	26,575	29,700
055	1500 GPM, 100 psi, 188 HP, 6" pump		.46		30,500	1,500	32,000	35,700
060	1500 GPM, 150 psi, 213 HP, 6" pump		.42		36,400	1,650	38,050	42,400
065	1500 GPM, 200 psi, 288 HP, 6" pump		.38		44,200	1,825	46,025	51,500
070	2000 GPM, 100 psi, 188 HP, 6" pump		.34		31,300	2,025	33,325	37,400
075	2000 GPM, 150 psi, 292 HP, 6"pump		.30		33,900	2,300	36,200	40,600
080	2500 GPM, 100 psi, 255 HP, 8" pump		.32		29,600	2,150	31,750	35,700
082	2500 GPM, 150 psi, 427 HP, 8" pump		.26		34,500	2,650	37,150	41,800
085	3000 GPM, 100 psi, 288 HP, 8"pump		.28		43,000	2,475	45,475	51,000
090	3000 GPM, 150 psi, 534 HP, 10" pump		.20		85,500	3,450	88,950	99,000
095	3500 GPM, 100 psi, 320 HP, 10" pump		.24		40,800	2,875	43,675	49,100
100	3500 GPM, 150 psi, 534 HP, 10" pump		.20		81,000	3,450	84,450	94,000
300	Electric							
310	250 GPM, 40 psi, 15 HP, 3550 RPM, 3"pump	Q-13	.70	Ea.	8,350	985	9,335	10,600
320	500 GPM, 50 psi, 30 HP, 3550 RPM, 4" pump		.68		8,975	1,025	10,000	11,400
325	500 GPM, 100 psi, 75 HP, 1770 RPM, 4" pump		.66		11,200	1,050	12,250	13,800
330	500 GPM, 125 psi, 100 HP, 1770 RPM, 4" pump		.62		11,500	1,125	12,625	14,300
335	750 GPM, 50 psi, 50 HP, 1770 RPM, 5" pump		.64		9,350	1,075	10,425	11,900
340	750 GPM, 100 psi, 100 HP, 1770 RPM, 5" pump		.58		12,100	1,200	13,300	15,000
345	750 GPM, 165 psi, 125 HP, 1770 RPM, 5" pump		.56		16,700	1,225	17,925	20,200
350	1000 GPM, 50 psi, 50 HP 1770 RPM, 5" pump		.60		10,100	1,150	11,250	12,800
355	1000 GPM, 100 psi, 125 HP, 1770 RPM, 5" pump		.54		13,100	1,275	14,375	16,300
360	1000 GPM, 150 psi, 200 HP, 1770 RPM, 6" pump		.50		22,900	1,375	24,275	27,200
365	1000 GPM, 200 psi, 250 HP, 1770 RPM, 6" pump		.36		26,500	1,925	28,425	31,900
370	1500 GPM, 50 psi, 75 HP, 1770 RPM, 6" pump		.50		11,900	1,375	13,275	15,100
375	1500 GPM, 100 psi, 150 HP, 1770 RPM, 6" pump		.46		14,600	1,500	16,100	18,200
380	1500 GPM, 150 psi, 200 HP, 1770 RPM, 6" pump		.36		23,700	1,925	25,625	28,900
385	1500 GPM, 200 psi, 250 HP, 1770 RPM, 6" pump		.32		27,400	2,150	29,550	33,300
390	2000 GPM, 100 psi, 200 HP, 1770 RPM, 6" pump		.34		16,700	2,025	18,725	21,300
395	2000 GPM, 150 psi, 300 HP, 1770 RPM, 6" pump		.28		21,700	2,475	24,175	27,500

		Description	CREW	DAILY OUTPUT	UNIT	BARE COSTS MAT.	BARE COSTS INST.	BARE COSTS TOTAL	TOTAL INCL O&P
40	400	2500 GPM, 100 psi, 250 HP, 1770 RPM, 8" pump	Q-13	.30	Ea.	22,700	2,300	25,000	28,300
	404	2500 GPM, 135 psi, 350 HP, 1770 RPM, 8" pump		.26		23,200	2,650	25,850	29,400
	410	3000 GPM, 100 psi, 250 HP, 1770 RPM, 8" pump		.28		25,900	2,475	28,375	32,100
	415	3000 GPM, 140 psi, 450 HP, 1770 RPM, 10" pump		.24		33,700	2,875	36,575	41,300
	420	3500 GPM, 100 psi, 300 HP, 1770 RPM, 10" pump		.26		25,900	2,650	28,550	32,400
	425	3500 GPM, 140 psi, 450 HP, 1770 RPM, 10" pump		.24		30,000	2,875	32,875	37,200
	500	For jockey pump 1", 7-1/2 HP, add	Q-12	2		1,375	165	1,540	1,750
	50-001	**FIRE VALVES**							
	002	Angle, combination pressure adjustable/restricting, rough brass							
	003	B8.2 -310 1-1/2"	1 Spri	12	Ea.	30	15.05	45.05	55
	004	2-1/2"	"	7		67	26	93	110
	005	B8.2 -320 For polished brass, add				20%			
	006	For polished chrome, add				29%			
	008	B8.2 -390 Wheel handle, 300 lb., 1-1/2"	1 Spri	12		22	15.05	37.05	46
	009	2-1/2"	"	7		49	26	75	91
	010	C8.2 -301 For polished brass, add				35%			
	011	For polished chrome, add				50%			
	100	C8.2 -302 Ball drip, automatic, rough brass, 1/2"	1 Spri	20		6	9.05	15.05	19.75
	101	3/4"	"	20		7.50	9.05	16.55	21
	110	Butterfly, 175 lb., sprinkler system, FM/UL, threaded, bronze							
	112	Slow close							
	115	1" size	1 Spri	19	Ea.	39	9.50	48.50	57
	116	1-1/4" size		15		44	12.05	56.05	66
	117	1-1/2" size		13		57	13.90	70.90	83
	118	2" size		11		74	16.45	90.45	105
	119	2-1/2" size	Q-12	15		105	22	127	145
	123	For supervisory switch kit, all sizes							
	124	One circuit, add	1 Spri	48	Ea.	41	3.77	44.77	51
	125	Two circuits, add	"	40	"	49	4.52	53.52	60
	128	Quarter turn for trim							
	130	1/2" size	1 Spri	22	Ea.	4.30	8.20	12.50	16.70
	131	3/4" size		20		6.60	9.05	15.65	20
	132	1" size		19		8.60	9.50	18.10	23
	133	1-1/4" size		15		14.30	12.05	26.35	33
	134	1-1/2" size		13		17.85	13.90	31.75	40
	135	2" size		11		22	16.45	38.45	48
	140	Caps, polished brass with chain, 3/4"				9.50		9.50	10.45M
	142	1"				17.50		17.50	19.25M
	144	1-1/2"				4.50		4.50	4.95M
	146	2-1/2"				7.50		7.50	8.25M
	148	3"				23		23	25M
	300	Gate, hose, wheel handle, NRS, rough brass, 1-1/2",	1 Spri	12		60	15.05	75.05	88
	304	2-1/2", 300 Lb.	"	7		78	26	104	125
	308	For polished brass, add				40%			
	309	For polished chrome, add				50%			
	380	Hydrant, screw type, crank handle, brass							
	384	2-1/2" size	Q-12	11	Ea.	145	30	175	205
	386	3" size	"	7		275	46	321	370
	388	For chrome, add				10%			
	420	Hydrolator, vent and draining, rough brass, 1-1/2"	1 Spri	12		31	15.05	46.05	56
	422	2-1/2"	"	7		61	26	87	105
	428	For polished brass, add				25%			
	429	For polished chrome, add				35%			
	500	Pressure restricting, adjustable rough brass, 1-1/2"	1 Spri	12		38	15.05	53.05	64
	502	2-1/2"	"	7		69	26	95	115
	508	For polished brass, add				15%			
	509	For polished chrome, add				28%			
	600	Roof manifold, horiz. or vert., brass, with valves & caps							
	604	2-1/2" x 2-1/2" x 4"	Q-12	4.80	Ea.	225	68	293	345

15.4 Fire Extinguishing Systems

Line	Ref.	Description	CREW	DAILY OUTPUT	UNIT	BARE COSTS MAT.	BARE COSTS INST.	BARE COSTS TOTAL	TOTAL INCL O&P
50 606		2-1/2" x 2-1/2" x 6"	Q-12	4.60	Ea.	275	71	346	405
608		2-1/2" x 2-1/2" x 2-1/2" x 4"		4.60		310	71	381	445
609		2-1/2" x 2-1/2" x 2-1/2" x 6"		4.60		375	71	446	515
700		Sprinkler line tester, cast brass				12.55		12.55	13.80M
800		Wye, leader line, ball type, swivel female x male x male							
804		2-1/2" x 1-1/2" x 1-1/2" polished brass			Ea.	185		185	205M
806		2-1/2" x 1-1/2" x 1-1/2" polished chrome			"	205		205	225M
60-001		**SPRINKLER SYSTEM COMPONENTS**							
060		Accelerator, including anti-flood device	1 Spri	8	Ea.	635	23	658	730
080	B8.2 -110	Air compressor for dry pipe system, automatic, complete							
082		200 gal. system capacity, 1/3 HP	1 Spri	1.30	Ea.	445	140	585	690
084	B8.2 -120	350 gal. system capacity, 1/2 HP		1.30		465	140	605	715
086		520 gal. system capacity, 1 HP		1.30		655	140	795	925
090	B8.2 -130	Air compressor, with starting switch for manual operation							
091		200 gal. system capacity, 1/3 HP	1 Spri	1.30	Ea.	410	140	550	655
092	B8.2 -140	350 gal. system capacity, 1/2 HP		1.30		430	140	570	675
093		520 gal. system capacity, 1 HP		1.30		515	140	655	770
096	B8.2 -150	Air pressure maintenance control		24		70	7.55	77.55	88
110		Alarm, electric pressure switch (circuit closer)		26		48	6.95	54.95	63
114	C8.2 -101	For explosion proof, max 20 PSI, contacts close or open		26		145	6.95	151.95	170
122		Water motor, complete with gong		4		85	45	130	160
140	C8.2 -102	Deluge system, pressured monitoring panel, 120V		18		270	10.05	280.05	310
142		Explosion proof panel, not incl. enclosure		16		200	11.30	211.30	235
160	C8.2 -103	Dehydrator package, incl. valves and nipples		12		145	15.05	160.05	180
162		For bowl guard, add				22		22	24M
180		Firecycle system, controls, includes panel,							
182		batteries, solenoid valves and pressure switches	Q-13	1	Ea.	4,250	690	4,940	5,675
198		Detector	1 Spri	16		145	11.30	156.30	175
200		Release, emergency, manual, for hydraulic or pneumatic system		12		49	15.05	64.05	76
206		Release, thermostatic, for hydraulic or pneumatic release line		20		105	9.05	114.05	130
220		Sprinkler cabinets, 6 head capacity		16		18.90	11.30	30.20	37
226		12 head capacity		16		20	11.30	31.30	38
234		Sprinkler head escutcheons, standard, brass, 2" size		40		1.51	4.52	6.03	8.25
236		Chrome, 2" size		40		1.76	4.52	6.28	8.50
240		Recessed type, brass		40		2.04	4.52	6.56	8.80
244		Chrome or white enamel		40		1.68	4.52	6.20	8.40
260		Sprinkler heads, not including supply piping							
264		Dry, pendent or upright, 1/2" orifice, 3/4" or 1" NPT							
266		3" to 4-3/4" length	1 Spri	8	Ea.	17	23	40	52
267		5" to 6-3/4" length		8		17	23	40	52
268		7" to 8-3/4" length		8		18.10	23	41.10	53
269		9" to 10-3/4" length		8		19.50	23	42.50	54
270		11" to 12-3/4" length		8		19.50	23	42.50	54
271		13" to 14-3/4" length		7		21	26	47	61
272		15" to 16-3/4" length		7		21	26	47	61
273		17" to 18-3/4" length		7		24	26	50	64
274		19" to 20-3/4" length		7		24	26	50	64
275		21" to 22-3/4" length		7		25	26	51	65
276		23" to 24-3/4" length		7		25	26	51	65
278		25" to 26-3/4" length		6		27	30	57	74
279		27" to 28-3/4" length		6		28	30	58	75
280		For each inch or fraction over 29", add				.53		.53	.58M
290		For recessed fitting, add				1.26		1.26	1.38M
360		Foam-water, pendent or upright, 1/2" NPT	1 Spri	8		22	23	45	57
363		On-off (automatic), 165°-212°F							
364		Pendent, brass	1 Spri	7	Ea.	31	26	57	72
365		Chrome on brass		7		33	26	59	74
366		Recessed, chrome on brass		7		49	26	75	91
370		Standard spray, pendent or upright, brass, 135°-286°F							

15.4 Fire Extinguishing Systems

			CREW	DAILY OUTPUT	UNIT	BARE COSTS MAT.	BARE COSTS INST.	BARE COSTS TOTAL	TOTAL INCL O&P
60	372	1/2" NPT, 3/8" orifice	1 Spri	10	Ea.	2.75	18.10	20.85	29
	373	1/2" NPT, 7/16" orifice		10		2.75	18.10	20.85	29
	374	1/2" NPT, 1/2" orifice		10		2.10	18.10	20.20	29
	376	1/2" NPT, 17/32" orifice		10		2.47	18.10	20.57	29
	378	3/4" NPT, 17/32" orifice		10		4.68	18.10	22.78	31
	380	For open sprinklers, deduct				.60		.60	.66M
	384	For chrome, add			Ea.	.66		.66	.72M
	386	For wax and lead coating, add				.96		.96	1.05M
	388	For wax coating, add				.44		.44	.48M
	390	For lead coating, add				.53		.53	.58M
	392	For 360°F, same cost							
	393	For 400°F, add				2.12		2.12	2.33M
	394	For 500°F, add				2.12		2.12	2.33M
	420	Sidewall, vertical, brass, 135°-286°F							
	422	1/2" NPT, 3/8" orifice	1 Spri	10	Ea.	3.45	18.10	21.55	30
	423	1/2" NPT, 7/16" orifice		10		3.45	18.10	21.55	30
	424	1/2" NPT, 1/2" orifice		10		2.80	18.10	20.90	29
	426	1/2" NPT, 17/32" orifice		10		3.17	18.10	21.27	30
	428	3/4" NPT, 17/32" orifice		10		4.72	18.10	22.82	31
	436	For satin chrome, add				.66		.66	.72M
	440	For 360°F, same cost							
	441	For 400°F, add				2.12		2.12	2.33M
	442	For 500°F, add				2.12		2.12	2.33M
	450	Sidewall, horizontal, brass, 135°-286°F							
	452	1/2" NPT, 1/2" orifice	1 Spri	10	Ea.	2.25	18.10	20.35	29
	454	For 360°F, same cost							
	455	For 400°F, add				2.12		2.12	2.33M
	456	For 500°F, add				2.12		2.12	2.33M
	480	Recessed pendent, brass, 135°-286°F							
	482	1/2" NPT, 3/8" orifice	1 Spri	8	Ea.	3.25	23	26.25	36
	483	1/2" NPT, 7/16" orifice		8		3.25	23	26.25	36
	484	1/2" NPT, 1/2" orifice		8		2.70	23	25.70	36
	486	1/2" NPT, 17/32" orifice		8		3.01	23	26.01	36
	490	For satin chrome, add				.30		.30	.33M
	500	Recessed-vertical sidewall, brass, 135°-286°F							
	502	1/2" NPT, 3/8" orifice	1 Spri	8	Ea.	3	23	26	36
	503	1/2" NPT, 7/16" orifice		8		3	23	26	36
	504	1/2" NPT, 1/2" orifice		8		2.45	23	25.45	36
	506	1/2" NPT, 17/32" orifice		8		2.77	23	25.77	36
	510	For bright nickel, add				.30		.30	.33M
	560	Concealed, complete with cover plate							
	562	1/2" NPT, 1/2" orifice, 135°F to 212°F	1 Spri	7	Ea.	3.30	26	29.30	41
	580	Window, brass, 1/2" NPT, 1/4" orifice		10		7.50	18.10	25.60	35
	581	1/2" NPT, 5/16" orifice		10		7.50	18.10	25.60	35
	582	1/2" NPT, 3/8" orifice		10		7.50	18.10	25.60	35
	583	1/2" NPT, 7/16" orifice		10		7.50	18.10	25.60	35
	584	1/2" NPT, 1/2" orifice		10		7.95	18.10	26.05	35
	586	For polished chrome, add				1.14		1.14	1.25M
	588	3/4" NPT, 5/8" orifice	1 Spri	10		8.30	18.10	26.40	35
	589	3/4 NPT, 3/4" orifice	"	10		8.30	18.10	26.40	35
	600	Sprinkler head guards, red				3.12		3.12	3.43M
	602	Chrome				4.44		4.44	4.88M
	610	Sprinkler head wrenches, standard head				5		5	5.50M
	612	Recessed head				6.15		6.15	6.75M
	620	Valves							
	621	Alarm, includes							
	622	retard chamber, trim, gauges, alarm line strainer							
	626	2-1/2" size	Q-12	3	Ea.	420	110	530	620
	628	4" size	"	2		450	165	615	730
	630	6" size	Q-13	4		480	175	655	780

15

15.4	Fire Extinguishing Systems	CREW	DAILY OUTPUT	UNIT	BARE COSTS MAT.	BARE COSTS INST.	BARE COSTS TOTAL	TOTAL INCL O&P
60 632	8" size	Q-13	3	Ea.	565	230	795	955
650	Check, swing, C.I. body, brass fittings							
652	4" size	Q-12	3	Ea.	110	110	220	280
654	6" size	Q-13	4	↓	175	175	350	445
658	8" size	"	3	↓	425	230	655	800
680	Check, wafer, butterfly type, C.I. body, bronze fittings							
682	4" size	Q-12	4	Ea.	125	81	206	255
684	6" size	Q-13	5.50	↓	225	125	350	430
686	8" size	↓	5	↓	370	140	510	610
688	10" size	↓	4.50	↓	590	155	745	870
690	12" size	↓	4	↓	830	175	1,005	1,175
700	Deluge, assembly, incl. trim, pressure							
702	operated relief, emergency release, gauges							
704	2" size	Q-12	2	Ea.	675	165	840	980
706	3" size	↓	1.50	↓	785	215	1,000	1,175
708	4" size	↓	1	↓	980	325	1,305	1,550
710	6" size	Q-13	1.40	↓	1,300	495	1,795	2,150
730	Detector check, valve only, flanged,							
732	4" size, painted	Q-12	4	Ea.	360	81	441	515
734	6" size, painted	Q-13	5.50	↓	605	125	730	850
736	8" size, painted	↓	5	↓	945	140	1,085	1,250
738	10" size, painted	↓	4.50	↓	1,975	155	2,130	2,400
740	4" size, galvanized	Q-12	4	↓	425	81	506	585
742	6" size, galvanized	Q-13	5.50	↓	785	125	910	1,050
744	8" size, galvanized	↓	5	↓	1,175	140	1,315	1,500
746	10" size, galvanized	↓	4.50	↓	2,200	155	2,355	2,650
756	By-pass trim, not including meter							
758	For 5/8" meter	1 Spri	6	Ea.	36	30	66	83
760	For 1 meter	↓	5.50	↓	53	33	86	105
762	For 1-1/2" meter	↓	5	↓	93	36	129	155
764	For 2" meter	↓	4.50	↓	130	40	170	200
780	Diaphragm by-pass valve, bronze, required on all							
782	pneumatic release systems, any size deluge	1 Spri	18	Ea.	92	10.05	102.05	115
800	Dry pipe air check valve, incl. trim, gauges, 3" size	Q-12	2	↓	460	165	625	745
820	Dry pipe valve, incl. trim and gauges, 3" size	↓	2	↓	600	165	765	895
822	4" size	↓	1	↓	675	325	1,000	1,225
824	6" size	Q-13	2	↓	785	345	1,130	1,375
828	For accelerator trim with gauges, add	1 Spri	8	↓	210	23	233	265
840	Firecycle package, includes swing check							
842	and flow control valves with required trim							
844	2" size	Q-12	2	Ea.	675	165	840	980
846	3" size	↓	1.50	↓	1,075	215	1,290	1,500
848	4" size	↓	1	↓	1,225	325	1,550	1,825
850	6" size	Q-13	1.40	↓	1,725	495	2,220	2,625
880	Flow control valve, includes trim and gauges, 2" size	Q-12	2	↓	620	165	785	920
882	3" size	"	1.50	↓	720	215	935	1,100
884	4" size	Q-13	2.80	↓	905	245	1,150	1,350
886	6" size	"	2	↓	1,150	345	1,495	1,775
900	Magnetic by-pass valve, normally open circuit	1 Spri	16	↓	450	11.30	461.30	510
904	Normally closed detection circuit	↓	16	↓	500	11.30	511.30	565
920	Pressure operated relief valve, brass body	↓	18	↓	86	10.05	96.05	110
930	Pressure regulating valve, brass body	↓	20	↓	355	9.05	364.05	405
940	Pressure relieving device, anti-water columning	↓	12	↓	310	15.05	325.05	365
960	Waterflow indicator, with recycling retard and							
961	two single pole retard switches, 2" thru 8" pipe size	1 Spri	8	Ea.	97	23	120	140

15

15.5 Heating		CREW	DAILY OUTPUT	UNIT	BARE COSTS MAT.	BARE COSTS INST.	BARE COSTS TOTAL	TOTAL INCL O&P
02-001	**AVERAGE** Square foot and percent of total							
002	job cost, see division 17.1							
04-001	**BOILERS, GENERAL** Prices do not include flue piping, elec. wiring,							
002	gas or oil piping, boiler base, pad, or tankless unless noted							
010	C8.3 -001 Boiler horsepower: 10 KW = 34 lbs/steam/hr = 33,475 BTU/hr.							
015	To convert SFR to BTU rating: Hot water, 150 x SFR;							
016	C8.3 -002 C8.3 -003 Forced hot water, 180 x SFR; steam, 240 x SFR							
06-001	**BOILERS, ELECTRIC, ASME** Standard controls and trim							
100	Steam, 6 KW, 20.5 MBH	Q-19	1.20	Ea.	2,450	415	2,865	3,300
102	B8.3 -110 9 KW, 30.7 MBH		1.20		2,475	415	2,890	3,325
104	12 KW, 40.9 MBH		1.20		2,500	415	2,915	3,350
106	B8.3 -120 18 KW, 61.4 MBH		1.20		2,525	415	2,940	3,375
108	24 KW, 81.8 MBH		1.10		2,650	455	3,105	3,575
110	30 KW, 102 MBH		1.10		2,675	455	3,130	3,600
112	36 KW, 123 MBH		1.10		2,775	455	3,230	3,700
114	45 KW, 153 MBH		1		2,850	500	3,350	3,850
116	60 KW, 205 MBH		1		3,650	500	4,150	4,725
118	75 KW, 256 MBH		.90		3,875	555	4,430	5,050
120	105 KW, 358 MBH		.80		5,675	620	6,295	7,150
122	120 KW, 409 MBH		.75		5,825	665	6,490	7,375
124	150 KW, 512 MBH		.65		6,550	765	7,315	8,300
126	180 KW, 614 MBH		.60		6,800	830	7,630	8,675
128	210 KW, 716 MBH		.55		7,625	905	8,530	9,700
130	240 KW, 819 MBH		.45		10,500	1,100	11,600	13,100
132	300 KW, 1023 MBH		.40		11,200	1,250	12,450	14,100
134	360 KW, 1228 MBH		.35		12,500	1,425	13,925	15,800
136	420 KW, 1433 MBH		.30		14,200	1,650	15,850	18,000
138	510 KW, 1740 MBH	Q-21	.36		15,700	1,875	17,575	20,000
140	600 KW, 2047 MBH		.34		15,900	2,000	17,900	20,400
142	720 KW, 2456 MBH		.32		17,500	2,125	19,625	22,300
144	810 KW, 2764 MBH		.30		18,800	2,250	21,050	23,900
146	900 KW, 3070 MBH		.28		20,300	2,425	22,725	25,800
148	1080 KW, 3685 MBH		.25		22,900	2,700	25,600	29,100
150	1260 KW, 4300 MBH		.22		26,000	3,075	29,075	33,000
152	1620 KW, 5527 MBH		.20		37,700	3,375	41,075	46,300
154	1800 KW, 6141 MBH		.19		39,300	3,550	42,850	48,400
156	2070 KW, 7063 MBH		.18		45,200	3,750	48,950	55,000
158	2250 KW, 7677 MBH		.17		46,800	3,975	50,775	57,000
160	2340 KW, 7984 MBH		.16		47,700	4,225	51,925	58,500
162	2430 KW, 8291 MBH		.14		48,900	4,825	53,725	61,000
164	2520 KW, 8598 MBH		.12		50,500	5,625	56,125	63,500
200	Hot water, 12 KW, 41 MBH	Q-19	1.30		2,350	385	2,735	3,125
202	15 KW, 52 MBH		1.30		2,375	385	2,760	3,175
204	24 KW, 82 MBH		1.20		2,525	415	2,940	3,375
206	30 KW, 103 MBH		1.20		2,550	415	2,965	3,400
207	36 KW, 123 MBH		1.20		2,650	415	3,065	3,525
208	45 KW, 154 MBH		1.10		2,700	455	3,155	3,625
210	60 KW, 205 MBH		1.10		3,125	455	3,580	4,100
212	90 KW, 308 MBH		1		3,600	500	4,100	4,675
214	120 KW, 410 MBH		.90		3,925	555	4,480	5,125
216	150 KW, 510 MBH		.75		5,350	665	6,015	6,850
218	180 KW, 615 MBH		.65		5,775	765	6,540	7,450
220	210 KW, 716 MBH		.60		6,200	830	7,030	8,025
222	240 KW, 820 MBH		.55		6,550	905	7,455	8,500
224	270 KW, 922 MBH		.50		8,250	995	9,245	10,500
226	300 KW, 1024 MBH		.45		8,950	1,100	10,050	11,400
228	360 KW, 1228 MBH		.40		9,425	1,250	10,675	12,200
230	420 KW, 1432 MBH		.35		10,300	1,425	11,725	13,400
232	480 KW, 1636 MBH	Q-21	.46		11,100	1,475	12,575	14,300

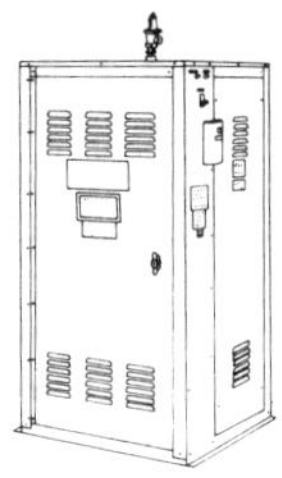

15

15.5 Heating

		CREW	DAILY OUTPUT	UNIT	BARE COSTS MAT.	BARE COSTS INST.	BARE COSTS TOTAL	TOTAL INCL O&P
06 234	510 KW, 1739 MBH	Q-21	.44	Ea.	11,500	1,525	13,025	14,900
236	570 KW, 1944 MBH		.43		12,400	1,575	13,975	15,900
238	630 KW, 2148 MBH		.42		13,300	1,600	14,900	17,000
240	690 KW, 2353 MBH		.40		14,200	1,700	15,900	18,100
242	720 KW, 2452 MBH		.39		14,600	1,725	16,325	18,600
244	810 KW, 2764 MBH		.38		16,200	1,775	17,975	20,400
246	900 KW, 3071 MBH		.37		17,400	1,825	19,225	21,800
248	1020 KW, 3480 MBH		.36		18,800	1,875	20,675	23,400
250	 1200 KW, 4095 MBH		.34		20,300	2,000	22,300	25,200
252	1320 KW, 4505 MBH		.33		23,800	2,050	25,850	29,100
254	1440 KW, 4915 MBH		.32		24,800	2,125	26,925	30,300
256	1560 KW, 5323 MBH		.31		26,000	2,175	28,175	31,700
258	1680 KW, 5733 MBH		.30		27,000	2,250	29,250	33,000
260	1800 KW, 6143 MBH		.29		28,500	2,325	30,825	34,700
262	1980 KW, 6757 MBH		.28		30,800	2,425	33,225	37,400
264	2100 KW, 7167 MBH		.27		31,900	2,500	34,400	38,700
266	2220 KW, 7576 MBH		.26		33,200	2,600	35,800	40,300
268	2400 KW, 8191 MBH		.25		34,600	2,700	37,300	42,000
270	2610 KW, 8905 MBH		.24		36,400	2,825	39,225	44,100
272	2790 KW, 9519 MBH		.23		38,100	2,950	41,050	46,200
274	2970 KW, 10133 MBH		.21		39,400	3,225	42,625	48,000
276	3150 KW, 10748 MBH		.19		41,000	3,550	44,550	50,000
278	3240 KW, 11055 MBH		.18		41,800	3,750	45,550	51,500
280	3420 KW, 11669 MBH		.17		42,800	3,975	46,775	53,000
282	3600 KW, 12,283 MBH		.16		44,600	4,225	48,825	55,000
08-001	**BOILERS, GAS FIRED** Natural or propane, standard controls							
100	Cast iron, with insulated jacket							
200	Steam, gross output, 81 MBH	Q-7	1.40	Ea.	955	485	1,440	1,750
202	102 MBH		1.30		1,100	525	1,625	1,975
204	122 MBH		1		1,200	680	1,880	2,300
206	163 MBH		.90		1,450	760	2,210	2,700
208	203 MBH		.90		1,650	760	2,410	2,900
210	240 MBH		.85		1,850	800	2,650	3,200
212	280 MBH		.80		2,000	850	2,850	3,425
214	320 MBH		.70		2,200	975	3,175	3,825
216	360 MBH		.65		2,525	1,050	3,575	4,300
218	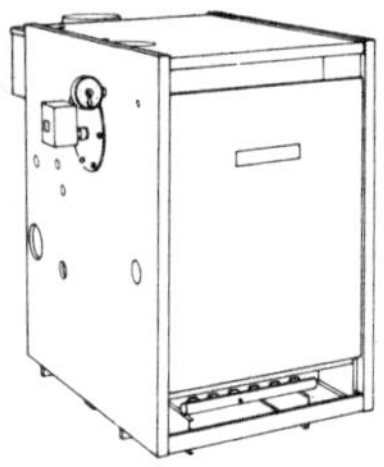 400 MBH		.60		2,700	1,125	3,825	4,600
220	440 MBH		.55		2,900	1,250	4,150	4,975
222	544 MBH		.50		3,825	1,375	5,200	6,175
224	816 MBH		.45		5,175	1,525	6,700	7,875
226	952 MBH		.40		5,900	1,700	7,600	8,950
228	1360 MBH		.36		8,275	1,900	10,175	11,800
230	1632 MBH		.31		9,725	2,200	11,925	13,900
232	2000 MBH		.28		11,600	2,425	14,025	16,300
234	2312 MBH		.26		13,300	2,625	15,925	18,400
236	2856 MBH		.24		16,100	2,850	18,950	21,800
238	3264 MBH		.23		17,900	2,975	20,875	24,000
240	3808 MBH		.18		20,900	3,800	24,700	28,500
242	4488 MBH		.16		24,600	4,250	28,850	33,200
244	5032 MBH		.14		27,300	4,875	32,175	37,100
246	5660 MBH		.13		28,300	5,250	33,550	38,700
248	6100 MBH		.12		33,600	5,675	39,275	45,200
250	6390 MBH		.10		35,000	6,825	41,825	48,400
252	6680 MBH		.09		36,100	7,575	43,675	50,500
254	6970 MBH		.08		37,400	8,525	45,925	53,500
300	Hot water, gross output, 80 MBH		1.46		845	465	1,310	1,600
302	100 MBH		1.35		990	505	1,495	1,825
304	122 MBH		1.10		1,100	620	1,720	2,100
306	163 MBH		1		1,350	680	2,030	2,475

15

15.5 Heating

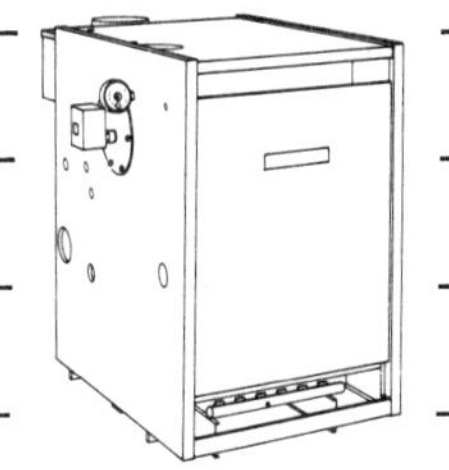

	15.5 Heating	CREW	DAILY OUTPUT	UNIT	BARE COSTS MAT.	BARE COSTS INST.	BARE COSTS TOTAL	TOTAL INCL O&P
08 308	203 MBH	Q-7	1	Ea.	1,550	680	2,230	2,700
310	240 MBH		.95		1,750	720	2,470	2,950
312	280 MBH		.90		2,025	760	2,785	3,325
314	320 MBH		.80		2,200	850	3,050	3,650
316	360 MBH		.75		2,525	910	3,435	4,100
318	400 MBH		.70		2,700	975	3,675	4,375
320	440 MBH		.65		2,900	1,050	3,950	4,700
322	544 MBH		.60		3,650	1,125	4,775	5,650
324	816 MBH		.55		5,025	1,250	6,275	7,325
326	1088 MBH		.50		6,450	1,375	7,825	9,075
328	1360 MBH		.46		8,125	1,475	9,600	11,100
330	1632 MBH		.42		9,550	1,625	11,175	12,900
332	2000 MBH		.38		11,400	1,800	13,200	15,100
334	2312 MBH		.36		13,900	1,900	15,800	18,000
336	2856 MBH		.33		16,700	2,075	18,775	21,400
338	3264 MBH		.30		18,600	2,275	20,875	23,700
340	3808 MBH		.26		22,100	2,625	24,725	28,100
342	4488 MBH		.22		25,800	3,100	28,900	32,900
344	5032 MBH		.18		27,200	3,800	31,000	35,400
346	5520 MBH		.14		30,100	4,875	34,975	40,100
348	6100 MBH		.12		35,400	5,675	41,075	47,200
350	6390 MBH		.10		36,700	6,825	43,525	50,000
352	6680 MBH		.09		37,900	7,575	45,475	52,500
354	6970 MBH		.08		39,200	8,525	47,725	55,500
400	Steel, insulating jacket							
450	Steam, not including burner, gross output							
550	1440 MBH	Q-6	.35	Ea.	7,575	1,425	9,000	10,400
552	1630 MBH		.30		7,850	1,675	9,525	11,000
554	1800 MBH		.26		8,125	1,925	10,050	11,700
556	2520 MBH		.20		10,100	2,500	12,600	14,700
558	3065 MBH		.18		10,800	2,775	13,575	15,900
560	3600 MBH		.16		11,500	3,125	14,625	17,200
562	5400 MBH		.14		16,000	3,575	19,575	22,800
564	7200 MBH	Q-7	.18		19,800	3,800	23,600	27,300
566	10,800 MBH		.17		24,700	4,000	28,700	33,000
568	12,600 MBH		.15		27,400	4,550	31,950	36,700
570	16,190 MBH		.12		34,700	5,675	40,375	46,400
572	17,990 MBH		.10		37,000	6,825	43,825	50,500
600	Hot water, including burner & one zone valve, gross output							
601	51.2 MBH	Q-6	2	Ea.	945	250	1,195	1,400
602	72 MBH		2		1,100	250	1,350	1,575
604	86 MBH		1.90		1,150	265	1,415	1,650
606	101 MBH		1.80		1,275	280	1,555	1,800
608	132 MBH		1.70		1,475	295	1,770	2,050
610	150 MBH		1.50		1,700	335	2,035	2,350
611	186 MBH		1.40		2,050	355	2,405	2,775
612	236 MBH		1.30		2,600	385	2,985	3,425
614	292 MBH		1.20		2,925	415	3,340	3,825
616	400 MBH		.80		3,875	625	4,500	5,175
618	480 MBH		.70		4,450	715	5,165	5,925
620	640 MBH		.60		5,275	835	6,110	7,000
622	800 MBH		.50		6,150	1,000	7,150	8,200
624	960 MBH		.45		7,625	1,100	8,725	10,000
626	1200 MBH		.40		9,400	1,250	10,650	12,100
628	1440 MBH		.35		10,600	1,425	12,025	13,700
630	1960 MBH		.30		14,100	1,675	15,775	17,900
632	2400 MBH		.20		17,200	2,500	19,700	22,500
634	3000 MBH		.15		21,300	3,325	24,625	28,200
700	For tankless water heater on smaller gas units, add				10%			
705	For additional zone valves up to 312 MBH add				64		64	70M

15

15.5 Heating

		CREW	DAILY OUTPUT	UNIT	BARE COSTS MAT.	BARE COSTS INST.	BARE COSTS TOTAL	TOTAL INCL O&P
08 800	Pulse combustion 44,000 BTU	Q-5	1.50	Ea.	1,450	215	1,665	1,900
805	88,000 BTU		1.40		1,675	230	1,905	2,175
808	134,000 BTU		1.20		2,375	270	2,645	3,000
10-001	**BOILERS, OIL FIRED** Standard controls, flame retention burner							
100	Cast iron, with insulated flush jacket							
200	Steam, gross output, 109 MBH	Q-7	1.20	Ea.	1,100	570	1,670	2,025
202	144 MBH		1.10		1,275	620	1,895	2,300
204	173 MBH		1		1,450	680	2,130	2,575
206	207 MBH		.90		1,575	760	2,335	2,825
208	236 MBH		.85		1,800	800	2,600	3,150
210	336 MBH		.70		2,725	975	3,700	4,400
212	480 MBH		.50		3,350	1,375	4,725	5,650
214	625 MBH		.48		4,150	1,425	5,575	6,625
216	720 MBH		.45		4,700	1,525	6,225	7,350
218	940 MBH		.42		5,550	1,625	7,175	8,450
220	1160 MBH		.38		6,300	1,800	8,100	9,525
222	1600 MBH		.31		8,100	2,200	10,300	12,100
224	2040 MBH		.28		9,700	2,425	12,125	14,200
226	2480 MBH		.25		12,200	2,725	14,925	17,400
228	2920 MBH		.23		13,900	2,975	16,875	19,600
230	3350 MBH		.22		15,500	3,100	18,600	21,500
232	3770 MBH		.19		17,600	3,600	21,200	24,500
234	4360 MBH		.17		19,300	4,000	23,300	27,000
236	4940 MBH		.15		21,100	4,550	25,650	29,800
238	5520 MBH		.13		22,900	5,250	28,150	32,800
240	6100 MBH		.12		27,600	5,675	33,275	38,600
242	6390 MBH		.10		28,800	6,825	35,625	41,500
244	6680 MBH		.09		30,000	7,575	37,575	43,900
246	6970 MBH		.08		31,300	8,525	39,825	46,700
300	Hot water, same price as steam							
400	For tankless coil in smaller sizes, add			Ea.	15%			
500	Steel, insulated jacket, burner							
600	Steam, full water leg construction, gross output							
602	144 MBH	Q-6	1.60	Ea.	1,250	310	1,560	1,825
604	198 MBH		1.40		1,375	355	1,730	2,025
606	252 MBH		1.30		1,725	385	2,110	2,450
608	324 MBH		1.20		1,950	415	2,365	2,750
610	396 MBH		.90		2,125	555	2,680	3,150
612	468 MBH		.80		2,350	625	2,975	3,475
614	648 MBH		.60		2,925	835	3,760	4,425
616	792 MBH		.50		3,450	1,000	4,450	5,250
618	1008 MBH		.45		4,250	1,100	5,350	6,275
620	1260 MBH		.40		4,725	1,250	5,975	7,000
622	1512 MBH		.35		5,650	1,425	7,075	8,275
624	1800 MBH		.33		6,200	1,525	7,725	9,000
626	2100 MBH		.26		7,300	1,925	9,225	10,800
628	2400 MBH		.22		8,150	2,275	10,425	12,200
640	Larger sizes are same as steel, gas fired.							
700	Hot water, gross output, 103 MBH	Q-6	1.90	Ea.	1,300	265	1,565	1,800
702	122 MBH		1.80		1,325	280	1,605	1,850
704	137 MBH		1.60		1,375	310	1,685	1,975
706	168 MBH		1.50		1,675	335	2,010	2,325
708	225 MBH		1.40		1,975	355	2,330	2,700
710	315 MBH		1.10		2,900	455	3,355	3,850
712	420 MBH		.80		3,275	625	3,900	4,500
714	525 MBH		.65		4,150	770	4,920	5,675
716	630 MBH		.60		4,600	835	5,435	6,275
718	735 MBH		.55		5,625	910	6,535	7,500
720	840 MBH		.50		6,425	1,000	7,425	8,500

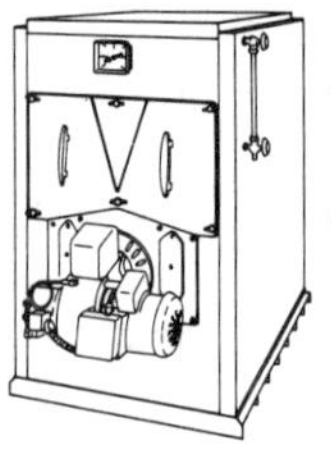

15

15.5 Heating

		CREW	DAILY OUTPUT	UNIT	BARE COSTS MAT.	BARE COSTS INST.	BARE COSTS TOTAL	TOTAL INCL O&P
10 722	1050 MBH	Q-6	.42	Ea.	7,800	1,200	9,000	10,300
724	1365 MBH		.37		9,975	1,350	11,325	12,900
726	1680 MBH		.33		11,700	1,525	13,225	15,100
728	2310 MBH		.24		15,100	2,075	17,175	19,600
730	2835 MBH		.17		18,700	2,950	21,650	24,800
732	3150 MBH		.13		21,000	3,850	24,850	28,700
734	For tankless coil in steam or hot water, add				7%			
12-001	**BOILERS, GAS/OIL** Combination with burners and controls							
100	Cast Iron with insulated jacket							
200	Steam, gross output, 720 MBH	Q-7	.40	Ea.	6,625	1,700	8,325	9,750
202	940 MBH		.37		7,500	1,850	9,350	10,900
204	1160 MBH		.33		8,500	2,075	10,575	12,300
206	1380 MBH		.30		9,750	2,275	12,025	14,000
208	1600 MBH		.26		11,000	2,625	13,625	15,900
210	2040 MBH		.23		12,900	2,975	15,875	18,500
212	2480 MBH		.20		15,000	3,400	18,400	21,400
214	2700 MBH		.19		15,700	3,600	19,300	22,500
216	2920 MBH		.18		16,700	3,800	20,500	23,800
218	3350 MBH		.17		18,500	4,000	22,500	26,100
220	3770 MBH		.15		22,700	4,550	27,250	31,500
222	4070 MBH		.14		23,700	4,875	28,575	33,100
224	4650 MBH		.13		25,300	5,250	30,550	35,400
226	5230 MBH		.11		28,800	6,200	35,000	40,600
228	5520 MBH		.10		29,800	6,825	36,625	42,600
230	5810 MBH		.09		31,500	7,575	39,075	45,600
232	6100 MBH		.08		35,000	8,525	43,525	51,000
234	6390 MBH		.07		36,400	9,750	46,150	54,000
236	6680 MBH		.06		37,500	11,400	48,900	57,500
238	6970 MBH		.05		38,800	13,600	52,400	62,500
300	Hot water, gross output, 584 MBH		.54		7,900	1,275	9,175	10,500
302	876 MBH		.50		11,800	1,375	13,175	15,000
304	1168 MBH		.48		15,600	1,425	17,025	19,200
306	1460 MBH		.45		19,500	1,525	21,025	23,600
308	2044 MBH		.36		27,200	1,900	29,100	32,700
310	2628 MBH		.32		35,000	2,125	37,125	41,600
312	3210 MBH		.28		42,700	2,425	45,125	50,500
314	3796 MBH		.24		50,500	2,850	53,350	59,500
316	4088 MBH		.22		54,500	3,100	57,600	64,500
318	4672 MBH		.18		62,000	3,800	65,800	73,500
320	5256 MBH		.14		67,000	4,875	71,875	80,500
322	6000 MBH, 179 BHP		.10		68,500	6,825	75,325	85,000
324	7130 MBH, 213 BHP		.07		69,500	9,750	79,250	90,500
326	9800 MBH, 286 BHP		.05		71,000	13,600	84,600	98,000
328	10,900 MBH, 325.6 BHP		.04		72,000	17,000	89,000	104,000
329	12,200 MBH, 364.5 BHP		.03		78,500	22,700	101,200	119,000
330	13,500 MBH, 403.3 BHP		.02		88,500	34,100	122,600	146,500
400	Steel, insulated jacket, skid base, tubeless							
410	Steam, 15 psi, gross output 517 MBH, 15 BHP	Q-6	.81	Ea.	6,500	615	7,115	8,050
411	690 MBH, 20 BHP		.78		6,775	640	7,415	8,375
412	862 MBH, 25 BHP		.61		7,250	820	8,070	9,150
413	1004 MBH, 30 BHP		.59		7,550	845	8,395	9,525
414	1340 MBH, 40 BHP		.52		8,600	960	9,560	10,800
415	1675 MBH, 50 BHP		.48		9,700	1,050	10,750	12,200
416	2175 MBH, 65 BHP		.42		10,600	1,200	11,800	13,400
417	2510 MBH, 75 BHP		.34		11,300	1,475	12,775	14,600
418	3015 MBH, 90 BHP		.28		13,600	1,775	15,375	17,500
419	3350 MBH, 100 BHP	Q-7	.32		16,200	2,125	18,325	20,900
420	3880 MBH, 116 BHP		.29		17,700	2,350	20,050	22,900
421	4685 MBH, 140 BHP		.27		19,000	2,525	21,525	24,500

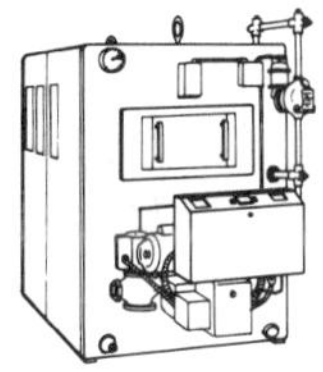

15.5 Heating

		CREW	DAILY OUTPUT	UNIT	BARE COSTS MAT.	BARE COSTS INST.	BARE COSTS TOTAL	TOTAL INCL O&P
12 422	5025 MBH, 150 BHP	Q-7	.26	Ea.	19,900	2,625	22,525	25,700
423	5550 MBH, 166 BHP		.17		26,300	4,000	30,300	34,700
424	6700 MBH, 200 BHP		.14		32,500	4,875	37,375	42,800
425	7760 MBH, 232 BHP		.13		35,000	5,250	40,250	46,100
426	8870 MBH, 265 BHP		.12		36,400	5,675	42,075	48,300
427	11,100 MBH, 322 BHP		.09		41,100	7,575	48,675	56,000
428	13,400 MBH, 400 BHP		.08		47,300	8,525	55,825	64,500
429	17,750 MBH, 530 BHP		.06		59,500	11,400	70,900	82,000
430	20,750 MBH, 620 BHP		.05		67,000	13,600	80,600	93,500
450	Steam, 150 psi gross output, 335 MBH, 10 BHP	Q-6	.65		6,575	770	7,345	8,350
452	502 MBH, 15 BHP		.55		8,075	910	8,985	10,200
456	670 MBH, 20 BHP		.45		8,425	1,100	9,525	10,900
460	1005 MBH, 30 BHP		.36		10,700	1,400	12,100	13,800
464	1339 MBH, 40 BHP		.28		13,400	1,775	15,175	17,300
466	1674 MBH, 50 BHP		.24		14,700	2,075	16,775	19,200
468	2009 MBH, 60 BHP		.21		15,800	2,375	18,175	20,800
472	2511 MBH, 75 BHP		.17		17,500	2,950	20,450	23,500
500	Hot water, gross output, 525 MBH		.60		6,100	835	6,935	7,925
502	630 MBH		.55		6,550	910	7,460	8,525
504	735 MBH		.50		7,575	1,000	8,575	9,775
506	840 MBH		.45		8,375	1,100	9,475	10,800
508	1050 MBH		.38		9,750	1,325	11,075	12,600
510	1365 MBH		.33		12,200	1,525	13,725	15,600
512	1680 MBH		.29		13,900	1,725	15,625	17,800
514	2310 MBH		.20		18,200	2,500	20,700	23,600
516	2835 MBH		.13		21,800	3,850	25,650	29,500
518	3150 MBH		.09		24,100	5,550	29,650	34,500
13-001	**BOILERS, SOLID FUEL**							
002	Wood, coal, wood/paper waste, peat. Steel, 3" diam. horizontal							
004	fire tubes, low pressure steam or hot water, natural or							
006	mechanical draft. Not including dry base.							
100	Natural draft, gross output, 1445 MBH	Q-6	.45	Ea.	5,575	1,100	6,675	7,725
105	1635 MBH		.43		5,825	1,150	6,975	8,075
110	1805 MBH		.41		6,075	1,225	7,300	8,450
115	2160 MBH		.38		6,525	1,325	7,850	9,075
120	2520 MBH		.32		7,875	1,550	9,425	10,900
125	3065 MBM	Q-7	.40		8,475	1,700	10,175	11,800
130	3600 MBH		.38		9,075	1,800	10,875	12,600
135	4500 MBH		.29		11,800	2,350	14,150	16,400
140	5400 MBH		.26		12,700	2,625	15,325	17,800
145	6295 MBH		.23		14,700	2,975	17,675	20,500
150	7200 MBH		.21		16,600	3,250	19,850	23,000
155	9000 MBH		.19		18,000	3,600	21,600	25,000
160	10,800 MBH		.17		20,600	4,000	24,600	28,500
165	12,600 MBH		.15		22,600	4,550	27,150	31,400
170	14,390 MBH		.13		27,200	5,250	32,450	37,500
175	16,190 MBH		.12		29,600	5,675	35,275	40,800
180	17,990 MBH		.10		32,900	6,825	39,725	46,000
300	Stoker fired (coal) cast iron with flush jacket and							
308	insulation, steam or water, gross output, 148 MBH	Q-6	1.50	Ea.	2,075	335	2,410	2,775
310	200 MBH		1.30		2,375	385	2,760	3,175
312	247 MBH		1.20		2,625	415	3,040	3,500
314	300 MBH		1		2,900	500	3,400	3,900
316	410 MBH		.96		3,225	520	3,745	4,300
318	495 MBH		.88		3,550	570	4,120	4,725
320	580 MBH		.80		3,875	625	4,500	5,175
340	1280 MBH		.36		9,450	1,400	10,850	12,400
342	1460 MBH		.30		10,100	1,675	11,775	13,500
344	1640 MBH		.28		10,800	1,775	12,575	14,500

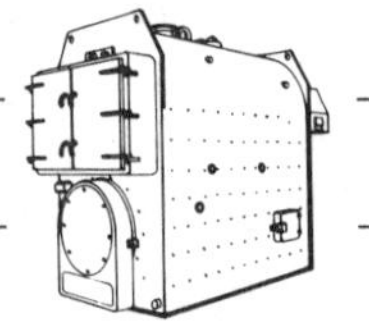

		CREW	DAILY OUTPUT	UNIT	BARE COSTS MAT.	BARE COSTS INST.	BARE COSTS TOTAL	TOTAL INCL O&P
13 346	1820 MBH	Q-6	.26	Ea.	11,500	1,925	13,425	15,400
348	2000 MBH		.25		12,300	2,000	14,300	16,400
350	2360 MBH		.23		13,600	2,175	15,775	18,100
352	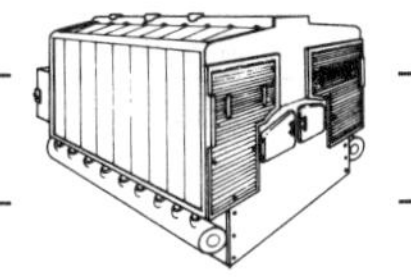 2540 MBH		.21		14,400	2,375	16,775	19,300
354	2725 MBH		.20		15,100	2,500	17,600	20,200
380	2950 MBH	Q-7	.16		26,000	4,250	30,250	34,800
382	3210 MBH		.15		27,200	4,550	31,750	36,500
384	3480 MBH		.14		28,500	4,875	33,375	38,400
386	3745 MBH		.14		29,600	4,875	34,475	39,600
388	4000 MBH		.13		30,900	5,250	36,150	41,600
390	4200 MBH		.13		32,200	5,250	37,450	43,000
392	4400 MBH		.12		33,800	5,675	39,475	45,400
394	4600 MBH		.12		35,500	5,675	41,175	47,300
500	Wood or coal and oil combination, circulator,							
505	mixing valve, controls							
510	Output (oil), 60 MBH, without burner	Q-5	.84	Ea.	815	380	1,195	1,450
515	80 MBH, without burner		.80		960	400	1,360	1,625
525	100 MBH, with burner		.73		1,750	440	2,190	2,550
530	130 MBH, with burner		.69		1,950	465	2,415	2,825
535	160 MBH, with burner		.65		2,125	495	2,620	3,050
14-001	**BOILERS, SCOTCH MARINE** Steam or hot water							
100	Packaged fire tube, #2 oil, gross output							
102	3348 MBH	Q-7	.12	Ea.	22,200	5,675	27,875	32,600
104	6696 MBH		.08		33,400	8,525	41,925	49,100
106	10,044 MBH		.06		44,500	11,400	55,900	65,500
108	16,740 MBH		.04		60,400	17,000	77,400	91,000
110	23,435 MBH		.03		74,300	22,700	97,000	114,500
112	To fire #6, add		.83		3,150	820	3,970	4,650
114	To fire #6, and gas, add		.42		4,625	1,625	6,250	7,425
116	For high pressure, add				3,100		3,100	3,400M
118	For duplex package feed system							
120	To 3348 MBH boiler, add	Q-7	.54	Ea.	3,100	1,275	4,375	5,225
122	To 6696 MBH boiler, add		.41		4,325	1,675	6,000	7,150
124	To 10,044 MBH boiler, add		.38		5,575	1,800	7,375	8,725
126	To 16,740 MBH boiler, add		.28		5,575	2,425	8,000	9,650
128	To 23,435 MBH boiler, add		.25		6,800	2,725	9,525	11,400
200	Packaged water tube, #2 oil, gross output							
204	1200 MBH	Q-7	.50	Ea.	8,975	1,375	10,350	11,800
206	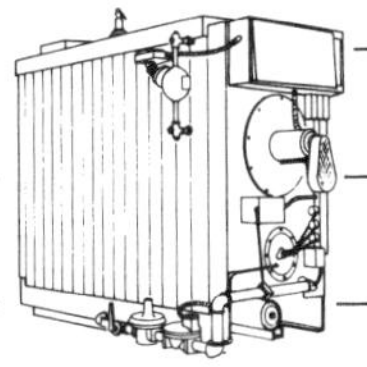 1600 MBH		.40		10,100	1,700	11,800	13,600
208	2400 MBH		.30		12,400	2,275	14,675	16,900
210	3200 MBH		.25		14,900	2,725	17,625	20,300
212	4800 MBH		.20		19,100	3,400	22,500	25,900
214	for gas fired, add		.40		3,100	1,700	4,800	5,875
17-001	**BOILER BLOWDOWN SYSTEMS**							
101	Boiler blowdown, auto/manual to 2000 MBH	Q-5	3.75	Ea.	1,400	86	1,486	1,675
102	7300 MBH	"	3	"	1,700	105	1,805	2,025
19-001	**BURNERS**							
099	Residential, conversion, gas fired, LP or natural							
100	Gun type, atmospheric input 35 to 180 MBH			Ea.	140		140	155M
102	50 to 240 MBH				160		160	175M
104	200 to 400 MBH				300		300	330M
200	Commercial and industrial, gun type, gas/oil, input							
205	400 to 980 MBH			Ea.	1,700		1,700	1,875M
208	Oil side pilot, second redundant valve, add				175		175	195M
211	Low High Off firing, add				400		400	440M
214	750 to 2100 MBH				2,650		2,650	2,925M
217	Low firing start, 2nd redundant switch, add				505		505	555M
220	1750 to 3080 MBH				3,200		3,200	3,525M

15.5 Heating

		CREW	DAILY OUTPUT	UNIT	BARE COSTS MAT.	BARE COSTS INST.	BARE COSTS TOTAL	TOTAL INCL O&P
19 223	Self check valve, 2nd redundant valve, add			Ea.	605		605	665M
226	3000 to 4000 MBH				4,825		4,825	5,300M
229	2nd redundant valve, low fire valve, add				660		660	725M
232	4000 to 6300 MBH				4,825		4,825	5,300M
235	Low fire start, self check valve, gas firing				455		455	500M
250	Impinged jet, rectangular, burner only, 3-1/2" WC,							
252	Gas fired, input, 420 to 640 MBH			Ea.	275		275	305M
254	560 to 860 MBH				395		395	435M
256	630 to 950 MBH				415		415	455M
258	950 to 1200 MBH				590		590	650M
260	1120 to 1700 MBH				785		785	865M
262	1700 to 1900 MBH				885		885	975M
264	1400 to 2100 MBH				980		980	1,075M
266	1600 to 2400 MBH				1,050		1,050	1,150M
268	1700 to 2500 MBH				1,175		1,175	1,300M
270	1880 to 2880 MBH				1,250		1,250	1,375M
272	2260 to 3680 MBH				1,450		1,450	1,600M
274	2590 to 4320 MBH				1,775		1,775	1,950M
276	3020 to 5040 MBH				2,050		2,050	2,250M
278	3450 to 5760 MBH				2,350		2,350	2,575M
285	Burner pilot				7.30		7.30	8.05M
286	Thermocouple				7.50		7.50	8.25M
287	Thermcouple & pilot bracket				20		20	22M
290	For 7" W.C. pressure, increase MBH 40%							
300	Flame retention oil fired assembly, input							
302	.50 to 2.25 GPH			Ea.	225		225	250M
304	2.0 to 5.0 GPH				255		255	280M
306	3.0 to 7.0 GPH				390		390	430M
308	6.0 to 12.0 GPH				615		615	675M
802	Coal stoker, automatic, 1000 MBH boiler	Q-7	1		5,475	680	6,155	7,000
804	2000 MBH boiler		.50		10,900	1,375	12,275	14,000
806	7300 MBH boiler		.20		38,300	3,400	41,700	47,100
22-001	**DRAFT CONTROLS, BAROMETRIC**							
100	Gas fired system only, 6" size for 5" and 6" pipes	1 Shee	20	Ea.	18.10	8.90	27	33
102	7" size, for 6" and 7" pipes		19		21	9.35	30.35	37
104	8" size, for 7" and 8" pipes		18		24	9.85	33.85	41
106	9" size, for 8" and 9" pipes		16		27	11.10	38.10	46
200	All fuel, oil, oil/gas, coal							
202	10" for 9" and 10" pipes	1 Shee	15	Ea.	41	11.85	52.85	62
204	12" for 11" and 12" pipes		15		54	11.85	65.85	77
206	14" for 13" and 14" pipes		14		67	12.70	79.70	92
208	16" for 15" and 16" pipes		13		94	13.65	107.65	125
210	18" for 17" and 18" pipes		12		120	14.80	134.80	155
212	20" for 19" and 21" pipes		10		145	17.75	162.75	185
214	24" for 22" and 25" pipes	Q-9	12		200	27	227	260
216	28" for 26" and 30" pipes		10		315	32	347	395
218	32" for 31" and 34" pipes		8		425	40	465	525
326	For thermal switch for above, add	1 Shee	24		23	7.40	30.40	36
500	Vent damper, bi-metal, gas	Q-9	16		49	20	69	83
510	Electric, automatic, gas, 6"		8		92	40	132	160
512	Automatic, oil, 8"		8		135	40	175	205
27-001	**DUCT FURNACES** Includes burner, controls, stainless steel							
002	heat exchanger. Gas fired, electric ignition							
003	Indoor installation							
008	100 MBH output	Q-5	5	Ea.	655	64	719	815
010	120 MBH output		4		755	80	835	945
013	200 MBH output		2.70		1,050	120	1,170	1,325
014	240 MBH output		2.30		1,125	140	1,265	1,450
016	280 MBH output		2		1,275	160	1,435	1,625

15

15.5 Heating

		CREW	DAILY OUTPUT	UNIT	BARE COSTS MAT.	BARE COSTS INST.	BARE COSTS TOTAL	TOTAL INCL O&P
27 018	320 MBH output	Q-5	1.60	Ea.	1,375	200	1,575	1,800
030	For powered venter and adapter, add			"	155		155	170M
050	For required flue pipe, see division 15.5-92							
100	Outdoor installation, with vent cap							
102	75 MBH output	Q-5	4	Ea.	660	80	740	840
104	94 MBH output		4		700	80	780	885
106	120 MBH output		4		795	80	875	990
108	157 MBH output		3.50		950	92	1,042	1,175
110	187 MBH output		3		1,075	105	1,180	1,325
112	225 MBH output		2.50		1,125	130	1,255	1,425
114	300 MBH output		1.80		1,400	180	1,580	1,800
116	375 MBH output		1.60		2,375	200	2,575	2,900
118	450 MBH output		1.40		2,500	230	2,730	3,075
120	600 MBH output		1		3,050	320	3,370	3,825
130	Aluminized exchanger, subtract				15%			
140	For powered venter, add				10%			
150	For two stage gas valve, add				150		150	165M
29-001	**DUCT HEATERS** Electric, slip-in. Includes blast coil, controls, fused							
002	transformer, air flow switch, thermal cutouts, contactors							
010	120 volt, 1 phase, 0.5 KW, 8" wide x 8" high	Q-20	14	Ea.	200	29	229	260
012	1 KW, 10" wide x 10" high		12		205	34	239	275
014	2 KW, 14" wide x 14" high		11		215	37	252	290
016	5 KW, 16" wide x 16" high		10		240	41	281	325
100	208 volt, 1 phase, 0.5 KW, 8" wide, 8" high		14		205	29	234	270
102	1 KW, 10" wide x 10" high		12		210	34	244	280
104	2 KW, 14" wide x 14" high		11		220	37	257	295
106	5 KW, 16" wide x 16" high		10		250	41	291	335
108	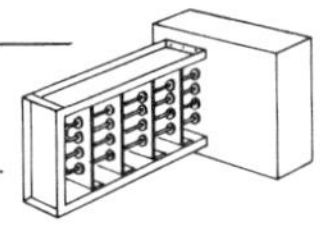 7.5 KW, 20" wide x 20" high		9		280	45	325	375
110	10 KW, 30" wide x 30" high		8		395	51	446	510
112	15 KW, 30" wide x 30" high		8		435	51	486	555
114	20 KW, 30" wide x 30" high		8		540	51	591	670
116	25 KW, 42" wide x 30" high		7		665	58	723	815
200	208 or 240 volt, 3 phase, 1.5 KW, 8" wide x 8" high		12		230	34	264	300
202	5 KW, 16" wide x 16" high		10		260	41	301	345
204	10 KW, 30" wide x 30" high		8		340	51	391	450
206	15 KW, 30" wide x 30" high		8		400	51	451	515
208	20 KW, 30" wide x 30" high		8		505	51	556	630
210	30 KW, 60" wide x 30" high		6		630	68	698	790
212	50 KW, 60" wide, x 30" high		6		880	68	948	1,075
214	75 KW, 72" wide x 30" high		4		1,250	100	1,350	1,525
215	85 KW, 72" wide x 30" high		4		1,425	100	1,525	1,725
216	100 KW, 72" wide x 30" high		4		1,625	100	1,725	1,925
300	480 volt, 3 phase, 110 KW, 72" wide x 30" high		2		1,500	205	1,705	1,950
302	125 KW, 72" wide x 30" high		2		1,650	205	1,855	2,100
304	150 KW, 72" wide x 30" high		2		1,850	205	2,055	2,325
350	To obtain BTU, multiply KW by 3413							
32-001	**FUEL OIL SPECIALTIES**							
002	Foot valve, single poppet, metal to metal construction,							
004	bevel seat, 1/2" diameter	1 Stpi	20	Ea.	27	8.90	35.90	43
006	3/4" diameter		18		28	9.90	37.90	45
008	1" diameter		16		30	11.15	41.15	49
010	1-1/4" diameter		15		40	11.90	51.90	61
012	1-1/2" diameter		13		46	13.70	59.70	70
014	2" diameter		11		60	16.20	76.20	89
100	Oil filters, 3/8" IPT., 20 gal. per hour		20		9.95	8.90	18.85	24
102	32 gal. per hour		18		28	9.90	37.90	45
104	40 gal. per hour		16		30	11.15	41.15	49
106	50 gal. per hour		14		32	12.75	44.75	54

	15.5 Heating		CREW	DAILY OUTPUT	UNIT	BARE COSTS MAT.	BARE COSTS INST.	BARE COSTS TOTAL	TOTAL INCL O&P
32	180	Pump and motor sets							
	181	Light fuel and diesel oils, 100 PSI							
	182	25 GPH 1/4 HP	Q-5	6	Ea.	320	54	374	430
	183	45 GPH, 1/4 HP		6		325	54	379	435
	184	90 GPH, 1/4 HP		5		330	64	394	455
	185	160 GPH, 1/3 HP		4		540	80	620	710
	186	325 GPH, 3/4 HP		4		645	80	725	825
	187	700 GPH, 1-1/2 HP		3		1,050	105	1,155	1,300
	188	1000 GPH, 2 HP		3		1,075	105	1,180	1,325
	189	1800 GPH, 5 HP		1.80		1,700	180	1,880	2,125
	200	Remote tank gauging system, self contained							
	202	30' transmission line							
	210	30" pointer travel	1 Stpi	2	Ea.	845	89	934	1,050
	212	5" pointer travel	"	2.50		550	71	621	710
	250	For each additional 10' transmission line, add				10		10	11M
	300	Valve, ball check, globe type, 3/8" diameter	1 Stpi	24		3.15	7.45	10.60	14.20
	350	Fusible, 3/8" diameter		24		2.95	7.45	10.40	14
	360	1/2" diameter		24		6.40	7.45	13.85	17.80
	361	3/4" diameter		20		10.65	8.90	19.55	25
	362	1" diameter		19		38	9.40	47.40	55
	400	Nonfusible, 3/8" diameter		24		4.45	7.45	11.90	15.65
	450	Shutoff, gate type, lever handle, spring-fusible kit							
	452	1/4" diameter	1 Stpi	14	Ea.	6.90	12.75	19.65	26
	454	3/8" diameter		12		6.70	14.85	21.55	29
	456	1/2" diameter		10		8.80	17.85	26.65	35
	457	3/4" diameter		8		11.05	22	33.05	44
	458	Lever handle, requires weight and fusible kit							
	460	1" diameter	1 Stpi	9	Ea.	80	19.80	99.80	115
	462	1-1/4" diameter		8		92	22	114	135
	464	1-1/2" diameter		7		110	25	135	160
	466	2" diameter		6		195	30	225	255
	468	For fusible link, weight and braided wire, add				5%			
	500	Vent alarm, whistling signal				14.10		14.10	15.50M
	550	Vent protector/breather, 1-1/4" diameter	1 Stpi	32		3	5.60	8.60	11.35
	552	1-1/2" diameter		32		3.75	5.60	9.35	12.20
	554	2" diameter		32		5.25	5.60	10.85	13.85
	556	3" diameter		28		22	6.35	28.35	33
	558	4" diameter		24		29	7.45	36.45	43
	800	Fuel oil and tank heaters							
	802	Electric, capacity rated at 230 volts							
	804	Immersion element in steel manifold							
	806	96 GPH at 50°F rise	Q-5	6.40	Ea.	510	50	560	635
	807	128 GPH at 50°F rise		6.20		525	52	577	650
	808	160 GPH at 50°F rise		5.90		555	54	609	690
	809	192 GPH at 50°F rise		5.50		585	58	643	730
	810	240 GPH at 50°F rise		5.10		630	63	693	785
	811	288 GPH at 50°F rise		4.60		705	70	775	875
	812	384 GPH at 50°F rise		3.10		1,025	105	1,130	1,275
	813	480 GPH at 50°F rise		2.30		1,450	140	1,590	1,800
	814	576 GPH at 50°F rise		2.10		1,550	155	1,705	1,925
	830	Suction stub, immersion type							
	832	75" long, 750 watts	1 Stpi	14	Ea.	375	12.75	387.75	430
	833	99" long, 2000 watts		12		410	14.85	424.85	470
	834	123" long, 3000 watts		10		440	17.85	457.85	510
	840	Electro/steam, electric capacity							
	842	rated at 230 volts/steam at 5 PSIG							
	844	28/50 GPH	Q-5	3.50	Ea.	1,075	92	1,167	1,325
	845	64/135 GPH		3		1,300	105	1,405	1,575
	846	80/257 GPH		2.50		1,600	130	1,730	1,950
	847	96/303 GPH		2.10		1,750	155	1,905	2,150

15

15.5 Heating

		CREW	DAILY OUTPUT	UNIT	BARE COSTS MAT.	BARE COSTS INST.	BARE COSTS TOTAL	TOTAL INCL O&P
32 848	120/384 GPH	Q-6	3	Ea.	1,875	165	2,040	2,300
849	144/720 GPH		1.80		3,275	280	3,555	4,000
850	192/800 GPH		1.60		3,700	310	4,010	4,525
851	240/1000 GPH		1.30		4,450	385	4,835	5,450
866	Steam, cross flow, rated at 5 PSIG							
868	42 GPH	Q-5	7	Ea.	510	46	556	625
869	73 GPH		6.70		550	48	598	675
870	112 GPH		6.20		575	52	627	705
871	158 GPH		5.80		615	55	670	755
872	187 GPH		4		925	80	1,005	1,125
873	270 GPH		3.60		990	89	1,079	1,225
874	365 GPH	Q-6	4.90		1,150	100	1,250	1,400
875	635 GPH		3.70		1,475	135	1,610	1,825
876	845 GPH		2.50		2,175	200	2,375	2,675
877	1420 GPH		1.60		3,525	310	3,835	4,325
878	2100 GPH		1.10		4,825	455	5,280	5,975
886	Steam or water, safety type							
888	Water capacity at 180°F/steam at 5 PSIG							
890	70/84 GPH	Q-5	3.70	Ea.	875	87	962	1,100
891	90/108 GPH		3.10		1,025	105	1,130	1,275
892	120/144 GPH		2.70		1,200	120	1,320	1,500
893	150/180 GPH		2.30		1,400	140	1,540	1,750
894	180/216 GPH		1.80		1,750	180	1,930	2,175
895	250/300 GPH		1.40		2,200	230	2,430	2,750
896	320/384 GPH	Q-6	1.90		2,650	265	2,915	3,300
897	400/449 GPH		1.50		3,250	335	3,585	4,050
898	480/536 GPH		1.30		3,775	385	4,160	4,700
899	600/670 GPH		1.10		4,775	455	5,230	5,900
910	Suction bell type							
912	160 GPH, 17" high	Q-5	8	Ea.	660	40	700	785
914	240 GPH, 24" high	"	7	"	825	46	871	975
920	Tank heating coil							
924	73" high by 15-3/8" diameter	Q-5	5	Ea.	1,375	64	1,439	1,600
928	Tank outlet heater (large capacity tank)							
930	rated for steam at 5/125 PSIG							
932	95/200 GPH	Q-5	4	Ea.	895	80	975	1,100
933	190/400 GPH	"	2.70		1,250	120	1,370	1,550
934	360/810 GPH	Q-6	3.20		1,750	155	1,905	2,150
935	490/1070 GPH		2.60		2,200	190	2,390	2,700
936	830/1850 GPH		2		2,825	250	3,075	3,475
937	1000/2200 GPH		1.70		3,175	295	3,470	3,925
938	1310/2920 GPH		1.30		4,325	385	4,710	5,325
939	1500/3300 GPH		1.10		4,725	455	5,180	5,850
940	2280/505o GPH		.83		7,725	600	8,325	9,375
941	2500/5550 GPH		.80		7,925	625	8,550	9,625
942	3000/6660 GPH		.72		8,725	695	9,420	10,600
943	4100/9150 GPH		.60		9,225	835	10,060	11,400
944	5450/12,100 GPH		.50		11,300	1,000	12,300	13,900
945	7000/15,500 GPH		.40		14,000	1,250	15,250	17,200
946	8700/19,300 GPH		.33		16,900	1,525	18,425	20,800
947	10,400/23,200 GPH		.29		18,700	1,725	20,425	23,100
35-001	**FURNACES** Hot air heating, blowers, standard controls,							
002	not including gas, oil or flue piping.							
100	Electric, UL listed, heat staging, 240 volt							
102	30 MBH	Q-20	4	Ea.	260	100	360	435
104	47 MBH		4		420	100	520	610
106	61 MBH		3.80		445	105	550	645
108	76 MBH		3.60		505	115	620	720
110	91 MBH		3.40		525	120	645	750

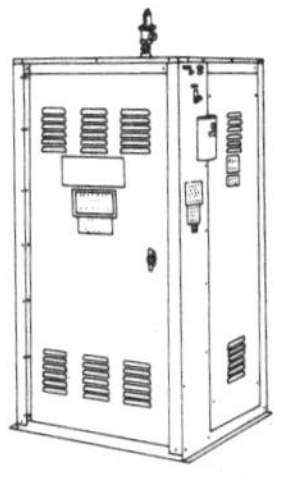

15

15.5 Heating

		CREW	DAILY OUTPUT	UNIT	BARE COSTS MAT.	INST.	TOTAL	TOTAL INCL O&P
35 112	112.7 MBH	Q-20	3.20	Ea.	650	130	780	900
114	131.2 MBH		3.10		675	130	805	935
116	140.8 MBH		3		690	135	825	955
150	For starter plenum, add	Q-9	16		43	20	63	76
200	For external filter rack, add		16		14.30	20	34.30	45
250	For electronic air filter, add		10		275	32	307	350
300	Gas, AGA certified, direct drive models							
302	42 MBH output	Q-9	4	Ea.	320	80	400	470
304	63 MBH output		3.80		335	84	419	490
306	79 MBH output		3.60		350	89	439	515
308	84 MBH output		3.40		380	94	474	555
310	105 MBH output		3.20		425	100	525	615
312	126 MBH output		3		550	105	655	760
313	160 MBH output		2.80		910	115	1,025	1,175
314	200 MBH output		2.60		1,275	125	1,400	1,575
316	300 MBH output		2.30		1,400	140	1,540	1,750
318	400 MBH output		2		1,650	160	1,810	2,050
328	For belt drive models, add				15%			
329	For starter plenum, add	Q-9	16		30	20	50	62
330	For external filter rack, add		16		18	20	38	49
331	For electronic air filter, add		10		280	32	312	355
340	Pulse combustion 38 MBH output		4		825	80	905	1,025
342	55 MBH output		3.80		840	84	924	1,050
345	72 MBH output		3.60		865	89	954	1,075
600	Oil, UL listed, atomizing gun type burner							
602	55 MBH output	Q-9	3.60	Ea.	585	89	674	775
603	84 MBH output		3.50		600	91	691	795
604	99 MBH output		3.40		650	94	744	850
606	125 MBH output		3.20		740	100	840	960
608	152 MBH output		3		935	105	1,040	1,175
610	200 MBH output		2.60		1,175	125	1,300	1,475
612	300 MBH output		2.30		1,250	140	1,390	1,575
614	400 MBH output		2		1,500	160	1,660	1,875
750	For plenum, add		14		44	23	67	82
800	For electronic air filter, add		10		275	32	307	350
850	Wood, coal and oil combination complete with burner							
852	112 MBH output (oil)	Q-9	3	Ea.	2,300	105	2,405	2,675
854	140 MBH output (oil)		3		2,300	105	2,405	2,675
856	150 MBH output (oil)		3		2,300	105	2,405	2,675
858	170 MBH output (oil)		2.50		2,300	130	2,430	2,725
860	For wood, coal and gas combination, add		14		93	23	116	135
37-001	**FURNACES, COMBINATION SYSTEMS** Heating, cooling, electric air							
002	cleaner, humidification, dehumidification.							
200	Gas fired, 80 MBH heat output, 24 MBH cooling	Q-9	1.20	Ea.	2,275	265	2,540	2,900
202	80 MBH heat output, 36 MBH cooling		1.20		2,450	265	2,715	3,075
204	100 MBH heat output, 29 MBH cooling		1		2,475	320	2,795	3,200
206	100 MBH heat output, 36 MBH cooling		1		2,600	320	2,920	3,325
208	100 MBH heat output, 47 MBH cooling		.90		2,875	355	3,230	3,675
210	120 MBH heat output, 29 MBH cooling	Q-10	1.30		2,600	385	2,985	3,425
212	120 MBH heat output, 42 MBH cooling		1.30		2,950	385	3,335	3,800
214	120 MBH heat output, 47 MBH cooling		1.20		2,975	415	3,390	3,875
216	120 MBH heat output, 55 MBH cooling		1.10		3,150	450	3,600	4,125
218	144 MBH heat output, 42 MBH cooling		1.20		3,000	415	3,415	3,900
220	144 MBH heat output, 47 MBH cooling		1.20		3,025	415	3,440	3,925
222	144 MBH heat output, 58 MBH cooling		1		3,175	495	3,670	4,225
250	For precharged tubing with connection							
252	15 feet			Ea.	83		83	91M
254	25 feet				110		110	120M
256	35 feet				140		140	155M

15

		CREW	DAILY OUTPUT	UNIT	BARE COSTS MAT.	BARE COSTS INST.	BARE COSTS TOTAL	TOTAL INCL O&P
37 300	Oil fired, 84 MBH heat output, 24 MBH cooling	Q-9	1.20	Ea.	2,425	265	2,690	3,050
302	84 MBH heat output, 36 MBH cooling		1.20		2,600	265	2,865	3,250
304	95.2 MBH heat output, 29 MBH cooling		1		2,625	320	2,945	3,350
306	95.2 MBH heat output, 36 MBH cooling		1		2,725	320	3,045	3,475
308	95.2 MBH heat output, 47 MBH cooling		.90		3,000	355	3,355	3,825
310	112 MBH heat output, 29 MBH cooling	Q-10	1.30		2,750	385	3,135	3,575
312	112 MBH heat output, 42 MBH cooling		1.30		3,050	385	3,435	3,925
314	112 MBH heat output, 47 MBH cooling		1.20		3,075	415	3,490	3,975
316	112 MBH heat output, 58 MBH cooling		1.10		3,250	450	3,700	4,225
318	151.2 MBH heat output, 42 MBH cooling		1.20		3,200	415	3,615	4,125
320	151.2 MBH heat output, 47 MBH cooling		1.20		3,250	415	3,665	4,175
324	151.2 MBH heat output, 58 MBH cooling		1		3,375	495	3,870	4,425
350	For precharged tubing with connection, add							
352	15 feet			Ea.	83		83	91M
354	25 feet				110		110	120M
356	35 feet				140		140	155M
40-001	**HEAT EXCHANGERS** 4 pass, 3/4" O.D. copper tubes,							
002	C.I. heads, C.I. tube sheet, steel shell							
010	Hot water 40°F to 180°F, by steam at 10 PSI							
012	8 GPM	Q-5	6	Ea.	455	54	509	580
014	10 GPM		5		700	64	764	865
016	40 GPM		4		1,100	80	1,180	1,325
018	64 GPM		2		1,625	160	1,785	2,025
020	96 GPM		1		2,150	320	2,470	2,825
022	120 GPM	Q-6	1.50		3,150	335	3,485	3,950
024	168 GPM		1		3,875	500	4,375	4,975
026	240 GPM		.80		6,050	625	6,675	7,550
050	For bronze head and tube sheet, add				50%			
100	Hot water 40°F to 140°F, by water at 200°F							
102	7 GPM	Q-5	6	Ea.	545	54	599	675
104	16 GPM		5		790	64	854	960
106	34 GPM		4		1,225	80	1,305	1,475
108	55 GPM		3		1,750	105	1,855	2,075
110	74 GPM		1.50		2,250	215	2,465	2,775
112	86 GPM		1.40		3,350	230	3,580	4,025
114	112 GPM	Q-6	2		4,250	250	4,500	5,025
116	126 GPM		1.80		5,300	280	5,580	6,225
118	152 GPM		1		6,775	500	7,275	8,175
42-001	**HEAT TRANSFER PACKAGES** Complete, controls,							
002	expansion tank, converter, air separator							
100	Hot water, 180°F enter, 200°F leaving, 15# steam							
101	One pump system, 28 GPM	Q-6	.75	Ea.	5,550	665	6,215	7,075
102	35 GPM		.70		6,250	715	6,965	7,900
104	55 GPM		.65		6,900	770	7,670	8,700
106	130 GPM		.55		8,950	910	9,860	11,200
108	255 GPM		.40		11,800	1,250	13,050	14,800
110	550 GPM		.30		15,900	1,675	17,575	19,900
112	800 GPM		.25		18,600	2,000	20,600	23,300
122	Two pump system, 28 GPM		.70		7,500	715	8,215	9,275
124	35 GPM		.65		9,150	770	9,920	11,200
126	55 GPM		.60		9,400	835	10,235	11,500
128	130 GPM		.50		12,100	1,000	13,100	14,800
130	255 GPM		.35		15,600	1,425	17,025	19,200
132	550 GPM		.25		20,300	2,000	22,300	25,200
134	800 GPM		.20		24,000	2,500	26,500	30,000
45-001	**HEATING & VENTILATING UNITS** Classroom							
002	Includes filter, heating/cooling coils, standard controls							
008	750 CFM, 2 tons cooling	Q-6	2	Ea.	1,750	250	2,000	2,275
010	1000 CFM, 2-1/2 tons cooling	"	1.60	"	2,075	310	2,385	2,725

15.5 Heating

		CREW	DAILY OUTPUT	UNIT	BARE COSTS MAT.	BARE COSTS INST.	BARE COSTS TOTAL	TOTAL INCL O&P
45 012	1250 CFM, 3 tons cooling	Q-6	1.40	Ea.	2,150	355	2,505	2,875
014	1500 CFM, 4 tons cooling		.80		2,300	625	2,925	3,425
016	2000 CFM 5 tons cooling		.50		2,750	1,000	3,750	4,475
050	For electric heat, add				35%			
100	For no cooling, deduct				25%	10%		
48-001	**HEAT RECOVERY PACKAGES**							
010	Air to air							
200	Kitchen exhaust, commercial, heat pipe exchanger							
204	Combined supply/exhaust air volume							
208	2.5 to 6.0 MCFM	Q-10	2.80	MCFM	2,950	180	3,130	3,500
212	6 to 16 MCFM		5		1,800	99	1,899	2,125
216	16 to 22 MCFM		6		1,400	83	1,483	1,650
800	Ventilating systems, moderate heat, paper plates							
801	and fins exchanger							
802	Standard capacity							
806	30 CFM	1 Shee	16	Ea.	155	11.10	166.10	185
810	57 CFM		14		330	12.70	342.70	380
814	71 CFM		12		385	14.80	399.80	445
824	212 CFM		8		1,600	22	1,622	1,800
830	277 CFM	Q-9	9		1,775	36	1,811	2,000
850	1175 CFM		4		12,300	80	12,380	13,600
860	2355 CFM		3		21,700	105	21,805	24,000
880	For intermediate capacities, use multiples							
51-001	**HYDRONIC HEATING** Terminal units, not including pipe to main supply							
100	Radiation							
110	Panel, baseboard, C.I., including supports, no covers	Q-5	20	L.F.	17.65	16.05	33.70	43
115	Fin tube, wall hung, 14" slope top cover, with damper							
120	1-1/4" copper tube, 4-1/4" alum. fin	Q-5	16	L.F.	23	20	43	54
125	1-1/4" steel tube, 4-1/4" steel fin		16		21	20	41	52
130	2" steel tube, 4-1/4" steel fin		13		23	25	48	61
131	Baseboard, pkgd, 1/2" copper tube, alum. fin, 7" h.		26		5.60	12.35	17.95	24
132	3/4" copper tube, alum. fin, 7" high		24		5.85	13.40	19.25	26
134	1" copper tube, alum. fin, 8-7/8" high		24		10.25	13.40	23.65	31
136	1-1/4" copper tube, alum. fin, 8-7/8" high		24		14.40	13.40	27.80	35
138	1-1/4" IPS steel tube with steel fins		24		16.30	13.40	29.70	37
150	Note: fin tube may also require corners, caps, etc.							
199	Convector unit, floor recessed, flush, with trim							
200	for under large glass wall areas, no damper	Q-5	20	L.F.	19.75	16.05	35.80	45
210	For unit with damper, add			"	8.65		8.65	9.50M
300	Radiators, cast iron							
310	Free standing or wall hung, 6 tube, 25" high	Q-5	96	Section	16.60	3.35	19.95	23
315	4 tube 25" high		96		12.45	3.35	15.80	18.55
320	4 tube, 19" high		96		11.55	3.35	14.90	17.55
325	Adj. brackets, 2 per wall radiator up to 30 sections	1 Stpi	32	Ea.	25	5.60	30.60	36
350	Recessed, 20" high x 5" deep, without grille	Q-5	60	Section	14	5.35	19.35	23
360	For inlet grille, add			"	.99		.99	1.08M
395	Unit heaters, propeller, 1 speed, 200° EWT							
400	Horizontal, 14.7 MBH	Q-5	12	Ea.	190	27	217	250
402	26.9 MBH		10		210	32	242	275
404	33.5 MBH		8		230	40	270	310
406	44.8 MBH		8		260	40	300	345
408	55.4 MBH		7.50		285	43	328	375
410	72.7 MBH		7		295	46	341	390
412	88.3 MBH		6.50		310	49	359	410
414	106.4 MBH		6		315	54	369	425
416	143.7 MBH		5		365	64	429	495
418	160.9 MBH		4		435	80	515	595
420	207.3 MBH		3		590	105	695	805
422	254.5 MBH		2.50		650	130	780	900

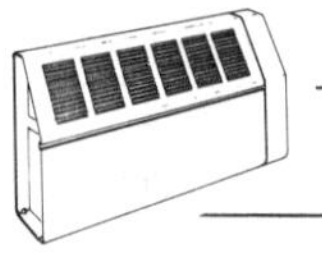

15

15.5 Heating

			CREW	DAILY OUTPUT	UNIT	BARE COSTS MAT.	BARE COSTS INST.	BARE COSTS TOTAL	TOTAL INCL O&P
51	424	292.5 MBH	Q-5	2	Ea.	710	160	870	1,025
	430	For vertical diffuser, add				76		76	84M
	500	Vertical flow, 52.4 MBH	Q-5	11		260	29	289	330
	502	71.4 MBH		8		285	40	325	370
	504	88.4 MBH		7		315	46	361	415
	506	123 MBH		6		365	54	419	480
	508	140 MBH		4		415	80	495	575
	510	177.4 MBH		3		465	105	570	665
	512	220 MBH		2.20		545	145	690	810
	514	296.7 MBH		2		830	160	990	1,150
	516	408 MBH	Q-6	1.80		1,050	280	1,330	1,550
	518	520 MBH	"	1.40		1,250	355	1,605	1,900
	530	For capacity of unit heaters using 2 psi							
	532	steam, multiply MBH by 1.43							
	600	Valance units, complete with 1/2" cooling coil, enclosure							
	602	2 tube	Q-5	18	L.F.	21	17.85	38.85	49
	604	3 tube		16		26	20	46	58
	606	4 tube		16		30	20	50	62
	608	5 tube		15		36	21	57	71
	610	6 tube		15		40	21	61	75
	612	8 tube		14		58	23	81	97
	620	For 3/4" cooling coil, add				10%			
	950	To convert SFR to BTU rating: Hot water, 150 x SFR							
	951	Forced hot water, 180 x SFR; steam, 240 x SFR							
	54-001	**HUMIDIFIERS**							
	003	Centrifugal atomizing							
	005	5 lb. per hour	Q-5	12	Ea.	745	27	772	860
	010	10 lb. per hour		10		960	32	992	1,100
	012	24 lb. per hour		8		1,475	40	1,515	1,675
	052	Steam, room or duct, filter, regulators, automatic controls, 220Volt							
	054	10 lb. per hour	Q-5	6	Ea.	1,850	54	1,904	2,100
	056	17 lb. per hour		5		1,900	64	1,964	2,175
	058	30 lb. per hour		4		1,925	80	2,005	2,225
	060	60 lb. per hour		4		2,775	80	2,855	3,175
	062	90 lb. per hour		3		2,850	105	2,955	3,300
	064	120 lb. per hour		2.50		4,850	130	4,980	5,525
	066	180 lb. per hour		2		5,075	160	5,235	5,825
	070	With blower							
	072	10 lb. per hour	Q-5	5.50	Ea.	1,975	58	2,033	2,250
	074	17 lb. per hour		4.75		2,025	68	2,093	2,325
	076	30 lb. per hour		3.75		2,075	86	2,161	2,400
	078	60 lb. per hour		3.50		2,550	92	2,642	2,950
	080	90 lb. per hour		2.75		2,950	115	3,065	3,425
	082	120 lb. per hour		2		4,475	160	4,635	5,150
	084	180 lb. per hour		1.50		6,150	215	6,365	7,075
	57-001	**INDUCED DRAFT FANS**							
	100	Breeching installation							
	180	Hot gas, 600°F,variable pitch pulley and motor							
	184	6" diam. inlet, 1/4 H.P., 1 phase, 400 CFM	Q-9	6	Ea.	720	53	773	870
	185	7" diam. inlet, 1/4 H.P., 1 phase, 800 CFM		5		765	64	829	935
	186	8" diam. inlet, 1/4 H.P., 1phase, 1120 CFM		4		1,150	80	1,230	1,375
	187	9" diam. inlet, 3/4 H.P., 1 phase, 1440 CFM		3.60		1,315	89	1,404	1,575
	188	10" diam. inlet, 3/4 H.P., 1 phase, 2000 CFM		3.30		1,535	97	1,632	1,825
	190	12" diam. inlet, 3/4 H.P., 3 phase, 2960 CFM		3		1,575	105	1,680	1,900
	191	14" diam. inlet, 1 H.P., 3 phase, 4160 CFM		2.60		1,890	125	2,015	2,250
	192	16" diam. inlet, 2 H.P., 3 phase, 6720 CFM		2.30		2,075	140	2,215	2,475
	194	18" diam. inlet, 3 H.P., 3 phase, 9120 CFM		2		2,575	160	2,735	3,075
	195	20" diam. inlet, 3 H.P., 3 phase, 9760 CFM		1.50		2,825	215	3,040	3,425
	196	22" diam. inlet, 5 H.P., 3 phase, 13,360 CFM		1		3,315	320	3,635	4,125

15.5 Heating

	15.5 Heating	CREW	DAILY OUTPUT	UNIT	BARE COSTS MAT.	BARE COSTS INST.	BARE COSTS TOTAL	TOTAL INCL O&P
57 198	24" diam. inlet, 7-1/2 H.P., 3 phase, 17,760 CFM	Q-9	.80	Ea.	4,050	400	4,450	5,050
230	For multi-blade damper at fan inlet, add				20%			
360	Chimneytop installation							
370	6" size	1 Shee	8	Ea.	240	22	262	295
374	8" size		7		245	25	270	305
378	13" size		6		340	30	370	415
388	For speed control switch, add				40		40	44M
392	For thermal fan control, add				17.25		17.25	18.95M
550	Flue shutter damper for draft control,							
551	parallel blades with circular collar and locking quadrant							
555	8" size	Q-9	8	Ea.	325	40	365	415
556	9" size		7.50		355	43	398	455
557	10" size		7		390	46	436	495
558	12" size		6.50		425	49	474	540
559	14" size		6		480	53	533	605
560	16" size		5.50		525	58	583	660
561	18" size		5		590	64	654	740
562	20" size		4.50		660	71	731	830
563	22" size		4		730	80	810	920
564	24" size		3.50		780	91	871	990
565	27" size		3		820	105	925	1,050
566	30" size		2.50		880	130	1,010	1,150
567	32" size		2		935	160	1,095	1,250
568	36" size		1.50		970	215	1,185	1,375
62-001	**INFRA-RED UNIT**							
002	Gas fired, unvented, electric ignition, 100% shutoff. Piping and							
003	wiring not included							
006	Input, 15 MBH	Q-5	7	Ea.	280	46	326	375
010	30 MBH		6		325	54	379	435
012	45 MBH		5		380	64	444	510
014	50 MBH		4.50		390	71	461	530
016	60 MBH		4		450	80	530	610
018	75 MBH		3		500	105	605	705
020	90 MBH		2.50		560	130	690	800
022	105 MBH		2		645	160	805	940
024	120 MBH		2		685	160	845	985
050	For deep parabolic reflector, add				50		50	55M
100	For opt. screen, req'd. for hangar or garage install., add				8%			
200	Electric, single or three phase							
205	6 KW, 20,478 BTU	Q-5	6	Ea.	300	54	354	405
210	13.5 KW, 40,956 BTU		5		405	64	469	540
215	24 KW, 81,912 BTU		3		770	105	875	1,000
300	Oil fired, two stage pump, controls, solenoid valve, venter							
305	117,000 BTU, one burner	Q-5	2.50	Ea.	1,975	130	2,105	2,350
308	234,000 BTU, two burner		2.25		3,450	145	3,595	4,000
311	351,000 BTU, three burner		2		4,925	160	5,085	5,650
314	468,000 BTU four burner		1.50		6,375	215	6,590	7,325
320	Free air inlet per burner, add				99		99	110M
325	Prepurger per unit, add				24		24	26M
65-001	**INSULATION**							
010	Rule of thumb, as a percentage of total mechanical costs						10%	
100	Boiler, 1-1/2" calcium silicate, 1/2" cement finish	Q-14	50	S.F.	2.80	6.40	9.20	12.50
102	2" fiberglass	"	80	"	1.80	4	5.80	7.85
200	Breeching, 2" calcium silicate with 1/2" cement finish, no lath							
202	Rectangular	Q-14	50	S.F.	3.90	6.40	10.30	13.70
204	Round	"	40	"	3.90	8	11.90	16.05
230	Calcium silicate block, + 200° to + 1200°F							
234	1" thick	Q-14	30	S.F.	2.20	10.65	12.85	18.10
236	1-1/2" thick	"	25	"	2.40	12.80	15.20	21

15

15.5 Heating

		CREW	DAILY OUTPUT	UNIT	BARE COSTS MAT.	BARE COSTS INST.	BARE COSTS TOTAL	TOTAL INCL O&P
65 238	2" thick	Q-14	22	S.F.	3.05	14.55	17.60	25
240	3" thick	"	18	"	4.85	17.75	22.60	31
290	Domestic water heater wrap kit							
292	1-1/2" with vinyl jacket, 20-60 gal.	1 Plum	8	Ea.	21	22	43	55
300	Ductwork							
302	Blanket type, fiberglass, flexible							
303	Fire resistant liner, black coating one side							
305	1/2" thick, 2 lb. density	Q-14	380	S.F.	.25	.84	1.09	1.51
306	1" thick, 1-1/2 lb. density		350		.31	.91	1.22	1.69
307	1-1/2" thick, 1-1/2 lb. density		320		.34	1	1.34	1.85
308	2" thick, 1-1/2 lb. density		300		.45	1.07	1.52	2.06
314	FRK vapor barrier wrap, .75 lb. density							
316	1" thick	Q-14	350	S.F.	.18	.91	1.09	1.54
317	1-1/2" thick		320		.21	1	1.21	1.70
318	2" thick		300		.27	1.07	1.34	1.87
319	3" thick		260		.39	1.23	1.62	2.24
321	Vinyl jacket, same as FRK							
328	Unfaced, 1 lb. density							
331	1" thick	Q-14	360	S.F.	.22	.89	1.11	1.55
332	1-1/2" thick		330		.33	.97	1.30	1.79
333	2" thick		310		.45	1.03	1.48	2.01
349	Board type, fiberglass, 3 lb. density							
350	Fire resistant, black pigmented, 1 side							
352	1" thick	Q-14	150	S.F.	.66	2.13	2.79	3.86
354	1-1/2" thick		130		.69	2.46	3.15	4.38
356	2" thick		120		.81	2.66	3.47	4.81
360	FRK vapor barrier							
362	1" thick	Q-14	150	S.F.	.97	2.13	3.10	4.21
363	1-1/2" thick		130		1.10	2.46	3.56	4.83
364	2" thick		120		1.35	2.66	4.01	5.40
368	No finish							
370	1" thick	Q-14	170	S.F.	.56	1.88	2.44	3.39
371	1-1/2" thick		140		.74	2.28	3.02	4.18
372	2" thick		130		.98	2.46	3.44	4.70
375	Finishes							
380	1/2" cement over 1" wire mesh, incl. corner bead	Q-14	116	S.F.	.89	2.76	3.65	5.05
382	Canvas, 8 oz. pasted on		246		1.50	1.30	2.80	3.56
390	Weatherproof, non-metallic, 2 lb. per S.F.		100		2.20	3.20	5.40	7.15
400	Pipe covering							
404	Air cell, with cover							
410	3 ply, 1/2" iron pipe size	Q-14	145	L.F.	.78	2.20	2.98	4.10
413	3/4" iron pipe size		145		.88	2.20	3.08	4.21
414	1" iron pipe size		145		.93	2.20	3.13	4.27
415	1-1/4" iron pipe size		140		1.04	2.28	3.32	4.51
416	1-1/2" iron pipe size		140		1.14	2.28	3.42	4.62
417	2" iron pipe size		135		1.24	2.37	3.61	4.85
418	2-1/2" iron pipe size		135		1.40	2.37	3.77	5.05
419	3" iron pipe size		130		1.55	2.46	4.01	5.35
420	3-1/2" iron pipe size		130		1.71	2.46	4.17	5.50
421	4" iron pipe size		125		2.02	2.56	4.58	6
422	5" iron pipe size		120		2.33	2.66	4.99	6.50
423	6" iron pipe size		115		2.48	2.78	5.26	6.80
424	8" iron pipe size		105		3.42	3.04	6.46	8.25
425	10" iron pipe size		95		4.04	3.37	7.41	9.40
426	12" iron pipe size		90		4.66	3.55	8.21	10.35
430	4 ply, 1/2" iron pipe size		115		1.09	2.78	3.87	5.30
433	3/4" iron pipe size		115		1.23	2.78	4.01	5.45
434	1" iron pipe size		115		1.30	2.78	4.08	5.50
435	1-1/4" iron pipe size		110		1.45	2.91	4.36	5.90
436	1-1/2" iron pipe size		110		1.59	2.91	4.50	6.05

15.5 Heating

		CREW	DAILY OUTPUT	UNIT	BARE COSTS MAT.	BARE COSTS INST.	BARE COSTS TOTAL	TOTAL INCL O&P
65 437	2" iron pipe size	Q-14	105	L.F.	1.74	3.04	4.78	6.40
438	2-1/2" iron pipe size		105		1.96	3.04	5	6.65
439	3" iron pipe size		100		2.17	3.20	5.37	7.10
440	3-1/2" iron pipe size		100		2.39	3.20	5.59	7.35
441	4" iron pipe size		95		2.83	3.37	6.20	8.05
442	5" iron pipe size		90		3.26	3.55	6.81	8.80
443	6" iron pipe size		90		3.48	3.55	7.03	9.05
444	8" iron pipe size		85		4.78	3.76	8.54	10.80
445	10" iron pipe size		75		5.65	4.26	9.91	12.50
446	12" iron pipe size		70		6.50	4.57	11.07	13.90
490	Calcium silicate, with cover							
510	1" wall, 1/2" iron pipe size	Q-14	170	L.F.	1.54	1.88	3.42	4.46
513	3/4" iron pipe size		170		1.62	1.88	3.50	4.55
514	1" iron pipe size		170		1.62	1.88	3.50	4.55
515	1-1/4" iron pipe size		165		1.65	1.94	3.59	4.67
516	1-1/2" iron pipe size		165		1.69	1.94	3.63	4.71
517	2" iron pipe size		160		1.72	2	3.72	4.83
518	2-1/2" iron pipe size		160		2.14	2	4.14	5.30
519	3" iron pipe size		150		2.29	2.13	4.42	5.65
520	4" iron pipe size		140		2.62	2.28	4.90	6.25
521	5" iron pipe size		135		2.90	2.37	5.27	6.70
522	6" iron pipe size		130		3.17	2.46	5.63	7.10
528	1-1/2" wall, 1/2" iron pipe size		150		1.86	2.13	3.99	5.20
531	3/4" iron pipe size		150		1.88	2.13	4.01	5.20
532	1" iron pipe size		150		2.06	2.13	4.19	5.40
533	1-1/4" iron pipe size		145		2.23	2.20	4.43	5.70
534	1-1/2" iron pipe size		145		2.32	2.20	4.52	5.80
535	2" iron pipe size		140		2.54	2.28	4.82	6.15
536	2-1/2" iron pipe size		140		2.78	2.28	5.06	6.40
537	3" iron pipe size		135		2.90	2.37	5.27	6.70
538	4" iron pipe size		125		3.37	2.56	5.93	7.45
539	5" iron pipe size		120		3.77	2.66	6.43	8.05
540	6" iron pipe size		110		3.93	2.91	6.84	8.60
546	2" wall, 1/2" iron pipe size		135		2.92	2.37	5.29	6.70
549	3/4" iron pipe size		135		3.09	2.37	5.46	6.90
550	1" iron pipe size		135		3.20	2.37	5.57	7
551	1-1/4" iron pipe size		130		3.46	2.46	5.92	7.45
552	1-1/2" iron pipe size		130		3.61	2.46	6.07	7.60
553	2" iron pipe size		125		3.78	2.56	6.34	7.90
554	2-1/2" iron pipe size		125		4.08	2.56	6.64	8.25
555	3" iron pipe size		120		4.37	2.66	7.03	8.75
556	4" iron pipe size		115		5.15	2.78	7.93	9.75
557	5" iron pipe size		110		5.85	2.91	8.76	10.70
558	6" iron pipe size		105		6.55	3.04	9.59	11.70
560	Calcium silicate, no cover							
572	1" wall, 1/2" iron pipe size	Q-14	180	L.F.	1.50	1.78	3.28	4.27
574	3/4" iron pipe size		180		1.50	1.78	3.28	4.27
575	1" iron pipe size		180		1.50	1.78	3.28	4.27
576	1-1/4" iron pipe size		175		1.51	1.83	3.34	4.35
577	1-1/2" iron pipe size		175		1.55	1.83	3.38	4.40
578	2" iron pipe size		170		1.86	1.88	3.74	4.82
579	2-1/2" iron pipe size		170		1.97	1.88	3.85	4.94
580	3" iron pipe size		160		2.12	2	4.12	5.25
581	4" iron pipe size		150		2.42	2.13	4.55	5.80
582	5" iron pipe size		145		2.67	2.20	4.87	6.20
583	6" iron pipe size		140		2.91	2.28	5.19	6.55
590	1-1/2" wall, 1/2" iron pipe size		160		1.74	2	3.74	4.86
592	3/4" iron pipe size		160		1.74	2	3.74	4.86
593	1" iron pipe size		160		1.92	2	3.92	5.05
594	1-1/4" iron pipe size		155		2.07	2.06	4.13	5.30

15.5 Heating

			CREW	DAILY OUTPUT	UNIT	BARE COSTS MAT.	BARE COSTS INST.	BARE COSTS TOTAL	TOTAL INCL O&P
65	595	1-1/2" iron pipe size	Q-14	155	L.F.	2.16	2.06	4.22	5.40
	596	2" iron pipe size		150		2.38	2.13	4.51	5.75
	597	2-1/2" iron pipe size		150		2.59	2.13	4.72	6
	598	3" iron pipe size		145		2.71	2.20	4.91	6.25
	599	4" iron pipe size		135		3.15	2.37	5.52	6.95
	600	5" iron pipe size		130		3.52	2.46	5.98	7.50
	601	6" iron pipe size		120		3.65	2.66	6.31	7.95
	602	7" iron pipe size		115		4.90	2.78	7.68	9.50
	603	8" iron pipe size		105		4.92	3.04	7.96	9.90
	604	9" iron pipe size		100		5.90	3.20	9.10	11.20
	605	10" iron pipe size		95		6.55	3.37	9.92	12.15
	606	12" iron pipe size		90		7.75	3.55	11.30	13.75
	607	14" iron pipe size		85		8.80	3.76	12.56	15.20
	608	16" iron pipe size		80		9.85	4	13.85	16.70
	609	18" iron pipe size		75		10.85	4.26	15.11	18.20
	612	2" wall, 1/2" iron pipe size		145		2.75	2.20	4.95	6.25
	614	3/4" iron pipe size		145		2.90	2.20	5.10	6.45
	615	1" iron pipe size		145		3.01	2.20	5.21	6.55
	616	1-1/4" iron pipe size		140		3.25	2.28	5.53	6.95
	617	1-1/2" iron pipe size		140		3.40	2.28	5.68	7.10
	618	2" iron pipe size		135		3.57	2.37	5.94	7.40
	619	2-1/2" iron pipe size		135		3.84	2.37	6.21	7.70
	620	3" iron pipe size		130		4.13	2.46	6.59	8.15
	621	4" iron pipe size		125		4.87	2.56	7.43	9.10
	622	5" iron pipe size		120		5.55	2.66	8.21	10.05
	623	6" iron pipe size		115		6.20	2.78	8.98	10.90
	624	7" iron pipe size		110		6.90	2.91	9.81	11.85
	625	8" iron pipe size		105		7.40	3.04	10.44	12.60
	626	9" iron pipe size		100		8.90	3.20	12.10	14.50
	627	10" iron pipe size		95		9.30	3.37	12.67	15.20
	628	12" iron pipe size		90		10.35	3.55	13.90	16.60
	629	14" iron pipe size		85		11.40	3.76	15.16	18.10
	630	16" iron pipe size		80		12.55	4	16.55	19.70
	631	18" iron pipe size		75		13.65	4.26	17.91	21
	632	20" iron pipe size		65		15.80	4.92	20.72	25
	633	22" iron pipe size		60		16.95	5.35	22.30	26
	634	24" iron pipe size		55		17.75	5.80	23.55	28
	636	3" wall, 1/2" iron pipe size		115		3.81	2.78	6.59	8.30
	638	3/4" iron pipe size		115		4.37	2.78	7.15	8.90
	639	1" iron pipe size		115		4.51	2.78	7.29	9.05
	640	1-1/4" iron pipe size		110		4.67	2.91	7.58	9.40
	641	1-1/2" iron pipe size		110		4.96	2.91	7.87	9.75
	642	2" iron pipe size		105		5.25	3.04	8.29	10.25
	643	2-1/2" iron pipe size		105		5.65	3.04	8.69	10.70
	644	3" iron pipe size		100		6.15	3.20	9.35	11.45
	645	4" iron pipe size		95		8.05	3.37	11.42	13.80
	646	5" iron pipe size		90		9.35	3.55	12.90	15.50
	647	6" iron pipe size		90		9.85	3.55	13.40	16.05
	648	7" iron pipe size		85		12	3.76	15.76	18.75
	649	8" iron pipe size		85		11.70	3.76	15.46	18.40
	650	9" iron pipe size		80		14.20	4	18.20	22
	651	10" iron pipe size		75		14.10	4.26	18.36	22
	652	12" iron pipe size		70		15.60	4.57	20.17	24
	653	14" iron pipe size		65		17.50	4.92	22.42	26
	654	16" iron pipe size		60		19.55	5.35	24.90	29
	655	18" iron pipe size		55		21	5.80	26.80	32
	656	20" iron pipe size		50		24	6.40	30.40	36
	657	22" iron pipe size		45		26	7.10	33.10	39
	658	24" iron pipe size	▼	40	▼	28	8	36	43
	660	Fiberglass, with all service jacket							

15

15.5 Heating

		CREW	DAILY OUTPUT	UNIT	BARE COSTS MAT.	BARE COSTS INST.	BARE COSTS TOTAL	TOTAL INCL O&P
65 664	1/2" wall, 1/2" iron pipe size	Q-14	250	L.F.	.60	1.28	1.88	2.54
666	3/4" iron pipe size		240		.68	1.33	2.01	2.71
667	1" iron pipe size		230		.72	1.39	2.11	2.84
668	1-1/4" iron pipe size		220		.80	1.45	2.25	3.02
669	1-1/2" iron pipe size		220		.91	1.45	2.36	3.14
670	2" iron pipe size		210		.98	1.52	2.50	3.32
671	2-1/2" iron pipe size		200		1.08	1.60	2.68	3.54
672	3" iron pipe size		190		1.19	1.68	2.87	3.79
673	3-1/2" iron pipe size		180		1.33	1.78	3.11	4.08
674	4" iron pipe size		160		1.58	2	3.58	4.68
675	5" iron pipe size		150		1.80	2.13	3.93	5.10
676	6" iron pipe size		120		1.94	2.66	4.60	6.05
684	1" wall, 1/2" iron pipe size		240		.75	1.33	2.08	2.79
686	3/4" iron pipe size		230		.86	1.39	2.25	2.99
687	1" iron pipe size		220		.91	1.45	2.36	3.14
688	1-1/4" iron pipe size		210		1.01	1.52	2.53	3.35
689	1-1/2" iron pipe size		210		1.12	1.52	2.64	3.47
690	2" iron pipe size		200		1.22	1.60	2.82	3.70
691	2-1/2" iron pipe size		190		1.36	1.68	3.04	3.97
692	3" iron pipe size		180		1.51	1.78	3.29	4.28
693	3-1/2" iron pipe size		170		1.66	1.88	3.54	4.60
694	4" iron pipe size		150		1.97	2.13	4.10	5.30
695	5" iron pipe size		140		2.26	2.28	4.54	5.85
696	6" iron pipe size		120		2.42	2.66	5.08	6.60
697	7" iron pipe size		110		3.04	2.91	5.95	7.60
698	8" iron pipe size		100		3.51	3.20	6.71	8.55
699	9" iron pipe size		90		3.84	3.55	7.39	9.45
700	10" iron pipe size		90		4.14	3.55	7.69	9.80
701	12" iron pipe size		80		4.72	4	8.72	11.10
702	14" iron pipe size		80		5.50	4	9.50	11.95
703	16" iron pipe size		70		6.55	4.57	11.12	13.95
704	18" iron pipe size		70		7	4.57	11.57	14.45
705	20" iron pipe size		60		7.80	5.35	13.15	16.45
706	24" iron pipe size		60		9.50	5.35	14.85	18.30
708	1-1/2" wall, 1/2" iron pipe size		230		1.60	1.39	2.99	3.81
710	3/4" iron pipe size		220		1.66	1.45	3.11	3.97
711	1" iron pipe size		210		1.77	1.52	3.29	4.19
712	1-1/4" iron pipe size		200		1.93	1.60	3.53	4.48
713	1-1/2" iron pipe size		200		2.03	1.60	3.63	4.59
714	2" iron pipe size		190		2.23	1.68	3.91	4.93
715	2-1/2" iron pipe size		180		2.42	1.78	4.20	5.30
716	3" iron pipe size		170		2.52	1.88	4.40	5.55
717	3-1/2" iron pipe size		160		2.75	2	4.75	5.95
718	4" iron pipe size		140		2.88	2.28	5.16	6.55
719	5" iron pipe size		130		3.24	2.46	5.70	7.20
720	6" iron pipe size		110		3.33	2.91	6.24	7.95
721	7" iron pipe size		100		3.86	3.20	7.06	8.95
722	8" iron pipe size		90		4.16	3.55	7.71	9.80
723	9" iron pipe size		85		4.79	3.76	8.55	10.80
724	10" iron pipe size		80		5.30	4	9.30	11.70
725	12" iron pipe size		75		5.80	4.26	10.06	12.65
726	14" iron pipe size		70		6.65	4.57	11.22	14.05
727	16" iron pipe size		65		7.50	4.92	12.42	15.50
728	18" iron pipe size		60		8.60	5.35	13.95	17.30
729	20" iron pipe size		55		9.10	5.80	14.90	18.55
730	24" iron pipe size		50		10.65	6.40	17.05	21
732	2" wall, 1/2" iron pipe size		220		2.57	1.45	4.02	4.97
734	3/4" iron pipe size		210		2.67	1.52	4.19	5.20
735	1" iron pipe size		200		2.87	1.60	4.47	5.50
736	1-1/4" iron pipe size	↓	190	↓	3.05	1.68	4.73	5.85

15.5 Heating

		Description	Crew	Daily Output	Unit	Bare Costs Mat.	Bare Costs Inst.	Bare Costs Total	Total Incl O&P
65	737	1-1/2" iron pipe size	Q-14	190	L.F.	3.18	1.68	4.86	6
	738	2" iron pipe size		180		3.33	1.78	5.11	6.30
	739	2-1/2" iron pipe size		170		3.59	1.88	5.47	6.70
	740	3" iron pipe size		160		3.83	2	5.83	7.15
	741	3-1/2" iron pipe size		150		4.14	2.13	6.27	7.70
	742	4" iron pipe size		130		4.44	2.46	6.90	8.50
	743	5" iron pipe size		120		5.05	2.66	7.71	9.50
	744	6" iron pipe size		100		5.15	3.20	8.35	10.35
	745	7" iron pipe size		90		5.95	3.55	9.50	11.80
	746	8" iron pipe size		80		6.40	4	10.40	12.95
	747	9" iron pipe size		75		7.05	4.26	11.31	14.05
	748	10" iron pipe size		70		7.65	4.57	12.22	15.15
	749	12" iron pipe size		65		8.50	4.92	13.42	16.60
	750	14" iron pipe size		60		9.55	5.35	14.90	18.35
	751	16" iron pipe size		55		10.50	5.80	16.30	20
	752	18" iron pipe size		50		11.45	6.40	17.85	22
	753	20" iron pipe size		45		12.70	7.10	19.80	24
	754	24" iron pipe size		40		14.25	8	22.25	27
	760	For fiberglass with standard canvas jacket, deduct				5%			
	766	For fittings, add 3 L.F. for each fitting							
	768	plus 4 L.F. for each flange of the fitting							
	770	For equipment or congested areas, add			L.F.		20%		
	772	Finishes, for .006" aluminum jacket, add	Q-14	120	S.F.	.17	2.66	2.83	4.11
	774	For .016" aluminum jacket, add		120		.29	2.66	2.95	4.24
	776	For .010" stainless steel, add		100		.23	3.20	3.43	4.96
	778	For single layer of felt, add				10%			
	780	For roofing paper, 45 lb. to 55 lb., add				25%			
	786	Rubber tubing, flexible closed cell foam							
	788	3/8" wall, 1/4" iron pipe size	1 Asbe	120	L.F.	.16	1.48	1.64	2.35
	790	3/8" iron pipe size		120		.19	1.48	1.67	2.39
	791	1/2" iron pipe size		115		.21	1.54	1.75	2.50
	792	3/4" iron pipe size		115		.24	1.54	1.78	2.54
	793	1" iron pipe size		110		.27	1.61	1.88	2.67
	794	1-1/4" iron pipe size		110		.34	1.61	1.95	2.75
	795	1-1/2" iron pipe size		110		.37	1.61	1.98	2.78
	796	2" iron pipe size		105		.49	1.69	2.18	3.03
	797	2-1/2" iron pipe size		100		.69	1.78	2.47	3.37
	798	3" iron pipe size		100		.77	1.78	2.55	3.46
	799	3-1/2" iron pipe size		100		1.08	1.78	2.86	3.80
	810	1/2" wall, 1/4" iron pipe size		90		.25	1.97	2.22	3.18
	812	3/8" iron pipe size		90		.28	1.97	2.25	3.21
	813	1/2" iron pipe size		89		.31	2	2.31	3.28
	814	3/4" iron pipe size		89		.34	2	2.34	3.31
	815	1" iron pipe size		88		.38	2.02	2.40	3.39
	816	1-1/4" iron pipe size		87		.44	2.04	2.48	3.49
	817	1-1/2" iron pipe size		87		.53	2.04	2.57	3.59
	818	2" iron pipe size		86		.64	2.07	2.71	3.74
	819	2-1/2" iron pipe size		86		.85	2.07	2.92	3.97
	820	3" iron pipe size		85		.88	2.09	2.97	4.04
	821	3-1/2" iron pipe size		85		1.18	2.09	3.27	4.37
	822	4" iron pipe size		80		1.32	2.22	3.54	4.72
	823	5" iron pipe size		80		1.95	2.22	4.17	5.40
	830	3/4" wall, 1/4" iron pipe size		90		.37	1.97	2.34	3.31
	832	3/8" iron pipe size		90		.43	1.97	2.40	3.38
	833	1/2" iron pipe size		89		.51	2	2.51	3.50
	834	3/4" iron pipe size		89		.62	2	2.62	3.62
	835	1" iron pipe size		88		.71	2.02	2.73	3.75
	836	1-1/4" iron pipe size		87		.95	2.04	2.99	4.05
	837	1-1/2" iron pipe size		87		1.08	2.04	3.12	4.19
	838	2" iron pipe size		86		1.27	2.07	3.34	4.44

15.5 Heating		CREW	DAILY OUTPUT	UNIT	BARE COSTS MAT.	BARE COSTS INST.	BARE COSTS TOTAL	TOTAL INCL O&P
65 839	2-1/2" iron pipe size	1 Asbe	86	L.F.	1.69	2.07	3.76	4.90
840	3" iron pipe size		85		1.73	2.09	3.82	4.98
841	3-1/2" iron pipe size		85		2.11	2.09	4.20	5.40
842	4" iron pipe size		80		2.42	2.22	4.64	5.95
843	5" iron pipe size		80		2.91	2.22	5.13	6.45
844	6" iron pipe size		80		4.12	2.22	6.34	7.80
868	Rubber insulation tape 1/8"x2"x30'			ROLL	5.50		5.50	6.05M
880	Urethane, with ASJ, -60°F to +225°F							
896	1" wall, 1/2" iron pipe size	Q-14	240	L.F.	1.02	1.33	2.35	3.08
898	3/4" iron pipe size		230		1.02	1.39	2.41	3.17
899	1" iron pipe size		220		1.58	1.45	3.03	3.88
900	1-1/4" iron pipe size		210		1.58	1.52	3.10	3.98
901	1-1/2" iron pipe size		210		1.74	1.52	3.26	4.16
902	2" iron pipe size		200		2.04	1.60	3.64	4.60
903	2-1/2" iron pipe size		190		1.93	1.68	3.61	4.60
904	3" iron pipe size		180		2.10	1.78	3.88	4.93
905	4" iron pipe size		150		2.64	2.13	4.77	6.05
906	5" iron pipe size		140		2.81	2.28	5.09	6.45
907	6" iron pipe size		120		3.15	2.66	5.81	7.40
912	1-1/2" wall, 1/2" iron pipe size		230		1.27	1.39	2.66	3.44
914	3/4" iron pipe size		220		1.39	1.45	2.84	3.67
915	1" iron pipe size		210		1.73	1.52	3.25	4.14
916	1-1/4" iron pipe size		200		1.73	1.60	3.33	4.26
917	1-1/2" iron pipe size		200		1.93	1.60	3.53	4.48
918	2" iron pipe size		190		2.28	1.68	3.96	4.99
919	2-1/2" iron pipe size		180		2.48	1.78	4.26	5.35
920	3" iron pipe size		170		2.86	1.88	4.74	5.90
921	4" iron pipe size		140		3.31	2.28	5.59	7
922	5" iron pipe size		130		3.57	2.46	6.03	7.55
923	6" iron pipe size		110		3.91	2.91	6.82	8.60
924	8" iron pipe size		90		5.05	3.55	8.60	10.80
929	Urethane, with ultraviolet cover							
931	1" wall, 1/2" pipe size	Q-14	216	L.F.	1.12	1.48	2.60	3.41
932	3/4" pipe size		207		1.12	1.54	2.66	3.51
933	1" pipe size		198		1.63	1.61	3.24	4.17
934	1-1/4" pipe size		189		1.63	1.69	3.32	4.28
935	1-1/2" pipe size		185		1.88	1.73	3.61	4.61
936	2" pipe size		180		2.25	1.78	4.03	5.10
940	1-1/2" wall, 1/2" pipe size		207		1.34	1.54	2.88	3.75
941	3/4" pipe size		198		1.51	1.61	3.12	4.04
942	1" pipe size		189		1.85	1.69	3.54	4.53
943	1-1/4" pipe size		180		1.85	1.78	3.63	4.65
944	1-1/2" pipe size		175		2.10	1.83	3.93	5
945	2" pipe size		171		2.44	1.87	4.31	5.45
950	Urethane, with ultraviolet cover, fittings							
951	90° & 45° elbows, 1" wall thickness							
952	1/2" pipe size	Q-14	108	Ea.	3.11	2.96	6.07	7.80
953	3/4" pipe size		104		3.27	3.07	6.34	8.10
954	1" pipe size		99		3.72	3.23	6.95	8.85
955	1-1/4" pipe size		95		4.08	3.37	7.45	9.45
956	1-1/2" pipe size		93		4.85	3.44	8.29	10.40
957	2" pipe size		90		6	3.55	9.55	11.85
965	90° & 45° elbows, 1-1/2" wall thickness							
966	1/2" pipe size	Q-14	104	Ea.	3.64	3.07	6.71	8.55
967	3/4" pipe size		99		3.89	3.23	7.12	9.05
968	1" pipe size		95		4.59	3.37	7.96	10
969	1-1/4" pipe size		90		4.71	3.55	8.26	10.40
970	1-1/2" pipe size		88		5	3.63	8.63	10.85
971	2" pipe size		86		6.55	3.72	10.27	12.70
980	Tees & valves, 1" wall thickness							

15

15.5 Heating		CREW	DAILY OUTPUT	UNIT	BARE COSTS MAT.	BARE COSTS INST.	BARE COSTS TOTAL	TOTAL INCL O&P
65 982	1/2" pipe size	Q-14	72	Ea.	3.72	4.44	8.16	10.65
983	3/4" pipe size		69		3.83	4.63	8.46	11.05
984	1" pipe size		66		4.59	4.84	9.43	12.20
985	1-1/4" pipe size		63		4.85	5.05	9.90	12.80
986	1-1/2" pipe size		62		5.55	5.15	10.70	13.70
987	2" pipe size		60		7.05	5.35	12.40	15.60
992	Tees & valves, 1-1/2" wall thickness							
993	1/2" pipe size	Q-14	69	Ea.	4.63	4.63	9.26	11.90
994	3/4" pipe size		66		4.88	4.84	9.72	12.50
995	1" pipe size		63		6.30	5.05	11.35	14.40
996	1-1/4" pipe size		60		6.45	5.35	11.80	14.95
997	1-1/2" pipe size		58		6.70	5.50	12.20	15.50
998	2" pipe size		57		8.90	5.60	14.50	18.05
68-001	**MAKE-UP AIR UNIT**							
002	Indoor suspension, natural/LP gas, direct fired,							
003	standard control. For flue see Division 15.5-92.							
004	70°F temperature rise, MBH is input							
010	2000 CFM, 168 MBH	Q-6	3	Ea.	2,800	165	2,965	3,325
012	3000 CFM, 252 MBH		2		2,950	250	3,200	3,600
014	4000 CFM, 336 MBH		1.80		3,050	280	3,330	3,750
016	6000 CFM, 502 MBH		1.50		4,025	335	4,360	4,900
018	8000 CFM, 670 MBH		1.40		4,125	355	4,480	5,050
020	10,000 CFM, 838 MBH		1.20		4,450	415	4,865	5,500
022	12,000 CFM, 1005 MBH		1		4,625	500	5,125	5,800
024	14,000 CFM, 1180 MBH		.94		4,750	530	5,280	6,000
026	16,000 CFM, 1340 MBH		.86		5,350	580	5,930	6,725
028	20,000 CFM, 1675 MBH		.76		5,600	655	6,255	7,100
030	24,000 CFM, 2007 MBH	Q-7	1		6,500	680	7,180	8,125
032	30,000 CFM, 2510 MBH		.96		7,150	710	7,860	8,900
034	35,000 CFM, 2930 MBH		.92		8,425	740	9,165	10,300
036	40,000 CFM, 3350 MBH		.88		8,850	775	9,625	10,900
038	45,000 CFM, 3770 MBH		.84		9,425	810	10,235	11,500
040	50,000 CFM, 4180 MBH		.80		9,825	850	10,675	12,000
042	55,000 CFM, 4600 MBH		.75		10,500	910	11,410	12,900
044	60,000 CFM, 5020 MBH		.70		10,900	975	11,875	13,400
046	65,000 CFM, 5435 MBH		.60		11,100	1,125	12,225	13,900
048	75,000 CFM, 6275 MBH		.50		12,100	1,375	13,475	15,300
060	For discharge louver assembly, add				15%			
070	For filters, add				5%			
080	For air shut-off damper section, add				10%			
090	For vertical unit, add				10%			
100	Rooftop unit, natural gas, gravity vent, S.S. exchanger							
101	70°F temperature rise, MBH is input							
102	995 CFM, 75 MBH	Q-6	4	Ea.	1,725	125	1,850	2,075
104	1245 CFM, 93.7 MBH		3.60		1,750	140	1,890	2,125
106	1590 CFM, 120 MBH		3.30		1,975	150	2,125	2,400
108	2090 CFM, 157.5 MBH		3		2,250	165	2,415	2,725
110	2490 CFM, 187.5 MBH		2.60		2,475	190	2,665	3,000
112	2985 CFM, 225 MBH		2.30		2,575	215	2,790	3,150
114	3975 CFM, 300 MBH		1.90		3,225	265	3,490	3,925
116	4970 CFM, 375 MBH		1.40		4,950	355	5,305	5,950
118	5950 CFM, 450 MBH		1.20		5,125	415	5,540	6,250
120	7940 CFM, 600 MBH		1		6,125	500	6,625	7,450
122	9930 CFM, 750 MBH		.80		7,900	625	8,525	9,600
124	11,900 CFM, 900 MBH		.60		9,100	835	9,935	11,200
126	15,880 CFM, 1200 MBH		.30		11,100	1,675	12,775	14,600
150	For power vent, add				5%			
160	For cleanable filters, add				65		65	72M
170	For electric modulating gas control, add				575		575	635M

15

Line	Ref.	15.5 Heating	Crew	Daily Output	Unit	Bare Costs Mat.	Bare Costs Inst.	Bare Costs Total	Total Incl O&P
68 180		For aluminized steel exchanger, subtract			Ea.	10%			
77-001		**SOLAR ENERGY**							
002		System/Package prices, not including connecting							
003		pipe, insulation, or special heating/plumbing fixtures							
010		For solar air conditioning, see Division 15.7-93							
015		For solar ultraviolet pipe insulation see Division 15.5-65-929							
050		Hot water, standard package, low temperature							
054	B8.3 -610	2 collectors, circulator, fittings, no tank	Q-1	.50	Ea.	915	640	1,555	1,925
058		2 collectors, circulator, fittings, 120 gal. tank		.40		1,625	800	2,425	2,950
062	B8.3 -620	3 collectors, circulator, fittings, 120 gal. tank		.40		2,000	800	2,800	3,350
070		Medium temperature package							
074	B8.3 -630	2 collectors, circulator, fittings, 80 gal. tank	Q-1	.40	Ea.	1,750	800	2,550	3,075
078		3 collectors, circulator, fittings, 120 gal. tank	"	.30	"	2,200	1,075	3,275	3,950
090	B8.3 -640	Commercial/process							
094		10 med. temp. collectors, fittings, 120 gal. tank	Q-2	.15	Ea.	6,675	3,325	10,000	12,100
098	B8.3 -650	For each additional 120 gal. tank, add			"	685		685	755M
130		Solar assist package, for space heating and domestic							
134	B8.3 -660	hot water, 10 collectors, fittings	Q-2	.13	Ea.	6,475	3,825	10,300	12,700
144		For heating, complete with heat pump, 9 collectors		.12		8,675	4,150	12,825	15,500
148	B8.3 -710	12 collectors		.10		10,100	4,975	15,075	18,300
154		18 collectors		.08		13,300	6,225	19,525	23,600
200	B8.3 -720	Seasonal pool heating package, fittings, 10 collectors		.30		2,675	1,650	4,325	5,350
220		Window mounted solar heater, complete with blower	1 Plum	16		400	11.10	411.10	455
225	B8.3 -730	Controller, liquid temperature	"	5		52	36	88	110
230		Circulators, air							
231	B8.3 -740	Blowers							
232		30 to 100 S.F. system, 1/20 HP	Q-9	16	Ea.	43	20	63	76
233		100-300 S.F. system, 1/10 HP		16		73	20	93	110
234		300-500 S.F. system, 1/5 HP		15		100	21	121	140
235		Two speed, 100-300 S.F., 1/10 HP		14		73	23	96	115
240		Reversible fan, 20" diameter, 2 speed		18		150	17.75	167.75	190
245		Room to room fan, 225 CFM		12		80	27	107	125
248		Shutter mounted fan, 12" diameter, 650 CFM		14		87	23	110	130
252		Space & DHW system, less duct work		.50		2,000	640	2,640	3,125
255		Booster fan 5" diameter, 70 CFM		16		20	20	40	51
257		6" diameter, 90 CFM		16		20	20	40	51
258		8" diameter, 150 CFM		16		25	20	45	57
259		8" diameter, 250 CFM		14		35	23	58	72
260		8" diameter, 310 CFM		14		37	23	60	74
265		Rheostat		32		9.95	10	19.95	26
266		Shutter/damper		12		23	27	50	64
267		Shutter motor		16		52	20	72	86
275		Circulators, liquid, 1/100 HP, 2 GPM	Q-1	14		64	23	87	105
277		1/100 HP, 3 GPM		14		75	23	98	115
280		1/25 HP, 5.3 GPM		14		79	23	102	120
282		1/20 HP, 17 GPM		12		70	27	97	115
285		1/20 HP, 17 GPM, stainless steel		12		145	27	172	200
287		1/12 HP, 30 GPM		10		105	32	137	160
300		Collector panels, air with aluminum absorber plate							
301		Wall or roof mount							
304		Flat black, plastic glazing							
308		4' x 8'	Q-9	6	Ea.	350	53	403	465
310		5' x 9'		5	"	500	64	564	645
315		Mounting, head clamp		138	L.F.	4.65	2.32	6.97	8.50
316		Sill clamp		151		3.85	2.12	5.97	7.30
317		T batten clamp		172		2.60	1.86	4.46	5.55
320		Flush roof mount, 10' to 16' x 22" wide		96		32	3.33	35.33	40
321		Manifold, by L.F. width of collectors		160		11	2	13	15
330		Collector panels, liquid with copper absorber plate							

15

15.5 Heating

77

		CREW	DAILY OUTPUT	UNIT	BARE COSTS MAT.	BARE COSTS INST.	BARE COSTS TOTAL	TOTAL INCL O&P
332	Black chrome, tempered glass glazing							
333	Alum. frame, 3' x 8', 5/32" single glazing	Q-1	9.50	Ea.	285	34	319	360
336	Alum. frame, 3' x 8', 5/32" double glazing		9		315	36	351	400
339	Alum. frame, 4' x 8', 3/16" glazing		6		630	53	683	770
344	Flat black							
345	Alum. frame, 3' x 8', 5/32" single glazing	Q-1	9	Ea.	260	36	296	335
350	Alum. frame 3' x 8', 5/32" double glazing		5.50		285	58	343	400
352	Alum. frame, 3' x 10', plastic glazing		10		255	32	287	325
354	Alum. frame, 3' x 8', 1/8" tempered glass		5		350	64	414	475
355	Liquid with fin tube absorber plate							
356	Alum. frame 4' x 8' tempered glass	Q-1	10	Ea.	375	32	407	460
358	Liquid with vacuum tubes, 4' x 6'-10"		9		695	36	731	815
360	Liquid, full wetted, plastic, alum. frame, 3' x 10'		5		100	64	164	200
365	Collector panel mounting, flat roof or ground rack		7		110	46	156	185
367	Roof clamps		70	Set	3	4.57	7.57	9.90
370	Roof strap, teflon	1 Plum	205	L.F.	3.15	.87	4.02	4.72
390	Differential controller with two sensors							
393	Thermostat, hard wired	1 Plum	8	Ea.	53	22	75	90
395	Line cord and receptacle		12		61	14.80	75.80	88
400	External adjustment		12		78	14.80	92.80	105
405	Pool valve system, 2" pipe size (plastic)		2.50		235	71	306	360
407	1-1/2" pipe size (copper)		2		305	89	394	465
408	Pool pump system, 2" pipe size		6		140	30	170	195
410	Six station with digital read-out		3		230	59	289	340
415	Sensors							
419	Brass plug, 1/4" MPT	1 Plum	32	Ea.	12.95	5.55	18.50	22
420	Brass plug, 1/2" MPT		32		12.95	5.55	18.50	22
421	Brass plug, reversed		32		12.95	5.55	18.50	22
422	Freeze prevention		32		9.95	5.55	15.50	18.95
423	Freeze recycle		32		9.95	5.55	15.50	18.95
424	Screw attached		32		8	5.55	13.55	16.80
425	Brass, immersion		32		18.95	5.55	24.50	29
430	Heat exchanger							
431	Fluid to air coil							
433	Up flow, 45 MBH	Q-1	4	Ea.	88	80	168	210
438	70 MBH		3.50		215	91	306	370
440	80 MBH		3		315	105	420	500
449	Horizontal, 110 MBH		2		180	160	340	430
452	Fluid to air package includes coil, blower, circulator							
453	and controller for DHW and space hot air	Q-1	2	Ea.	880	160	1,040	1,200
458	Fluid to fluid package includes two circulating pumps							
459	expansion tank, check valve, relief valve							
460	controller, high temperature cutoff and sensors	Q-1	2.50	Ea.	555	130	685	795
465	Heat transfer fluid							
470	Propylene glycol, inhibited anti-freeze	1 Plum	28	Gal.	5.80	6.35	12.15	15.55
480	Solar storage tanks, knocked down							
481	Air, galvanized steel clad, double wall, 4" fiberglass							
482	insulation, 20 Mil PVC lining							
486	4' high, 3' x 3', = 36 C.F./250 gallons	Q-9	2.60	Ea.	885	125	1,010	1,150
487	4' x 4', = 64 C.F./480 gallons		2		1,200	160	1,360	1,550
488	5' x 5' = 100 C.F./750 gallons		1.50		1,550	215	1,765	2,025
489	6' x 6' = 144 C.F./1000 gallons		1.30		1,975	245	2,220	2,525
490	7' x 7' = 196 C.F./1450 gallons		1		2,725	320	3,045	3,475
501	6'-3" high, 7' x 7' = 306 C.F./2000 gallons	Q-10	1.20		4,025	415	4,440	5,025
502	7' x 10'-6" = 459 C.F./3000 gallons		.80		5,050	620	5,670	6,450
503	7' x 14' = 613 C.F./4000 gallons		.60		6,075	830	6,905	7,900
504	10'-6" x 10'-6" = 689 C.F./4500 gallons		.50		6,475	995	7,470	8,575
505	10'-6" x 14' = 919 C.F./6000 gallons		.40		7,750	1,250	9,000	10,300
506	14' x 14' = 1225 C.F./8000 gallons	Q-11	.40		9,300	1,700	11,000	12,700
507	14' x 17'-6" = 1531 C.F./10,000 gallons	"	.30		10,500	2,275	12,775	14,800

15.5 Heating		CREW	DAILY OUTPUT	UNIT	BARE COSTS MAT.	BARE COSTS INST.	BARE COSTS TOTAL	TOTAL INCL O&P
77 508	17'-6" x 17'-6" = 1914 C.F./12,500 gallons	Q-11	.25	Ea.	12,000	2,725	14,725	17,200
509	17'-6" x 21' = 2297 C.F./15,000 gallons		.20		13,800	3,400	17,200	20,100
510	21' x 21' = 2756 C.F./18,000 gallons	↓	.18	↓	15,800	3,775	19,575	22,900
512	30 Mil reinforced Chemflex lining,							
513	4' high, 3' x 3' = 36 C.F./250 gallons	Q-9	2.60	Ea.	980	125	1,105	1,250
514	4' x 4' = 64 C.F./480 gallons		2		1,300	160	1,460	1,675
515	5' x 5' = 100 C.F./750 gallons		1.50		1,725	215	1,940	2,200
516	6' x 6' = 144 C.F./1000 gallons		1.30		2,175	245	2,420	2,750
517	7' x 7' = 196 C.F./1450 gallons	↓	1		2,725	320	3,045	3,475
519	6'-3" high, 7' x 7' = 306 C.F./2000 gallons	Q-10	1.20		4,375	415	4,790	5,425
520	7' x 10'-6" = 459 C.F./3000 gallons		.80		5,450	620	6,070	6,900
521	7' x 14' = 613 C.F./4000 gallons		6		6,525	83	6,608	7,300
522	10'-6" x 10'-6" = 689 C.F./4500 gallons		.50		6,950	995	7,945	9,100
523	10'-6" x 14' = 919 C.F./6000 gallons	↓	4		8,325	125	8,450	9,350
524	14' x 14' = 1225 C.F./8000 gallons	Q-11	.40		9,975	1,700	11,675	13,400
525	14' x 17'-6" = 1531 C.F./10,000 gallons		.30		11,200	2,275	13,475	15,600
526	17'-6" x 17'-6" = 1914 C.F./12,500 gallons		.25		12,900	2,725	15,625	18,100
527	17'-6" x 21' = 2297 C.F./15,000 gallons		.20		14,700	3,400	18,100	21,100
528	21' x 21' = 2756 C.F./18,000 gallons	↓	.18	↓	16,700	3,775	20,475	23,900
529	30 Mil reinforced Hypalon lining, add				.05%			
700	Solar control valves and vents							
705	Air purger, 1" pipe size	1 Plum	12	Ea.	9.95	14.80	24.75	32
707	Air eliminator, automatic 3/4" size		32		5.50	5.55	11.05	14.05
709	Air vent, automatic, 1/8" fitting		32		4.77	5.55	10.32	13.25
710	Manual, 1/8" NPT		32		1.27	5.55	6.82	9.40
712	Backflow preventer, 1/2" pipe size		16		32	11.10	43.10	51
713	3/4" pipe size		16		32	11.10	43.10	51
715	Balancing valve, 3/4" pipe size		20		9.50	8.90	18.40	23
718	Draindown valve, 1" pipe size		9		80	19.75	99.75	115
720	Flow control valve, 1/2" pipe size		22		17.35	8.05	25.40	31
722	Expansion tank, up to 5 gal.		32		28	5.55	33.55	39
724	Over 5 gal.		32		34	5.55	39.55	45
725	Hydronic controller		8		18.75	22	40.75	53
740	Pressure gauge, 2" dial		32		4.95	5.55	10.50	13.45
745	Relief valve, temp. and pressure 3/4" pipe size	↓	30	↓	6.95	5.90	12.85	16.20
750	Solenoid valve, normally closed							
752	Brass, 3/4" NPT, 24V	1 Plum	9	Ea.	45	19.75	64.75	78
753	1" NPT, 24V		9		59	19.75	78.75	93
755	1-1/2" NPT, 24V		8		75	22	97	115
756	2" NPT, 24V		7		78	25	103	120
760	Plastic 1-1/2" NPT, 24V		9		58	19.75	77.75	92
761	2" NPT, 24V	↓	8	↓	71	22	93	110
765	Solenoid valve, normally open							
767	Brass, 1-1/2" NPT, 24V	1 Plum	7	Ea.	78	25	103	120
768	2" NPT, 24V		6.50		82	27	109	130
770	Plastic, 1-1/2" NPT, 24V		9		67	19.75	86.75	100
771	2" NPT, 24V		8		76	22	98	115
775	Vacuum relief valve, 3/4" pipe size		32		14	5.55	19.55	23
777	Vent and vacuum breaker valve, 1/2" pipe size	↓	32	↓	15	5.55	20.55	25
780	Thermometers							
782	Digital temperature monitoring, 6 locations	1 Plum	2.50	Ea.	160	71	231	280
783	8 locations		2.30		270	77	347	410
784	10 locations		2		300	89	389	460
787	Indoor, outdoor		8		22	22	44	56
789	In line, dial, 1/2" NPT		8		20	22	42	54
790	Upright, 1/2" NPT		8		11.75	22	33.75	45
795	Maximum - minimum		16		16.95	11.10	28.05	35
797	Remote probe, 2" dial		8		23	22	45	57
799	Stem, 2" dial, 9" stem	↓	16	↓	12.95	11.10	24.05	30
825	Water storage tank with heat exchanger and electric element							

15

	15.5 Heating	CREW	DAILY OUTPUT	UNIT	BARE COSTS MAT.	BARE COSTS INST.	BARE COSTS TOTAL	TOTAL INCL O&P
77 826	66 gal. with 2" x 1/2 lb. density insulation	1 Plum	1.60	Ea.	365	110	475	560
827	66 gal. with 2" x 2 lb. density insulation		1.60		390	110	500	590
828	80 gal. with 2" x 1/2 lb. density insulation		1.60		385	110	495	585
830	80 gal. with 2" x 2 lb. density insulation		1.60		415	110	525	615
835	120 gal. with 2" x 1/2 lb. density insulation		1.40		580	125	705	820
838	120 gal. with 2" x 2 lb. density insulation		1.40		610	125	735	855
840	120 gal. with 2" x 2 lb. density insul., 40 S.F. heat coil		1.40		655	125	780	905
850	Water storage module, plastic							
860	Tubular, 12" diameter, 4' high	1 Carp	48	Ea.	43	3.27	46.27	52
861	12" diameter, 8' high		40		59	3.92	62.92	71
862	18" diameter, 5' high		38		75	4.13	79.13	88
863	18" diameter, 10' high		32		98	4.90	102.90	115
864	58" diameter, 5' high	F-2	32		250	10.25	260.25	290
865	Cap, 12" diameter				5.95		5.95	6.55M
866	18" diameter				6.95		6.95	7.65M
80-001	**SPACE HEATERS** Cabinet, grilles, fan, controls, burner,							
002	thermostat, no piping. For flue see division 15.5-92							
100	Gas fired, floor mounted							
110	60 MBH output	Q-5	10	Ea.	340	32	372	420
112	80 MBH output		9		360	36	396	450
114	100 MBH output		8		390	40	430	485
116	120 MBH output		7		420	46	466	530
118	180 MBH output		6		520	54	574	650
150	Rooftop mounted, gravity vent, stainless steel exchanger							
152	75 MBH output	Q-6	4	Ea.	1,425	125	1,550	1,750
154	93.7 MBH output		3.60		1,500	140	1,640	1,850
156	120 MBH output		3.30		1,725	150	1,875	2,125
158	157.5 MBH output		3		1,925	165	2,090	2,350
160	187.5 MBH output		2.60		2,100	190	2,290	2,600
162	225 MBH output		2.30		2,200	215	2,415	2,725
164	300 MBH output		1.90		2,700	265	2,965	3,350
166	375 MBH output		1.40		4,725	355	5,080	5,725
168	450 MBH output		1.20		5,175	415	5,590	6,300
170	600 MBH output		1		5,450	500	5,950	6,725
172	750 MBH output		.80		7,550	625	8,175	9,200
174	900 MBH output		.60		8,300	835	9,135	10,300
176	1200 MBH output		.30		10,100	1,675	11,775	13,500
180	For power vent, add				10%			
190	For aluminized steel exchanger, subtract				14%			
200	Suspension mounted, propeller fan, 36 MBH output	Q-5	8		315	40	355	405
202	48 MBH output		7.50		330	43	373	425
204	60 MBH output		7		360	46	406	460
206	84 MBH output		6		405	54	459	525
208	100 MBH output		5.50		470	58	528	600
210	120 MBH output		5		495	64	559	635
212	140 MBH output		4.50		545	71	616	705
214	160 MBH output		4		595	80	675	770
216	180 MBH output		3.50		645	92	737	840
218	200 MBH output		3		695	105	800	920
220	240 MBH output		2.70		825	120	945	1,075
222	280 MBH output		2.30		975	140	1,115	1,275
224	320 MBH output		2		1,100	160	1,260	1,450
250	For powered venter and adapter, add				155		155	170M
300	Suspension mounted, blower type, 40 MBH output	Q-5	6.80		535	47	582	655
302	60 MBH output		6.60		565	49	614	690
304	84 MBH output		5.80		590	55	645	730
306	104 MBH output		5.20		630	62	692	780
308	140 MBH output		4.30		730	75	805	910
310	180 MBH output		3.30		840	97	937	1,075

15

15.5 Heating

		CREW	DAILY OUTPUT	UNIT	BARE COSTS MAT.	BARE COSTS INST.	BARE COSTS TOTAL	TOTAL INCL O&P
80 312	240 MBH output	Q-5	2.50	Ea.	975	130	1,105	1,250
314	280 MBH output	"	2	"	1,200	160	1,360	1,550
400	Suspension mounted, sealed combustion system,							
402	aluminized steel exchanger, powered vent							
404	100 MBH output	Q-5	5	Ea.	1,150	64	1,214	1,350
406	120 MBH output		4.70		1,225	68	1,293	1,450
408	160 MBH output		3.70		1,375	87	1,462	1,650
410	200 MBH output		2.90		1,650	110	1,760	1,975
412	240 MBH output		2.50		1,725	130	1,855	2,075
414	320 MBH output		1.70		2,025	190	2,215	2,500
500	Wall furnace, 17 MBH output		6		185	54	239	280
502	24 MBH output		5		230	64	294	345
504	49 MBH output		4		325	80	405	475
600	Oil fired, suspension mounted, 94 MBH output		4		935	80	1,015	1,150
604	140 MBH output		3		995	105	1,100	1,250
606	184 MBH output		3		1,075	105	1,180	1,325
608	250 MBH output	↓	2		1,650	160	1,810	2,050
610	308 MBH output	Q-6	2.60	↓	1,775	190	1,965	2,225
83-001	**STEAM GENERATORS** See division 15.5-04							
86-001	**SWIMMING POOL HEATERS** Not including wiring, external							
002	piping, base or pad,							
006	Gas fired, gross output, 50 MBH	Q-6	3	Ea.	325	165	490	600
010	80 MBH		2		355	250	605	750
016	120 MBH		1.50		410	335	745	930
018	145 MBH		1.30		475	385	860	1,075
020	170 MBH		1		490	500	990	1,250
022	200 MBH		.70		530	715	1,245	1,625
024	280 MBH		.60		625	835	1,460	1,900
026	450 MBH		.50		1,525	1,000	2,525	3,125
028	500 MBH		.40		1,575	1,250	2,825	3,525
030	600 MBH		.35		1,675	1,425	3,100	3,900
032	750 MBH		.33		1,800	1,525	3,325	4,175
034	890 MBH		.26		2,025	1,925	3,950	5,000
036	965 MBH		.22		2,075	2,275	4,350	5,575
037	1080 MBH		.21		2,250	2,375	4,625	5,900
038	1312 MBH		.19		2,650	2,625	5,275	6,725
040	1440 MBH		.14		3,050	3,575	6,625	8,500
041	2000 MBH		.13		4,425	3,850	8,275	10,400
042	2450 MBH		.11		5,250	4,550	9,800	12,300
044	3000 MBH		.09		5,450	5,550	11,000	14,000
100	Oil fired, gross output, 97 MBH		1.70		1,150	295	1,445	1,700
102	118 MBH		1.70		1,175	295	1,470	1,725
104	134 MBH		1.60		1,250	310	1,560	1,825
106	161 MBH		1.30		1,525	385	1,910	2,225
108	187 MBH		1.20		1,600	415	2,015	2,350
110	214 MBH		1.10		1,825	455	2,280	2,675
112	262 MBH		.89		2,050	560	2,610	3,075
114	340 MBH		.68		2,725	735	3,460	4,050
116	420 MBH		.58		3,125	860	3,985	4,675
118	525 MBH		.46		4,000	1,075	5,075	5,975
120	630 MBH		.41		4,475	1,225	5,700	6,675
122	735 MBH		.35		5,500	1,425	6,925	8,100
124	840 MBH		.33		6,300	1,525	7,825	9,125
126	1050 MBH		.27		7,675	1,850	9,525	11,100
128	1365 MBH		.21		9,850	2,375	12,225	14,300
130	1680 MBH		.18		11,600	2,775	14,375	16,800
132	2310 MBH		.13		14,900	3,850	18,750	21,900
134	2835 MBH		.11		18,600	4,550	23,150	27,000
136	3150 MBH	↓	.09	↓	20,900	5,550	26,450	31,000

15

15.5 Heating

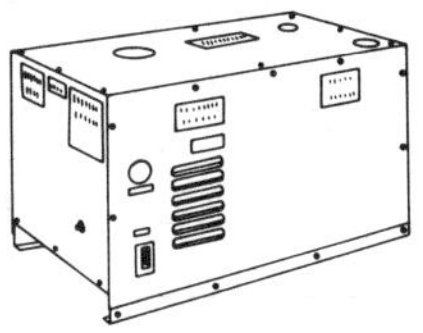

		CREW	DAILY OUTPUT	UNIT	BARE COSTS MAT.	BARE COSTS INST.	BARE COSTS TOTAL	TOTAL INCL O&P
86 200	Electric, 12 KW, 4800 gallon pool	Q-19	3	Ea.	860	165	1,025	1,175
202	18 KW, 7200 gallon pool		2.80		885	180	1,065	1,225
204	24 KW, 9600 gallon pool		2.40		940	205	1,145	1,325
206	30 KW, 12,000 gallon pool		2		980	250	1,230	1,425
208	36 KW, 14,400 gallon pool		1.60		1,075	310	1,385	1,625
210	54 KW, 24,000 gallon pool		1.20		1,275	415	1,690	2,000
212	300 KW, 120,000 gallon pool		1		7,800	500	8,300	9,300
214	600 KW, 240,000 gallon pool		.80		12,900	620	13,520	15,100
216	1200 KW, 480,000 gallon pool	↓	.60	↓	20,300	830	21,130	23,500
900	To select pool heater: 12 BTUH x S.F. pool area							
901	X temperature differential =required output							
905	For electric, KW = gallons x 2.5 divided by 1000							
910	For family home type pool, double the							
911	Rated gallon capacity = 1/2°F rise per hour							
89-001	**TANKS**							
002	Fiberglass, underground, U.L. listed, not including							
003	manway or hold-down strap							
004	550 gallon capacity	Q-6	3	Ea.	1,100	165	1,265	1,450
010	1000 gallon capacity	"	2		1,400	250	1,650	1,900
014	2000 gallon capacity	Q-7	2		2,250	340	2,590	2,975
016	4000 gallon capacity		1.30		2,975	525	3,500	4,025
018	6000 gallon capacity		1		3,750	680	4,430	5,100
020	8000 gallon capacity		.80		4,375	850	5,225	6,050
022	10,000 gallon capacity		.60		5,000	1,125	6,125	7,150
024	12,000 gallon capacity		.50		6,075	1,375	7,450	8,650
026	15,000 gallon capacity		.40		8,200	1,700	9,900	11,500
028	20,000 gallon capacity		.30		10,600	2,275	12,875	14,900
030	25,000 gallon capacity		.20		16,700	3,400	20,100	23,300
032	30,000 gallon capacity		.20		21,300	3,400	24,700	28,400
034	40,000 gallon capacity		.10		25,300	6,825	32,125	37,700
036	48,000 gallon capacity	↓	.10		29,900	6,825	36,725	42,700
050	For manway, fittings and hold-downs, add				20%	15%		
100	For helical heating coil, add	Q-5	2.50	↓	575	130	705	820
150	Plastic, corrosion resistant, see Div. 15.2-60							
200	Steel, liquid expansion, painted, 8 gallon capacity	Q-5	20	Ea.	44	16.05	60.05	72
202	12 gallon capacity		18		50	17.85	67.85	81
204	15 gallon capacity		17		53	18.90	71.90	86
206	18 gallon capacity		16		58	20	78	93
208	24 gallon capacity		14		67	23	90	105
210	30 gallon capacity		12		77	27	104	125
212	40 gallon capacity		10		105	32	137	160
300	Steel ASME expansion, rubber diaphragm, 19 gallon capacity		17		515	18.90	533.90	595
302	31 gallon capacity		12		665	27	692	770
304	61 gallon capacity		9		1,025	36	1,061	1,175
306	79 gallon capacity		8		1,200	40	1,240	1,375
308	119 gallon capacity		8		1,350	40	1,390	1,550
310	158 gallon capacity		7		1,950	46	1,996	2,200
312	211 gallon capacity		6		2,200	54	2,254	2,500
314	317 gallon capacity		5		2,775	64	2,839	3,150
316	422 gallon capacity		4		4,150	80	4,230	4,675
318	528 gallon capacity	↓	3	↓	4,575	105	4,680	5,175
400	Steel, storage, above ground, including supports, coating,							
402	fittings, not including mat, pumps or piping							
404	275 gallon capacity	Q-5	5	Ea.	155	64	219	265
406	550 gallon capacity	"	4		585	80	665	760
408	1000 gallon capacity	Q-7	4		815	170	985	1,150
410	1500 gallon capacity		3.70		1,125	185	1,310	1,500
412	2000 gallon capacity		3		1,150	225	1,375	1,600
414	5000 gallon capacity	↓	1	↓	2,950	680	3,630	4,225
500	Steel underground, coated, set in place, incl. hold-down bars.							

15.5 Heating

		CREW	DAILY OUTPUT	UNIT	BARE COSTS MAT.	BARE COSTS INST.	BARE COSTS TOTAL	TOTAL INCL O&P
89 550	Excavation, pad, pumps and piping not included							
552	1000 gallon capacity, 7 gauge shell	Q-7	4	Ea.	725	170	895	1,050
554	5000 gallon capacity, 1/4" thick shell		1		2,625	680	3,305	3,875
556	10,000 gallon capacity, 1/4" thick shell		.70		3,875	975	4,850	5,675
558	10,000 gallon capacity, 5/16" thick shell		.60		4,650	1,125	5,775	6,750
560	20,000 gallon capacity, 5/16" thick shell		.30		7,650	2,275	9,925	11,700
562	30,000 gallon capacity, 3/8" thick shell		.20		12,700	3,400	16,100	18,900
600	For 20 year warranty galvanic protection,							
602	add to tank cost							
604	1000 gallon tank			Ea.	610		610	670M
606	2000 gallon tank				795		795	875M
608	5000 gallon tank				1,250		1,250	1,375M
610	10,000 gallon tank				1,625		1,625	1,800M
612	20,000 gallon tank				2,575		2,575	2,825M
614	30,000 gallon tank				3,575		3,575	3,925M
92-001	**VENT CHIMNEY** Prefab metal, U.L. listed							
002	Gas, double wall, galvanized steel							
008	3" diameter	Q-9	72	V.L.F.	2	4.44	6.44	8.65
010	4" diameter		68		2.43	4.70	7.13	9.55
012	5" diameter		64		2.88	5	7.88	10.45
014	6" diameter		60		3.38	5.35	8.73	11.50
016	7" diameter		56		4.59	5.70	10.29	13.35
018	8" diameter		52		5.10	6.15	11.25	14.55
020	10" diameter		48		10.80	6.65	17.45	22
022	12" diameter		44		14.40	7.25	21.65	26
024	14" diameter		42		24	7.60	31.60	38
026	16" diameter		40		32	8	40	47
028	18" diameter		38		40	8.40	48.40	56
030	20" diameter	Q-10	36		48	13.80	61.80	73
032	22" diameter		34		60	14.65	74.65	87
034	24" diameter		32		74	15.55	89.55	105
036	26" diameter		31		89	16.05	105.05	120
038	28" diameter		30		94	16.60	110.60	130
040	30" diameter		28		99	17.75	116.75	135
042	32" diameter		27		110	18.40	128.40	150
044	34" diameter		26		125	19.15	144.15	165
046	36" diameter		25		135	19.90	154.90	175
048	38" diameter		24		145	21	166	190
050	40" diameter		23		165	22	187	215
052	42" diameter		22		170	23	193	220
054	44" diameter		21		190	24	214	245
056	46" diameter		20		210	25	235	265
058	48" diameter		19		230	26	256	290
060	For 4", 5" and 6" oval, add				50%			
065	Gas, double wall, galvanized steel, fittings							
066	Elbow 45°, 3" diameter	Q-9	36	Ea.	4.90	8.90	13.80	18.35
067	4" diameter		34		5.75	9.40	15.15	20
068	5" diameter		32		6.75	10	16.75	22
069	6" diameter		30		8.10	10.65	18.75	24
070	7" diameter		28		11.55	11.40	22.95	29
071	8" diameter		26		12.30	12.30	24.60	31
072	10" diameter		24		31	13.30	44.30	54
073	12" diameter		22		39	14.55	53.55	64
074	14" diameter		21		55	15.20	70.20	83
075	16" diameter		20		69	16	85	99
076	18" diameter		19		87	16.85	103.85	120
077	20" diameter	Q-10	18		99	28	127	150
078	22" diameter		17		150	29	179	210
079	24" diameter		16		190	31	221	255

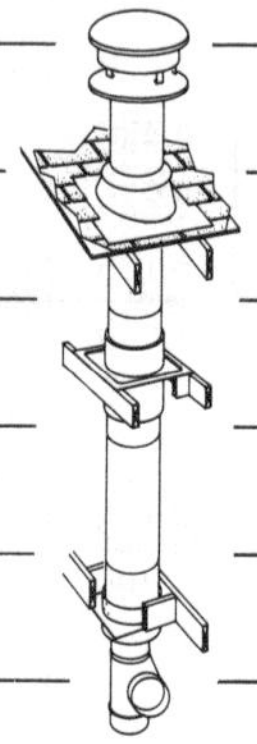

15

15.5 Heating

		CREW	DAILY OUTPUT	UNIT	BARE COSTS MAT.	BARE COSTS INST.	BARE COSTS TOTAL	TOTAL INCL O&P
92 080	26" diameter	Q-10	16	Ea.	370	31	401	450
081	28" diameter		15		380	33	413	465
082	30" diameter		14		400	36	436	490
083	32" diameter		14		435	36	471	530
084	34" diameter		13		475	38	513	580
085	36" diameter		12		490	41	531	600
086	38" diameter		12		530	41	571	645
087	40" diameter		11		555	45	600	675
088	42" diameter		11		590	45	635	715
089	44" diameter		10		625	50	675	760
090	46" diameter		10		670	50	720	810
091	48" diameter		10		695	50	745	835
095	Elbow 90°, adjustable, 3" diameter	Q-9	36		8.85	8.90	17.75	23
096	4" diameter		34		10	9.40	19.40	25
097	5" diameter		32		11.55	10	21.55	27
098	6" diameter		30		13.95	10.65	24.60	31
099	7" diameter		28		20	11.40	31.40	39
101	8" diameter		26		21	12.30	33.30	41
104	Roof flashing, 3" diameter		36		3.46	8.90	12.36	16.75
105	4" diameter		34		4.75	9.40	14.15	18.95
106	5" diameter		32		5.25	10	15.25	20
107	6" diameter		30		7.35	10.65	18	24
108	7" diameter		28		8.35	11.40	19.75	26
109	8" diameter		26		9.10	12.30	21.40	28
110	10" diameter		24		25	13.30	38.30	47
111	12" diameter		22		32	14.55	46.55	56
112	14" diameter		20		39	16	55	66
113	16" diameter		18		48	17.75	65.75	79
114	18" diameter		16		62	20	82	97
115	20" diameter	Q-10	18		80	28	108	130
116	22" diameter		14		100	36	136	160
117	24" diameter		12		125	41	166	200
120	Tee, 3" diameter	Q-9	27		12	11.85	23.85	30
121	4" diameter		26		12.55	12.30	24.85	32
122	5" diameter		25		13.25	12.80	26.05	33
123	6" diameter		24		14.75	13.30	28.05	36
124	7" diameter		23		18.25	13.90	32.15	40
125	8" diameter		22		20	14.55	34.55	43
126	10" diameter		21		56	15.20	71.20	84
127	12" diameter		20		67	16	83	97
128	14" diameter		18		110	17.75	127.75	145
129	16" diameter		16		155	20	175	200
130	18" diameter		14		185	23	208	235
131	20" diameter	Q-10	17		250	29	279	320
132	22" diameter		13		310	38	348	395
133	24" diameter		12		360	41	401	455
134	26" diameter		12		510	41	551	620
135	28" diameter		11		575	45	620	700
136	30" diameter		10		625	50	675	760
137	32" diameter		10		695	50	745	835
138	34" diameter		9		705	55	760	855
139	36" diameter		9		835	55	890	1,000
140	38" diameter		8		840	62	902	1,025
141	40" diameter		7		920	71	991	1,125
142	42" diameter		6		1,000	83	1,083	1,225
143	44" diameter		5		1,100	99	1,199	1,350
144	46" diameter		5		1,200	99	1,299	1,475
145	48" diameter		5		1,275	99	1,374	1,550
146	Tee cap, 3" diameter	Q-9	45		.91	7.10	8.01	11.35
147	4" diameter	"	42		1.12	7.60	8.72	12.35

15

15.5 Heating

92

		CREW	DAILY OUTPUT	UNIT	BARE COSTS MAT.	BARE COSTS INST.	BARE COSTS TOTAL	TOTAL INCL O&P
148	5" diameter	Q-9	40	Ea.	1.23	8	9.23	13
149	6" diameter		37		1.44	8.65	10.09	14.20
150	7" diameter		35		2.14	9.15	11.29	15.65
151	8" diameter		34		2.52	9.40	11.92	16.50
152	10" diameter		32		4.97	10	14.97	20
153	12" diameter		30		6.60	10.65	17.25	23
154	14" diameter		28		13.75	11.40	25.15	32
155	16" diameter		25		15.95	12.80	28.75	36
156	18" diameter		24		17.10	13.30	30.40	38
157	20" diameter	Q-10	27		18.75	18.40	37.15	47
158	22" diameter		22		21	23	44	56
159	24" diameter		21		24	24	48	61
160	26" diameter		20		41	25	66	81
161	28" diameter		19		48	26	74	91
162	30" diameter		18		55	28	83	100
163	32" diameter		17		62	29	91	110
164	34" diameter		16		69	31	100	120
165	36" diameter		15		77	33	110	135
166	38" diameter		14		84	36	120	145
167	40" diameter		14		91	36	127	150
168	42" diameter		13		98	38	136	165
169	44" diameter		12		105	41	146	175
170	46" diameter		12		115	41	156	185
171	48" diameter		11		125	45	170	205
175	Top, 3" diameter	Q-9	46		4.48	6.95	11.43	15.05
176	4" diameter		44		4.76	7.25	12.01	15.85
177	5" diameter		42		6.50	7.60	14.10	18.25
178	6" diameter		40		8.35	8	16.35	21
179	7" diameter		38		11.55	8.40	19.95	25
180	8" diameter		36		14.60	8.90	23.50	29
181	10" diameter		34		34	9.40	43.40	51
182	12" diameter		32		53	10	63	73
183	14" diameter		30		60	10.65	70.65	82
184	16" diameter		28		105	11.40	116.40	130
185	18" diameter		26		175	12.30	187.30	210
186	20" diameter	Q-10	28		195	17.75	212.75	240
187	22" diameter		22		265	23	288	325
188	24" diameter		20		335	25	360	405
300	All fuel, double wall, stainless steel, 6" diameter	Q-9	60	V.L.F.	13.95	5.35	19.30	23
302	7" diameter		56		17.25	5.70	22.95	27
304	8" diameter		52		21	6.15	27.15	32
306	10" diameter		48		29	6.65	35.65	42
308	12" diameter		44		38	7.25	45.25	52
310	14" diameter		42		50	7.60	57.60	66
800	All fuel, double wall, stainless steel fittings							
801	Roof support 6" diameter	Q-9	30	Ea.	35	10.65	45.65	54
802	7" diameter		28		39	11.40	50.40	60
803	8" diameter		26		42	12.30	54.30	64
804	10" diameter		24		54	13.30	67.30	79
805	12" diameter		22		65	14.55	79.55	93
806	14" diameter		21		81	15.20	96.20	110
810	Elbow 15°, 6" diameter		30		30	10.65	40.65	49
812	7" diameter		28		34	11.40	45.40	54
814	8" diameter		26		39	12.30	51.30	61
816	10" diameter		24		49	13.30	62.30	73
818	12" diameter		22		59	14.55	73.55	86
820	14" diameter		21		70	15.20	85.20	99
830	Insulated tee with insulated tee cap, 6" diameter		30		56	10.65	66.65	77
834	7" diameter		28		73	11.40	84.40	97
836	8" diameter		26		83	12.30	95.30	110

15

15.5 Heating		CREW	DAILY OUTPUT	UNIT	BARE COSTS MAT.	BARE COSTS INST.	BARE COSTS TOTAL	TOTAL INCL O&P
92 838	10" diameter	Q-9	24	Ea.	115	13.30	128.30	145
840	12" diameter		22		160	14.55	174.55	195
842	14" diameter		21		210	15.20	225.20	255
850	Joist shield, 6" diameter		30		16.45	10.65	27.10	34
851	7" diameter		28		17.95	11.40	29.35	36
852	8" diameter		26		18.75	12.30	31.05	39
853	10" diameter		24		22	13.30	35.30	44
854	12" diameter		22		32	14.55	46.55	56
855	14" diameter		21		33	15.20	48.20	58
860	Round top, 6" diameter		30		20	10.65	30.65	38
862	7" diameter		28		26	11.40	37.40	45
864	8" diameter		26		36	12.30	48.30	58
866	10" diameter		24		62	13.30	75.30	88
868	12" diameter		22		88	14.55	102.55	120
870	14" diameter		21		115	15.20	130.20	150
880	Adjustable roof flashing, 6" diameter		30		63	10.65	73.65	85
882	7" diameter		28		71	11.40	82.40	95
884	8" diameter		26		74	12.30	86.30	99
886	10" diameter		24		92	13.30	105.30	120
888	12" diameter		22		140	14.55	154.55	175
890	14" diameter		21		190	15.20	205.20	230
900	High temp. (2000°F), steel jacket, acid resist. refractory lining							
901	11 ga. galvanized jacket, U.L. listed							
902	Straight section, 48" long, 10" diameter	Q-10	13.30	V.L.F.	250	37	287	330
903	12" diameter		11.20		270	44	314	360
904	18" diameter		7.40		375	67	442	510
905	24" diameter		4.60		510	110	620	720
906	30" diameter		3.70		620	135	755	880
907	36" diameter		2.70		770	185	955	1,125
908	42" diameter	Q-11	3.10		1,050	220	1,270	1,475
909	48" diameter		2.70		1,250	250	1,500	1,750
910	54" diameter		2.20		1,450	310	1,760	2,050
911	60" diameter		2		1,750	340	2,090	2,425
912	Tee section, 10" diameter	Q-10	4.40	Ea.	415	115	530	620
913	12" diameter		3.70		450	135	585	690
914	18" diameter		2.40		625	205	830	990
915	24" diameter		1.50		875	330	1,205	1,450
916	30" diameter		1.20		1,175	415	1,590	1,900
917	36" diameter		.80		1,900	620	2,520	3,000
918	42" diameter	Q-11	1		2,600	680	3,280	3,850
919	48" diameter		.90		3,050	755	3,805	4,450
920	54" diameter		.70		4,000	970	4,970	5,825
921	60" diameter		.60		4,700	1,125	5,825	6,825
922	Cleanout pier section, 10" diameter	Q-10	3.50		435	140	575	685
923	12" diameter		2.50		465	200	665	800
924	18" diameter		2		630	250	880	1,050
925	24" diameter		1.30		780	385	1,165	1,425
926	30" diameter		1		920	495	1,415	1,725
927	36" diameter		.75		1,100	665	1,765	2,175
928	42" diameter	Q-11	.90		1,450	755	2,205	2,700
929	48" diameter		.75		1,650	905	2,555	3,125
930	54" diameter		.60		1,925	1,125	3,050	3,775
931	60" diameter		.50		2,250	1,350	3,600	4,450
932	For drain, add				53%			
933	Elbow, 30° and 45°, 10" diameter	Q-10	6.60		240	75	315	375
934	12" diameter		5.60		275	89	364	430
935	18" diameter		3.70		425	135	560	665
936	24" diameter		2.30		665	215	880	1,050
937	30" diameter		1.85		895	270	1,165	1,375
938	36" diameter		1.30		1,150	385	1,535	1,825

15.5 Heating

			CREW	DAILY OUTPUT	UNIT	BARE COSTS MAT.	INST.	TOTAL	TOTAL INCL O&P
92	939	42" diameter	Q-11	1.60	Ea.	1,925	425	2,350	2,725
	940	48" diameter		1.35		2,400	505	2,905	3,375
	941	54" diameter		1.10		2,800	615	3,415	3,975
	942	60" diameter		1		3,275	680	3,955	4,600
	943	For 60° and 90° elbow, add				112%			
	944	End cap, 10" diameter	Q-10	26		495	19.15	514.15	570
	945	12" diameter		22		535	23	558	620
	946	18" diameter		15		730	33	763	850
	947	24" diameter		9		940	55	995	1,125
	948	30" diameter		7.30		1,225	68	1,293	1,450
	949	36" diameter		5		1,400	99	1,499	1,675
	950	42" diameter	Q-11	6.30		1,775	110	1,885	2,100
	951	48" diameter		5.30		2,125	130	2,255	2,525
	952	54" diameter		4.40		2,575	155	2,730	3,050
	953	60" diameter		3.90		3,100	175	3,275	3,675
	954	Increaser (1 diameter), 10" diameter	Q-10	6.60		350	75	425	495
	955	12" diameter		5.60		375	89	464	540
	956	18" diameter		3.70		510	135	645	755
	957	24" diameter		2.30		690	215	905	1,075
	958	30" diameter		1.85		845	270	1,115	1,325
	959	36" diameter		1.30		1,075	385	1,460	1,750
	960	For expansion joints, add to straight section				7%			
	961	For 1/4" hot rolled steel jacket, add				157%			
	962	For 2950°F very high temperature, add				89%			
	963	26 ga. aluminized jacket, straight section, 48" long							
	964	10" diameter	Q-10	15.30	V.L.F.	115	33	148	175
	965	12" diameter		12.90		125	39	164	195
	966	18" diameter		8.50		200	59	259	305
	967	24" diameter		5.30		275	94	369	440
	968	30" diameter		4.30		380	115	495	585
	969	36" diameter		3.10		455	160	615	735
	970	Accessories (all models)							
	971	Guy band, 10" diameter	Q-10	32	Ea.	36	15.55	51.55	62
	972	12" diameter		30		44	16.60	60.60	73
	973	18" diameter		26		56	19.15	75.15	89
	974	24" diameter		24		115	21	136	155
	975	30" diameter		20		125	25	150	175
	976	36" diameter		18		145	28	173	200
	977	42" diameter	Q-11	22		575	31	606	675
	978	48" diameter		20		615	34	649	725
	979	54" diameter		16		680	42	722	810
	980	60" diameter		12		755	57	812	915
	981	Draw band, 11 gauge, 10" diameter	Q-10	32		28	15.55	43.55	53
	982	12" diameter		30		30	16.60	46.60	57
	983	18" diameter		26		45	19.15	64.15	77
	984	24" diameter		24		60	21	81	96
	985	30" diameter		20		79	25	104	125
	986	36" diameter		18		95	28	123	145
	987	42" diameter	Q-11	22		115	31	146	170
	988	48" diameter		20		135	34	169	200
	989	54" diameter		16		150	42	192	225
	990	60" diameter		12		165	57	222	265
	991	Draw band, 26 gauge, 10" diameter	Q-10	32		11.65	15.55	27.20	35
	992	12" diameter		30		11.65	16.60	28.25	37
	993	18" diameter		26		13.30	19.15	32.45	43
	994	24" diameter		24		15	21	36	47
	995	30" diameter		20		16.65	25	41.65	55
	996	36" diameter		15		16.65	33	49.65	67

15.6 HVAC Piping Specialties

		CREW	DAILY OUTPUT	UNIT	BARE COSTS MAT.	BARE COSTS INST.	BARE COSTS TOTAL	TOTAL INCL O&P
04-001	**AUTOMATIC AIR VENT**							
002	Cast iron body, stainless steel internals							
006	1/2" NPT inlet, 300 psi	1 Stpi	12	Ea.	37	14.85	51.85	62
014	3/4" NPT inlet, 300 psi		12		37	14.85	51.85	62
018	1/2" NPT inlet, 250 psi		10		120	17.85	137.85	160
022	3/4" NPT inlet, 250 psi		10		120	17.85	137.85	160
026	1" NPT inlet, 250 psi		10		175	17.85	192.85	220
034	1-1/2" NPT inlet, 250 psi	Q-5	12		360	27	387	435
038	2" NPT inlet, 250 psi	"	12		360	27	387	435
060	Forged steel body, stainless steel internals							
064	1/2" NPT, 750 psi	1 Stpi	12	Ea.	365	14.85	379.85	425
068	3/4" NPT inlet, 750 psi		12		365	14.85	379.85	425
076	3/4" NPT inlet, 1000 psi		10		545	17.85	562.85	625
080	1" NPT inlet, 1000 psi	Q-5	12		545	27	572	640
088	1-1/2" NPT inlet, 1000 psi		10		1,550	32	1,582	1,750
092	2" NPT inlet, 1000 psi		10		1,550	32	1,582	1,750
110	Formed steel body, non corrosive							
111	1/8" NPT inlet 35 psi	1 Stpi	32	Ea.	5.95	5.60	11.55	14.60
112	1/4" NPT inlet 150 psi		32		15.05	5.60	20.65	25
113	3/4" NPT inlet 150 psi		32		15.05	5.60	20.65	25
130	Chrome plated brass, automatic/manual, for radiators							
131	1/8" NPT inlet, nickel plated brass	1 Stpi	32	Ea.	1.75	5.60	7.35	10
08-001	**AIR CONTROL** With strainer							
004	2" diameter	Q-5	6	Ea.	290	54	344	395
008	2-1/2" diameter		5		320	64	384	445
010	3" diameter		4		495	80	575	660
012	4" diameter		3		710	105	815	935
014	6" diameter	Q-6	3.40		1,075	145	1,220	1,400
016	8" diameter		3		1,625	165	1,790	2,025
018	10" diameter		2.20		2,475	225	2,700	3,050
020	12" diameter		1.70		3,575	295	3,870	4,350
12-001	**CIRCUIT SENSOR** Flow meter,							
006	2-1/2" pipe size	Q-5	12	Ea.	70	27	97	115
010	3" pipe size		11		80	29	109	130
014	4" pipe size		8		95	40	135	165
018	5" pipe size		7.30		115	44	159	190
022	6" pipe size		6.40		135	50	185	220
026	8" pipe size	Q-6	5.30		190	94	284	345
028	10" pipe size		4.60		225	110	335	405
036	12" pipe size		4.20		325	120	445	530
16-001	**CIRCUIT SETTER** Balance valve							
002	3/4" pipe size	1 Stpi	20	Ea.	31	8.90	39.90	47
004	1" pipe size		18		39	9.90	48.90	57
006	1-1/2" pipe size		12		64	14.85	78.85	92
008	2" pipe size		10		94	17.85	111.85	130
010	2-1/2" pipe size	Q-5	15		180	21	201	230
012	3" pipe size		10		270	32	302	345
014	4" pipe size		3		385	105	490	580
020	For differential meter, add				440		440	485M
20-001	**EXPANSION JOINTS** Bellows type, neoprene cover, flanged spool							
010	6" face to face, 1/2" diameter	1 Stpi	14	Ea.	155	12.75	167.75	190
011	3/4" diameter		14		155	12.75	167.75	190
012	1" diameter		13		160	13.70	173.70	195
014	1-1/4" diameter		11		165	16.20	181.20	205
016	1-1/2" diameter		10.60		170	16.85	186.85	210
018	2" diameter	Q-5	13.30		185	24	209	240
020	3" diameter	"	11.40		205	28	233	265

15.6 HVAC Piping Specialties

		CREW	DAILY OUTPUT	UNIT	BARE COSTS MAT.	BARE COSTS INST.	BARE COSTS TOTAL	TOTAL INCL O&P
20 048	10" face to face, 2" diameter	Q-5	13	Ea.	225	25	250	285
050	2-1/2" diameter		12		230	27	257	290
052	3" diameter		11		250	29	279	315
054	4" diameter		8		260	40	300	345
056	5" diameter		7		300	46	346	395
058	6" diameter		6		325	54	379	435
060	8" diameter		5		375	64	439	505
062	10" diameter		4.60		425	70	495	570
064	12" diameter		4		490	80	570	655
066	14" diameter		3.80		545	85	630	720
24-001	**EXPANSION COUPLINGS** Hydronic							
010	Copper to copper, sweat							
012	3/4" diameter	1 Stpi	20	Ea.	14.85	8.90	23.75	29
014	1" diameter		19		18.65	9.40	28.05	34
016	1-1/4" diameter		15		26	11.90	37.90	46
100	Baseboard riser fitting, 5" stub by coupling 12" long							
102	1/2" diameter	1 Stpi	24	Ea.	6.90	7.45	14.35	18.35
104	3/4" diameter		20		9.50	8.90	18.40	23
106	1" diameter		19		13.15	9.40	22.55	28
108	1-1/4" diameter		15		16.40	11.90	28.30	35
118	9" Stub by tubing 8" long							
120	1/2" diameter	1 Stpi	24	Ea.	6.20	7.45	13.65	17.55
122	3/4" diameter		20		7.70	8.90	16.60	21
124	1" diameter		19		11.30	9.40	20.70	26
126	1-1/4" diameter		15		14.30	11.90	26.20	33
28-001	**FLEXIBLE METAL HOSE** Connectors, standard lengths							
010	Bronze braided, bronze ends							
012	3/8" diameter x 12"	1 Stpi	26	Ea.	14.95	6.85	21.80	26
014	1/2" diameter x 12"		24		18.60	7.45	26.05	31
016	3/4" diameter x 12"		20		26	8.90	34.90	41
018	1" diameter x 18"		19		43	9.40	52.40	61
020	1-1/2" diameter x 18"		13		64	13.70	77.70	90
022	2" diameter x 18"		11		90	16.20	106.20	120
100	Carbon steel ends							
102	1/4" diameter x 12"	1 Stpi	28	Ea.	11.60	6.35	17.95	22
104	3/8" diameter x 12"		26		13.25	6.85	20.10	24
106	1/2" diameter x 12"		24		16.85	7.45	24.30	29
108	1/2" diameter x 24"		24		24	7.45	31.45	37
110	1/2" diameter x 36"		24		31	7.45	38.45	45
112	3/4" diameter x 12"		20		23	8.90	31.90	38
114	3/4" diameter x 24"		20		34	8.90	42.90	50
116	3/4" diameter x 36"		20		45	8.90	53.90	62
118	1" diameter x 18"		19		34	9.40	43.40	51
120	1" diameter x 30"		19		48	9.40	57.40	66
122	1" diameter x 36"		19		55	9.40	64.40	74
124	1-1/4" diameter x 18"		15		41	11.90	52.90	62
126	1-1/4" diameter x 36"		15		67	11.90	78.90	91
128	1-1/2" diameter x 18"		13		49	13.70	62.70	74
130	1-1/2" diameter x 36"		13		80	13.70	93.70	110
132	2" diameter x 24"		11		82	16.20	98.20	115
134	2" diameter x 36"		11		110	16.20	126.20	145
136	2-1/2" diameter x 24"		9		100	19.80	119.80	140
138	2-1/2" diameter x 36"		9		135	19.80	154.80	175
140	3" diameter x 24"		7		125	25	150	175
142	3" diameter x 36"		7		165	25	190	220
200	Carbon steel braid, carbon steel solid ends							
210	1/2" diameter x 12"	1 Stpi	24	Ea.	14	7.45	21.45	26
212	3/4" diameter x 12"		20		17	8.90	25.90	32
214	1" diameter x 12"		19		21	9.40	30.40	37

15.6 HVAC Piping Specialties

		CREW	DAILY OUTPUT	UNIT	BARE COSTS MAT.	INST.	TOTAL	TOTAL INCL O&P
28 216	1-1/4" diameter x 12"	1 Stpi	15	Ea.	25	11.90	36.90	45
218	1-1/2" diameter x 12"	"	13	"	29	13.70	42.70	52
300	Stainless steel braid, welded on carbon steel ends							
310	1/2" diameter x 12"	1 Stpi	24	Ea.	21	7.45	28.45	34
312	3/4" diameter x 12"		20		25	8.90	33.90	40
314	3/4" diameter x 24"		20		37	8.90	45.90	54
316	3/4" diameter x 36"		20		48	8.90	56.90	66
318	1" diameter x 12"		19		31	9.40	40.40	48
320	1" diameter x 24"		19		46	9.40	55.40	64
322	1" diameter x 36"		19		62	9.40	71.40	82
324	1-1/4" diameter x 12"		15		37	11.90	48.90	58
326	1-1/4" diameter x 24"		15		55	11.90	66.90	78
328	1-1/4" diameter x 36"		15		73	11.90	84.90	97
330	1-1/2" diameter x 12"		13		41	13.70	54.70	65
332	1-1/2" diameter x 24"		13		62	13.70	75.70	88
334	1-1/2" diameter x 36"	↓	13	↓	83	13.70	96.70	110
32-001	**HEATING CONTROL VALVES**							
005	Hot water, nonelectric, thermostatic							
010	Radiator supply, 1/2" diameter	1 Stpi	24	Ea.	62	7.45	69.45	79
012	3/4" diameter		20		66	8.90	74.90	85
014	1" diameter		19		79	9.40	88.40	100
016	1-1/4" diameter	↓	15		87	11.90	98.90	115
050	For low pressure steam, add			↓	25%			
100	Manual, radiator supply							
101	1/2" pipe size, angle union	1 Stpi	24	Ea.	11.15	7.45	18.60	23
102	3/4" pipe size, angle union		20		15.45	8.90	24.35	30
103	1" pipe size, angle union	↓	19	↓	19.90	9.40	29.30	35
110	Radiator, balancing							
111	1/2" pipe size, angle union	1 Stpi	24	Ea.	11.60	7.45	19.05	24
112	3/4" pipe size, angle union		20		13.45	8.90	22.35	28
113	1" pipe size, angle union	↓	19	↓	21	9.40	30.40	37
120	Steam, radiator, supply							
121	1/2" pipe size, angle union	1 Stpi	24	Ea.	14.95	7.45	22.40	27
122	3/4" pipe size, angle union		20		16.55	8.90	25.45	31
123	1" pipe size, angle union		19		19.95	9.40	29.35	36
124	1-1/4" pipe size, angle union	↓	15	↓	25	11.90	36.90	45
800	System balancing and shut-off							
802	Butterfly, quarter turn, calibrated, threaded or solder							
804	Bronze, -30°F to +350°F, pressure to 175 psi							
806	1/2" size	1 Stpi	22	Ea.	6.85	8.10	14.95	19.25
807	3/4" size		20		8.95	8.90	17.85	23
808	1" size		19		10.85	9.40	20.25	26
809	1-1/4" size		15		16.65	11.90	28.55	36
810	1-1/2" size		13		19.85	13.70	33.55	42
811	2" size	↓	11	↓	24	16.20	40.20	50
36-001	**MIXING VALVE** automatic water tempering							
005	3/4" size	1 Stpi	18	Ea.	140	9.90	149.90	170
010	1" size		16		200	11.15	211.15	235
012	1-1/4" size		13		290	13.70	303.70	340
014	1-1/2" size		10		350	17.85	367.85	410
016	2" size		8		445	22	467	520
018	3" size	↓	4	↓	1,000	45	1,045	1,175
40-001	**PRESSURE REDUCING VALVE,** Steam, pilot operated							
010	Threaded, iron body							
012	1/2" size	1 Stpi	18	Ea.	375	9.90	384.90	425
014	3/4" size		14		405	12.75	417.75	465
016	1" size		13		450	13.70	463.70	515
018	1-1/4" size	↓	10	↓	510	17.85	527.85	585

	15.6 HVAC Piping Specialties	CREW	DAILY OUTPUT	UNIT	BARE COSTS MAT.	BARE COSTS INST.	BARE COSTS TOTAL	TOTAL INCL O&P
40 020	1-1/2" size	1 Stpi	8	Ea.	585	22	607	675
022	2" size	"	5	"	705	36	741	825
100	Flanged, iron body, 125 lb. flanges							
102	2" size	1 Stpi	8	Ea.	780	22	802	890
104	2-1/2" size	"	4		970	45	1,015	1,125
106	3" size	Q-5	4.50		1,175	71	1,246	1,400
108	4" size	"	3		1,375	105	1,480	1,675
150	For 250 lb. flanges, add				5%			
44-001	**PRESSURE REGULATOR**							
005	Oil, threaded, 1/4" size for #2 oil	1 Stpi	24	Ea.	53	7.45	60.45	69
010	1/2" size for #4,5,6 oil		24		110	7.45	117.45	130
012	3/4" size for #4,5,6 oil		20		125	8.90	133.90	150
014	1" size for #4,5,6 oil		19		150	9.40	159.40	180
100	Process steam, threaded							
150	Bronze body, 1/2" size	1 Stpi	24	Ea.	95	7.45	102.45	115
152	3/4" size		20		115	8.90	123.90	140
154	1" size		19		175	9.40	184.40	205
200	Iron body, 1/2" size		24		67	7.45	74.45	84
202	3/4" size		20		81	8.90	89.90	100
204	1" size		19		105	9.40	114.40	130
206	1-1/4" size		15		145	11.90	156.90	175
208	1-1/2" size		13		165	13.70	178.70	200
210	2" size		11		235	16.20	251.20	280
300	Steam, high capacity, iron body, stainless steel trim							
302	Threaded, 1/2" diameter	1 Stpi	24	Ea.	225	7.45	232.45	260
303	3/4" diameter		24		225	7.45	232.45	260
304	1" diameter		19		255	9.40	264.40	295
306	1-1/4" diameter		15		280	11.90	291.90	325
308	1-1/2" diameter		13		320	13.70	333.70	370
310	2" diameter		11		390	16.20	406.20	450
312	2-1/2" diameter	Q-5	12		490	27	517	580
314	3" diameter	"	11		560	29	589	660
350	Flanged connection, 125 lb. W.S.P.							
352	3" diameter	Q-5	11	Ea.	610	29	639	715
354	4" diameter	"	5	"	770	64	834	940
900	For water pressure regulators, see Div. 15.1-80							
50-001	**SEPARATORS** Entrainment eliminator, steel body, 150 PSIG							
010	1/4" size	1 Stpi	24	Ea.	80	7.45	87.45	99
012	1/2" size		24		85	7.45	92.45	105
014	3/4" size		20		90	8.90	98.90	110
016	1" size		19		135	9.40	144.40	160
018	1-1/4" size		15		155	11.90	166.90	190
020	1-1/2" size		13		175	13.70	188.70	210
022	2" size		11		195	16.20	211.20	240
024	2-1/2" size	Q-5	15		685	21	706	785
026	3" size		13		775	25	800	890
028	4" size		10		875	32	907	1,000
030	5" size		6		1,075	54	1,129	1,250
032	6" size		3		1,300	105	1,405	1,575
034	8" size	Q-6	4.40		1,600	115	1,715	1,925
036	10" size		4		2,250	125	2,375	2,650
038	12" size		3		3,775	165	3,940	4,400
040	14" size		2		5,025	250	5,275	5,900
100	For 300 PSIG, add				15%			
55-001	**STEAM CONDENSATE METER**							
005	250 lb. per hour	1 Stpi	16	Ea.	905	11.15	916.15	1,000
010	500 lb. per hour		14		1,025	12.75	1,037.75	1,150
012	750 lb. per hour		11		1,100	16.20	1,116.20	1,225
014	1500 lb. per hour		7		1,275	25	1,300	1,450
016	3000 lb. per hour		5		2,200	36	2,236	2,475

15.6 HVAC Piping Specialties

		CREW	DAILY OUTPUT	UNIT	BARE COSTS MAT.	BARE COSTS INST.	BARE COSTS TOTAL	TOTAL INCL O&P
55 018	6500 lb. per hour	1 Stpi	3	Ea.	2,525	59	2,584	2,875
020	12,000 lb. per hour	Q-5	3.50	″	3,075	92	3,167	3,525
60-001	**STEAM TRAP** Cast Iron							
004	Inverted bucket							
005	1/2″ NPT	1 Stpi	12	Ea.	48	14.85	62.85	74
007	3/4″ NPT		10		80	17.85	97.85	115
010	1″ NPT		9		125	19.80	144.80	165
012	1-1/4″ NPT		8		185	22	207	235
100	Float & thermostatic, 15 psig							
101	3/4″ pipe size	1 Stpi	16	Ea.	56	11.15	67.15	78
102	1″ pipe size		15		68	11.90	79.90	92
103	1-1/4″ pipe size		13		85	13.70	98.70	115
104	1-1/2″ pipe size		9		125	19.80	144.80	165
106	2″ pipe size		6		225	30	255	290
130	Thermostatic, angle union, 15 psi							
131	1/2″ pipe size	1 Stpi	24	Ea.	17.65	7.45	25.10	30
132	3/4″ pipe size		20		30	8.90	38.90	46
133	1″ pipe size		19		41	9.40	50.40	59
64-001	**STRAINERS, BASKET TYPE** Perforated stainless steel basket							
010	Brass or monel available							
200	Simplex style							
230	Bronze body							
232	Screwed, 3/8″ pipe size	1 Stpi	22	Ea.	89	8.10	97.10	110
234	1/2″ pipe size		20		89	8.90	97.90	110
236	3/4″ pipe size		17		140	10.50	150.50	170
238	1″ pipe size		15		140	11.90	151.90	170
240	1-1/4″ pipe size		13		190	13.70	203.70	230
242	1-1/2″ pipe size		12		210	14.85	224.85	250
244	2″ pipe size		10		385	17.85	402.85	450
246	2-1/2″ pipe size	Q-5	15		540	21	561	625
248	3″ pipe size	″	14		645	23	668	745
260	Flanged, 2″ pipe size	1 Stpi	6		445	30	475	530
262	2-1/2″ pipe size	Q-5	4.50		695	71	766	870
264	3″ pipe size		3.50		845	92	937	1,050
266	4″ pipe size		3		1,525	105	1,630	1,825
268	5″ pipe size	Q-6	3.40		2,025	145	2,170	2,450
270	6″ pipe size	″	3		2,625	165	2,790	3,125
360	Iron body							
370	Screwed, 3/8″ pipe size	1 Stpi	22	Ea.	51	8.10	59.10	68
372	1/2″ pipe size		20		51	8.90	59.90	69
374	3/4″ pipe size		17		67	10.50	77.50	89
376	1″ pipe size		15		67	11.90	78.90	91
378	1-1/4″ pipe size		13		93	13.70	106.70	120
380	1-1/2″ pipe size		12		110	14.85	124.85	140
382	2″ pipe size		10		145	17.85	162.85	185
384	2-1/2″ pipe size	Q-5	15		190	21	211	240
386	3″ pipe size	″	14		260	23	283	320
400	Flanged, 2″ pipe size	1 Stpi	6		220	30	250	285
402	2-1/2″ pipe size	Q-5	4.50		280	71	351	410
404	3″ pipe size		3.50		340	92	432	505
406	4″ pipe size		3		585	105	690	800
408	5″ pipe size	Q-6	3.40		685	145	830	965
410	6″ pipe size		3		830	165	995	1,150
412	8″ pipe size		2.50		1,800	200	2,000	2,275
414	10″ pipe size		2.20		3,100	225	3,325	3,750
416	12″ pipe size		1.70		4,025	295	4,320	4,850
418	14″ pipe size		1.40		5,300	355	5,655	6,350
420	16″ pipe size		1		5,500	500	6,000	6,775
600	Cast steel body							
640	Screwed, 1″ pipe size	1 Stpi	15	Ea.	135	11.90	146.90	165

15.6 HVAC Piping Specialties

		CREW	DAILY OUTPUT	UNIT	BARE COSTS MAT.	BARE COSTS INST.	BARE COSTS TOTAL	TOTAL INCL O&P
64 642	1-1/2" pipe size	1 Stpi	12	Ea.	225	14.85	239.85	270
644	2" pipe size	"	10		375	17.85	392.85	440
646	2-1/2" pipe size	Q-5	15		550	21	571	635
648	3" pipe size	"	14		660	23	683	760
656	Flanged, 2" pipe size	1 Stpi	6		525	30	555	620
658	2-1/2" pipe size	Q-5	4.50		755	71	826	935
660	3" pipe size		3.50		815	92	907	1,025
662	4" pipe size		3		1,175	105	1,280	1,450
664	6" pipe size	Q-6	3		2,425	165	2,590	2,900
666	8" pipe size	"	2.50		4,100	200	4,300	4,800
700	Stainless steel body							
720	Screwed, 1" pipe size	1 Stpi	15	Ea.	310	11.90	321.90	360
722	1-1/2" pipe size		12		400	14.85	414.85	460
724	2" pipe size		10		615	17.85	632.85	700
726	2-1/2" pipe size	Q-5	15		955	21	976	1,075
728	3" pipe size	"	14		1,450	23	1,473	1,625
740	Flanged, 2" pipe size	1 Stpi	6		775	30	805	895
742	2-1/2" pipe size	Q-5	4.50		1,225	71	1,296	1,450
744	3" pipe size		3.50		1,800	92	1,892	2,125
746	4" pipe size		3		2,550	105	2,655	2,950
748	6" pipe size	Q-6	3		3,525	165	3,690	4,125
750	8" pipe size	"	2.50		7,625	200	7,825	8,675
810	Duplex style							
820	Bronze body							
824	Screwed, 3/4" pipe size	1 Stpi	16	Ea.	470	11.15	481.15	535
826	1" pipe size		14		470	12.75	482.75	535
828	1-1/4" pipe size		12		895	14.85	909.85	1,000
830	1-1/2" pipe size		11		895	16.20	911.20	1,000
832	2" pipe size		9		1,400	19.80	1,419.80	1,575
834	2-1/2" pipe size	Q-5	14		1,400	23	1,423	1,575
842	Flanged, 2" pipe size	1 Stpi	6		1,600	30	1,630	1,800
844	2-1/2" pipe size	Q-5	4.50		1,625	71	1,696	1,900
846	3" pipe size		3.50		2,250	92	2,342	2,600
848	4" pipe size		3		3,425	105	3,530	3,925
850	5" pipe size	Q-6	3.40		6,000	145	6,145	6,800
852	6" pipe size	"	3		7,725	165	7,890	8,750
870	Iron body							
874	Screwed, 3/4" pipe size	1 Stpi	16	Ea.	265	11.15	276.15	310
876	1" pipe size		14		265	12.75	277.75	310
878	1-1/4" pipe size		12		460	14.85	474.85	525
880	1-1/2" pipe size		11		460	16.20	476.20	530
882	2" pipe size		9		750	19.80	769.80	855
884	2-1/2" pipe size	Q-5	14		750	23	773	860
900	Flanged, 2" pipe size	1 Stpi	6		790	30	820	910
902	2-1/2" pipe size	Q-5	4.50		810	71	881	995
904	3" pipe size		3.50		1,050	92	1,142	1,300
906	4" pipe size		3		1,650	105	1,755	1,975
908	5" pipe size	Q-6	3.40		2,625	145	2,770	3,100
910	6" pipe size		3		3,325	165	3,490	3,900
912	8" pipe size		2.50		6,275	200	6,475	7,200
914	10" pipe size		2.20		9,525	225	9,750	10,800
916	12" pipe size		1.70		12,200	295	12,495	13,800
918	16" pipe size		1		17,500	500	18,000	20,000
930	Cast steel body							
934	Screwed, 1" pipe size	1 Stpi	14	Ea.	655	12.75	667.75	740
936	1-1/2" pipe size		11		970	16.20	986.20	1,100
938	2" pipe size		9		1,625	19.80	1,644.80	1,825
946	Flanged, 2" pipe size		6		1,725	30	1,755	1,950
948	2-1/2" pipe size	Q-5	4.50		1,775	71	1,846	2,050
950	3" pipe size	"	3.50		2,375	92	2,467	2,750

15

15.6 HVAC Piping Specialties

		CREW	DAILY OUTPUT	UNIT	BARE COSTS MAT.	BARE COSTS INST.	BARE COSTS TOTAL	TOTAL INCL O&P
64 952	4" pipe size	Q-5	3	Ea.	3,500	105	3,605	4,000
954	6" pipe size	Q-6	3		6,000	165	6,165	6,850
956	8" pipe size	"	2.50		11,900	200	12,100	13,400
970	Stainless steel body							
974	Screwed, 1" pipe size	1 Stpi	14	Ea.	1,200	12.75	1,212.75	1,350
976	1-1/2" pipe size		11		1,825	16.20	1,841.20	2,025
978	2" pipe size		9		2,950	19.80	2,969.80	3,275
986	Flanged, 2" pipe size		6		3,050	30	3,080	3,400
988	2-1/2" pipe size	Q-5	4.50		3,150	71	3,221	3,575
990	3" pipe size		3.50		4,975	92	5,067	5,600
992	4" pipe size		3		6,050	105	6,155	6,800
994	6" pipe size	Q-6	3		10,700	165	10,865	12,000
996	8" pipe size	"	2.50		21,300	200	21,500	23,700
65-001	**STRAINERS, BOWL TYPE,** Bronze screen							
010	Bronze body, 1/8" pipe size	1 Stpi	24	Ea.	17.10	7.45	24.55	30
012	1/4" pipe size		24		17.10	7.45	24.55	30
014	3/8" pipe size		24		27	7.45	34.45	40
016	1/2" pipe size		22		27	8.10	35.10	41
018	3/4" pipe size		20		33	8.90	41.90	49
100	Iron body, 1/8" pipe size		24		14.90	7.45	22.35	27
102	1/4" pipe size		24		14.90	7.45	22.35	27
104	3/8" pipe size		23		19.15	7.75	26.90	32
105	1/2" pipe size		23		19.15	7.75	26.90	32
106	3/4" pipe size		20		21	8.90	29.90	36
70-001	**STRAINERS, Y TYPE** Bronze body							
005	Screwed, 150 lb., 1/4" pipe size	1 Stpi	24	Ea.	13.20	7.45	20.65	25
007	3/8" pipe size		24		13.20	7.45	20.65	25
010	1/2" pipe size		20		15.60	8.90	24.50	30
012	3/4" pipe size		19		18.60	9.40	28	34
014	1" pipe size		17		27	10.50	37.50	45
015	1-1/4" pipe size		15		34	11.90	45.90	55
016	1-1/2" pipe size		14		48	12.75	60.75	71
018	2" pipe size		13		78	13.70	91.70	105
020	2-1/2" pipe size	Q-5	17		160	18.90	178.90	205
022	3" pipe size		16		315	20	335	375
024	4" pipe size		15		720	21	741	825
050	For 300 lb. rating, add				15%			
100	Flanged, 150 lb., 1-1/2" pipe size	1 Stpi	11		205	16.20	221.20	250
102	2" pipe size	"	8		265	22	287	325
104	3" pipe size	Q-5	4.50		505	71	576	660
106	4" pipe size	"	3		765	105	870	995
108	5" pipe size	Q-6	3.40		1,125	145	1,270	1,450
110	6" pipe size		3		1,450	165	1,615	1,825
112	8" pipe size		2.50		4,600	200	4,800	5,350
150	For 300 lb. rating, add				40%			
75-001	**STRAINERS, Y TYPE** Iron body							
005	Screwed, 250 lb., 1/4" pipe size	1 Stpi	20	Ea.	6.30	8.90	15.20	19.80
007	3/8" pipe size		20		6.30	8.90	15.20	19.80
010	1/2" pipe size		20		6.30	8.90	15.20	19.80
012	3/4" pipe size		18		7.20	9.90	17.10	22
014	1" pipe size		16		9.60	11.15	20.75	27
015	1-1/4" pipe size		15		12.60	11.90	24.50	31
016	1-1/2" pipe size		12		16.20	14.85	31.05	39
018	2" pipe size		8		27	22	49	62
020	2-1/2" pipe size	Q-5	12		42	27	69	85
022	3" pipe size		11		61	29	90	110
024	4" pipe size		5		225	64	289	340

15.6 HVAC Piping Specialties		CREW	DAILY OUTPUT	UNIT	BARE COSTS MAT.	BARE COSTS INST.	BARE COSTS TOTAL	TOTAL INCL O&P
75 050	For galvanized body, add			Ea.	50%			
100	Flanged, 125 lb., 1-1/2" pipe size	1 Stpi	11		65	16.20	81.20	95
102	2" pipe size	"	8		65	22	87	105
103	2-1/2" pipe size	Q-5	5		66	64	130	165
104	3" pipe size		4.50		77	71	148	190
105	3-1/2" pipe size		4		150	80	230	280
106	4" pipe size		3		150	105	255	320
108	5" pipe size	Q-6	3.40		235	145	380	470
110	6" pipe size		3		285	165	450	555
112	8" pipe size		2.50		485	200	685	820
114	10" pipe size		2		930	250	1,180	1,375
116	12" pipe size		1.70		1,375	295	1,670	1,925
117	14" pipe size		1.30		2,175	385	2,560	2,950
118	16" pipe size		1		3,075	500	3,575	4,100
150	For 250 lb. rating, add				20%			
200	For galvanized body, add				50%			
250	For steel body, add				40%			
80-001	**THERMOFLO INDICATOR** For balancing							
100	Sweat connections, 1-1/4" pipe size	1 Stpi	12	Ea.	205	14.85	219.85	245
102	1-1/2" pipe size		10		210	17.85	227.85	255
104	2" pipe size		8		220	22	242	275
106	2-1/2" pipe size		7		325	25	350	395
200	Flange connections, 3" pipe size	Q-5	5		390	64	454	520
202	4" pipe size		4		475	80	555	640
203	5" pipe size		3.50		590	92	682	780
204	6" pipe size		3		640	105	745	860
206	8" pipe size		2		745	160	905	1,050
85-001	**VENTURI FLOW** Measuring device							
005	1/2" diameter	1 Stpi	24	Ea.	60	7.45	67.45	77
010	3/4" diameter		20		60	8.90	68.90	79
012	1" diameter		19		68	9.40	77.40	88
014	1-1/4" diameter		15		74	11.90	85.90	99
016	1-1/2" diameter		13		82	13.70	95.70	110
018	2" diameter		11		99	16.20	115.20	130
020	2-1/2" diameter	Q-5	16		135	20	155	180
022	3" diameter		14		165	23	188	215
024	4" diameter		11		210	29	239	275
026	5" diameter	Q-6	4		265	125	390	470
028	6" diameter		3.50		310	145	455	545
030	8" diameter		3		465	165	630	750
032	10" diameter		2		1,150	250	1,400	1,625
050	For meter, add				660		660	725M
90-001	**WATER LEVEL CONTROLS**							
100	Electric water feeder	1 Stpi	12	Ea.	105	14.85	119.85	135
200	Feeder cut-off combination							
210	Steam system up to 5000 sq. ft.	1 Stpi	12	Ea.	195	14.85	209.85	235
220	Steam system above 5000 sq. ft.		12		250	14.85	264.85	295
230	Steam and hot water, high pressure		10		285	17.85	302.85	340
300	Low water cut-off for hot water boiler, 50 psi maximum							
310	1" top & bottom equalizing pipes, manual reset	1 Stpi	14	Ea.	110	12.75	122.75	140
320	1" top & bottom equalizing pipes		14		110	12.75	122.75	140
330	2-1/2" side connection for nipple-to-boiler		14		96	12.75	108.75	125
400	Low water cut-off for low pressure steam with quick hook-up ftgs.							
410	For installation in gauge glass tappings	1 Stpi	16	Ea.	87	11.15	98.15	110
420	Built-in type, 2-1/2" tap - 3-1/8" insertion		16		66	11.15	77.15	89
430	Built-in type 2-1/2" tap - 1-3/4" insertion		16		70	11.15	81.15	93
440	Side connection to 2-1/2" tapping		16		78	11.15	89.15	100
500	Pump control, low water cut-off and alarm switch		14		160	12.75	172.75	195

15.6 HVAC Piping Specialties

		CREW	DAILY OUTPUT	UNIT	BARE COSTS MAT.	BARE COSTS INST.	BARE COSTS TOTAL	TOTAL INCL O&P
95-001	**WELD END BALL JOINTS** Steel							
005	2-1/2" diameter	Q-17	13	Ea.	185	26	211	240
010	3" diameter		12		220	28	248	280
012	4" diameter		11		330	31	361	405
014	5" diameter	Q-18	14		520	37	557	625
016	6" diameter		12		675	43	718	805
018	8" diameter		9		975	58	1,033	1,150
020	10" diameter		8		1,175	65	1,240	1,375
022	12" diameter		6		1,700	86	1,786	2,000
024	14" diameter		4		2,300	130	2,430	2,725

15.7 Air Conditioning & Ventilating

		CREW	DAILY OUTPUT	UNIT	BARE COSTS MAT.	BARE COSTS INST.	BARE COSTS TOTAL	TOTAL INCL O&P
01-001	**AVERAGE** Square foot and percent of total							
010	job cost, see division 17.1							
04-001	**ABSORPTION COLD GENERATORS** Water chiller							
002	Steam or hot water, water cooled							
005	101 ton			Ea.	43,000	12,900	55,900	66,000
010	148 ton				55,500	16,600	72,100	85,000
020	200 ton				62,000	18,500	80,500	94,500
030	354 ton				85,000	25,500	110,500	130,500
040	420 ton				95,000	28,600	123,600	146,000
050	665 ton				133,500	41,000	174,500	206,000
060	750 ton				151,500	45,400	196,900	232,500
070	850 ton				162,500	48,700	211,200	249,000
080	955 ton				174,500	52,500	227,000	268,000
090	1125 ton				200,500	60,000	260,500	307,500
100	1250 ton				221,500	66,500	288,000	340,000
110	1465 ton				262,000	78,500	340,500	401,500
120	1660 ton				308,500	92,500	401,000	473,000
200	For two stage unit, add				80%	25%		
300	Gas fired, air cooled							
315	2 ton	Q-5	2.10	Ea.	1,475	155	1,630	1,850
318	3 ton		1.30		1,700	245	1,945	2,225
320	4 ton		.90		2,250	355	2,605	3,000
322	5 ton		.60		2,500	535	3,035	3,525
325	8 ton		.50		5,800	640	6,440	7,300
327	10 ton		.40		6,350	805	7,155	8,150
07-001	C8.4 -002 **AIR CONDITIONING REQUIREMENTS** For various types of buildings							
10-001	**AIR FILTERS**							
005	Activated charcoal type, full flow			MCFM	400		400	440M
050	Chemical media filtration type							
110	Industrial air fume & odor scrubber unit w/pump & motor							
111	corrosion resistant PVC construction							
112	Single pack filter, horizontal type							
113	500 CFM	Q-9	14	Ea.	2,425	23	2,448	2,700
114	1000 CFM		11		2,850	29	2,879	3,175
115	2000 CFM		8		3,450	40	3,490	3,850
116	3000 CFM		7		4,000	46	4,046	4,475
117	5000 CFM		5		5,200	64	5,264	5,825
118	8000 CFM		4		7,275	80	7,355	8,125
119	12,000 CFM		3		9,650	105	9,755	10,800
120	16,000 CFM		2.50		11,900	130	12,030	13,300

15.7 Air Conditioning & Ventilating

10

		CREW	DAILY OUTPUT	UNIT	BARE COSTS MAT.	BARE COSTS INST.	BARE COSTS TOTAL	TOTAL INCL O&P
121	20,000 CFM	Q-9	2	Ea.	14,900	160	15,060	16,600
122	26,000 CFM	"	1.50		17,200	215	17,415	19,200
123	30,000 CFM	Q-10	2		19,500	250	19,750	21,800
124	40,000 CFM	"	1.50		25,500	330	25,830	28,500
125	50,000 CFM	Q-11	2		31,100	340	31,440	34,700
126	55,000 CFM		1.80		33,900	375	34,275	37,800
127	60,000 CFM		1.50		36,600	455	37,055	40,900
130	Double pack filter, horizontal type							
131	500 CFM	Q-9	10	Ea.	3,150	32	3,182	3,500
132	1000 CFM		8		3,775	40	3,815	4,200
133	2000 CFM		6		4,550	53	4,603	5,075
134	3000 CFM		5		5,450	64	5,514	6,100
135	5000 CFM		4		6,900	80	6,980	7,700
136	8000 CFM		3		9,150	105	9,255	10,200
137	12,000 CFM		2.50		12,400	130	12,530	13,800
138	16,000 CFM		2		16,200	160	16,360	18,100
139	20,000 CFM		1.50		19,600	215	19,815	21,900
140	26,000 CFM		1		22,700	320	23,020	25,400
141	30,000 CFM	Q-10	1.50		26,600	330	26,930	29,700
142	40,000 CFM	"	1.30		34,100	385	34,485	38,100
143	50,000 CFM	Q-11	1.50		42,600	455	43,055	47,500
144	55,000 CFM		1.30		46,600	520	47,120	52,000
145	60,000 CFM		1		50,500	680	51,180	56,500
150	Single pack filter, vertical type							
151	500 CFM	Q-9	24	Ea.	2,300	13.30	2,313.30	2,550
152	1000 CFM		18		2,725	17.75	2,742.75	3,025
153	2000 CFM		12		3,275	27	3,302	3,650
154	3000 CFM		9		3,825	36	3,861	4,250
155	5000 CFM		6		5,100	53	5,153	5,700
156	8000 CFM		4		7,275	80	7,355	8,125
157	12,000 CFM		3		9,400	105	9,505	10,500
158	16,000 CFM		2		11,500	160	11,660	12,900
159	20,000 CFM		1.80		14,600	180	14,780	16,300
160	24,000 CFM		1.60		16,300	200	16,500	18,200
165	Double pack filter, vertical type							
166	500 CFM	Q-9	22	Ea.	3,000	14.55	3,014.55	3,325
167	1000 CFM		16		3,575	20	3,595	3,950
168	2000 CFM		10		4,300	32	4,332	4,775
169	3000 CFM		7		5,050	46	5,096	5,625
170	5000 CFM		5		6,625	64	6,689	7,375
171	8000 CFM		3		9,075	105	9,180	10,100
172	12,000 CFM		2.50		12,300	130	12,430	13,700
173	16,000 CFM		2		15,400	160	15,560	17,200
174	20,000 CFM		1.50		19,200	215	19,415	21,400
175	24,000 CFM		1		21,400	320	21,720	24,000
180	Inlet or outlet transition, horizontal							
181	Single pack to 12,000 CFM, add				4%			
182	Single pack to 30,000 CFM, add				6%			
183	Single pack to 60,000 CFM, add				8%			
184	Double pack to 12,000 CFM, add				3%			
185	Double pack to 30,000 CFM, add				5%			
186	Double pack to 60,000 CFM, add				6%			
187	Inlet or outlet transition, vertical							
188	Single pack to 5000 CFM, add				2%			
189	Single pack to 24,000 CFM, add				3%			
190	Double pack to 24,000 CFM, add				2%			
200	Electronic air cleaner, self-contained							
205	185 CFM	1 Shee	2.40	Ea.	185	74	259	310
210	200 CFM		2.30		365	77	442	515
215	500 CFM		2.30		520	77	597	685

15

15.7 Air Conditioning & Ventilating

Line	Description	CREW	DAILY OUTPUT	UNIT	BARE COSTS MAT.	BARE COSTS INST.	BARE COSTS TOTAL	TOTAL INCL O&P
10 220	1000 CFM	1 Shee	2.20	Ea.	715	81	796	905
225	1200 CFM		2.10		1,050	85	1,135	1,275
230	2500 CFM		2		1,200	89	1,289	1,450
240	In-line installation, duct or plenum							
242	Power pack, one required for each							
243	Installation, one or two filter cells	1 Shee	2	Ea.	195	89	284	345
244	Filter cell, 2" thick, 1" perimeter offset		4		190	44	234	275
245	20" x 20"		4		150	44	194	230
246	20" x 16"		4		135	44	179	215
247	25" x 20"		3.50		170	51	221	260
248	25" x 16"		3		150	59	209	250
295	Mechanical media filtration units							
300	High efficiency type, with frame, non-supported			MCFM	35		35	39M
310	Supported type				47		47	52M
400	Medium efficiency, extended surface				5.25		5.25	5.75M
450	Permanent washable				20		20	22M
500	Renewable disposable roll				70		70	77M
550	Throwaway glass or paper media type			Ea.	1.87		1.87	2.05M
13-001	**ANTI-FREEZE** Inhibited							
090	Ethylene glycol concentrated							
100	55 gallon drums, small quantities			Gal.	3.80		3.80	4.18M
120	Large quantities				2.45		2.45	2.69M
200	Propylene glycol, for solar heat, small quantities				5.80		5.80	6.40M
210	Large quantities				4.30		4.30	4.73M
16-001	**BALANCING, AIR** (Subcontract including material & labor)							
090	Heating and ventilating equipment							
100	Centrifugal fans, utility sets			Ea.	155		155	170
110	Heating and ventilating unit				235		235	260
120	In-line fan				235		235	260
130	Propeller and wall fan				44		44	48
140	Roof exhaust fan				105		105	115
200	Air conditioning equipment, central station				335		335	370
210	Built-up low pressure unit				310		310	340
220	Built-up high pressure unit				365		365	400
230	Built-up high pressure dual duct				570		570	625
240	Built-up variable volume				675		675	745
250	Multi-zone A.C. and heating unit				235		235	260
260	For each zone over one, add				52		52	57
270	Package A.C. unit				130		130	145
280	Rooftop heating and cooling unit				180		180	200
300	Supply, return, exhaust, registers & diffusers, avg. height ceiling				31		31	34
310	High ceiling				47		47	52
320	Floor height				26		26	29
330	Off mixing box				21		21	23
350	Induction unit				34		34	37
360	Lab fume hood				155		155	170
370	Linear supply				78		78	86
380	Linear supply high				91		91	100
400	Linear return				26		26	29
410	Light troffers				31		31	34
420	Moduline - master				31		31	34
430	Moduline - slaves				16		16	17.60
440	Regenerators				205		205	225
450	Taps into ceiling plenums				39		39	43
460	Variable volume boxes				31		31	34
19-001	**BALANCING, WATER** (Subcontract, including material & labor)							
005	Air cooled condenser			Ea.	90		90	99
010	Cabinet unit heater				31		31	34
020	Chiller				220		220	240

15.7 Air Conditioning & Ventilating

		CREW	DAILY OUTPUT	UNIT	BARE COSTS MAT.	BARE COSTS INST.	BARE COSTS TOTAL	TOTAL INCL O&P
19 030	Convector			Ea.	26		26	29
040	Converter				130		130	145
050	Cooling tower				170		170	185
060	Fan coil unit, unit ventilator				46		46	51
070	Fin tube and radiant panels				52		52	57
080	Main and duct re-heat coils				48		48	53
090	Main balancing cocks				39		39	43
100	Pumps				115		115	125
110	Unit heater			↓	36		36	40
22-001	**CENTRAL STATION AIR-HANDLING UNIT, CHILL WATER**							
100	Modular, capacity at 700 FPM face velocity							
110	1507 CFM	Q-5	1.20	Ea.	1,275	270	1,545	1,800
120	2196 CFM		1.10		1,450	290	1,740	2,025
130	3749 CFM	↓	.80		1,600	400	2,000	2,350
140	5440 CFM	Q-6	.80		2,500	625	3,125	3,650
150	8050 CFM		.60		3,250	835	4,085	4,775
160	12,110 CFM		.46		4,200	1,075	5,275	6,200
170	18,410 CFM		.33		6,125	1,525	7,650	8,925
180	22,330 CFM		.29		6,825	1,725	8,550	10,000
190	33,670 CFM	↓	.19		10,800	2,625	13,425	15,700
200	52,500 CFM	Q-7	.16		18,100	4,250	22,350	26,100
210	63,000 CFM	"	.13		22,000	5,250	27,250	31,800
220	For hot water heating coil, add				10%	5%		
300	Packaged, 2370-3950 CFM, 7.5 ton	Q-6	1.70		1,225	295	1,520	1,775
310	3210-5350 CFM, 10 ton		1.40		1,625	355	1,980	2,300
320	5070-9300 CFM, 15 ton		1		2,200	500	2,700	3,150
330	5410-9920 CFM, 20 ton		.80		2,650	625	3,275	3,825
340	8070-14,800 CFM, 30 ton	↓	.50		3,675	1,000	4,675	5,475
350	For hot water heat coil, add			↓	20%	5%		
25-001	**COILS, FLANGED**							
050	Chilled water cooling, 6 rows, 24" x 48"	Q-5	2	Ea.	1,700	160	1,860	2,100
100	Direct expansion cooling, 6 rows, 24" x 48"		2		1,850	160	2,010	2,275
150	Hot water heating, 1 row, 24" x 48"		3		670	105	775	890
200	Steam heating, 1 row, 24" x 48"	↓	3	↓	960	105	1,065	1,200
28-001	**COMPRESSORS**							
099	Refrigeration, reciprocating hermetic, switches & protective devices							
100	(Ratings are ARI standard 515 group IV using R-22) 2 ton	Q-5	1	Ea.	1,950	320	2,270	2,600
110	15 ton	Q-6	.72		3,875	695	4,570	5,275
120	20 ton		.64		4,400	780	5,180	5,975
130	40 ton		.44		4,825	1,125	5,950	6,950
140	50 ton	↓	.20		10,200	2,500	12,700	14,800
150	75 ton	Q-7	.27		13,100	2,525	15,625	18,100
160	130 ton	"	.21	↓	15,700	3,250	18,950	22,000
500	Air, diaphragm, w/skid, motor, drive, valves, switches & starter							
503	Single stage, 220 psig, 0.47 CFM	Q-5	3.30	Ea.	17,000	97	17,097	18,800
504	1.7 CFM		2		21,000	160	21,160	23,300
505	3.5 CFM		1.50		23,000	215	23,215	25,600
506	8.5 CFM	↓	1		33,000	320	33,320	36,800
510	14 CFM	Q-6	1		42,000	500	42,500	46,900
512	28 CFM		.60		57,000	835	57,835	64,000
513	56 CFM	↓	.30		90,000	1,675	91,675	101,500
515	Two stage, 3600 psig, 0.47 CFM	Q-5	2.50		26,000	130	26,130	28,800
516	1.7 CFM		1.50		32,000	215	32,215	35,500
517	3.5 CFM	↓	1.10		39,000	290	39,290	43,300
518	8.5 CFM	Q-6	1.10		53,000	455	53,455	59,000
519	14 CFM		.60		62,000	835	62,835	69,500
520	28 CFM	↓	.30	↓	91,000	1,675	92,675	102,500
525	Reciprocating air cooled heavy duty base mounted							

	15.7 Air Conditioning & Ventilating	CREW	DAILY OUTPUT	UNIT	BARE COSTS MAT.	BARE COSTS INST.	BARE COSTS TOTAL	TOTAL INCL O&P
28 535	2 stage motor & drive, 9 CFM at 150 psi 2 HP	Q-5	4	Ea.	1,500	80	1,580	1,775
536	12 CFM at 150 psi 3 HP		3.80		1,550	85	1,635	1,825
537	14 CFM at 150 psi 3 HP		3.40		1,700	94	1,794	2,000
538	21 CFM at 150 psi 5 HP		3.20		1,750	100	1,850	2,075
539	23 CFM at 150 psi 5 HP		3		2,100	105	2,205	2,475
540	34 CFM at 75 psi 7-1/2 HP		2.80		2,275	115	2,390	2,675
541	36 CFM at 150 psi 7-1/2 HP		2.30		2,925	140	3,065	3,425
542	49 CFM at 150 psi 10 HP		2.20		3,000	145	3,145	3,500
543	67 CFM at 150 psi 15 HP		2		3,800	160	3,960	4,400
544	78 CFM 150 psi 15 HP		1.60		4,225	200	4,425	4,950
545	104 CFM at 125 psi 20 HP		1.40		4,900	230	5,130	5,725
546	117 CFM at 150 psi 25 HP		1.20		5,575	270	5,845	6,525
547	132 CFM at 125 psi 30 HP		1		5,775	320	6,095	6,825
560	2 stage pkg. motor starter aftercooler recvr. drain							
565	115 CFM at 100 psi 25 HP	Q-5	.70	Ea.	8,675	460	9,135	10,200
567	180 CFM at 100 psi 40 HP		.60		11,300	535	11,835	13,200
568	230 CFM at 100 psi 50 HP		.50		14,500	640	15,140	16,900
569	360 CFM at 100 psi 75 HP		.40		19,700	805	20,505	22,800
580	Single stage motor & starter, low pressure							
585	631 CFM at 50 psi 100 HP	Q-6	.60	Ea.	31,400	835	32,235	35,700
586	738 CFM at 50 psi 125 HP	″	.50	″	35,700	1,000	36,700	40,700
600	Reciprocating water cooled heavy duty slow speed 100 psi							
605	Lubricated, 73 CFM by 20 HP motor	Q-5	.80	Ea.	10,600	400	11,000	12,200
606	100 CFM by 25 HP motor		.60		11,100	535	11,635	13,000
607	132 CFM by 30 HP motor		.50		13,500	640	14,140	15,800
608	185 CFM by 40 HP motor		.40		16,000	805	16,805	18,800
609	210 CFM by 50 HP motor	Q-6	.60		17,700	835	18,535	20,700
610	246 CFM by 50 HP motor		.50		21,000	1,000	22,000	24,500
611	330 CFM by 75 HP motor		.30		25,100	1,675	26,775	30,000
612	428 CFM by 100 HP motor	Q-7	.40		30,100	1,700	31,800	35,600
613	557 CFM by 125 HP motor	″	.30		35,900	2,275	38,175	42,800
620	Nonlubricated, 73 CFM by 20 HP motor	Q-5	.80		12,700	400	13,100	14,600
621	100 CFM by 25 HP motor		.60		13,300	535	13,835	15,400
622	132 CFM by 30 HP motor		.50		15,000	640	15,640	17,400
623	185 CFM by 40 HP motor		.40		18,000	805	18,805	21,000
625	210 CFM by 50 HP motor	Q-6	.60		21,000	835	21,835	24,300
626	246 CFM by 50 HP motor		.50		22,400	1,000	23,400	26,100
627	330 CFM by 75 HP motor		.30		28,300	1,675	29,975	33,500
628	428 CFM by 100 HP motor	Q-7	.40		33,800	1,700	35,500	39,600
629	557 CFM by 125 HP motor	″	.30		40,100	2,275	42,375	47,400
650	Rotary screw motor & starter aftercooler heavy duty,							
655	Air cooled single stage, 100 CFM at 115 psi 25 HP	Q-5	2.60	Ea.	6,775	125	6,900	7,625
656	120 CFM at 115 psi 30 HP		2		7,875	160	8,035	8,900
657	160 CFM at 115 psi 40 HP		1.40		9,075	230	9,305	10,300
658	215 CFM at 115 psi 50 HP in enclosure		1		13,100	320	13,420	14,900
659	260 CFM at 115 psi 60 HP in enclosure		.80		15,300	400	15,700	17,400
660	330 CFM at 115 psi 75 HP in enclosure	Q-6	1		17,200	500	17,700	19,600
661	490 CFM at 115 psi 100 HP		.90		18,900	555	19,455	21,600
662	600 CFM at 115 psi 125 HP		.80		22,500	625	23,125	25,700
663	710 CFM at 115 psi 150 HP		.60		24,700	835	25,535	28,400
664	950 CFM at 100 psi 200 HP	Q-7	.80		29,900	850	30,750	34,100
665	1000 CFM at 115 psi 250 HP		.70		38,600	975	39,575	43,900
666	1250 CFM at 115 psi 300 HP		.60		44,200	1,125	45,325	50,500
667	1500 CFM at 115 psi 350 HP		.50		53,000	1,375	54,375	60,500
668	2000 CFM at 100 psi 400 HP		.40		64,500	1,700	66,200	73,500
669	2000 CFM at 115 psi 450 HP		.30		70,000	2,275	72,275	80,500
680	Water cooled single stage, 100 CFM at 115 psi 25 HP	Q-5	1.60		6,775	200	6,975	7,750
681	120 CFM at 115 psi 30 HP		1.40		7,875	230	8,105	9,000
682	180 CFM at 100 psi 40 HP in enclosure		1		11,600	320	11,920	13,200
683	235 CFM at 100 psi 50 HP in enclosure		.80		12,800	400	13,200	14,700

15.7 Air Conditioning & Ventilating		CREW	DAILY OUTPUT	UNIT	BARE COSTS MAT.	BARE COSTS INST.	BARE COSTS TOTAL	TOTAL INCL O&P
28 684	285 CFM at 100 psi 60 HP in enclosure	Q-5	.60	Ea.	15,000	535	15,535	17,300
685	355 CFM at 100 psi 75 HP in enclosure	Q-6	1		17,000	500	17,500	19,400
686	515 CFM at 100 psi 100 HP		.80		18,400	625	19,025	21,100
687	640 CFM at 100 psi 125 HP		.70		22,000	715	22,715	25,200
688	750 CFM at 100 psi 150 HP		.60		23,900	835	24,735	27,500
690	950 CFM at 100 psi 200 HP	Q-7	.60		27,300	1,125	28,425	31,700
691	1000 CFM at 115 psi 250 HP		.50		35,600	1,375	36,975	41,100
692	1250 CFM at 115 psi 300 HP		.50		49,600	1,375	50,975	56,500
693	1500 CFM at 115 psi 350 HP		.40		49,300	1,700	51,000	56,500
694	2000 CFM at 100 psi 400 HP		.40		58,500	1,700	60,200	67,000
695	2000 CFM at 115 psi 450 HP		.30		62,000	2,275	64,275	71,500
715	Tanks with mounted compressor add							
718	60 gallon				300		300	330M
719	80 gallon				345		345	380M
720	120 gallon				375		375	415M
721	200 gallon				740		740	815M
722	240 gallon				835		835	920M
750	Vacuum pump, centrifugal motor coupling & guard base mounted							
753	Single stage, 0" to 25" Hg.							
760	15 CFM	Q-5	5	Ea.	1,500	64	1,564	1,750
761	35 CFM		4		1,600	80	1,680	1,875
762	65 CFM		3		2,375	105	2,480	2,775
763	110 CFM		2.80		2,625	115	2,740	3,050
764	115 CFM		2.50		2,775	130	2,905	3,250
765	250 CFM		1.60		4,675	200	4,875	5,425
766	310 CFM	Q-6	2.10		5,275	240	5,515	6,150
767	575 CFM		1.60		7,975	310	8,285	9,225
768	700 CFM		1.20		9,025	415	9,440	10,500
769	1050 CFM		1		13,500	500	14,000	15,600
775	Two stage, up to 29" Hg.							
780	15 CFM	Q-5	4	Ea.	1,600	80	1,680	1,875
781	35 CFM		3.40		1,800	94	1,894	2,125
782	50 CFM		3		1,875	105	1,980	2,225
783	75 CFM		2.80		1,950	115	2,065	2,300
784	100 CFM		2.50		2,825	130	2,955	3,300
785	140 CFM		2		3,100	160	3,260	3,650
786	200 CFM		1.60		4,500	200	4,700	5,250
787	310 CFM	Q-6	2		5,125	250	5,375	6,000
788	600 CFM		1		9,275	500	9,775	10,900
789	800 CFM		.80		12,400	625	13,025	14,500
790	1000 CFM		.60		15,500	835	16,335	18,300
15 29-001	**COMPRESSOR ACCESSORIES**							
010	Aftercooler							
013	Air cooled, based on 100 psig and 250°F							
014	Tank mounted single stage, 20 SCFM	Q-5	7.50	Ea.	450	43	493	555
015	46 SCFM		7		530	46	576	650
016	77 SCFM		6.50		880	49	929	1,050
017	Floor mounted horizontal draft, 36 SCFM		6.50		450	49	499	565
018	78 SCFM		6		560	54	614	695
019	190 SCFM		5.70		865	56	921	1,025
020	260 SCFM		5.20		975	62	1,037	1,150
021	440 SCFM		4		1,275	80	1,355	1,525
022	650 SCFM		3.50		1,400	92	1,492	1,675
023	Floor mounted vertical draft, 850 SCFM		2.50		2,425	130	2,555	2,850
026	2060 SCFM		2		3,800	160	3,960	4,400
028	3640 SCFM		1.70		9,300	190	9,490	10,500
030	Separator with automatic float trap							
033	600 lb/hr drain cap. 125 psi 3/4" conn.	Q-5	7.50	Ea.	225	43	268	310
034	1200 lb/hr drain capacity 125 psi 1" conn	"	7	"	235	46	281	325

15.7 Air Conditioning & Ventilating

29

		CREW	DAILY OUTPUT	UNIT	BARE COSTS MAT.	BARE COSTS INST.	BARE COSTS TOTAL	TOTAL INCL O&P
035	1200 lb/hr drain cap. 125 psi 1-1/2" conn	Q-5	6	Ea.	265	54	319	370
036	1200 lb/hr drain capacity 125 psi 2" conn		5		280	64	344	400
037	1500 CFM at 200 psi 2-1/2" conn		4		690	80	770	875
038	4000 CFM at 200 psi 4" conn		3		730	105	835	960
040	Water cooled, fixed tube at 250°F							
043	52 CFM	Q-5	5	Ea.	260	64	324	380
044	104 CFM		4		305	80	385	450
045	182 CFM		3.50		485	92	577	665
046	312 CFM		3		530	105	635	740
048	494 CFM		2.80		1,175	115	1,290	1,450
049	702 CFM		2.40		1,575	135	1,710	1,925
050	806 CFM		2.20		1,875	145	2,020	2,275
051	1092 CFM		2		2,300	160	2,460	2,750
052	1430 CFM		1.50		2,725	215	2,940	3,300
060	Separator, horizontal							
061	100 CFM	Q-5	15	Ea.	150	21	171	195
062	350 CFM		10		215	32	247	285
063	900 CFM		7		510	46	556	625
064	1500 CFM		5		565	64	629	715
065	2800 CFM		3.50		830	92	922	1,050
070	Dryers, air, deliquescent at 90 psi							
073	21CFM	Q-5	14	Ea.	460	23	483	540
074	63 CFM		12		875	27	902	1,000
075	140 CFM		8		1,475	40	1,515	1,675
076	335 CFM		5.50		2,125	58	2,183	2,425
077	525 CFM		5		2,825	64	2,889	3,200
078	760 CFM		3		3,975	105	4,080	4,525
079	1050 CFM		2.50		4,950	130	5,080	5,625
080	1340 CFM		2		5,975	160	6,135	6,800
081	1700 CFM		1.50		7,475	215	7,690	8,525
082	2550 CFM	Q-6	2		10,000	250	10,250	11,400
083	3550 CFM		1.50		12,500	335	12,835	14,200
084	4700 CFM		1		14,800	500	15,300	17,000
085	6100 CFM	Q-7	1		17,300	680	17,980	20,000
086	8400 CFM		.80		21,500	850	22,350	24,900
087	10200 CFM		.60		23,400	1,125	24,525	27,400
088	12100 CFM		.50		25,000	1,375	26,375	29,500
089	14200 CFM		.40		26,400	1,700	28,100	31,500
095	Desiccant 33 lb. bag				21		21	23M
096	55 lb. bag				33		33	36M
098	2200 lb. bulk bag 1 or 2				1,075		1,075	1,175M
099	2200 lb. bulk bag 3 or 4				970		970	1,075M
125	Refrigeration, 50°F pressure dewpoint							
128	20 CFM	Q-5	7	Ea.	920	46	966	1,075
129	60 CFM		6		1,550	54	1,604	1,775
130	130 CFM		5.50		2,150	58	2,208	2,450
131	150 CFM		5		2,775	64	2,839	3,150
132	200 CFM		4.50		2,925	71	2,996	3,325
133	280 CFM		4		3,575	80	3,655	4,050
134	350 CFM		3.50		3,875	92	3,967	4,400
135	450 CFM		3		4,600	105	4,705	5,225
136	520 CFM		2.80		5,100	115	5,215	5,775
137	650 CFM		2.40		5,725	135	5,860	6,500
138	800 CFM		2.20		6,650	145	6,795	7,525
139	1000 CFM		2		8,125	160	8,285	9,175
140	1400 CFM		1.50		9,750	215	9,965	11,000
141	2000 CFM		1.20		12,200	270	12,470	13,800
142	2400 CFM		1		16,300	320	16,620	18,400
143	3000 CFM	Q-6	1.30		22,100	385	22,485	24,900
144	3600 CFM	"	1.20		26,300	415	26,715	29,500

15

15.7 Air Conditioning & Ventilating

		CREW	DAILY OUTPUT	UNIT	BARE COSTS MAT.	BARE COSTS INST.	BARE COSTS TOTAL	TOTAL INCL O&P
29 145	4800 CFM	Q-6	1	Ea.	33,500	500	34,000	37,600
146	6000 CFM		.80		37,400	625	38,025	42,000
147	7200 CFM		.50		41,200	1,000	42,200	46,800
150	Package includes dryer & filtration systems							
153	150 CFM	Q-5	4.20	Ea.	6,175	76	6,251	6,900
154	200 CFM		3.90		6,500	82	6,582	7,275
155	280 CFM		3.50		7,350	92	7,442	8,225
156	350 CFM		3.20		7,625	100	7,725	8,525
157	450 CFM		3		9,500	105	9,605	10,600
158	520 CFM		2.50		10,500	130	10,630	11,700
159	650 CFM		2.20		10,800	145	10,945	12,100
160	800 CFM		2		11,400	160	11,560	12,800
161	1000 CFM		1.80		14,800	180	14,980	16,500
162	1200 CFM		1.40		16,500	230	16,730	18,500
163	1400 CFM		1.20		18,000	270	18,270	20,200
164	2000 CFM		1		20,800	320	21,120	23,300
165	2400 CFM		.70		26,200	460	26,660	29,500
200	Regenerative adsorption with desiccant							
203	50 CFM	Q-5	5	Ea.	4,400	64	4,464	4,925
204	75 CFM		4.50		4,650	71	4,721	5,225
205	100 CFM		4.20		5,100	76	5,176	5,725
206	125 CFM		4		5,400	80	5,480	6,050
207	170 CFM		3.70		5,800	87	5,887	6,500
208	240 CFM		3.40		7,000	94	7,094	7,825
209	350 CFM		3		7,700	105	7,805	8,625
210	520 CFM		2.50		8,725	130	8,855	9,775
211	650 CFM		2		9,950	160	10,110	11,200
212	800 CFM		1.50		12,500	215	12,715	14,100
213	1000 CFM		1.20		14,200	270	14,470	16,000
214	1200 CFM		1		15,000	320	15,320	17,000
215	1600 CFM		.80		20,300	400	20,700	22,900
216	2000 CFM		.70		21,500	460	21,960	24,300
217	2500 CFM		.60		27,600	535	28,135	31,100
218	3000 CFM	Q-6	.90		29,800	555	30,355	33,600
219	3500 CFM		.70		39,000	715	39,715	43,900
220	4000 CFM		.50		44,000	1,000	45,000	49,800
230	Desiccant, 50 lb. bag				85		85	94M
233	350 lb. drum				600		600	660M
300	Filters Air							
310	Adsorptive, odors by glass fiber & carbon, 90 psi							
314	110 CFM	Q-5	8	Ea.	545	40	585	660
315	190 CFM		7.50		620	43	663	745
316	290 CFM		7		770	46	816	915
317	380 CFM		6		1,050	54	1,104	1,225
318	570 CFM		5		1,475	64	1,539	1,725
319	870 CFM		4.50		1,675	71	1,746	1,950
320	1160 CFM		4		2,150	80	2,230	2,475
321	1450 CFM		3.50		2,525	92	2,617	2,900
322	1740 CFM		3		2,800	105	2,905	3,225
326	Coalescent system, liquids & solids 90 psi high purity							
330	110 CFM	Q-5	7	Ea.	2,050	46	2,096	2,325
331	190 CFM		6.70		2,200	48	2,248	2,500
332	290 CFM		6.50		2,350	49	2,399	2,650
333	380 CFM		6		2,750	54	2,804	3,100
334	570 CFM		4.50		3,450	71	3,521	3,900
335	870 CFM		4		3,650	80	3,730	4,125
336	1160 CFM		3.50		4,150	92	4,242	4,700
337	1450 CFM		3		4,675	105	4,780	5,300
338	1740 CFM		2.50		4,950	130	5,080	5,625
339	2030 CFM		2		5,125	160	5,285	5,875

15

15.7 Air Conditioning & Ventilating

	15.7 Air Conditioning & Ventilating	CREW	DAILY OUTPUT	UNIT	BARE COSTS MAT.	BARE COSTS INST.	BARE COSTS TOTAL	TOTAL INCL O&P
31-001	**COMPUTER ROOM UNITS**							
100	Air cooled, includes remote condenser but not							
102	interconnecting tubing or refrigerant							
108	3 ton	Q-5	.50	Ea.	6,100	640	6,740	7,650
112	B8.4 -810 5 ton		.45		6,700	715	7,415	8,400
116	6 ton		.30		10,500	1,075	11,575	13,100
120	8 ton		.27		11,500	1,200	12,700	14,400
124	10 ton		.25		12,600	1,275	13,875	15,700
128	15 ton		.22		14,700	1,450	16,150	18,300
132	20 ton	Q-6	.29		16,800	1,725	18,525	21,000
136	23 ton	"	.28		17,800	1,775	19,575	22,200
220	Chilled water, for connection to							
222	existing chiller system of adequate capacity							
226	5 ton	Q-5	.74	Ea.	7,000	435	7,435	8,325
228	6 ton		.52		10,000	620	10,620	11,900
230	8 ton		.50		11,000	640	11,640	13,000
232	10 ton		.49		12,100	655	12,755	14,300
236	15 ton		.48		13,600	670	14,270	15,900
240	20 ton		.46		15,700	700	16,400	18,300
244	23 ton		.42		16,800	765	17,565	19,600
400	Glycol system, complete except for interconnecting tubing							
406	3 ton	Q-5	.40	Ea.	8,600	805	9,405	10,600
410	5 ton		.38		9,200	845	10,045	11,300
412	6 ton		.25		13,600	1,275	14,875	16,800
414	8 ton		.23		14,700	1,400	16,100	18,200
416	10 ton		.21		15,700	1,525	17,225	19,500
420	15 ton	Q-6	.26		18,900	1,925	20,825	23,600
424	20 ton		.24		21,000	2,075	23,075	26,100
428	23 ton		.22		22,000	2,275	24,275	27,500
800	Water cooled system, not including condenser,							
802	water supply or cooling tower							
806	3 ton	Q-5	.62	Ea.	6,075	520	6,595	7,425
810	5 ton		.54		7,000	595	7,595	8,550
812	6 ton		.35		10,800	915	11,715	13,200
814	8 ton		.33		11,400	975	12,375	13,900
816	10 ton		.31		12,100	1,025	13,125	14,800
820	15 ton		.27		14,100	1,200	15,300	17,200
824	20 ton	Q-6	.38		15,400	1,325	16,725	18,800
828	23 ton	"	.34		17,200	1,475	18,675	21,000
34-001	**CONDENSERS** Ratings are for 30°F TD, R-22							
008	Air cooled, belt drive, propeller fan							
010	20 ton	Q-5	1	Ea.	3,425	320	3,745	4,225
018	30 ton	"	.60		3,550	535	4,085	4,675
020	B8.4 -110 37 ton	Q-6	.60		4,425	835	5,260	6,075
022	41 ton		.54		4,650	925	5,575	6,450
024	B8.4 -240 48 ton		.48		5,450	1,050	6,500	7,500
026	51 ton		.40		5,600	1,250	6,850	7,975
028	B8.4 -250 60 ton		.32		6,200	1,550	7,750	9,075
030	66 ton		.30		6,975	1,675	8,650	10,100
032	72 ton		.29		7,600	1,725	9,325	10,800
034	77 ton		.28		8,325	1,775	10,100	11,700
036	82 ton		.27		8,875	1,850	10,725	12,400
038	90 ton		.26		9,775	1,925	11,700	13,500
040	100 ton	Q-7	.30		13,100	2,275	15,375	17,700
050	157 ton		.20		21,200	3,400	24,600	28,200
060	235 ton		.12		32,600	5,675	38,275	44,100
070	314 ton		.09		43,800	7,575	51,375	59,000
080	471 ton		.06		51,000	11,400	62,400	72,500
090	630 ton		.05		82,000	13,600	95,600	110,000

15

	15.7 Air Conditioning & Ventilating	CREW	DAILY OUTPUT	UNIT	BARE COSTS MAT.	BARE COSTS INST.	BARE COSTS TOTAL	TOTAL INCL O&P
34 100	840 ton	Q-7	.04	Ea.	110,000	17,000	127,000	145,500
150	May be specified single or multi-circuit							
155	Air cooled, direct drive, propeller fan							
160	1-1/2 ton	Q-5	3.60	Ea.	325	89	414	485
162	2 ton		3.20		390	100	490	575
163	3 ton		2.40		585	135	720	835
164	5 ton		2		1,025	160	1,185	1,350
165	8 ton		1.80		1,225	180	1,405	1,600
166	10 ton		1.40		1,650	230	1,880	2,150
167	12 ton		1.30		1,775	245	2,020	2,300
168	14 ton		1.20		1,950	270	2,220	2,525
169	16 ton		1.10		2,100	290	2,390	2,725
170	21 ton		1		2,425	320	2,745	3,125
172	26 ton		.74		3,000	435	3,435	3,925
174	30 ton		.60		3,450	535	3,985	4,575
176	41 ton	Q-6	.54		4,650	925	5,575	6,450
178	52 ton		.40		5,550	1,250	6,800	7,900
180	63 ton		.32		6,700	1,550	8,250	9,625
182	76 ton		.28		7,700	1,775	9,475	11,000
184	86 ton		.27		9,700	1,850	11,550	13,300
186	97 ton		.25		10,900	2,000	12,900	14,900
188	105 ton	Q-7	.30		12,600	2,275	14,875	17,100
200	Centrifugal fan, integral receiver							
210	3 ton	Q-5	1.40	Ea.	1,850	230	2,080	2,375
220	7.5 ton		.90		2,675	355	3,030	3,450
230	10 ton		.70		3,450	460	3,910	4,450
240	15 ton		.60		3,775	535	4,310	4,925
250	20 ton		.50		5,400	640	6,040	6,875
260	30 ton	Q-6	.55		7,150	910	8,060	9,175
270	40 ton		.28		12,700	1,775	14,475	16,500
280	50 ton		.24		16,100	2,075	18,175	20,700
290	60 ton		.21		18,400	2,375	20,775	23,700
340	Evaporative, copper coil, pump, fan motor							
344	10 ton	Q-5	.54	Ea.	3,000	595	3,595	4,150
346	15 ton		.50		3,125	640	3,765	4,375
348	20 ton		.47		3,250	685	3,935	4,550
350	25 ton		.45		3,475	715	4,190	4,850
352	30 ton		.42		3,700	765	4,465	5,175
354	40 ton	Q-6	.49		4,900	1,025	5,925	6,875
356	50 ton		.39		6,125	1,275	7,400	8,600
358	65 ton		.35		6,800	1,425	8,225	9,550
360	80 ton		.33		7,450	1,525	8,975	10,400
362	90 ton		.29		8,125	1,725	9,850	11,400
364	100 ton	Q-7	.36		8,850	1,900	10,750	12,500
366	110 ton		.33		9,700	2,075	11,775	13,700
368	125 ton		.30		10,600	2,275	12,875	14,900
370	135 ton		.28		11,400	2,425	13,825	16,100
372	150 ton		.25		12,500	2,725	15,225	17,700
374	165 ton		.23		13,800	2,975	16,775	19,500
376	185 ton		.22		14,600	3,100	17,700	20,500
386	For fan damper control, add	Q-5	2		310	160	470	575
37-001	**CONDENSING UNITS**							
003	Air cooled, compressor, standard controls							
005	1.5 ton	Q-5	2.50	Ea.	515	130	645	750
010	2 ton		2.10		565	155	720	840
020	2.5 ton		1.70		670	190	860	1,000
030	3 ton		1.30		845	245	1,090	1,275
040	4 ton		.90		1,150	355	1,505	1,775
050	5 ton		.60		2,425	535	2,960	3,450

15.7	Air Conditioning & Ventilating	CREW	DAILY OUTPUT	UNIT	BARE COSTS MAT.	BARE COSTS INST.	BARE COSTS TOTAL	TOTAL INCL O&P
37 055	7.5 ton	Q-5	.55	Ea.	2,400	585	2,985	3,475
060	10 ton	"	.50		3,225	640	3,865	4,475
070	20 ton	Q-6	.40		5,750	1,250	7,000	8,125
080	40 ton		.20		11,000	2,500	13,500	15,700
090	75 ton		.14		26,100	3,575	29,675	33,900
100	80 ton		.12		27,900	4,175	32,075	36,700
110	100 ton	↓	.09	↓	33,700	5,550	39,250	45,100
200	Water cooled, compressor, heat exchanger, controls							
210	5 ton	Q-5	.70	Ea.	1,875	460	2,335	2,725
220	15 ton	"	.50		3,550	640	4,190	4,825
230	20 ton	Q-6	.40		5,175	1,250	6,425	7,500
240	40 ton		.20		8,450	2,500	10,950	12,900
250	100 ton	↓	.11	↓	22,800	4,550	27,350	31,600
40-001	**CONTROLS COMPONENTS**							
070	Controller, receiver							
073	Panel mount, single input	1 Plum	8	Ea.	145	22	167	190
074	With conversion mounting bracket		8		150	22	172	195
075	Dual input, with control point adjustment	↓	7		160	25	185	215
085	Electric, single snap switch	1 Elec	4		205	44	249	290
086	Dual snap switches	"	3	↓	270	59	329	380
100	Enthalpy control, boiler water temperature control							
101	governed by outdoor temperature, with timer	1 Elec	3	Ea.	125	59	184	220
200	Gauges, pressure or vacuum							
210	2" diameter dial	1 Stpi	32	Ea.	8.35	5.60	13.95	17.25
220	2-1/2" diameter dial		32		9.85	5.60	15.45	18.90
230	3-1/2" diameter dial		32		12.65	5.60	18.25	22
240	4-1/2" diameter dial	↓	32	↓	16.15	5.60	21.75	26
270	Flanged iron case, black ring							
280	3-1/2" diameter dial	1 Stpi	32	Ea.	46	5.60	51.60	59
290	4-1/2" diameter dial		32		52	5.60	57.60	65
300	6" diameter dial		32		86	5.60	91.60	105
310	8-1/2" diameter dial		30		86	5.95	91.95	105
320	12" diameter dial	↓	28		210	6.35	216.35	240
330	For compound pressure-vacuum,add			↓	18%			
335	Humidistat							
339	Electric operated	1 Shee	8	Ea.	32	22	54	68
340	Relays							
343	Pneumatic/electric	1 Plum	16	Ea.	105	11.10	116.10	130
344	Pneumatic proportioning		8		45	22	67	82
345	Pneumatic switching		12		47	14.80	61.80	73
346	Selector, 3 point		6		48	30	78	96
347	Time delay	↓	8	↓	60	22	82	98
350	Sensor, air operated							
352	Humidity	1 Plum	16	Ea.	160	11.10	171.10	190
354	Pressure		16		64	11.10	75.10	86
356	Temperature	↓	12	↓	105	14.80	119.80	135
360	Electric operated							
362	Humidity	1 Elec	8	Ea.	32	22	54	67
365	Pressure		8		375	22	397	445
368	Temperature	↓	10	↓	48	17.70	65.70	78
400	Thermometers							
410	Dial type, 3-1/2" diameter, vapor type, union connection	1 Stpi	32	Ea.	82	5.60	87.60	98
412	Liquid type, union connection		32		115	5.60	120.60	135
450	Stem type, 6-1/2" case, 2" stem, 1/2" NPT		32		24	5.60	29.60	34
452	4" stem, 1/2" NPT		32		32	5.60	37.60	43
460	9" case, 3-1/2" stem, 3/4" NPT		28		53	6.35	59.35	68
462	6" stem, 3/4" NPT		28		57	6.35	63.35	72
464	8" stem, 3/4" NPT		28		68	6.35	74.35	84
466	12" stem, 1" NPT	↓	26	↓	82	6.85	88.85	100

15

15.7 Air Conditioning & Ventilating

		CREW	DAILY OUTPUT	UNIT	BARE COSTS MAT.	BARE COSTS INST.	BARE COSTS TOTAL	TOTAL INCL O&P
40 500	Thermostats							
503	1 set back, manual	1 Shee	8	Ea.	40	22	62	76
504	1 set back, electric, timed		8		63	22	85	100
505	2 set back, electric, timed		8		64	22	86	105
510	Locking cover				9.15		9.15	10.05M
520	24 hour, automatic, clock	1 Shee	8		71	22	93	110
522	Electric, 2 wire	1 Elec	13		9.90	13.60	23.50	30
523	3 wire	″	10		12.50	17.70	30.20	39
525	Pneumatic, 2 pipe	1 Plum	10		86	17.75	103.75	120
530	Transmitter, pneumatic							
532	Temperature averaging element	Q-1	8	Ea.	175	40	215	250
535	Pressure differential	1 Plum	7		340	25	365	410
537	Humidity, duct		8		115	22	137	160
538	Room		12		115	14.80	129.80	150
539	Temperature, with averaging element		6		77	30	107	125
542	Electric operated, humidity	1 Elec	8		23	22	45	57
543	DPST	″	8		27	22	49	61
600	Valves, motorized zone							
610	Sweat connections, 1/2" C x C	1 Stpi	20	Ea.	31	8.90	39.90	47
611	3/4" C x C		20		31	8.90	39.90	47
612	1" C x C		19		35	9.40	44.40	52
614	1/2" C x C, with end switch, 2 wire		20		36	8.90	44.90	52
615	3/4" C x C, with end switch, 2 wire		20		36	8.90	44.90	52
616	1" C x C, with end switch, 2 wire		19		40	9.40	49.40	58
618	1-1/4" C x C, w/end switch, 2 wire		15		44	11.90	55.90	66
700	Threaded connections, 1/2" NPT		20		32	8.90	40.90	48
701	3/4" NPT		20		32	8.90	40.90	48
709	Valves, motor controlled							
710	Electric motor actuated							
720	Brass, two way, screwed							
721	1/2" pipe size	L-6	36	Ea.	67	7.40	74.40	84
722	3/4" pipe size		30		72	8.85	80.85	92
723	1" pipe size		28		80	9.50	89.50	100
724	1-1/2" pipe size		19		105	14	119	135
725	2" pipe size		16		130	16.65	146.65	165
735	Brass, three way, screwed							
736	1/2" pipe size	L-6	33	Ea.	83	8.05	91.05	105
737	3/4" pipe size		27		105	9.85	114.85	130
738	1" pipe size		25.50		120	10.45	130.45	145
739	1-1/2" pipe size		17		145	15.65	160.65	180
740	2" pipe size		14		180	19	199	225
755	Iron body, two way, flanged							
756	2-1/2" pipe size	L-6	4	Ea.	215	67	282	330
757	3" pipe size		3		280	89	369	435
758	4" pipe size		2		530	135	665	775
759	6" pipe size		1.50		1,750	175	1,925	2,175
785	Iron body, three way, flanged							
786	2-1/2" pipe size	L-6	3	Ea.	495	89	584	670
787	3" pipe size		2.50		680	105	785	900
788	4" pipe size		2		815	135	950	1,100
789	6" pipe size		1.50		2,050	175	2,225	2,500
800	Pneumatic, air operated							
805	Brass, two way, screwed							
806	1/2" pipe size	1 Plum	24	Ea.	220	7.40	227.40	255
807	3/4" pipe size		20		230	8.90	238.90	265
808	1" pipe size		19		350	9.35	359.35	400
809	1-1/4" pipe size		15		390	11.85	401.85	445
810	1-1/2" pipe size		13		420	13.65	433.65	480
811	2" pipe size		11		615	16.15	631.15	700
818	Brass, three way, screwed							

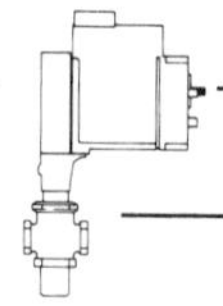

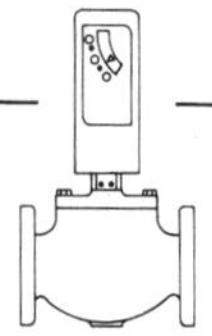

15

	15.7 Air Conditioning & Ventilating	CREW	DAILY OUTPUT	UNIT	BARE COSTS MAT.	BARE COSTS INST.	BARE COSTS TOTAL	TOTAL INCL O&P
40 819	1/2" pipe size	1 Plum	22	Ea.	115	8.05	123.05	140
820	3/4" pipe size		18		115	9.85	124.85	140
821	1" pipe size		17		130	10.45	140.45	160
822	1-1/2" pipe size		11		160	16.15	176.15	200
823	2" pipe size		9		215	19.75	234.75	265
845	Iron body, two way, flanged							
846	2-1/2" pipe size	Q-1	5	Ea.	275	64	339	395
847	3" pipe size		4.50		400	71	471	545
848	4" pipe size		3		770	105	875	1,000
849	6" pipe size	Q-2	3		1,200	165	1,365	1,550
856	Iron body, three way, flanged							
857	2-1/2" pipe size	Q-1	4.50	Ea.	615	71	686	780
858	3" pipe size		4		780	80	860	975
859	4" pipe size		2.50		1,200	130	1,330	1,500
860	6" pipe size	Q-2	3		1,575	165	1,740	1,975
43-001	**CONTROL SYSTEMS, PNEUMATIC** (Subcontract includes material & labor)							
010	Heating and Ventilating, split system							
020	Mixed air control, economizer cycle, panel readout, tubing							
022	Up to 10 tons			Ea.	2,375		2,375	2,625
024	For 10 to 20 tons, add				15%			
026	For over 20 tons, add				25%			
027	Enthalpy cycle, up to 10 tons				2,925		2,925	3,225
028	For 10 to 20 tons, add				15%			
029	For over 20 tons, add				25%			
030	Heating coil, hot water, 3 way valve,							
032	freezestat, limit control on discharge readout			Ea.	1,800		1,800	1,975
050	Cooling coil, chilled water, room							
052	thermostat, 3 way valve			Ea.	695		695	765
060	Cooling tower, fan cycle, damper control,							
062	condenser, water readout in/out at panel			Ea.	1,825		1,825	2,000
100	Unit ventilator, day/night operation,							
110	freezestat, ASHRAE, cycle 2			Ea.	1,750		1,750	1,925
200	Compensated hot water from boiler, valve control,							
210	readout and reset at panel, up to 60 GPM			Ea.	2,400		2,400	2,650
212	For 120 GPM, add				7%			
214	For 240 GPM, add				12%			
300	Boiler room combustion air, damper, controls				850		850	935
350	Fan coil, heating and cooling valves, 4 pipe system				765		765	840
390	Multizone control (one per zone), includes thermostat, damper							
391	motor and reset of discharge temperature			Ea.	1,500		1,500	1,650
400	Pneumatic thermostat, controlling room radiator valve				230		230	255
404	Program energy saving optimizer				3,225		3,225	3,550
408	Reheat coil				545		545	600
450	Air supply for pneumatic control system							
460	Tank mounted duplex compressor, starter, alternator,							
462	piping, dryer, PRV station and filter							
463	1/2 HP			Ea.	5,475		5,475	6,025
464	3/4 HP				5,700		5,700	6,275
465	1 HP				5,825		5,825	6,400
466	1-1/2 HP				7,175		7,175	7,900
468	3 HP				7,975		7,975	8,775
469	5 HP				13,100		13,100	14,400
480	Main air supply, includes 3/8" copper main and labor			C.L.F.	230		230	255
481	If poly tubing used, deduct				30%			
700	Static pressure control for air handling unit, includes pressure							
701	sensor, receiver controller, readout and damper motors			Ea.	2,600		2,600	2,850
702	If return air fan requires control, add				70%			
860	VAV boxes, incl. thermostat, damper motor, reheat coil & tubing				800		800	880
861	If no reheat coil, deduct				110		110	120

15

15.7 Air Conditioning & Ventilating

		CREW	DAILY OUTPUT	UNIT	BARE COSTS MAT.	BARE COSTS INST.	BARE COSTS TOTAL	TOTAL INCL O&P
46-001	**CONTROL SYSTEMS, ELECTRONIC**							
002	Pneumatic, costs, add to division 15.7-43			Ea.	15%			
49-001	**COOLING TOWERS**							
008	Draw thru, single flow							
010	B8.4 -120 Belt drive, 60 tons	Q-6	16	Ton	58	31	89	110
015	90 tons		20		53	25	78	94
020	B8.4 -230 100 tons		20		50	25	75	91
100	For higher capacities, use multiples							
150	Induced air, double flow							
180	Gear drive, 125 ton	Q-6	22	Ton	49	23	72	87
190	150 ton		22		46	23	69	83
200	300 ton		24		43	21	64	77
210	600 ton		26		34	19.20	53.20	65
215	840 ton		28		28	17.85	45.85	57
220	Up to 1000 tons		28		27	17.85	44.85	55
300	For higher capacities, use multiples							
350	For pumps and piping, add	Q-6	38	Ton	31.90	13.15	45.05	54
400	For absorption systems, add				75%	75%		
450	For rigging, add per 5 floors of building height (subcontract)					38	38	55L
50-001	**CURBS/PADS PREFABRICATED**							
600	Pad, fiberglass reinforced concrete with polystyrene foam core							
605	Condenser, 2" thick, 20" x 54"	1 Shee	8	Ea.	13.90	22	35.90	48
607	24" x 24"		16		7.40	11.10	18.50	24
609	24" x 36"		12		11.10	14.80	25.90	34
611	24" x 48"		8		14.80	22	36.80	49
615	26" x 56"	Q-9	8		18.70	40	58.70	79
617	28" x 38"	1 Shee	8		13.70	22	35.70	47
619	30" x 30"		12		11.60	14.80	26.40	34
622	30" x 36"		8		13.90	22	35.90	48
624	30" x 48"	Q-9	8		18.50	40	58.50	79
626	30" x 56"		7		21	46	67	90
628	36" x 36"		8		16.60	40	56.60	77
630	36" x 48"		7		22	46	68	91
632	36" x 60"		7		28	46	74	97
634	36" x 72"		6		33	53	86	115
636	40" x 42"		7		22	46	68	91
640	40" x 72"		6		37	53	90	120
643	45" x 50"		7		29	46	75	98
646	45" x 58"		6		34	53	87	115
649	26" round		10		8.80	32	40.80	56
655	30" round		9		11.10	36	47.10	64
660	36" round		8		16.60	40	56.60	77
670	For 3" thick pad, add				20%	5%		
750	Heat pump							
755	16" x 28", 4" thick	1 Shee	8	Ea.	16.30	22	38.30	50
760	16" x 28", 8" thick		6		21	30	51	66
765	22" x 36", 4" thick		6		23	30	53	68
770	24" round, 6" thick		8		18	22	40	52
880	Stand, concrete coated polystyrene foam							
885	Heat pump, 6" high	1 Shee	32	Ea.	2.80	5.55	8.35	11.15
887	10" high	"	32	"	3.35	5.55	8.90	11.80
51-001	**DEHUMIDIFIERS**							
600	Selfcontained with filters and standard controls							
604	1 lb./hr.	1 Plum	8	Ea.	1,600	22	1,622	1,800
606	3 lb./hr.	Q-1	12		4,000	27	4,027	4,450
607	12 lb/hr		5		7,000	64	7,064	7,800
608	3-40 lb/hr		4		12,000	80	12,080	13,300
609	10-60 lb/hr		3		18,000	105	18,105	20,000
610	20-120 lb/hr		2		26,000	160	26,160	28,800

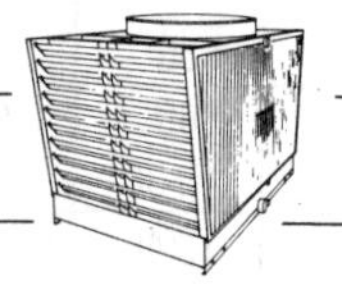

15

15.7 Air Conditioning & Ventilating

Line	Description	Crew	Daily Output	Unit	Bare Costs Mat.	Bare Costs Inst.	Bare Costs Total	Total Incl O&P
51 611	40-300 lb/hr	Q-1	1.50	Ea.	41,000	215	41,215	45,400
52-001	**DIFFUSERS** Aluminum, opposed blade damper unless noted							
010	Ceiling, linear, also for sidewall							
012	C8.4 -006 C8.4 -007 2" wide	1 Shee	32	L.F.	14.40	5.55	19.95	24
014	3" wide		30		17.10	5.90	23	27
016	4" wide		26		18.55	6.85	25.40	30
018	6" wide		24		23	7.40	30.40	36
020	8" wide		22		36	8.05	44.05	51
022	10" wide		20		40	8.90	48.90	57
024	12" wide		18		45	9.85	54.85	64
026	For floor or sill application, add				15%			
050	Perforated, 24" x 24", panel size 6" x 6"	1 Shee	16	Ea.	41	11.10	52.10	61
052	8" x 8"		15		43	11.85	54.85	65
054	10" x 10"		14		45	12.70	57.70	68
056	12" x 12"		12		46	14.80	60.80	72
058	15" x 15"		11		50	16.15	66.15	79
060	18" x 18"		10		53	17.75	70.75	84
100	Rectangular, 1 to 4 way blow, 6" x 6"		16		25	11.10	36.10	44
102	12" x 6"		15		32	11.85	43.85	52
104	12" x 9"		14		39	12.70	51.70	61
106	12" x 12"		12		44	14.80	58.80	70
108	18" x 12"		11		58	16.15	74.15	87
110	24" x 12"		10		73	17.75	90.75	105
112	15" x 15"		10		63	17.75	80.75	95
114	18" x 15"		9		73	19.75	92.75	110
116	21" x 21"		8		105	22	127	150
118	24" x 24"		7		130	25	155	180
150	Round, butterfly damper, 6" diameter		18		10.70	9.85	20.55	26
152	8" diameter		16		11.80	11.10	22.90	29
154	10" diameter		14		13.90	12.70	26.60	34
156	12" diameter		12		18	14.80	32.80	41
158	14" diameter		10		19.55	17.75	37.30	47
160	18" diameter		9		98	19.75	117.75	135
162	22" diameter		8		130	22	152	175
200	T bar mounting, 24" x 24" lay-in frame, 6" x 6"		16		43	11.10	54.10	63
202	9" x 9"		14		49	12.70	61.70	72
204	12" x 12"		12		63	14.80	77.80	91
206	15" x 15"		11		82	16.15	98.15	115
208	18" x 18"		10		81	17.75	98.75	115
250	Combination supply and return							
252	21" x 21" supply, 15" x 15" return	Q-9	10	Ea.	205	32	237	270
254	24" x 24" supply, 18" x 18" return		9.50		235	34	269	310
256	27" x 27" supply, 18" x 18" return		9		275	36	311	355
258	30" x 30" supply, 21" x 21" return		8.50		330	38	368	420
260	33" x 33" supply, 24" x 24" return		8		335	40	375	425
262	36" x 36" supply, 24" x 24" return		7.50		405	43	448	510
300	Baseboard, white enameled steel							
310	18" long	1 Shee	20	Ea.	6.20	8.90	15.10	19.75
312	24" long		18		9.15	9.85	19	24
314	48" long		16		14.25	11.10	25.35	32
330	For connector, add				1.40		1.40	1.54M
340	For matching return, deduct				10%			
400	Floor,steel,adjustable pattern							
410	2"x 10"	1 Shee	34	Ea.	5.95	5.20	11.15	14.15
412	2" x 12"		32		6.55	5.55	12.10	15.30
414	2" x 14"		30		7.45	5.90	13.35	16.80
420	4" x 10"		28		6.85	6.35	13.20	16.80
422	4" x 12"		26		7.65	6.85	14.50	18.35
424	4" x 14"		25		8.65	7.10	15.75	19.85

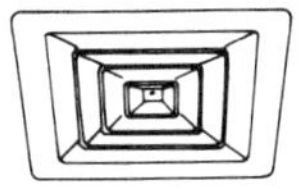

15.7 Air Conditioning & Ventilating

		CREW	DAILY OUTPUT	UNIT	BARE COSTS MAT.	BARE COSTS INST.	BARE COSTS TOTAL	TOTAL INCL O&P
52 426	6" x 10"	1 Shee	26	Ea.	10.35	6.85	17.20	21
428	6" x 12"		24		10.75	7.40	18.15	23
430	6" x 14"		22		11.35	8.05	19.40	24
500	Sidewall, aluminum, 3 way dispersion							
510	8" x 4"	1 Shee	24	Ea.	4.70	7.40	12.10	15.95
511	8" x 6"		24		4.60	7.40	12	15.85
512	10" x 4"		22		4.30	8.05	12.35	16.50
513	10" x 6"		22		4.60	8.05	12.65	16.85
514	10" x 8"		20		5.80	8.90	14.70	19.30
516	12" x 4"		18		4.60	9.85	14.45	19.45
518	12" x 8"		16		6.45	11.10	17.55	23
520	14" x 4"		15		4.95	11.85	16.80	23
522	14" x 6"		14		5.55	12.70	18.25	25
524	14" x 8"		13		6.80	13.65	20.45	27
600	For steel diffusers instead of aluminum, deduct				10%			
55-001	**GRILLES**							
002	Aluminum							
010	Air supply, single deflection, adjustable							
012	8" x 4"	1 Shee	30	Ea.	9.50	5.90	15.40	19.10
014	8" x 8"		28		11.55	6.35	17.90	22
016	10" x 4"		24		10.55	7.40	17.95	22
018	10" x 10"		23		14.70	7.70	22.40	27
020	12" x 6"		23		12.65	7.70	20.35	25
022	12" x 12"		22		16.95	8.05	25	30
024	14" x 8"		23		15.90	7.70	23.60	29
026	14" x 14"		22		21	8.05	29.05	35
028	18" x 10"		23		21	7.70	28.70	34
030	18" x 18"		21		34	8.45	42.45	50
032	20" x 12"		22		24	8.05	32.05	38
034	20" x 20"		21		39	8.45	47.45	55
036	24" x 14"		18		31	9.85	40.85	48
038	24" x 24"		15		50	11.85	61.85	72
040	30" x 8"		20		28	8.90	36.90	44
042	30" x 10"		19		31	9.35	40.35	48
044	30" x 12"		18		34	9.85	43.85	52
046	30" x 16"		17		39	10.45	49.45	58
048	30" x 18"		17		51	10.45	61.45	71
050	30" x 30"		14		70	12.70	82.70	95
052	36" x 12"		17		40	10.45	50.45	59
054	36" x 16"		16		47	11.10	58.10	68
056	36" x 18"		15		61	11.85	72.85	84
058	36" x 24"		14		71	12.70	83.70	97
060	36" x 28"		13		78	13.65	91.65	105
062	36" x 30"		12		82	14.80	96.80	110
064	36" x 32"		12		86	14.80	100.80	115
066	36" x 34"		11		100	16.15	116.15	135
068	36" x 36"		11		105	16.15	121.15	140
070	For double deflecting, add				70%			
100	Air return, 6" x 6"	1 Shee	26		5.45	6.85	12.30	15.95
102	10" x 6"		24		5.85	7.40	13.25	17.20
104	14" x 6"		23		7.15	7.70	14.85	19.10
106	10" x 8"		23		6.35	7.70	14.05	18.25
108	16" x 8"		22		8.40	8.05	16.45	21
110	12" x 12"		22		8.40	8.05	16.45	21
112	24" x 12"		18		15.30	9.85	25.15	31
114	30" x 12"		16		17.90	11.10	29	36
116	14" x 14"		22		10.30	8.05	18.35	23
118	16" x 16"		22		12.65	8.05	20.70	26
120	18" x 18"		21		16.50	8.45	24.95	30

15.7 Air Conditioning & Ventilating

		CREW	DAILY OUTPUT	UNIT	BARE COSTS MAT.	BARE COSTS INST.	BARE COSTS TOTAL	TOTAL INCL O&P
55 122	24" x 18"	1 Shee	16	Ea.	19.95	11.10	31.05	38
124	36" x 18"		15		30	11.85	41.85	50
126	24" x 24"		15		27	11.85	38.85	47
128	36" x 24"		14		43	12.70	55.70	66
130	48" x 24"		12		60	14.80	74.80	88
132	48" x 30"		11		78	16.15	94.15	110
134	36" x 36"		13		75	13.65	88.65	100
136	48" x 36"		11		95	16.15	111.15	130
138	48" x 48"		8		135	22	157	180
200	Door grilles, 12" x 12"		22		14.80	8.05	22.85	28
202	18" x 12"		22		19.15	8.05	27.20	33
204	24" x 12"		18		22	9.85	31.85	39
206	18" x 18"		18		28	9.85	37.85	45
208	24" x 18"		16		35	11.10	46.10	55
210	24" x 24"		15		48	11.85	59.85	70
300	Filter grille with filter, 12" x 12"		24		14.10	7.40	21.50	26
302	18" x 12"		20		19.80	8.90	28.70	35
304	24" x 18"		18		34	9.85	43.85	52
306	24" x 24"		16		46	11.10	57.10	67
308	30" x 24"		14		59	12.70	71.70	83
310	30" x 30"		13		76	13.65	89.65	105
395	Eggcrate, framed, 6" x 6" opening		26		8.05	6.85	14.90	18.80
396	10" x 10" opening		23		10.10	7.70	17.80	22
397	12" x 12" opening		22		12.35	8.05	20.40	25
398	14" x 14" opening		22		15	8.05	23.05	28
399	18" x 18" opening		21		23	8.45	31.45	38
402	22" x 22" opening		17		33	10.45	43.45	52
404	24" x 24" opening		15		39	11.85	50.85	60
405	48" x 24" opening		12		88	14.80	102.80	120
406	Eggcrate, lay-in, T-bar system							
407	48" x 24" sheet	1 Shee	40	Ea.	26	4.44	30.44	35
500	Transfer grille, vision proof, 8" x 4"		30		13.10	5.90	19	23
502	8" x 6"		28		16.25	6.35	22.60	27
504	8" x 8"		26		18.65	6.85	25.50	30
506	10" x 6"		24		16	7.40	23.40	28
508	10" x 10"		23		20	7.70	27.70	33
510	12" x 10"		22		22	8.05	30.05	36
512	14" x 10"		22		24	8.05	32.05	38
514	16" x 8"		21		24	8.45	32.45	39
516	18" x 12"		20		32	8.90	40.90	48
518	20" x 12"		20		34	8.90	42.90	50
520	20" x 20"		19		53	9.35	62.35	72
522	24" x 12"		17		42	10.45	52.45	61
524	24" x 24"		16		80	11.10	91.10	105
526	30" x 6"		15		37	11.85	48.85	58
528	30" x 8"		15		43	11.85	54.85	65
530	30" x 12"		14		54	12.70	66.70	78
532	30" x 16"		13		64	13.65	77.65	90
534	30" x 20"		12		81	14.80	95.80	110
536	30" x 24"		11		100	16.15	116.15	135
538	30" x 30"		10		120	17.75	137.75	160
600	For steel grilles instead of aluminum in above, deduct				10%			
620	Plastic, eggcrate, lay-in, T-bar system							
621	48" x 24" sheet	1 Shee	50	Ea.	7.70	3.55	11.25	13.65
625	Steel door louver							
627	With fire link, steel only							
635	12" x 18"	1 Shee	22	Ea.	70	8.05	78.05	89
641	12" x 24"		21		74	8.45	82.45	94
647	18" x 18"		20		81	8.90	89.90	100
653	18" x 24"		19		85	9.35	94.35	105

15

15.7 Air Conditioning & Ventilating

		CREW	DAILY OUTPUT	UNIT	BARE COSTS MAT.	BARE COSTS INST.	BARE COSTS TOTAL	TOTAL INCL O&P
55 660	18" x 36"	1 Shee	17	Ea.	115	10.45	125.45	140
666	24" x 24"		15		97	11.85	108.85	125
672	24" x 30"		15		115	11.85	126.85	145
678	30" x 30"		13		145	13.65	158.65	180
684	30" x 36"		12		165	14.80	179.80	205
690	36" x 36"		10		185	17.75	202.75	230
58-001	**REGISTERS**							
098	Air supply							
100	Ceiling/wall, O.B. damper, anodized aluminum							
101	One or two way deflection, adj. curved face bars							
102	8" x 4"	1 Shee	26	Ea.	22	6.85	28.85	34
104	8" x 8"		24		25	7.40	32.40	38
106	10" x 6"		20		27	8.90	35.90	43
108	10" x 10"		19		31	9.35	40.35	48
110	12" x 6"		19		30	9.35	39.35	47
112	12" x 12"		18		37	9.85	46.85	55
114	14" x 8"		17		40	10.45	50.45	59
116	14" x 14"		18		44	9.85	53.85	63
118	18" x 8"		18		42	9.85	51.85	61
120	18" x 18"		17		58	10.45	68.45	79
122	20" x 4"		19		38	9.35	47.35	55
124	20" x 6"		18		42	9.85	51.85	61
126	20" x 8"		18		47	9.85	56.85	66
128	20" x 20"		17		88	10.45	98.45	110
130	24" x 4"		17		42	10.45	52.45	61
132	24" x 6"		16		48	11.10	59.10	69
134	24" x 8"		13		53	13.65	66.65	78
136	24" x 24"		11		115	16.15	131.15	150
138	30" x 4"		16		52	11.10	63.10	73
140	30" x 6"		15		58	11.85	69.85	81
142	30" x 8"		14		64	12.70	76.70	89
144	30" x 24"		12		135	14.80	149.80	170
146	30" x 30"		10		185	17.75	202.75	230
150	For 3 or 4 way deflection, adj. curved face bars, add				20%			
198	One way deflection, adj. vert. or horiz. face bars							
200	8" x 4"	1 Shee	26	Ea.	18	6.85	24.85	30
202	8" x 8"		24		21	7.40	28.40	34
204	10" x 6"		20		22	8.90	30.90	37
206	10" x 10"		19		26	9.35	35.35	42
208	12" x 6"		19		24	9.35	33.35	40
210	12" x 12"		18		32	9.85	41.85	50
212	14" x 8"		17		30	10.45	40.45	48
214	14" x 12"		18		35	9.85	44.85	53
216	14" x 14"		18		38	9.85	47.85	56
218	16" x 6"		18		30	9.85	39.85	47
220	16" x 12"		18		39	9.85	48.85	57
222	16" x 16"		17		44	10.45	54.45	64
224	18" x 8"		18		36	9.85	45.85	54
226	18" x 12"		17		42	10.45	52.45	61
228	18" x 18"		16		57	11.10	68.10	79
230	20" x 10"		18		42	9.85	51.85	61
232	20" x 16"		16		52	11.10	63.10	73
234	20" x 20"		16		81	11.10	92.10	105
236	24" x 12"		15		52	11.85	63.85	74
238	24" x 16"		14		60	12.70	72.70	84
240	24" x 20"		12		93	14.80	107.80	125
242	24" x 24"		10		100	17.75	117.75	135
244	30" x 12"		13		62	13.65	75.65	88
246	30" x 16"		13		77	13.65	90.65	105
248	30" x 24"		11		120	16.15	136.15	155

15.7	Air Conditioning & Ventilating	CREW	DAILY OUTPUT	UNIT	BARE COSTS MAT.	BARE COSTS INST.	BARE COSTS TOTAL	TOTAL INCL O&P
58 250	30" x 30"	1 Shee	9	Ea.	155	19.75	174.75	200
252	36" x 12"		11		73	16.15	89.15	105
254	36" x 24"		10		140	17.75	157.75	180
256	36" x 36"		8		215	22	237	270
260	For 2 way deflect., adj. vert. or horiz. face bars, add				40%			
270	Above registers in steel instead of aluminum, deduct				10%			
300	Baseboard, hand adj. damper,enameled steel							
302	10" x 6"	1 Shee	24	Ea.	5.10	7.40	12.50	16.40
304	12" x 5"		23		12	7.70	19.70	24
306	12" x 6"		23		5.45	7.70	13.15	17.25
308	12" x 8"		22		14.05	8.05	22.10	27
310	14" x 6"		20		6	8.90	14.90	19.55
400	Floor, toe operated damper, enameled steel							
402	4" x 6"	1 Shee	32	Ea.	16.65	5.55	22.20	26
404	4" x 12"		26		9.90	6.85	16.75	21
406	6" x 8"		28		9.95	6.35	16.30	20
408	6" x 14"		22		11.90	8.05	19.95	25
410	8" x 10"		22		10.45	8.05	18.50	23
412	8" x 16"		20		35	8.90	43.90	51
414	10" x 10"		20		26	8.90	34.90	42
416	10" x 16"		18		39	9.85	48.85	57
418	12" x 12"		18		15.20	9.85	25.05	31
420	12" x 24"		16		77	11.10	88.10	100
422	14" x 14"		16		22	11.10	33.10	40
424	14" x 20"		15		74	11.85	85.85	99
426	16" x 16"		15		78	11.85	89.85	105
498	Air return							
500	Ceiling or wall, fixed 45° face blades,							
501	Adjustable O.B. damper, anodized aluminum							
502	8" x 4"	1 Shee	26	Ea.	8.05	6.85	14.90	18.80
504	8" x 6"		24		8.80	7.40	16.20	20
506	10" x 6"		19		9.60	9.35	18.95	24
508	16" x 6"		18		12.35	9.85	22.20	28
510	10" x 8"		19		10.45	9.35	19.80	25
512	12" x 8"		16		11.40	11.10	22.50	29
514	10" x 10"		18		11.25	9.85	21.10	27
518	18" x 12"		18		19.05	9.85	28.90	35
520	30" x 12"		12		32	14.80	46.80	57
522	16" x 16"		16		21	11.10	32.10	39
524	18" x 18"		17		26	10.45	36.45	44
526	36" x 18"		10		53	17.75	70.75	84
528	24" x 24"		11		44	16.15	60.15	72
530	36" x 24"		8		77	22	99	115
532	48" x 24"		6		105	30	135	160
600	Above registers in steel instead of aluminum, deduct				10%			
61-001	**DUCT ACCESSORIES**							
005	Air extractors, 12" x 4"	1 Shee	24	Ea.	7.15	7.40	14.55	18.65
010	8" x 6"		22		6.45	8.05	14.50	18.85
012	12" x 6"		21		9.65	8.45	18.10	23
014	16" x 6"		20		12.10	8.90	21	26
016	24" x 6"		18		17.35	9.85	27.20	33
018	12" x 8"		20		10.70	8.90	19.60	25
020	20" x 8"		16		15.40	11.10	26.50	33
024	18" x 10"		14		15.65	12.70	28.35	36
026	24" x 10"		12		21	14.80	35.80	45
028	24" x 12"		10		22	17.75	39.75	50
030	30" x 12"		8		28	22	50	63
100	Duct access door, insulated, 6" x 8"		14		10.70	12.70	23.40	30
102	10" x 12"		11		13.90	16.15	30.05	39

15

15.7 Air Conditioning & Ventilating		CREW	DAILY OUTPUT	UNIT	BARE COSTS MAT.	INST.	TOTAL	TOTAL INCL O&P
61 104	12" x 16"	1 Shee	10	Ea.	20	17.75	37.75	48
106	16" x 24"		9		31	19.75	50.75	63
108	24" x 24"		8		30	22	52	65
150	With insulated plexiglass window, 12" x 16"		10		41	17.75	58.75	71
200	Fabrics for flexible connections, with metal edge		100	L.F.	1.03	1.78	2.81	3.72
210	Without metal edge		160	"	.58	1.11	1.69	2.26
300	Fire damper, curtain type, vertical, 8" x 4"		24	Ea.	18.15	7.40	25.55	31
302	12" x 4"		22		18.15	8.05	26.20	32
304	10" x 6"		22		15.40	8.05	23.45	29
306	20" x 6"		18		17.60	9.85	27.45	34
308	12" x 8"		22		15.40	8.05	23.45	29
310	24" x 8"		16		19.80	11.10	30.90	38
312	12" x 10"		21		16.50	8.45	24.95	30
314	24" x 10"		15		21	11.85	32.85	40
316	36" x 10"		12		31	14.80	45.80	56
318	14" x 12"		20		17.60	8.90	26.50	32
320	24" x 12"		13		21	13.65	34.65	43
322	48" x 12"		10		42	17.75	59.75	72
324	16" x 14"		18		19.80	9.85	29.65	36
326	24" x 14"		12		25	14.80	39.80	49
328	30" x 14"		11		31	16.15	47.15	58
330	18" x 16"		17		19.80	10.45	30.25	37
332	24" x 16"		11		25	16.15	41.15	51
334	36" x 16"		10		34	17.75	51.75	63
336	24" x 18"		10		29	17.75	46.75	58
338	48" x 18"		8		54	22	76	92
340	24" x 20"		8		29	22	51	64
342	36" x 20"		7		39	25	64	80
344	24" x 22"		7		31	25	56	71
346	30" x 22"		6		40	30	70	87
348	26" x 24"		7		33	25	58	73
350	48" x 24"	Q-9	12		63	27	90	110
352	28" x 26"		13		39	25	64	79
354	30" x 28"		12		44	27	71	87
356	48" x 30"		11		74	29	103	125
358	48" x 36"		10		83	32	115	140
360	44" x 44"		10		90	32	122	145
362	48" x 48"		8		105	40	145	175
370	U.L. label included in above							
380	For horizontal operation, add			Ea.	20%			
390	For cap for blades out of air stream, add				30%			
400	For 10" 22 ga., U.L. approved sleeve, add				35%			
410	For 10" 22 ga., U.L. sleeve, 100% free area, add				65%			
420	For oversize openings group dampers							
450	Fire/smoke combination damper, louver type, UL							
451	8" x 8"	1 Shee	22	Ea.	53	8.05	61.05	70
452	16" x 8"		20		55	8.90	63.90	73
454	18" x 8"		18		58	9.85	67.85	78
456	20" x 8"		16		58	11.10	69.10	80
458	10" x 10"		21		53	8.45	61.45	71
460	24" x 10"		15		62	11.85	73.85	85
462	30" x 10"		12		72	14.80	86.80	100
464	12" x 12"		20		69	8.90	77.90	89
466	18" x 12"		18		73	9.85	82.85	95
468	24" x 12"		13		78	13.65	91.65	105
470	30" x 12"		11		90	16.15	106.15	125
472	14" x 14"		17		83	10.45	93.45	105
474	16" x 14"		18		83	9.85	92.85	105
476	20" x 14"		14		86	12.70	98.70	115
478	24" x 14"		12		89	14.80	103.80	120

	15.7 Air Conditioning & Ventilating	CREW	DAILY OUTPUT	UNIT	BARE COSTS MAT.	BARE COSTS INST.	BARE COSTS TOTAL	TOTAL INCL O&P
61 480	30" x 14"	1 Shee	10	Ea.	105	17.75	122.75	140
482	16" x 16"		16		83	11.10	94.10	105
484	20" x 16"		14		86	12.70	98.70	115
486	24" x 16"		11		89	16.15	105.15	120
488	30" x 16"		8		105	22	127	150
490	18" x 18"		15		95	11.85	106.85	120
500	24" x 18"		10		97	17.75	114.75	135
502	36" x 18"		7		120	25	145	170
504	20" x 20"		13		95	13.65	108.65	125
506	24" x 20"		8		97	22	119	140
508	30" x 20"		8		110	22	132	155
510	36" x 20"		7		120	25	145	170
512	24" x 24"		8		105	22	127	150
514	36" x 24"	Q-9	12		130	27	157	180
518	Mixing box, includes electric or pneumatic motor.							
519	Recommend use with attenuator, see 15.7-61-900							
520	Constant volume, 150 to 250 CFM	Q-9	12	Ea.	320	27	347	390
521	250 to 400 CFM		11		330	29	359	405
522	400 to 700 CFM		9		350	36	386	435
523	700 to 1000 CFM		9		370	36	406	460
524	1000 to 1500 CFM		8		390	40	430	485
525	1500 to 2200 CFM		6		420	53	473	540
550	Variable volume, 100 to 300 CFM		12		270	27	297	335
551	250 to 550 CFM		11		275	29	304	345
552	550 to 1150 CFM		9		290	36	326	370
553	900 to 1700 CFM		9		300	36	336	380
554	1200 to 2200 CFM		8		320	40	360	410
555	1500 to 3000 CFM		6		325	53	378	435
600	Multi-blade dampers, opposed blade, 12" x 12"	1 Shee	21		7.65	8.45	16.10	21
602	12" x 18"		18		10.60	9.85	20.45	26
604	18" x 24"		12		20	14.80	34.80	44
606	18" x 30"		10		22	17.75	39.75	50
608	24" x 24"		8		25	22	47	60
610	24" x 30"		6		29	30	59	75
612	30" x 30"	Q-9	11		35	29	64	81
614	36" x 36"		10		59	32	91	110
616	44" x 30"		8		58	40	98	120
618	48" x 36"		7		78	46	124	150
620	56" x 36"		7		98	46	144	175
622	60" x 36"		6		98	53	151	185
624	60" x 44"		5		115	64	179	220
700	Splitter damper assembly, self-locking, 1 ft. rod	1 Shee	24		8.35	7.40	15.75	19.95
702	3 ft. rod		22		10.95	8.05	19	24
704	4 ft. rod		20		12.15	8.90	21.05	26
706	6 ft. rod		18		14.55	9.85	24.40	30
750	Variable volume modulating motorized damper							
752	12" x 12"	1 Shee	12	Ea.	92	14.80	106.80	125
754	18" x 12"		10		98	17.75	115.75	135
756	24" x 12"		8		110	22	132	155
758	28" x 16"		6		130	30	160	185
760	30" x 18"		4		125	44	169	200
770	For thermostat, add		8		37	22	59	73
780	For transformer 40 VA capacity, add		16		18.75	11.10	29.85	37
800	Volume control, dampers							
810	8" x 8"	1 Shee	24	Ea.	14.75	7.40	22.15	27
812	12" x 8"		22		15.95	8.05	24	29
814	16" x 10"		20		21	8.90	29.90	36
816	18" x 12"		18		26	9.85	35.85	43
818	22" x 12"		15		30	11.85	41.85	50
820	24" x 16"		11		40	16.15	56.15	68

15

15.7 Air Conditioning & Ventilating

		CREW	DAILY OUTPUT	UNIT	BARE COSTS MAT.	BARE COSTS INST.	BARE COSTS TOTAL	TOTAL INCL O&P
61 822	28" x 16"	1 Shee	10	Ea.	46	17.75	63.75	76
824	30" x 16"		8		49	22	71	86
826	30" x 18"		7		54	25	79	96
900	Silencers, noise control for air flow, duct			MCFM	36		36	40M
910	Louvers, galvanized				50		50	55M
912	Louvers, mill finish aluminum				70		70	77M
920	Plenums, measured by panel surface			S.F.	8		8	8.80M
930	Vent, air transfer			Ea.	135		135	150M
64-001	**DUCTWORK**							
002	Fabricated rectangular, includes fittings, joints, supports,							
003	allowance for flexible connections, no insulation							
010	Aluminum, alloy 3003-H14, under 300 lb.	Q-10	75	Lb.	2.08	6.65	8.73	11.95
011	300 to 500 lb.		80		1.81	6.20	8.01	11.05
012	500 to 1000 lb.		95		1.70	5.25	6.95	9.50
014	1000 to 2000 lb.		120		1.54	4.14	5.68	7.75
015	2000 to 10,000 lb.		130		1.47	3.83	5.30	7.20
016	Over 10,000 lb.		145		1.41	3.43	4.84	6.55
050	Galvanized steel, under 400 lb.		235		1.19	2.12	3.31	4.39
052	C8.4-005 400 to 1000 lb.		255		.76	1.95	2.71	3.68
054	1000 to 2000 lb.		265		.58	1.88	2.46	3.37
056	C8.4-009 2000 to 5000 lb.		275		.34	1.81	2.15	3.01
057	5000 to 10,000 lb.		285		.34	1.74	2.08	2.92
058	Over 10,000 lb.		300		.34	1.66	2	2.79
060	For large quantities special prices available							
100	Stainless steel, type 304, under 400 lb.	Q-10	165	Lb.	2.77	3.01	5.78	7.45
102	400 to 1000 lb.		185		2.25	2.69	4.94	6.40
104	1000 to 2000 lb.		200		1.46	2.49	3.95	5.25
105	2000 to 10,000 lb.		225		1.36	2.21	3.57	4.72
106	Over 10,000 lb.		235		1.30	2.12	3.42	4.51
110	For medium pressure ductwork, add					15%		
120	For high pressure ductwork, add					40%		
130	Flexible, vinyl coated spring steel or aluminum,							
140	pressure to 10" (WG) UL-181.							
150	Non-insulated, 3" diameter	Q-9	400	L.F.	.30	.80	1.10	1.50
152	4" diameter		360		.31	.89	1.20	1.64
154	5" diameter		320		.38	1	1.38	1.87
156	6" diameter		280		.41	1.14	1.55	2.12
158	7" diameter		240		.48	1.33	1.81	2.47
160	8" diameter		200		.60	1.60	2.20	2.99
162	9" diameter		180		.67	1.78	2.45	3.33
164	10" diameter		160		.75	2	2.75	3.74
166	12" diameter		120		.94	2.66	3.60	4.92
168	14" diameter		80		1.11	4	5.11	7.05
170	16" diameter		60		1.33	5.35	6.68	9.25
180	For adjustable clamps, add			Ea.	.51		.51	.56M
190	Insulated, 4" diameter	Q-9	340	L.F.	.99	.94	1.93	2.46
192	5" diameter		300		1.15	1.07	2.22	2.82
194	6" diameter		260		1.39	1.23	2.62	3.32
196	7" diameter		220		1.51	1.45	2.96	3.78
198	8" diameter		180		1.68	1.78	3.46	4.44
200	9" diameter		160		1.85	2	3.85	4.95
202	10" diameter		140		2.03	2.28	4.31	5.55
204	12" diameter		100		2.40	3.20	5.60	7.30
206	14" diameter		80		2.84	4	6.84	8.95
208	16" diameter		60		3.64	5.35	8.99	11.75
210	18" diameter		45		4.84	7.10	11.94	15.70
212	20" diameter	Q-10	65		4.96	7.65	12.61	16.60
250	Fiberglass, aluminumized jacket, 1-1/2" blanket							
252	4" diameter	Q-9	340	L.F.	.87	.94	1.81	2.33
254	5" diameter	"	300	"	.96	1.07	2.03	2.61

15.7 Air Conditioning & Ventilating

			CREW	DAILY OUTPUT	UNIT	BARE COSTS MAT.	BARE COSTS INST.	BARE COSTS TOTAL	TOTAL INCL O&P
64	256	6" diameter	Q-9	260	L.F.	1.03	1.23	2.26	2.93
	258	7" diameter		220		1.16	1.45	2.61	3.39
	260	8" diameter		180		1.26	1.78	3.04	3.97
	262	9" diameter		160		1.39	2	3.39	4.44
	264	10" diameter		140		1.49	2.28	3.77	4.97
	266	12" diameter		100		1.86	3.20	5.06	6.70
	268	14" diameter		80		2.28	4	6.28	8.35
	270	16" diameter		60		2.65	5.35	8	10.70
	272	18" diameter		45		3.20	7.10	10.30	13.90
	300	Rigid fiberglass, round, .003" foil scrim jacket							
	310	4" diameter	Q-9	310	L.F.	1.55	1.03	2.58	3.21
	312	5" diameter		275		1.77	1.16	2.93	3.64
	314	6" diameter		240		1.96	1.33	3.29	4.10
	316	7" diameter		220		2.28	1.45	3.73	4.63
	318	8" diameter		180		2.60	1.78	4.38	5.45
	320	9" diameter		160		3.01	2	5.01	6.20
	322	10" diameter		140		3.18	2.28	5.46	6.85
	324	12" diameter		100		4.06	3.20	7.26	9.15
	326	14" diameter		80		4.78	4	8.78	11.10
	328	16" diameter		60		6	5.35	11.35	14.35
	330	18" diameter		45		8.40	7.10	15.50	19.60
	332	20" diameter	Q-10	65		10.25	7.65	17.90	22
	334	22" diameter		60		11.85	8.30	20.15	25
	336	24" diameter		55		14	9.05	23.05	29
	338	26" diameter		50		16.30	9.95	26.25	32
	340	28" diameter		45		18.55	11.05	29.60	37
	342	30" diameter		40		21	12.45	33.45	41
	350	Rectangular, 1" thick, alum. faced, (FRK), std. weight		350	S.F.Surf	.43	1.42	1.85	2.54
	352	Heavy weight		350	"	.62	1.42	2.04	2.75
	400	Rigid plastic, corrosive fume resistant							
	402	Straight, 6" diameter	Q-9	220	L.F.	4.62	1.45	6.07	7.20
	403	7" diameter		160		6.65	2	8.65	10.25
	404	8" diameter		160		6.65	2	8.65	10.25
	405	9" diameter		140		8.45	2.28	10.73	12.60
	406	10" diameter		120		8.45	2.66	11.11	13.20
	407	12" diameter		100		9.85	3.20	13.05	15.50
	408	14" diameter		70		12.75	4.57	17.32	21
	409	16" diameter		62		14.75	5.15	19.90	24
	410	18" diameter		58		23	5.50	28.50	33
	411	20" diameter	Q-10	75		47	6.65	53.65	61
	412	22" diameter		60		52	8.30	60.30	69
	413	24" diameter		55		57	9.05	66.05	76
	414	26" diameter		42		62	11.85	73.85	85
	415	28" diameter		35		66	14.20	80.20	93
	416	30" diameter		28		71	17.75	88.75	105
	417	36" diameter		23		85	22	107	125
	418	42" diameter		20		100	25	125	145
	420	48" diameter		18		115	28	143	165
	425	Coupling, 6" diameter	Q-9	88	Ea.	4.51	3.63	8.14	10.25
	426	7" diameter		82		4.93	3.90	8.83	11.10
	427	8" diameter		78		4.93	4.10	9.03	11.40
	428	9" diameter		74		5.40	4.32	9.72	12.25
	429	10" diameter		70		5.40	4.57	9.97	12.60
	430	12" diameter		55		5.85	5.80	11.65	14.90
	431	14" diameter		44		7.10	7.25	14.35	18.40
	432	16" diameter		40		8.40	8	16.40	21
	433	18" diameter		38		9.70	8.40	18.10	23
	434	20" diameter	Q-10	55		16.85	9.05	25.90	32
	435	22" diameter		50		19.45	9.95	29.40	36
	436	24" diameter		45		22	11.05	33.05	40

15.7	Air Conditioning & Ventilating	CREW	DAILY OUTPUT	UNIT	BARE COSTS MAT.	INST.	TOTAL	TOTAL INCL O&P
64 437	26" diameter	Q-10	39	Ea.	23	12.75	35.75	44
438	28" diameter		33		24	15.05	39.05	48
439	30" diameter		27		26	18.40	44.40	55
440	36" diameter		25		42	19.90	61.90	75
441	42" diameter		23		47	22	69	83
442	48" diameter		21		55	24	79	95
447	Elbow, 90°, 6" diameter	Q-9	44		42	7.25	49.25	57
448	7" diameter		36		43	8.90	51.90	60
449	8" diameter		28		45	11.40	56.40	66
450	9" diameter		22		59	14.55	73.55	86
451	10" diameter		18		72	17.75	89.75	105
452	12" diameter		15		100	21	121	140
453	14" diameter		11		135	29	164	190
454	16" diameter		10		165	32	197	230
455	18" diameter	Q-10	15		170	33	203	235
456	20" diameter		14		190	36	226	260
457	22" diameter		13		240	38	278	320
458	24" diameter		12		270	41	311	355
459	26" diameter		11		310	45	355	405
460	28" diameter		10		375	50	425	485
461	30" diameter		9		430	55	485	555
462	36" diameter		7		600	71	671	765
463	42" diameter		5		815	99	914	1,050
464	48" diameter		4		1,075	125	1,200	1,375
475	Elbow 45°, use 90° and deduct				45%			
504	Spiral preformed, aluminum, straight, 18 ga., 3" diameter	Q-9	400	L.F.	1.06	.80	1.86	2.33
505	4" diameter		360		1.41	.89	2.30	2.85
506	5" diameter		320		1.77	1	2.77	3.40
507	6" diameter		280		2.12	1.14	3.26	4
508	7" diameter		240		2.48	1.33	3.81	4.67
509	8" diameter		200		2.83	1.60	4.43	5.45
510	9" diameter		180		3.16	1.78	4.94	6.05
511	10" diameter		160		3.52	2	5.52	6.80
512	12" diameter		120		4.22	2.66	6.88	8.55
513	14" diameter		80		4.93	4	8.93	11.25
514	16" diameter		60		5.65	5.35	11	14
515	18" diameter		45		6.35	7.10	13.45	17.35
516	20" diameter	Q-10	65		7.05	7.65	14.70	18.90
518	24" diameter		55		8.45	9.05	17.50	22
520	30" diameter		45		10.60	11.05	21.65	28
522	36" diameter		40		12.70	12.45	25.15	32
524	40" diameter		35		14.10	14.20	28.30	36
526	48" diameter		30		16.95	16.60	33.55	43
540	Steel, galvanized, straight lengths							
541	4" diameter, 26 ga.	Q-9	360	L.F.	.63	.89	1.52	1.99
542	6" diameter, 26 ga.		280		.95	1.14	2.09	2.71
543	8" diameter, 26 ga.		200		1.27	1.60	2.87	3.73
544	10" diameter, 24 ga.		160		2.07	2	4.07	5.20
545	.12" diameter, 24 ga.		120		2.48	2.66	5.14	6.60
546	14" diameter, 24 ga.		80		2.89	4	6.89	9
548	16" diameter, 22 ga.		60		3.86	5.35	9.21	12
550	20" diameter, 22 ga.	Q-10	65		4.95	7.65	12.60	16.60
552	24" diameter, 20 ga.		55		7	9.05	16.05	21
554	30" diameter, 18 ga.		45		11.65	11.05	22.70	29
560	36" diameter, 18 ga.		40		13.95	12.45	26.40	33
580	Connector, 4" diameter	Q-9	100	Ea.	2.97	3.20	6.17	7.95
582	6" diameter		88		3.47	3.63	7.10	9.10
584	8" diameter		78		3.92	4.10	8.02	10.30
586	10" diameter		70		4.37	4.57	8.94	11.45
588	12" diameter		50		6.05	6.40	12.45	16

15

15.7 Air Conditioning & Ventilating

			CREW	DAILY OUTPUT	UNIT	BARE COSTS MAT.	BARE COSTS INST.	BARE COSTS TOTAL	TOTAL INCL O&P
64	590	14" diameter	Q-9	44	Ea.	6.90	7.25	14.15	18.20
	592	16" diameter		40		7.50	8	15.50	19.90
	594	20" diameter		34		8.85	9.40	18.25	23
	596	24" diameter		28		10.70	11.40	22.10	28
	598	30" diameter		22		13.80	14.55	28.35	36
	600	36" diameter		18		23	17.75	40.75	51
	630	Elbow, 45°, 4" diameter		60		9.90	5.35	15.25	18.65
	632	6" diameter		44		13.85	7.25	21.10	26
	634	8" diameter		28		22	11.40	33.40	41
	636	10" diameter		18		35	17.75	52.75	64
	638	12" diameter		13		47	25	72	88
	640	14" diameter		11		55	29	84	105
	642	16" diameter		10		63	32	95	115
	644	20" diameter	Q-10	14		82	36	118	140
	646	24" diameter		12		105	41	146	175
	648	30" diameter		9		135	55	190	230
	650	36" diameter		7		185	71	256	305
	660	Elbow, 90° use 45° and add				90%	20%		
	680	Reducing coupling, 6" x 4"	Q-9	46		13.05	6.95	20	24
	682	8" x 6"		40		14.90	8	22.90	28
	684	10" x 8"		32		18.75	10	28.75	35
	686	12" x 10"		24		25	13.30	38.30	47
	688	14" x 12"		20		31	16	47	57
	690	16" x 14"		18		31	17.75	48.75	60
	692	18" x 16"		16		37	20	57	70
	694	20" x 18"	Q-10	24		40	21	61	74
	696	24" x 22"		22		52	23	75	90
	698	30" x 28"		18		69	28	97	115
	700	36" x 34"		16		105	31	136	160
	740	Steel, PVC coated both sides, straight lengths							
	742	6" diameter, 26 ga.	Q-9	280	L.F.	1.76	1.14	2.90	3.60
	744	8" diameter, 26 ga.		200		2.31	1.60	3.91	4.87
	746	10" diameter, 24 ga.		160		3.82	2	5.82	7.10
	748	12" diameter, 24 ga.		120		4.56	2.66	7.22	8.90
	750	14" diameter, 24 ga.		80		5.40	4	9.40	11.75
	752	16" diameter, 22 ga.		60		7.65	5.35	13	16.20
	754	18" diameter, 22 ga.		45		8.45	7.10	15.55	19.65
	756	20" diameter, 22 ga.	Q-10	65		9.45	7.65	17.10	22
	758	24" diameter, 20 ga.		55		13.75	9.05	22.80	28
	760	30" diameter, 18 ga.		45		26	11.05	37.05	45
	762	36" diameter, 18 ga.		40		29	12.45	41.45	50
	790	Connector, 6" diameter	Q-9	88	Ea.	2.30	3.63	5.93	7.85
	792	8" diameter		78		2.96	4.10	7.06	9.25
	794	10" diameter		70		5.20	4.57	9.77	12.40
	796	12" diameter		50		6.20	6.40	12.60	16.15
	798	14" diameter		44		7.25	7.25	14.50	18.55
	800	16" diameter		40		8.10	8	16.10	21
	802	18" diameter		37		9.05	8.65	17.70	23
	804	20" diameter		34		9.80	9.40	19.20	24
	806	24" diameter		28		11.50	11.40	22.90	29
	808	30" diameter		22		14.95	14.55	29.50	38
	810	36" diameter		18		22	17.75	39.75	50
	840	Elbow, 45°, 6" diameter		44		3.60	7.25	10.85	14.55
	842	8" diameter		28		4.77	11.40	16.17	22
	844	10" diameter		18		8.05	17.75	25.80	35
	846	12" diameter		13		12	25	37	49
	848	14" diameter		11		19.85	29	48.85	64
	850	16" diameter		10		29	32	61	79
	852	18" diameter		9		37	36	73	92
	854	20" diameter	Q-10	13		45	38	83	105

15

15.7 Air Conditioning & Ventilating

		CREW	DAILY OUTPUT	UNIT	BARE COSTS MAT.	BARE COSTS INST.	BARE COSTS TOTAL	TOTAL INCL O&P
64 856	24" diameter	Q-10	11	Ea.	60	45	105	130
858	30" diameter		9		92	55	147	180
860	36" diameter		7		120	71	191	235
880	Elbow, 90° use 45° and add				90%	20%		
900	Reducing coupling, tapered offset							
902	Diameter x any smaller diameter							
904	6" x	Q-9	46	Ea.	5.75	6.95	12.70	16.45
906	8" x		40		7	8	15	19.35
908	10" x		32		8.90	10	18.90	24
910	12" x		24		12.70	13.30	26	33
912	14" x		20		21	16	37	46
914	16" x		18		31	17.75	48.75	60
916	18" x		16		40	20	60	73
918	20" x	Q-10	24		50	21	71	85
920	24" x		22		71	23	94	110
922	30" x		18		105	28	133	155
924	36" x		16		140	31	171	200
940	Lateral 45°, straight barrel							
942	6" size	Q-9	30	Ea.	9.45	10.65	20.10	26
944	8" size		19		11.65	16.85	28.50	37
946	10" size		12		16.85	27	43.85	57
948	12" size		9		21	36	57	75
950	14" size		7		39	46	85	110
952	16" size		6.70		56	48	104	130
954	18" size		6		83	53	136	170
956	20" size	Q-10	8.70		92	57	149	185
958	24" size		7		125	71	196	240
960	30" size		6		175	83	258	315
962	36" size		5		225	99	324	390
964	For reducing lateral, add				40%			
970	Register boot, elbow or perimeter							
971	6" duct diameter	Q-9	60	Ea.	7.80	5.35	13.15	16.35
972	8" duct diameter		56		8.55	5.70	14.25	17.75
973	10" duct diameter		45		10.55	7.10	17.65	22
974	12" duct diameter		36		12.85	8.90	21.75	27
980	Steel, stainless, straight lengths							
981	3" diameter	Q-9	400	L.F.	2.83	.80	3.63	4.28
982	4" diameter		360		3.77	.89	4.66	5.45
983	5" diameter		320		4.74	1	5.74	6.65
984	6" diameter		280		5.70	1.14	6.84	7.95
985	7" diameter		240		6.65	1.33	7.98	9.25
986	8" diameter		200		7.60	1.60	9.20	10.70
987	9" diameter		180		8.55	1.78	10.33	12
988	10" diameter		160		9.50	2	11.50	13.35
989	12" diameter		120		11.40	2.66	14.06	16.40
990	14" diameter		80		13.30	4	17.30	20
991	16" diameter		60		15.20	5.35	20.55	24
992	18" diameter		45		17.10	7.10	24.20	29
993	20" diameter	Q-10	65		19.05	7.65	26.70	32
994	24" diameter		55		23	9.05	32.05	38
995	30" diameter		45		29	11.05	40.05	48
996	36" diameter		40		34	12.45	46.45	56
997	40" diameter		35		38	14.20	52.20	63
66-001	**EVAPORATORS,** DX coils, remote compressors not included							
100	Coolers, reach-in type, above freezing temperatures							
130	Shallow depth, wall mount, 7 fins per inch, air defrost							
131	600 BTUH, 8" fan, rust-proof core	Q-5	3.80	Ea.	140	85	225	275
132	900 BTUH, 8" fan, rust-proof core		3.30		155	97	252	310
133	1200 BTUH, 8" fan, rust-proof core		3		175	105	280	345

15.7 Air Conditioning & Ventilating

			CREW	DAILY OUTPUT	UNIT	BARE COSTS MAT.	BARE COSTS INST.	BARE COSTS TOTAL	TOTAL INCL O&P
66	134	1800 BTUH, 8" fan, rust-proof core	Q-5	2.40	Ea.	220	135	355	435
	135	2500 BTUH, 10" fan, rust-proof core		2.20		265	145	410	500
	136	3500 BTUH, 10" fan		2		280	160	440	540
	137	4500 BTUH, 10" fan		1.90		335	170	505	615
	139	For pan drain, add				21		21	23M
	160	Undercounter refrigerators, ceiling or wall mount,							
	161	8 fins per inch, air defrost, rust-proof core							
	163	800 BTUH, one 6" fan	Q-5	5	Ea.	105	64	169	210
	164	1300 BTUH, two 6" fans		4		130	80	210	260
	165	1700 BTUH, two 6" fans		3.60		145	89	234	290
	200	Coolers, reach-in and walk-in types, above freezing temperatures							
	260	Two-way discharge, ceiling mount, 150-4100 CFM,							
	261	above 34°F applications, air defrost							
	263	900 BTUH, 7 fins per inch, 8" fan	Q-5	4	Ea.	130	80	210	260
	266	2500 BTUH, 7 fins per inch, 10" fan		2.60		250	125	375	455
	269	5500 BTUH, 7 fins per inch, 12" fan		1.70		385	190	575	695
	272	8500 BTUH, 8 fins per inch, 16" fan		1.20		535	270	805	975
	275	15,000 BTUH, 7 fins per inch, 18" fan		1.10		760	290	1,050	1,250
	277	24,000 BTUH, 7 fins per inch, two 16" fans		1		1,150	320	1,470	1,725
	279	30,000 BTUH, 7 fins per inch, two 18" fans		.90		1,425	355	1,780	2,075
	285	Two-way discharge, low profile, ceiling mount,							
	286	8 fins per inch, 200-570 CFM, air defrost							
	288	800 BTUH, one 6" fan	Q-5	4	Ea.	130	80	210	260
	289	1300 BTUH, two 6" fans		3.50		155	92	247	305
	290	1800 BTUH, two 6" fans		3.30		165	97	262	320
	291	2700 BTUH, three 6" fans		2.70		195	120	315	385
	300	Coolers, walk-in type, above freezing temperatures							
	330	General use, ceiling mount, 108-2080 CFM							
	332	600 BTUH, 7 fins per inch, 6" fan	Q-5	5.20	Ea.	100	62	162	200
	334	1200 BTUH, 7 fins per inch, 8" fan		3.70		140	87	227	280
	336	1800 BTUH, 7 fins per inch, 10" fan		2.70		200	120	320	390
	338	3500 BTUH, 7 fins per inch, 12" fan		2.40		270	135	405	490
	340	5500 BTUH, 8 fins per inch, 12" fan		2		325	160	485	590
	343	8500 BTUH, 8 fins per inch, 16" fan		1.40		485	230	715	865
	346	15,000 BTUH, 7 fins per inch, two 16" fans		1.10		780	290	1,070	1,275
	364	Low velocity, high latent load, ceiling mount,							
	365	1050-2420 CFM, air defrost, 6 fins per inch							
	367	6700 BTUH, two 10" fans	Q-5	1.30	Ea.	610	245	855	1,025
	368	10,000 BTUH, three 10" fans		1		790	320	1,110	1,325
	369	13,500 BTUH, three 10" fans		1		940	320	1,260	1,500
	370	18,000 BTUH, four 10" fans		1		1,200	320	1,520	1,775
	371	26,500 BTUH, four 10" fans		.80		1,475	400	1,875	2,200
	373	For electric defrost, add				17%			
	500	Freezers and coolers, reach-in type, above 34°F							
	503	to sub-freezing temperature range, low latent load,							
	505	air defrost, 7 fins per inch							
	507	1200 BTUH, 8" fan, rustproof core	Q-5	4.40	Ea.	120	73	193	235
	508	1500 BTUH, 8" fan, rustproof core		4.20		130	76	206	255
	509	1800 BTUH, 8" fan, rustproof core		3.70		140	87	227	280
	510	2500 BTUH, 10" fan, rustproof core		2.90		180	110	290	360
	511	3500 BTUH, 10" fan, rustproof core		2.50		210	130	340	415
	512	4500 BTUH, 12" fan, rustproof core		2.10		255	155	410	500
	600	Freezers and coolers, walk-in type							
	605	1960-18,000 CFM, medium profile							
	606	Standard motor, 4 fins per inch, to -30°F							
	608	10,500 BTUH, air defrost, one 18" fan	Q-5	1.40	Ea.	605	230	835	995
	609	12,500 BTUH, air defrost, one 18" fan		1.40		640	230	870	1,025
	610	15,000 BTUH, air defrost, one 20" fan		1.30		860	245	1,105	1,300
	611	21,000 BTUH, air defrost, two 18" fans		1		1,075	320	1,395	1,650
	612	25,000 BTUH, air defrost, two 18" fans		1		1,150	320	1,470	1,725

Line	15.7 Air Conditioning & Ventilating	CREW	DAILY OUTPUT	UNIT	BARE COSTS MAT.	BARE COSTS INST.	BARE COSTS TOTAL	TOTAL INCL O&P
66 613	30,000 BTUH, air defrost, two 20" fans	Q-6	1.30	Ea.	1,175	385	1,560	1,850
614	37,300 BTUH, air defrost, two 18" fans		1.25		1,250	400	1,650	1,950
615	44,800 BTUH, air defrost, two 20" fans		1		1,650	500	2,150	2,525
616	60,700 BTUH, elec defrost, three 20" fans		1		2,650	500	3,150	3,625
617	91,000 BTUH, elec defrost, four 20" fans		.80		3,750	625	4,375	5,025
618	108,900 BTUH, elec defrost, four 24" fans		.60		4,375	835	5,210	6,025
633	Hot gas defrost, standard motor units, add				45%			
637	Electric defrost, 230V, standard motor, add				27%			
641	460V, standard or high capacity motor, add				61%			
660	Six fins per inch, increases BTUH 45%							
661	Standard capacity, add			Ea.	10%			
663	Hot gas defrost, standard capacity, add				57%			
665	Electric defrost, standard capacity, add				37%			
668	Electric defrost, 460V, either capacity				73%			
680	Eight fins per inch, increases BTUH 75%							
681	Standard capacity, add			Ea.	17%			
683	Hot gas & electric defrost not recommended							
690	Aluminum tube coils. add			Ea.	25%			
700	580-3750 CFM, 10" fans							
701	Four fins per inch							
703	3400 BTUH, 1 fan	Q-5	2.80	Ea.	185	115	300	370
704	4000 BTUH, 2 fans		2.40		215	135	350	430
705	5300 BTUH, 2 fans		1.90		270	170	440	540
706	6800 BTUH, 2 fans		1.70		340	190	530	645
707	8000 BTUH, 3 fans		1.60		385	200	585	715
708	11,100 BTUH, 3 fans		1.40		480	230	710	860
709	13,600 BTUH, 4 fans		1.30		615	245	860	1,025
710	17,000 BTUH, 5 fans		1.20		720	270	990	1,175
711	20,400 BTUH, 6 fans		1		885	320	1,205	1,450
715	Five fins per inch increases BTUH 21%, add				3%			
716	Six fins per inch increases BTUH 35%, add				5%			
718	Eight fins per inch increases BTUH 53%,							
719	not recommended for below 35°F, add			Ea.	8%			
721	For 12°F temperature differential, add				20%			
722	For aluminum tube core, add				25%			
723	For adjustable thermostat control, add				41		41	45M
724	For hot gas defrost, except 8 fin models							
725	on applications below 35°F, add			Ea.	101%			
726	For electric defrost, except 8 fin models							
727	on applications below 35°F, add			Ea.	61%			
800	Freezers, pass-thru door uprights, walk-in storage							
830	Low temperature, thin profile, electric defrost							
831	138-1800 CFM, 5 fins per inch							
832	900 BTUH, one 6" fan	Q-5	3.30	Ea.	170	97	267	330
834	1500 BTUH, two 6" fans		2.20		255	145	400	490
836	2600 BTUH, three 6" fans		1.90		350	170	520	630
837	3300 BTUH, four 6" fans		1.70		405	190	595	720
838	4400 BTUH, five 6" fans		1.40		485	230	715	865
840	7000 BTUH, three 10" fans		1.30		635	245	880	1,050
841	8700 BTUH, four 10" fans		1.10		730	290	1,020	1,225
845	For air defrost, deduct				115		115	125M
67-001	**EVAPORATIVE COOLERS** Ducted, not incl. duct.							
010	Side draft style, capacities at .25" S.P.							
012	1785 CFM, 1/3 HP, 115 V	Q-9	5	Ea.	235	64	299	350
014	2740 CFM, 1/3 HP, 115 V		4.50		285	71	356	415
016	3235 CFM, 1/2 HP, 115 V		4		295	80	375	440
018	3615 CFM, 1/2 HP, 230 V		3.60		365	89	454	530
020	4215 CFM, 3/4 HP, 230 V		3.20		380	100	480	565
022	5255 CFM. 1 HP, 115/230 V		3		625	105	730	845

	15.7 Air Conditioning & Ventilating	CREW	DAILY OUTPUT	UNIT	BARE COSTS MAT.	BARE COSTS INST.	BARE COSTS TOTAL	TOTAL INCL O&P
67 024	6090 CFM, 1 HP, 230/460 V	Q-9	2.80	Ea.	895	115	1,010	1,150
026	8300 CFM, 2 HP, 230/460 V		2.60		905	125	1,030	1,175
028	8360 CFM, 1-1/2" HP, 230/460 V		2.20		1,250	145	1,395	1,575
030	9725 CFM, 2 HP, 230/460 V		1.80		1,275	180	1,455	1,650
032	11,715 CFM, 3 HP, 230/460 V		1.40		1,300	230	1,530	1,775
034	14,410 CFM, 5 HP, 230/460 V		1		1,325	320	1,645	1,925
040	For two-speed motor, add				5%			
050	For down discharge style, add				10%			
100	High capacities at .4" S.P.							
104	12,100 CFM,2 HP, 230/460 V	Q-9	.54	Ea.	3,500	590	4,090	4,725
106	14,000 CFM, 3HP, 230/460 V		.50		3,600	640	4,240	4,900
108	17,100 CFM, 5HP, 230/460 V		.48		3,725	665	4,390	5,075
110	22,500 CFM, 7.5 HP, 230/460 V	Q-10	.54		5,350	920	6,270	7,225
112	25,100 CFM, 10 HP, 230/460 V		.50		5,675	995	6,670	7,700
114	29,500 CFM, 15 HP, 230/460 V		.46		5,950	1,075	7,025	8,125
68-001	**EXHAUST SYSTEMS**							
050	Engine exhaust,garage,in-floor system							
051	B8.5 -110 Single tube outlet assemblies							
052	For transite pipe ducting, self-storing tube							
053	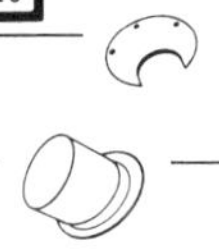 3" tubing adapter plate	1 Shee	16	Ea.	175	11.10	186.10	210
054	4" tubing adapter plate		16		175	11.10	186.10	210
055	5" tubing adapter plate		16		180	11.10	191.10	215
060	For vitrified tile ducting							
061	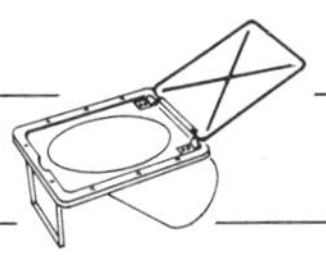 3" tubing adapter plate, self-storing tube	1 Shee	16	Ea.	160	11.10	171.10	190
062	4" tubing adapter plate, self-storing tube		16		160	11.10	171.10	190
065	4" tube size opening, non-storing tube		16		105	11.10	116.10	130
066	5" tubing adapter plate, self-storing tube		16		165	11.10	176.10	200
070	For vitrified tile & transite pipe ducting							
071	3" tube size opening, self-storing tube	1 Shee	16	Ea.	145	11.10	156.10	175
072	4" tube size opening, self-storing tube	"	16	"	145	11.10	156.10	175
080	Two tube outlet assemblies							
081	For transite pipe ducting, self-storing tube							
082	 3" tubing, dual exhaust adapter plate	1 Shee	16	Ea.	190	11.10	201.10	225
085	For vitrified tile ducting							
086	3" tubing, dual exhaust, self-storing tube	1 Shee	16	Ea.	175	11.10	186.10	210
087	3" tubing, double outlet, non-storing tubes	"	16	"	155	11.10	166.10	185
090	Accessories for metal tubing, (overhead systems also)							
091	Adapters, for metal tubing end							
092	3" tail pipe type			Ea.	36		36	40M
093	4" tail pipe type				38		38	42M
094	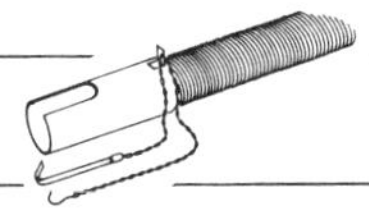 5" tail pipe type				39		39	43M
096	4" bail type				16.50		16.50	18.15M
097	5" bail type				17.50		17.50	19.25M
099	5" diesel stack type				44		44	48M
100	6" diesel stack type				63		63	69M
110	Bullnose (guide) required for in-floor assemblies							
111	 3" tubing size			Ea.	12		12	13.20M
112	4" tubing size				14.50		14.50	15.95M
113	5" tubing size				15.50		15.50	17.05M
115	Plain rings, for tubing end							
116	3" tubing size			Ea.	6		6	6.60M
117	4" tubing size			"	11.70		11.70	12.85M
120	Tubing, galvanized, flexible, (for overhead sys. also)							
121	3" ID			L.F.	3.55		3.55	3.90M
122	4" ID				4.75		4.75	5.20M
123	5" ID				6.60		6.60	7.25M
124	6" ID				9.65		9.65	10.60M
125	Stainless steel, flexible, (for overhead sys, also)							
126	3" ID			L.F.	11.25		11.25	12.35M

15.7 Air Conditioning & Ventilating

	15.7 Air Conditioning & Ventilating	CREW	DAILY OUTPUT	UNIT	BARE COSTS MAT.	BARE COSTS INST.	BARE COSTS TOTAL	TOTAL INCL O&P
68 127	4″ ID			L.F.	15.70		15.70	17.25M
128	5″ ID				23		23	25M
129	6″ ID				28		28	31M
150	Engine exhaust, garage, overhead components, for neoprene tubing							
151	Alternate metal tubing & accessories see above							
155	Adapters, for neoprene tubing end							
156	3″ tail pipe, adjustable, neoprene			Ea.	16		16	17.60M
157	3″ tail pipe, heavy wall neoprene				13		13	14.30M
158	4″ tail pipe, heavy wall neoprene				30		30	33M
159	5″ tail pipe, heavy wall neoprene				35		35	39M
160	4″ truck stack, aluminum				41		41	45M
165	Connectors, tubing							
166	3″ interior, aluminum			Ea.	9.25		9.25	10.15M
167	4″ interior, aluminum				10.50		10.50	11.55M
171	5″ interior, neoprene				20		20	22M
175	3″ spiralock, neoprene				4.45		4.45	4.89M
176	4″ spiralock, neoprene				7.25		7.25	7.95M
178	Y for 3″ ID tubing, neoprene, dual exhaust				20		20	22M
179	Y for 4″ ID tubing, aluminum, dual exhaust				29		29	32M
185	Elbows, aluminum, splice into tubing for strap							
186	3″ neoprene tubing size			Ea.	14.90		14.90	16.40M
187	4″ neoprene tubing size			″	22		22	24M
190	Flange assemblies, connect tubing to overhead duct							
191	For galvanized & stainless tubing							
192	3″ cast aluminum			Ea.	12		12	13.20M
193	4″ cast aluminum				14.50		14.50	15.95M
194	5″ cast aluminum				15.50		15.50	17.05M
195	6″ cast aluminum				16.50		16.50	18.15M
196	For neoprene tubing							
197	3″ neoprene			Ea.	12.75		12.75	14M
198	4″ cast aluminum			″	14		14	15.40M
200	Hardware and accessories							
202	Cable, vinyl coated, 5/32″ diameter			L.F.	.35		.35	.38M
204	Cleat, tie down cable or rope			Ea.	.90		.90	.99M
206	Pulley				2.50		2.50	2.75M
208	Pulley hook, universal				1.85		1.85	2.03M
210	Rope, manila, 5/16″ diameter			L.F.	.25		.25	.27M
212	Winch, 1″ diameter			Ea.	45		45	50M
215	Lifting strap, mounts on neoprene							
216	3″ tubing size			Ea.	8		8	8.80M
217	4″ tubing size				8		8	8.80M
218	5″ tubing size				9		9	9.90M
219	6″ tubing size				15		15	16.50M
220	Tubing, neoprene, 11′ lengths							
221	3″ ID			L.F.	5.45		5.45	6M
222	4″ ID				10.10		10.10	11.10M
223	5″ ID				13.15		13.15	14.45M
250	Engine exhaust, thru-door outlet							
251	3″ for up to 1/4″ thick door	1 Carp	16	Ea.	41	9.80	50.80	59
252	3″ for over 1/4″ thick door		16		44	9.80	53.80	63
253	4″ for up to 1/4″ thick door		16		46	9.80	55.80	65
750	Welding fume elimination accessories for garage exhaust systems							
760	Cut off (blast gate)							
761	3″ tubing size, 3″ x 6″ opening	1 Shee	24	Ea.	21	7.40	28.40	34
762	4″ tubing size, 4″ x 8″ opening		24		24	7.40	31.40	37
763	5″ tubing size, 5″ x 10″ opening		24		29	7.40	36.40	43
770	Hoods, magnetic, with handle & screen							
771	3″ tubing size, 3″ x 6″ opening	1 Shee	24	Ea.	115	7.40	122.40	135
772	4″ tubing size, 4″ x 8″ opening		24		115	7.40	122.40	135
773	5″ tubing size, 5″ x 10″ opening		24		125	7.40	132.40	150

15.7 Air Conditioning & Ventilating

		CREW	DAILY OUTPUT	UNIT	BARE COSTS MAT.	BARE COSTS INST.	BARE COSTS TOTAL	TOTAL INCL O&P
70-001	**FANS**							
002	Air conditioning and process air handling							
003	C8.4 -004 Axial flow, compact, low sound, 2.5" S.P.							
005	3800 CFM, 5 HP	Q-20	3.40	Ea.	1,925	120	2,045	2,300
008	C8.4 -007 6400 CFM, 5 HP		2.80		2,150	145	2,295	2,575
010	10,500 CFM, 7-1/2 HP		2.40		2,675	170	2,845	3,200
012	15,600 CFM, 10 HP		1.60		3,350	255	3,605	4,050
014	23,000 CFM, 15 HP		.70		5,175	585	5,760	6,550
016	28,000 CFM, 20 HP		.40		5,775	1,025	6,800	7,825
020	In-line centrifugal, supply/exhaust booster,							
022	aluminum wheel/hub, disconnect switch, 1/4" S.P.							
024	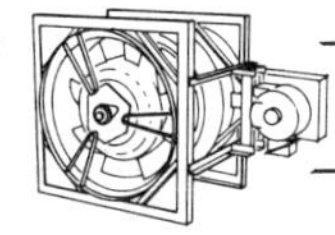 500 CFM, 10" diameter connection	Q-20	3	Ea.	380	135	515	615
026	1380 CFM, 12" diameter connection		2		535	205	740	885
028	1520 CFM, 16" diameter connection		2		575	205	780	930
030	2560 CFM, 18" diameter connection		1		765	410	1,175	1,425
032	3480 CFM, 20" diameter connection		.80		925	510	1,435	1,750
150	Vane-axial, low pressure, 2000 CFM, 1/2 HP		3.60		1,300	115	1,415	1,600
152	4000 CFM, 1 HP		3.20		1,400	130	1,530	1,725
154	8000 CFM, 2 HP		2.80		1,700	145	1,845	2,075
156	16,000 CFM, 5 HP		2.40		2,100	170	2,270	2,550
200	Blowers, direct drive with motor, complete							
202	1030 CFM @ .5" S.P., 1/6 HP	Q-20	18	Ea.	130	23	153	175
204	1150 CFM @ .5" S.P., 1/6 HP		18		130	23	153	175
206	1640 CFM @ 1.0" S.P., 1/3 HP		18		155	23	178	205
208	1720 CFM @ 1.0" S.P., 1/3 HP		18		155	23	178	205
209	4 speed							
210	600 to 1160 CFM @ .5" S.P., 1/5 HP	Q-20	16	Ea.	175	26	201	230
212	740 to 1700 CFM @ 1.0" S.P., 1/3 HP	"	14	"	235	29	264	300
250	Ceiling fan, right angle, extra quiet, 0.10" S.P.							
252	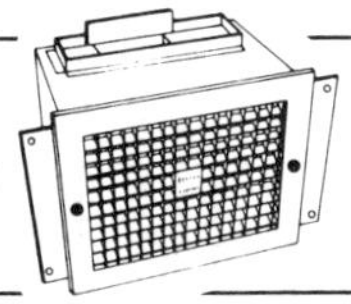 95 CFM	Q-20	20	Ea.	110	20	130	150
254	210 CFM		19		120	21	141	165
256	385 CFM		18		150	23	173	200
258	885 CFM		16		290	26	316	355
260	1650 CFM		13		380	31	411	465
262	2960 CFM		11		515	37	552	620
264	For wall or roof cap, add	1 Shee	16		67	11.10	78.10	90
266	For straight thru fan, add				10%			
268	For speed control switch, add	1 Elec	16		37	11.05	48.05	57
300	Paddle blade air circulator, 2 speed switch							
302	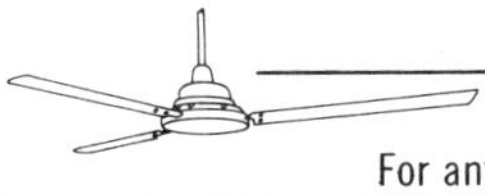 36", 4000 CFM high, 3000 CFM low	Q-20	6	Ea.	125	68	193	235
304	52", 7000 CFM high, 4000 CFM low	"	4		140	100	240	300
310	For antique white motor, add				22		22	24M
320	For brass plated motor, add				37		37	41M
330	For light adaptor kit, add				17.25		17.25	18.95M
350	Centrifugal, airfoil, motor and drive, complete							
352	1000 CFM, 1/2 HP	Q-20	2.50	Ea.	1,100	165	1,265	1,450
354	2000 CFM, 1 HP		2		1,200	205	1,405	1,625
356	4000 CFM, 3 HP		1.80		1,600	225	1,825	2,100
358	8000 CFM, 7-1/2 HP		1.40		2,600	290	2,890	3,275
360	12,000 CFM, 10 HP		1		3,200	410	3,610	4,125
400	Single width, belt drive, not incl. motor, capacities							
402	at 2000 FPM, 2.5" S.P. for indicated motor							
404	6900 CFM, 5 HP	Q-9	2.40	Ea.	1,500	135	1,635	1,850
406	10,340 CFM, 7-1/2" HP		2.20		2,050	145	2,195	2,475
408	15,320 CFM, 10 HP		2		2,425	160	2,585	2,900
410	22,780 CFM, 15 HP		1.80		3,600	180	3,780	4,225
412	33,840 CFM, 20 HP		1.60		4,800	200	5,000	5,575
414	41,400 CFM, 25 HP		1.40		6,375	230	6,605	7,350
416	50,100 CFM, 30 HP		.80		8,100	400	8,500	9,500
420	Double width wheel, 12,420 CFM, 7.5 HP		2.20		2,100	145	2,245	2,525

15

15.7 Air Conditioning & Ventilating

		CREW	DAILY OUTPUT	UNIT	BARE COSTS MAT.	INST.	TOTAL	TOTAL INCL O&P
70 422	18,620 CFM, 15 HP	Q-9	2	Ea.	2,875	160	3,035	3,400
424	27,580 CFM, 20 HP		1.80		3,525	180	3,705	4,125
426	40,980 CFM, 25 HP		1.50		5,525	215	5,740	6,400
428	60,920 CFM, 40 HP		1		7,450	320	7,770	8,650
430	74,520 CFM, 50 HP		.80		8,800	400	9,200	10,300
432	90,160 CFM, 50 HP		.70		12,100	455	12,555	14,000
434	110,300 CFM, 60 HP		.50		16,400	640	17,040	19,000
436	134,960 CFM, 75 HP		.40		21,600	800	22,400	24,900
450	Corrosive fume resistant, plastic							
460	roof ventilators, centrifugal, V belt drive, motor							
462	1/4" S.P., 250 CFM, 1/4 HP	Q-20	6	Ea.	1,525	68	1,593	1,775
464	895 CFM, 1/3 HP		5		1,650	82	1,732	1,925
466	1630 CFM, 1/2 HP		4		1,950	100	2,050	2,300
468	2240 CFM, 1 HP		3		2,025	135	2,160	2,425
470	3810 CFM, 2 HP		2		2,250	205	2,455	2,775
472	11760 CFM, 5 HP		1		3,800	410	4,210	4,775
474	18810 CFM, 10 HP		.70		5,550	585	6,135	6,950
480	For intermediate capacity, motors may be varied							
481	For explosion proof motor, add			Ea.	15%			
500	Utility set, centrifugal, V belt drive, motor							
502	1/4" S.P., 1900 CFM, 1/4 HP	Q-20	6	Ea.	2,700	68	2,768	3,075
504	2170 CFM, 1/3 HP		5		2,725	82	2,807	3,125
506	2680 CFM, 1/2 HP		4		2,725	100	2,825	3,150
508	3020 CFM, 3/4 HP		3		2,750	135	2,885	3,225
510	1/2" S.P., 3195 CFM, 1 HP		2		2,775	205	2,980	3,350
512	3610 CFM, 1-1/2 HP		1.60		2,800	255	3,055	3,450
514	4120 CFM, 2 HP		1.40		2,850	290	3,140	3,550
516	7850 CFM, 5 HP		1.30		3,425	315	3,740	4,225
518	10,200 CFM, 7-1/2 HP		1.20		3,475	340	3,815	4,325
520	For explosion proof motor, add				15%			
525	Vane-axial, direct drive, 1/4" S.P.							
530	615 CFM, 1/4 HP	Q-20	3	Ea.	1,400	135	1,535	1,750
532	1320 CFM, 1/4 HP		2.60		1,750	155	1,905	2,150
534	1550 CFM, 1/4 HP		2		1,750	205	1,955	2,225
536	2840 CFM, 1/2 HP		1		1,800	410	2,210	2,575
538	For explosion proof motor, add				15%			
550	Industrial exhauster, for air which may contain granular material							
552	1000 CFM, 1-1/2 HP	Q-20	2.50	Ea.	1,500	165	1,665	1,875
554	2000 CFM, 3 HP		2		1,800	205	2,005	2,275
556	4000 CFM, 7-1/2 HP		1.80		2,500	225	2,725	3,075
558	8000 CFM, 15 HP		1.40		4,200	290	4,490	5,050
560	12,000 CFM, 30 HP		1		6,700	410	7,110	7,975
600	Propeller exhaust, wall shutter, 1/4" S.P.							
602	Direct drive, two speed							
610	375 CFM, 1/10 HP	Q-20	10	Ea.	135	41	176	210
612	730 CFM, 1/7 HP		9		165	45	210	245
614	1000 CFM, 1/8 HP		8		225	51	276	320
616	1890 CFM, 1/4 HP		7		230	58	288	340
618	3275 CFM, 1/2 HP		6		275	68	343	400
620	4720 CFM, 1 HP		5		435	82	517	595
630	V-belt drive, 3 phase							
632	6175 CFM, 3/4 HP	Q-20	5	Ea.	440	82	522	605
634	7500 CFM, 3/4 HP		5		490	82	572	660
636	10,100 CFM, 1 HP		4.50		625	91	716	820
638	14,300 CFM, 1-1/2 HP		4		770	100	870	995
640	19,800 CFM, 2 HP		3		1,050	135	1,185	1,350
642	26,250 CFM, 3 HP		2.60		1,375	155	1,530	1,750
644	38,500 CFM, 5 HP		2.20		1,700	185	1,885	2,150
646	46,000 CFM, 7-1/2 HP		2		1,875	205	2,080	2,350
648	51,500 CFM, 10 HP		1.80		2,075	225	2,300	2,600

15

15.7 Air Conditioning & Ventilating

		CREW	DAILY OUTPUT	UNIT	BARE COSTS MAT.	BARE COSTS INST.	BARE COSTS TOTAL	TOTAL INCL O&P
70 649	V-belt drive, 115V., residential, whole house							
650	Ceiling-wall, 5200 CFM, 1/4 HP, 30" x 30"	1 Shee	6	Ea.	180	30	210	240
653	13,200 CFM, 1/3 HP, 48"x48"		4		225	44	269	310
654	15,445 CFM, 1/2 HP, 48"x48"		4		250	44	294	340
655	17,025 CFM, 1/2 HP, 54"x54"		4		475	44	519	585
656	For two speed motor, add				20%			
657	Shutter, automatic, ceiling/wall							
658	30"x30"	1 Shee	8	Ea.	45	22	67	82
659	36"x36"		8		48	22	70	85
660	42"x42"		8		54	22	76	92
661	48"x48"		7		66	25	91	110
662	54"x54"		6		76	30	106	125
663	Timer, shut off, to 12 Hr.		20		19	8.90	27.90	34
665	Residential, bath exhaust, grille, back draft damper							
666	50 CFM	Q-20	24	Ea.	28	17	45	56
667	110 CFM		22		55	18.55	73.55	87
668	Light combination, squirrel cage, 100 watt, 70 CFM		24		61	17	78	92
670	Light/heater combination, ceiling mounted							
671	70 CFM, 1450 watt	Q-20	24	Ea.	96	17	113	130
680	Heater combination, recessed, 70 CFM		24		56	17	73	86
682	With 2 infrared bulbs		23		71	17.75	88.75	105
690	Kitchen exhaust, grille, complete, 160 CFM		22		47	18.55	65.55	79
692	344 CFM		18		97	23	120	140
700	Roof exhauster, centrifugal, aluminum housing, 12" galvanized							
702	curb, bird screen, back draft damper, 1/4" S.P.							
710	Direct drive, 420 CFM, 8" sq. damper	Q-20	7	Ea.	320	58	378	435
712	675 CFM, 12" sq. damper		6		450	68	518	595
714	770 CFM, 16" sq. damper		5		665	82	747	850
716	1870 CFM, 20" sq. damper		4.20		925	97	1,022	1,150
718	2150 CFM, 20" sq. damper		4		1,075	100	1,175	1,325
720	V-belt drive, 1660 CFM, 12" sq. damper		6		535	68	603	685
722	2830 CFM, 14" sq. damper		5		650	82	732	835
724	4600 CFM, 20" sq. damper		4		855	100	955	1,100
726	8750 CFM, 26" sq. damper		3		945	135	1,080	1,225
728	12,500 CFM, 32" sq. damper		2		1,650	205	1,855	2,100
730	21,600 CFM, 40" sq. damper		1		2,750	410	3,160	3,625
732	For 2 speed winding, add				15%			
734	For explosion-proof motor, add				220		220	240M
736	For belt drive, top discharge, add				15%			
750	Utility set, steel construction, pedestal, 1/4" S.P.							
752	Direct drive, 150 CFM, 1/8 HP	Q-20	6.40	Ea.	345	64	409	470
754	485 CFM, 1/6 HP		5.80		435	70	505	580
756	1950 CFM, 1/2 HP		4.80		515	85	600	690
758	2410 CFM, 3/4 HP		4.40		945	93	1,038	1,175
760	3328 CFM, 1-1/2 HP		3		1,050	135	1,185	1,350
768	V-belt drive, drive cover, 3 phase							
770	800 CFM, 1/4 HP	Q-20	6	Ea.	330	68	398	460
772	1300 CFM, 1/3 HP		5		370	82	452	525
774	2000 CFM, 1 HP		4.60		390	89	479	560
776	2900 CFM, 3/4 HP		4.20		600	97	697	800
778	3600 CFM, 3/4 HP		4		685	100	785	900
780	4800 CFM, 1 HP		3.50		855	115	970	1,100
782	6700 CFM, 1-1/2 HP		3		930	135	1,065	1,225
784	11,000 CFM, 3 HP		2		1,800	205	2,005	2,275
786	13,000 CFM, 3 HP		1.60		2,100	255	2,355	2,675
788	15,000 CFM, 5 HP		1		2,150	410	2,560	2,950
790	17,000 CFM, 7-1/2 HP		.80		2,200	510	2,710	3,150
792	20,000 CFM, 7-1/2 HP		.80		2,225	510	2,735	3,200
800	Ventilation, residential							
802	Attic, roof type							

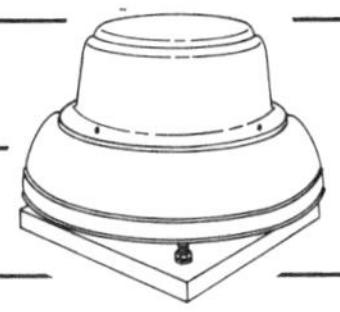

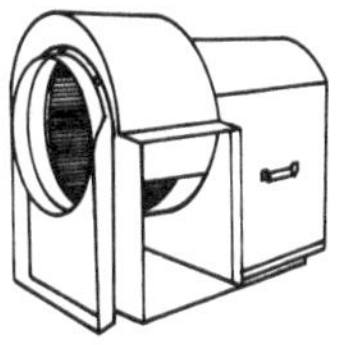

15

15.7	Air Conditioning & Ventilating	CREW	DAILY OUTPUT	UNIT	BARE COSTS MAT.	BARE COSTS INST.	BARE COSTS TOTAL	TOTAL INCL O&P
70 803	Aluminum dome, damper & curb							
804	6" diameter, 300 CFM	1 Elec	16	Ea.	120	11.05	131.05	150
805	7" diameter, 450 CFM		15		140	11.80	151.80	170
806	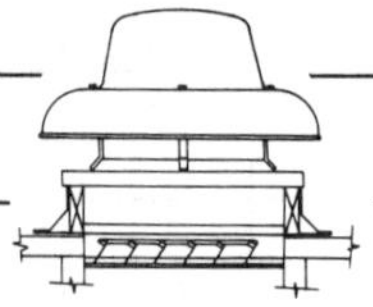 9" diameter, 900 CFM		14		250	12.65	262.65	295
808	12" diameter, 1000 CFM (gravity)		10		180	17.70	197.70	225
809	16" diameter, 1500 CFM (gravity)		9		220	19.65	239.65	270
810	20" diameter, 2500 CFM (gravity)		8		270	22	292	330
811	26" diameter, 4000 CFM (gravity)		7		340	25	365	410
812	32" diameter, 6500 CFM (gravity)		6		465	29	494	555
813	38" diameter, 8000 CFM (gravity)		5		640	35	675	755
814	50" diameter, 13,000 CFM (gravity)		4		1,025	44	1,069	1,200
816	Plastic, ABS dome							
818	930 CFM	1 Elec	14	Ea.	66	12.65	78.65	91
820	1600 CFM	"	12	"	92	14.75	106.75	120
824	Attic, wall type, with shutter, one speed							
825	12" diameter, 1000 CFM	1 Elec	14	Ea.	94	12.65	106.65	120
826	14" diameter, 1500 CFM		12		110	14.75	124.75	140
827	16" diameter, 2000 CFM		9		135	19.65	154.65	175
829	Whole house, wall type, with shutter, one speed							
830	 30" diameter, 4800 CFM	1 Elec	7	Ea.	265	25	290	330
831	36" diameter, 7000 CFM		6		310	29	339	385
832	42" diameter, 10,000 CFM		5		370	35	405	460
833	48" diameter, 16,000 CFM		4		445	44	489	555
834	For two speed, add				10		10	11M
835	Whole house, lay-down type, with shutter, one speed							
836	30" diameter, 4500 CFM	1 Elec	8	Ea.	275	22	297	335
837	36" diameter, 6500 CFM		7		325	25	350	395
838	42" diameter, 9000 CFM		6		380	29	409	460
839	48" diameter, 12,000 CFM		5		455	35	490	550
844	For two speed, add				10		10	11M
845	For 12 hour timer switch, add	1 Elec	32		15	5.55	20.55	24
850	Wall exhausters, centrifugal, auto damper, 1/8" S.P.							
852	Direct drive , 635 CFM, 1/20 HP	Q-20	14	Ea.	165	29	194	225
854	845 CFM, 1/12 HP		13		190	31	221	255
856	1005 CFM, 1/6 HP		12		275	34	309	350
858	1220 CFM, 1/6 HP		12		305	34	339	385
860	1575 CFM, 1/4 HP		11		340	37	377	430
862	2065 CFM, 1/4 HP		10		405	41	446	505
864	2660 CFM, 1/2 HP		9		460	45	505	570
866	3260 CFM, 3/4 HP		8		530	51	581	655
950	V-belt drive, 3 phase							
952	2800 CFM, 1/4 HP	Q-20	9	Ea.	685	45	730	820
954	3740 CFM, 1/2 HP		8		705	51	756	850
956	4400 CFM, 3/4 HP		7		725	58	783	880
958	5700 CFM, 1-1/2" HP		6		795	68	863	975
73-001	**FAN COIL AIR CONDITIONING** Cabinet mounted, filters							
010	Chilled water, 1/2 ton cooling	Q-5	8	Ea.	720	40	760	850
012	1 ton cooling		6		810	54	864	970
014	1.5 ton cooling		5.50		815	58	873	980
016	2.5 ton cooling		5		1,200	64	1,264	1,425
018	3 ton cooling		4		1,475	80	1,555	1,750
020	10 ton cooling	Q-6	2.80		1,650	180	1,830	2,075
022	15 ton cooling		1.50		2,075	335	2,410	2,775
024	20 ton cooling		.80		2,975	625	3,600	4,175
026	30 ton cooling		.60		4,600	835	5,435	6,275
100	Direct expansion, air cooled condensing, 5 ton cooling	Q-5	3		525	105	630	730
102	7.5 ton cooling	"	3		1,250	105	1,355	1,525
104	10 ton cooling	Q-6	2.60		1,525	190	1,715	1,950
106	20 ton cooling	"	.70		2,350	715	3,065	3,625

15

15.7 Air Conditioning & Ventilating		CREW	DAILY OUTPUT	UNIT	BARE COSTS MAT.	BARE COSTS INST.	BARE COSTS TOTAL	TOTAL INCL O&P
73 108	30 ton cooling	Q-6	.60	Ea.	3,250	835	4,085	4,775
110	40 ton cooling		.45		5,150	1,100	6,250	7,275
112	50 ton cooling		.38		6,200	1,325	7,525	8,725
150	For hot water coil, add				15%	10%		
76-001	HEAT PUMPS							
100	Air to air, split system, not including curbs or pads							
101	For curbs/pads see division 15.7-50							
102	2 ton cooling, 8.5 MBH heat @ 0°F	Q-5	1.20	Ea.	1,250	270	1,520	1,750
104	3 ton cooling, 13 MBH heat @0°F		.80		1,725	400	2,125	2,475
106	5 ton cooling, 27 MBH heat @0°F		.50		2,925	640	3,565	4,150
108	7 ton cooling, 33 MBH heat @0°F		.30		4,725	1,075	5,800	6,750
110	10 ton cooling, 50 MBH heat @0°F	Q-6	.38		6,200	1,325	7,525	8,725
112	15 ton cooling, 64 MBH heat @0°F	"	.26		8,550	1,925	10,475	12,200
114	25 ton cooling, 119 MBH heat @0°F	Q-7	.20		13,900	3,400	17,300	20,200
116	30 ton cooling, 163 MBH heat @0°F		.17		20,000	4,000	24,000	27,800
118	40 ton cooling, 193 MBH heat @0°F		.12		22,000	5,675	27,675	32,400
130	For supplementary electric heat coil, add				5%			
150	Single package, not including curbs, pads, or plenums							
152	2 ton cooling, 6.5 MBH heat @0°F	Q-5	1.50	Ea.	1,200	215	1,415	1,625
154	2.5 ton cooling, 8 MBH heat @0°F		1.40		1,525	230	1,755	2,000
156	3 ton cooling, 10 MBH heat @0°F		1.20		1,725	270	1,995	2,275
158	4 ton cooling, 13 MBH heat @0°F		.96		2,000	335	2,335	2,675
160	4.5 ton cooling, 14.1 MBH heat @0°F		.80		2,150	400	2,550	2,950
162	5 ton cooling, 27 MBH heat @0°F		.65		2,575	495	3,070	3,550
164	7 ton cooling, 35 MBH heat @0°F		.40		3,625	805	4,430	5,150
166	15 ton cooling, 56 MBH heat @0°F	Q-6	.30		9,450	1,675	11,125	12,800
167	20 ton cooling, 100 MBH heat @0°F	"	.20		12,100	2,500	14,600	16,900
168	25 ton cooling, 120 MBH heat @0°F	Q-7	.23		14,400	2,975	17,375	20,100
169	30 ton cooling, 163 MBH heat @0°F	"	.19		18,400	3,600	22,000	25,400
170	For supplementary electric heat coil, add				5%			
200	Water source to air, single package							
210	1 ton cooling, 13 MBH heat @75°F	Q-5	2	Ea.	845	160	1,005	1,150
212	1.5 ton cooling, 17 MBH heat @75°F		1.80		890	180	1,070	1,225
214	2 ton cooling, 19 MBH heat @75°F		1.70		950	190	1,140	1,325
216	2.5 ton cooling, 25 MBH heat @75°F		1.60		1,075	200	1,275	1,475
218	3 ton cooling, 27 MBH heat @75°F		1.40		1,150	230	1,380	1,600
220	4 ton cooling, 31 MBH heat @75°F		1.20		1,500	270	1,770	2,025
222	5 ton cooling, 29 MBH heat @75°F		.90		2,025	355	2,380	2,750
224	7.5 ton cooling, 35 MBH heat @75°F		.60		2,575	535	3,110	3,600
226	10 ton cooling, 50 MBH heat @75°F		.53		3,200	605	3,805	4,400
228	15 ton cooling, 64 MBH heat @75°F	Q-6	.47		5,600	1,075	6,675	7,700
230	20 ton cooling, 100 MBH heat @75°F	"	.41		6,400	1,225	7,625	8,800
396	For supplementary heat coil, add				10%			
400	For increase in capacity thru use							
402	of solar collector, size boiler at 60%							
79-001	LOUVERS							
010	Aluminum, extruded, with screen, mill finish							
100	Brick vent, (see also division 4.1-80)							
110	Standard, 4" deep, 8" wide, 5" high	1 Shee	24	Ea.	18.85	7.40	26.25	32
120	Modular, 4" deep, 7-3/4" wide, 5" high		24		20	7.40	27.40	33
130	Speed brick, 4" deep, 11-5/8" wide, 3-7/8" high		24		20	7.40	27.40	33
140	Fuel oil brick, 4" deep, 8" wide, 5" high		24		35	7.40	42.40	49
200	Cooling tower and mechanical equip., screens, light weight		40	S.F.	7.55	4.44	11.99	14.80
202	Standard weight		35		25	5.05	30.05	35
250	Dual combination, automatic, intake or exhaust		20		31	8.90	39.90	47
252	Manual operation		20		25	8.90	33.90	40
254	Electric or pneumatic operation		20		25	8.90	33.90	40
256	Motor, for electric or pneumatic		14	Ea.	115	12.70	127.70	145
300	Fixed blade, continuous line							

15.7 Air Conditioning & Ventilating

		CREW	DAILY OUTPUT	UNIT	BARE COSTS MAT.	BARE COSTS INST.	BARE COSTS TOTAL	TOTAL INCL O&P
79 310	Mullion type, stormproof	1 Shee	28	S.F.	23	6.35	29.35	35
320	Stormproof		28		25	6.35	31.35	37
330	Vertical line		28		31	6.35	37.35	43
350	For damper to use with above, add				50%	30%		
352	Motor, for damper, electric or pneumatic	1 Shee	14	Ea.	115	12.70	127.70	145
400	Operating, 45°, manual, electric or pneumatic		24	S.F.	25	7.40	32.40	38
410	Motor, for electric or pneumatic		14	Ea.	115	12.70	127.70	145
420	Penthouse, roof		56	S.F.	12.60	3.17	15.77	18.50
430	Walls		40		31	4.44	35.44	41
500	Thinline, under 4" thick, fixed blade		40		10.05	4.44	14.49	17.55
501	Finishes, applied by mfr. at additional cost, available in colors							
502	Prime coat only, add			S.F.	1.33		1.33	1.46M
504	Baked enamel finish coating, add				2.65		2.65	2.91M
506	Anodized finish, add				3.18		3.18	3.49M
508	Duranodic finish, add				6		6	6.60M
510	Fluoropolymer finish coating, add				9.30		9.30	10.25M
998	For small orders (under 10 pieces), add				20%			
82-001	**PACKAGED TERMINAL AIR CONDITIONER** Cabinet, wall sleeve,							
010	louver, electric heat, thermostat, manual changeover, 208 V							
020	6000 BTUH cooling, 8800 BTU heat	Q-5	6	Ea.	630	54	684	770
022	9,000 BTUH cooling, 13,900 BTU heat		5		655	64	719	815
024	12,000 BTUH cooling, 13,900 BTU heat		4		700	80	780	885
026	15,000 BTUH cooling, 13,900 BTU heat		3		730	105	835	960
028	19,000 BTUH cooling, 28,000 BTU heat		2.40		990	135	1,125	1,275
030	23,400 BTUH cooling, 28,000 BTU heat		1.90		1,025	170	1,195	1,375
032	31,000 BTUH cooling, 35,000 BTU heat		1.40		1,250	230	1,480	1,700
034	35,600 BTUH cooling, 35,000 BTU heat		1.25		1,300	255	1,555	1,800
036	42,500 BTUH cooling, 36,000 BTU heat		1		1,625	320	1,945	2,250
038	46,500 BTUH cooling, 54,000 BTU heat		.90		1,800	355	2,155	2,500
050	For hot water coil, increase heat by 10%, add				5%	10%		
100	For steam, increase heat output by 30%, add				8%	10%		
85-001	**ROOF TOP AIR CONDITIONERS** Standard controls, curb							
100	Single zone, electric cool, gas heat							
110	B8.4 -210 — 2 ton cooling, 60 MBH heating	Q-5	1.50	Ea.	2,075	215	2,290	2,600
112	4 ton cooling, 95 MBH heating		1.10		3,150	290	3,440	3,875
114	5 ton cooling, 112 MBH heating		.56		3,350	575	3,925	4,525
116	10 ton cooling, 200 MBH heating	Q-6	.46		6,475	1,075	7,550	8,700
118	15 ton cooling, 270 MBH heating	"	.31		8,625	1,600	10,225	11,800
120	20 ton cooling, 360 MBH heating	Q-7	.32		12,200	2,125	14,325	16,500
122	30 ton cooling, 540 MBH heating		.22		17,800	3,100	20,900	24,100
124	40 ton cooling, 675 MBH heating		.16		24,100	4,250	28,350	32,700
126	50 ton cooling, 810 MBH heating		.13		28,300	5,250	33,550	38,700
130	For hot water heat coil, deduct				6%			
140	For steam heat coil, deduct				2%			
150	For electric heat, deduct				3%	5%		
158	With variable volume distribution, 20 ton	Q-7	.32		11,900	2,125	14,025	16,200
160	30 ton		.14		16,900	4,875	21,775	25,600
162	50 ton		.08		23,000	8,525	31,525	37,600
170	Gas cool, gas heat							
172	3 ton cooling, 90 MBH heating	Q-5	1.40	Ea.	2,375	230	2,605	2,950
173	4 ton cooling, 112.5 MBH heating		1.10		3,150	290	3,440	3,875
175	5 ton cooling, 135 MBH heating		.56		3,475	575	4,050	4,650
200	Multizone, electric cool, gas heat, economizer							
210	B8.4 -220 — 15 ton cooling, 360 MBH heating	Q-7	.22	Ea.	20,800	3,100	23,900	27,400
212	20 ton cooling, 360 MBH heating		.21		21,800	3,250	25,050	28,700
214	25 ton cooling, 450 MBH heating		.18		24,600	3,800	28,400	32,500
216	28 ton cooling, 450 MBH heating		.16		27,400	4,250	31,650	36,300
218	30 ton cooling, 540 MBH heating		.15		29,600	4,550	34,150	39,100
220	37 ton cooling, 540 MBH heating		.13		34,200	5,250	39,450	45,200

15.7 Air Conditioning & Ventilating

		Crew	Daily Output	Unit	Bare Costs Mat.	Bare Costs Inst.	Bare Costs Total	Total Incl O&P
85 222	70 ton cooling, 1500 MBH heating	Q-7	.09	Ea.	48,600	7,575	56,175	64,500
224	80 ton cooling, 1500 MBH heating		.08		55,500	8,525	64,025	73,500
226	90 ton cooling, 1500 MBH heating		.07		62,500	9,750	72,250	83,000
228	105 ton cooling, 1500 MBH heating		.06		73,000	11,400	84,400	96,500
240	For hot water heat coil, deduct				5%			
250	For steam heat coil, deduct				2%			
260	For electric heat, deduct				3%	5%		
88-001	**SELF-CONTAINED SINGLE PACKAGE**							
010	Air cooled, for free blow or duct, not including remote condenser							
020	B8.4-240 3 ton cooling	Q-5	1	Ea.	2,450	320	2,770	3,150
021	4 ton cooling	"	.80		2,675	400	3,075	3,525
022	5 ton cooling	Q-6	1.20		3,100	415	3,515	4,000
024	10 ton cooling	Q-7	1		5,475	680	6,155	7,000
026	20 ton cooling		.90		8,975	760	9,735	11,000
028	30 ton cooling		.80		11,900	850	12,750	14,300
030	40 ton cooling		.60		16,800	1,125	17,925	20,100
032	50 ton cooling		.40		20,700	1,700	22,400	25,200
034	60 ton cooling	Q-8	.40		24,100	1,750	25,850	29,000
040	For duct mounting, no price change							
050	For hot water or steam heating coils, add			Ea.	10%	10%		
100	Water cooled for free blow or duct, not including tower							
110	B8.4-230 3 ton cooling	Q-6	1	Ea.	2,125	500	2,625	3,050
112	5 ton cooling	"	1		2,525	500	3,025	3,500
114	10 ton cooling	Q-7	.90		5,125	760	5,885	6,725
116	20 ton cooling		.80		8,400	850	9,250	10,500
118	30 ton cooling		.70		10,600	975	11,575	13,100
120	40 ton cooling		.40		14,300	1,700	16,000	18,200
122	50 ton cooling		.30		17,600	2,275	19,875	22,600
124	60 ton cooling	Q-8	.30		20,800	2,325	23,125	26,200
130	For hot water or steam heat coils, add				10%	10%		
91-001	**SHOCK AND VIBRATION ABSORBERS**							
010	Hangers, neoprene flex							
020	10-120 lb. capacity			Ea.	4.35		4.35	4.78M
022	75-550 lb. capacity				5.35		5.35	5.90M
024	250-1100 lb. capacity				13		13	14.30M
026	1000-4000 lb. capacity				37		37	41M
050	Spring flex, 60-450 lb. capacity				15.05		15.05	16.55M
052	85-450 lb. capacity				23		23	25M
054	600-900 lb. capacity				29		29	32M
056	1100-1300 lb. capacity				32		32	35M
100	Mounts, neoprene, 135-380 lb. capacity				5.35		5.35	5.90M
102	250-1100 lb. capacity				13		13	14.30M
104	1000-4000 lb. capacity				32		32	35M
110	Spring flex, 60-450 lb. capacity				15.80		15.80	17.40M
112	85-900 lb. capacity				24		24	26M
114	1100-1300 lb. capacity				27		27	30M
116	900-1800 lb. capacity				40		40	44M
118	2200-2600 lb. capacity				44		44	48M
120	1800-3600 lb. capacity				68		68	75M
122	4400-5200 lb. capacity				75		75	83M
124	3150-6300 lb. capacity				98		98	110M
126	5400-10,800 lb. capacity				175		175	195M
128	9600-14,400 lb. capacity				340		340	375M
130	12,000-18,000 lb. capacity				380		380	420M
132	17,600-20,800 lb. capacity				465		465	510M
134	22,000-26,000 lb. capacity				525		525	580M
200	Pads, cork rib, 18" x 18" x 1", 10-50 psi				56		56	62M
202	18" x 36" x 1", 10-50 psi				100		100	110M

15.7 Air Conditioning & Ventilating

Line	Description	Crew	Daily Output	Unit	Bare Costs Mat.	Bare Costs Inst.	Bare Costs Total	Total Incl O&P
91 210	Shear flexible pads, 18" x 18" x 3/8", 20-70 psi			Ea.	33		33	36M
212	18" x 36" x 3/8", 20-70 psi			"	64		64	70M
300	Note overlap in capacities due to deflections							
93-001	**SOLAR AIR CONDITIONING**							
010	Absorption chiller, standard controls							
020	3 ton	Q-5	.80	Ea.	1,550	400	1,950	2,275
022	25 ton	Q-6	.20	"	18,400	2,500	20,900	23,800
050	Fan coil filter unit, horizontal, chilled-hot water							
052	24 MBH cool, 60 MBH heat	Q-5	5	Ea.	565	64	629	715
054	36 MBH cool, 90 MBH heat		4		570	80	650	745
056	48 MBH cool, 109 MBH heat		3.70		710	87	797	905
058	60 MBH cool, 136 MBH heat		3.20		800	100	900	1,025
070	For fan delay switch, add				14		14	15.40M
96-001	**VENTILATORS** Base, damper & bird screen, CFM in 5 MPH wind							
050	Rotary syphon, galvanized, 6" neck diameter, 185 CFM	Q-9	16	Ea.	33	20	53	65
052	8" neck diameter, 215 CFM		14		36	23	59	73
054	10" neck diameters, 260 CFM		12		41	27	68	84
056	12" neck diameter, 310 CFM		10		45	32	77	96
058	14" neck diameter, 500 CFM		10		59	32	91	110
060	16" neck diameter, 635 CFM		9		68	36	104	125
062	18" neck diameter, 835 CFM		9		78	36	114	140
064	20" neck diameter, 1080 CFM		8		105	40	145	175
066	24" neck diameter, 1530 CFM		8		160	40	200	235
068	30" neck diameter, 2500 CFM		7		235	46	281	325
070	36" neck diameter, 3800 CFM		6		345	53	398	455
072	42" neck diameter, 4500 CFM		4		565	80	645	740
080	For aluminum, add				300%			
082	For stainless steel, add				600%			
128	Spinner ventilators, wind driven, galvanized							
130	4" neck diameter, 180 CFM	Q-9	20	Ea.	34	16	50	61
132	5" neck diameter, 210 CFM		18		38	17.75	55.75	68
134	6" neck diameter, 250 CFM		16		38	20	58	71
136	8" neck diameter, 360 CFM		14		41	23	64	78
138	10" neck diameter, 540 CFM		12		48	27	75	92
140	12" neck diameter, 770 CFM		10		56	32	88	110
144	16" neck diameter, 1200 CFM		9		85	36	121	145
146	18" neck diameter, 1700 CFM		9		96	36	132	155
148	20" neck diameter, 2100 CFM		8		125	40	165	195
150	24" neck diameter, 3100 CFM		8		195	40	235	275
152	30" neck diameter, 4500 CFM		7		310	46	356	410
154	36" neck diameter, 5500 CFM		6		440	53	493	560
160	For aluminum, add				300%			
162	For stainless steel, add				600%			
200	Stationary, gravity, syphon, galvanized							
210	3" diameter, 40 CFM	Q-9	24	Ea.	24	13.30	37.30	46
212	4" neck diameter, 50 CFM		20		24	16	40	50
214	5" neck diameter, 58 CFM		18		25	17.75	42.75	53
216	6" neck diameter, 66 CFM		16		25	20	45	57
218	7" neck diameter, 86 CFM		15		27	21	48	61
220	8" neck diameter, 110 CFM		14		28	23	51	64
222	10" neck diameter, 140 CFM		12		30	27	57	72
224	12" neck diameter, 160 CFM		10		39	32	71	90
226	14" neck diameter, 250 CFM		10		48	32	80	99
228	16" neck diameter, 380 CFM		9		58	36	94	115
230	18" neck diameter, 500 CFM		9		65	36	101	125
232	20" neck diameter, 625 CFM		8		88	40	128	155
234	24" neck diameter, 900 CFM		8		145	40	185	220
236	30" neck diameter, 1375 CFM		7		235	46	281	325
238	36" neck diameter, 2000 CFM		6		325	53	378	435

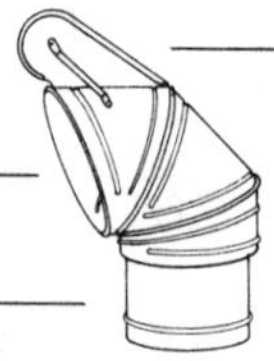

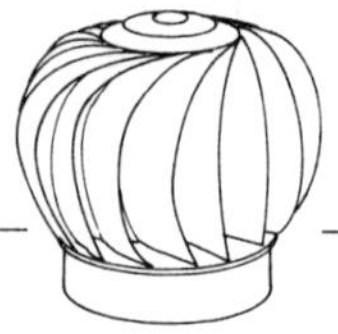

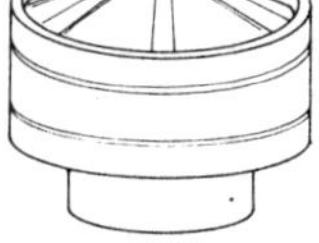

15

15.7 Air Conditioning & Ventilating

		CREW	DAILY OUTPUT	UNIT	BARE COSTS MAT.	BARE COSTS INST.	BARE COSTS TOTAL	TOTAL INCL O&P
96 240	42" neck diameter, 3000 CFM	Q-9	4	Ea.	425	80	505	585
250	For aluminum, add				300%			
252	For stainless steel, add				600%			
300	Rotating chimney cap, galvanized, 4" neck diameter	Q-9	20		26	16	42	52
302	5" neck diameter 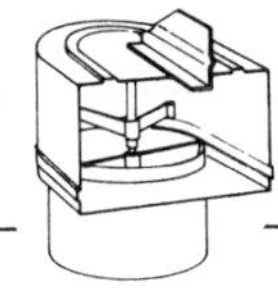		18		27	17.75	44.75	56
304	6" neck diameter		16		28	20	48	60
306	7" neck diameter		15		29	21	50	63
308	8" neck diameter		14		31	23	54	67
310	10" neck diameter		12		41	27	68	84
360	Stationary chimney rain cap, galvanized, 3" neck diameter		24		10.90	13.30	24.20	31
362	4" neck diameter		20		11	16	27	35
364	6" neck diameter		16		11.30	20	31.30	42
368	8" neck diameter		14		11.55	23	34.55	46
370	10" neck diameter		12		18.50	27	45.50	59
372	12" neck diameter		10		23	32	55	72
374	14" neck diameter		10		26	32	58	75
376	16" neck diameter		9		31	36	67	86
420	Stationary mushroom, aluminum, 16" orifice diameter		10		260	32	292	335
422	24" orifice diameter		9		325	36	361	410
424	36" orifice diameter		8		530	40	570	640
426	50" orifice diameter	↓	7	↓	825	46	871	975
500	Relief vent							
550	Rectangular, aluminum, galvanized curb							
551	intake/exhaust, 0.05" SP							
560	500 CFM, 12" x 16"	Q-9	8	Ea.	330	40	370	420
562	750 CFM, 12" x 20"		7.20		365	44	409	465
564	1000 CFM, 12" x 24"		6.60		390	48	438	500
566	1500 CFM, 12" x 36" 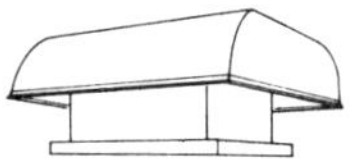		5.80		470	55	525	595
568	3000 CFM, 20" x 42"		4		670	80	750	855
570	6000 CFM, 20" x 84"		2.60		1,025	125	1,150	1,300
572	8000 CFM, 24" x 96"		2.30		1,175	140	1,315	1,500
574	10,000 CFM, 48" x 60"		1.80		1,500	180	1,680	1,900
576	12,500 CFM, 48" x 72"		1.60		1,700	200	1,900	2,150
578	15,000 CFM, 48" x 96"		1.30		1,975	245	2,220	2,525
580	20,000 CFM, 48" x 120"		1.20		2,250	265	2,515	2,875
582	25,000 CFM, 60" x 120"		.90		3,000	355	3,355	3,825
584	30,000 CFM, 72" x 120"		.70		3,950	455	4,405	5,000
586	40,000 CFM, 96" x 120"		.60		5,075	535	5,610	6,350
588	50,000 CFM, 96" x 144"	↓	.50	↓	5,375	640	6,015	6,850
700	Note: sizes based on exhaust. Intake, with 0.125" SP							
710	loss, approximately twice listed capacity.							
750	For power open, spring return damper							
760	motor for each S.F. of throat area	Q-9	100	S.F.	36	3.20	39.20	44
98-001	**WATER CHILLERS**							
002	Centrifugal liquid chiller, water cooled							
003	not including water tower							
010	Open drive, 2000 ton			Ea.	324,000	103,500	427,500	506,000
024	Packaged hermetic, 200 ton				53,000	16,900	69,900	82,500
028	400 ton				73,000	23,400	96,400	114,000
032	1000 ton				149,000	47,800	196,800	233,000
034	1300 ton				181,000	58,000	239,000	283,000
036	1500 ton				204,500	65,500	270,000	319,500
050	Reciprocating, air cooled, 20 ton cooling	Q-7	.35		10,600	1,950	12,550	14,500
052	40 ton cooling		.24		19,200	2,850	22,050	25,200
054	65 ton cooling		.14		25,600	4,875	30,475	35,200
060	100 ton cooling	↓	.10		38,500	6,825	45,325	52,000
062	110 ton cooling	Q-8	.09		40,000	7,775	47,775	55,000
063	125 ton cooling	"	.08	↓	42,500	8,750	51,250	59,500
068	Water cooled, single compressor, semi-hermetic							

15

15.7 Air Conditioning & Ventilating

	Air Conditioning & Ventilating	CREW	DAILY OUTPUT	UNIT	BARE COSTS MAT.	BARE COSTS INST.	BARE COSTS TOTAL	TOTAL INCL O&P
98 070	2 to 5 ton cooling	Q-5	.44	Ea.	3,925	730	4,655	5,375
072	6 ton cooling		.38		4,500	845	5,345	6,175
074	8 ton cooling		.27		6,100	1,200	7,300	8,425
076	10 ton cooling	Q-6	.36		7,050	1,400	8,450	9,750
078	15 ton cooling	″	.33		7,700	1,525	9,225	10,700
080	20 ton cooling	Q-7	.38		10,300	1,800	12,100	13,900
082	30 ton cooling		.28		12,600	2,425	15,025	17,400
084	35 ton cooling		.27		13,000	2,525	15,525	17,900
098	Water cooled, multiple compressors, semi-hermetic							
100	15 ton cooling	Q-6	.30	Ea.	8,500	1,675	10,175	11,800
102	20 ton cooling	Q-7	.36		9,500	1,900	11,400	13,200
104	25 ton cooling		.32		11,500	2,125	13,625	15,700
106	30 ton cooling		.27		12,800	2,525	15,325	17,700
108	40 ton cooling		.24		14,100	2,850	16,950	19,600
110	50 ton cooling		.21		16,400	3,250	19,650	22,700
112	60 ton cooling		.19		17,500	3,600	21,100	24,400
114	80 ton cooling	Q-8	.15		24,000	4,675	28,675	33,100
116	100 ton cooling		.12		29,000	5,825	34,825	40,300
118	120 ton cooling		.10		33,800	7,000	40,800	47,200
120	140 ton cooling		.09		38,700	7,775	46,475	53,500
130	Water cooled, single compressor, direct drive							
132	40 ton cooling,	Q-8	.19	Ea.	18,800	3,675	22,475	26,000
134	50 ton cooling		.16		22,500	4,375	26,875	31,000
136	60 ton cooling		.15		24,200	4,675	28,875	33,300
138	80 ton cooling		.12		30,500	5,825	36,325	41,900
145	Water cooled, dual compressors, direct drive							
150	80 ton cooling	Q-8	.10	Ea.	34,500	7,000	41,500	48,000
152	100 ton cooling		.09		38,000	7,775	45,775	53,000
154	120 ton cooling		.08		44,000	8,750	52,750	61,000
156	135 ton cooling		.07		49,200	10,000	59,200	68,500
158	150 ton cooling		.07		53,500	10,000	63,500	73,000
160	175 ton cooling		.06		61,500	11,700	73,200	84,500
162	200 ton cooling		.05		65,500	14,000	79,500	92,000
164	225 ton cooling		.05		71,000	14,000	85,000	98,000
166	250 ton cooling		.04		74,500	17,500	92,000	107,000
200	Packaged chiller, condenserless for remote condenser							
202	40 ton cooling	Q-7	.27	Ea.	16,100	2,525	18,625	21,400
204	60 ton cooling		.21		23,200	3,250	26,450	30,200
206	80 ton cooling		.16		27,000	4,250	31,250	35,900
208	100 ton cooling		.11		35,800	6,200	42,000	48,300
210	120 ton cooling		.10		38,900	6,825	45,725	52,500
212	160 ton cooling		.09		45,800	7,575	53,375	61,500
99-001	**WINDOW UNIT AIR CONDITIONERS**							
400	Portable, 15 amp 125V grounded receptacle required							
402	Standard models							
405	4000 BTUH, 1 speed fan	1 Carp	8	Ea.	195	19.60	214.60	245
406	4000 BTUH, 2 speed fan		8		225	19.60	244.60	275
408	5000 BTUH, 2 speed fan		8		255	19.60	274.60	310
410	For high efficiency (EER rating), add				40		40	44M
425	Semi-permanent installation, 3 speed fan							
426	15 amp 125V grounded receptacle required							
428	Standard models, 2 way air direction							
432	5000 BTUH	1 Carp	8	Ea.	275	19.60	294.60	330
434	6000 BTUH	″	8	″	310	19.60	329.60	370
440	High efficiency models							
445	5900 BTUH, 2 way air direction	1 Carp	8	Ea.	360	19.60	379.60	425
448	8000 BTUH, 4 way high thrust air		6		440	26	466	520
450	10,000 BTUH, 4 way high thrust air		6		440	26	466	520
452	12,000 BTUH, 4 way high thrust air	L-2	8		540	35	575	645
454	14,000 BTUH, 4 way high thrust air	″	8		600	35	635	710

15.7 Air Conditioning & Ventilating

		CREW	DAILY OUTPUT	UNIT	BARE COSTS MAT.	BARE COSTS INST.	BARE COSTS TOTAL	TOTAL INCL O&P
99 460	15 amp 250V grounded receptacle required							
470	High efficiency, 4 way high thrust air							
474	15,000 BTUH	L-2	6	Ea.	570	46	616	695
478	18,000 BTUH	"	6	"	660	46	706	795
482	20 amp 250V grounded receptacle required							
484	High efficiency, 4 way high thrust air							
486	21,000 BTUH	L-2	6	Ea.	700	46	746	835
490	30 amp 250V grounded receptacle required							
491	High efficiency, 4 way high thrust air							
494	25,000 BTUH	L-2	4	Ea.	830	70	900	1,025
496	29,000 BTUH	"	4	"	860	70	930	1,050

16.3 Starters, Boards & Switches

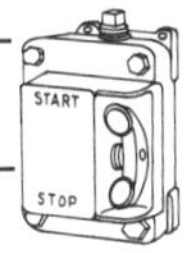

		CREW	DAILY OUTPUT	UNIT	BARE COSTS MAT.	BARE COSTS INST.	BARE COSTS TOTAL	TOTAL INCL O&P
15-001	CONTROL STATIONS							
005	NEMA 1, heavy duty, stop/start	1 Elec	8	Ea.	34	22	56	69
010	Stop/start, pilot light		6.20		81	29	110	130
020	Hand/off/automatic		6.20		34	29	63	78
040	Stop/start/reverse		5.30		62	33	95	115
050	NEMA 7, heavy duty, stop/start		6		98	29	127	150
060	Stop/start, pilot light	↓	4	↓	225	44	269	310
27-001	MOTOR CONNECTIONS							
002	Flexible conduit and fittings, 1 HP motor	1 Elec	8	Ea.	3	22	25	35
005	5 HP motor		6.50		3.50	27	30.50	43
010	7-1/2 HP motor		5.50		4.15	32	36.15	51
012	10 HP motor		4.20		4.15	42	46.15	65
015	15 HP motor		3.30		5.35	54	59.35	83
020	25 HP motor		2.70		10	65	75	105
040	50 HP motor		2.20		28	80	108	145
060	100 HP motor	↓	1.50	↓	115	120	235	295
30-001	MOTOR STARTERS & CONTROLS							
005	Magnetic, FVNR, with enclosure and heaters, 480 volt							
008	2 HP, size 00	1 Elec	3.50	Ea.	95	51	146	175
010	5 HP, size 0		2.30		115	77	192	235
020	10 HP, size 1		1.60		125	110	235	295
030	25 HP, size 2		1.10		240	160	400	495
040	50 HP, size 3		.90		390	195	585	710
050	100 HP, size 4		.60		865	295	1,160	1,375
060	200 HP, size 5		.45		2,025	395	2,420	2,800
061	400 HP, size 6		.40		5,700	440	6,140	6,900
062	NEMA 7, 5 HP, size 0		1.60		435	110	545	635
063	10 HP, size 1		1.10		450	160	610	725
064	25 HP, size 2		.90		725	195	920	1,075
065	50 HP, size 3		.60		1,075	295	1,370	1,600
066	100 HP, size 4		.45		1,725	395	2,120	2,450
067	200 HP, size 5		.25		4,125	705	4,830	5,550
070	Combination, with motor circuit protectors, 5 HP, size 0		1.80		370	98	468	550
080	10 HP, size 1		1.30		380	135	515	615
090	25 HP, size 2		1		535	175	710	845
100	50 HP, size 3		.66		775	270	1,045	1,225
120	100 HP, size 4		.40		1,675	440	2,115	2,475
122	NEMA 7, 5 HP, size 0	↓	1.30	↓	765	135	900	1,025

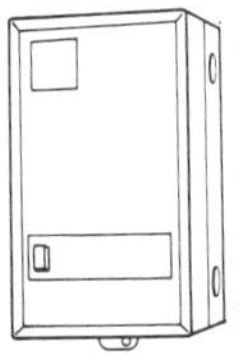

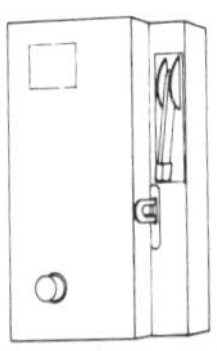

For expanded coverage of these items see *Means' Electrical Cost Data 1985*

16.3 Starters, Boards & Switches

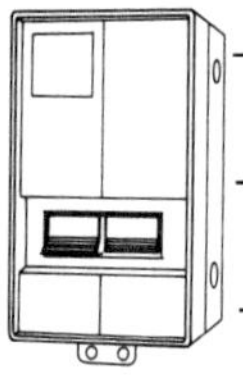

16.3		CREW	DAILY OUTPUT	UNIT	BARE COSTS MAT.	BARE COSTS INST.	BARE COSTS TOTAL	TOTAL INCL O&P
30 123	10 HP, size 1	1 Elec	1	Ea.	785	175	960	1,125
124	25 HP, size 2		.66		1,050	270	1,320	1,550
125	50 HP, size 3		.40		1,725	440	2,165	2,525
126	100 HP, size 4		.30		2,675	590	3,265	3,800
127	200 HP, size 5		.20		5,800	885	6,685	7,650
140	Combination, with fused switch, 5 HP, size 0		1.80		295	98	393	465
160	10 HP, size 1		1.30		310	135	445	535
180	25 HP, size 2		1		475	175	650	775
200	50 HP, size 3		.66		800	270	1,070	1,275
220	100 HP, size 4		.40		1,500	440	1,940	2,275
350	Magnetic FVNR with NEMA 12, enclosure & heaters, 480 volt							
360	5 HP, size 0	1 Elec	2.20	Ea.	145	80	225	275
370	10 HP, size 1		1.50		160	120	280	345
380	25 HP, size 2		1		300	175	475	585
390	50 HP, size 3		.80		465	220	685	830
400	100 HP, size 4		.50		1,100	355	1,455	1,725
410	200 HP, size 5		.40		2,650	440	3,090	3,550
420	Combination with motor circuit protectors, 5 HP, size 0		1.70		435	105	540	630
430	10 HP, size 1		1.20		450	145	595	705
440	25 HP, size 2		.90		630	195	825	975
450	50 HP, size 3		.60		900	295	1,195	1,425
460	100 HP, size 4		.37		2,025	480	2,505	2,925
470	Combination with fused switch, 5 HP, size 0		1.70		360	105	465	545
480	10 HP, size 1		1.20		375	145	520	625
490	25 HP, size 2		.90		570	195	765	910
500	50 HP, size 3		.60		925	295	1,220	1,450
510	100 HP, size 4		.37		1,850	480	2,330	2,725
520	Factory installed controls, adders to size 0 thru 5							
530	Start-stop push button	1 Elec	32	Ea.	23	5.55	28.55	33
540	Hand-off-auto-selector switch		32		23	5.55	28.55	33
550	Pilot light		32		42	5.55	47.55	54
560	Start-stop-pilot		32		65	5.55	70.55	79
570	Auxiliary contact, NO or NC		32		31	5.55	36.55	42
580	NO-NC		32		62	5.55	67.55	76

16.8 Special Systems

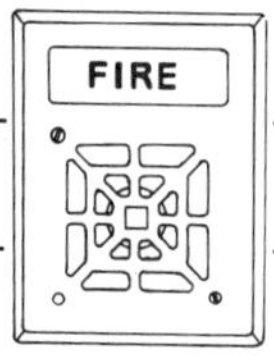

16.8		CREW	DAILY OUTPUT	UNIT	BARE COSTS MAT.	BARE COSTS INST.	BARE COSTS TOTAL	TOTAL INCL O&P
15-001	**DETECTION SYSTEMS**							
010	Burglar alarm, battery operated, mechanical trigger	1 Elec	4	Ea.	125	44	169	200
020	Electrical trigger		4		155	44	199	235
040	For outside key control, add		8		21	22	43	55
060	For remote signaling circuitry, add		8		37	22	59	72
080	Card reader, flush type, standard		2.70		540	65	605	690
100	Multi-code		2.70		675	65	740	835
360	Fire, sprinkler & standpipe alarm, control panel, 4 zone		2		660	88	748	855
380	8 zone		1		950	175	1,125	1,300
400	12 zone		.66		1,325	270	1,595	1,850
402	Alarm device		8		85	22	107	125
405	Actuating device		8		225	22	247	280
420	Battery and rack		4		530	44	574	645
440	Automatic charger		8		340	22	362	405
460	Signal bell		8		32	22	54	67
480	Trouble buzzer or manual station		8		21	22	43	55
500	Detector, rate of rise		8		12	22	34	45
510	Fixed temperature		8		12	22	34	45

16

For expanded coverage of these items see *Means' Electrical Cost Data 1985*

16.8 Special Systems

		CREW	DAILY OUTPUT	UNIT	BARE COSTS MAT.	BARE COSTS INST.	BARE COSTS TOTAL	TOTAL INCL O&P
15 520	Smoke detector, ceiling type	1 Elec	6.20	Ea.	64	29	93	110
540	Duct type		3.20		185	55	240	285
560	Light and horn		5.30		75	33	108	130
580	Fire alarm horn		6.70		21	26	47	61
600	Door holder, electro-magnetic		4		48	44	92	115
620	Combination holder and closer		3.20		320	55	375	430
640	Code transmitter		4		530	44	574	645
660	Drill switch		8		32	22	54	67
680	Master box		2.70		1,150	65	1,215	1,350
700	Break glass station		8		27	22	49	61
780	Remote annunciator, 8 zone lamp		1.80		160	98	258	315
800	12 zone lamp		1.30		210	135	345	425
820	16 zone lamp		1.10		265	160	425	520
33-001	ELECTRIC HEATING							
020	Snow melting for paved surface embedded mat heaters & controls	1 Elec	130	S.F.	4.10	1.36	5.46	6.45
040	Cable heating, radiant heat plaster, no controls, in South		130		4.40	1.36	5.76	6.80
060	In North		90		4.40	1.96	6.36	7.65
080	Cable on 1/2" board, not incl. controls, tract housing		90		3.90	1.96	5.86	7.10
100	Custom housing		80		4.40	2.21	6.61	8
110	Rule of thumb: Baseboard units, including control		4.40	KW	42	40	82	105
120	Duct heaters, including controls		5.30	"	36	33	69	88
130	Baseboard heaters, 2' long, 375 watt		8	Ea.	28	22	50	63
140	3' long, 500 watt		8		35	22	57	70
160	4' long, 750 watt		6.70		43	26	69	85
180	5' long, 935 watt		5.70		57	31	88	105
200	6' long, 1125 watt		5		64	35	99	120
220	7' long, 1310 watt		4.40		79	40	119	145
240	8' long, 1500 watt		4		86	44	130	160
260	9' long, 1680 watt		3.60		98	49	147	180
280	10' long, 1875 watt		3.30		98	54	152	185
295	Wall heaters with fan, 120 to 277 volt							
297	surface mounted, residential, 750 watt	1 Elec	7	Ea.	45	25	70	86
298	1000 watt		7		48	25	73	89
299	1250 watt		6		65	29	94	115
300	1500 watt		5		68	35	103	125
301	2000 watt		5		69	35	104	125
304	2250 watt		4		85	44	129	155
305	2500 watt		4		105	44	149	180
307	4000 watt		3.50		105	51	156	190
308	Commercial, 750 watt		7		47	25	72	88
309	1000 watt		7		84	25	109	130
310	1250 watt		6		84	29	113	135
311	1500 watt		5		130	35	165	195
312	2000 watt		5		130	35	165	195
313	2500 watt		4.50		140	39	179	210
314	3000 watt		4		140	44	184	220
315	4000 watt		3.50		150	51	201	240
316	Recessed, residential, 750 watt		6		55	29	84	105
317	1000 watt		6		55	29	84	105
318	1250 watt		5		83	35	118	140
319	1500 watt		4		83	44	127	155
321	2000 watt		4		85	44	129	155
323	2500 watt		3.50		105	51	156	190
324	3000 watt		3		105	59	164	200
325	4000 watt		2.70		105	65	170	210
326	Commercial, 750 watt		6		68	29	97	115
327	1000 watt		6		79	29	108	130
328	1250 watt		5		84	35	119	145
329	1500 watt		4		83	44	127	155

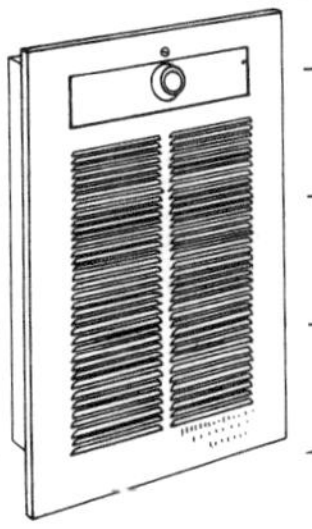

For expanded coverage of these items see *Means' Electrical Cost Data 1985*

16.8 Special Systems

		CREW	DAILY OUTPUT	UNIT	BARE COSTS MAT.	BARE COSTS INST.	BARE COSTS TOTAL	TOTAL INCL O&P
33 330	2000 watt	1 Elec	4	Ea.	85	44	129	155
331	2500 watt		3.50		130	51	181	215
332	3000 watt		3		130	59	189	230
333	4000 watt		2.70		105	65	170	210
360	Thermostats, integral		16		15	11.05	26.05	32
380	Line voltage, 1 pole		8		14	22	36	47
381	2 pole		8		18	22	40	52
382	Low voltage, 1 pole		8		14	22	36	47
500	Radiant heating ceiling panels, 2' x 4', 500 watt		16		115	11.05	126.05	140
505	750 watt		16		105	11.05	116.05	130
520	For recessed plaster frame, add		32		25	5.55	30.55	35
530	Infra-red quartz heaters, 120 volts, 1000 watts		6.70		76	26	102	120
535	1500 watt		5		76	35	111	135
540	240 volts, 1500 watt		5		76	35	111	135
545	2000 watt		4		76	44	120	145
550	3000 watt		3		93	59	152	185
555	4000 watt		2.60		110	68	178	220
557	Modulating control		.80		31	220	251	350
560	Unit heaters, heavy duty, with fan & mounting bracket							
565	Single phase, 208-240-277 volt, 3 KW	1 Elec	3.20	Ea.	140	55	195	235
570	4 KW		2.80		140	63	203	245
575	5 KW		2.40		160	74	234	280
580	7 KW		1.90		255	93	348	415
585	10 KW		1.30		305	135	440	530
590	13 KW		1		420	175	595	715
595	15 KW		.90		410	195	605	735
600	480 volt, 3KW		3.30		230	54	284	330
602	4 KW		3		230	59	289	340
604	5 KW		2.60		230	68	298	350
606	7 KW		2		265	88	353	420
608	10 KW		1.40		365	125	490	585
610	13 KW		1.10		375	160	535	645
612	15 KW		1		435	175	610	735
614	20 KW		.90		520	195	715	855
630	3 phase, 208-240 volt, 5 KW		2.40		175	74	249	300
632	7 KW		1.90		250	93	343	410
634	10 KW		1.30		305	135	440	530
636	15 KW		.90		400	195	595	720
638	20 KW		.70		480	255	735	890
640	25 KW		.50		585	355	940	1,150
650	480 volt, 5 KW		2.60		240	68	308	360
652	7 KW		2		265	88	353	420
654	10 KW		1.40		365	125	490	585
656	13 KW		1.10		375	160	535	645
658	15 KW		1		435	175	610	735
660	20 KW		.90		525	195	720	860
662	25 KW		.60		585	295	880	1,075
663	30 KW		.70		860	255	1,115	1,300
664	40 KW		.60		1,075	295	1,370	1,600
665	50 KW		.50		1,275	355	1,630	1,900
680	Vertical discharge heaters, with fan							
682	Single phase, 208-240-277 volt, 10 KW	1 Elec	1.30	Ea.	490	135	625	735
684	15 KW		.90		575	195	770	915
690	3 phase, 208-240 volt, 10 KW		1.30		440	135	575	680
692	15 KW		.90		500	195	695	830
694	20 KW		.70		720	255	975	1,150
696	25 KW		.50		740	355	1,095	1,325
698	30 KW		.40		870	440	1,310	1,600
700	40 KW		.36		1,025	490	1,515	1,825
702	50 KW		.32		1,200	555	1,755	2,125

For expanded coverage of these items see *Means' Electrical Cost Data 1985*

16.8 Special Systems

		CREW	DAILY OUTPUT	UNIT	BARE COSTS MAT.	BARE COSTS INST.	BARE COSTS TOTAL	TOTAL INCL O&P
33 710	480 volt, 10 KW	1 Elec	1.40	Ea.	515	125	640	750
712	15 KW		1		610	175	785	925
714	20 KW		.90		700	195	895	1,050
716	25 KW		.60		800	295	1,095	1,300
718	30 KW		.50		945	355	1,300	1,550
720	40 KW		.40		1,125	440	1,565	1,875
722	50 KW		.35		1,250	505	1,755	2,100
790	Cabinet convector heaters, 240 volt							
792	2 ft. long, 1000 watt	1 Elec	5.30	Ea.	200	33	233	270
794	1500 watt		5.30		210	33	243	280
796	2000 watt		5.30		225	33	258	295
798	3 ft. long, 1500 watt		4.60		230	38	268	310
800	2250 watt		4.60		240	38	278	320
802	3000 watt		4.60		255	38	293	335
804	4 ft. long, 2000 watt		4		260	44	304	350
806	3000 watt		4		275	44	319	365
808	4000 watt		4		290	44	334	385
810	Available also in 208 or 277 volt							
820	Cabinet unit heaters, 120 to 277 volt, 1 pole,							
822	wall mounted, 2000 watt	1 Elec	4.60	Ea.	650	38	688	770
823	3000 watt		4.60		675	38	713	800
824	4000 watt		4.40		720	40	760	850
825	5000 watt		4.40		745	40	785	875
826	6000 watt		4.20		765	42	807	900
827	8000 watt		4		780	44	824	920
828	10,000 watt		3.80		805	47	852	950
829	12,000 watt		3.50		920	51	971	1,075
830	13,500 watt		2.90		1,075	61	1,136	1,275
831	16,000 watt		2.70		1,075	65	1,140	1,275
832	20,000 watt		2.30		1,175	77	1,252	1,400
833	24,000 watt		1.90		1,275	93	1,368	1,525
835	Recessed, 2000 watt		4.40		710	40	750	840
837	3000 watt		4.40		735	40	775	865
838	4000 watt		4.20		780	42	822	920
839	5000 watt		4.20		805	42	847	945
840	6000 watt		4		825	44	869	970
841	8000 watt		3.80		840	47	887	990
842	10,000 watt		3.50		865	51	916	1,025
843	12,000 watt		2.90		980	61	1,041	1,175
844	13,500 watt		2.70		1,125	65	1,190	1,325
845	16,000 watt		2.30		1,150	77	1,227	1,375
846	20,000 watt		1.90		1,250	93	1,343	1,500
847	24,000 watt		1.60		1,350	110	1,460	1,650
849	Ceiling mounted, 2000 watt		3.20		655	55	710	800
851	3000 watt		3.20		680	55	735	825
852	4000 watt		3		720	59	779	875
853	5000 watt		3		745	59	804	905
854	6000 watt		2.80		765	63	828	930
855	8000 watt		2.40		780	74	854	965
856	10,000 watt		2.20		805	80	885	1,000
857	12,000 watt		2		920	88	1,008	1,150
858	13,500 watt		1.50		1,075	120	1,195	1,350
859	16,000 watt		1.30		1,075	135	1,210	1,375
860	20,000 watt		.90		1,175	195	1,370	1,575
861	24,000 watt		.60		1,275	295	1,570	1,825
863	208 to 480 volt, 3 pole							
865	Wall mounted, 2000 watt	1 Elec	4.60	Ea.	650	38	688	770
867	3000 watt		4.60		675	38	713	800
868	4000 watt		4.40		720	40	760	850
869	5000 watt		4.40		745	40	785	875

For expanded coverage of these items see *Means' Electrical Cost Data 1985*

16.8 Special Systems

		CREW	DAILY OUTPUT	UNIT	BARE COSTS MAT.	BARE COSTS INST.	BARE COSTS TOTAL	TOTAL INCL O&P
33 870	6000 watt	1 Elec	4.20	Ea.	765	42	807	900
871	8000 watt		4		780	44	824	920
872	10,000 watt		3.80		805	47	852	950
873	12,000 watt		3.50		920	51	971	1,075
874	13,500 watt		2.90		1,075	61	1,136	1,275
875	16,000 watt		2.70		1,075	65	1,140	1,275
876	20,000 watt		2.30		1,175	77	1,252	1,400
877	24,000 watt		1.90		1,275	93	1,368	1,525
879	Recessed, 2000 watt		4.40		710	40	750	840
881	3000 watt		4.40		735	40	775	865
882	4000 watt		4.20		780	42	822	920
883	5000 watt		4.20		805	42	847	945
884	6000 watt		4		825	44	869	970
885	8000 watt		3.80		840	47	887	990
886	10,000 watt		3.50		865	51	916	1,025
887	12,000 watt		2.90		980	61	1,041	1,175
888	13,500 watt		2.70		1,125	65	1,190	1,325
889	16,000 watt		2.30		1,150	77	1,227	1,375
890	20,000 watt		1.90		1,250	93	1,343	1,500
892	24,000 watt		1.60		1,350	110	1,460	1,650
894	Ceiling mount, 2000 watt		3.20		650	55	705	795
895	3000 watt		3.20		675	55	730	820
896	4000 watt		3		720	59	779	875
897	5000 watt		3		745	59	804	905
898	6000 watt		2.80		765	63	828	930
899	8000 watt		2.40		780	74	854	965
900	10,000 watt		2.20		805	80	885	1,000
901	12,000 watt		2		920	88	1,008	1,150
902	13,500 watt		1.50		1,075	120	1,195	1,350
903	16,000 watt		1.30		1,075	135	1,210	1,375
904	20,000 watt		.90		1,200	195	1,395	1,600
906	24,000 watt	↓	.60	↓	1,275	295	1,570	1,825
81-001	**WIRING, RESIDENTIAL** 20' average runs							
018	Air conditioning receptacle							
020	Using non-metallic sheathed cable	1 Elec	10	Ea.	6	17.70	23.70	32
024	BX cable		8.30		10	21	31	42
026	EMT conduit		6.70		10	26	36	49
028	Aluminum conduit	↓	5	↓	17.40	35	52.40	70
098	Exhaust fan wiring							
100	Using non-metallic sheathed cable	1 Elec	10	Ea.	5	17.70	22.70	31
102	BX cable		8.30		8	21	29	39
106	EMT conduit		6.70		8	26	34	47
108	Aluminum conduit		5		16	35	51	68
110	Galvanized steel conduit	↓	4.70	↓	15	38	53	71
118	Fire alarm smoke detector & horn							
120	Using non-metallic sheathed cable	1 Elec	10	Ea.	36	17.70	53.70	65
122	BX cable		8.30		39	21	60	74
124	EMT conduit		6.70		39	26	65	81
128	Aluminum conduit		5		46	35	81	100
130	Galvanized steel conduit	↓	4.70	↓	45	38	83	105
158	Furnace circuit and switch							
160	Using non-metallic sheathed cable	1 Elec	6	Ea.	7.10	29	36.10	50
162	BX cable		5		11	35	46	63
164	EMT conduit		4		11	44	55	76
168	Aluminum conduit		3		18.60	59	77.60	105
169	Galvanized steel conduit	↓	2.80	↓	17.60	63	80.60	110
179	Heater circuits							
180	Using non-metallic sheathed cable	1 Elec	8	Ea.	5	22	27	37
182	BX cable		6.60		8	27	35	47
184	EMT conduit	↓	5.40	↓	8	33	41	56

For expanded coverage of these items see *Means' Electrical Cost Data 1985*

16.8 Special Systems

		CREW	DAILY OUTPUT	UNIT	BARE COSTS MAT.	BARE COSTS INST.	BARE COSTS TOTAL	TOTAL INCL O&P
81 186	Aluminum conduit	1 Elec	4	Ea.	16	44	60	81
188	Galvanized steel conduit	"	3.80	"	15	47	62	83
358	Water heater circuit							
360	Using non-metallic sheathed cable	1 Elec	5	Ea.	7	35	42	59
362	BX cable		4.20		12	42	54	74
364	EMT conduit		3.40		12	52	64	88
366	Aluminum conduit		2.50		18	71	89	120
368	Galvanized steel conduit		2.40		17	74	91	125

17.1 S.F., C.F. and % of Total Costs

		UNIT	UNIT COSTS 1/4	UNIT COSTS MEDIAN	UNIT COSTS 3/4	% OF TOTAL 1/4	% OF TOTAL MEDIAN	% OF TOTAL 3/4
01-001	**APARTMENTS** Low Rise (1 to 3 story)	S.F.	31	39	52			
002	Total project cost	C.F.	2.73	3.54	4.42			
272	C14.1 -000 Plumbing	S.F.	2.40	3.16	4.01	6.80%	9%	10.20%
277	Heating, ventilating, air conditioning		1.54	1.90	2.75	4.20%	5.80%	7.60%
290	Electrical		1.80	2.41	3.34	5.30%	6.70%	8.90%
310	Total: Mechanical & Electrical		5.35	6.55	8.45	16%	18.30%	22.40%
900	Per apartment unit, total cost	Apt.	24,800	35,900	53,600			
950	Total: Mechanical & Electrical	"	4,425	6,425	8,875			
02-001	**APARTMENTS** Mid Rise (4 to 7 story)	S.F.	42	49	61			
002	Total project costs	C.F.	3.22	4.49	5.30			
272	Plumbing	S.F.	2.41	3.02	4.08	6.20%	7.40%	8.90%
290	Electrical		2.75	3.69	4.43	6.30%	6.90%	8.80%
310	Total: Mechanical & Electrical		7.65	9.35	11	17.10%	19.50%	20.80%
900	Per apartment unit, total cost	Apt.	37,600	49,300	64,900			
950	Total: Mechanical & Electrical	"	9,100	10,500	15,500			
03-001	**APARTMENTS** High Rise (8 to 24 story)	S.F.	46.20	53.40	61.85			
002	Total project costs	C.F.	3.77	4.96	6.23			
272	Plumbing	S.F.	2.99	4.06	5.03	6.70%	9.10%	10.40%
290	Electrical		3.39	4.14	5.51	7%	7.60%	9.10%
310	Total: Mechanical & Electrical		9.60	11.70	14.45	19.70%	22%	24.50%
900	Per apartment unit, total cost	Apt.	41,850	50,200	55,500			
950	Total: Mechanical & Electrical	"	10,400	12,000	12,700			
04-001	**AUDITORIUMS**	S.F.	45.60	64.70	83.45			
002	Total project costs	C.F.	2.86	4.18	5.80			
272	Plumbing	S.F.	2.98	3.96	5.14	5.70%	6.80%	8.40%
277	Heating, ventilating, air conditioning		6.20	15	17.40	6.90%	16%	19.80%
290	Electrical		3.68	5.25	6.90	6.80%	8.80%	10.90%
310	Total: Mechanical & Electrical		7.75	9.90	16	14.40%	17.80%	23.30%
05-001	**AUTOMOTIVE SALES**		32.30	40.90	53.30			
002	Total project costs	C.F.	2.39	2.84	3.68			
272	Plumbing	S.F.	1.62	2.27	3.10	2.80%	5.90%	6.40%
277	Heating, ventilating, air conditioning		2.36	3.92	5.65	6.30%	10.20%	10.70%
290	Electrical		2.79	4.90	6.45	7.30%	9.90%	12.30%
310	Total: Mechanical & Electrical		5.90	11.20	13.10	15.40%	19.20%	30.30%
06-001	**BANKS**		71.30	89.45	117.20			
002	Total project costs	C.F.	5.04	6.85	8.95			
272	Plumbing	S.F.	2.35	3.34	4.95	2.90%	4%	5%
277	Heating, ventilating, air conditioning		4.50	6.15	8.75	5.20%	7.40%	8.70%
290	Electrical		6.90	9.10	12.20	8.40%	10.30%	12.30%
310	Total: Mechanical & Electrical		11.55	16	22.75	14.20%	18%	23.50%
13-001	**CHURCHES**		47.70	59.35	73.70			
002	Total project costs	C.F.	3.03	3.77	4.89			

For expanded coverage of these items see *Means' Square Foot Costs Data 1985*

17.1 S.F., C.F. and % of Total Costs

		UNIT	UNIT COSTS 1/4	UNIT COSTS MEDIAN	UNIT COSTS 3/4	% OF TOTAL 1/4	% OF TOTAL MEDIAN	% OF TOTAL 3/4
13 272	Plumbing	S.F.	1.90	2.72	3.96	3.60%	4.90%	6.40%
277	Heating, ventilating, air conditioning		4.48	5.85	8.20	7.80%	10%	12.10%
290	Electrical		3.92	5.28	6.93	7.30%	8.80%	10.90%
310	Total: Mechanical & Electrical		8.52	12.15	16.15	15.90%	21.80%	26.60%
15-001	**CLUBS, COUNTRY**	↓	47.85	57.40	75.10			
002	Total project costs	C.F.	3.98	4.95	6.80			
272	Plumbing	S.F.	2.95	4.09	7.40	5.40%	8.90%	10%
277	Heating, ventilating, air conditioning		2.85	6.10	9.10	6.70%	10%	10.70%
290	Electrical		3.44	5.60	7.45	7.80%	9.30%	11%
310	Total: Mechanical & Electrical		9.35	12.65	20.25	17.20%	23.70%	30.90%
17-001	**CLUBS, SOCIAL** Fraternal	↓	41.05	56.50	73.80			
002	Total project costs	C.F.	2.43	3.85	4.65			
272	Plumbing	S.F.	1.94	3.05	3.48	4.90%	6.70%	7.80%
277	Heating, ventilating, air conditioning		3.64	5.40	6.10	8.20%	10.90%	14.40%
290	Electrical		2.90	4.96	5.60	6.70%	9.50%	11.40%
310	Total: Mechanical & Electrical		8.90	11.95	15.75	18.50%	28.70%	31.90%
18-001	**CLUBS, Y.M.C.A.**	↓	48.15	60.70	73			
002	Total project costs	C.F.	2.35	3.87	4.64			
272	Plumbing	S.F.	3.44	5.20	7.20	6%	8.40%	11%
290	Electrical		3.35	4.74	6.15	6%	7.80%	9%
310	Total: Mechanical & Electrical	↓	9.60	14.85	22.50	17.20%	27.90%	32.10%
19-001	**COLLEGES** Classrooms & Administration	S.F.	59.90	78.25	101			
002	Total project costs	C.F.	4.12	5.90	8.70			
272	Plumbing	S.F.	3.10	4.18	6.65	4%	6.40%	8%
277	Heating, ventilating, air conditioning		6.23	9.90	14.70	8.70%	12.20%	14.60%
290	Electrical		4.52	7.40	11.15	7.80%	9.80%	11.50%
310	Total: Mechanical & Electrical		13	21.75	29.75	18.70%	28%	35.10%
21-001	**COLLEGES** Science, Engineering, Laboratories	↓	67.80	91.55	110			
002	Total project costs	C.F.	5.05	6.90	8			
272	Plumbing	S.F.	3.33	4.98	6.95	5.90%	6.80%	8%
277	Heating, ventilating, air conditioning		5.20	10.65	12.60	8.20%	14.40%	19.10%
290	Electrical		6.05	8.65	12.95	8.10%	10.10%	11.40%
310	Total: Mechanical & Electrical		19	27.40	34.95	25.40%	34.90%	38.70%
23-001	**COLLEGES** Student Unions	↓	59.85	84.30	97.15			
002	Total project costs	C.F.	3.41	4.32	5.53			
272	Plumbing	S.F.	3.75	4.85	6.20	4.20%	5.20%	8.60%
277	Heating, ventilating, air conditioning		9.15	10.65	15.50	10.90%	14.10%	17.40%
290	Electrical		4.85	7.30	9.70	7.90%	9.70%	10.70%
310	Total: Mechanical & Electrical		15.90	22.40	27.15	23.40%	26.20%	31.70
25-001	**COMMUNITY CENTERS**	↓	50	61.40	77.60			
002	Total project costs	C.F.	3.06	4.38	5.65			
272	Plumbing	S.F.	2.60	4.22	5.70	5.40%	7%	9.10%
277	Heating, ventilating, air conditioning		4.17	5.95	7.95	7.60%	10.20%	12.90%
290	Electrical		4.17	5.35	7.60	7.50%	9.10%	10.80%
310	Total: Mechanical & Electrical		10.20	15.10	21.45	19.20%	26%	30.70%
28-001	**COURT HOUSES**	↓	67	81.75	101			
002	Total project costs	C.F.	5.40	6.10	7.10			
272	Plumbing	S.F.	3.46	4.67	5.48	3.90%	6.30%	7.40%
277	Heating, ventilating, air conditioning		9.10	10.25	14.90	10.30%	11.90%	14.80%
290	Electrical		6.35	7.85	12.60	8.40%	9.80%	12%
310	Total: Mechanical & Electrical		14.70	20.75	29.90	20.10%	25.70%	28.70%
30-001	**DEPARTMENT STORES**	↓	27.30	36.40	41.25			
002	Total project costs	C.F.	1.22	1.78	2.36			
272	Plumbing	S.F.	.88	1.23	1.60	2.80%	3.90%	5.30%
277	Heating, ventilating, air conditioning		2.84	3.77	5.70	8.30%	11.80%	14.80%
290	Electrical		3.12	4.20	5.20	10.40%	11.90%	13.60%
310	Total: Mechanical & Electrical		5.90	10.25	11.85	22.10%	26.70%	32.30%
31-001	**DORMITORIES** Low Rise (1 to 3 story)	↓	43.70	59.80	75.75			
002	Total project costs	C.F.	3.29	5.35	6.90			

For expanded coverage of these items see *Means' Square Foot Costs Data 1985*

17.1 S.F., C.F. and % of Total Costs		UNIT	UNIT COSTS 1/4	UNIT COSTS MEDIAN	UNIT COSTS 3/4	% OF TOTAL 1/4	% OF TOTAL MEDIAN	% OF TOTAL 3/4
31 272	Plumbing	S.F.	3.07	3.85	5.15	8%	8.90%	9.60%
277	Heating, ventilating, air conditioning		3.24	3.73	5.20	4.60%	7.60%	9.90%
290	Electrical		3.14	4.31	5.20	6.40%	7.80%	9.50%
310	Total: Mechanical & Electrical		7.15	11.10	16.25	18.20%	20.90%	28.40%
900	Per bed, total cost	Bed	9,170	14,680	26,180			
32-001	**DORMITORIES** Mid Rise (4 to 8 story)	S.F.	64.55	75.10	90.50			
002	Total project costs	C.F.	5.65	6.90	7.60			
272	Plumbing	S.F.	5	5.05	7.40	6.40%	6.60%	10.30%
290	Electrical		5.15	7.30	8.80	8.20%	8.90%	9.70%
310	Total: Mechanical & Electrical		13.30	16.60	23.90	17.70%	22.20%	25.70%
900	Per bed, total cost	Bed	8,780	15,150	18,560			
34-001	**FACTORIES**	S.F.	22.65	31.90	54.60			
002	Total project costs	C.F.	2	2.15	3.20			
272	Plumbing	S.F.	1.76	2.44	2.67	4.10%	4.40%	4.70%
277	Heating, ventilating, air conditioning		3.90	3.95	4.80	7%	9.20%	10.40%
290	Electrical		2.95	3.65	5.70	6.20%	7.70%	11%
310	Total: Mechanical & Electrical		8.10	10.55	12.30	18.80%	20.80%	27.20%
36-001	**FIRE STATIONS**		46.80	63.05	77.85			
002	Total project costs	C.F.	3	3.95	5.10			
272	Plumbing	S.F.	3.04	4.76	6.27	5.90%	7.20%	9.40%
277	Heating, ventilating, air conditioning		2.62	4.18	6.64	4.90%	7.30%	9.20%
290	Electrical		3.42	5.90	8.55	7.20%	9.70%	12.10%
310	Total: Mechanical & Electrical		8.90	13.90	19.20	17.40%	22.50%	27.50%
37-001	**FRATERNITY HOUSES** And Sorority Houses		47.10	55.75	62.20			
002	Total project costs	C.F.	4.50	5.45	6			
272	Plumbing	S.F.	3.55	4.10	5.45	5.90%	8%	10.80%
290	Electrical		3.10	4.06	7.35	6.50%	8.80%	10.40%
310	Total: Mechanical & Electrical		8.80	12.35	14.90	14.60%	20.70%	24.20%
38-001	**FUNERAL HOMES**	S.F.	44.45	55.85	81.85			
002	Total project costs	C.F.	3.14	4.52	5.36			
272	Plumbing	S.F.	1.76	2.44	2.67	4.10%	4.40%	4.70%
277	Heating, ventilating, air conditioning		3.91	3.95	4.79	7%	9.20%	10.40%
290	Electrical		2.95	3.65	5.70	6.20%	7.70%	11%
310	Total: Mechanical & Electrical		8.10	10.55	12.30	18.80%	20.80%	27.20%
39-001	**GARAGES, COMMERCIAL**		25.75	42.50	56.40			
002	Total project costs	C.F.	1.70	2.47	3.52			
272	Plumbing	S.F.	1.72	2.71	5.47	4.90%	7.30%	11%
273	Heating & ventilating		2.55	3.51	4.23	5.20%	7.20%	9.50%
290	Electrical		2.38	3.99	5.58	7.10%	9%	11.80%
310	Total: Mechanical & Electrical		5.35	9.90	14.30	15.70%	21.90%	28.40%
40-001	**GARAGES, MUNICIPAL**		28.65	41.55	62.50			
002	Total project costs	C.F.	2.05	2.78	3.72			
272	Plumbing	S.F.	1.79	3.37	5.50	5.20%	6.90%	8.70%
273	Heating & ventilating		2.12	3.52	4.04	4.60%	7.80%	11.90%
290	Electrical		2.53	3.87	5.35	6.60%	8.10%	10.10%
310	Total: Mechanical & Electrical		5.75	9.65	15.60	15.50%	23.50%	31.50%
41-001	**GARAGES, PARKING**		15	18.95	30.20			
002	Total project costs	C.F.	1.33	1.75	2.89			
272	Plumbing	S.F.	.28	.47	.77	2.10%	2.70%	3.80%
290	Electrical		.65	.94	1.53	4.20%	5.20%	6.30%
310	Total: Mechanical & Electrical		.97	1.47	1.93	6.80%	8.30%	9.50%
900	Per car, total cost	Car	4,930	6,720	9,460			
950	Total: Mechanical & Electrical	"	355	530	650			
43-001	**GYMNASIUMS**	S.F.	41.80	54.75	69.70			
002	Total project costs	C.F.	2.02	2.61	3.58			
272	Plumbing	S.F.	2.52	3.40	4.33	4.80%	7.20%	8.50%
277	Heating, ventilating, air conditioning	"	2.65	4.59	7.53	7.40%	9.70%	14%

17.1 S.F., C.F. and % of Total Costs

		UNIT	UNIT COSTS 1/4	UNIT COSTS MEDIAN	UNIT COSTS 3/4	% OF TOTAL 1/4	% OF TOTAL MEDIAN	% OF TOTAL 3/4
43 290	Electrical	S.F.	3.43	4.26	6.23	6.20%	9.30%	10.70%
310	Total: Mechanical & Electrical		6.45	11.40	15	16.60%	21.70%	27%
46-001	**HOSPITALS**	↓	90.20	111	146			
002	Total project costs	C.F.	6.60	7.85	10.55			
272	Plumbing	S.F.	7.45	9.85	13.60	7.20%	9.10%	10.70%
277	Heating, ventilating, air conditioning		10.70	14.25	19.95	11.50%	15.80%	16.90%
290	Electrical		10.45	13.45	19.75	11.20%	12.90%	15.20%
310	Total: Mechanical & Electrical	↓	29	37.15	54.45	28.20%	38.50%	40.30%
900	Per bed or person, total cost	Bed	26,600	41,700	60,600			
48-001	**HOUSING** For the Elderly	S.F.	43.42	54.70	69.25			
002	Total project costs	C.F.	3.06	4.25	5.52			
272	Plumbing	S.F.	3.30	4.57	6.78	8.50%	9.70%	11%
273	Heating, ventilating, air conditioning		1.26	2.17	3.10	3%	5.60%	6.80%
290	Electrical		3.19	4.48	6.35	7.50%	9.10%	10.80%
291	Electrical incl. electric heat		3.65	6.87	8.15	9.60%	11%	13.30%
310	Total: Mechanical & Electrical	↓	8	11.20	14.60	18.60%	22%	24.70%
900	Per rental unit, total cost	Unit	38,800	45,500	50,200			
950	Total: Mechanical & Electrical	"	7,680	9,630	11,300			
50-001	**HOUSING** Public (low-rise)	S.F.	33.10	45.15	62.30			
002	Total project costs	C.F.	2.77	3.62	4.54			
272	Plumbing	S.F.	2.41	3.24	4.26	7.10%	9.50%	11.60%
273	Heating, ventilating, air conditioning		1.10	2.27	2.64	4.20%	5.10%	6.40%
290	Electrical		2.11	3.04	4.27	5%	6.50%	8.20%
310	Total: Mechanical & Electrical	↓	5.95	8.90	12.15	15.70%	19.20%	22.20%
900	Per apartment, total cost	Apt.	36,760	41,420	51,540			
950	Total: Mechanical & Electrical	"	6,110	8,470	10,520			
51-001	**ICE SKATING RINKS**	S.F.	31.10	43.45	71.30			
002	Total project costs	C.F.	1.80	2.25	2.65			
272	Plumbing	S.F.	.94	1.38	2.12	3.10%	3.20%	4.60%
290	Electrical		2.45	3.24	4.46	5.70%	7%	10.10%
310	Total: Mechanical & Electrical	↓	4.45	6.30	9.50	12.40%	16.40%	25.90%
52-001	**JAILS**	S.F.	101	113	128			
002	Total project costs	C.F.	7.40	9.25	11.60			
272	Plumbing	S.F.	5.85	10.10	12.10	7%	8.90%	12%
277	Heating, ventilating, air conditioning		5.85	8.35	13.65	6.30%	8%	10.40%
290	Electrical		8.55	11.35	13.45	7.80%	9.80%	12.40%
310	Total: Mechanical & Electrical		22.10	31.75	40.60	22.20%	28.80%	35.30%
53-001	**LIBRARIES**	↓	55.95	68.90	87.10			
002	Total project costs	C.F.	3.93	4.70	5.90			
272	Plumbing	S.F.	2.33	3.20	4.29	3.60%	4.50%	5.80%
277	Heating, ventilating, air conditioning		5.70	8.05	10.50	8.70%	11%	13.20%
290	Electrical		5.80	7.05	9.25	8.40%	10.90%	12.10%
310	Total: Mechanical & Electrical		12.10	16.95	24	19.40%	24.50%	29.60%
55-001	**MEDICAL CLINICS**	↓	54.20	66.65	82.50			
002	Total project costs	C.F.	4.02	5.25	7			
272	Plumbing	S.F.	3.74	5.15	7.10	6.10%	8.40%	10.20%
277	Heating, ventilating, air conditioning		4.55	5.90	8.55	6.70%	9%	11.70%
290	Electrical		5.10	6.55	8.60	8.10%	10%	12%
310	Total: Mechanical & Electrical		11.85	15.10	21.05	19%	24.20%	30.10%
57-001	**MEDICAL OFFICES**	↓	50.85	62.85	75.75			
002	Total project costs	C.F.	3.85	5.10	6.75			
272	Plumbing	S.F.	3.15	4.59	6.26	5.70%	7%	9.40%
277	Heating, ventilating, air conditioning		3.61	5.40	6.85	6.50%	8%	10.40%
290	Electrical		4.34	6.15	8.05	7.90%	9.80%	11.80%
310	Total: Mechanical & Electrical		9.85	14	18.60	17.20%	22.80%	27.70%
59-001	**MOTELS**	↓	46.85	48.70	71.80			
002	Total project costs	C.F.	2.87	5.40	6.60			

For expanded coverage of these items see *Means' Square Foot Costs Data 1985*

17.1 S.F., C.F. and % of Total Costs

		UNIT	UNIT COSTS 1/4	UNIT COSTS MEDIAN	UNIT COSTS 3/4	% OF TOTAL 1/4	% OF TOTAL MEDIAN	% OF TOTAL 3/4
59 272	Plumbing	S.F.	2.17	3.93	6.70	3.80%	8.90%	12.60%
290	Electrical		2.56	4.63	6.95	4.70%	9.50%	10.90%
310	Total: Mechanical & Electrical	↓	8.15	13.70	16	17.30%	27.70%	33.30%
900	Per rental unit, total cost	Unit	13,170	22,120	31,480			
950	Total: Mechanical & Electrical	"	3,655	4,900	5,380			
60-001	**NURSING HOMES**	S.F.	51.05	66.45	82.10			
002	Total project costs	C.F.	4.05	5.20	6.95			
272	Plumbing	S.F.	4.43	5.40	8.15	8.30%	10.30%	14.10%
277	Heating, ventilating, air conditioning		4.79	6.85	8.15	10.60%	11.70%	11.80%
290	Electrical		5.15	6.50	8	9.70%	11%	12.50%
310	Total: Mechanical & Electrical	↓	11.65	16.02	22.90	22%	28.10%	33.20%
900	Per bed or person, total cost	Bed	19,850	25,630	31,880			
61-001	**OFFICES** Low-Rise (1 to 4 story)	S.F.	41.10	52.75	69.80			
002	Total project costs	C.F.	2.99	4.21	5.65			
272	Plumbing	S.F.	1.57	2.37	3.38	3.60%	4.50%	6.10%
277	Heating, ventilating, air conditioning		3.34	4.74	6.90	7.20%	10.40%	11.90%
290	Electrical		3.50	4.84	6.65	7.60%	9.70%	11.20%
310	Total: Mechanical & Electrical		7.10	10.60	15.60	14.50%	20.50%	27.10%
62-001	**OFFICES** Mid-Rise (5 to 10 story)	↓	46.75	58.15	79.05			
002	Total project costs	C.F.	3.22	4.27	5.94			
272	Plumbing	S.F.	1.39	2.12	3.07	2.80%	3.60%	4.50%
277	Heating, ventilating, air conditioning		3.46	4.95	7.90	7.60%	9.30%	11%
290	Electrical		2.95	4.33	7.10	6.60%	8.30%	10%
310	Total: Mechanical & Electrical		7.45	10.95	18	14.50%	21.80%	25.50%
63-001	**OFFICES** High-Rise (11 to 20 story)	↓	56.65	69.85	92.35			
002	Total project costs	C.F.	3.40	4.94	7.10			
290	Electrical	S.F.	2.49	4.06	6.15	6.20%	8%	10.50%
310	Total: Mechanical & Electrical		9.30	12.75	23.10	17.90%	21.70%	29.40%
64-001	**POLICE STATIONS**	↓	66.50	87.75	113			
002	Total project costs	C.F.	5.05	6.50	8.05			
272	Plumbing	S.F.	3.87	5.45	8.95	5.70%	6.80%	10.60%
277	Heating, ventilating, air conditioning		5.50	7.35	10.60	7%	10.50%	11.90%
290	Electrical		6.80	9.60	13.75	9.30%	11.40%	14.50%
310	Total: Mechanical & Electrical		18.25	22.25	29.70	22.20%	27.40%	33%
65-001	**POST OFFICES**	↓	56	66.40	86.55			
002	Total project costs	C.F.	3.14	3.80	4.45			
272	Plumbing	S.F.	2.38	3.06	2.79	4.10%	5.30%	5.60%
277	Heating, ventilating, air conditioning		3.49	4.60	6.95	6.70%	8%	9.80%
290	Electrical		4.38	6.26	7.33	7.80%	9.70%	11%
310	Total: Mechanical & Electrical		9.65	13.15	17.10	16.50%	21.40%	26.30%
66-001	**POWER PLANTS**	↓	262	437	570			
002	Total project costs	C.F.	6.30	15.55	22.05			
290	Electrical	S.F.	19.15	38.45	70.55	9.50%	12.70%	17.20%
810	Total: Mechanical & Electrical		43.65	717	262	25.50%	32.50%	47.70%
67-001	**RELIGIOUS EDUCATION**	↓	41.75	48.95	60.70			
002	Total project costs	C.F.	2.45	3.30	4.55			
272	Plumbing	S.F.	1.80	2.60	3.75	4.10%	5.10%	7.10%
277	Heating, ventilating, air conditioning		4.15	4.80	6.30	8.10%	9.90%	11.20%
290	Electrical		3.12	4.18	5.40	6.90%	8.60%	10.10%
310	Total: Mechanical & Electrical		6.90	9.40	13.80	14.70%	19.50%	24.60%
69-001	**RESEARCH** Laboratories and facilities	↓	57.85	88.45	133			
002	Total project costs	C.F.	4.07	6.60	10.10			
272	Plumbing	S.F.	6.05	8.30	11.75	5.20%	8.30%	10.80%
277	Heating, ventilating, air conditioning		5.90	19.35	22.60	7.20%	16.50%	17.70%
290	Electrical		7	10.45	22.35	9.10%	11.50%	16.60%
310	Total: Mechanical & Electrical		14.95	29.80	58.90	20.30%	29.50%	39.60%
70-001	**RESTAURANTS**	↓	59.05	77.85	100			
002	Total project costs	C.F.	5.25	6.75	8.50			

17

For expanded coverage of these items see *Means' Square Foot Costs Data 1985*

17.1 S.F., C.F. and % of Total Costs

		UNIT	UNIT COSTS 1/4	UNIT COSTS MEDIAN	UNIT COSTS 3/4	% OF TOTAL 1/4	% OF TOTAL MEDIAN	% OF TOTAL 3/4
70 272	Plumbing	S.F.	4.91	6.20	8.40	6%	8.20%	9.30%
277	Heating, ventilating, air conditioning		6.80	9.20	11.80	9.60%	12.30%	13.60%
290	Electrical		6.35	8.15	10.80	8.30%	10.40%	12%
310	Total: Mechanical & Electrical	↓	14.15	19.15	25.70	18.30%	23.20%	32.20%
900	Per seat unit, total cost	Seat	1,914	2,767	3,633			
950	Total: Mechanical & Electrical	"	429	560	869			
72-001	**RETAIL STORES**	S.F.	27.50	37.90	49.65			
002	Total project costs	C.F.	1.93	2.73	3.81			
272	Plumbing	S.F.	1.08	1.76	3.10	3.20%	4.50%	7%
277	Heating, ventilating, air conditioning		3.30	3.14	4.90	6.80%	8.70%	10.20%
290	Electrical		2.65	3.55	5.20	7.50%	10.10%	12.20%
310	Total: Mechanical & Electrical		4.66	7.14	10.38	14.70%	18%	24.20%
74-001	**SCHOOLS** Elementary	↓	46.70	57.95	68.80			
002	Total project costs	C.F.	3.14	3.96	5			
272	Plumbing	S.F.	2.75	3.94	5.30	5.60%	7.10%	9.20%
273	Heating & ventilating		4.07	6.56	9.05	8.10%	10.80%	15.10%
290	Electrical		4.29	5.50	7.15	8.40%	10%	11.80%
310	Total: Mechanical & Electrical	↓	9.80	14.20	18.35	19.80%	25.90%	32%
900	Per pupil, total cost	Pupil	4,000	6,360	8,660			
950	Total: Mechanical & Electrical	"	1,250	1,810	13,150			
76-001	**SCHOOLS** Junior High & Middle	S.F.	49.15	57.35	69.25			
002	Total project costs	C.F.	3.08	3.77	4.48			
272	Plumbing	S.F.	2.87	3.49	4.54	5.40%	6.90%	8.10%
277	Heating, ventilating, air conditioning		3.55	6.75	9.60	8.70%	11.50%	17.40%
290	Electrical		4.76	5.70	6.70	8%	9.60%	10.60%
310	Total: Mechanical & Electrical	↓	10.20	14.15	20	19.20%	27%	32.20%
900	Per pupil, total cost	Pupil	4,410	6,540	8,600			
78-001	**SCHOOLS** Senior High	S.F.	48.55	56.20	76.60			
002	Total project costs	C.F.	2.96	3.63	4.52			
272	Plumbing	S.F.	2.59	4.31	6.90	5%	6.50%	8%
277	Heating, ventilating, air conditioning		5.85	6.65	9.50	8.90%	11.50%	14.20%
290	Electrical		4.90	6.15	8.70	8.30%	10.50%	13.60%
310	Total: Mechanical & Electrical	↓	9.80	15.50	20.30	16.10%	25.60%	30.10%
900	Per pupil, total cost	Pupil	5,970	8,130	10,500			
80-001	**SCHOOLS** Vocational	S.F.	40.75	54.45	72.60			
002	Total project costs	C.F.	2.55	3.47	4.81			
272	Plumbing	S.F.	2.71	4	5.75	5.40%	7%	8.50%
277	Heating, ventilating, air conditioning		5.15	7.15	11.80	9.30%	12.50%	17%
290	Electrical		4.42	5.95	8.91	9.50%	11.80%	13.90%
310	Total: Mechanical & Electrical	↓	10.20	14.65	22.45	21.10%	29.70%	36%
900	Per pupil, total cost	Pupil	2,480	14,930	22,100			
950	Total: Mechanical & Electrical	"	840	2,020	5,200			
83-001	**SPORTS ARENAS**	S.F.	35.80	44.70	57			
002	Total project costs	C.F.	1.94	3.51	4.49			
272	Plumbing	S.F.	1.81	3.11	5.55	4.30%	6.30%	8.50%
277	Heating, ventilating, air conditioning		3.80	5.30	6.90%	5.80%	10.20%	13.50%
290	Electrical		2.95	4.90	6.10	7.10%	9.70%	12.20%
310	Total: Mechanical & Electrical		6.48	11.60	14.90	13.40%	22.50%	30.80%
85-001	**SUPERMARKETS**	↓	32.45	37.40	43.35			
002	Total project costs	C.F.	1.81	2.13	2.73			
272	Plumbing	S.F.	1.85	2.33	2.71	5%	6%	6.90%
277	Heating, ventilating, air conditioning		2.23	3.21	3.74	8.50%	8.60%	9.50%
290	Electrical		3.82	4.66	5.60	10.40%	12.50%	13.40%
310	Total: Mechanical & Electrical		6.07	8.70	10.90	17.80%	21.70%	27.60%
86-001	**SWIMMING POOLS**	↓	59.15	64.50	90.90			
002	Total project costs	C.F.	4.35	5.05	5.90			
272	Plumbing	S.F.	3.49	5.73	8.03	4.60%	9.60%	12.40%
290	Electrical	"	3.82	5.36	8.10	6.50%	7.60%	7.90%

17

For expanded coverage of these items see *Means' Square Foot Costs Data 1985*

17.1 S.F., C.F. and % of Total Costs

			UNIT	UNIT COSTS 1/4	UNIT COSTS MEDIAN	UNIT COSTS 3/4	% OF TOTAL 1/4	% OF TOTAL MEDIAN	% OF TOTAL 3/4
86	310	Total: Mechanical & Electrical	S.F.	8.80	15.80	31.25	17.50%	24.90%	31%
87	001	**TELEPHONE EXCHANGES**	S.F.	74.05	99.15	131			
	002	Total project costs	C.F.	4.38	6.50	9.35			
	272	Plumbing	S.F.	2.50	4.30	6.40	3.50%	5.70%	6.60%
	277	Heating, ventilating, air conditioning		6.10	14.05	17.45	11.70%	16%	18.40%
	290	Electrical		6.60	11.30	20.55	10.70%	14%	17.80%
	310	Total: Mechanical & Electrical		13.80	20.30	39.65	19.80%	27.40%	34.70%
89	001	**TERMINALS** Bus		32.80	43.45	59.60			
	002	Total project costs	C.F.	1.63	2.81	3.59			
	272	Plumbing	S.F.	1.10	2.47	3.80	2.30%	7.20%	8.80%
	290	Electrical		.77	2.26	5.49	3.70%	7.50%	10.50%
	310	Total: Mechanical & Electrical		1.73	4.18	8.45	8.30%	16.40%	19.60%
91	001	**THEATERS**	S.F.	42.85	53.10	82.15			
	002	Total project costs	C.F.	2.18	3.06	4.49			
	272	Plumbing	S.F.	1.39	1.61	4.82	2.90%	4.60%	6.10%
	277	Heating, ventilating, air conditioning		3.87	5.15	6	7.30%	11.60%	13.30%
	290	Electrical		3.93	5.30	8.15	8%	9.30%	11.20%
	310	Total: Mechanical & Electrical		7.80	10.95	20.10	13.70%	24.90%	27.40%
94	001	**TOWN HALLS** City Halls & Municipal Buildings		50.20	63.60	82.95			
	002	Total project costs	C.F.	3.38	5	6.40			
	272	Plumbing	S.F.	1.77	3.45	6	4.20%	5.90%	7.90%
	277	Heating, ventilating, air conditioning		3.77	7.50	8.55	7%	9%	13.20%
	290	Electrical		3.84	6.10	8.49	7.90%	9.40%	11.60%
	310	Total: Mechanical & Electrical		7.95	12.95	20.70	15.60%	21.10%	30%
97	001	**WAREHOUSES** And Storage Buildings		18.15	25.15	38.85			
	002	Total project costs	C.F.	.97	1.52	2.49			
99	001	**WAREHOUSE & OFFICES** Combination	S.F.	21.45	29.05	39.90			
	002	Total project costs	C.F.	1.14	1.70	2.47			
	272	Plumbing	S.F.	.87	1.49	2.30	3.60%	4.60%	6.20%
	277	Heating, ventilating, air conditioning		1.39	2.17	3.03	5%	5.60%	9.50%
	290	Electrical		1.53	2.25	3.58	6%	7.90%	9.90%
	310	Total: Mechanical & Electrical		2.92	4.87	7.69	11.70%	16.20%	21.30%

For expanded coverage of these items see *Means' Square Foot Costs Data 1985*

18.1 Repair & Remodeling

			CREW	DAILY OUTPUT	UNIT	BARE COSTS EQUIP.	BARE COSTS LABOR	BARE COSTS TOTAL	TOTAL INCL O&P
12	001	**CUTOUTS** Openings to 5 S.F., interior walls, not incl. re-framing							
	601								
	610	C18.1 -100 Drywall to 5/8" thick	F-1	24	Ea.	.29	6.55	6.84	9.75
	620	Paneling to 3/4" thick		20		.35	7.85	8.20	11.70
	630	Plaster on gypsum lath		20		.35	7.85	8.20	11.70
	634	On wire lath		14		.50	11.20	11.70	16.70
	700	Wood frame, openings to 5 S.F., not including re-framing							
	720	Floors, incl. subfloor, underlayment, wood or resilient flooring							
	721	up to 2" thick	F-1	5	Ea.	1.40	31	32.40	47
	730	Roofs, sheathing to 1" thick		6		1.17	26	27.17	39
	740	Walls, sheathing to 1" thick		7		1	22	23	33
						BARE COSTS MAT.	**BARE COSTS INST.**	**BARE COSTS TOTAL**	
20	001	**FACTORS** To adjust figures in other sections of this							
	002	book for repair and remodeling projects:							
	050	C18.1 -100 Cut & patch to match existing construction, add, minimum			Costs	2%	3%		
	055	Maximum			"	5%	9%		

For expanded coverage of these items see Means' *Repair & Remodeling Cost Data 1985*

18.1 Repair & Remodeling

	18.1 Repair & Remodeling	CREW	DAILY OUTPUT	UNIT	BARE COSTS MAT.	BARE COSTS INST.	BARE COSTS TOTAL	TOTAL INCL O&P
20 080	Dust protection, add, minimum			Costs	1%	2%		
085	Maximum				4%	11%		
110	Equipment usage curtailment, add, minimum				1%	1%		
115	Maximum				3%	10%		
140	Material handling & storage limitation, add, minimum				1%	1%		
145	Maximum				6%	7%		
170	Protection of existing work, add, minimum				2%	2%		
175	Maximum				5%	7%		
200	Shift work requirements, add, minimum					5%		
205	Maximum					30%		
230	Temporary shoring and bracing, add, minimum				2%	5%		
235	Maximum			↓	5%	12%		
25-001	**CLEANING MASONRY** No staging included							
020	Chemical cleaning, brush and wash, average	D-1	1,000	S.F.	.04	.29	.33	.45
040	High pressure, water only, average	B-9	3,000			.26	.26	.36
080	Water and chemical, average	"	1,250	↓	.04	.62	.66	.90
30-001	**CAULKING MASONRY** No staging included							
002	Re-caulk only, 3/8" deep x 5/8" wide							
005	joints, 82 L.F. per 12 tubes, oil base	1 Bric	225	L.F.	.20	.71	.91	1.23
010	Butyl		205		.45	.78	1.23	1.61
020	Polysulfide		200		.38	.80	1.18	1.56
030	Silicone rubber	↓	195	↓	.81	.82	1.63	2.06
35-001	**NEEDLE MASONRY** Including shoring							
040	Block, concrete, 8" thick	B-9	7.10	Ea.	33	110	143	190
080	Brick, 4" thick with 8" backup block		5.70		33	135	168	225
100	Brick solid, 8" thick	↓	6.20	↓	33	125	158	210
45-001	**POINTING MASONRY**							
002								
030	Cut and repoint brick, common bond, hard mortar	1 Bric	80	S.F.	.15	2.01	2.16	3.01
060	Soft old mortar		100	"	.15	1.60	1.75	2.44
070	Stone work, hard mortar		140	L.F.	.23	1.15	1.38	1.88
072	Soft old mortar		160	"	.23	1	1.23	1.68
100	Repoint, mask and grout method common bond	↓	90	S.F.	.24	1.78	2.02	2.79
190								
200	Scrub coat, sand grout on walls, minimum	1 Bric	120	S.F.	.25	1.34	1.59	2.17
202	Maximum	"	100	"	.28	1.60	1.88	2.59
55-001	**INTERIOR CLEANUP**							
010	Window sill repair, to 1.25 S.F. area	1 Pord	32	Ea.	.56	4.71	5.27	7.30
050	Floor, clean and touchup		1,600	S.F.		.09	.09	.13L
100	Wax and polish		2,400		.05	.06	.11	.14
150	Wallpaper removal, 3 layer, minimum		400		.04	.38	.42	.58
200	Maximum		560		.04	.27	.31	.43
250	Wash walls and ceilings	↓	2,000	↓	.01	.08	.09	.12
60-001	**REMOVE EXISTING LEAD PAINT**							
002	Refinish with 2 coats of paint							
005	Baseboard, to 6" wide	1 Pord	190	L.F.	.13	.79	.92	1.27
010	Balustrades, one side		150	S.F.	.28	1.01	1.29	1.74
030	Cabinets, simple design		85		.13	1.77	1.90	2.66
100	Ornate design		40		.15	3.77	3.92	5.50
150	Cornice, simple design		65		.13	2.32	2.45	3.44
200	Doors, one side, flush		125		.13	1.21	1.34	1.86
250	Four panel		95		.13	1.59	1.72	2.40
300	Fence, picket, one side		80		.14	1.89	2.03	2.83
350	Grilles, one side, simple design		95	↓	.15	1.59	1.74	2.42
400	Pipes, to 4" diameter		200	L.F.	.13	.75	.88	1.21
450	To 8" diameter		90	"	.25	1.68	1.93	2.65
500	Siding	↓	170	S.F.	.13	.89	1.02	1.40
550	Windows, one side only, double hung							
600	1/1 light, 24" x 48" high	1 Pord	12	Ea.	1.01	12.55	13.56	18.95

For expanded coverage of these items see Means' *Repair & Remodeling Cost Data 1985*

18.1 Repair & Remodeling

		CREW	DAILY OUTPUT	UNIT	BARE COSTS MAT.	BARE COSTS INST.	BARE COSTS TOTAL	TOTAL INCL O&P
60 650	36" x 72" high	1 Pord	8	Ea.	2.30	18.85	21.15	29
700	Colonial window, 6/6 light, 24" x 48" high		7		1.30	22	23.30	32
750	36" x 72" high		4		2.55	38	40.55	56
800	12/12 light, 24" x 48" high		5		1.90	30	31.90	45
70-001	**SCRAPE AND SEAL AFTER FIRE DAMAGE**							
005	Boards, floor, 1" x 4"	1 Pord	145	L.F.	.07	1.04	1.11	1.55
010	1" x 6"		110		.08	1.37	1.45	2.03
015	Wall, 1" x 4"		116		.07	1.30	1.37	1.92
020	1" x 6"		88		.08	1.71	1.79	2.52
030	Framing, 2" x 4"		110		.08	1.37	1.45	2.03
100	2" x 6"		90		.11	1.68	1.79	2.50
150	2" x 8"		70		.12	2.15	2.27	3.19
200	Heavy framing 3" x 4"		110		.08	1.37	1.45	2.03
250	4" x 6"		90		.12	1.68	1.80	2.51
300	4" x 10"		50		.16	3.02	3.18	4.46
350	For sealing only, minimum		889	S.F.	.08	.17	.25	.33
400	Maximum		470	"	.08	.32	.40	.54
75-001	**REMOVE AND RESET DOORS**							
030	Combination, storm and screen	F-2	14	Ea.		23	23	33
160	Entrance door, flush birch, solid core		10			33	33	47
400	Interior passage door, solid core		13			25	25	36
460	Hollow core		14			23	23	33
80-001	**RETROFIT GLASS**							
002	Additional plate glass in existing metal sash							
010	Minimum	1 Glaz	50	S.F.	4.30	3.18	7.48	9.25
020	Maximum	"	25	"	6.30	6.35	12.65	15.95
85-001	**ROOF REPAIR**							
010	Clay, mission tile, extensive replacement	1 Rots	1	Sq.	465	150	615	735
015	Individual pieces		19	Ea.	2.81	7.80	10.61	14.90
020	Slate, Pennsylvania tile, extensive replacement		1	Sq.	570	150	720	850
025	Individual pieces		19	Ea.	4.05	7.80	11.85	16.30
030	Wood shingles, 16" no. 1 red cedar, extensive replacement		1.50	Sq.	88	99	187	245
035	Individual pieces		40	Ea.	.58	3.70	4.28	6.25
90-001	**WINDOW GLASS** Replace broken window lites							
010	1/8" float glass (9 S.F. maximum size)	1 Glaz	48	S.F.	1.88	3.31	5.19	6.80
020	3/16" float glass (16 S.F. maximum size)	"	48	"	1.95	3.31	5.26	6.85
95-001	**WOOD FRAMING**							
010	Sister, joist or rafter, 2" x 4"	F-2	940	L.F.	.25	.35	.60	.77
020	2" x 6"		800		.37	.41	.78	.99
030	2" x 8"		640		.56	.51	1.07	1.35

Description		CREW	DAILY OUTPUT	UNIT	BARE COSTS MAT.	BARE COSTS INST.	BARE COSTS TOTAL	TOTAL INCL O&P

ITEM LOCATION - SYSTEMS FORMAT - PART B & C

The Systems Format part of the book is divided into Sections B and C.

Section B is made up of systems whose costs are computer generated using the items in the Unit Cost part of this book. The prices used include Overhead and Profit

Section C is made up of mechanical and electrical related items, cost modifiers, technical data, codes, sizing information and general data.

Section B

Section C

8

B8.1-010 Sample Estimate

Example of Plumbing Cost Calculations: The bathroom system includes the individual fixtures such as bathtub, lavatory, shower and water closet. These fixtures are listed below as separate items merely as a checklist.

8.1-010 Plumbing Systems 20 Unit, 2 Story Apartment Building

	FIXTURE	SYSTEM	LINE	QUANTITY	UNIT	COST EACH		
						MAT.	INST.	TOTAL
0440	Bathroom	B8.1-630	3640	20	Ea.	17,700	26,700	44,400
0480	Bathtub							
0520	Booster pump [1]	not req'd.						
0560	Drinking fountain							
0600	Garbage disposal [1]	not incl.						
0660								
0680	Grease interceptor							
0720	Water heater	B8.1-170	2140	1	Ea.	3100	1450	4550
0760	Kitchen sink	B8.1-431	1960	20	Ea.	7125	11,300	18,425
0800	Laundry sink	B8.1-432	1840	4	Ea.	1625	1800	3425
0840	Lavatory							
0900								
0920	Roof drain, 1 floor	B8.1-310	4200	2	Ea.	345	395	740
0960	Roof drain, add'l floor	B8.1-310	4240	20	L.F.	115	210	325
1000	Service sink	B8.1-434	4300	1	Ea.	410	605	1015
1040	Sewage ejector [1]	not req'd.						
1080	Shower							
1100								
1160	Sump pump							
1200	Urinal							
	Water closet							
1320								
1360	**SUB TOTAL**					30,400	42,500	72,900
1480	Water controls	C8.1-031		10% [2]		3,040	4,250	7,290
1520	Pipe & fittings [3]	C8.1-031		30% [2]		9,120	12,750	21,870
1560	Other							
	Quality/complexity	C8.1-031		15% [2]		4,560	6,375	10,935
1680								
1720	**TOTAL**					47,120	65,875	112,995

[1] **Note:** Cost for items such as booster pumps, backflow preventers, sewage ejectors, water meters, etc., may be obtained from Sections 15.1, 15.2 and 15.3. Water controls, pipe and fittings, and the Quality/Complexity factors come from Table C8.1-031.

[2] Percentage of sub total.

[3] Long easily discernible runs of pipe would be more accurately priced from Section 15.1. If this is done, reduce the miscellaneous percentage in proportion.

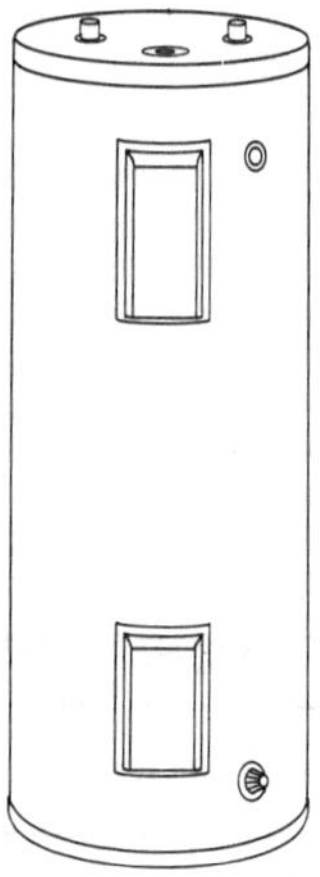

Installation includes piping and fittings within 10′ of heater. Electric water heaters do not require venting.

1 Kilowatt hour will raise:			
Gals. of Water	Degrees F	Gals. of Water	Degrees F
4.1 Gals.	100°F	6.8 Gals.	60°F
4.5	90	8.2	50
5.1	80	10.0	40
5.9	70		

System Components	QUANTITY	UNIT	COST EACH MAT.	INST.	TOTAL
SYSTEM 08.1-110-1780					
ELECTRIC WATER HEATER, RESIDENTIAL, 100° F RISE					
10 GALLON TANK, 7 GPH					
Water heater, residential electric, glass lined tank, 10 Gal	1.000	Ea.	115.50	109.50	225
Copper tubing, type L, 50/50 solder joint, hanger 10′ OC 1/2″ diam	30.000	L.F.	25.20	145.80	171
Wrought copper 90° elbow for solder joints 1/2″ diam	4.000	Ea.	.56	51.24	51.80
Wrought copper Tee for solder joints, 1/2″ diam	2.000	Ea.	.48	39.52	40
Wrought copper union for soldered joints, 1/2″ diam	2.000	Ea.	2.80	27	29.80
Valve, gate, bronze, 125 lb, NRS, soldered 1/2″ diam	2.000	Ea.	23	21	44
Relief valve, bronze, press & temp, self-close, 3/4″ IPS	1.000	Ea.	31.90	9.10	41
Wrought copper adapter, CTS to MPT 3/4″ IPS	1.000	Ea.	.55	12.20	12.75
Copper tubing, type L, 50/50 solder joints, 3/4″ diam	1.000	L.F.	1.22	5.83	7.05
Wrought copper 90° elbow for solder joints 3/4″ diam	1.000	Ea.	.33	13.52	13.85
TOTAL			201.54	434.71	636.25

8.1-110	Electric Water Heaters - Residential Systems	COST EACH MAT.	INST.	TOTAL
1760	Electric water heater, residential, 100°F rise			
1780	10 gallon tank, 7 GPH	200	435	635
1820	C8.1-102 20 gallon tank, 7 GPH	250	490	740
1860	30 gallon tank, 7 GPH	275	515	790
1900	40 gallon tank, 8 GPH	315	585	900
1940	52 gallon tank, 10 GPH	345	595	940
1980	66 gallon tank, 13 GPH	505	700	1205
2020	80 gallon tank, 16 GPH	550	735	1285
2060	120 gallon tank, 23 GPH	880	860	1740

8

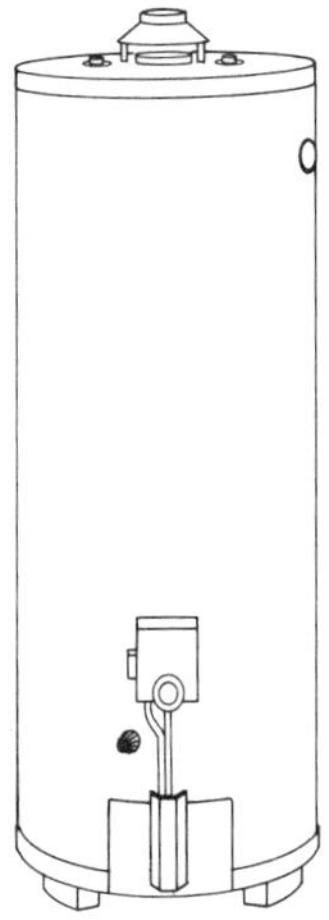

Installation includes piping and fittings within 10′ of heater. Gas heaters require vent piping (not included with these units).

System Components	QUANTITY	UNIT	COST EACH MAT.	COST EACH INST.	COST EACH TOTAL
SYSTEM 08.1-120-2220					
GAS FIRED WATER HEATER, RESIDENTIAL, 100° F RISE					
20 GALLON, 25 GPH					
Water heater, residential, gas, glass lined tank, 20 Gal	1.000	Ea.	154	121	275
Copper tubing, type L, 50/50 solder joint, hanger 10′ OC 3/4″ diam	32.000	L.F.	39.04	186.56	225.60
Wrought copper 90° elbow for solder joints 3/4″ diam	5.000	Ea.	1.65	67.60	69.25
Wrought copper tee for solder joints, 3/4″ diam	2.000	Ea.	1.22	42.78	44
Wrought copper union for soldered joints, 3/4″ diam	2.000	Ea.	3.98	28.52	32.50
Valve, bronze, 125 lb, NRS, soldered 3/4″ diam	2.000	Ea.	29.82	26.18	56
Relief valve, press & temp, bronze, self-close, 3/4″ IPS	1.000	Ea.	31.90	9.10	41
Wrought copper, adapter, CTS to MPT 3/4″ IPS	1.000	Ea.	.55	12.20	12.75
Pipe steel black, schedule 40, threaded, 1/2″ diam	10.000	L.F.	11.90	40.60	52.50
Pipe, 90° elbow, malleable iron black, 150 lb, threaded, 1/2″ diam	2.000	Ea.	1.24	34.16	35.40
Pipe, union with brass seat, malleable iron black, 1/2″ diam	1.000	Ea.	2.46	18.54	21
Valve, gas stop w/o check, brass, 1/2″ IPS	1.000	Ea.	7.32	10.68	18
TOTAL			285.08	597.92	883

8.1-120	Gas Fired Water Heaters - Residential Systems	COST EACH MAT.	COST EACH INST.	COST EACH TOTAL
2200	Gas fired water heater, residential, 100° F rise			
2220	20 gallon tank, 25 GPH	285	600	885
2260	C8.1-102 30 gallon tank, 32 GPH	295	615	910
2300	40 gallon tank, 32 GPH	340	705	1045
2340	50 gallon tank, 63 GPH	390	710	1100
2380	75 gallon tank, 63 GPH	645	810	1455
2420	100 gallon tank, 63 GPH	905	850	1755

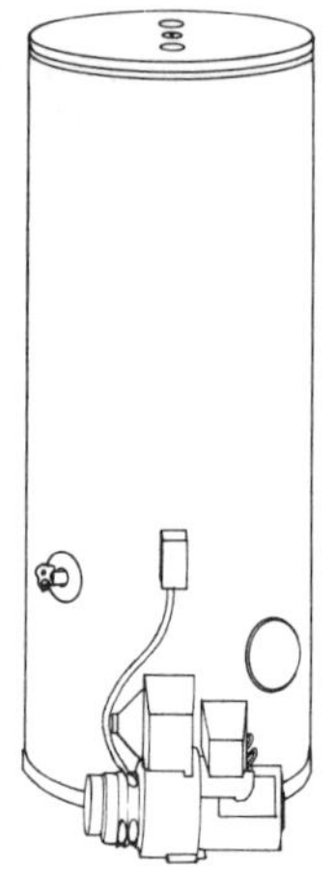

Installation includes piping and fittings within 10′ of heater. Oil fired heaters require vent piping (not included in these prices).

8

System Components	QUANTITY	UNIT	COST EACH MAT.	INST.	TOTAL
SYSTEM 08.1-130-2220					
OIL FIRED WATER HEATER, RESIDENTIAL, 100° F RISE					
30 GALLON TANK, 103 GPH					
Water heater, residential, oil glass lined tank, 30 Gal	1.000	Ea.	742.50	127.50	870
Copper tubing, type L, 50/50 solder joint, hanger 10′ O.C. 3/4″ diam	33.000	L.F.	40.26	192.39	232.65
Wrought copper 90° elbow for solder joints 3/4″ diam	5.000	Ea.	1.65	67.60	69.25
Wrought copper Tee for solder joints, 3/4″ diam	2.000	Ea.	1.22	42.78	44
Wrought copper union for soldered joints, 3/4″ diam	2.000	Ea.	3.98	28.52	32.50
Valve, gate, bronze, 125 lb, NRS, soldered 3/4″ diam	2.000	Ea.	29.82	26.18	56
Relief valve, bronze, press & temp, self-close, 3/4″ IPS	1.000	Ea.	31.90	9.10	41
Wrought copper adapter, CTS to MPT, 3/4″ IPS	1.000	Ea.	.55	12.20	12.75
Copper tubing, type L, 50/50 solder joint, hanger 10′ OC 3/8″ diam	10.000	L.F.	7	46	53
Wrought copper 90° elbow for solder joints 3/8″ diam	2.000	Ea.	1.52	23.28	24.80
Valve, globe, fusible, 3/8″ diam	1.000	Ea.	3.25	10.75	14
TOTAL			863.65	586.30	1449.95

8.1-130	Oil Fired Water Heaters - Residential Systems	COST EACH MAT.	INST.	TOTAL
2200	Oil fired water heater, residential, 100° F rise			
2220	30 gallon tank, 103 GPH	865	585	1450
2260	C8.1 -102 50 gallon tank, 145 GPH	1125	680	1805
2300	70 gallon tank, 164 GPH	1275	780	2055
2340	85 gallon tank, 181 GPH	1700	805	2505

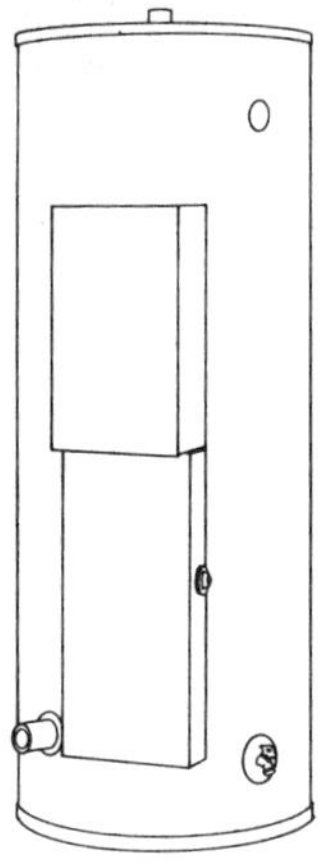

Systems below include piping and fittings within 10' of heater. Electric water heaters do not require venting.

System Components	QUANTITY	UNIT	COST EACH MAT.	COST EACH INST.	COST EACH TOTAL
SYSTEM 08.1-160-1820					
ELECTRIC WATER HEATER, COMMERCIAL, 100° F RISE					
50 GALLON TANK, 9 KW, 37 GPH					
Water heater, commercial, electric, 50 Gal, 9 KW, 37 GPH	1.000	Ea.	962.50	137.50	1100
Copper tubing, type L, 50/50 solder joint, hanger 10' OC 3/4" diam	34.000	L.F.	41.48	198.22	239.70
Wrought copper 90° elbow for solder joints 3/4" diam	5.000	Ea.	1.65	67.60	69.25
Wrought copper Tee for solder joints, 3/4" diam	2.000	Ea.	1.22	42.78	44
Wrought copper union for soldered joints, 3/4" diam	2.000	Ea.	3.98	28.52	32.50
Valve, gate, bronze, 125 lb, NRS, soldered 3/4" diam	2.000	Ea.	29.82	26.18	56
Relief valve, bronze, press & temp, self-close, 3/4" IPS	1.000	Ea.	31.90	9.10	41
Wrought copper adapter, copper tubing to male, 3/4" IPS	1.000	Ea.	.55	12.20	12.75
TOTAL			1073.10	522.10	1595.20

8

8.1-160	Electric Water Heaters - Commercial Systems	COST EACH MAT.	COST EACH INST.	COST EACH TOTAL
1800	Electric water heater, commercial, 100° F rise			
1820	50 gallon tank, 9 KW 37 GPH	1075	520	1595
1860	C8.1 -101 80 gal, 12 KW 49 GPH	1425	680	2105
1900	36 KW 147 GPH	1925	740	2665
1940	C8.1 -102 120 gal, 36 KW 147 GPH	2075	785	2860
1980	150 gal, 120 KW 490 GPH	6575	830	7405
2020	200 gal, 120 KW 490 GPH	6875	860	7735
2060	250 gal, 150 KW 615 GPH	7700	995	8695
2100	300 gal, 180 KW 738 GPH	8375	1050	9425
2140	350 gal, 30 KW 123 GPH	5900	1100	7000
2180	180 KW 738 GPH	8575	1125	9700
2220	500 gal, 30 KW 123 GPH	7575	1275	8850
2260	240 KW 984 GPH	11,900	1250	13,150
2300	700 gal, 30 KW 123 GPH	9200	1450	10,650
2340	300 KW 1230 GPH	14,500	1450	15,950
2380	1000 gal, 60 KW 245 GPH	11,300	1950	13,250
2420	480 KW 1970 GPH	19,700	1950	21,650
2460	1500 gal, 60 KW 245 GPH	16,500	2450	18,950
2500	480 KW 1970 GPH	23,300	2425	25,725
2540	2000 gal, 60 KW 245 GPH	20,700	3525	24,225
2580	480 KW 1970 GPH	28,500	3550	32,050

B8.1-170 Gas Fired Water Heaters - Comm.

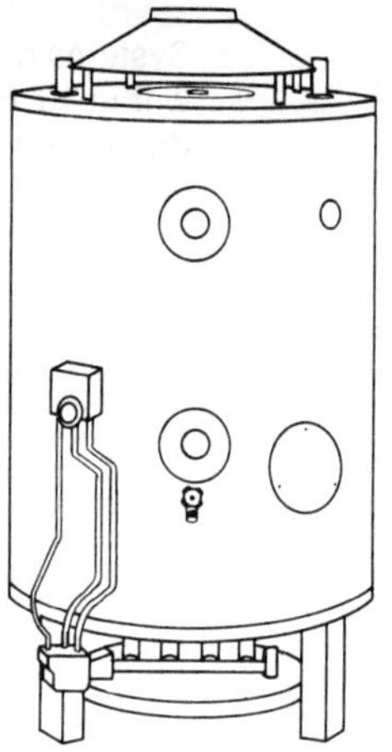

Units may be installed in multiples for increased capacity.

Included below is the heater with self-energizing gas controls, safety pilots, insulated jacket, hi-limit aquastat and pressure relief valve.

Installation includes piping and fittings within 10′ of heater. Gas heaters require vent piping (not included in these prices).

System Components	QUANTITY	UNIT	COST EACH MAT.	INST.	TOTAL
SYSTEM 08.1-170-1780					
GAS FIRED WATER HEATER, COMMERCIAL, 100° F RISE					
75.5 MBH INPUT, 63 GPH					
Water heater, commercial, gas, 75.5 MBH, 63 GPH	1.000	Ea.	704	181	885
Copper tubing, type L, 50/50 solder joint, hanger 10′ OC, 1-1/4″ diam	30.000	L.F.	68.70	265.80	334.50
Wrought copper 90° elbow for solder joints 1-1/4″ diam	4.000	Ea.	6.08	68.32	74.40
Wrought copper Tee for solder joints, 1-1/4″ diam	2.000	Ea.	6.86	57.14	64
Wrought copper union for soldered joints, 1-1/4″ diam	2.000	Ea.	9.50	36.50	46
Valve, gate, bronze, 125 lb, NRS, soldered 1-1/4″ diam	2.000	Ea.	48.40	33.60	82
Relief valve, bronze, press & temp, self-close, 3/4″ IPS	1.000	Ea.	31.90	9.10	41
Copper tubing, type L, 50/50 solder joints, 3/4″ diam	8.000	L.F.	9.76	46.64	56.40
Wrought copper 90° elbow for solder joints 3/4″ diam	1.000	Ea.	.33	13.52	13.85
Wrought copper, adapter, CTS to MPT, 3/4″ IPS	1.000	Ea.	.55	12.20	12.75
Pipe steel black, schedule 40, threaded, 3/4″ diam	10.000	L.F.	13.60	41.90	55.50
Pipe, 90° elbow, malleable iron black, 150 lb threaded, 3/4″ diam	2.000	Ea.	1.44	36.66	38.10
Pipe, union with brass seat, malleable iron black, 3/4″ diam	1.000	Ea.	2.52	19.48	22
Valve, gas stop w/o check, brass, 3/4″ IPS	1.000	Ea.	8.42	11.58	20
TOTAL			912.06	833.44	1745.50

8.1-170		Gas Fired Water Heaters - Commercial Systems	COST EACH MAT.	INST.	TOTAL
1760		Gas fired water heater, commercial, 100° F rise			
1780		75.5 MBH input, 63 GPH	910	835	1745
1860	C8.1 -101	100 MBH input, 91 GPH	1125	870	1995
1980		155 MBH input, 150 GPH	1525	995	2520
2060	C8.1 -102	200 MBH input, 192 GPH	1975	1175	3150
2140		300 MBH input, 278 GPH	3100	1450	4550
2180		390 MBH input, 374 GPH	3475	1450	4925
2220		500 MBH input, 480 GPH	4600	1550	6150
2260		600 MBH input, 576 GPH	5225	1675	6900
2300		800 MBH input, 768 GPH	6125	1825	7950
2340		1000 MBH input, 960 GPH	7200	1850	9050
2420		1500 MBH input, 1440 GPH	11,100	2350	13,450
2460		1800 MBH input, 1730 GPH	12,400	2600	15,000
2500		2450 MBH input, 2350 GPH	16,300	2975	19,275
2540		3000 MBH input, 2880 GPH	19,700	3575	23,275
2580		3750 MBH input, 3600 GPH	24,300	3675	27,975

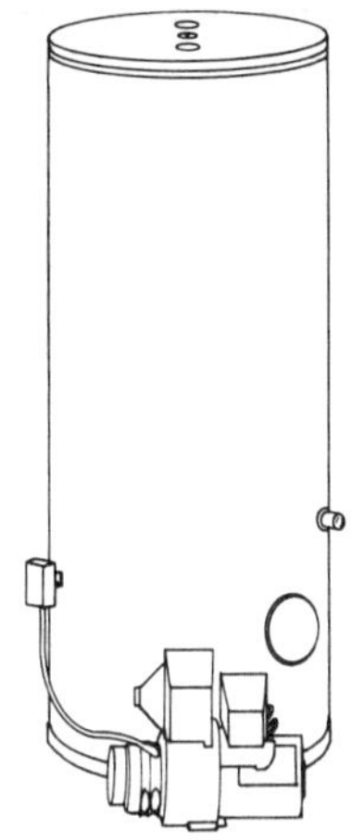

Units may be installed in multiples for increased capacity.

Included below is the heater, wired-in flame retention burners, cadmium cell primary controls, hi-limit controls, ASME pressure relief valves, draft controls, and insulated jacket.

Oil fired water heater systems include piping and fittings within 10' of heater. Oil fired heaters require vent piping (not included in these systems).

Systems Components	QUANTITY	UNIT	COST EACH MAT.	INST.	TOTAL
SYSTEM 08.1-180-1820					
OIL FIRED WATER HEATER, COMMERCIAL, 100° F RISE					
97 MBH INPUT, 116 GPH					
Water heater, commercial, gas, 95 MBH	1.000	Ea.	1265	235	1500
Copper tubing, type L, 50/50 solder joint, hanger 10' OC, 3/4" diam	34.000	L.F.	41.48	198.22	239.70
Wrought copper 90° elbow for solder joints 3/4" diam	5.000	Ea.	1.65	67.60	69.25
Wrought copper Tee for solder joints, 3/4" diam	2.000	Ea.	1.22	42.78	44
Wrought copper union for soldered joints, 3/4" diam	2.000	Ea.	3.98	28.52	32.50
Valve, bronze, 125 lb, NRS, soldered 3/4" diam	2.000	Ea.	29.82	26.18	56
Relief valve, bronze, press & temp, self-close, 3/4" IPS	1.000	Ea.	31.90	9.10	41
Wrought copper adapter, copper tubing to male, 3/4" IPS	1.000	Ea.	.55	12.20	12.75
Copper tubing, type L, 50/50 solder joint, hanger 10' OC, 3/8" diam	10.000	L.F.	7	46	53
Wrought copper 90° elbow for solder joints 3/8" diam	2.000	Ea.	1.52	23.28	24.80
Valve, globe, fusible, 3/8" IPS	1.000	Ea.	3.25	10.75	14
TOTAL			1387.37	699.63	2087

8

8.1-180	Oil Fired Water Heaters - Commercial Systems	COST EACH MAT.	INST.	TOTAL
1800	Oil fired water heater, commercial, 100° F rise			
1820	97 MBH output, 116 GPH	1375	700	2075
1900	C8.1-101 134 MBH output, 161 GPH	1575	925	2500
1940	161 MBH output, 192 GPH	1875	1025	2900
1980	C8.1-102 187 MBH output, 224 GPH	2000	1200	3200
2060	262 MBH output, 315 GPH	2525	1350	3875
2100	341 MBH output, 409 GPH	3350	1425	4775
2140	420 MBH output, 504 GPH	3775	1525	5300
2180	525 MBH output, 630 GPH	4800	1725	6525
2220	630 MBH output, 756 GPH	5325	1725	7050
2260	735 MBH output, 880 GPH	6550	1975	8525
2300	840 MBH output, 1000 GPH	7425	1950	9375
2340	1050 MBH output, 1260 GPH	8950	2025	10,975
2380	1365 MBH output, 1640 GPH	11,400	2300	13,700
2420	1680 MBH output, 2000 GPH	13,600	2900	16,500
2460	2310 MBH output, 2780 GPH	17,300	3450	20,750
2500	2835 MBH output, 3400 GPH	21,300	3500	24,800
2540	3150 MBH output, 3780 GPH	23,800	4675	28,475

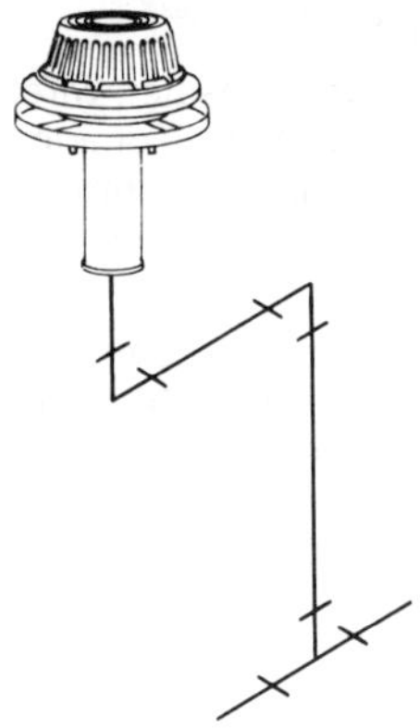

Design Assumptions: Vertical conductor size is based on a maximum rate of rainfall of 4" per hour. To convert roof area to other rates multiply rate by four and divide by local rate, then multiply the resultant factor by the area shown in table.

Basic cost is for roof drain, 10' of vertical leader and 10' of horizontal, plus connection to the main.

PIPE DIA.	MAX. S.F. ROOF AREA	GALLONS PER MIN.
2"	544	23
3"	1610	67
4"	3460	144
5"	6280	261
6"	10,200	424
8"	22,000	913

System Components	QUANTITY	UNIT	COST EACH MAT.	INST.	TOTAL
SYSTEM 081-310-1880					
ROOF DRAIN, DWV PVC PIPE, 2" DIAM., 10' HIGH					
Drain, roof, main, PVC, dome type 2" pipe size	1.000	Ea.	35.20	32.80	68
Clamp, roof drain, underdeck	1.000	Ea.	9.30	19.70	29
Pipe, Tee, PVC DWV, schedule 40, 2" pipe size	1.000	Ea.	2.20	26.80	29
Pipe, PVC, DWV, schedule 40, 2" diam.	20.000	L.F.	30.40	156.60	187
Pipe, elbow, PVC schedule 40, 2" diam.	2.000	Ea.	1.70	33	34.70
TOTAL			78.80	268.90	347.70

8.1-310	Roof Drain Systems	COST PER SYSTEM MAT.	INST.	TOTAL
1880	Roof drain, DWV PVC, 2" diam., piping, 10' high	79	270	349
1920	For each additional foot add	1.52	7.85	9.37
1960	3" diam., 10' high	110	325	435
2000	For each additional foot add	2.73	8.70	11.43
2040	4" diam., 10' high	150	365	515
2080	For each additional foot add	3.80	9.60	13.40
2120	5" diam., 10' high	300	390	690
2160	For each additional foot add	6.10	10.75	16.85
2200	6" diam., 10' high	400	460	860
2240	For each additional foot add	8.25	11.75	20
2280	8" diam., 10' high	650	755	1405
2320	For each additional foot add	13.60	15.40	29
3940	C.I., soil, single hub, service wt., 2" diam. piping, 10' high	165	315	480
3980	For each additional foot add	3.10	9.25	12.35
4120	3" diam., 10' high	205	345	550
4160	For each additional foot add	4.41	9.65	14.06
4200	4" diam., 10' high	250	375	625
4240	For each additional foot add	5.85	10.50	16.35
4280	5" diam., 10' high	325	410	735
4320	For each additional foot add	7.70	11.60	19.30
4360	6" diam., 10' high	455	430	885
4400	For each additional foot add	11.20	11.80	23
4440	8" diam., 10' high	735	875	1610
4480	For each additional foot add	17.10	19.90	37
6040	Steel galv. sch 40 threaded, 2" diam. piping, 10' high	200	270	470
6080	For each additional foot add	4.54	7.20	11.74

8.1-310	Roof Drain Systems	COST PER SYSTEM		
		MAT.	INST.	TOTAL
6120	3" diam., 10' high	335	390	725
6160	For each additional foot add	9.45	10.55	20
6200	4" diam., 10' high	475	500	975
6240	For each additional foot add	14.30	12.70	27
6280	5" diam., 10' high	965	630	1595
6320	For each additional foot add	35	17.45	52
6360	6" diam, 10' high	1175	775	1950
6400	For each additional foot add	40	23	63
6440	8" diam., 10' high	1750	1075	2825
6480	For each additional foot add	56	27	83

B8.1-410 Bathtub Systems

Systems are complete with trim and rough-in (supply, waste and vent) to connect to supply branches and waste mains.

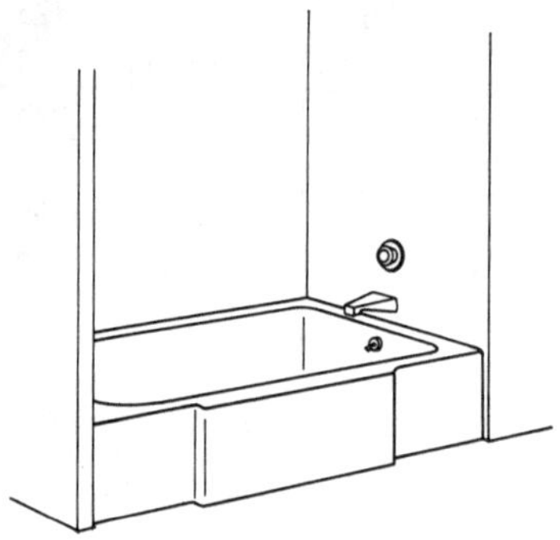

Recessed Bathtub

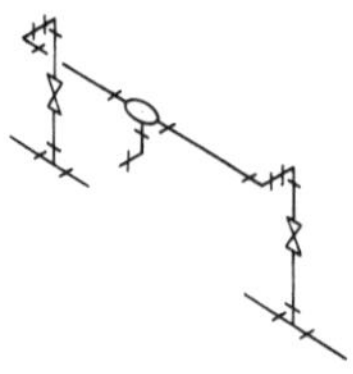

Supply

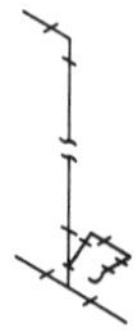

Waste/Vent

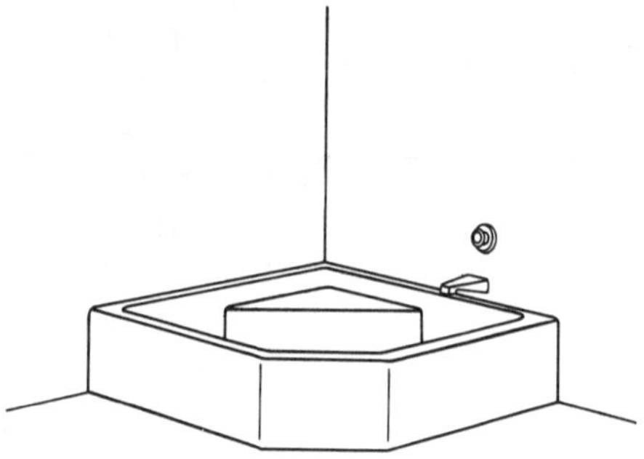

Corner Bathtub

System Components

System Components	QUANTITY	UNIT	COST EACH MAT.	INST.	TOTAL
SYSTEM 08.1-410-1960					
BATHTUB RECESSED, PORCELAIN ENAMEL ON CAST IRON, 42" X 37"					
Bath tub, porcelain enamel on cast iron, w/fittings, 42" x 37"	1.000	Ea.	484	91	575
Pipe, steel, galvanized, schedule 40, threaded, 1-1/4" diam	4.000	L.F.	11.32	20.68	32
Pipe, CI, soil, no hub, w/coupling 10' OC, hanger 5' OC, 4" diam	3.000	L.F.	17.07	26.13	43.20
Combination Y and 1/8 bend for C.I. soil pipe, no hub, 4" pipe size	1.000	Ea.	11.90		11.90
1/8 bend, CI, soil, no hub, 2" pipe size, no labor	1.000	Ea.	2.20		2.20
Drum trap, 3" x 5", copper, 1-1/2" diam	1.000	Ea.	11.50	16.50	28
Copper tubing, type L, 50/50 solder joints, hangers 10' OC 1/2" diam	10.000	L.F.	8.40	48.60	57
Wrought copper 90° elbow for solder joints, 1/2" diam	6.000	Ea.	.84	76.86	77.70
Wrought copper Tee for solder joints, 1/2" diam	2.000	Ea.	.48	39.52	40
Stop, angle supply 1/2" diam	2.000	Ea.	23.88	24.12	48
Copper tubing type DWV, 50/50 solder joint, hanger 10'OC 1-1/2" diam	4.000	L.F.	9.16	36.64	45.80
Pipe coupling, standard, C.I. soil no hub, 4" pipe size	6.000	Ea.	19.80	124.20	144
TOTAL			600.55	504.25	1104.80

8.1-410	Bathtub Systems	COST EACH MAT.	INST.	TOTAL
1960	Bathtub, recessed, P.E. on CI., 42" x 37"	600	505	1105
2000	48" x 42"	865	530	1395
2040	C8.1 -401 72" x 36"	890	570	1460
2080	Mat bottom, 5' long	385	520	905
2120	5'-6" long	725	530	1255
2160	Corner, 48" x 42"	980	520	1500
2200	Formed steel, enameled, 4'-6" long	310	490	800
2240	5' long	300	495	795

B8.1-420 Drinking Fountain Systems

Systems are complete with trim and rough-in (supply, waste and vent) to connect to supply branches and waste mains.

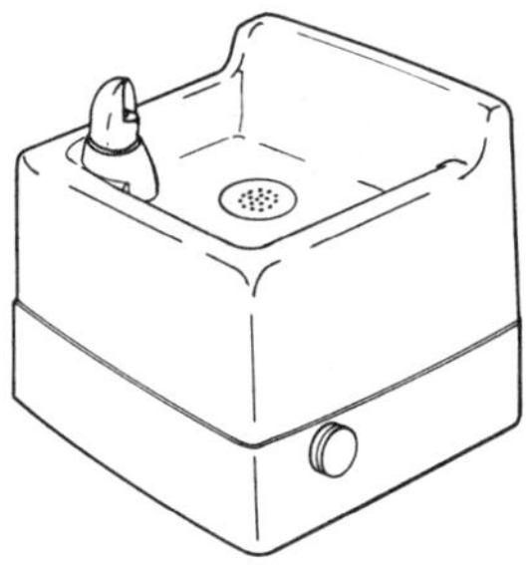

Wall Mounted, No Back

Supply

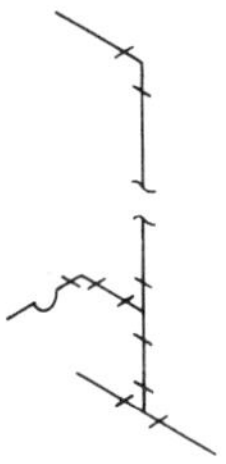

Waste/Vent

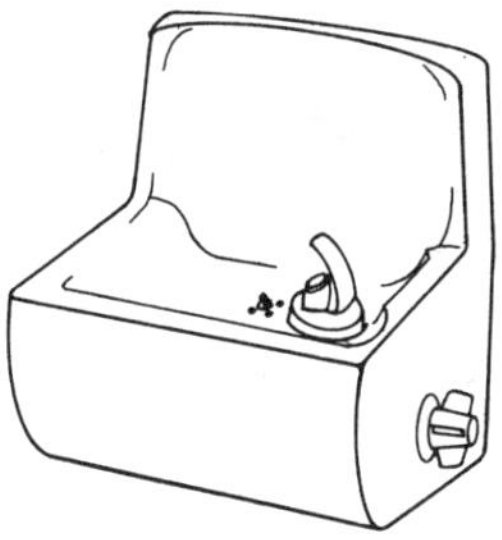

Wall Mounted, Low Back

System Components	QUANTITY	UNIT	COST EACH MAT.	COST EACH INST.	COST EACH TOTAL
SYSTEM 08.1-420-1800					
DRINKING FOUNTAIN, ONE BUBBLER, WALL MOUNTED					
NON RECESSED, BRONZE, NO BACK					
Drinking fountain, wall mount, bronze, 1 bubbler	1.000	Ea.	550	65	615
Copper tubing, type L, 50/50 solder joint, hanger 10' OC 3/8" diam	10.000	L.F.	7	46	53
Stop, supply, straight, chrome, 3/8" diam	1.000	Ea.	10.73	10.27	21
Wrought copper 90° elbow for solder joints 3/8" diam	3.000	Ea.	2.28	34.92	37.20
Wrought copper Tee for solder joints, 3/8" diam	1.000	Ea.	1.18	18.32	19.50
Copper tubing, type DWV, 50/50 solder jnt, hngr 10' OC 1-1/4" diam	5.000	L.F.	9.25	42.75	52
P trap, standard, copper drainage, 1-1/4" diam	1.000	Ea.	9.57	14.43	24
Wrought copper, DWV, Tee, sanitary, 2" diam	1.000	Ea.	4.94	37.06	42
TOTAL			594.95	268.75	863.70

8.1-420	Drinking Fountain Systems	COST EACH MAT.	COST EACH INST.	COST EACH TOTAL
1740	Drinking fountain, one bubbler, wall mounted			
1760	Non recessed			
1800	C8.1 -401 Bronze, no back	595	270	865
1840	Cast iron, enameled, low back	300	265	565
1880	Fiberglass, 12" back	270	270	540
1920	Stainless steel, no back	590	270	860
1960	Semi-recessed, poly marble	335	265	600
2040	Stainless steel	375	270	645
2080	Vitreous china	375	270	645
2120	Full recessed, poly marble	480	270	750
2200	Stainless steel	430	270	700
2240	Floor mounted, pedestal type, aluminum	320	335	655
2320	Bronze	715	335	1050
2360	Stainless steel	730	330	1060

B8.1-431 Kitchen Sink Systems

Systems are complete with trim and rough-in (supply, waste and vent) to connect to supply branches and waste mains.

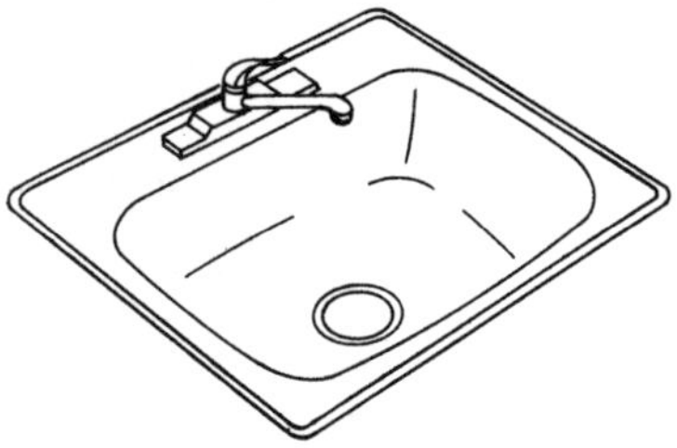

Countertop
Single Bowl

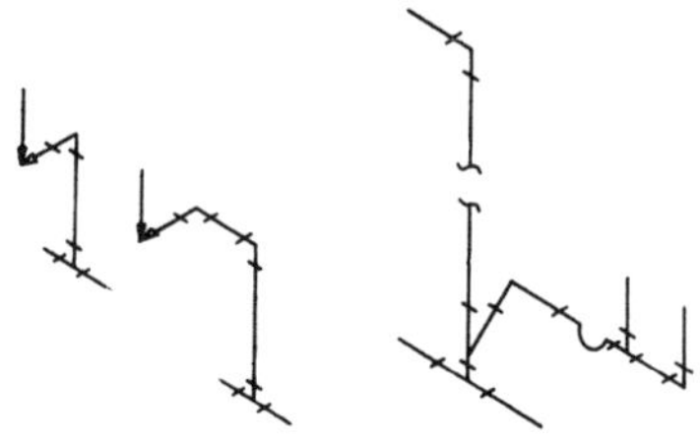

Supply

Waste/Vent

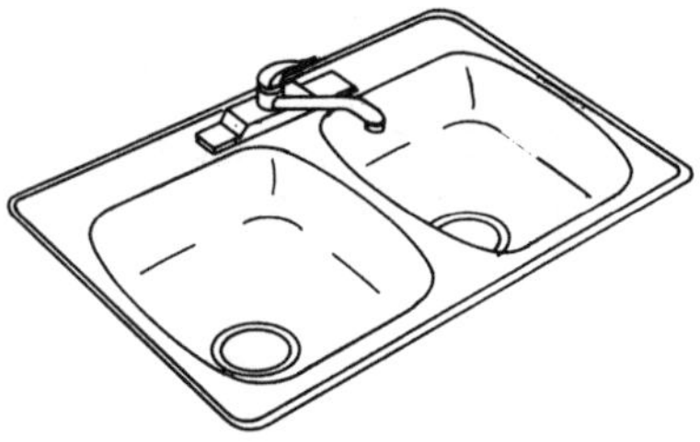

Countertop
Double Bowl

System Components	QUANTITY	UNIT	COST EACH MAT.	COST EACH INST.	COST EACH TOTAL
SYSTEM 08.1-431-1720					
KITCHEN SINK W/TRIM, COUNTERTOP. P E ON C I , 24"X21", SINGLE BOWL					
Kitchen sink, countertop, PE on CI, 1 bowl, 24" x 21" OD	1.000	Ea.	121	144	265
Pipe, steel, galvanized, schedule 40, threaded, 1-1/4" diam	10.000	L.F.	28.30	51.70	80
Copper tubing, type DWV, 50/50 solder, hangers 10' OC 1-1/4" diam	6.000	L.F.	11.10	51.30	62.40
Wrought copper, DWV, Tee, sanitary, 3" diam	1.000	Ea.	11.22	65.78	77
P trap, standard, copper, 1-1/4" diam	1.000	Ea.	9.57	14.43	24
Copper tubing, type L, 50/50 solder joints, hangers 10' OC 1/2" diam	10.000	L.F.	8.40	48.60	57
Wrought copper 90° elbow for solder joints 1/2" diam	6.000	Ea.	.84	76.86	77.70
Wrought copper Tee for solder joints, 1/2" diam	2.000	Ea.	.48	39.52	40
Stop, angle supply, chrome 1/2' diam	2.000	Ea.	23.88	24.12	48
TOTAL			214.79	516.31	731.10

8.1-431		Kitchen Sink Systems	COST EACH MAT.	COST EACH INST.	COST EACH TOTAL
1720		Kitchen sink w/trim, countertop, PE on CI, 24"x21", single bowl	215	515	730
1760		30" x 21" single bowl	235	515	750
1800	C8.1 -102	32" x 21" double bowl	255	565	820
1840		42" x 21" double bowl	365	575	940
1880	C8.1 -201	Stainless steel, 19" x 18" single bowl	315	515	830
1920		25" x 22" single bowl	335	515	850
1960	C8.1 -202	33" x 22" double bowl	355	565	920
2000		43" x 22" double bowl	385	575	960
2040		44" x 22" triple bowl	495	615	1110
2080		44" x 24" corner double bowl	400	575	975
2120		Steel, enameled, 24" x 21" single bowl	160	515	675
2160		32" x 21" double bowl	180	565	745
2240		Raised deck, PE on CI, 32" x 21", dual level, double bowl	340	685	1025
2280		42" x 21" dual level, triple bowl	415	795	1210

Laundry Sink Systems

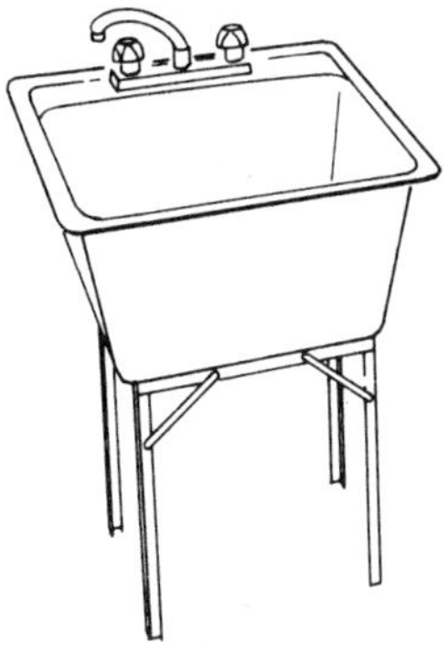

Single Compartment Sink

Systems are complete with trim and rough-in (supply, waste and vent) to connect to supply branches and waste mains.

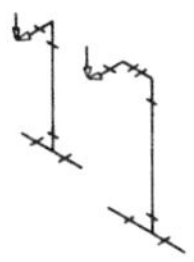

Supply

Waste/Vent

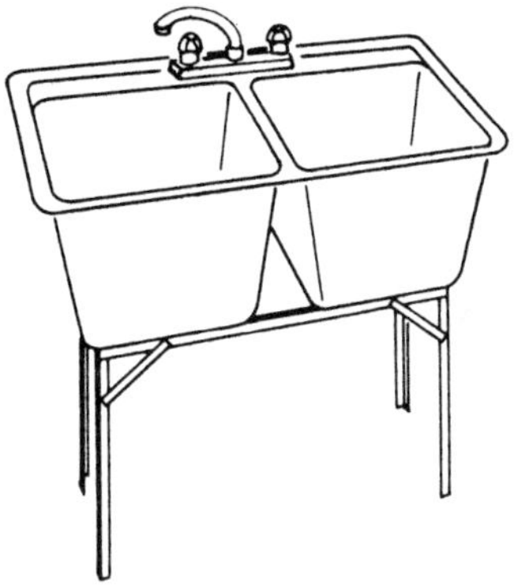

Double Compartment Sink

System Components	QUANTITY	UNIT	COST EACH MAT.	INST.	TOTAL
SYSTEM 08.1-432-1760					
LAUNDRY SINK W/TRIM, PE ON CI, BLACK IRON FRAME					
24" X 20" OD, SINGLE COMPARTMENT					
Laundry sink PE on CI w/trim & frame, 24" x 20" OD, 1 compartment	1.000	Ea.	176	79	255
Pipe, steel, galvanized, schedule 40, threaded, 1-1/4" diam	10.000	L.F.	28.30	51.70	80
Copper tubing, type DWV, 50/50 solder joint, hanger 10' OC 1-1/4"diam	6.000	L.F.	11.10	51.30	62.40
Wrought copper, DWV, Tee, sanitary, 2" diam	1.000	Ea.	4.94	37.06	42
P trap, standard, copper, 1-1/4" diam	1.000	Ea.	9.57	14.43	24
Copper tubing type L, 50/50 solder jnt, hngr 10' OC 1/2" diam	10.000	L.F.	8.40	48.60	57
Wrought copper 90° elbow for solder joints 1/2" diam	6.000	Ea.	.84	76.86	77.70
Wrought copper Tee for solder joints, 1/2" diam	2.000	Ea.	.48	39.52	40
Stop, angle supply, 1/2" diam	2.000	Ea.	23.88	24.12	48
TOTAL			263.51	422.59	686.10

8

8.1-432	Laundry Sink Systems	COST EACH MAT.	INST.	TOTAL
1740	Laundry sink w/trim, PE on CI, black iron frame			
1760	24" x 20", single compartment	265	425	690
1800	C8.1 -401 24" x 23" single compartment	290	420	710
1840	48" x 21" double compartment	405	450	855
1920	Molded stone, on wall, 22" x 21" single compartment	185	420	605
1960	45"x 21" double compartment	255	450	705
2040	Plastic, on wall or legs, 18" x 23" single compartment	145	410	555
2080	20" x 24" single compartment	165	415	580
2120	36" x 23" double compartment	175	445	620
2160	40" x 24" double compartment	225	445	670
2200	Stainless steel, countertop, 22" x 17" single compartment	235	420	655
2240	19" x 22" single compartment	245	420	665
2280	33" x 22" double compartment	335	455	790

B8.1-433 Lavatory Systems

Systems are complete with trim and rough-in (supply, waste and vent) to connect to supply branches and waste mains.

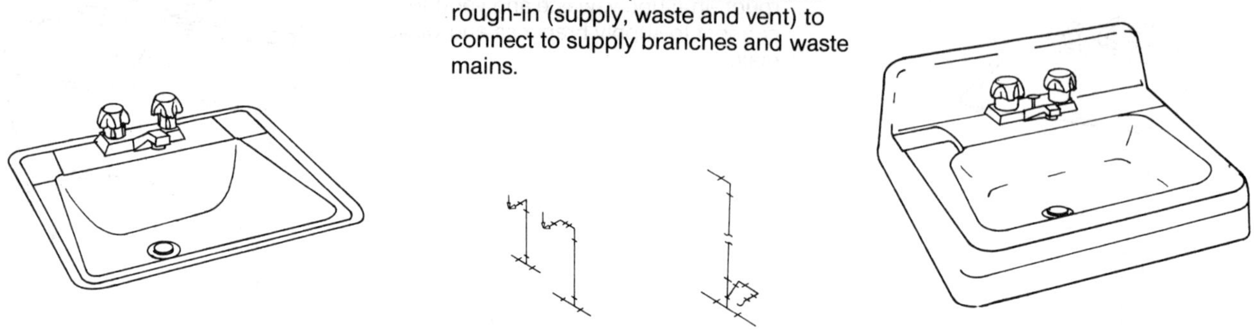

Vanity Top   Supply   Waste/Vent   Wall Hung

System Components

System Components	QUANTITY	UNIT	COST EACH MAT.	COST EACH INST.	COST EACH TOTAL
SYSTEM 08.1-433-1560					
LAVATORY W/TRIM, VANITY TOP, PE ON CI, 20"X18"					
Lavatory w/trim, PE on CI, white, vanity top, 20" x 18" oval	1.000	Ea.	110	70	180
Pipe, steel, galvanized, schedule 40, threaded, 1-1/4" diam	4.000	L.F.	11.32	20.68	32
Copper tubing type DWV, 50/50 solder joint, hanger 10'OC 1-1/4" diam	4.000	L.F.	7.40	34.20	41.60
Wrought copper DWV, Tee, sanitary, 2" diam	1.000	Ea.	4.94	37.06	42
P trap w/cleanout, chromed brass, 20 ga, 1-1/4" diam	1.000	Ea.	11	14	25
Copper tubing type L, 50/50 solder joint, hanger 10' OC 1/2" diam	10.000	L.F.	8.40	48.60	57
Wrought copper 90° elbow for solder joints 1/2" diam	6.000	Ea.	.84	76.86	77.70
Wrought copper Tee for solder joints, 1/2" diam	2.000	Ea.	.48	39.52	40
Stop, chrome, angle supply, 1/2" diam	2.000	Ea.	23.88	24.12	48
TOTAL			178.26	365.04	543.30

8.1-433 Lavatory Systems

C8.1 -401

	8.1-433 Lavatory Systems	COST EACH MAT.	COST EACH INST.	COST EACH TOTAL
1560	Lavatory w/trim, vanity top, PE on CI, 20" x 18"	180	365	545
1600	19" x 16" oval	170	365	535
1640	18" round	170	370	540
1680	Cultured marble, 19" x 17"	140	370	510
1720	25" x 19"	155	365	520
1760	Stainless, self-rimming, 25" x 22"	205	370	575
1800	17" x 22"	195	370	565
1840	Steel enameled, 20" x 17"	135	375	510
1880	19" round	130	375	505
1920	Vitreous china, 20" x 16"	235	380	615
1960	19" x 16"	185	380	565
2000	22" x 13"	180	380	560
2040	Wall hung, PE on CI, 18" x 15"	315	395	710
2080	19" x 17"	280	400	680
2120	20" x 18"	240	395	635
2160	Vitreous china, 18" x 15"	245	405	650
2200	19" x 17"	190	405	595
2240	24" x 20"	295	405	700

B8.1-434 Laboratory Sink Systems

Corrosion resistant laboratory sink systems are complete with trim and rough-in (supply, waste and vent) to connect to supply branches and waste mains.

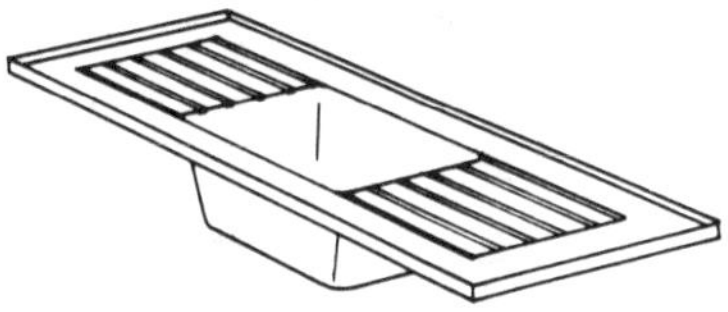

Laboratory Sink

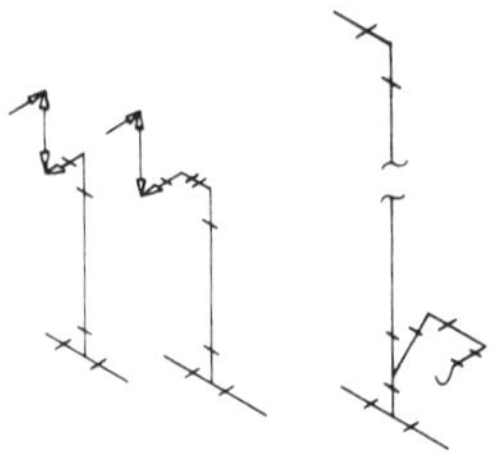

Supply Waste/Vent

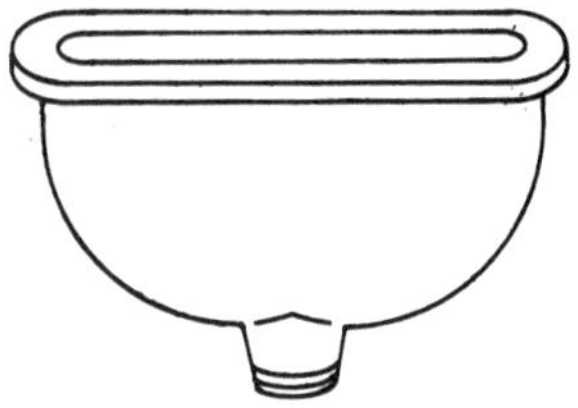

Polypropylene Cup Sink

System Components

System Components	QUANTITY	UNIT	COST EACH MAT.	COST EACH INST.	COST EACH TOTAL
SYSTEM 08.1-434-1600					
LABORATORY SINK W/TRIM, POLYETHYLENE, SINGLE BOWL					
DOUBLE DRAINBOARD, 54" X 24" OD					
Sink w/trim, polyethylene, 1 bowl, 2 drainboards 54" x 24" OD	1.000	Ea.	539	156	695
Pipe, polypropylene, schedule 40, acid resistant 1-1/2" diam	16.000	L.F.	25.76	108.64	134.40
Tee, sanitary, polypropylene, acid resistant, 3" diam	1.000	Ea.	28.60	42.40	71
P trap, polypropylene, acid resistant, 1-1/2" diam	1.000	Ea.	11	15	26
Copper tubing type L, 50/50 solder joint, hanger 10' OC 1/2" diam	10.000	L.F.	8.40	48.60	57
Wrought copper, 90° elbow for solder joints 1/2" diam	6.000	Ea.	.84	76.86	77.70
Wrought copper, Tee for solder joints, 1/2" diam	2.000	Ea.	.48	39.52	40
Stop, angle supply, chrome, 1/2" diam	2.000	Ea.	23.88	24.12	48
TOTAL			637.96	511.14	1149.10

8.1-434	Laboratory Sink Systems	COST EACH MAT.	COST EACH INST.	COST EACH TOTAL
1580	Laboratory sink w/trim, polyethylene, single bowl,			
1600	Double drainboard, 54" x 24" O.D.	640	510	1150
1640	Single drainboard, 47" x 24"O.D.	595	510	1105
1680	70" x 24" O.D.	685	510	1195
1760	Flanged, 14-1/2" x 14-1/2" O.D.	235	475	710
1800	18-1/2" x 18-1/2" O.D.	260	470	730
1840	23-1/2" x 20-1/2" O.D.	300	470	770
1920	Polypropylene, cup sink, oval, 7" x 4" O.D.	135	430	565
1960	10" x 4-1/2" O.D.	155	435	590

Service sink systems are complete with trim and rough-in (supply, waste and vent) to connect to supply branches and waste mains.

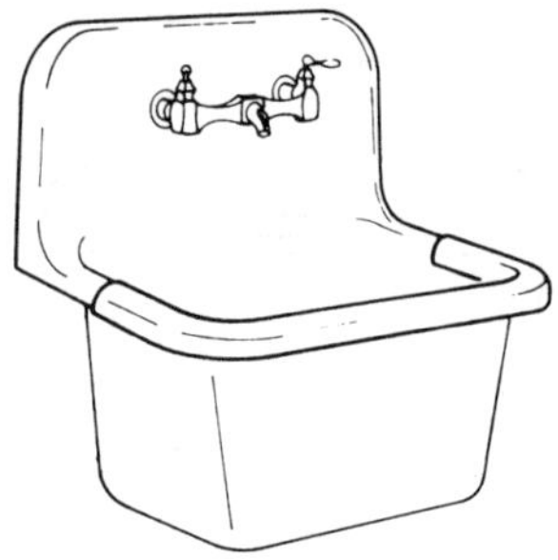

Wall Hung

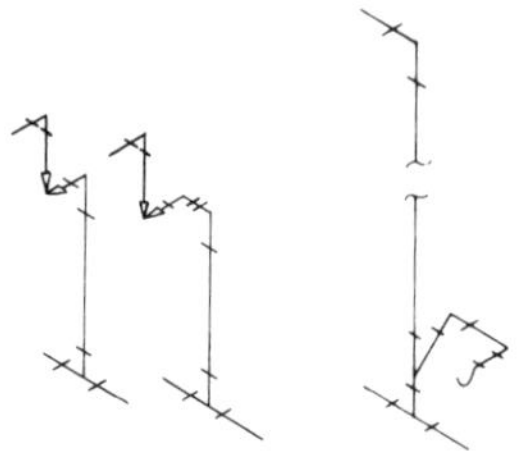

Supply Waste/Vent

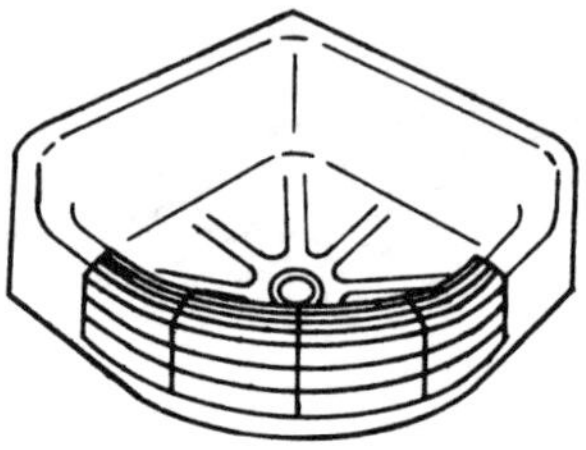

Corner, Floor

System Components	QUANTITY	UNIT	COST EACH MAT.	INST.	TOTAL
SYSTEM 08.1-434-4260					
SERVICE SINK, PE ON CI, CORNER FLOOR, 28"X28", W/RIM GUARD & TRIM					
Service sink, corner, floor, PE on CI, 28" x 28", w/rim guard & trim	1.000	Ea.	346.50	113.50	460
Pipe, steel, galvanized, schedule 40, threaded, 1-1/4" diam	10.000	L.F.	28.30	51.70	80
Copper tubing type DWV, 50/50 solder joint, hanger 10'OC 1-1/4" diam	6.000	L.F.	11.10	51.30	62.40
Copper tubing type DWV, 50/50 solder joint, hanger 10'OC 2" diam	6.000	L.F.	17.76	64.14	81.90
Wrought copper DWV, Tee, sanitary, 3" diam	1.000	Ea.	11.22	65.78	77
P trap with cleanout & slip joint, copper 2" diam	1.000	Ea.	16.78	17.22	34
Copper tubing, type L, 50/50 solder joint, hanger 10' OC 1/2" diam	10.000	L.F.	8.40	48.60	57
Wrought copper 90° elbow for solder joints 1/2" diam	6.000	Ea.	.84	76.86	77.70
Wrought copper Tee for solder joints, 1/2" diam	2.000	Ea.	.48	39.52	40
Stop, angle supply, chrome, 1/2" diam	2.000	Ea.	23.88	24.12	48
TOTAL			465.26	552.74	1018

8.1-434	Service Sink Systems	COST EACH MAT.	INST.	TOTAL
4260	Service sink w/trim, PE on CI, corner floor, 28" x 28", w/rim guard	465	555	1020
4300	Wall hung w/rim guard, 22" x 18"	430	670	1100
4340	24" x 20"	460	670	1130
4380	Vitreous china, wall hung 22" x 20"	450	670	1120

PLUMBING | **B8.1-440** | # Shower Systems

Systems are complete with trim and rough-in (supply, waste and vent) for connection to supply branches and waste mains.

Three Wall

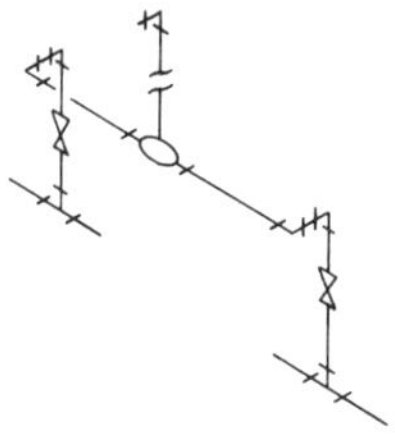

Supply

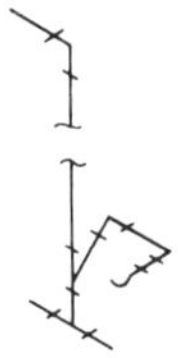

Waste/Vent

Corner Angle

System Components

System Components	QUANTITY	UNIT	COST EACH MAT.	COST EACH INST.	COST EACH TOTAL
SYSTEM 08.1-440-1560					
SHOWER, STALL, BAKED ENAMEL, MOLDED STONE RECEPTOR, 30" SQUARE					
Shower stall, enameled steel, molded stone receptor, 30" square	1.000	Ea.	286	229	515
Pipe, steel, galvanized, schedule 40, threaded, 1-1/4" diam	6.000	L.F.	16.98	31.02	48
Copper tubing type DWV, 50/50 solder jnt, hngr 10'OC 1-1/4" diam	6.000	L.F.	11.10	51.30	62.40
Wrought copper DWV, Tee, sanitary, 3" diam	1.000	Ea.	11.22	65.78	77
Trap, standard, copper 1-1/4" diam	1.000	Ea.	9.57	14.43	24
Copper tubing type L, 50/50 solder joint, hanger 10' OC 1-1/4" diam	16.000	L.F.	13.44	77.76	91.20
Wrought copper 90° elbow for solder joints 1/2" diam	6.000	Ea.	.84	76.86	77.70
Wrought copper Tee for solder joints, 1/2" diam	2.000	Ea.	.48	39.52	40
Stop and waste, straightway, bronze, solder joint 1/2" diam	2.000	Ea.	3.40	10.70	14.10
TOTAL			353.03	596.37	949.40

8.1-440	Shower Systems	COST EACH MAT.	COST EACH INST.	COST EACH TOTAL
1560	Shower, stall, baked enamel, molded stone receptor, 30" square	355	595	950
1600	32" square	355	595	950
1640	C8.1 -401 Terrazzo receptor, 32" square	575	595	1170
1680	36" square	605	625	1230
1720	36" corner angle	605	625	1230
1800	Fiberglass one piece, three walls, 32" square	300	560	860
1840	36" square	330	560	890
1880	Polypropylene, molded stone receptor, 30" square	385	600	985
1920	32" square	385	600	985
1960	Built-in, head, arm, bypass, stops and handles	175	440	615

B8.1-450 Urinal Systems

Systems are complete with trim, flush valve and rough-in (supply, waste and vent) for connection to supply branches and waste mains.

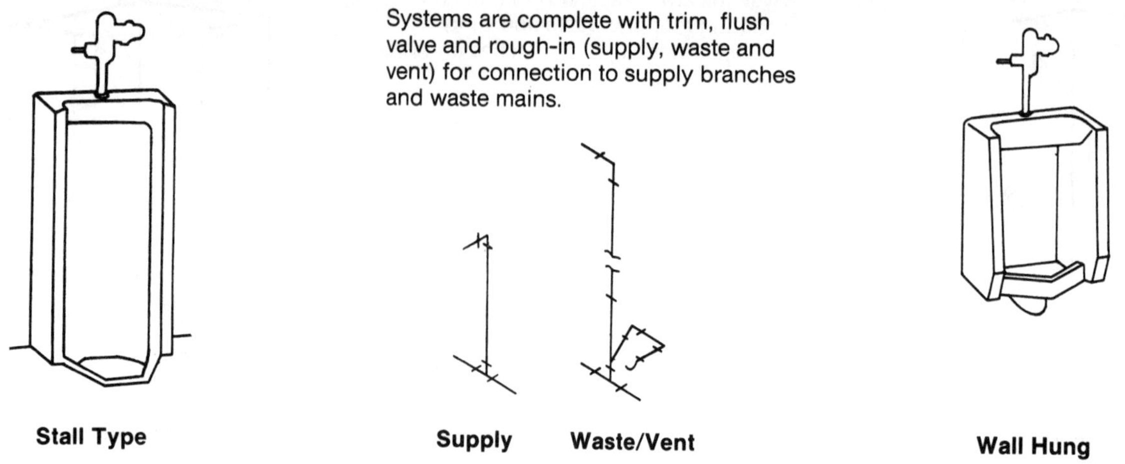

System Components	QUANTITY	UNIT	COST EACH MAT.	INST.	TOTAL
SYSTEM 08.1-450-2000					
URINAL, VITREOUS CHINA, WALL HUNG					
Urinal, wall hung, vitreous china, incl hanger	1.000	Ea.	302.50	152.50	455
Pipe, steel, galvanized, schedule 40, threaded, 1-1/2" diam	6.000	L.F.	19.86	34.74	54.60
Copper tubing type DWV, 50/50 solder joint, hanger 10'OC 1-1/2" diam	10.000	L.F.	22.90	91.60	114.50
Combination Y & 1/8 bend for CI soil pipe, no hub, 3" diam	1.000	Ea.	6.25		6.25
Pipe, CI, no hub, cplg 10' OC, hanger 5' OC, 3" diam	4.000	L.F.	17.60	31.80	49.40
Pipe coupling standard, CI soil, no hub, 3" diam	4.000	Ea.	11.44	76.56	88
Copper tubing type L, 50/50 solder joint, hanger 10' OC 1-1/2" diam	6.000	L.F.	7.32	34.98	42.30
Wrought copper 90° elbow for solder joints 3/4" diam	3.000	Ea.	.99	40.56	41.55
Wrought copper Tee for solder joints, 3/4" diam	1.000	Ea.	.61	21.39	22
TOTAL			389.47	484.13	873.60

8.1-450	Urinal Systems	COST EACH MAT.	INST.	TOTAL
2000	Urinal, vitreous china, wall hung	390	485	875
2040	Stall type	480	530	1010
	C8.1 -401			

PLUMBING | B8.1-460 | Water Cooler Systems

Systems are complete with trim and rough-in (supply, waste and vent) for connection to supply branches and waste mains.

Wall Hung

Supply

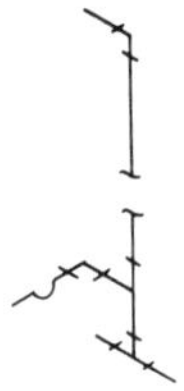

Waste/Vent

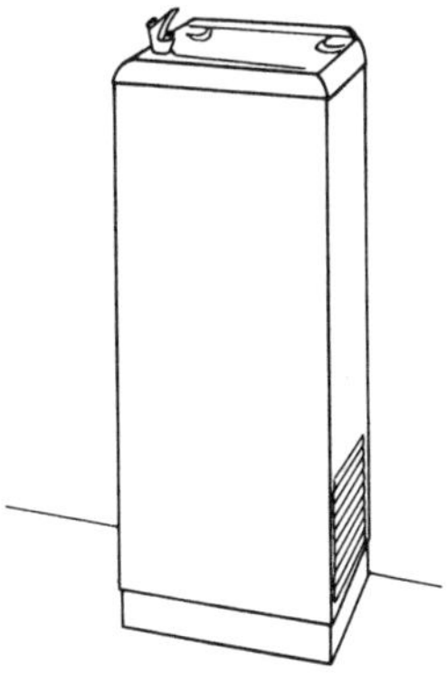

Floor Mounted

System Components	QUANTITY	UNIT	COST EACH MAT.	INST.	TOTAL
SYSTEM 08.1-460-1840					
WATER COOLER, ELECTRIC, SELF CONTAINED, WALL HUNG, 8.2 GPH					
Water cooler, wall mounted, 8.2 GPH	1.000	Ea.	368.50	116.50	485
Copper tubing type DWV, 50/50 solder joint, hanger 10'OC 1-1/4" diam	5.000	L.F.	9.25	42.75	52
Wrought copper DWV, Tee, sanitary 2" diam	1.000	Ea.	4.94	37.06	42
P trap, copper drainage, 1-1/4" diam	1.000	Ea.	9.57	14.43	24
Copper tubing type L, 50/50 solder joint, hanger 10' OC 3/8" diam	10.000	L.F.	7	46	53
Wrought copper 90° elbow for solder joints 3/8" diam	3.000	Ea.	2.28	34.92	37.20
Wrought copper Tee for solder joints, 3/8" diam	1.000	Ea.	1.18	18.32	19.50
Stop and waste, straightway, bronze, solder, 3/8" diam	1.000	Ea.	3.81	10.69	14.50
TOTAL			406.53	320.67	727.20

8.1-460	Water Cooler Systems	COST EACH MAT.	INST.	TOTAL
1840	Water cooler, electric, wall hung, 8.2 GPH	405	320	725
1880	Dual height, 14.3 GPH	550	330	880
1920	C8.1 -401 Wheelchair type, 7.5 G.P.H.	990	330	1320
1960	Semi recessed, 8.1 G.P.H.	515	320	835
2000	C8.1 -403 Full recessed, 8 G.P.H.	845	335	1180
2040	Floor mounted, 14.3 G.P.H.	430	290	720
2080	Dual height, 14.3 G.P.H.	595	335	930
2120	Refrigerated compartment type, 1.5 G.P.H.	775	290	1065
2160	Cafeteria type, dual glass fillers, 27 G.P.H.	1425	380	1805

B8.1-470 Water Closet Systems

Systems are complete with trim, seat and rough-in (supply, waste and vent) for connection to supply branches and waste mains.

One Piece Wall Hung

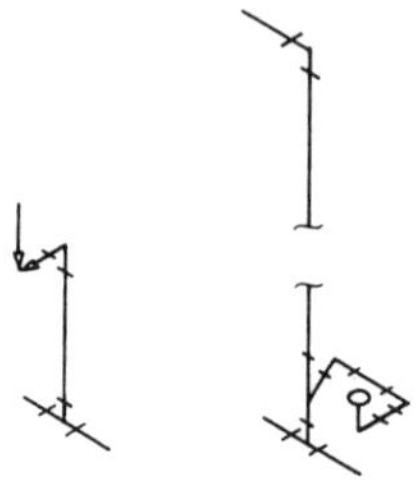

Supply Waste/Vent

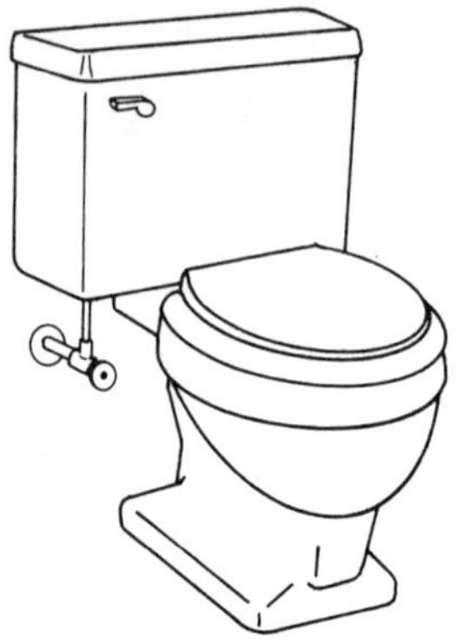

Floor Mount

System Components	QUANTITY	UNIT	COST EACH MAT.	INST.	TOTAL
SYSTEM 08.1-470-1840					
WATER CLOSET, VITREOUS CHINA, ELONGATED					
TANK TYPE, WALL HUNG, ONE PIECE					
Water closet tank type vit china wall hung 1 piece w/seat supply & stop	1.000	Ea.	467.50	87.50	555
Pipe steel galvanized, schedule 40, threaded, 2" diam	4.000	L.F.	18.16	28.84	47
Pipe, CI soil, no hub, cplg 10' OC, hanger 5' OC, 4" diam	3.000	L.F.	17.07	26.13	43.20
Combination Y & 1/8 bend for CI soil pipe, no hub, 4" diam	1.000	Ea.	11.90		11.90
Pipe, coupling, standard coupling, CI soil, no hub, 4" diam	5.000	Ea.	16.50	103.50	120
Copper tubing type L, 50/50 solder joint, hanger 10' OC 1/2" diam	6.000	L.F.	5.04	29.16	34.20
Wrought copper 90° elbow for solder joints 1/2" diam	3.000	Ea.	.42	38.43	38.85
Wrought copper Tee for solder joints 1/2" diam	1.000	Ea.	.24	19.76	20
Support/carrier, for water closet, siphon jet, horiz, single, 4" waste	1.000	Ea.	88	42	130
TOTAL			624.83	375.32	1000.15

8.1-470	Water Closet Systems	COST EACH MAT.	INST.	TOTAL
1800	Water closet, vitreous china, elongated			
1840	Tank type, wall hung, one piece	625	375	1000
1880	C8.1 -401 Close coupled two piece	445	375	820
1920	Floor mount, one piece	505	370	875
1960	One piece low profile	725	370	1095
2000	Two piece close coupled	215	370	585
2040	Bowl only with flush valve			
2080	Wall hung	410	395	805
2120	Floor mount	325	390	715

B8.1-510 Water Closets Group

Systems are complete with trim, seat, flush valve and rough-in (supply, waste and vent) for connection to supply branches and waste mains.

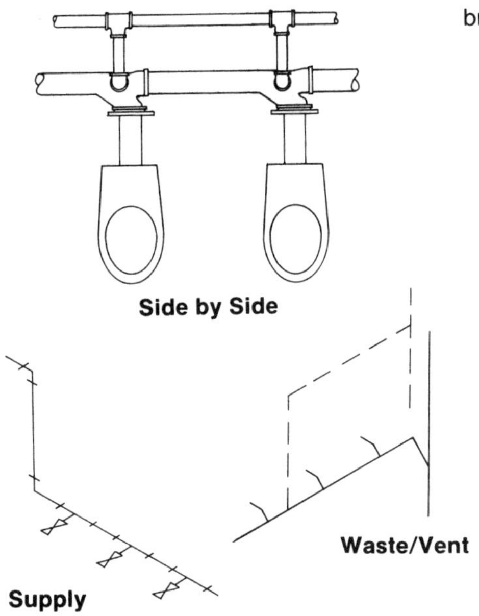

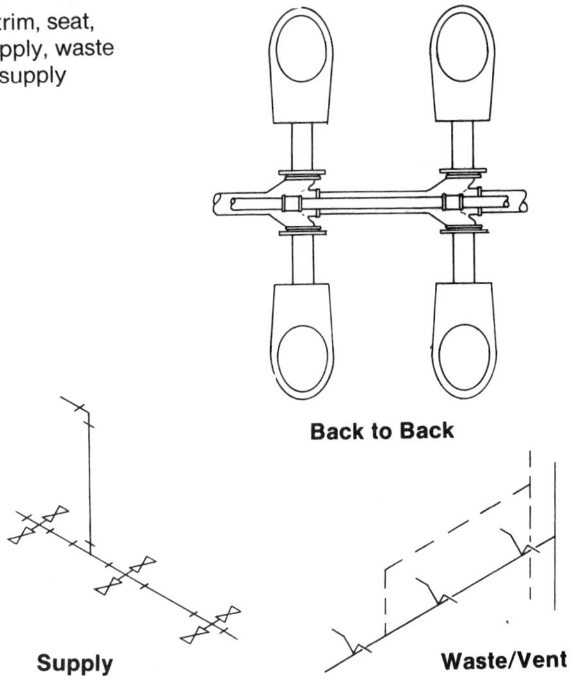

System Components	QUANTITY	UNIT	COST EACH MAT.	INST.	TOTAL
SYSTEM 08.1-510-1760					
WATER CLOSETS, BATTERY MOUNT, WALL HUNG, SIDE BY SIDE, FIRST CLOSET					
Water closet, bowl only w/flush valve, seat, wall hung	1.000	Ea.	242	78	320
Cleanout Tee, C.I. with countersunk plug, 5" diam	1.000	Ea.	49.50	85.50	135
Pipe, CI soil, no hub, cplg 10' OC, hanger 5' OC, 5" diam	4.000	L.F.	33.88	38.92	72.80
Combination Y & 1/8 bend for CI soil pipe, no hub, 5" diam	1.000	Ea.	23		23
Coupling, standard, CI, soil, no hub, 5" diam	5.000	Ea.	40.15	129.85	170
Copper tubing type L, 50/50 solder joint, hanger 10' OC 1" diam	6.000	L.F.	10.62	42.78	53.40
Copper tubing, type DWV, 50/50 solder joint, hanger 10' OC 2" diam	5.000	L.F.	14.80	53.45	68.25
Wrought copper 90° elbow for solder joints 1" diam	3.000	Ea.	2.58	48.12	50.70
Wrought copper Tee for solder joints, 1" diam	1.000	Ea.	2.37	25.63	28
Support/carrier, siphon jet, horiz, adjustable single, 4" pipe	1.000	Ea.	88	42	130
Valve, gate, bronze, 125 lb, NRS, soldered 1" diam	1.000	Ea.	18.04	13.96	32
Wrought copper, DWV, 90° elbow, 2" diam	1.000	Ea.	3.36	25.64	29
TOTAL			528.30	583.85	1112.15

8.1-510	Water Closets Group	COST EACH MAT.	INST.	TOTAL
1760	Water closets, battery mount, wall hung, side by side, first closet	530	585	1115
1800	Each additional water closet, add	385	300	685
3000	C8.1 -401 Back to back, 2 closets	855	760	1615
3100	Each additional pair of closets, back to back	710	475	1185

Systems are complete with trim and rough-in (supply, waste and vent) for connection to supply branches and waste mains.

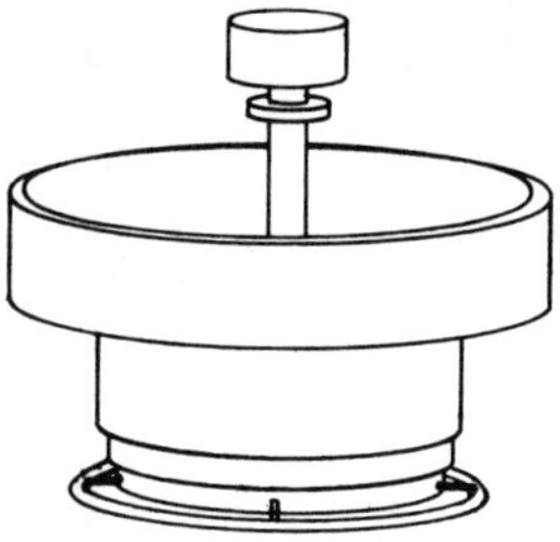

Circular Fountain

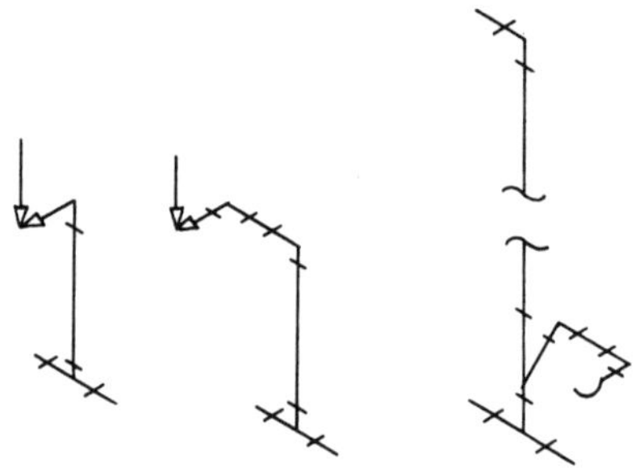

Supply

Waste/Vent

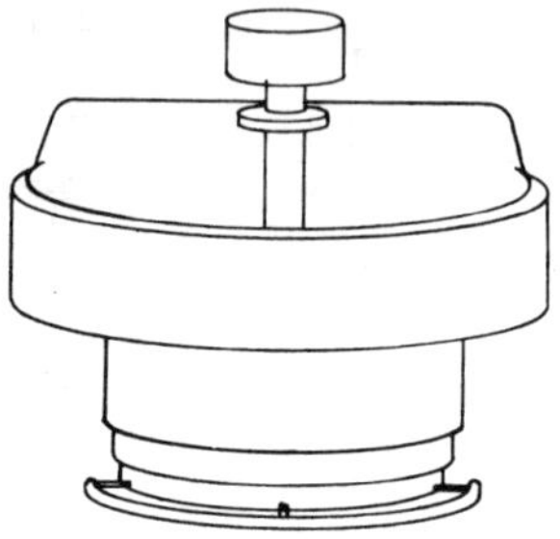

Semi-Circular Fountain

System Components	QUANTITY	UNIT	COST EACH MAT.	INST.	TOTAL
SYSTEM 08.1-560-1760					
GROUP WASH FOUNTAIN, PRECAST TERRAZZO					
CIRCULAR, 36" DIAMETER					
Wash fountain, group, precast terrazzo, foot control 36" diam	1.000	Ea.	918.50	231.50	1150
Copper tubing type DWV, 50/50 solder joint, hanger 10' OC 2" diam	10.000	L.F.	29.60	106.90	136.50
P trap, standard, copper, 2" diam	1.000	Ea.	16.78	17.22	34
Wrought copper, Tee, sanitary, 3" diam	1.000	Ea.	11.22	65.78	77
Copper tubing type L, 50/50 solder joint, hanger 10' OC 3/4" diam	20.000	L.F.	24.40	116.60	141
Wrought copper 90° elbow for solder joints 3/4" diam	6.000	Ea.	1.98	81.12	83.10
Wrought copper Tee for solder joints, 3/4" diam	2.000	Ea.	1.22	42.78	44
TOTAL			1003.70	661.90	1665.60

8.1-560	Group Wash Fountain Systems	COST EACH MAT.	INST.	TOTAL
1740	Group wash fountain, precast terrazzo			
1760	Circular, 36" diameter	1000	660	1660
1800	54" diameter	1150	710	1860
1840	Semi-circular, 36" diameter	925	665	1590
1880	54" diameter	1125	720	1845
1960	Stainless steel, circular, 36" diameter	1175	630	1805
2000	54" diameter	1525	675	2200
2040	Semi-circular, 36" diameter	1025	635	1660
2080	54" diameter	1325	695	2020
2160	Thermoplastic, circular, 36" diameter	1150	645	1795
2200	54" diameter	1450	680	2130
2240	Semi-circular, 36" diameter	1000	640	1640
2280	54" diameter	1275	700	1975

B8.1-620 Two Fixture Bathrooms

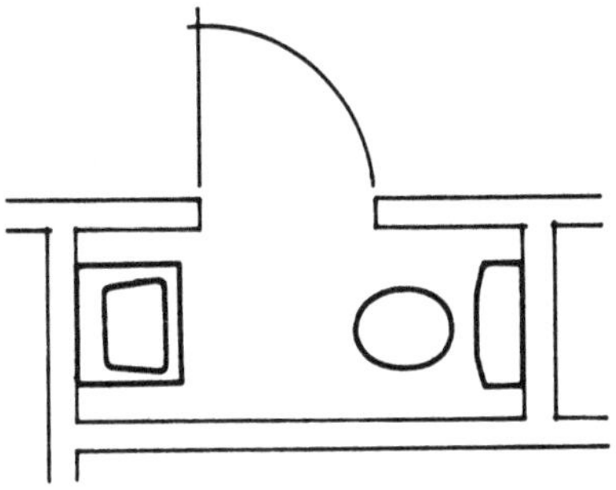

Two fixture bathroom systems consisting of a lavatory, water closet and service piping.
- Prices for plumbing and fixtures only.

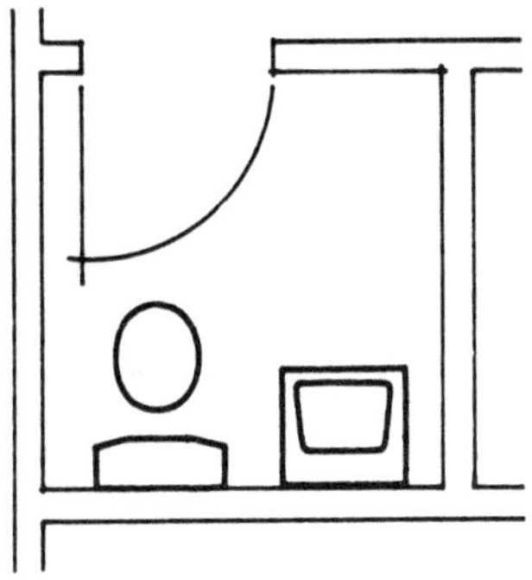

* Common wall is with an adjacent bathroom

System Components	QUANTITY	UNIT	COST EACH MAT.	COST EACH INST.	COST EACH TOTAL
SYSTEM 08.1-620-1180					
BATHROOM, LAVATORY & WATER CLOSET, TWO WALL PLUMBING, STAND ALONE					
Water closet, 2 Pc close cpld vit china flr mntd w/seat, supply & stop	1.000	Ea.	115.50	89.50	205
Water closet, rough-in waste & vent	1.000	Set	100.31	264.69	365
Lavatory w/ftngs, wall hung, white, PE on CI, 20" x 18"	1.000	Ea.	126.50	58.50	185
Lavatory, rough-in waste & vent	1.000	Set	121.81	338.19	460
Copper tubing type L, 50/50 solder joint, hanger 10' OC 1/2" diam	10.000	L.F.	8.40	48.60	57
Pipe, steel, galvanized, schedule 40, threaded, 2" diam	12.000	L.F.	54.48	86.52	141
Pipe, CI soil, no hub, coupling 10' OC, hanger 5' OC, 4" diam	7.000	L.F.	41.09	73.36	114.45
TOTAL			568.09	959.36	1527.45

8.1-620	Two Fixture Bathroom, Two Wall Plumbing	COST EACH MAT.	COST EACH INST.	COST EACH TOTAL
1180	Bathroom, lavatory & water closet, two wall plumbing, stand alone	570	960	1530
1200	Share common plumbing wall*	514.29	838.66	1352

8.1-620	Two Fixture Bathroom, One Wall Plumbing	COST EACH MAT.	COST EACH INST.	COST EACH TOTAL
2220	Bathroom, lavatory & water closet, one wall plumbing, stand alone	530	870	1400
2240	Share common plumbing wall*	465	750	1215

8

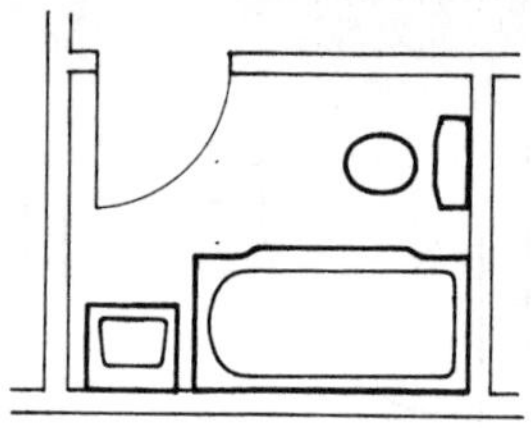

Three fixture bathroom systems consisting of a lavatory, water closet, bathtub or shower and service piping.
- Prices for plumbing and fixtures only.

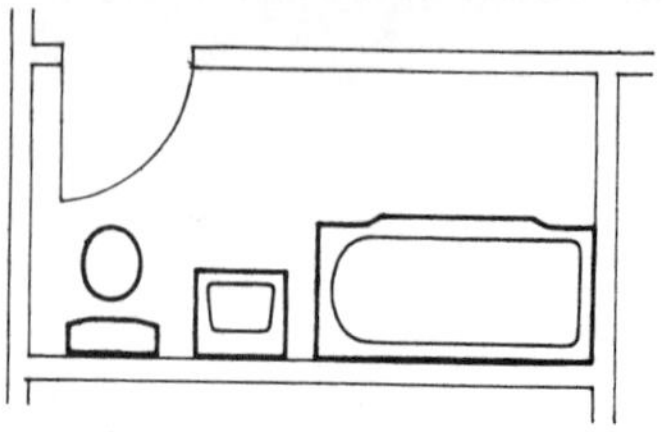

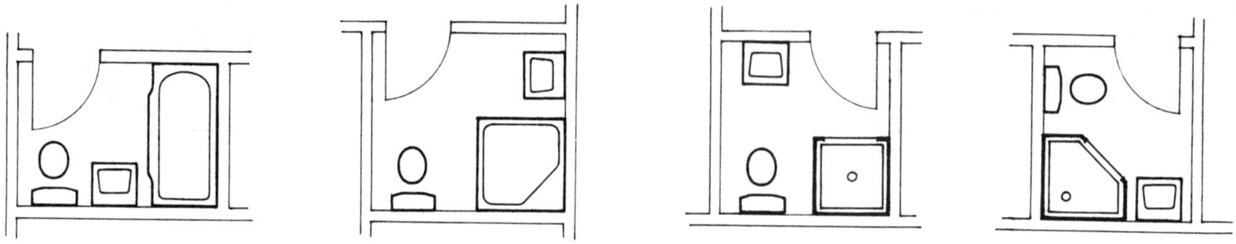

* Common wall is with an adjacent bathroom

System Components	QUANTITY	UNIT	COST EACH MAT.	COST EACH INST.	COST EACH TOTAL
SYSTEM 08.1-630-1170					
BATHROOM, LAVATORY, WATER CLOSET & BATHTUB					
ONE WALL PLUMBING, STAND ALONE					
Water closet, 2 piece close cpld vit china flr mntd w/seat, supply & stop	1.000	Ea.	115.50	89.50	205
Water closet, rough-in waste & vent	1.000	Set	100.31	264.69	365
Lavatory w/ftngs, wall hung, white, PE on CI, 20" x 18"	1.000	Ea.	126.50	58.50	185
Lavatory, rough-in waste & vent	1.000	Set	121.81	338.19	460
Bathtub, white PE on CI, w/ftgs, mat bottom, recessed, 5' long	1.000	Ea.	269.50	105.50	375
Baths, rough-in waste and vent	1.000	Set	86.98	354.02	441
TOTAL			820.60	1210.40	2031

8.1-630	Three Fixture Bathroom, One Wall Plumbing	COST EACH MAT.	COST EACH INST.	COST EACH TOTAL
1150	Bathroom, three fixture, one wall plumbing			
1160	Lavatory, water closet & bathtub			
1170	Stand alone	820	1200	2020
1180	Share common plumbing wall *	713.41	912.79	1626

8.1-630	Three Fixture Bathroom, Two Wall Plumbing	COST EACH MAT.	COST EACH INST.	COST EACH TOTAL
2130	Bathroom, three fixture, two wall plumbing			
2140	Lavatory, water closet & bathtub			
2160	Stand alone	825	1225	2050
2180	Long plumbing wall common *	745	1000	1745
3610	Lavatory, bathtub & water closet			
3620	Stand alone	895	1400	2295
3640	Long plumbing wall common *	840	1275	2115
4660	Water closet, corner bathtub & lavatory			
4680	Stand alone	1425	1250	2675
4700	Long plumbing wall common *	1325	855	2180
6100	Water closet, stall shower & lavatory			
6120	Stand alone	900	1500	2400
6140	Long plumbing wall common *	855	1400	2255
7060	Lavatory, corner stall shower & water closet			
7080	Stand alone	1075	1375	2450
7100	Short plumbing wall common *	975	920	1895

B8.1-640 Four Fixture Bathrooms

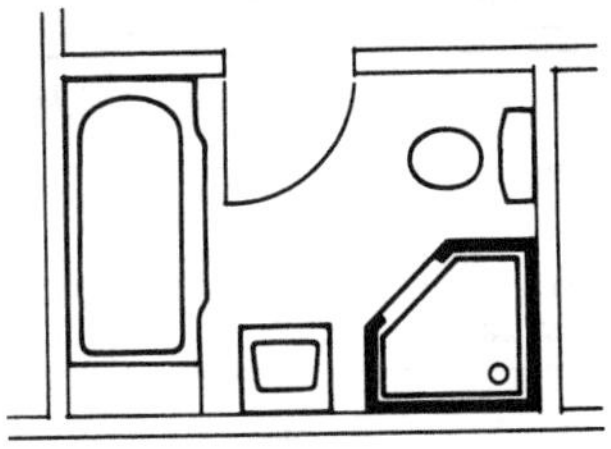

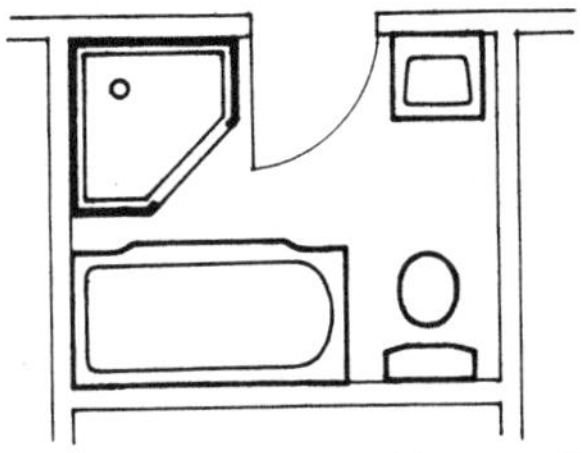

Four fixture bathroom systems consisting of a lavatory, water closet, bathtub, shower and service piping.

- Prices for plumbing and fixtures only.

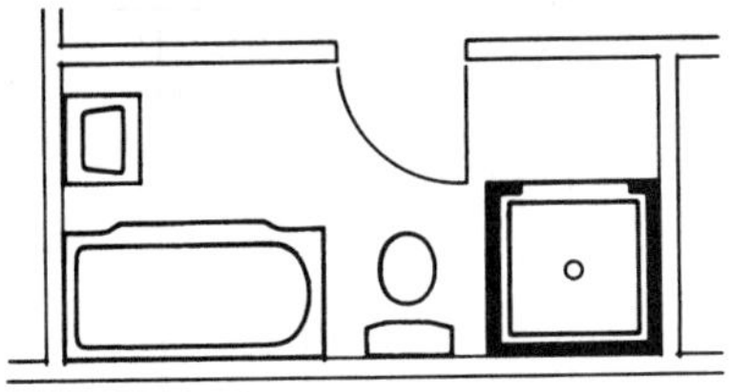

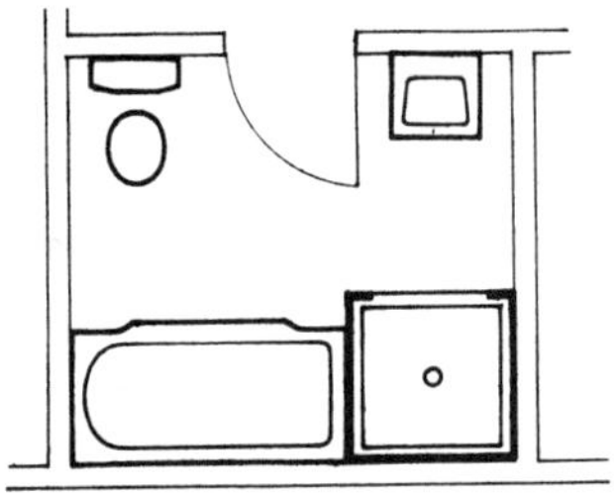

* Common wall is with an adjacent bathroom

System Components	QUANTITY	UNIT	COST EACH MAT.	COST EACH INST.	COST EACH TOTAL
SYSTEM 08.1-640-1160					
BATHROOM, BATHTUB, WATER CLOSET, STALL SHOWER & LAVATORY					
TWO WALL PLUMBING, STAND ALONE					
Water closet, 2 piece close cpld vit china flr mntd w/seat, supply & stop	1.000	Ea.	115.50	89.50	205
Water closet, rough-in waste & vent	1.000	Set	100.31	264.69	365
Lavatory w/ftngs, wall hung, white PE on CI 20" x 18"	1.000	Ea.	126.50	58.50	185
Lavatory, rough-in waste & vent	1.000	Set	12.18	33.82	46
Bathtub, white PE on CI, w/ftgs, mat bottom, recessed, 5' long	1.000	Ea.	269.50	105.50	375
Baths, rough-in waste and vent	1.000	Set	96.64	393.36	490
Shower stall, baked enamel, molded stone receptor, door & trim 32" sq	1.000	Ea.	286	229	515
Shower stall, rough-in supply, waste & vent	1.000	Set	76.40	373.60	450
TOTAL			1083.03	1547.97	2631

8.1-640	Four Fixture Bathroom, Two Wall Plumbing	COST EACH MAT.	COST EACH INST.	COST EACH TOTAL
1140	Bathroom, four fixture, two wall plumbing			
1150	Bathtub, water closet, stall shower & lavatory			
1160	Stand alone	1075	1550	2625
1180	Long plumbing wall common *	985	1150	2135
2260	Bathtub, lavatory, corner stall shower & water closet			
2280	Stand alone	1325	1575	2900
2320	Long plumbing wall common *	1250	1175	2425
3620	Bathtub, stall shower, lavatory & water closet			
3640	Stand alone	1200	1850	3050
3660	Long plumbing wall (opp. door) common *	1096.02	1458.98	2555

8.1-640	Four Fixture Bathroom, Three Wall Plumbing	COST EACH MAT.	COST EACH INST.	COST EACH TOTAL
4680	Bathroom, four fixture, three wall plumbing			
4700	Bathtub, stall shower, lavatory & water closet			
4720	Stand alone	1500	2025	3525
4760	Long plumbing wall (opposite door) common *	1450	1900	3350

B8.1-650 Five Fixture Bathrooms

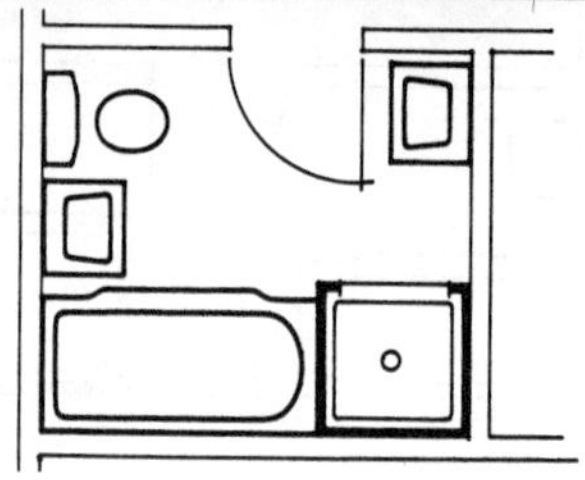

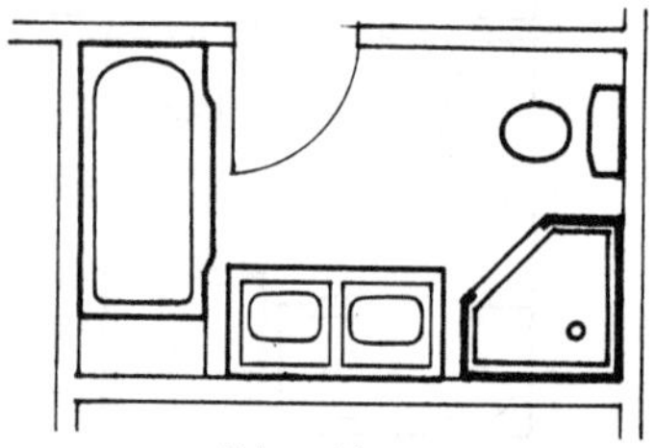

Five fixture bathroom systems consisting of two lavatories, a water closet, bathtub, shower and service piping.

- Prices for plumbing and fixtures only.

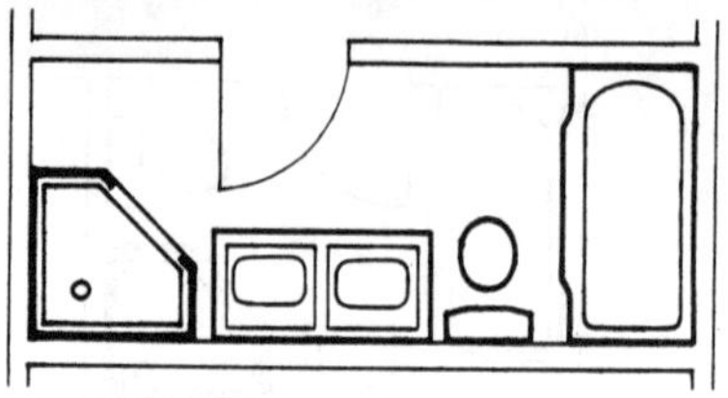

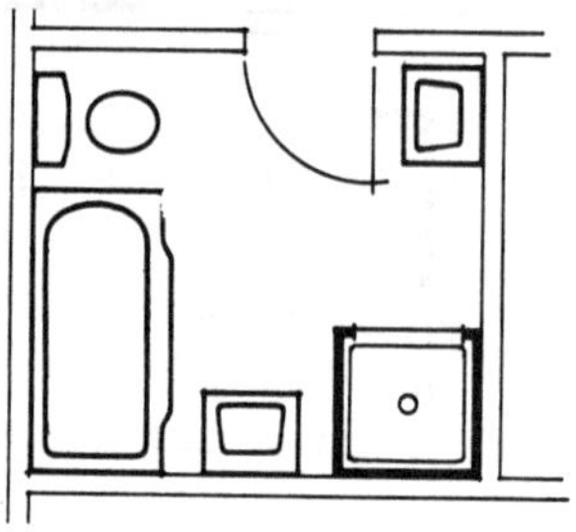

* Common wall is with an adjacent bathroom

System Components	QUANTITY	UNIT	COST EACH MAT.	COST EACH INST.	COST EACH TOTAL
SYSTEM 08.1-650-1360					
BATHROOM, BATHTUB, WATER CLOSET, STALL SHOWER & TWO LAVATORIES					
TWO WALL PLUMBING, STAND ALONE					
Water closet, 2 piece close cpld vit china flr mntd incl seat, supply & stop	1.000	Ea.	115.50	89.50	205
Water closet rough-in waste & vent	1.000	Set	100.31	264.69	365
Lavatory w/ftngs, wall hung, white PE on CI, 20" x 18"	2.000	Ea.	253	117	370
Lavatory rough-in waste & vent	2.000	Set	243.62	676.38	920
Bathtub, white PE on CI, w/ftgs, mat bottom, recessed, 5' long	1.000	Ea.	269.50	105.50	375
Bath rough-in waste and vent	1.000	Set	96.64	393.36	490
Shower stall, baked enamel, molded stone receptor, door & ftng, 32" sq	1.000	Ea.	286	229	515
Shower stall rough-in supply, waste & vent,	1.000	Set	76.40	373.60	450
TOTAL			1440.97	2249.03	3690

8.1-650	Five Fixture Bathroom, Two Wall Plumbing	COST EACH MAT.	COST EACH INST.	COST EACH TOTAL
1320	Bathroom, five fixture, two wall plumbing			
1340	Bathtub, water closet, stall shower & two lavatories			
1360	Stand alone	1450	2250	3700
1400	One short plumbing wall common *	1350	1850	3200
1440				
1500	Bathtub, two lavatories, corner stall shower & water closet			
1520	Stand alone	1700	2275	3975
1540	Long plumbing wall common*	1550	1850	3400

8.1-650	Five Fixture Bathroom, Three Wall Plumbing	COST EACH MAT.	COST EACH INST.	COST EACH TOTAL
2360	Bathroom, five fixture, three wall plumbing			
2380	Water closet, bathtub, two lavatories & stall shower			
2400	Stand alone	1700	2275	3975
2440	One short plumbing wall common *	1600	1875	3475

8.1-650	Five Fixture Bathroom, One Wall Plumbing	COST EACH MAT.	COST EACH INST.	COST EACH TOTAL
4080	Bathroom, five fixture, one wall plumbing			
4100	Bathtub, two lavatories, corner stall shower & water closet			
4120	Stand alone	1625	2075	3700
4160	Share common wall *	1375	1425	2800

B8.2-110 Wet Pipe Sprinkler Systems

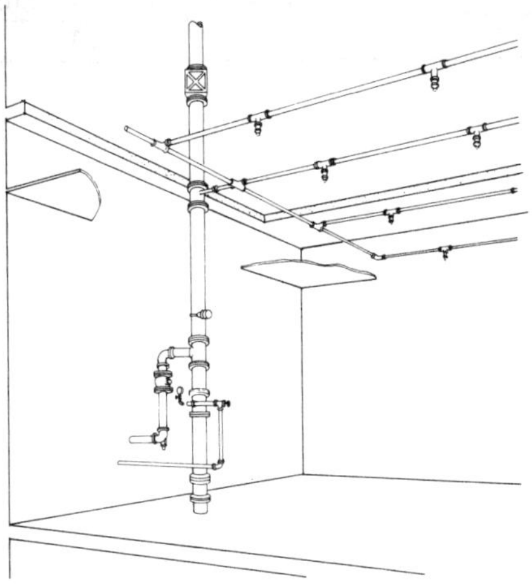

Wet pipe system. A system employing automatic sprinklers attached to a piping system containing water and connected to a water supply so that water discharges immediately from sprinklers opened by heat from a fire.

Costs for large jobs and those using prefabrications will run 15 to 25% less than these.

All areas are assumed to be open.

System Components	QUANTITY	UNIT	COST EACH MAT.	INST.	TOTAL
SYSTEM 08.2-110-0580					
WET PIPE SPRINKLER, STEEL, BLACK, SCH. 40 PIPE					
LIGHT HAZARD, ONE FLOOR, 2000 S.F.					
Valve, gate, iron body, 125 lb, OS&Y, flanged, 4" diam	1.000	Ea.	176	154	330
Valve, swing check, bronze, 125 lb, regrinding disc, 2-1/2" pipe size	1.000	Ea.	85.80	29.20	115
Valve, angle, bronze, 150 lb, rising stem, threaded, 2" diam	1.000	Ea.	110	25	135
*Alarm valve, 2-1/2" pipe size	1.000	Ea.	462	158	620
Alarm, water motor, complete with gong	1.000	Ea.	93.50	66.50	160
Valve, swing check, w/balldrip CI with brass trim 4" pipe size	1.000	Ea.	121	159	280
Pipe, steel, black, schedule 40, 4" diam	10.000	L.F.	99.90	130.10	230
*Flow control valve, trim & gauges, 4" pipe size	1.000	Set	995.50	354.50	1350
Fire alarm horn, electric	1.000	Ea.	23.10	37.90	61
Pipe, steel, black, schedule 40, threaded, cplg & hngr 10'OC, 2-1/2" diam	20.000	L.F.	128.20	184.80	313
Pipe, steel, black, schedule 40, threaded, cplg & hngr 10'OC, 2" diam	12.500	L.F.	49.25	90.13	139.38
Pipe, steel, black, schedule 40, threaded, cplg & hngr 10'OC, 1-1/4" diam	37.500	L.F.	91.88	195	286.88
Pipe steel, black, schedule 40, threaded cplg & hngr 10'OC, 1" diam	112.000	L.F.	219.52	542.08	761.60
Pipe Tee, malleable iron black, 150 lb threaded, 4" pipe size	2.000	Ea.	70.40	229.60	300
Pipe Tee, malleable iron black, 150 lb threaded, 2-1/2" pipe size	2.000	Ea.	26.96	103.04	130
Pipe Tee, malleable iron black, 150 lb threaded, 2" pipe size	1.000	Ea.	6.05	41.95	48
Pipe Tee, malleable iron black, 150 lb threaded, 1-1/4" pipe size	5.000	Ea.	16.70	163.30	180
Pipe Tee, malleable iron black, 150 lb threaded, 1" pipe size	4.000	Ea.	8.24	127.76	136
Pipe, 90° elbow , malleable iron black, 150 lb threaded, 1" pipe size	6.000	Ea.	7.98	118.02	126
Sprinkler head, standard spray, brass 135°-286°F 1/2" NPT, 3/8" orifice	12.000	Ea.	36.36	311.64	348
Valve, gate, bronze, NRS, class 150, threaded, 1" pipe size	1.000	Ea.	18.26	13.74	32
*Standpipe connection, wall, single, flush w/plug & chain 2-1/2"x2-1/2"	1.000	Ea.	79.20	95.80	175
TOTAL			2925.80	3331.06	6256.86
COST PER S.F.			1.46	1.67	3.13

*Not included in systems under 2000 S.F.

8.2-110		Wet Pipe Sprinkler Systems	COST PER S.F. MAT.	INST.	TOTAL
0520		Wet pipe sprinkler systems, steel, black, sch. 40 pipe			
0530		Light hazard, one floor, 500 S.F.	.93	1.69	2.62
0560	C8.2 -101	1000 S.F.	1.36	1.74	3.10
0580		2000 S.F.	1.46	1.67	3.13
0600	C8.2 -102 C8.2 -103	5000 S.F.	.76	1.22	1.98
0620		10,000 S.F.	.54	1.03	1.57

8.2-110	Wet Pipe Sprinkler Systems	COST PER S.F.		
		MAT.	INST.	TOTAL
0640	50,000 S.F.	.45	.93	1.38
0660	Each additional floor, 500 S.F.	.56	1.41	1.97
0680	1000 S.F.	.51	1.29	1.80
0700	2000 S.F.	.47	1.14	1.61
0720	5000 S.F.	.36	1.01	1.37
0740	10,000 S.F.	.36	.92	1.28
0760	50,000 S.F.	.42	.91	1.33
1000	Ordinary hazard, one floor, 500 S.F.	1.11	1.86	2.97
1020	1000 S.F.	1.33	1.67	3
1040	2000 S.F.	1.56	1.80	3.36
1060	5000 S.F.	.84	1.32	2.16
1080	10,000 S.F.	.68	1.36	2.04
1100	50,000 S.F.	.95	1.49	2.44
1140	Each additional floor, 500 S.F.	.74	1.58	2.32
1160	1000 S.F.	.49	1.27	1.76
1180	2000 S.F.	.57	1.28	1.85
1200	5000 S.F.	.48	1.11	1.59
1220	10,000 S.F.	.50	1.26	1.76
1240	50,000 S.F.	.48	1.18	1.66
1500	Extra hazard, one floor, 500 S.F.	3.45	3.08	6.53
1520	1000 S.F.	2.13	2.49	4.62
1540	2000 S.F.	1.63	2.34	3.97
1560	5000 S.F.	.96	1.85	2.81
1580	10,000 S.F.	1.17	1.96	3.13
1600	50,000 S.F.	1.55	2.25	3.80
1660	Each additional floor, 500 S.F.	.82	1.91	2.73
1680	1000 S.F.	.80	1.82	2.62
1700	2000 S.F.	.71	1.82	2.53
1720	5000 S.F.	.61	1.64	2.25
1740	10,000 S.F.	.75	1.68	2.43
1760	50,000 S.F.	1.02	1.95	2.97
2020	Grooved steel, black sch. 40 pipe, light hazard, one floor, 2000 S.F.	1.54	1.45	2.99
2060	10,000 S.F.	.68	.93	1.61
2100	Each additional floor, 2000 S.F.	.55	.93	1.48
2150	10,000 S.F.	.39	.79	1.18
2200	Ordinary hazard, one floor, 2000 S.F.	1.57	1.56	3.13
2250	10,000 S.F.	.70	1.16	1.86
2300	Each additional floor, 2000 S.F.	.58	1.04	1.62
2350	10,000 S.F.	.52	1.06	1.58
2400	Extra hazard, one floor, 2000 S.F.	1.74	2.02	3.76
2450	10,000 S.F.	1.02	1.52	2.54
2500	Each additional floor, 2000 S.F.	.83	1.51	2.34
2550	10,000 S.F.	.70	1.36	2.06
3050	Grooved steel black sch. 10 pipe, light hazard, one floor, 2000 S.F.	1.52	1.44	2.96
3100	10,000 S.F.	.57	.88	1.45
3150	Each additional floor, 2000 S.F.	.53	.91	1.44
3200	10,000 S.F.	.39	.78	1.17
3250	Ordinary hazard, one floor, 2000 S.F.	1.56	1.55	3.11
3300	10,000 S.F.	.69	1.13	1.82
3350	Each additional floor, 2000 S.F.	.56	1.02	1.58
3400	10,000 S.F.	.51	1.03	1.54
3450	Extra hazard, one floor, 2000 S.F.	1.73	2.01	3.74
3500	10,000 S.F.	.91	1.49	2.40
3550	Each additional floor, 2000 S.F.	.82	1.50	2.32
3600	10,000 S.F.	.66	1.34	2
4050	Copper tubing, type M, light hazard, one floor, 2000 S.F.	1.40	1.82	3.22
4100	10,000 S.F.	.51	1.26	1.77
4150	Each additional floor, 2000 S.F.	.42	1.31	1.73

8.2-110	Wet Pipe Sprinkler Systems	COST PER S.F. MAT.	COST PER S.F. INST.	COST PER S.F. TOTAL
4200	10,000 S.F.	.33	1.16	1.49
4250	Ordinary hazard, one floor, 2000 S.F.	1.45	2.03	3.48
4300	10,000 S.F.	.59	1.52	2.11
4350	Each additional floor, 2000 S.F.	.41	1.26	1.67
4400	10,000 S.F.	.40	1.40	1.80
4450	Extra hazard, one floor, 2000 S.F.	1.58	2.60	4.18
4500	10,000 S.F.	1.02	2.16	3.18
4550	Each additional floor, 2000 S.F.	.67	2.09	2.76
4600	10,000 S.F.	.63	1.94	2.57
5050	Copper tubing, type M, T-drill system, light hazard, one floor			
5060	2000 S.F.	1.41	1.79	3.20
5100	10,000 S.F.	.49	1.14	1.63
5150	Each additional floor, 2000 S.F.	.43	1.28	1.71
5200	10,000 S.F.	.31	1.04	1.35
5250	Ordinary hazard, one floor, 2000 S.F.	1.42	1.84	3.26
5300	10,000 S.F.	.58	1.44	2.02
5350	Each additional floor, 2000 S.F.	.43	1.31	1.74
5400	10,000 S.F.	.40	1.34	1.74
5450	Extra hazard, one floor, 2000 S.F.	1.47	2.24	3.71
5500	10,000 S.F.	.87	1.78	2.65
5550	Each additional floor, 2000 S.F.	.61	1.73	2.34
5600	10,000 S.F.	.49	1.56	2.05

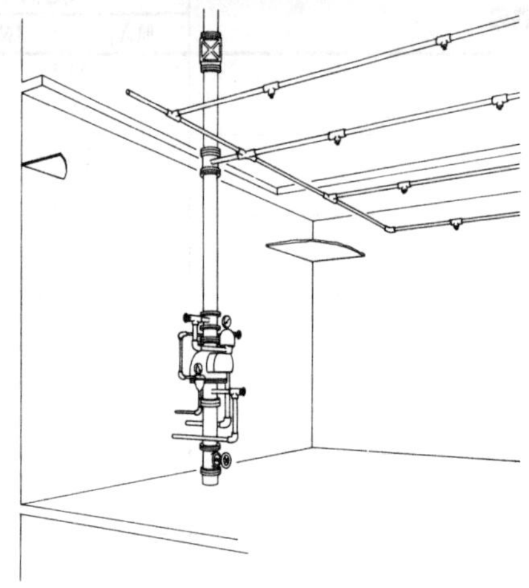

Dry Pipe System: A system employing automatic sprinklers attached to a piping system containing air under pressure, the release of which as from the opening of sprinklers permits the water pressure to open a valve known as a "dry pipe valve". The water then flows into the piping system and out the opened sprinklers.

Costs for large jobs and those using prefabrications will run 15 to 25% less than these.

All areas are assumed to be open.

System Components	QUANTITY	UNIT	COST EA. MAT.	INST.	TOTAL
SYSTEM 08.2-120-0580					
DRY PIPE SPRINKLER, STEEL BLACK, SCH. 40 PIPE					
LIGHT HAZARD, ONE FLOOR, 2000 S.F.					
Valve, gate, iron body 125 lb, OS&Y, flanged, 4" pipe size	1.000	Ea.	176	154	330
Valve, swing check, bronze, 125 lb, regrinding disc, 2-1/2" pipe size	1.000	Ea.	85.80	29.20	115
Valve, angle, bronze, 150 lb, rising stem, threaded, 2" pipe size	1.000	Ea.	110	25	135
*Alarm valve, 2-1/2" pipe size	1.000	Ea.	462	158	620
Alarm, water motor, complete with gong,	1.000	Ea.	93.50	66.50	160
Fire alarm horn, electric	1.000	Ea.	23.10	37.90	61
Valve swing check w/balldrip CI with brass trim, 4" pipe size	1.000	Ea.	121	159	280
Pipe, steel, black, schedule 40, 4" diam	10.000	L.F.	99.90	130.10	230
Dry pipe valve, trim & gauges, 4" pipe size	1.000	Ea.	742.50	482.50	1225
Pipe, steel, black, schedule 40, threaded, cplg & hngr 10'OC 2-1/2" diam	20.000	L.F.	128.20	184.80	313
Pipe, steel, black, schedule 40, threaded, cplg & hngr 10'OC 2" diam	12.500	L.F.	49.25	90.13	139.38
Pipe, steel, black, schedule 40, threaded, cplg & hngr 10'OC 1-1/4" diam	37.500	L.F.	91.88	195	286.88
Pipe, steel, black, schedule 40, threaded, cplg & hngr 10'OC 1" diam	112.000	L.F.	219.52	542.08	761.60
Pipe Tee, malleable iron black, 150 lb threaded, 4" pipe size	2.000	Ea.	70.40	229.60	300
Pipe Tee, malleable iron black, 150 lb threaded, 2-1/2" pipe size	2.000	Ea.	26.96	103.04	130
Pipe Tee, malleable iron black, 150 lb threaded, 2" pipe size	1.000	Ea.	6.05	41.95	48
Pipe Tee, malleable iron black, 150 lb threaded, 1-1/4" pipe size	5.000	Ea.	16.70	163.30	180
Pipe Tee, malleable iron black, 150 lb threaded, 1" pipe size	4.000	Ea.	8.24	127.76	136
Pipe, 90° elbow, malleable iron black, 150 lb threaded, 1" pipe size	6.000	Ea.	7.98	118.02	126
Sprinkler head dry 1/2" orifice 1" NPT, 3" to 4-3/4" length	12.000	Ea.	224.40	399.60	624
Air compressor, 200 Gal sprinkler system capacity, 1/3 HP	1.000	Ea.	489.50	200.50	690
*Standpipe connection, wall, flush, brs w/plug & chain 2-1/2"x2-1/2"	1.000	Ea.	79.20	95.80	175
Valve gate bronze, 300 psi, NRS, class 150, threaded, 1" pipe size	1.000	Ea.	18.26	13.74	32
TOTAL			3350.34	3747.52	7097.86
COST PER S.F.			1.68	1.87	3.55

*Not included in systems under 2000 S.F.

8.2-120	Dry Pipe Sprinkler Systems	COST PER S.F. MAT.	INST.	TOTAL
0520	Dry pipe sprinkler systems, steel, black, sch. 40 pipe			
0530	Light hazard, one floor, 500 S.F.	3.70	3.28	6.98
0560	C8.2 -101 1000 S.F.	1.95	1.95	3.90
0580	2000 S.F.	1.68	1.87	3.55
0600	C8.2 -102 C8.2 -103 5000 S.F.	.90	1.33	2.23
0620	10,000 S.F.	.65	1.10	1.75

8.2-120	Dry Pipe Sprinkler Systems	COST PER S.F. MAT.	COST PER S.F. INST.	COST PER S.F. TOTAL
0640	50,000 S.F.	.53	.98	1.51
0660	Each additional floor, 500 S.F.	.86	1.64	2.50
0680	1000 S.F.	.64	1.34	1.98
0700	2000 S.F.	.57	1.18	1.75
0720	5000 S.F.	.46	1.05	1.51
0740	10,000 S.F.	.44	.96	1.40
0760	50,000 S.F.	.56	1.48	2.04
1000	Ordinary hazard, one floor, 500 S.F.	3.70	3.28	6.98
1020	1000 S.F.	1.93	1.93	3.86
1040	2000 S.F.	1.80	2.02	3.82
1060	5000 S.F.	1.02	1.44	2.46
1080	10,000 S.F.	.83	1.45	2.28
1100	50,000 S.F.	1.08	1.56	2.64
1140	Each additional floor, 500 S.F.	.86	1.64	2.50
1160	1000 S.F.	.62	1.32	1.94
1180	2000 S.F.	.69	1.34	2.03
1200	5000 S.F.	.61	1.17	1.78
1220	10,000 S.F.	.63	1.32	1.95
1240	50,000 S.F.	.60	1.24	1.84
1500	Extra hazard, one floor, 500 S.F.	4.73	4.06	8.79
1520	1000 S.F.	2.75	2.96	5.71
1540	2000 S.F.	1.93	2.60	4.53
1560	5000 S.F.	1.17	1.99	3.16
1580	10,000 S.F.	1.36	2.04	3.40
1600	50,000 S.F.	1.72	2.34	4.06
1660	Each additional floor, 500 S.F.	1.01	2	3.01
1680	1000 S.F.	.99	1.91	2.90
1700	2000 S.F.	.90	1.91	2.81
1720	5000 S.F.	.76	1.70	2.46
1740	10,000 S.F.	.94	1.77	2.71
1760	50,000 S.F.	1.22	2.04	3.26
2020	Grooved steel, black, sch. 40 pipe, light hazard, one floor, 2000 S.F.	1.75	1.66	3.41
2060	10,000 S.F.	.69	.97	1.66
2100	Each additional floor, 2000 S.F.	.65	.97	1.62
2150	10,000 S.F.	.48	.83	1.31
2200	Ordinary hazard, one floor, 2000 S.F.	1.82	1.78	3.60
2250	10,000 S.F.	.85	1.25	2.10
2300	Each additional floor, 2000 S.F.	.71	1.09	1.80
2350	10,000 S.F.	.64	1.12	1.76
2400	Extra hazard, one floor, 2000 S.F.	2.05	2.27	4.32
2450	10,000 S.F.	1.23	1.64	2.87
2500	Each additional floor, 2000 S.F.	1.02	1.60	2.62
2550	10,000 S.F.	.89	1.44	2.33
3050	Grooved steel black sch. 10 pipe, light hazard, one floor, 2000 S.F.	1.73	1.64	3.37
3100	10,000 S.F.	.68	.95	1.63
3150	Each additional floor, 2000 S.F.	.62	.95	1.57
3200	10,000 S.F.	.48	.82	1.30
3250	Ordinary hazard, one floor, 2000 S.F.	1.80	1.77	3.57
3300	10,000 S.F.	.83	1.23	2.06
3350	Each additional floor, 2000 S.F.	.69	1.08	1.77
3400	10,000 S.F.	.63	1.09	1.72
3450	Extra hazard, one floor, 2000 S.F.	2.03	2.26	4.29
3500	10,000 S.F.	1.12	1.61	2.73
3550	Each additional floor, 2000 S.F.	1.01	1.59	2.60
3600	10,000 S.F.	.85	1.43	2.28
4050	Copper tubing, type M, light hazard, one floor, 2000 S.F.	1.61	2.03	3.64
4100	10,000 S.F.	.62	1.33	1.95
4150	Each additional floor, 2000 S.F.	.52	1.36	1.88

8.2-120	Dry Pipe Sprinkler Systems	MAT.	INST.	TOTAL
		COST PER S.F.		
4200	10,000 S.F.	.42	1.20	1.62
4250	Ordinary hazard, one floor, 2000 S.F.	1.70	2.26	3.96
4300	10,000 S.F.	.74	1.61	2.35
4350	Each additional floor, 2000 S.F.	.54	1.32	1.86
4400	10,000 S.F.	.53	1.46	1.99
4450	Extra hazard, one floor, 2000 S.F.	1.89	2.85	4.74
4500	10,000 S.F.	1.24	2.29	3.53
4550	Each additional floor, 2000 S.F.	.86	2.17	3.03
4600	10,000 S.F.	.82	2.03	2.85
5050	Copper tubing, type M, T-drill system, light hazard, one floor			
5060	2000 S.F.	1.62	1.99	3.61
5100	10,000 S.F.	.60	1.21	1.81
5150	Each additional floor, 2000 S.F.	.52	1.32	1.84
5200	10,000 S.F.	.39	1.08	1.47
5250	Ordinary hazard, one floor, 2000 S.F.	1.66	2.06	3.72
5300	10,000 S.F.	.72	1.53	2.25
5350	Each additional floor, 2000 S.F.	.56	1.37	1.93
5400	10,000 S.F.	.52	1.40	1.92
5450	Extra hazard, one floor, 2000 S.F.	1.78	2.49	4.27
5500	10,000 S.F.	1.10	1.90	3
5550	Each additional floor, 2000 S.F.	.75	1.81	2.56
5600	10,000 S.F.	.67	1.65	2.32

8

Preaction Sprinkler Systems

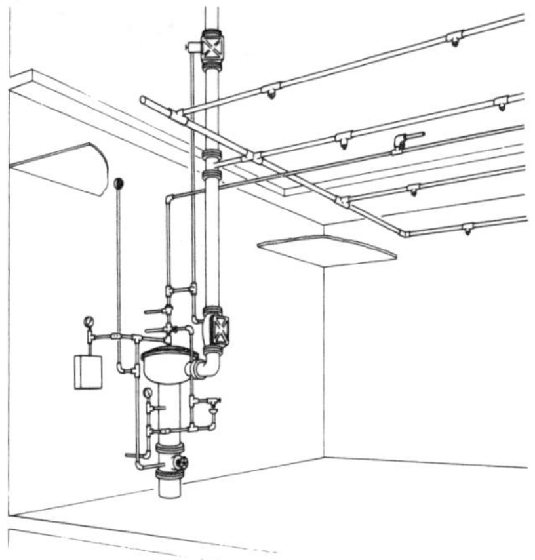

Pre-Action System: A system employing automatic sprinklers attached to a piping system containing air that may or may not be under pressure, with a supplemental heat responsive system of generally more sensitive characteristics than the automatic sprinklers themselves, installed in the same areas as the sprinklers; actuation of the heat responsive system, as from a fire, opens a valve which permits water to flow into the sprinkler piping system and to be discharged from those sprinklers which were opened by heat from the fire.

Costs for large jobs and those using prefabrications will run 15 to 25% less than these.

All areas are assumed to be open.

System Components	QUANTITY	UNIT	COST EACH MAT.	COST EACH INST.	COST EACH TOTAL
SYSTEM 08.2-130-0580					
PREACTION SPRINKLER SYSTEM, STEEL BLACK SCH. 40 PIPE					
LIGHT HAZARD, 1 FLOOR, 2000 S.F.					
Valve, gate, iron body 125 lb, OS&Y, flanged, 4" pipe size	1.000	Ea.	176	154	330
*Valve, swing check w/ball drip CI with brass trim 4" pipe size	1.000	Ea.	121	159	280
Valve, swing check, bronze, 125 lb, regrinding disc, 2-1/2" pipe size	1.000	Ea.	85.80	29.20	115
Valve, angle, bronze, 150 lb, rising stem, threaded, 2" pipe size	1.000	Ea.	110	25	135
*Alarm valve, 2-1/2" pipe size	1.000	Ea.	462	158	620
Alarm, water motor, complete with gong	1.000	Ea.	93.50	66.50	160
Fire alarm horn, electric	1.000	Ea.	23.10	37.90	61
Thermostatic release for release line	2.000	Ea.	231	29	260
Pipe, steel, black, shedule 40, 4" diam	10.000	L.F.	99.90	130.10	230
Dry pipe valve, trim & gauges, 4" pipe size	1.000	Ea.	742.50	482.50	1225
Pipe, steel, black, schedule 40, threaded, cplg & hngr 10'OC 2-1/2" diam	20.000	L.F.	128.20	184.80	313
Pipe steel black, schedule 40, threaded, cplg & hngr 10'OC 2" diam	12.500	L.F.	49.25	90.13	139.38
Pipe, steel, black, schedule 40, threaded, cplg & hngr 10'OC 1-1/4" diam	37.500	L.F.	91.88	195	286.88
Pipe, steel, black, schedule 40, threaded, cplg & hngr 10'OC 1" diam	112.000	L.F.	219.52	542.08	761.60
Pipe, Tee, malleable iron, black, 150 lb threaded, 4" diam	2.000	Ea.	70.40	229.60	300
Pipe, Tee, malleable iron, black, 150 lb threaded, 2-1/2" pipe size	2.000	Ea.	26.96	103.04	130
Pipe, Tee, malleable iron, black, 150 lb threaded, 2" pipe size	1.000	Ea.	6.05	41.95	48
Pipe, Tee, malleable iron, black, 150 lb threaded, 1-1/4" pipe size	5.000	Ea.	16.70	163.30	180
Pipe, Tee, malleable iron, black, 150 lb threaded, 1" pipe size	4.000	Ea.	8.24	127.76	136
Pipe, 90° elbow, malleable iron blk, 150 lb threaded, 1" pipe size	6.000	Ea.	7.98	118.02	126
Sprinkler head, std spray, brass 135°-286°F 1/2" NPT, 3/8" orifice	12.000	Ea.	36.36	311.64	348
Air compressor auto complete 200 Gal sprinkler sys cap, 1/3 HP	1.000	Ea.	489.50	200.50	690
*Standpipe connection, wall, flush, brs w/plug & chain 2-1/2" x 2-1/2"	1.000	Ea.	79.20	95.80	175
Valve, gate, bronze, 300 psi, NRS, class 150, threaded, 1" pipe size	1.000	Ea.	18.26	13.74	32
TOTAL			3393.30	3688.56	7081.86
COST PER S.F.			1.70	1.84	3.54

*Not included in systems under 2000 S.F.

8.2-130	Preaction Sprinkler Systems	COST PER S.F. MAT.	COST PER S.F. INST.	COST PER S.F. TOTAL
0520	Preaction sprinkler systems, steel, black, sch. 40 pipe			
0530	Light hazard, one floor, 500 S.F.	3.46	2.59	6.05
0560	C8.2 -101 C8.2 -102 C8.2 -103 1000 S.F.	1.94	1.95	3.89
0580	2000 S.F.	1.70	1.84	3.54

8.2-130	Preaction Sprinkler Systems	COST PER S.F.		
		MAT.	INST.	TOTAL
0600	5000 S.F.	.90	1.29	2.19
0620	10,000 S.F.	.64	1.07	1.71
0640	50,000 S.F.	.53	.95	1.48
0660	Each additional floor, 500 S.F.	.79	1.44	2.23
0680	1000 S.F.	.63	1.30	1.93
0700	2000 S.F.	.59	1.16	1.75
0720	5000 S.F.	.45	1.02	1.47
0740	10,000 S.F.	.44	.93	1.37
0760	50,000 S.F.	.49	.92	1.41
1000	Ordinary hazard, one floor, 500 S.F.	1.34	1.89	3.23
1020	1000 S.F.	1.92	1.89	3.81
1040	2000 S.F.	1.85	1.99	3.84
1060	5000 S.F.	.98	1.39	2.37
1080	10,000 S.F.	.79	1.40	2.19
1100	50,000 S.F.	1.03	1.51	2.54
1140	Each additional floor, 500 S.F.	.97	1.61	2.58
1160	1000 S.F.	.61	1.28	1.89
1180	2000 S.F.	.62	1.28	1.90
1200	5000 S.F.	.57	1.12	1.69
1220	10,000 S.F.	.58	1.27	1.85
1240	50,000 S.F.	.55	1.19	1.74
1500	Extra hazard, one floor, 500 S.F.	4.61	3.51	8.12
1520	1000 S.F.	2.60	2.64	5.24
1540	2000 S.F.	1.86	2.52	4.38
1560	5000 S.F.	1.09	1.92	3.01
1580	10,000 S.F.	1.25	1.96	3.21
1600	50,000 S.F.	1.59	2.26	3.85
1660	Each additional floor, 500 S.F.	1.05	1.94	2.99
1680	1000 S.F.	.91	1.83	2.74
1700	2000 S.F.	.82	1.84	2.66
1720	5000 S.F.	.70	1.65	2.35
1740	10,000 S.F.	.84	1.69	2.53
1760	50,000 S.F.	1.09	1.96	3.05
2020	Grooved steel, black, sch. 40 pipe, light hazard, one floor, 2000 S.F.	1.78	1.63	3.41
2060	10,000 S.F.	.68	.93	1.61
2100	Each additional floor of 2000 S.F.	.67	.94	1.61
2150	10,000 S.F.	.48	.80	1.28
2200	Ordinary hazard, one floor, 2000 S.F.	1.81	1.74	3.55
2250	10,000 S.F.	.80	1.20	2
2300	Each additional floor, 2000 S.F.	.70	1.05	1.75
2350	10,000 S.F.	.60	1.07	1.67
2400	Extra hazard, one floor, 2000 S.F.	1.97	2.20	4.17
2450	10,000 S.F.	1.14	1.57	2.71
2500	Each additional floor, 2000 S.F.	.95	1.52	2.47
2550	10,000 S.F.	.78	1.37	2.15
3050	Grooved steel, black, sch. 10 pipe light hazard, one floor, 2000 S.F.	1.76	1.62	3.38
3100	10,000 S.F.	.68	.92	1.60
3150	Each additional floor, 2000 S.F.	.65	.92	1.57
3200	10,000 S.F.	.47	.79	1.26
3250	Ordinary hazard, one floor, 2000 S.F.	1.79	1.73	3.52
3300	10,000 S.F.	.73	1.17	1.90
3350	Each additional floor, 2000 S.F.	.68	1.04	1.72
3400	10,000 S.F.	.59	1.04	1.63
3450	Extra hazard, one floor, 2000 S.F.	1.96	2.19	4.15
3500	10,000 S.F.	1.01	1.54	2.55
3550	Each additional floor, 2000 S.F.	.93	1.51	2.44
3600	10,000 S.F.	.74	1.35	2.09
4050	Copper tubing, type M, light hazard, one floor, 2000 S.F.	1.63	2	3.63

8.2-130	Preaction Sprinkler Systems	COST PER S.F. MAT.	INST.	TOTAL
4100	10,000 S.F.	.61	1.30	1.91
4150	Each additional floor, 2000 S.F.	.54	1.33	1.87
4200	10,000 S.F.	.35	1.16	1.51
4250	Ordinary hazard, one floor, 2000 S.F.	1.69	2.21	3.90
4300	10,000 S.F.	.70	1.56	2.26
4350	Each additional floor, 2000 S.F.	.53	1.27	1.80
4400	10,000 S.F.	.48	1.41	1.89
4450	Extra hazard, one floor, 2000 S.F.	1.81	2.77	4.58
4500	10,000 S.F.	1.12	2.21	3.33
4550	Each additional floor, 2000 S.F.	.79	2.10	2.89
4600	10,000 S.F.	.71	1.95	2.66
5050	Copper tubing, type M, T-drill system, light hazard, one floor			
5060	2000 S.F.	1.64	1.96	3.60
5100	10,000 S.F.	.59	1.18	1.77
5150	Each additional floor, 2000 S.F.	.54	1.29	1.83
5200	10,000 S.F.	.39	1.05	1.44
5250	Ordinary hazard, one floor, 2000 S.F.	1.65	2.01	3.66
5300	10,000 S.F.	.68	1.49	2.17
5350	Each additional floor, 2000 S.F.	.55	1.33	1.88
5400	10,000 S.F.	.48	1.35	1.83
5450	Extra hazard, one floor, 2000 S.F.	1.71	2.41	4.12
5500	10,000 S.F.	.98	1.82	2.80
5550	Each additional floor, 2000 S.F.	.68	1.74	2.42
5600	10,000 S.F.	.57	1.57	2.14

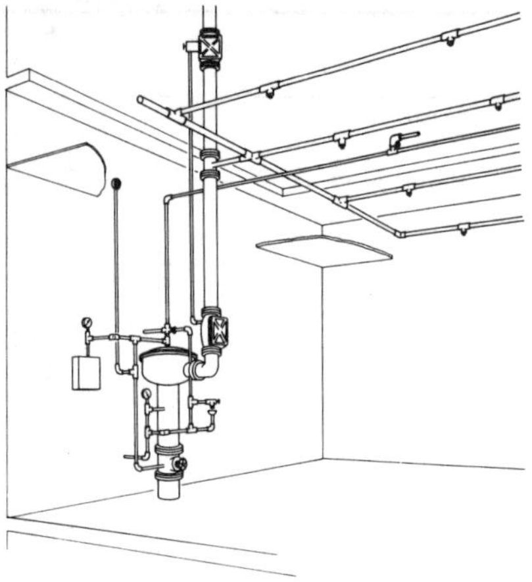

Deluge System: A system employing open sprinklers attached to a piping system connected to a water supply through a valve which is opened by the operation of a heat responsive system installed in the same areas as the sprinklers. When this valve opens, water flows into the piping system and discharges from all sprinklers attached thereto.

Costs for large jobs and those using prefabrications will run 15 to 25% less than these.

All areas are assumed to be open.

8

System Components	QUANTITY	UNIT	COST EA. MAT.	INST.	TOTAL
SYSTEM 08.2-140-0580					
DELUGE SPRINKLER SYSTEM, STEEL BLACK SCH. 40 PIPE					
LIGHT HAZARD, 1 FLOOR, 2000 S.F.					
Valve, gate, iron body 125 lb, OS&Y, flanged, 4" pipe size	1.000	Ea.	176	154	330
Valve, swing check w/ball drip, CI w/brass ftngs, 4" pipe size	1.000	Ea.	121	159	280
Valve, swing check, bronze, 125 lb, regrinding disc, 2-1/2" pipe size	1.000	Ea.	85.80	29.20	115
Valve, angle, bronze, 150 lb, rising stem, threaded, 2" pipe size	1.000	Ea.	110	25	135
*Alarm valve, 2-1/2" pipe size	1.000	Ea.	462	158	620
Alarm, water motor, complete with gong	1.000	Ea.	93.50	66.50	160
Fire alarm horn, electric	1.000	Ea.	23.10	37.90	61
Thermostatic release for release line	2.000	Ea.	231	29	260
Pipe, steel, black, schedule 40, 4" diam	10.000	L.F.	99.90	130.10	230
Deluge valve trim, pressure relief, emergency release, gauge, 4" pipe size	1.000	Ea.	1078	472	1550
Deluge system pressured monitoring panel, 120V	1.000	Ea.	297	13	310
Pipe, steel, black, schedule 40, threaded, cplg & hngr 10'OC 2-1/2" diam	20.000	L.F.	128.20	184.80	313
Pipe, steel, black, schedule 40, threaded, cplg & hngr 10'OC 2" diam	12.500	L.F.	49.25	90.13	139.38
Pipe, steel, black, schedule 40, threaded, cplg & hngr 10'OC 1-1/4" diam	37.500	L.F.	91.88	195	286.88
Pipe, steel, black, schedule 40, threaded, cplg & hngr 10'OC 1" diam	112.000	L.F.	219.52	542.08	761.60
Pipe, Tee, malleable iron, black, 150 lb threaded, 4" pipe size	2.000	Ea.	70.40	229.60	300
Pipe, Tee, malleable iron, black, 150 lb threaded, 2-1/2" pipe size	2.000	Ea.	26.96	103.04	130
Pipe, Tee, malleable iron, black, 150 lb threaded, 2" pipe size	1.000	Ea.	6.05	41.95	48
Pipe, Tee, malleable iron, black, 150 lb threaded, 1-1/4" pipe size	5.000	Ea.	16.70	163.30	180
Pipe, Tee, malleable iron, black, 150 lb threaded, 1" pipe size	4.000	Ea.	8.24	127.76	136
Pipe, 90° elbow, malleable iron, black, 150 lb threaded 1" pipe size	6.000	Ea.	7.98	118.02	126
Sprinkler head, std spray, brass 135°-286°F 1/2" NPT, 3/8" orifice	9.720	Ea.	29.45	252.43	281.88
Air compressor, auto, complete, 200 Gal sprinkler sys cap, 1/3 HP	1.000	Ea.	489.50	200.50	690
*Standpipe connection, wall, flush w/plug & chain 2-1/2"x2-1/2"	1.000	Ea.	79.20	95.80	175
Valve, gate, bronze, 300 psi, NRS, class 150, threaded, 1" pipe size	1.000	Ea.	18.26	13.74	32
TOTAL			4018.89	3631.85	7650.74
COST PER S.F.			2.01	1.82	3.83

*Not included in systems under 2000 S.F.

8.2-140	Deluge Sprinkler Systems	COST PER S.F. MAT.	INST.	TOTAL
0520	Deluge sprinkler systems, steel, black, sch. 40 pipe			
0530	Light hazard, one floor, 500 S.F.	4.22	2.58	6.80
0560	C8.2-101 C8.2-102 C8.2-103 1000 S.F.	2.31	1.84	4.15
0580	2000 S.F.	2.01	1.82	3.83

8.2-140	Deluge Sprinkler Systems	COST PER S.F. MAT.	COST PER S.F. INST.	COST PER S.F. TOTAL
0600	5000 S.F.	1.02	1.26	2.28
0620	10,000 S.F.	.70	1.04	1.74
0640	50,000 S.F.	.54	.92	1.46
0660	Each additional floor, 500 S.F.	.79	1.40	2.19
0680	1000 S.F.	.63	1.30	1.93
0700	2000 S.F.	.58	1.13	1.71
0720	5000 S.F.	.45	.99	1.44
0740	10,000 S.F.	.43	.91	1.34
0760	50,000 S.F.	.49	.89	1.38
1000	Ordinary hazard, one floor, 500 S.F.	4.64	2.90	7.54
1020	1000 S.F.	2.29	1.86	4.15
1040	2000 S.F.	2.16	1.95	4.11
1060	5000 S.F.	1.10	1.35	2.45
1080	10,000 S.F.	.84	1.36	2.20
1100	50,000 S.F.	1.04	1.47	2.51
1140	Each additional floor, 500 S.F.	.96	1.57	2.53
1160	1000 S.F.	.60	1.24	1.84
1180	2000 S.F.	.62	1.24	1.86
1200	5000 S.F.	.57	1.08	1.65
1220	10,000 S.F.	.64	1.39	2.03
1240	50,000 S.F.	.54	1.15	1.69
1500	Extra hazard, one floor, 500 S.F.	5.35	3.48	8.83
1520	1000 S.F.	3.15	2.68	5.83
1540	2000 S.F.	2.17	2.46	4.63
1560	5000 S.F.	1.22	1.87	3.09
1580	10,000 S.F.	1.32	1.92	3.24
1600	50,000 S.F.	1.66	2.22	3.88
1660	Each additional floor, 500 S.F.	1.04	1.88	2.92
1680	1000 S.F.	.91	1.77	2.68
1700	2000 S.F.	.82	1.78	2.60
1720	5000 S.F.	.69	1.60	2.29
1740	10,000 S.F.	.83	1.63	2.46
1760	50,000 S.F.	1.09	1.89	2.98
2000	Grooved steel, black, sch. 40 pipe, light hazard, one floor			
2020	2000 S.F.	2.09	1.60	3.69
2060	10,000 S.F.	.74	.91	1.65
2100	Each additional floor, 2,000 S.F.	.66	.91	1.57
2150	10,000 S.F.	.47	.77	1.24
2200	Ordinary hazard, one floor, 2000 S.F.	1.80	1.70	3.50
2250	10,000 S.F.	.86	1.16	2.02
2300	Each additional floor, 2000 S.F.	.69	1.01	1.70
2350	10,000 S.F.	.59	1.03	1.62
2400	Extra hazard, one floor, 2000 S.F.	2.28	2.14	4.42
2450	10,000 S.F.	1.21	1.53	2.74
2500	Each additional floor, 2000 S.F.	.94	1.46	2.40
2550	10,000 S.F.	.78	1.31	2.09
3000	Grooved steel, black, sch. 10 pipe, light hazard, one floor			
3050	2000 S.F.	1.86	1.52	3.38
3100	10,000 S.F.	.74	.89	1.63
3150	Each additional floor, 2000 S.F.	.64	.90	1.54
3200	10,000 S.F.	.47	.76	1.23
3250	Ordinary hazard, one floor, 2000 S.F.	2.10	1.69	3.79
3300	10,000 S.F.	.79	1.13	1.92
3350	Each additional floor, 2000 S.F.	.68	1	1.68
3400	10,000 S.F.	.58	1	1.58
3450	Extra hazard, one floor, 2000 S.F.	2.27	2.13	4.40
3500	10,000 S.F.	1.07	1.48	2.55
3550	Each additional floor, 2000 S.F.	.93	1.45	2.38

8

8.2-140	Deluge Sprinkler Systems	COST PER S.F. MAT.	INST.	TOTAL
3600	10,000 S.F.	.73	1.29	2.02
4000	Copper tubing, type M, light hazard, one floor			
4050	2000 S.F.	1.95	1.97	3.92
4100	10,000 S.F.	.67	1.27	1.94
4150	Each additional floor, 2000 S.F.	.53	1.30	1.83
4200	10,000 S.F.	.35	1.13	1.48
4250	Ordinary hazard, one floor, 2000 S.F.	2	2.17	4.17
4300	10,000 S.F.	.75	1.53	2.28
4350	Each additional floor, 2000 S.F.	.52	1.23	1.75
4400	10,000 S.F.	.48	1.37	1.85
4450	Extra hazard, one floor, 2000 S.F.	2.12	2.72	4.84
4500	10,000 S.F.	1.21	2.17	3.38
4550	Each additional floor, 2000 S.F.	.78	2.04	2.82
4600	10,000 S.F.	.70	1.89	2.59
5000	Copper tubing, type M, T-drill system, light hazard, one floor			
5050	2000 S.F.	1.95	1.94	3.89
5100	10,000 S.F.	.65	1.15	1.80
5150	Each additional floor, 2000 S.F.	.59	1.25	1.84
5200	10,000 S.F.	.38	1.02	1.40
5250	Ordinary hazard, one floor, 2000 S.F.	1.97	1.98	3.95
5300	10,000 S.F.	.74	1.45	2.19
5350	Each additional floor, 2000 S.F.	.54	1.29	1.83
5400	10,000 S.F.	.47	1.31	1.78
5450	Extra hazard, one floor, 2000 S.F.	2.02	2.36	4.38
5500	10,000 S.F.	1.03	1.76	2.79
5550	Each additional floor, 2000 S.F.	.67	1.68	2.35
5600	10,000 S.F.	.56	1.51	2.07

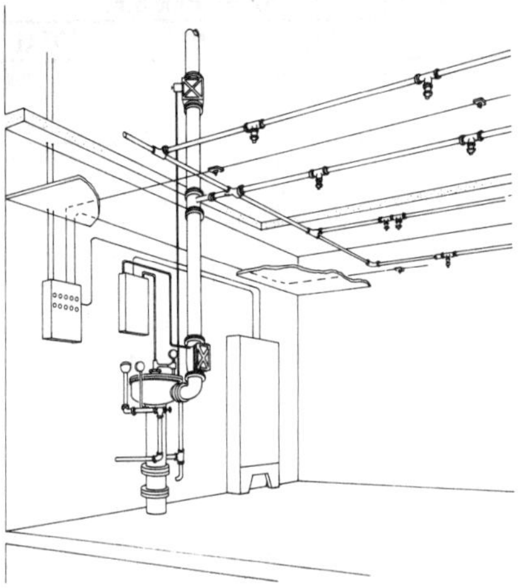

Firecycle is a fixed fire protection sprinkler system utilizing water as its extinguishing agent. It is a time delayed, recycling, preaction type which automatically shuts the water off when heat is reduced below the detector operating temperature and turns the water back on when that temperature is exceeded.

The system senses a fire condition through a closed circuit electrical detector system which controls water flow to the fire automatically. Batteries supply up to 90 hour emergency power supply for system operation. The piping system is dry (until water is required) and is monitored with pressurized air. Should any leak in the system piping occur, an alarm will sound, but water will not enter the system until heat is sensed by a Firecycle detector.

Costs for large jobs and those using prefabrications will run 15 to 25% less than these.

All areas are assumed to be open.

System Components	QUANTITY	UNIT	COST EACH MAT.	COST EACH INST.	COST EACH TOTAL
SYSTEM 08.2-150-0580					
FIRECYCLE SPRINKLER SYSTEM, STEEL BLACK SCH. 40 PIPE					
LIGHT HAZARD, ONE FLOOR, 2000 S.F.					
Valve, gate, iron body 125 lb, OS&Y, flanged, 4" pipe size	1.000	Ea.	176	154	330
Valve, angle, bronze, 150 lb, rising stem, threaded, 2" pipe size	1.000	Ea.	110	25	135
Valve, swing check, bronze, 125 lb, regrinding disc, 2-1/2" pipe size	1.000	Ea.	85.80	29.20	115
*Alarm valve, 2-1/2" pipe size	1.000	Ea.	462	158	620
Alarm, water motor, complete with gong	1.000	Ea.	93.50	66.50	160
Pipe, steel, black, schedule 40, 4" diam	10.000	L.F.	99.90	130.10	230
Fire alarm, horn, electric	1.000	Ea.	23.10	37.90	61
Pipe, steel, black, schedule 40, threaded, cplg & hngr 10'OC 2-1/2" diam	20.000	L.F.	128.20	184.80	313
Pipe, steel, black, schedule 40, threaded, cplg & hngr 10'OC 2" diam	12.500	L.F.	49.25	90.13	139.38
Pipe, steel, black, schedule 40, threaded, cplg & hngr 10'OC 1-1/4" diam	37.500	L.F.	91.88	195	286.88
Pipe, steel, black, schedule 40, threaded, cplg & hngr 10'OC 1" diam	112.000	L.F.	219.52	542.08	761.60
Pipe, Tee, malleable iron, black, 150 lb threaded, 4" pipe size	2.000	Ea.	70.40	229.60	300
Pipe, Tee, malleable iron, black, 150 lb threaded, 2-1/2" pipe size	2.000	Ea.	26.96	103.04	130
Pipe, Tee, malleable iron, black, 150 lb threaded, 2" pipe size	1.000	Ea.	6.05	41.95	48
Pipe, Tee, malleable iron, black, 150 lb threaded, 1-1/4" pipe size	5.000	Ea.	16.70	163.30	180
Pipe, Tee, malleable iron, black, 150 lb threaded, 1" pipe size	4.000	Ea.	8.24	127.76	136
Pipe, 90° elbow, malleable iron, black, 150 lb threaded, 1" pipe size	6.000	Ea.	7.98	118.02	126
Sprinkler head std spray, brass 135°-286°F 1/2" NPT, 3/8" orifice	12.000	Ea.	36.36	311.64	348
Firecycle controls, incls panel, battery, solenoid valves, press switches	1.000	Ea.	4675	1000	5675
Detector, firecycle system	2.000	Ea.	319	31	350
Firecycle pkg, swing check & flow control valves w/trim 4" pipe size	1.000	Ea.	1347.50	477.50	1825
Air compressor, auto, complete, 200 Gal sprinkler sys cap, 1/3 HP	1.000	Ea.	489.50	200.50	690
*Standpipe connection, wall, flush, brass w/plug & chain 2-1/2"x2-1/2"	1.000	Ea.	79.20	95.80	175
Valve, gate, bronze, 300 psi, NRS, class 150, threaded, 1" diam	1.000	Ea.	18.26	13.74	32
TOTAL			8640.30	4526.56	13,166.86
COST PER S.F.			4.32	2.26	6.58

*Not included in systems under 2000 S.F.

8.2-150	Firecycle Sprinkler Systems	COST PER S.F. MAT.	COST PER S.F. INST.	COST PER S.F. TOTAL
0520	Firecycle sprinkler systems, steel black sch. 40 pipe			
0530	Light hazard, one floor, 500 S.F.	14.20	5.05	19.25
0560	C8.2-101 C8.2-102 C8.2-103 1000 S.F.	7.30	3.17	10.47
0580	2000 S.F.	4.32	2.26	6.58

8.2-150	Firecycle Sprinkler Systems	COST PER S.F.		
		MAT.	INST.	TOTAL
0600	5000 S.F.	1.96	1.46	3.42
0620	10,000 S.F.	1.20	1.15	2.35
0640	50,000 S.F.	.67	.96	1.63
0660	Each additional floor of 500 S.F.	.88	1.44	2.32
0680	1000 S.F.	.67	1.30	1.97
0700	2000 S.F.	.55	1.15	1.70
0720	5000 S.F.	.49	1.02	1.51
0740	10,000 S.F.	.48	.94	1.42
0760	50,000 S.F.	.52	.92	1.44
1000	Ordinary hazard, one floor, 500 S.F.	14.40	5.20	19.60
1020	1000 S.F.	7.30	3.11	10.41
1040	2000 S.F.	4.41	2.40	6.81
1060	5000 S.F.	2.05	1.56	3.61
1080	10,000 S.F.	1.35	1.49	2.84
1100	50,000 S.F.	1.17	1.53	2.70
1140	Each additional floor, 500 S.F.	1.06	1.62	2.68
1160	1000 S.F.	.65	1.28	1.93
1180	2000 S.F.	.72	1.29	2.01
1200	5000 S.F.	.61	1.12	1.73
1220	10,000 S.F.	.63	1.27	1.90
1240	50,000 S.F.	.59	1.19	1.78
1500	Extra hazard, one floor, 500 S.F.	15.35	5.95	21.30
1520	1000 S.F.	7.90	3.85	11.75
1540	2000 S.F.	4.48	2.94	7.42
1560	5000 S.F.	2.16	2.09	4.25
1580	10,000 S.F.	1.80	2.04	3.84
1600	50,000 S.F.	1.69	2.25	3.94
1660	Each additional floor, 500 S.F.	1.14	1.94	3.08
1680	1000 S.F.	.96	1.83	2.79
1700	2000 S.F.	.87	1.84	2.71
1720	5000 S.F.	.73	1.65	2.38
1740	10,000 S.F.	.88	1.70	2.58
1760	50,000 S.F.	1.38	3.24	4.62
2020	Grooved steel, black, sch. 40 pipe, light hazard, one floor			
2030	2000 S.F.	4.40	2.05	6.45
2060	10,000 S.F.	1.26	1.04	2.30
2100	Each additional floor, 2000 S.F.	.71	.94	1.65
2150	10,000 S.F.	.52	.80	1.32
2200	Ordinary hazard, one floor, 2000 S.F.	4.43	2.16	6.59
2250	10,000 S.F.	1.46	1.36	2.82
2300	Each additional floor, 2000 S.F.	.74	1.05	1.79
2350	10,000 S.F.	.65	1.07	1.72
2400	Extra hazard, one floor, 2000 S.F.	4.60	2.62	7.22
2450	10,000 S.F.	1.68	1.64	3.32
2500	Each additional floor, 2000 S.F.	.99	1.52	2.51
2550	10,000 S.F.	.83	1.37	2.20
3050	Grooved steel, black, sch. 10 pipe light hazard, one floor,			
3060	2000 S.F.	4.38	2.03	6.41
3100	10,000 S.F.	1.24	1.01	2.25
3150	Each additional floor, 2000 S.F.	.69	.93	1.62
3200	10,000 S.F.	.52	.79	1.31
3250	Ordinary hazard, one floor, 2000 S.F.	4.41	2.15	6.56
3300	10,000 S.F.	1.35	1.26	2.61
3350	Each additional floor, 2000 S.F.	.72	1.04	1.76
3400	10,000 S.F.	.63	1.05	1.68
3450	Extra hazard, one floor, 2000 S.F.	4.58	2.61	7.19
3500	10,000 S.F.	1.57	1.61	3.18
3550	Each additional floor, 2000 S.F.	.98	1.51	2.49

8.2-150	Firecycle Sprinkler Systems	COST PER S.F. MAT.	COST PER S.F. INST.	COST PER S.F. TOTAL
3600	10,000 S.F.	.79	1.35	2.14
4060	Copper tubing, type M, light hazard, one floor, 2000 S.F.	4.26	2.42	6.68
4100	10,000 S.F.	1.17	1.39	2.56
4150	Each additional floor, 2000 S.F.	.58	1.33	1.91
4200	10,000 S.F.	.46	1.17	1.63
4250	Ordinary hazard, one floor, 2000 S.F.	4.31	2.63	6.94
4300	10,000 S.F.	1.26	1.65	2.91
4350	Each additional floor, 2000 S.F.	.57	1.27	1.84
4400	10,000 S.F.	.53	1.41	1.94
4450	Extra hazard, one floor, 2000 S.F.	4.44	3.19	7.63
4500	10,000 S.F.	1.73	2.31	4.04
4550	Each additional floor, 2000 S.F.	.83	2.10	2.93
4600	10,000 S.F.	.76	1.95	2.71
5060	Copper tubing, type M, T-drill system, light hazard,one floor 2000 S.F.	4.26	2.38	6.64
5100	10,000 S.F.	1.15	1.27	2.42
5150	Each additional floor, 2000 S.F.	.67	1.31	1.98
5200	10,000 S.F.	.43	1.05	1.48
5250	Ordinary hazard, one floor, 2000 S.F.	4.28	2.43	6.71
5300	10,000 S.F.	1.24	1.57	2.81
5350	Each additional floor, 2000 S.F.	.59	1.33	1.92
5400	10,000 S.F.	.52	1.35	1.87
5450	Extra hazard, one floor, 2000 S.F.	4.33	2.83	7.16
5500	10,000 S.F.	1.53	1.90	3.43
5550	Each additional floor, 2000 S.F.	.73	1.74	2.47
5600	10,000 S.F.	.61	1.57	2.18

8

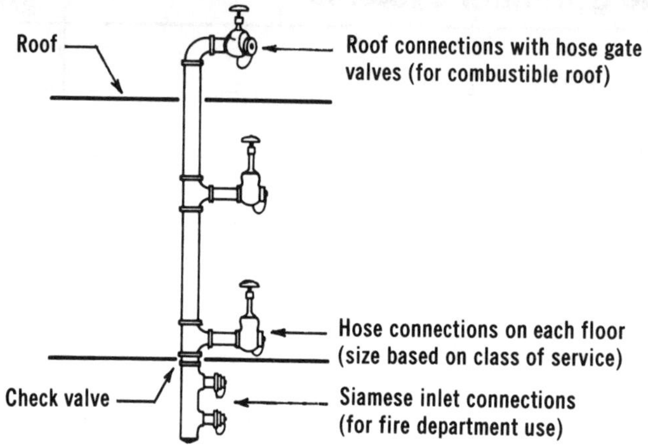

System Components	QUANTITY	UNIT	COST EACH MAT.	INST.	TOTAL
SYSTEM: 082-310-0560					
WET STANDPIPE RISER, CLASS I, STEEL, BLACK, SCH. 40 PIPE, 10' HEIGHT					
4" DIAMETER PIPE, ONE FLOOR					
Pipe, steel, black, schedule 40, threaded, 4" diam	20.000	L.F.	252	248	500
Pipe, Tee, malleable iron, black, 150 lb threaded, 4" pipe size	2.000	Ea.	70.40	229.60	300
Pipe, 90° elbow, malleable iron, black, 150 lb threaded 4" pipe size	1.000	Ea.	26.40	78.60	105
Pipe, nipple, steel, black, schedule 40, 2-1/2" pipe size x 3" long	2.000	Ea.	13.42	58.58	72
Fire valve, gate, 300 lb, brass, hose, handwheel, 2-1/2" pipe size	1.000	Ea.	85.80	39.20	125
Fire valve, pressure restricting, adj, rgh brs, 2-1/2" pipe size	1.000	Ea.	151.80	78.20	230
Valve, swing check, w/ball drip, CI w/brs ftngs 4" pipe size	1.000	Ea.	121	159	280
Standpipe conn wall dble flush brs w/plugs & chains 2-1/2"x2-1/2"x4"	1.000	Ea.	242	93	335
Valve, swing check, bronze, 125 lb, regrinding disc, 2-1/2" pipe size	1.000	Ea.	85.80	29.20	115
Roof manifold, fire, w/valves & caps, horiz/vert brs 2-1/2"x2-1/2"x4"	1.000	Ea.	247.50	97.50	345
Fire, hydrolator, vent & drain, 2-1/2" pipe size	1.000	Ea.	67.10	37.90	105
Valve, gate, iron body 125 lb, OS&Y, threaded, 4" pipe size	1.000	Ea.	247.50	47.50	295
TOTAL			1610.72	1196.28	2807

8

8.2-310	Wet Standpipe Risers, Class I		COST PER FLOOR MAT.	INST.	TOTAL
0550		Wet standpipe risers, Class I, steel black sch. 40, 10' height			
0560		4" diameter pipe, one floor	1600	1200	2800
0580	B8.2 -390	Additional floors	405	415	820
0600		6" diameter pipe, one floor	2750	1975	4725
0620	C8.2 -301	Additional floors	810	585	1395
0640		8" diameter pipe, one floor	3825	2350	6175
0660	C8.2 -302	Additional floors	1025	705	1730

8.2-310	Wet Standpipe Risers, Class II	COST PER FLOOR MAT.	INST.	TOTAL
1030	Wet standpipe risers, Class II, steel black sch. 40, 10' height			
1040	2" diameter pipe, one floor	560	435	995
1060	Additional floors	210	180	390
1080	2-1/2" diameter pipe, one floor	770	620	1390
1100	Additional floors	235	210	445

8.2-310	Wet Standpipe Risers, Class III	COST PER FLOOR MAT.	COST PER FLOOR INST.	COST PER FLOOR TOTAL
1530	Wet standpipe risers, Class III, steel black sch. 40, 10' height			
1540	4" diameter pipe, one floor	1650	1200	2850
1560	Additional floors	325	345	670
1580	6" diameter pipe, one floor	2775	1975	4750
1600	Additional floors	830	585	1415
1620	8" diameter pipe, one floor	3850	2350	6200
1640	Additional floors	1050	705	1755

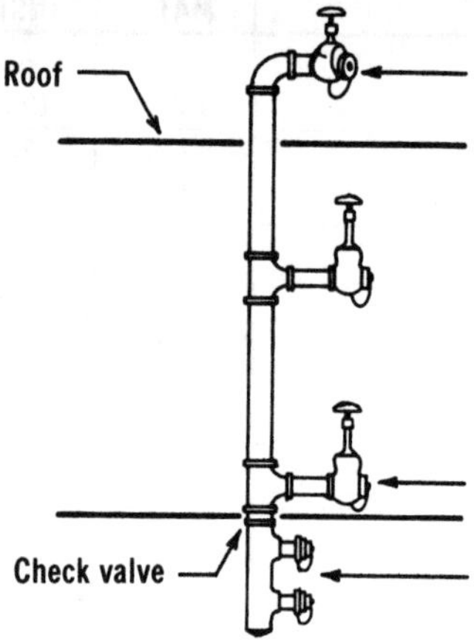

Costs for large jobs and those using prefabrications will run 15-25% less than these.

All areas are assumed to be open.

System Components	QUANTITY	UNIT	COST PER FLOOR		
			MAT.	INST.	TOTAL
SYSTEM 08.2-320-0540					
DRY STANDPIPE RISER, CLASS I, PIPE, STEEL, BLACK, SCH 40, 10' HEIGHT					
4" DIAMETER PIPE, ONE FLOOR					
Pipe, steel, black, schedule 40, threaded, 4" diam	20.000	L.F.	252	248	500
Pipe, Tee, malleable iron, black, 150 lb threaded, 4" pipe size	2.000	Ea.	70.40	229.60	300
Pipe, 90° elbow, malleable iron, black, 150 lb threaded 4" pipe size	1.000	Ea.	26.40	78.60	105
Pipe, nipple, steel, black, schedule 40, 2-1/2" pipe size x 3" long	2.000	Ea.	13.42	58.58	72
Fire valve, gate NRS 300 lb, brass, hose, handwheel, 2-1/2" pipe size	1.000	Ea.	85.80	39.20	125
Fire valve, pressure restricting, adj, rgh brs, 2-1/2" pipe size	1.000	Ea.	75.90	39.10	115
Standpipe conn, wall, dble flush, brs w/plugs & chains 2-1/2"x2-1/2"x4"	1.000	Ea.	242	93	335
Valve, swing check w/ball drip CI w/brs ftngs, 4" pipe size	1.000	Ea.	121	159	280
Roof manifold, fire, w/valves & caps, horiz/vert brs 2-1/2"x2-1/2"x4"	1.000	Ea.	247.50	97.50	345
TOTAL			1134.42	1042.58	2177

8.2-320	Dry Standpipe Risers, Class I	COST PER FLOOR		
		MAT.	INST.	TOTAL
0530	Dry standpipe risers, Class I, steel black sch. 40, 10' height			
0540	4" diameter pipe, one floor	1125	1050	2175
0560	Additional floors	335	375	710
0580	6" diameter pipe, one floor	2250	1625	3875
0600	Additional floors	745	550	1295
0620	8" diameter pipe, one floor	3125	1950	5075
0640	Additional floors	970	665	1635

8.2-320	Dry Standpipe Risers, Class II	COST PER FLOOR		
		MAT.	INST.	TOTAL
1030	Dry standpipe risers, Class II, steel black sch. 40, 10' height			
1040	2" diameter pipe, one floor	485	485	970
1060	Additional floor	175	160	335
1080	2-1/2" diameter pipe, one floor	575	565	1140
1100	Additional floors	200	190	390

8.2-320	Dry Standpipe Risers, Class III	COST PER FLOOR		
		MAT.	INST.	TOTAL
1530	Dry standpipe risers, Class III, steel black sch. 40, 10' height			
1540	4" diameter pipe, one floor	1150	1050	2200
1560	Additional floors	265	335	600
1580	6" diameter pipe, one floor	2275	1625	3900
1600	Additional floors	760	550	1310
1620	8" diameter pipe, one floor	3125	1950	5075
1640	Additional floor	985	665	1650

8.2-390	Standpipe Equipment	COST EACH		
		MAT.	INST.	TOTAL
0100	Adapters, reducing, 1 piece, FxM, hexagon, cast brass, 2-1/2" x 1-1/2"	32		32
0200	Pin lug, 1-1/2" x 1"	10.30		10.30
0250	3" x 2-1/2"	30		30
0300	For polished chrome, add	75%		
0400	Cabinets, D.S. glass in door, recessed, steel box, not equipped			
0500	Single extinguisher, steel door & frame	68	57	125
0550	Stainless steel door & frame	140	58	198
0600	Valve, 2-1/2" angle, steel door & frame	67	38	105
0650	Aluminum door & frame	77	38	115
0700	Stainless steel door & frame	140	38	178
0750	Hose rack assy, 2-1/2" x 1-1/2" valve & 100' hose, steel door & frame	110	80	190
0800	Aluminum door & frame	115	80	195
0850	Stainless steel door & frame	235	79	314
0900	Hose rack assy,& extinguisher, 2-1/2"x1-1/2" valve & hose, steel door & frame	120	94	214
0950	Aluminum	125	94	219
1000	Stainless steel	265	96	361
1550	Compressor, air, dry pipe system, automatic, 200 gal., 1/3 H.P.	490	200	690
1600	520 gal., 1 H.P.	565	205	770
1650	Alarm, electric pressure switch (circuit closer)	53	10.20	63.20
2500	Couplings, hose, rocker lug, cast brass, 1-1/2"	13.20		13.20
2550	2-1/2"	28		28
3000	Escutcheon plate, for angle valves, polished brass, 1-1/2"	11		11
3050	2-1/2"	13.75		13.75
3500	Fire pump, electric, w/controller, fittings, relief valve			
3550	4" pump, 30 H.P., 500 G.P.M.	9875	1525	11,400
3600	5" pump, 40 H.P., 1000 G.P.M.	11,100	1700	12,800
3650	5" pump, 100 H.P., 1000 G.P.M.	14,400	1900	16,300
3700	For jockey pump system, add	1525	240	1765
5000	Hose, per linear foot, synthetic jacket, lined,			
5100	300 lb. test, 1-1/2" diameter	1.13		1.13
5150	2-1/2" diameter	1.85		1.85
5200	500 lb. test, 1-1/2" diameter	1.34		1.34
5250	2-1/2" diameter	2.15		2.15
5500	Nozzle, plain stream, polished brass, 1-1/2" x 10"	16.15		16.15
5550	2-1/2" x 15" x 13/16" or 1-1/2"	47		47
5600	Heavy duty combination adjustable fog and straight stream w/handle 1-1/2"	180		180
5650	2-1/2" direct connection	235		235
6000	Rack, for 1-1/2" diameter hose 100 ft. long, steel	17.40	24	41.40
6050	Brass	64	23	87
6500	Reel, steel, for 50 ft. long 1-1/2" diameter hose	46	34	80
6550	For 75 ft. long 2-1/2" diameter hose	61	34	95
7050	Siamese, w/plugs & chains, polished brass, sidewalk, 4" x 2-1/2" x 2-1/2"	260	190	450
7100	6" x 2-1/2" x 2-1/2"	360	240	600
7200	Wall type, flush, 4" x 2-1/2" x 2-1/2"	240	93	333
7250	6" x 2-1/2" x 2-1/2"	280	105	385
7300	Projecting, 4" x 2-1/2" x 2-1/2"	205	97	302
7350	6" x 2-1/2" x 2-1/2"	235	105	340
7400	For chrome plate, add	15%		
8000	Valves, angle, wheel handle, 300 Lb., rough brass, 1-1/2"	24	22	46
8050	2-1/2"	54	37	91
8100	Combination pressure restricting, 1-1/2"	33	22	55
8150	2-1/2"	74	36	110
8200	Pressure restricting, adjustable, satin brass, 1-1/2"	42	22	64
8250	2-1/2"	76	39	115
8300	Hydrolator, vent and drain, rough brass, 1-1/2"	34	22	56
8350	2-1/2"	67	38	105
8400	Cabinet assy, incls. 2-1/2" valve, adapter, rack, hose, nozzle & hydrolator	415	180	595

B8.2-810 Halon Fire Suppression

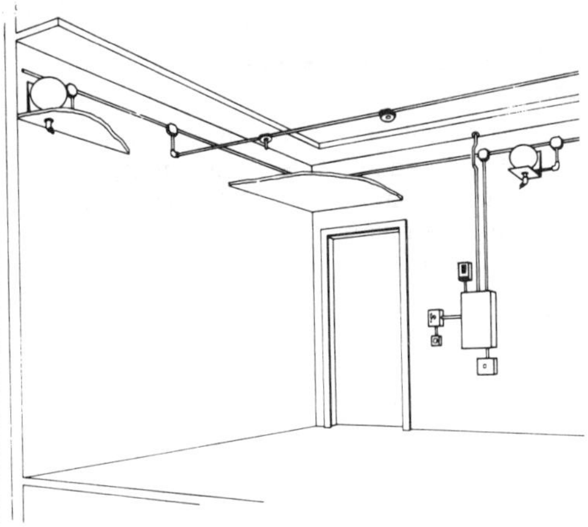

General: Automatic fire protection (suppression) systems other than water sprinklers may be desired for special environments, high risk areas, isolated locations or unusual hazards. Some typical applications would include:

- Paint dip tanks
- Securities vaults
- Electronic data processing
- Tape and data storage
- Transformer rooms
- Spray booths
- Petroleum storage
- High rack storage

Piping and wiring costs are dependent on the individual application and must be added to the component costs shown below.

Costs for large jobs and those using prefabrications will run 15 to 25% less than these.

All areas are assumed to be open.

8.2-810	Unit Components	COST EACH		
		MAT.	INST.	TOTAL
0020	Detectors with brackets			
0040	Fixed temperature heat detector	13.20	32	45.20
0060	Rate of temperature rise detector	13.20	32	45.20
0080	Ion detector (smoke) detector	70	40	110
0100				
0200	Extinguisher agent			
0240	200 lb halon, container	2850	115	2965
0280	75 lb carbon dioxide cylinder	735	78	813
0300				
0320	Dispersion nozzle			
0340	Halon 1-1/2" dispersion nozzle	69	18.70	87.70
0380	Carbon dioxide 3" x 5" dispersion nozzle	77	14	91
0400				
0420	Control station			
0440	Single zone control station with batteries	1150	245	1395
0470	Multizone (4) control station with batteries	2275	520	2795
0490				
0500	Electric mechanical release	175	130	305
0520				
0550	Manual pull station	29	42	71
0570				
0640	Battery standby power 10" x 10" x 17"	585	62	647
0700				
0740	Bell signalling device	35.20	31.80	67

8.2-810	Halon Systems	COST PER C.F.		
		MAT.	INST.	TOTAL
0820	Average halon system, minimum		.65	.65
0840	Maximum		1.05	1.05

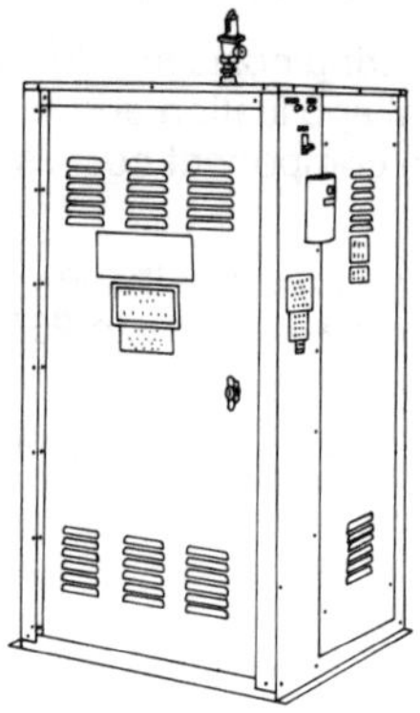

Boiler

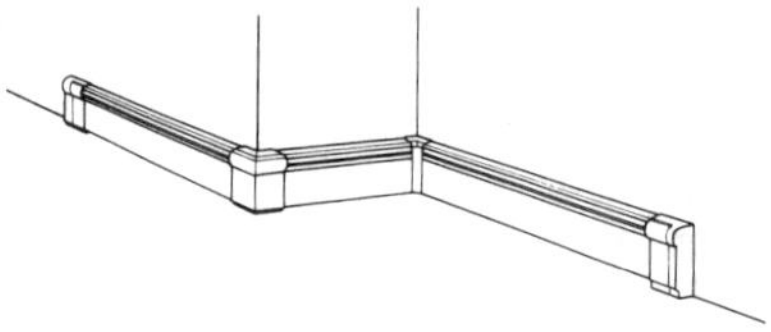

Baseboard Radiation

Small Electric Boiler Systems Considerations:

1. Terminal units are fin tube baseboard radiation rated at 720 BTU/hr with 200° water temperature or 820 BTU/hr steam.
2. Primary use being for residential or smaller supplementary areas,the floor levels are based on 7-1/2′ ceiling heights.
3. All distribution piping is copper for boilers through 205 MBH. All piping for larger systems is steel pipe.

8

System Components	QUANTITY	UNIT	COST EACH MAT.	COST EACH INST.	COST EACH TOTAL
SYSTEM 08.3-110-1120					
SMALL HEATING SYSTEMS, HYDRONIC, ELECTRIC BOILER					
1,480 S.F., 61 MBH, STEAM, 1 FLOOR					
Boiler, electric, steam, standard controls & trim, 18 KW, 61.4 MBH	1.000	Ea.	2777.50	597.50	3375
Copper tubing type L, 50/50 solder joint, hanger 10′OC, 1-1/4″ diam	160.000	L.F.	366.40	1417.60	1784
Radiation, 3/4″ copper tube w/alum fin baseboard pkg 7″ high	60.000	L.F.	386.40	1173.60	1560
Rough in baseboard panel or fin tube with valves & traps	10.000	Set	767.10	2432.90	3200
Boiler room fittings and valves	1.000	System	277.75	59.75	337.50
Pipe covering, calcium silicate w/cover 1″ wall 1-1/4″ diam	160.000	L.F.	291.20	456	747.20
Low water cut-off, quick hookup, in gage glass tappings	1.000	Ea.	95.70	14.30	110
TOTAL			4962.05	6151.65	11,113.70
COST PER S.F.			3.35	4.16	7.51

8.3-110	Small Heating Systems, Hydronic, Electric Boilers	COST PER S.F. MAT.	COST PER S.F. INST.	COST PER S.F. TOTAL
1100	Small heating systems, hydronic, electric boilers			
1120	Steam, 1 floor, 1480 S.F., 61 MBH	3.35	4.15	7.50
1160	C8.3-001 C8.3-002 3,000 S.F., 123 M.B.H.	2.27	3.65	5.92
1200	5,000 S.F., 205 M.B.H.	1.97	3.34	5.31
1240	C8.3-003 2 floors, 12,400 S.F., 512 M.B.H.	1.79	3.03	4.82
1280	3 floors, 24,800 S.F., 1023 M.B.H.	1.72	3	4.72
1320	34,750 S.F., 1,433 M.B.H.	1.61	2.93	4.54
1360	Hot water, 1 floor, 1,000 S.F., 41 M.B.H.	5.05	2.61	7.66
1400	2,500 S.F., 103 M.B.H.	3.24	4.11	7.35
1440	2 floors, 4,850 S.F., 205 M.B.H.	3.01	5.15	8.16
1480	3 floors, 9,700 S.F., 410 M.B.H.	2.88	5.40	8.28

Boiler

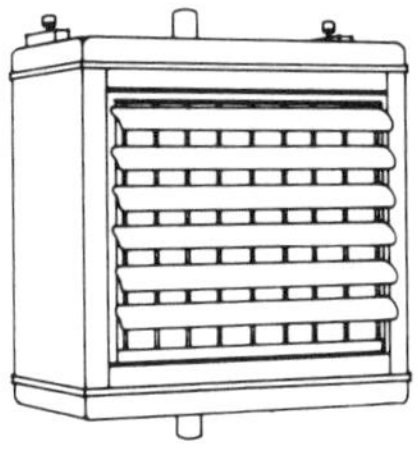

Unit Heater

Large Electric Boiler System Considerations:

1. Terminal units are all unit heaters of the same size. Quantities are varied to accommodate total requirements.
2. All air is circulated through the heaters a minimum of three times per hour.
3. As the capacities are adequate for commercial use, floor levels are based on 10′ ceiling heights.
4. All distribution piping is black steel pipe.

System Components	QUANTITY	UNIT	COST EACH MAT.	INST.	TOTAL
SYSTEM 08.3-120-1240					
LARGE HEATING SYSTEM, HYDRONIC, ELECTRIC BOILER					
9,280 S.F., 150 KW, 510 MBH, 1 FLOOR					
Boiler, electric hot water, standard controls & trim 150 KW, 510 MBH	1.000	Ea.	5885	965	6850
Boiler room fittings and valves	1.000	System	2942.50	482.50	3425
Expansion tank, painted steel, 60 Gal capacity ASME	1.000	Ea.	1127.50	47.50	1175
Circulating pump, CI, close cpld, 50 GPM, 2 HP, 2" pipe conn	1.000	Ea.	709.50	155.50	865
Unit heater, 1 speed propeller, horizontal, 200°F EWT, 72.7 MBH	7.000	Ea.	2271.50	458.50	2730
Unit heater piping hookup with controls	7.000	Set	1668.66	3826.34	5495
Pipe, steel, black, schedule 40, welded, 2-1/2" diam	380.000	L.F.	2112.80	3891.20	6004
Pipe covering, calcium silicate w/cover, 1" wall, 2-1/2" diam	380.000	L.F.	893	1121	2014
TOTAL			17,610.46	10,947.54	28,558
COST PER S.F.			1.90	1.18	3.08

8.3-120		Large Heating Systems, Hydronic, Electric Boilers	COST PER S.F. MAT.	INST.	TOTAL
1230		Large heating systems, hydronic, electric boilers			
1240		9,280 S.F., 150 KW, 510 MBH, 1 floor	1.90	1.18	3.08
1280		14,900 S.F., 240 K.W., 820 M.B.H., 2 floors	2.13	1.92	4.05
1320		18,600 S.F., 300 K.W., 1,024 M.B.H., 3 floors	2.30	2.11	4.41
1360	C8.3 -001	26,100 S.F., 420 K.W., 1,432 M.B.H., 4 floors	2.14	2.05	4.19
1400		39,100 S.F., 630 K.W., 2,148 M.B.H., 4 floors	1.82	1.71	3.53
1440	C8.3 -002	57,700 S.F., 900 K.W., 3,071 M.B.H., 5 floors	1.83	1.69	3.52
1480		111,700 S.F., 1,800 K.W., 6,148 M.B.H., 6 floors	1.58	1.45	3.03
1520	C8.3 -003	149,000 S.F., 2,400 K.W., 8,191 M.B.H., 8 floors	1.55	1.44	2.99
1560		223,300 S.F., 3,600 K.W., 12,283 M.B.H., 14 floors	1.61	1.64	3.25

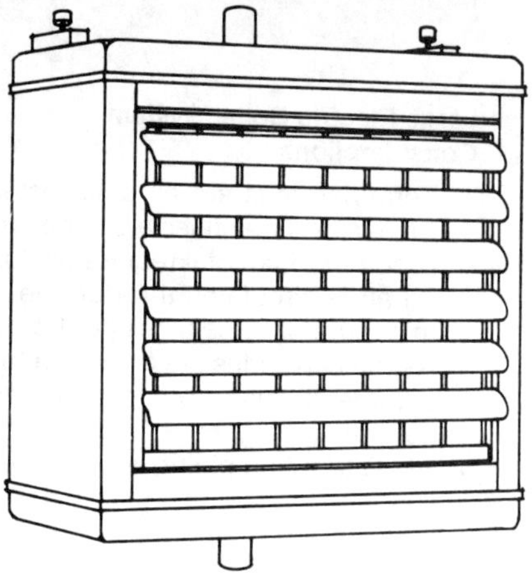
Unit Heater

Fossil Fuel Boiler System Considerations:

1. Terminal units are horizontal unit heaters. Quantities are varied to accommodate total heat loss per building.
2. Unit heater selection was determined by their capacity to circulate the building volume a minimum of three times per hour in addition to the BTU output.
3. Systems shown are forced hot water. Steam boilers cost slightly more than hot water boilers, however, this is compensated for by the smaller size or fewer terminal units required with steam.
4. Floor levels are based on 10′ story heights.
5. MBH requirements are gross boiler output.

System Components	QUANTITY	UNIT	COST EACH MAT.	INST.	TOTAL
SYSTEM 08.3-141-1280					
HEATING SYSTEM, HYDRONIC, FOSSIL FUEL, TERMINAL UNIT HEATERS					
CAST IRON BOILER, GAS, 80 MBH, 1,070 S.F. BUILDING					
Boiler, gas, hot water, CI, burner, controls & insulation 80 MBH	1.000	Ea.	929.50	670.50	1600
Pipe, steel, black, schedule 40, threaded, cplg & hngr 10′OC, 2″ diam	200.000	L.F.	788	1442	2230
Unit heater, 1 speed propeller, horizontal, 200° EWT, 72.7 MBH	2.000	Ea.	649	131	780
Unit heater piping hookup with controls	2.000	Set	476.76	1093.24	1570
Boiler breeching	1.000	System	46.48	33.52	80.
Expansion tank, painted steel, ASME, 18 Gal capacity	1.000	Ea.	566.50	28.50	595
Circulating pump, CI, flange connection, 1/12 HP	1.000	Ea.	126.50	78.50	205
Pipe covering, calcium silicate w/cover, 1″ wall, 2″ diam	200.000	L.F.	378	588	966
TOTAL			3960.74	4065.26	8026.
COST PER S.F.			3.70	3.80	7.50

8.3-141		Heating Systems, Hydronic, Fossil Fuel, Unit Heaters	COST PER S.F. MAT.	INST.	TOTAL
1260		Heating systems, hydronic, fossil fuel, terminal unit heaters,			
1280		Cast iron boiler, gas, 80 M.B.H., 1,070 S.F. bldg.	3.70	3.80	7.50
1320	C8.3 -001	163 M.B.H., 2,140 S.F. bldg.	2.56	2.58	5.14
1360		544 M.B.H., 7,250 S.F. bldg.	1.92	1.77	3.69
1400	C8.3 -002	1,088 M.B.H., 14,500 S.F. bldg.	1.77	1.68	3.45
1440		3,264 M.B.H., 43,500 S.F. bldg.	1.54	1.24	2.78
1480	C8.3 -003	5,032 M.B.H., 67,100 S.F. bldg.	1.61	1.31	2.92
1520		Oil, 109 M.B.H., 1,420 S.F. bldg.	3.72	3.35	7.07
1560		235 M.B.H., 3,150 S.F. bldg.	2.64	2.43	5.07
1600		940 M.B.H., 12,500 S.F. bldg.	1.99	1.59	3.58
1640		1,600 M.B.H., 21,300 S.F. bldg.	1.92	1.52	3.44
1680		2,480 M.B.H., 33,100 S.F. bldg.	1.88	1.39	3.27
1720		3,350 M.B.H., 44,500 S.F. bldg.	1.75	1.41	3.16
1760		Coal, 148 M.B.H., 1,975 S.F. bldg.	2.91	2.22	5.13
1800		300 M.B.H., 4,000 S.F. bldg.	2.25	1.72	3.97
1840		2,360 M.B.H., 31,500 S.F. bldg.	1.67	1.38	3.05
1880		Steel boiler, gas, 72 M.B.H., 1,020 S.F. bldg.	3.48	2.82	6.30
1920		240 M.B.H., 3,200 S.F. bldg.	2.60	2.19	4.79
1960		480 M.B.H., 6,400 S.F. bldg.	2.13	1.63	3.76
2000		800 M.B.H., 10,700 S.F. bldg.	1.87	1.44	3.31
2040		1,960 M.B.H., 26,100 S.F. bldg.	1.71	1.30	3.01
2080		3,000 M.B.H., 40,000 S.F. bldg.	1.67	1.33	3

8.3-141	Heating Systems, Hydronic, Fossil Fuel, Unit Heaters	COST PER S.F. MAT.	INST.	TOTAL
2120	Oil, 97 M.B.H., 1,300 S.F. bldg.	3.86	3.12	6.98
2160	315 M.B.H., 4,550 S.F. bldg.	2.31	1.62	3.93
2200	525 M.B.H., 7,000 S.F. bldg.	2.45	1.69	4.14
2240	1,050 M.B.H., 14,000 S.F. bldg.	2.21	1.66	3.87
2280	2,310 M.B.H. 30,800 S.F. bldg.	2.07	1.47	3.54
2320	3,150 M.B.H., 42,000 S.F. bldg.	1.97	1.50	3.47

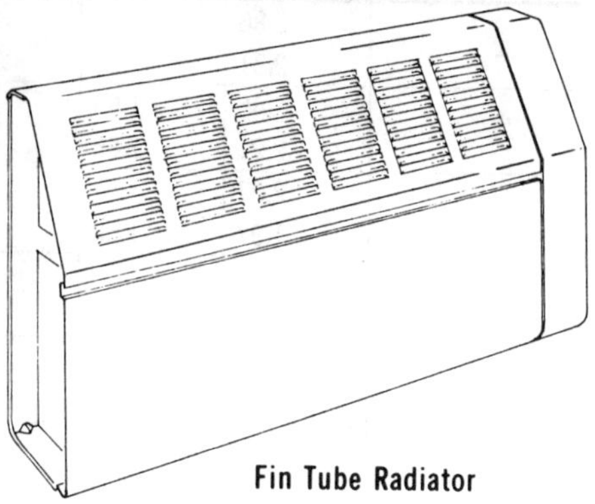

Fin Tube Radiator

Fossil Fuel Boiler System Considerations:

1. Terminal units are commercial steel fin tube radiation. Quantities are varied to accommodate total heat loss per building.
2. Systems shown are forced hot water. Steam boilers cost slightly more than hot water boilers, however, this is compensated for by the smaller size or fewer terminal units required with steam.
3. Floor levels are based on 10′ story heights.
4. MBH requirements are gross boiler output.

8

System Components	QUANTITY	UNIT	COST EACH MAT.	COST EACH INST.	COST EACH TOTAL
SYSTEM 08.3-142-3240					
HEATING SYSTEM, HYDRONIC, FOSSIL FUEL, FIN TUBE RADIATION					
CAST IRON BOILER, GAS, 80 MBH, 1,070 S.F. BLDG.					
Boiler, gas, hot water, CI, burner, controls & insulation, 80 MBH	1.000	Ea.	929.50	670.50	1600
Pipe, steel, black, schedule 40, threaded, 2" diam	200.000	L.F.	788	1442	2230
Radiation,steel 1-1/4" tube & 4-1/4" fin w/cover & damper, wall hung	80.000	L.F.	1848	2312	4160
Rough-in, wall hung steel fin radiation with supply & balance valves	8.000	Set	1343.76	2256.24	3600
Boiler breeching	1.000	System	46.48	33.52	80.
Expansion tank, painted steel, ASME, 18 Gal capacity	1.000	Ea.	566.50	28.50	595
Circulating pump, CI flanged, 1/12 HP	1.000	Ea.	126.50	78.50	205
Pipe covering, calcium silicate w/cover, 1" wall, 2" diam	200.000	L.F.	378	588	966
TOTAL			6026.74	7409.26	13,436.
COST PER S.F.			5.63	6.92	12.55

	8.3-142 Heating Systems, Hydronic, Fossil Fuel, Fin Tube Rad.	COST PER S.F. MAT.	COST PER S.F. INST.	COST PER S.F. TOTAL
3230	Heating systems, hydronic, fossil fuel, fin tube radiation			
3240	Cast iron boiler, gas, 80 MBH, 1,070 S.F. bldg.	5.65	6.90	12.55
3280	C8.3-001 169 M.B.H., 2,140 S.F. bldg.	3.62	4.38	8
3320	544 M.B.H., 7,250 S.F. bldg.	3.20	3.77	6.97
3360	C8.3-002 1,088 M.B.H., 14,500 S.F. bldg.	3.08	3.73	6.81
3400	3,264 M.B.H., 43,500 S.F. bldg.	2.90	3.34	6.24
3440	C8.3-003 5,032 M.B.H., 67,100 S.F. bldg.	2.97	3.42	6.39
3480	Oil, 109 M.B.H., 1,420 S.F. bldg.	6.50	7.60	14.10
3520	235 M.B.H., 3,150 S.F. bldg.	3.89	4.41	8.30
3560	940 M.B.H., 12,500 S.F. bldg.	3.32	3.66	6.98
3600	1,600 M.B.H., 21,300 S.F. bldg.	3.29	3.63	6.92
3640	2,480 M.B.H., 33,100 S.F. bldg.	3.25	3.50	6.75
3680	3,350 M.B.H., 44,500 S.F. bldg.	3.11	3.52	6.63
3720	Coal, 148 M.B.H., 1,975 S.F. bldg.	4.16	4.20	8.36
3760	300 M.B.H., 4,000 S.F. bldg.	3.50	3.69	7.19
3800	2,360 M.B.H., 31,500 S.F. bldg.	3.02	3.47	6.49
3840	Steel boiler, gas, 72 M.B.H., 1,020 S.F. bldg.	6.05	6.70	12.75
3880	240 M.B.H., 3,200 S.F. bldg.	3.92	4.26	8.18
3920	480 M.B.H., 6,400 S.F. bldg.	3.41	3.64	7.05
3960	800 M.B.H., 10,700 S.F. bldg.	3.51	3.97	7.48
4000	1,960 M.B.H., 26,100 S.F. bldg.	3.05	3.39	6.44
4040	3,000 M.B.H., 40,000 S.F. bldg.	3.01	3.41	6.42

8.3-142	Heating Systems, Hydronic, Fossil Fuel, Fin Tube Rad.	COST PER S.F.		
		MAT.	INST.	TOTAL
4080	Oil, 97 M.B.H., 1,300 S.F. bldg.	5.05	6.90	11.95
4120	315 M.B.H., 4,550 S.F. bldg.	3.49	3.47	6.96
4160	525 M.B.H., 7,000 S.F. bldg.	3.79	3.77	7.56
4200	1,050 M.B.H., 14,000 S.F. bldg.	3.55	3.73	7.28
4240	2,310 M.B.H., 30,800 S.F. bldg.	3.41	3.55	6.96
4280	3,150 M.B.H., 42,000 S.F. bldg.	3.32	3.58	6.90

B8.3-151 Fin Tube Radiation, Residential

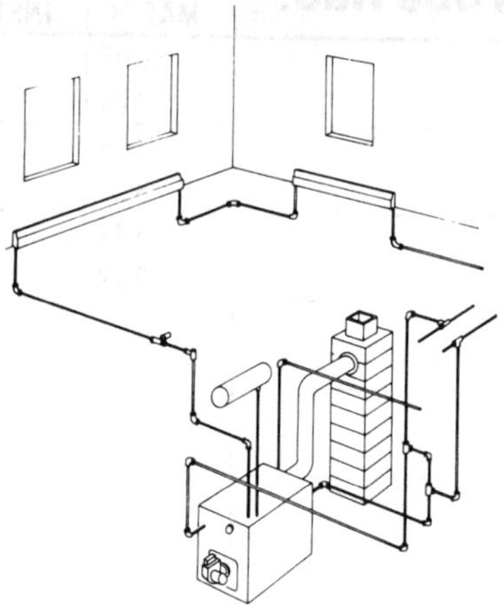

Basis for Heat Loss Estimate., Apartment Type Structures:

1. Masonry walls and flat roof are insulated. U factor is assumed at .08.
2. Window glass area taken as BOCA minimum, 1/10th of floor area. Double insulating glass with 1/4" air space, U = .65.
3. Infiltration = 0.3 cu. ft. per hour per sq. ft. of net wall.
4. Concrete floor loss is 2 BTUH per sq. ft.
5. Temperature difference taken as 70° F.
6. Ventilating or makeup air has not been included and must be added if desired.

System Components	QUANTITY	UNIT	COST EACH MAT.	COST EACH INST.	COST EACH TOTAL
SYSTEM 08.3-151-1760					
HEATING SYSTEM, FIN TUBE RADIATION, FORCED HOT WATER					
1,000 S.F. AREA, 10,000 C.F. VOLUME					
Boiler, oil fired, CI, burner, controls & insulation 109 MBH	1.000	Ea.	1210	815	2025
Circulating pump, CI flange connection, 1/12 HP	1.000	Ea.	126.50	78.50	205
Expansion tank, painted steel, ASME 18 Gal capacity	1.000	Ea.	566.50	28.50	595
Storage tank, steel, above ground, 275 Gal capacity w/supports	1.000	Ea.	170.50	94.50	265
Copper tubing type L, 50/50 solder joint, hanger 10' OC 3/4" diam	100.000	L.F.	122	583	705
Radiation, 3/4" copper tube w/alum fin baseboard pkg, 7" high	30.000	L.F.	193.20	586.80	780
Pipe covering, calcium silicate w/cover, 1" wall, 3/4" diam	100.000	L.F.	169	277	446
Boiler room fittings and valves	1.000	System	605	407.50	1012.50
Boiler breeching	1.000	System	60.50	40.75	101.25
Fuel oil piping system	1.000	System	242	163	405
TOTAL			3465.20	3074.55	6539.75
COST PER S.F.			3.47	3.07	6.54

8.3-151		Apartment Building Heating - Fin Tube Radiation	COST PER S.F. MAT.	COST PER S.F. INST.	COST PER S.F. TOTAL
1740		Heating systems, fin tube radiation, forced hot water			
1760		1,000 S.F. area, 10,000 C.F. volume	3.47	3.07	6.54
1800	C8.3 -001	10,000 S.F. area, 100,000 C.F. volume	1.03	2	3.03
1840		20,000 S.F. area, 200,000 C.F. volume	1.05	2.39	3.44
1880	C8.3 -002	30,000 S.F. area, 300,000 C.F. volume	1.04	2.33	3.37

B8.3-161 Fin Tube Radiation, Commercial

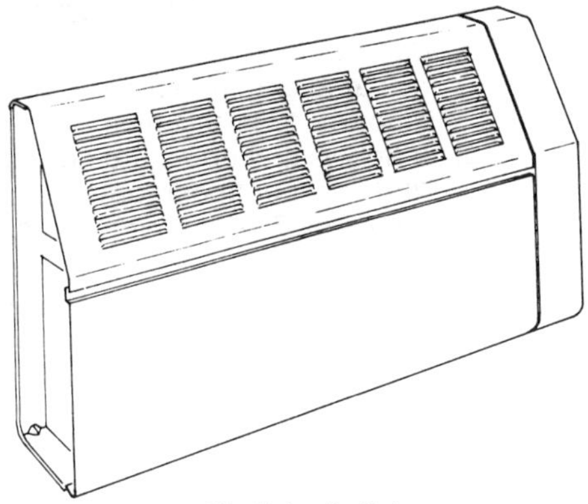

Fin Tube Radiator

Basis for Heat Loss Estimate, Factory or Commercial Type Structures:

1. Walls and flat roof are of lightly insulated concrete block or metal. U factor is assumed at .17.
2. Windows and doors are figured at 1-1/2 sq. ft. per linear foot of building perimeter. The U factor for flat single glass is 1.13.
3. Infiltration is approximately 5 cu. ft. per hour per sq. ft. of wall.
4. Concrete floor loss is 2 BTUH per sq. ft. of floor.
5. Temperature difference is assumed at 70° F.
6. Ventilation or makeup air has not been included and must be added if desired.

System Components	QUANTITY	UNIT	COST EACH MAT.	INST.	TOTAL
SYSTEM 08.3-161-1960					
HEATING SYSTEM, FIN TUBE RADIATION, FORCED HOT WATER					
1,000 S.F. BLDG., ONE FLOOR					
Boiler, oil fired, CI, burner, controls & insulation, 109 MBH	1.000	Ea.	1210	815	2025
Boiler room fittings and valves	1.000	System	605	407.50	1012.50
Boiler breeching	1.000	System	60.50	40.75	101.25
Fuel oil piping system	1.000	System	242	163	405
Expansion tank, painted steel, ASME 18 Gal capacity	1.000	Ea.	566.50	28.50	595
Storage tank, steel, above ground, 550 Gal capacity w/supports	1.000	Ea.	643.50	116.50	760
Circulating pump, CI flanged 3/4" to 1-1/2" pipe conn 1/8 HP	1.000	Ea.	198	77	275
Pipe, steel, black, schedule 40, threaded, cplg & hngr 10'OC 1-1/2" diam	260.000	L.F.	746.20	1502.80	2249
Pipe covering, calcium silicate w/cover , 1" wall, 1-1/2" diam	260.000	L.F.	483.60	741	1224.60
Radiation,steel 1-1/4" tube & 4-1/4" fin w/cover & damper, wall hung	42.000	L.F.	970.20	1213.80	2184
Rough in, steel fin tube radiation w/supply & balance valves	4.000	Set	671.88	1128.12	1800
TOTAL			6397.38	6233.97	12,631.35
COST PER S.F.			6.40	6.23	12.63

8.3-161		Commercial Building Heating - Fin Tube Radiation	COST PER S.F. MAT.	INST.	TOTAL
1940		Heating systems, fin tube radiation, forced hot water			
1960		1,000 S.F. bldg, one floor	6.40	6.25	12.65
2000	C8.3 -001	10,000 S.F., 100,000 C.F., total two floors	2.09	2.40	4.49
2040		100,000 S.F., 1,000,000 C.F., total three floors	.88	1.07	1.95
2080	C8.3 -002	1,000,000 S.F., 10,000,000 C.F., total five floors	.44	.58	1.02

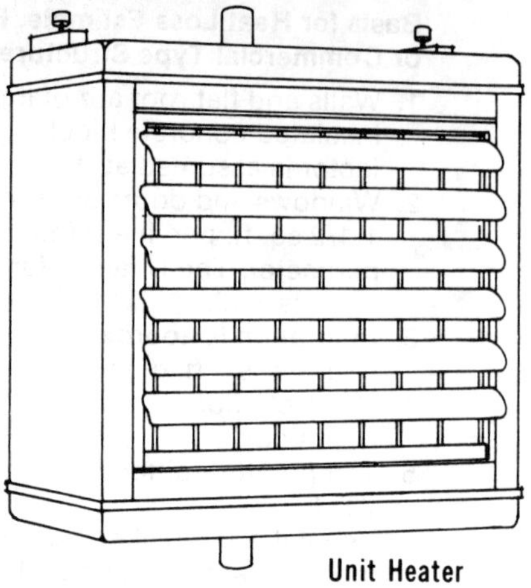
Unit Heater

Basis for Heat Loss Estimate, Factory or Commercial Type Structures:

1. Walls and flat roof are of lightly insulated concrete block or metal. U factor of .17 is assumed.
2. Windows and doors are figured at 1-1/2 sq. ft. per linear foot of building perimeter. The U factor for flat single glass is 1.13.
3. Infiltration is approximately 5 cu. ft. per hour per sq. ft. of wall.
4. Concrete floor loss is 2 BTUH per sq. ft. of floor.
5. Temperature difference is assumed at 70° F.
6. Ventilation or makeup air has not been included and must be added if desired.

System Components	QUANTITY	UNIT	COST EACH MAT.	COST EACH INST.	COST EACH TOTAL
SYSTEM 08.3-162-1880					
HEATING SYSTEM, TERMINAL UNIT HEATERS, FORCED HOT WATER					
1,000 S.F. BLDG., ONE FLOOR					
Boiler oil fired, CI, burner, controls & insulation 109 MBH	1.000	Ea.	1210	815	2025
Boiler room fittings and valves	1.000	System	605	407.50	1012.50
Boiler breeching	1.000	System	60.50	40.75	101.25
Fuel oil piping system	1.000	System	242	163	405
Expansion tank, painted steel, ASME 18 Gal capacity	1.000	Ea.	566.50	28.50	595
Storage tank, steel, above ground, 550 Gal capacity w/supports	1.000	Ea.	643.50	116.50	760
Circulating pump, CI flanged 1/8 HP	1.000	Ea.	198	77	275
Pipe, steel, black, schedule 40, threaded, cplg & hngr 10'OC 1-1/2" diam	260.000	L.F.	746.20	1502.80	2249
Pipe covering, calcium silicate w/cover, 1" wall, 1-1/2" diam	260.000	L.F.	483.60	741	1224.60
Unit heater, 1 speed propeller, horizontal, 200° EWT, 26.9 MBH	2.000	Ea.	462	88	550
Unit heater piping hookup with controls	2.000	Set	476.76	1093.24	1570
TOTAL			5694.06	5073.29	10,767.35
COST PER S.F.			5.69	5.07	10.76

8.3-162	Commercial Bldg. Heating - Terminal Unit Heaters		COST PER S.F. MAT.	COST PER S.F. INST.	COST PER S.F. TOTAL
1860		Heating systems, terminal unit heaters, forced hot water			
1880		1,000 S.F. bldg., one floor	5.70	5.05	10.75
1920	C8.3 -001	10,000 S.F. bldg., 100,000 C.F. total two floors	1.71	1.79	3.50
1960		100,000 S.F. bldg., 1,000,000 C.F. total three floors	.78	.87	1.65
2000	C8.3 -002	1,000,000 S.F. bldg., 10,000,000 C.F. total five floors	.48	.57	1.05

Many styles of active solar energy systems exist. Those shown on the following page represent the majority of systems now being installed in different regions of the country.

The five active domestic hot water (DHW) systems which follow are typically specified as two or three panel systems with additional variations being the type of glazing and size of the storage tanks. Various combinations have been costed for the user's evaluation and comparison. The basic specifications from which the following systems were developed satisfy the construction detail requirements specified in the HUD Intermediate Minimum Property Standards (IMPS). If these standards are not complied with, the renewable energy system's costs could be significantly lower than shown.

To develop the system's specifications and costs it was necessary to make a number of assumptions about the systems. Certain systems are more appropriate to one climatic region than another or the systems may require modifications to be usable in particular locations. Specific instances in which the systems are not appropriate throughout the country as specified, or in which modification will be needed include the following:

- The freeze protection mechanisms provided in the DHW systems vary greatly. In harsh climates, where freeze is a major concern, a closed-loop indirect collection system may be more appropriate than a direct collection system.
- The thermosyphon water heater system described cannot be used when temperatures drop below 32° F.
- In warm climates it may be necessary to modify the systems installed to prevent overheating.

For each renewable resource (solar) system a schematic diagram and descriptive summary of the system is provided along with a list of all the components priced as part of the system. The costs were developed based on these specifications.

Considerations affecting costs which may increase or decrease beyond the estimates presented here include the following:

- Special structural qualities (allowance for earthquake, futureexpansion, high winds, and unusual spans or shapes);
- Isolated building site or rough terrain that would affect the transportation of personnel, material, or equipment;
- Unusual climatic conditions during the construction process;
- Substitution of other materials or system components for those used in the system specifications.

B8.3-610 Closed Loop, Immersed Exchanger

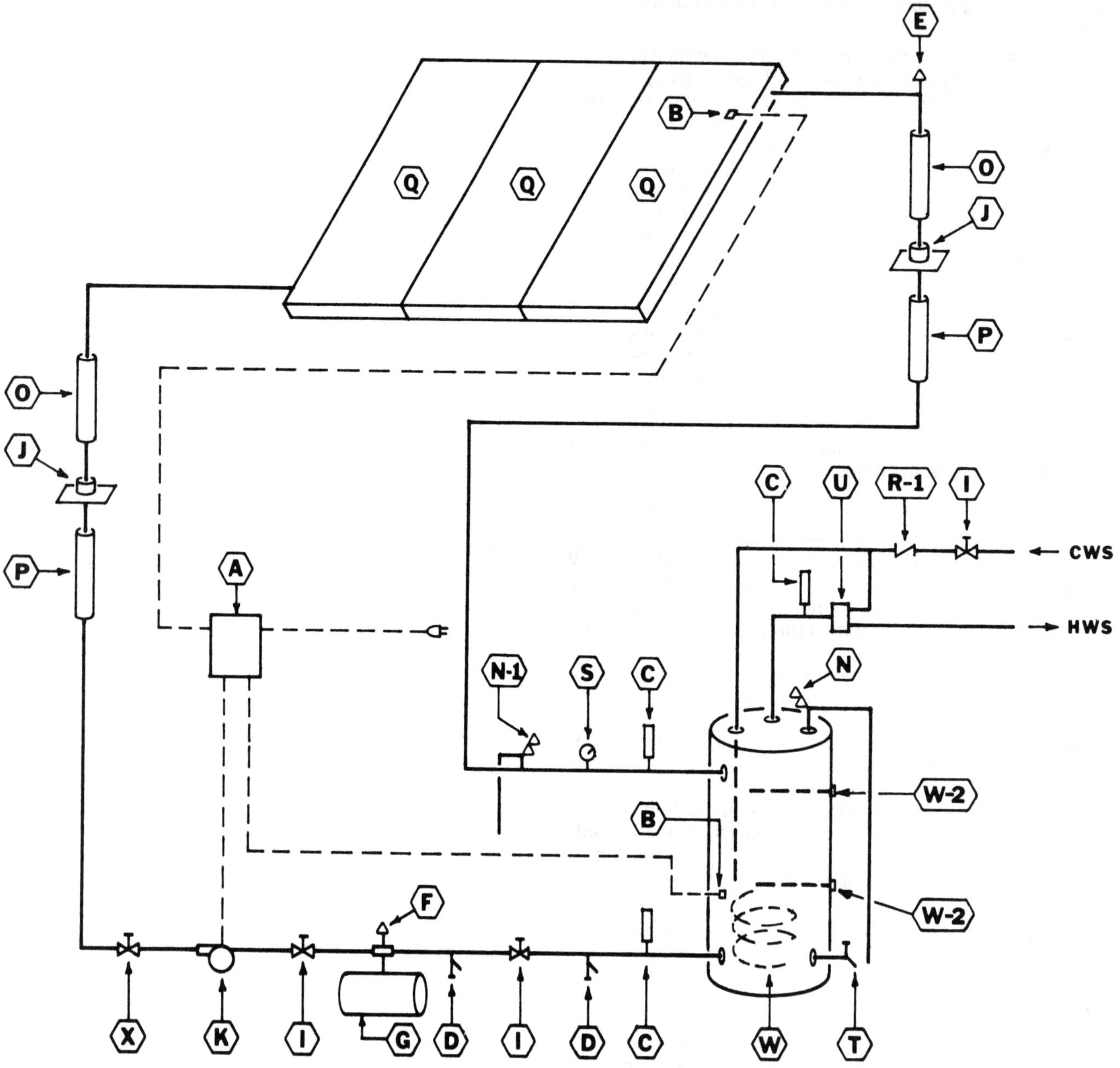

In this closed-loop indirect collection system, fluid with a low freezing temperature, such as propylene glycol, transports heat from the collectors to water storage. The transfer fluid is contained in a closed-loop consisting of collectors, supply and return piping, and a heat exchanger immersed in the storage tank. A typical two-or-three panel system contains 5 to 6 gallons of heat transfer fluid.

When the collectors become approximately 20°F warmer than the storage temperature, a controller activates the circulator. The circulator moves the fluid continuously through the collectors until the temperature difference between the collectors and storage is such that heat collection no longer occurs; at that point, the circulator shuts off. Since the heat transfer fluid has a very low freezing temperature, there is no need for it to be drained from the collectors between periods of collection.

B8.3-610 Closed Loop, Immersed Exchanger

System Components		QUANTITY	UNIT	COST EACH MAT.	COST EACH INST.	COST EACH TOTAL
	SYSTEM 08.3-610-2760					
	SOLAR, CLOSED LOOP, HOT WATER SYSTEM, IMMERSED HEAT EXCHANGER					
	3/4" TUBING, THREE 3'X7' BLACK CHROME COLLECTORS					
A, B	Differential controller, 2 sensors, thermostat, solar energy system	1.000	Ea.	67.10	20.90	88
C	Thermometer for solar energy system 2" dial	3.000	Ea.	42.75	47.25	90
D, T	Fill and drain valves, brass, 3/4" connection	3.000	Ea.	13.44	32.01	45.45
E	Air vent, manual, for solar energy system, 1/8" fitting	1.000	Ea.	1.40	8	9.40
F	Air purger	1.000	Ea.	10.95	21.05	32
G	Expansion tank, for solar energy system	1.000	Ea.	30.80	8.20	39
I	Valve, gate, bronze, NRS, soldered 3/4" diam	3.000	Ea.	44.73	39.27	84
J	Neoprene vent flashing	2.000	Ea.	8.96	25.64	34.60
K	Circulator, solar heated liquid, 1/25 HP	1.000	Ea.	86.90	33.10	120
N-1, N	Relief valve, temp & press 150 psi 210°F self-closing 3/4" IPS	2.000	Ea.	15.30	17.10	32.40
P	Pipe covering, urethane ultraviolet cover, 1" wall, 3/4" diam	20.000	L.F.	24.60	45.60	70.20
	Pipe covering, fiberglass, all service jacket, 1" wall, 3/4" diam	50.000	L.F.	47.50	102	149.50
	Roof clamps for solar energy collector panel	3.000	Set	9.90	19.80	29.70
Q	Collector panel solar blk chrome on copper, 1/8" temp glass, 3'x7'	3.000	Ea.	940.50	139.50	1080
R-1	Valve, swing check, bronze, regrinding disc, 3/4" diam, soldered	1.000	Ea.	13.70	13.30	27
S	Pressure gauge, 60 psi, 2-1/2" dial, for solar energy system	1.000	Ea.	5.45	8	13.45
U	Valve, water tempering, bronze, sweat connections, 3/4" diam	1.000	Ea.	22	13	35
W-2, W	Tank, water storage immersed heat exchr elec elem 2"x2# insul 120 Gal	1.000	Ea.	671	184	855
X	Valve, globe, bronze, rising stem, 3/4" diam, soldered	1.000	Ea.	20.24	12.76	33
	Copper tubing type L, 50/50 solder joint, hanger 10' OC 3/4" diam	20.000	L.F.	24.40	116.60	141
	Copper tubing, type M, 50/50 solder joint, hanger 10' OC 3/4" diam	70.000	L.F.	67.20	391.30	458.50
	Sensor wire, #22-2 conductor multistranded	.500	C.L.F.	4.40	12.60	17
	Solar energy heat transfer fluid, propylene glycol, anti-freeze	6.000	Gal.	38.28	55.02	93.30
	Wrought copper fittings & solder, 3/4" diam	76.000	Ea.	25.08	1027.52	1052.60
	TOTAL			2236.58	2393.52	4630.10

8.3-610	Solar, Closed Loop, Hot Water System	COST EACH MAT.	COST EACH INST.	COST EACH TOTAL
2550	Solar, closed loop, hot water system, immersed heat exchanger			
2560	3/8" tubing, 3 ea. 4' x 4'-4" vacuum tube collectors, 80 gal. tank	3350	2100	5450
2580	1/2" tubing, 4 ea. 4' x 4'-4" vacuum tube collectors, 80 gal. tank	4075	2275	6350
2600	120 gal. tank	4300	2300	6600
2640	C8.3 -601 2 ea. 3'x7' black chrome collectors, 80 gal. tank	1650	2150	3800
2660	120 gal. tank	1850	2175	4025
2700	2 ea. 3'x7' flat black collectors, 120 gal. tank	1800	2175	3975
2720	3 ea. 3'x7' flat black collectors, 120 gal. tank	2075	2225	4300
2760	3/4" tubing, 3 ea. 3'x7' black chrome collectors, 120 gal. tank	2225	2400	4625
2780	3 ea. 3'x7' flat black collectors, 120 gal. tank	2150	2400	4550
2800	2 ea. 4'x9' flat black w/plastic glazing collectors 120 gal. tank	1925	2425	4350
2840	1" tubing, 4 ea. 2'x9' plastic absorber & glazing collectors 120 gal. tank	3050	2750	5800
2860	4 ea. 3'x7' black chrome collectors, 120 gal. tank	2675	2750	5425
2880	4 ea. 3'x7' flat black absorber collectors, 120 gal. tank	2550	2750	5300

8

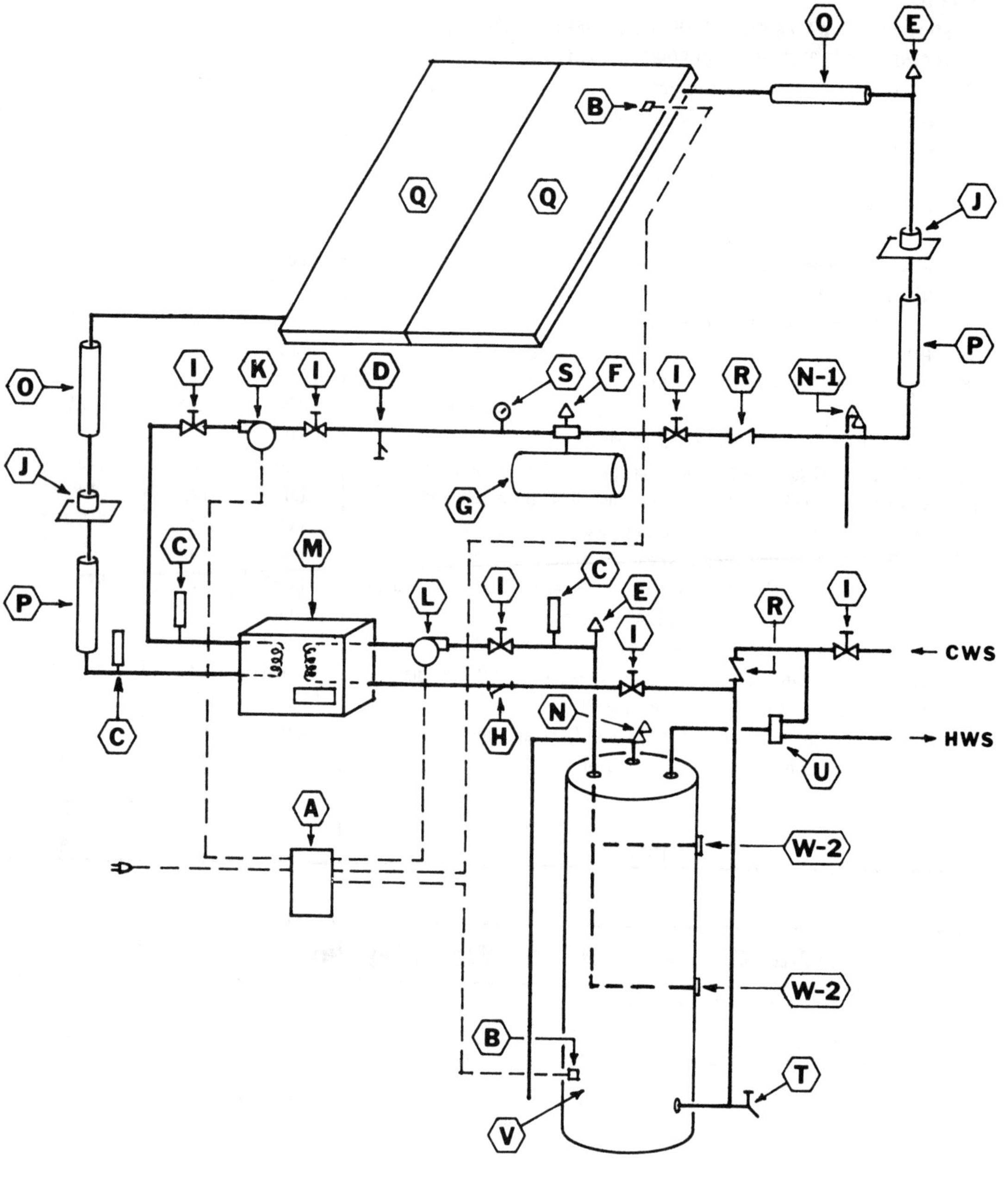

In this closed-loop indirect collection system, fluid with a low freezing temperature, such as propylene glycol, transports heat from the collectors to water storage. The transfer fluid is contained in a closed-loop consisting of collectors, supply and return piping, and a remote heat exchanger. The heat exchanger transfers heat energy from the fluid in the collector loop to potable water circulated in a storage loop. A typical two-or-three panel system contains 5 to 6 gallons of heat transfer fluid.

When the collectors become approximately 20°F warmer than the storage temperature, a controller activates the circulator on the collector and storage loops. The circulators will move the fluid and potable water through the heat exchanger until heat collection no longer occurs. At that point, the system shuts down. Since the heat transfer medium is a fluid with a very low freezing temperature, there is no need for it to be drained from the system between periods of collection.

System Components

System Components		QUANTITY	UNIT	COST EACH MAT.	COST EACH INST.	COST EACH TOTAL
	SYSTEM 08.3-620-2760 SOLAR, CLOSED LOOP, ADD-ON HOT WATER SYSTEM, EXTERNAL HEAT EXCHANGER 3/4" TUBING, TWO 3'X7' BLACK CHROME COLLECTORS					
A, B, G, L, K, M	Heat exchanger fluid-fluid pkg incl 2 circulators, expansion tank, check valve, relief valve, controller, hi temp cutoff, & 2 sensors	1.000	Ea.	610	185	795
C	Thermometer 2" dial	3.000	Ea.	42.75	47.25	90
D, T	Fill and drain valve, brass, 3/4" connection	1.000	Ea.	4.48	10.67	15.15
E	Air vent, manual, 1/8" diameter	2.000	Ea.	2.80	16	18.80
F	Air purger	1.000	Ea.	10.95	21.05	32
H	Strainer, Y type, bronze body, 3/4" IPS	1.000	Ea.	20.46	13.54	34
I	Valve, gate, bronze, NRS, soldered 3/4" diam	6.000	Ea.	89.46	78.54	168
J	Neoprene vent flashing	2.000	Ea.	8.96	25.64	34.60
N-1, N	Relief valve temp & press, 150 psi 210°F self-closing 3/4" IPS	1.000	Ea.	7.65	8.55	16.20
O	Pipe covering, urethane ultraviolet cover, 1" wall 3/4" diam	20.000	L.F.	24.60	45.60	70.20
P	Pipe covering, fiberglass, all service jacket, 1" wall, 3/4" diam	50.000	L.F.	47.50	102	149.50
Q	Collector panel solar energy blk chrome on copper, 1/8" temp glass 3'x7'	2.000	Ea.	627	93	720
	Roof clamps for solar energy collector panels	2.000	Set	6.60	13.20	19.80
R	Valve, swing check, bronze, regrinding disc, 3/4" diam	2.000	Ea.	27.40	26.60	54
S	Pressure gauge, 60 psi, 2-1/2" dial	1.000	Ea.	5.45	8	13.45
U	Valve, water tempering, bronze, sweat connections, 3/4" diam	1.000	Ea.	22	13	35
W-2, V	Tank water storage w/heating element, drain, relief valve, existing	1.000	Ea.			
	Copper tubing type L, 50/50 solder joint, hanger 10' OC 3/4" diam	20.000	L.F.	24.40	116.60	141
	Copper tubing, type M, 50/50 solder joint, hanger 10' OC 3/4" diam	70.000	L.F.	67.20	391.30	458.50
	Sensor wire, #22-2 conductor multistranded	.500	C.L.F.	4.40	12.60	17
	Solar energy heat transfer fluid, propylene glycol anti-freeze	6.000	Gal.	38.28	55.02	93.30
	Wrought copper fittings & solder, 3/4" diam	76.000	Ea.	25.08	1027.52	1052.60
	TOTAL			1715	2310	4025

8

8.3-620 Solar, Closed Loop, Add-on Hot Water System

8.3-620	Solar, Closed Loop, Add-on Hot Water System	COST EACH MAT.	COST EACH INST.	COST EACH TOTAL
2550	Solar, closed loop, add-on hot water system, external heat exchanger			
2570	3/8" tubing, 3 ea. 4' x 4'-4" vacuum tube collectors	3345	2090	5435
2580	1/2" tubing, 4 ea 4 x 4'-4" vacuum tube collectors (C8.3-601)	4075	2260	6335
2600	2 ea 3'x7' black chrome collectors	1640	2135	3775
2620	3 ea 3'x7' black chrome collectors	1955	2190	4145
2640	2 ea 3'x7' flat black collectors	1585	2140	3725
2660	3 ea 3'x7' flat black collectors	1875	2195	4070
2700	3/4" tubing, 3 ea 3'x7' black chrome collectors	2035	2365	4400
2720	3 ea 3'x7' flat black absorber plate collectors	1950	2370	4320
2740	2 ea 4'x9' flat black w/plastic glazing collectors	1715	2390	4105
2760	2 ea 3'x7' black chrome collectors	1715	2310	4025
2780	1" tubing,4 ea 2'x9' plastic absorber & glazing collectors	2880	2745	5625
2800	4 ea 3'x7' black chrome absorber collectors	2485	2740	5225
2820	4 ea 3'x7' flat black absorber collectors	2375	2750	5125

B8.3-630 Drainback, Immersed Exchanger

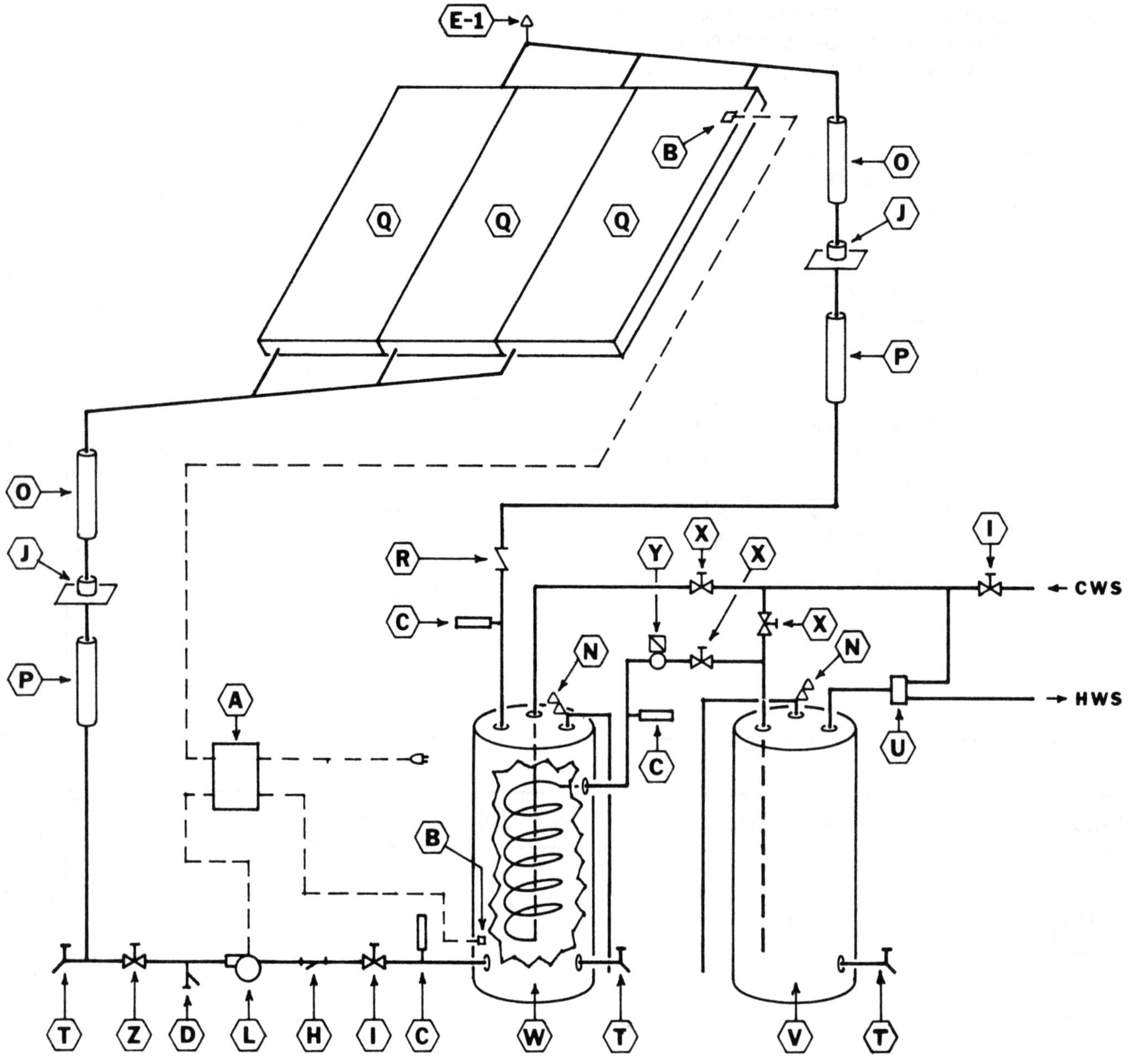

In the drainback indirect-collection system, the heat transfer fluid is distilled water contained in a loop consisting of collectors, supply and return piping, and an unpressurized holding tank. A large heat exchanger containing incoming potable water is immersed in the holding tank. When a controller activates solar collection, the distilled water is pumped through the collectors and heated and pumped back down to the holding tank. When the temperature differential between the water in the collectors and water in storage is such that collection no longer occurs, the pump turns off and gravity causes the distilled water in the collector loop to drain back to the holding tank. All the loop piping is pitched so that the water can drain out of the collectors and piping and not freeze there. As hot water is needed in the home, incoming water first flows through the holding tank with the immersed heat exchanger and is warmed and then flows through a conventional heater for any supplemental heating that is necessary.

System Components	QUANTITY	UNIT	COST EACH MAT.	INST.	TOTAL
SYSTEM 08.3-630-2760					
SOLAR, DRAINBACK, ADD ON, HOT WATER, IMMERSED HEAT EXCHANGER					
3/4" TUBING, THREE EA 3'X7' BLACK CHROME COLLECTOR					
A, B Differential controller 2 sensors, thermostat, solar energy system	1.000	Ea.	67.10	20.90	88
C Thermometer 2" dial	3.000	Ea.	42.75	47.25	90
D, T Fill and drain valve, brass, 3/4" connection	1.000	Ea.	4.48	10.67	15.15
E-1 Automatic air vent 1/8" diameter	1.000	Ea.	5.25	8	13.25
H Strainer,Y type, bronze body, 3/4" IPS	1.000	Ea.	20.46	13.54	34
I Valve, gate, bronze, NRS, soldered 3/4" diam	2.000	Ea.	29.82	26.18	56
J Neoprene vent flashing	2.000	Ea.	8.96	25.64	34.60
L Circulator, solar heated liquid, 1/20 HP	1.000	Ea.	77	38	115
N Relief valve temp & press 150 psi 210°F self-closing 3/4" IPS	1.000	Ea.	7.65	8.55	16.20
O Pipe covering, urethane ultraviolet cover, 1" wall, 3/4" diam	20.000	L.F.	24.60	45.60	70.20
P Pipe covering, fiberglass, all service jacket, 1" wall, 3/4" diam	50.000	L.F.	47.50	102	149.50
Q Collector panel solar energy blk chrome on copper, 1/8" temp glas 3'x7'	3.000	Ea.	940.50	139.50	1080
Roof clamps for solar energy collector panels	3.000	Set	9.90	19.80	29.70
R Valve, swing check, bronze, regrinding disc, 3/4" diam	1.000	Ea.	13.70	13.30	27
U Valve, water tempering, bronze sweat connections, 3/4" diam	1.000	Ea.	22	13	35
W Tank, water storage immersed heat exchr elec elem 2"x1/2# insul 120 gal	1.000	Ea.	638	182	820
V Tank, water storage w/heating element, drain, relief valve, existing	1.000	Ea.			
X Valve, globe, bronze, rising stem, 3/4" diam, soldered	3.000	Ea.	60.72	38.28	99
Y Flow control valve, 3/4" diam	1.000	Ea.	18	12	30
Z Valve, ball, bronze, solder 3/4" diam, solar loop flow control	1.000	Ea.	6.22	12.83	19.05
Copper tubing, type L, 50/50 solder joint, hanger 10' OC 3/4" diam	20.000	L.F.	24.40	116.60	141
Copper tubing, type M, 50/50 solder joint, hanger 10' OC 3/4" diam	70.000	L.F.	67.20	391.30	458.50
Sensor wire, #22-2 conductor, multistranded	.500	C.L.F.	4.40	12.60	17
Wrought copper fittings & solder, 3/4" diam	76.000	Ea.	25.08	1027.52	1052.60
TOTAL			2165	2325	4490

8

8.3-630	Solar, Drainback, Hot Water System	COST EACH MAT.	INST.	TOTAL
2550	Solar, drainback, hot water, immersed heat exchanger			
2560	3/8" tubing, 3 ea. 4' x 4'-4" vacuum tube collectors	3275	2025	5300
2580	C8.3 -601 1/2" tubing, 4 ea 4' x 4'-4" vacuum tube collectors, 80 gal tank	4000	2200	6200
2600	120 gal tank	4200	2225	6425
2640	2 ea. 3'x7' blk chrome collectors, 80 gal tank	1575	2075	3650
2660	3 ea. 3'x7' blk chrome collectors, 120 gal tank	2075	2150	4225
2700	2 ea. 3'x7' flat blk collectors, 120 gal tank	1700	2100	3800
2720	3 ea. 3'x7' flat blk collectors, 120 gal tank	1975	2150	4125
2760	3/4" tubing, 3 ea 3'x7' black chrome collectors, 120 gal tank	2165	2325	4490
2780	3 ea. 3'x7' flat black absorber collectors, 120 gal tank	2075	2325	4400
2800	2 ea. 4'x9' flat blk w/plastic glazing collectors 120 gal tank	1825	2350	4175
2840	1" tubing, 4 ea. 2'x9' plastic absorber & glazing collectors, 120 gal tank	3025	2700	5725
2860	4 ea. 3'x7' black chrome absorber collectors, 120 gal tank	2625	2700	5325
2880	4 ea. 3'x7' flat black absorber collectors, 120 gal tank	2510	2700	5210

Draindown, Direct Collection

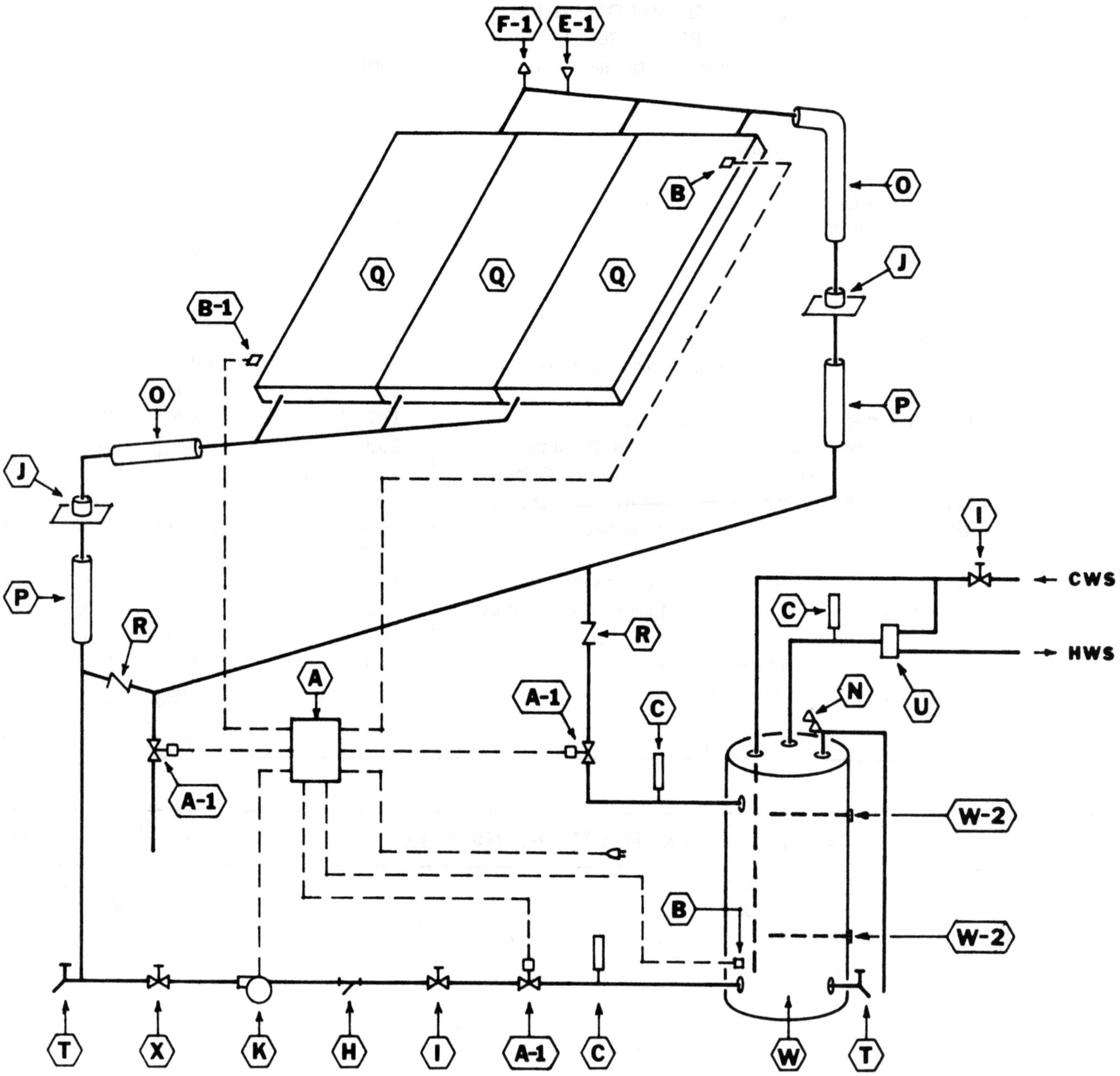

In the draindown direct-collection system, incoming domestic water is heated in the collectors. When the controller activates solar collection, domestic water is first heated as it flows through the collectors and is then pumped to storage. When conditons are no longer suitable for heat collection, the pump shuts off and the water in the loop drains down and out of the system by means of solenoid valves and properly pitched piping.

System Components

System Components	QUANTITY	UNIT	COST EACH MAT.	COST EACH INST.	COST EACH TOTAL
SYSTEM 08.3-640-2760					
SOLAR, DRAINDOWN, HOT WATER, DIRECT COLLECTION					
3/4" TUBING, THREE EA 3' X 7' BLACK CHROME COLLECTORS					
A, B Differential controller, 2 sensors, thermostat, solar energy system	1.000	Ea.	67.10	20.90	88
A-1 Solenoid valve, solar heating loop, brass, 3/4" diam, 24 volts	3.000	Ea.	148.50	85.50	234
B-1 Solar energy sensor, freeze prevention	1.000	Ea.	10.95	8	18.95
C Thermometer, 2" dial	3.000	Ea.	42.75	47.25	90
E-1 Vacuum relief valve for solar energy system, 3/4" diam	1.000	Ea.	15.40	7.60	23
F-1 Air vent, automatic, 1/8" diameter	1.000	Ea.	5.25	8	13.25
H Strainer,Y type, bronze body, 3/4" IPS	1.000	Ea.	20.46	13.54	34
I Valve, gate, bronze, NRS, soldered, 3/4" diam	2.000	Ea.	29.82	26.18	56
J Vent flashing neoprene	2.000	Ea.	8.96	25.64	34.60
K Circulator, solar heated liquid, 1/25 HP	1.000	Ea.	86.90	33.10	120
N Relief valve temp & press 150 psi 210°F self-closing 3/4" IPS	1.000	Ea.	7.65	8.55	16.20
O Pipe covering, urethane, ultraviolet cover, 1" wall, 3/4" diam	20.000	L.F.	24.60	45.60	70.20
P Pipe covering, fiberglass, all service jacket, 1" wall, 3/4" diam	50.000	L.F.	47.50	102	149.50
Roof clamps for solar energy collector panels	3.000	Set	9.90	19.80	29.70
Q Collector panel solar energy blk chrome on copper, 1/8" temp glass 3'x7'	3.000	Ea.	940.50	139.50	1080
R Valve, swing check, bronze, regrinding disc, 3/4" diam, soldered	2.000	Ea.	27.40	26.60	54
T Drain valve, brass, 3/4" connection	2.000	Ea.	8.96	21.34	30.30
U Valve, water tempering, bronze, sweat connections, 3/4" diam	1.000	Ea.	22	13	35
W-2, W Tank, water storage elec elem 2"x1/2# insul 120 gal	1.000	Ea.	638	182	820
X Valve, globe, bronze, rising stem, 3/4" diam, soldered	1.000	Ea.	20.24	12.76	33
Copper tubing, type L, 50/50 solder jnt, hngr 10' OC 3/4" diam	20.000	L.F.	24.40	116.60	141
Copper tubing, type M, 50/50 solder jnt, hngr 10' OC 3/4" diam	70.000	L.F.	67.20	391.30	458.50
Sensor wire, #22-2 conductor, multistranded	.500	C.L.F.	4.40	12.60	17
Wrought copper fittings & solder, 3/4" diam	76.000	Ea.	25.08	1027.52	1052.60
TOTAL			2303.92	2394.88	4698.80

8.3-640	Solar, Draindown, Hot Water System	COST EACH MAT.	COST EACH INST.	COST EACH TOTAL
2550	Solar, draindown, hot water			
2560	3/8" tubing, 3 ea. 4' x 4'-4" vacuum tube collectors, 80 gal tank	3400	2225	5625
2580	C8.3-601 1/2" tubing, 4 ea. 4' x 4'-4" vacuum tube collectors, 80 gal tank	4175	2275	6450
2600	120 gal tank	4350	2300	6650
2640	2 ea 3'x7' black chrome collectors, 80 gal tank	1725	2150	3875
2660	3 ea 3'x7' black chrome collectors, 120 gal tank	2225	2225	4450
2700	2 ea 3'x7' flat black collectors, 120 gal tank	1875	2175	4050
2720	3 ea 3'x7' flat black collectors, 120 gal tank	2150	2225	4375
2760	3/4" tubing, 3 ea 3'x7' black chrome collectors, 120 gal tank	2300	2400	4700
2780	3 ea 3'x7' flat collectors, 120 gal tank	2225	2400	4625
2800	2 ea. 4'x9' flat black & plastic glazing collectors, 120 gal tank	1975	2425	4400
2840	1" tubing, 4 ea. 2'x9' plastic absorber & glazing collectors, 120 gal tank	3200	2775	5975
2860	4 ea 3'x7' black chrome absorber collectors, 120 gal tank	2800	2775	5575
2880	4 ea 3'x7' flat black absorber collectors, 120 gal tank	2700	2775	5475

B8.3-650 Recirculation, Hot Water

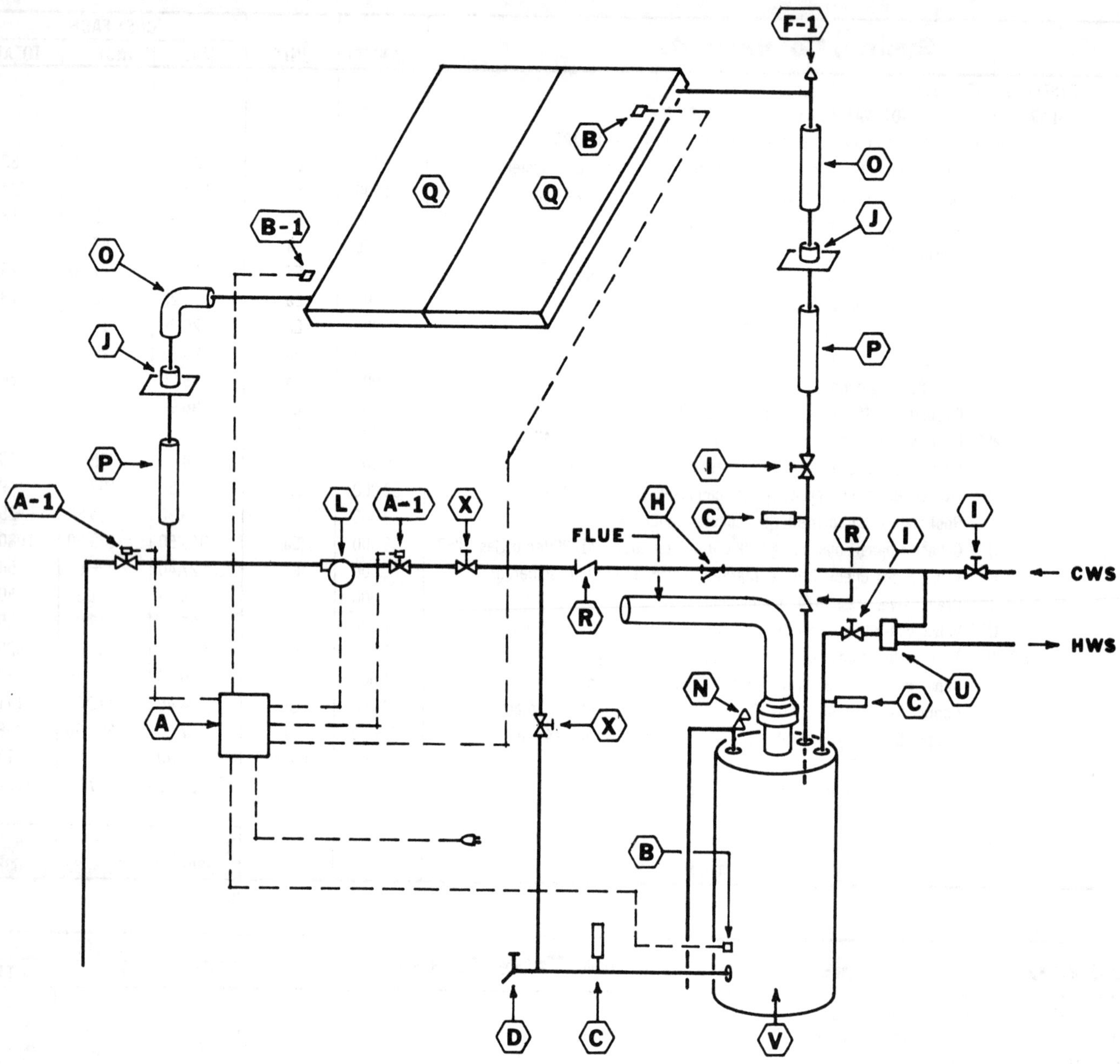

In the recirculation system, (a direct-collection system), incoming domestic water is heated in the collectors. When the controller activates solar collection, domestic water is heated as it flows through the collectors and then it flows back to storage. When conditions are not suitable for heat collection, the pump shuts off and the flow of the water stops. In this type of system, water remains in the collector loop at all times. A "frost sensor" at the collector activates the circulation of warm water from storage through the collectors when protection from freezing is required.

System Components		QUANTITY	UNIT	COST EACH MAT.	COST EACH INST.	COST EACH TOTAL
SYSTEM 08.3-650-2820						
SOLAR, RECIRCULATION, HOT WATER						
3/4" TUBING, TWO 3'X7' BLACK CHROME COLLECTORS						
A, B	Differential controller 2 sensors, thermostat, for solar energy system	1.000	Ea.	67.10	20.90	88
A-1	Solenoid valve, solar heating loop, brass, 3/4" IPS, 24 volts	2.000	Ea.	99	57	156
B-1	Solar energy sensor freeze prevention	1.000	Ea.	10.95	8	18.95
C	Thermometer, 2-1/2" dial	3.000	Ea.	42.75	47.25	90
D	Drain valve, brass, 3/4" connection	1.000	Ea.	4.48	10.67	15.15
F-1	Air vent, automatic, 1/8" diameter	1.000	Ea.	5.25	8	13.25
H	Strainer,Y type, bronze body, 3/4" IPS	1.000	Ea.	20.46	13.54	34
I	Valve, gate, bronze, 125 lb, soldered 3/4" diam	3.000	Ea.	44.73	39.27	84
J	Vent flashing, neoprene	2.000	Ea.	8.96	25.64	34.60
L	Circulator, solar heated liquid, 1/20 HP	1.000	Ea.	77	38	115
N	Relief valve, temp & press 150 psi 210°F self-closing 3/4" IPS	2.000	Ea.	15.30	17.10	32.40
O	Pipe covering, urethane, ultraviolet cover, 1" wall, 3/4" diam	20.000	L.F.	24.60	45.60	70.20
P	Pipe covering, fiberglass, all service jacket, 1" wall 3/4" diam	50.000	L.F.	47.50	102	149.50
Q	Collector panel solar energy blk chrome on copper, 1/8" temp glass 3'x7'	2.000	Ea.	627	93	720
	Roof clamps for solar energy collector panels	2.000	Set	6.60	13.20	19.80
R	Valve, swing check, bronze, 125 lb, regrinding disc, soldered 3/4" diam	2.000	Ea.	27.40	26.60	54
U	Valve, water tempering, bronze, sweat connections, 3/4" diam	1.000	Ea.	22	13	35
V	Tank, water storage, w/heating element, drain, relief valve, existing	1.000	Ea.			
X	Valve, globe, bronze, 125 lb, 3/4" diam	2.000	Ea.	40.48	25.52	66
	Copper tubing, type L, 50/50 solder jnt, hngr 10' OC 3/4" diam	20.000	L.F.	24.40	116.60	141
	Copper tubing, type M, 50/50 solder jnt, hngr 10' OC 3/4" diam	70.000	L.F.	67.20	391.30	458.50
	Wrought copper fittings & solder, 3/4" diam	76.000	Ea.	25.08	1027.52	1052.60
	Sensor wire, #22-2 conductor, multistranded	.500	C.L.F.	4.40	12.60	17
	TOTAL			1312.64	2152.31	3464.95

8

8.3-650	Solar Recirculation, Domestic Hot Water	COST EACH MAT.	COST EACH INST.	COST EACH TOTAL
2550	Solar recirculation, hot water			
2560	3/8" tubing, 3 ea. 4' x 4'-4" vacuum tube collectors	2950	1925	4875
2580	1/2" tubing, 4 ea. 4' x 4'-4" vacuum tube collectors	3675	2100	5775
2640	2 ea. 3'x7' black chrome collectors	1225	1975	3200
2660	3 ea. 3'x7' black chrome collectors	1550	2025	3575
2700	2 ea. 3'x7' flat black collectors	1175	1975	3150
2720	3 ea. 3'x7' flat black collectors	1475	2050	3525
2760	3/4" tubing, 3 ea. 3'x7' black chrome collectors	1625	2200	3825
2780	3 ea. 3'x7' flat black absorber plate collectors	1550	2225	3775
2800	2 ea. 4'x9' flat black w/plastic glazing collectors	1325	2225	3550
2820	2 ea. 3'x7' black chrome collectors	1325	2150	3475
2840	1" tubing, 4 ea. 2'x9' black plastic absorber & glazing collectors	2525	2575	5100
2860	4 ea. 3'x7' black chrome absorber collectors	2125	2575	4700
2880	4 ea. 3'x7' flat black absorber collectors	2025	2600	4625

C8.3 -601

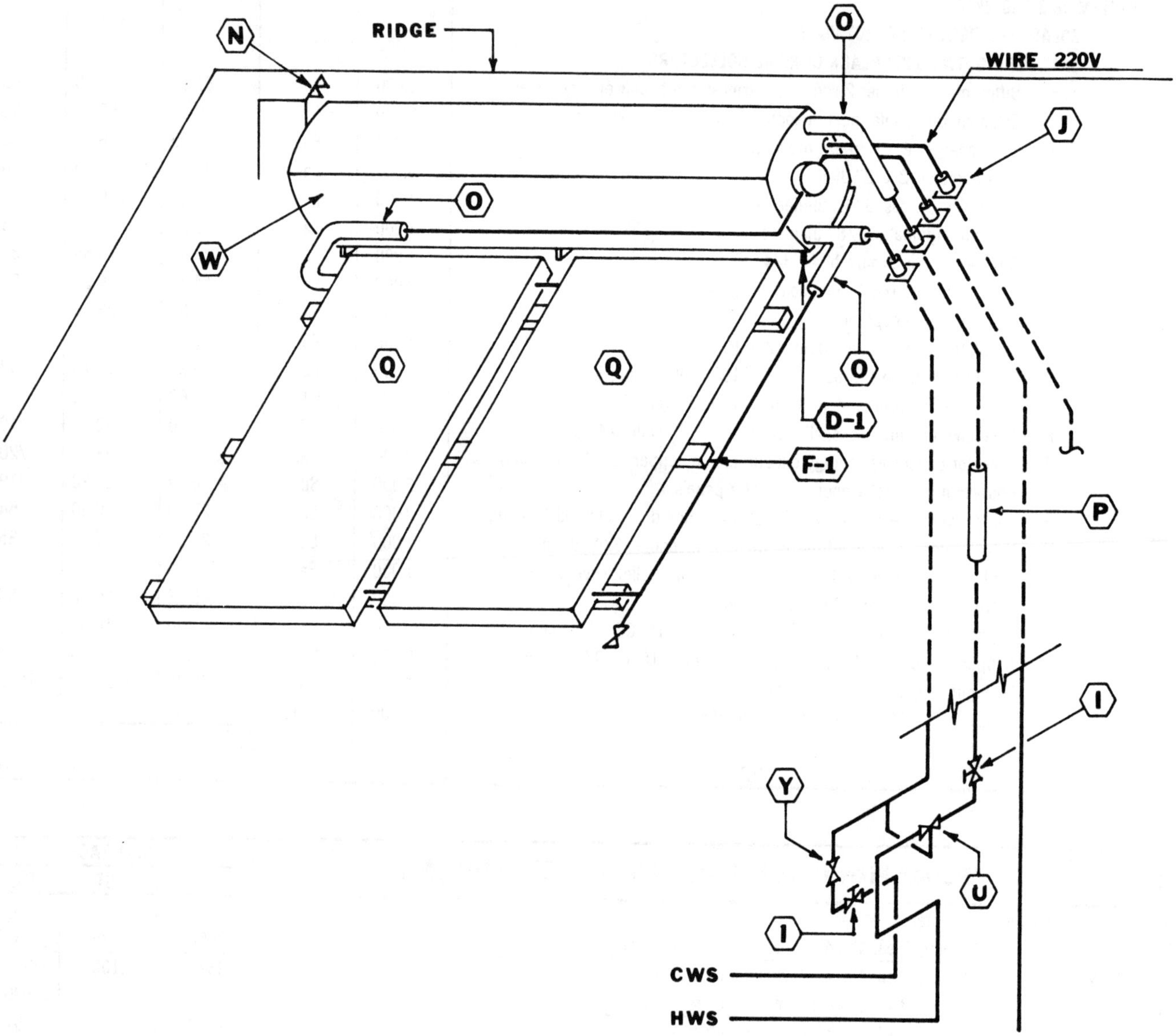

The thermosyphon domestic hot water system, a direct collection system, operates under city water pressure and does not require pumps for system operation. An insulated water storage tank is located above the collectors. As the sun heats the collectors, warm water in them rises by means of natural convection; the colder water in the storage tank flows into the collectors by means of gravity. As long as the sun is shining the water continues to flow through the collectors and to become warmer.

To prevent freezing, the system must be drained or the collectors covered with an insulated lid when the temperature drops below 32° F.

System Components		QUANTITY	UNIT	COST EACH MAT.	INST.	TOTAL
SYSTEM 08.3-660-0940						
SOLAR, THERMOSYPHON, WATER HEATER						
3/4" TUBING, TWO 3'X7' BLACK CHROME COLLECTORS						
D-1	Framing lumber, fir, 2" x 6" x 8', tank cradle	.008	M.B.F.	2.90	3.58	6.48
F-1	Framing lumber, fir, 2" x 4" x 24', sleepers	.016	M.B.F.	5.98	8.82	14.80
I	Valve, gate, bronze, 125 lb, soldered 1/2" diam	2.000	Ea.	23	21	44
J	Vent flashing, neoprene	4.000	Ea.	17.92	51.28	69.20
O	Pipe covering, urethane, ultraviolet cover, 1" wall, 1/2" diam	40.000	L.F.	49.20	87.20	136.40
P	Pipe covering fiberglass all service jacket 1" wall 1/2" diam	160.000	L.F.	132.80	313.60	446.40
Q	Collector panel solar energy blk chrome copper, 3/16" temp glass 3'-8"x6'	2.000	Ea.	572	98	670
Y	Flow control valve, globe, bronze, 125#, soldered, 1/2" diam	1.000	Ea.	16.12	10.88	27
U	Valve, water tempering, bronze, sweat connections, 1/2" diam	1.000	Ea.	19.09	10.91	30
W	Tank, water storage, solar energy system, 80 Gal, 2" x 1/2 lb insul	1.000	Ea.	423.50	161.50	585
	Copper tubing type L, 50/50 solder jnt, hngr 10' OC 1/2" diam	150.000	L.F.	126	729	855
	Copper tubing type M, 50/50 solder jnt, hngr 10' OC 1/2" diam	50.000	L.F.	33.50	229	262.50
	Sensor wire, #22-2 conductor, multistranded	.500	C.L.F.	4.40	12.60	17
	Wrought copper fittings & solder, 1/2" diam	75.000	Ea.	10.50	960.75	971.25
	TOTAL			1449.04	2717.34	4166.38

C8.3-601

Closed Loop, Space/Hot Water

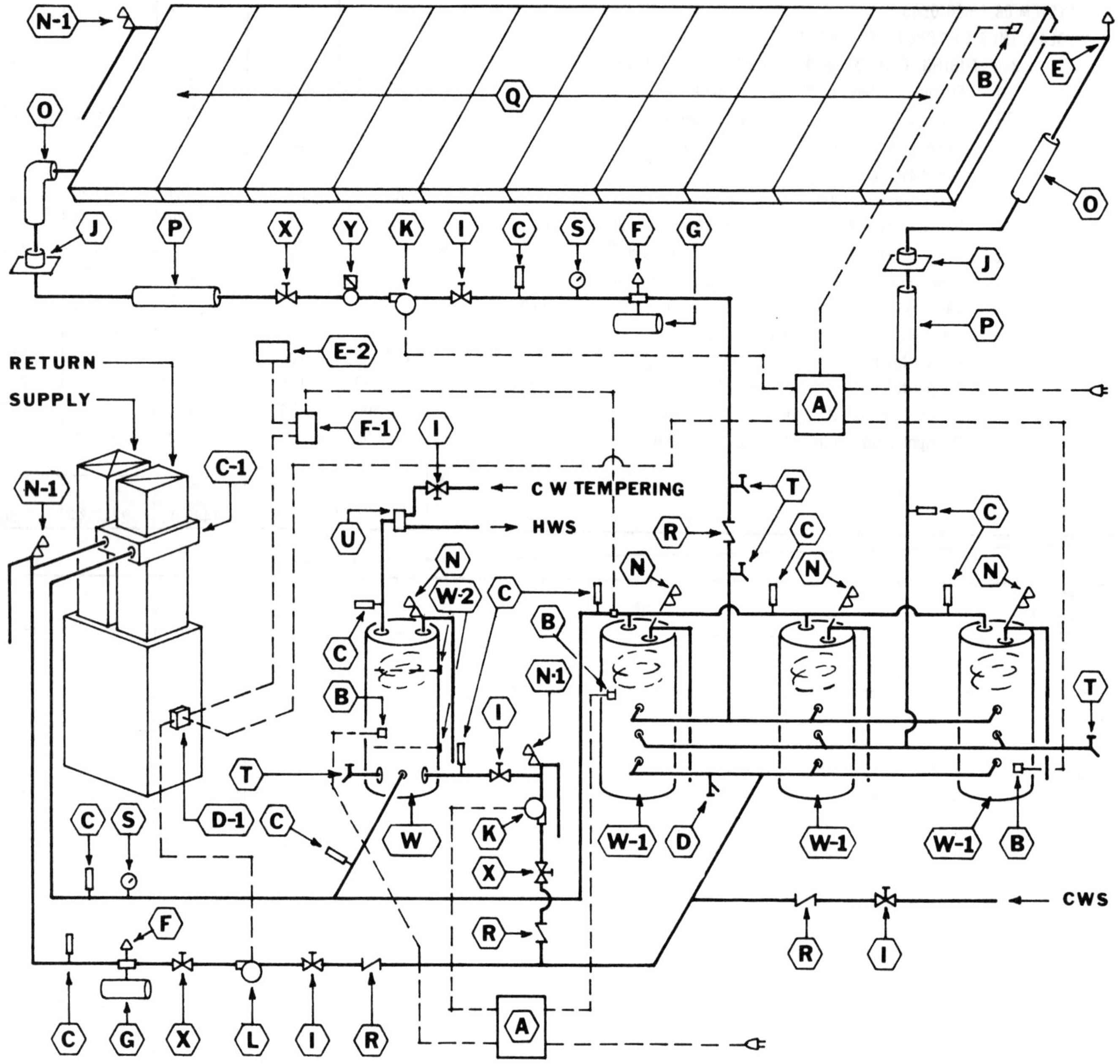

In this closed-loop indirect collection system, fluid with a low freezing temperature, such as propylene glycol, transports heat from the collectors to water storage. The transfer fluid is contained in a closed-loop consisting of collectors, supply and return piping, and a heat exchanger immersed in the storage tank.

When the collectors become approximately 20° F warmer than the storage temperature, the controller activates the circulator. The circulator moves the fluid continuously until the temperature difference between fluid in the collectors and storage is such that the collection will no longer occur and then the circulator turns off. Since the heat transfer fluid has a very low freezing temperature, there is no need for it to be drained from the collectors between periods of collection.

B8.3-710 Closed Loop, Space/Hot Water

System Components

System Components		QUANTITY	UNIT	COST EACH MAT.	COST EACH INST.	COST EACH TOTAL
SYSTEM 08.3-710-2750						
SOLAR, CLOSED LOOP, SPACE/HOT WATER						
1" TUBING, TEN 3'X7' BLACK CHROME ON COPPER ABSORBER COLLECTORS						
A, B	Differential controller 2 sensors, thermostat, solar energy system	2.000	Ea.	134.20	42	176
C	Thermometer, 2" dial	10.000	Ea.	142.50	157.50	300
C-1	Heat exchanger, solar energy system, fluid to air, up flow, 80 MBH	1.000	Ea.	346.50	153.50	500
D, T	Fill and drain valves, brass, 3/4" connection	5.000	Ea.	22.40	53.35	75.75
D-1	Fan center	1.000	Ea.	61.60	58.40	120
E	Air vent, manual, 1/8" diameter	1.000	Ea.	1.40	8	9.40
E-2	Thermostat, 2 stage for sensing room temperature	1.000	Ea.	70.40	34.60	105
F	Air purger	2.000	Ea.	21.90	42.10	64
F-1	Controller, liquid temperature, solar energy system	1.000	Ea.	57.20	52.80	110
G	Expansion tank	2.000	Ea.	61.60	16.40	78
I	Valve, gate, bronze, 125 lb, soldered, 1" diam	5.000	Ea.	90.20	69.80	160
J	Vent flashing, neoprene	2.000	Ea.	8.96	25.64	34.60
K	Circulator, solar heated liquid, 1/25 HP	2.000	Ea.	173.80	66.20	240
L	Circulator, solar heated liquid, 1/20 HP	1.000	Ea.	77	38	115
N	Relief valve, temp & pressure 150 psi 210°F self-closing	4.000	Ea.	30.60	34.20	64.80
N-1	Relief valve, pressure poppet, bronze, 30 psi, 3/4" IPS	3.000	Ea.	30.87	27.48	58.35
O	Pipe covering, urethane ultraviolet cover, 1" wall, 1" diam	50.000	L.F.	61.50	114	175.50
P	Pipe covering, fiberglass, all service jacket, 1" wall, 1" diam	60.000	L.F.	57	122.40	179.40
Q	Collector panel solar energy blk chrome/copr 1/8 temp glass 3'x7'	10.000	Ea.	3135	465	3600
	Roof clamps for solar energy collector panel	10.000	Set	33	66	99
R	Valve, swing check, bronze, 125 lb, regrinding disc, 3/4" & 1" diam	4.000	Ea.	71.72	52.28	124
S	Pressure gage, 0-60 psi	2.000	Ea.	10.90	16	26.90
U	Valve, water tempering, bronze, sweat connections, 3/4" diam	1.000	Ea.	22	13	35
W-2, W-1, W	Tank, waterstorage, immersed heat xchr elec elem 2"x1/2# insul 120 gal	4.000	Ea.	2552	728	3280
X	Valve, globe, bronze, 125 lb, rising stem, 1" diam	3.000	Ea.	99	42	141
Y	Valve, flow control, for solar energy system, 1" diam	1.000	Ea.	19.09	11.91	31
	Copper tubing type M 50/50 solder joint, hanger 10' OC, 1" diam	110.000	L.F.	152.90	760.10	913
	Copper tubing, type L, 50/50 solder joint, hanger 10' OC 3/4" diam	20.000	L.F.	24.40	116.60	141
	Wrought copper fittings & solder, 3/4" & 1" diam	121.000	Ea.	104.06	1940.84	2044.90
	Sensor, wire, #22-2 conductor, multistranded	.700	C.L.F.	6.16	17.64	23.80
	Ductwork, galvanized steel, for heat exchanger	8.000	Lb.	10.48	24.64	35.12
	Solar energy heat transfer, fluid, propylene glycol, anti-freeze	25.000	Gal.	159.50	229.25	388.75
	TOTAL			7842.47	5604.80	13,447.27

8

8.3-710	Solar, Closed Loop, Space/hot Water System	COST EACH MAT.	COST EACH INST.	COST EACH TOTAL
2540	Solar, closed loop, space/hot water			
2550	1/2" tubing, 12 ea. 4'x4'4" vacuum tube collectors	13,600	5050	18,650
2600	C8.3 -601 3/4" tubing, 12 ea. 4'x4'4" vacuum tube collectors	13,800	5300	19,100
2650	10 ea. 3' x 7' black chrome absorber collectors	7725	5150	12,875
2700	10 ea. 3' x 7' flat black absorber collectors	7450	5175	12,625
2750	1" tubing, 10 ea. 3' x 7' black chrome absorber collectors	7850	5600	13,450
2800	10 ea. 3' x 7' flat black absorber collectors	7575	5625	13,200
2850	6 ea. 4' x 9' flat black w/plastic glazing collectors	6575	5625	12,200
2900	12 ea. 2' x 9' plastic absorber and glazing collectors	9675	5725	15,400

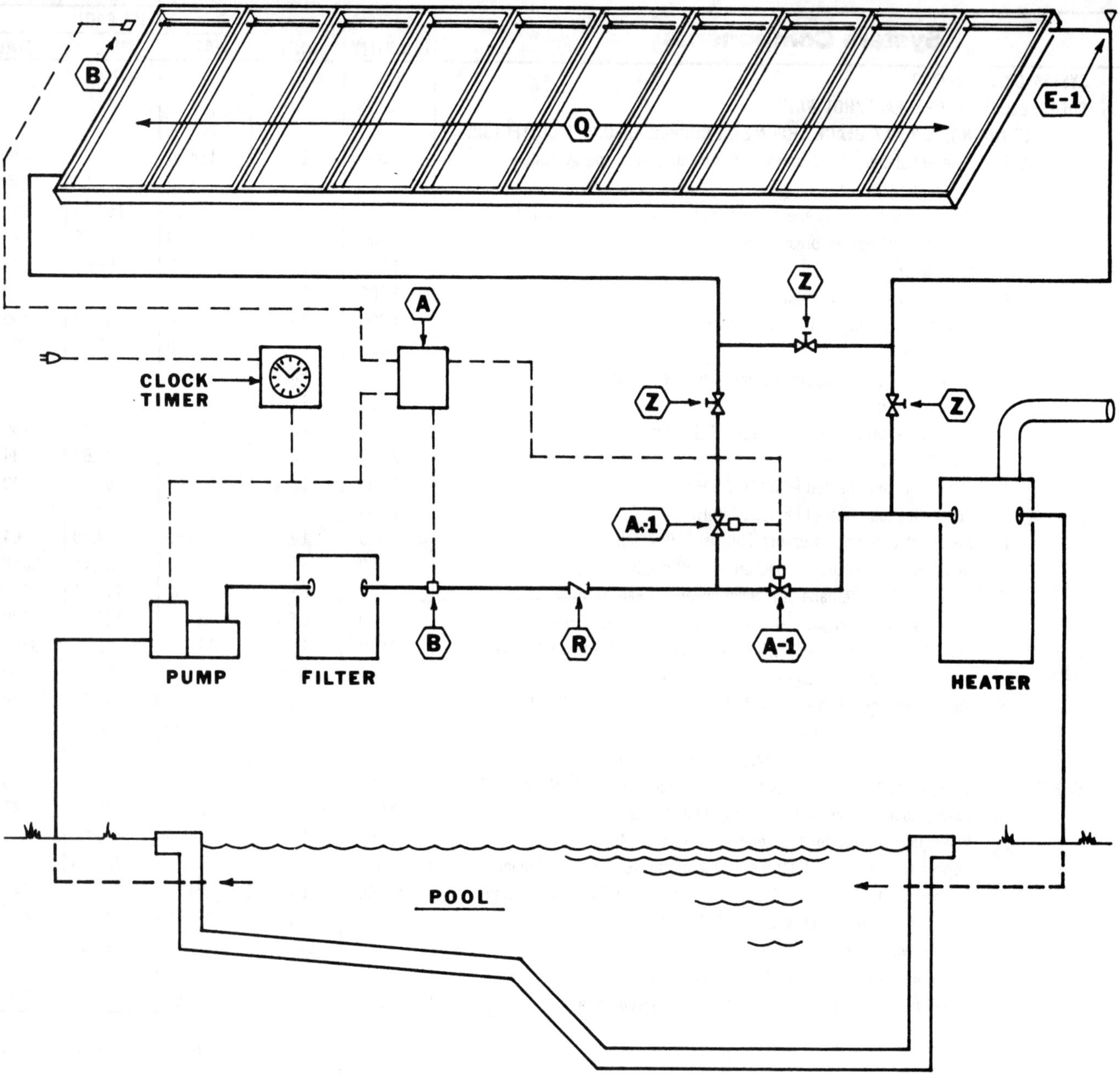

This draindown pool system uses a differential thermostat similar to those used in solar domestic hot water and space heating applications. To heat the pool, the pool water passes through the conventional pump-filter loop and then flows through the collectors. When collection is not possible, or when the desired pool temperature is reached, all water drains from the solar loop back to the pool through the existing piping. The modes are controlled by solenoid valves or other automatic valves in conjunction with a vacuum breaker relief valve, which facilitates draindown.

System Components	QUANTITY	UNIT	COST EACH MAT.	INST.	TOTAL
SYSTEM 08.3-720-2640					
SOLAR SWIMMING POOL HEATER, ROOF MOUNTED COLLECTORS					
TEN 4' X 10' FULLY WETTED UNGLAZED PLASTIC ABSORBERS					
A Differential thermostat/controller, 110V, adj pool pump system	1.000	Ea.	335.50	129.50	465
A-1 Solenoid valve, PVC, normally 1 open 1 closed, included	2.000	Ea.			
B Sensor, thermistor type, included	2.000	Ea.			
E-1 Valve, vacuum relief, for solar energy system	1.000	Ea.	15.40	7.60	23
Q Collector panel, solar energy, plastic, liquid full wetted, 4' x 10'	10.000	Ea.	1100	900	2000
R Valve, ball check, PVC, socket, 1-1/2" diam	1.000	Ea.	40.70	13.30	54
Z Valve, ball, PVC, socket, 1-1/2" diam	3.000	Ea.	89.10	39.90	129
Pipe, PVC, sch 40, 1-1/2" diam	80.000	L.F.	96.80	571.20	668
Pipe fittings, PVC sch 40, socket joint, 1-1/2" diam	10.000	Ea.	5.40	160.10	165.50
Sensor wire, #22-2 conductor, multistranded	.500	C.L.F.	4.40	12.60	17
Roof clamps for solar energy collector panels	10.000	Set	33	66	99
Roof strap, teflon for solar energy collector panels	26.000	L.F.	90.22	32.50	122.72
TOTAL			1810.52	1932.70	3743.22

8.3-720	Solar Swimming Pool Heater Systems	COST EACH MAT.	INST.	TOTAL
2530	Solar swimming pool heater systems, roof mounted collectors			
2540	10 ea. 3'x7' black chrome absorber, 1/8" temp. glass	3850	1500	5350
2560	C8.3 -601 10 ea. 4'x8' black chrome absorber, 3/16" temp. glass	7650	1800	9450
2580	10 ea. 3'8"x6' flat black absorber, 3/16" temp. glass	3575	1525	5100
2600	10 ea. 4'x9' flat black absorber, plastic glazing	3850	1900	5750
2620	10 ea. 2'x9' rubber absorber, plastic glazing	9000	2050	11,050
2640	10 ea. 4'x10' fully wetted unglazed plastic absorber	1800	1925	3725
2660	Ground mounted collectors			
2680	10 ea. 3'x7' black chrome absorber, 1/8" temp. glass	4175	1725	5900
2700	10 ea. 4'x8' black chrome absorber, 3/16" temp glass	7950	2050	10,000
2720	10 ea. 3'8"x6' flat blk absorber, 3/16" temp. glass	3900	1750	5650
2740	10 ea. 4'x9' flat blk absorber, plastic glazing	4175	2125	6300
2760	10 ea. 2'x9' rubber absorber, plastic glazing	9275	2225	11,500
2780	10 ea. 4'x10' fully wetted unglazed plastic absorber	2125	2175	4300

B8.3-730 Air To Water Heat Exchange

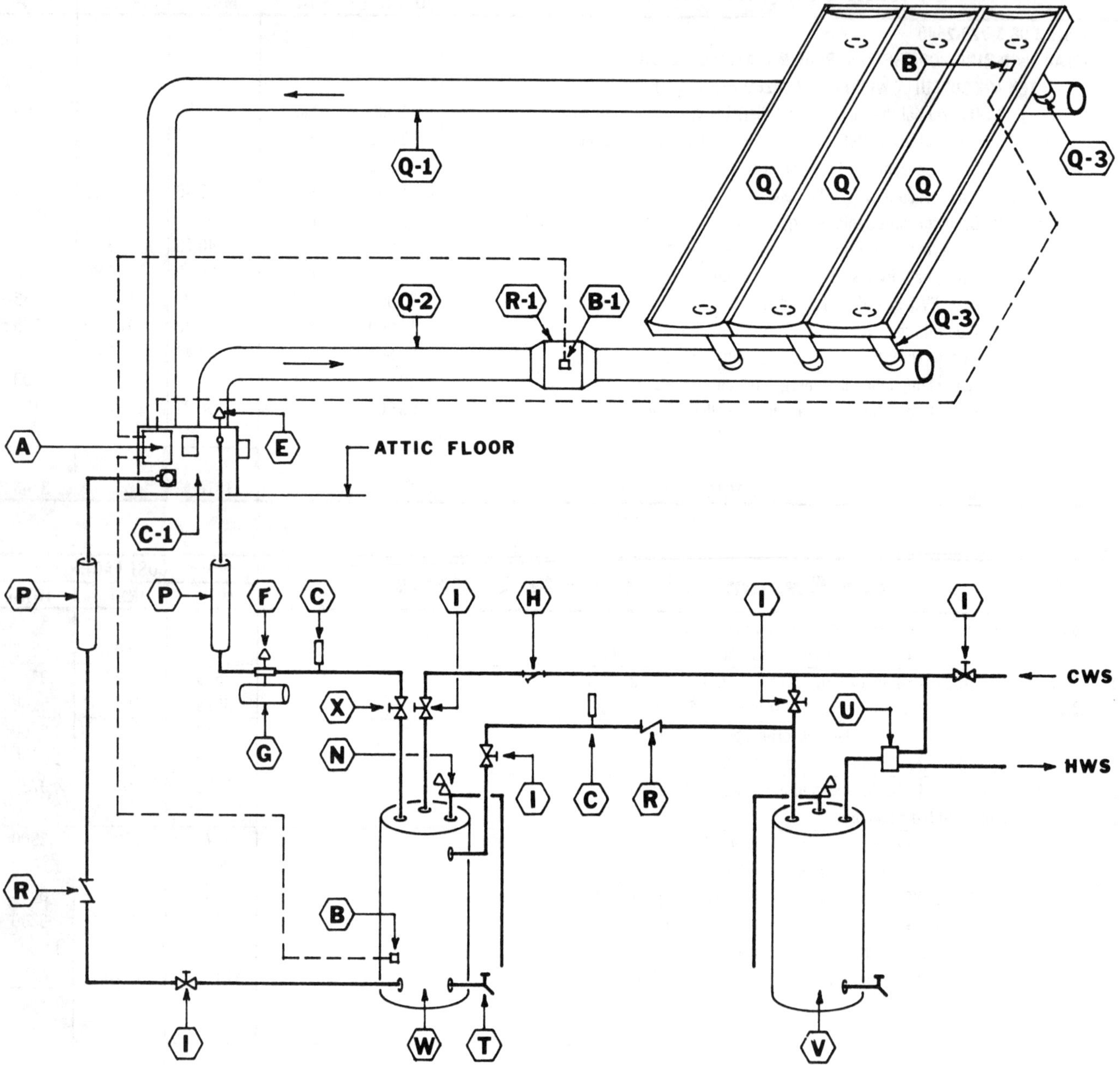

This domestic hot water pre-heat system includes heat exchanger with a circulating pump, blower, air-to-water, coil and controls, mounted in the upper collector manifold. Heat from the hot air coming out of the collectors is transferred through the heat exchanger. For each degree of DHW preheating gained, one degree less heating is needed from the fuel fired water heater. The system is simple, inexpensive to operate and can provide a substantial portion of DHW requirements for modest additional cost.

System Components		QUANTITY	UNIT	COST EACH MAT.	INST.	TOTAL
	SYSTEM 08.3-730-2560					
	SOLAR, HOT WATER, AIR TO WATER HEAT EXCHANGE					
	THREE COLLECTORS, OPTICAL BLACK ON ALUMINUM, 10'X2',80GAL TANK					
A, B	Differential controller, 2 sensors, thermostat, solar energy system	1.000	Ea.	67.10	20.90	88
C	Thermometer, 2" dial	2.000	Ea.	28.50	31.50	60
C-1	Heat exchanger, solar energy system, air to fluid, up flow 70 MBH	1.000	Ea.	236.50	133.50	370
E	Air vent, manual, 1/8" diameter	1.000	Ea.	1.40	8	9.40
F	Air purger	1.000	Ea.	10.95	21.05	32
G	Expansion tank	1.000	Ea.	30.80	8.20	39
H	Strainer,Y type, bronze body, 1/2" IPS	1.000	Ea.	17.16	12.84	30
I	Valve, gate, bronze, 125 lb, NRS, soldered 1/2" diam	5.000	Ea.	57.50	52.50	110
N	Relief valve, temp & pressure solar 150 psi 210°F self-closing	1.000	Ea.	7.65	8.55	16.20
P	Pipe covering, fiberglass, all service jacket, 1" wall, 1/2" diam	60.000	L.F.	49.80	117.60	167.40
Q	Collector panel solar energy, air, black on alum plate, flush mount, 10'x2'	30.000	L.F.	1056	144	1200
B-1, R-1	Shutter damper with shutter motor	1.000	Ea.	82.50	67.50	150
R	Backflow preventer for solar energy system, 1/2" diam	2.000	Ea.	70.40	31.60	102
T	Drain valve, brass, 3/4" connection	1.000	Ea.	4.48	10.67	15.15
U	Valve, water tempering, bronze, sweat connections, 1/2" diam	1.000	Ea.	19.09	10.91	30
W	Tank, water storage, solar energy system, 80 Gal, 2" x 2 lb insul	1.000	Ea.	456.50	158.50	615
V	Tank, water storage, w/heating element, drain, relief valve, existing	1.000	System			
X	Valve, globe, bronze, 125 lb, rising stem, 1/2" diam	1.000	Ea.	16.12	10.88	27
	Copper tubing type M, 50/50 solder jnt, hngr 10' OC 1/2" diam	50.000	L.F.	33.50	229	262.50
	Copper tubing type L, 50/50 solder jnt, hngr 10' OC 1/2" diam	10.000	L.F.	8.40	48.60	57
	Wrought copper fittings & solder, 1/2" diam	10.000	Ea.	1.40	128.10	129.50
	Sensor wire, #22-2 conductor, multistranded	.500	C.L.F.	4.40	12.60	17
Q-1, Q-2	Ductwork, fiberglass, aluminized jacket, 1-1/2" thick, 8" diam	32.000	L.F.	44.48	82.56	127.04
Q-3	Manifold for flush mount solar energy collector panels, air	6.000	L.F.	72.60	17.40	90
	TOTAL			2388.18	1374.96	3763.14

8.3-730	Solar Hot Water, Air To Water Heat Exchange Systems	COST EACH MAT.	INST.	TOTAL
2550	Solar hot water, air to water heat exchange			
2560	Three collectors, optical black on aluminum, 10'x2', 80 Gal tank	2400	1375	3775
2580	C8.3 -601 Four collectors, optical black on aluminum, 10'x2', 80 Gal tank	2775	1450	4225
2600	Four collectors, optical black on aluminum, 10'x2', 120 Gal tank	2975	1475	4450

8

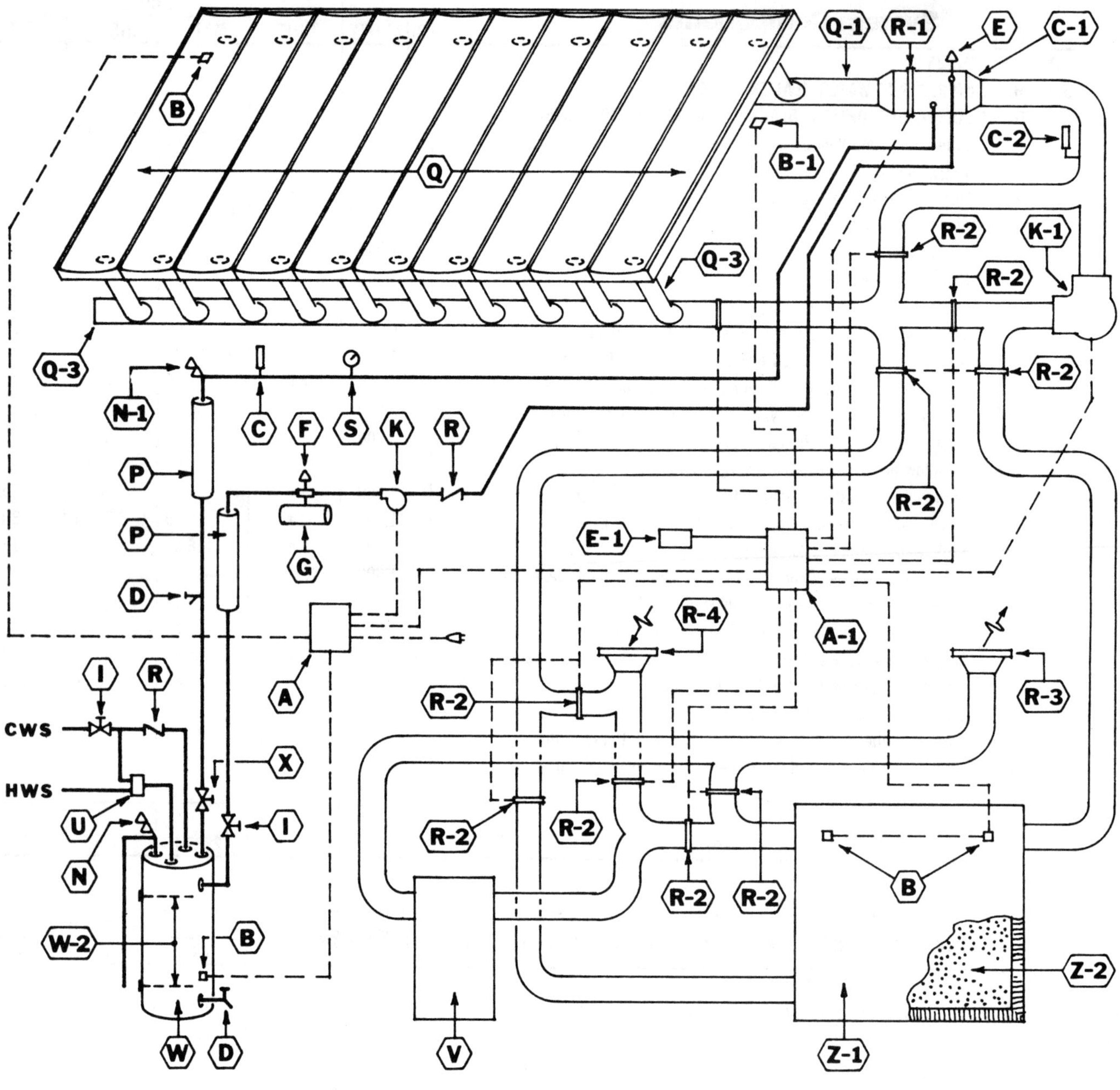

The complete Solar Air Heating System provides maximum savings of conventional fuel with both space heating and year-round domestic hot water heating. It allows for the home air conditioning to operate simultaneously and independently from the solar domestic water heating in summer. The system's modes of operation are:

MODE 1: The building is heated directly from the collectors with air circulated by the Solar Air Mover.

MODE 2: When heat is not needed in the building, the dampers change within the air mover to circulate the air from the collectors to the rock storage bin.

MODE 3: When heat is not available from the collector array and is available in rock storage, the air mover draws heated air from rock storage and directs it into the building. When heat is not available from the collectors or the rock storage bin, the auxiliary heating unit will provide heat for the building. The size of the collector array is typically 25% the size of the main floor area.

System Components		QUANTITY	UNIT	COST EACH MAT.	COST EACH INST.	COST EACH TOTAL
SOLAR, SPACE/HOT WATER, AIR TO WATER HEAT EXCHANGE						
A, B	Differential controller 2 sensors thermostat, solar energy sys liquid loop	1.000	Ea.	67.10	20.90	88
A-1, B	Differential controller 2 sensors 6 station solar energy sys air loop	1.000	Ea.	253	87	340
B-1	Solar energy sensor, freeze prevention	1.000	Ea.	10.95	8	18.95
C	Thermometer for solar energy system, 2" dial	2.000	Ea.	28.50	31.50	60
C-1	Heat exchanger, solar energy system, air to fluid, up flow, 70 MBH	1.000	Ea.	236.50	133.50	370
D	Drain valve, brass, 3/4" connection	2.000	Ea.	8.96	21.34	30.30
E	Air vent, manual, for solar energy system 1/8" fitting	1.000	Ea.	1.40	8	9.40
E-1	Thermostat, 2 stage for sensing room temperature	1.000	Ea.	70.40	34.60	105
F	Air purger	1.000	Ea.	10.95	21.05	32
G	Expansion tank, for solar energy system	1.000	Ea.	30.80	8.20	39
I	Valve, gate, bronze, 125 lb, NRS, soldered 3/4" diam	2.000	Ea.	29.82	26.18	56
K	Circulator, solar heated liquid, 1/25 HP	1.000	Ea.	86.90	33.10	120
N	Relief valve temp & press 150 psi 210°F self-closing, 3/4" IPS	1.000	Ea.	7.65	8.55	16.20
N-1	Relief valve, pressure, poppet, bronze, 30 psi, 3/4" IPS	1.000	Ea.	10.29	9.16	19.45
P	Pipe covering, fiberglass, all service jacket, 1" wall, 3/4" diam	60.000	L.F.	57	122.40	179.40
Q-3	Manifold for flush mount solar energy collector panels	20.000	L.F.	242	58	300
Q	Collector panel solar energy, air, black on alum. plate, flush mount 10'x2'	100.000	L.F.	3520	480	4000
R	Valve, swing check, bronze, 125 lb, regrinding disc, 3/4" diam	2.000	Ea.	27.40	26.60	54
S	Pressure gage, 2" dial, for solar energy system	1.000	Ea.	5.45	8	13.45
U	Valve, water tempering, bronze, sweat connections, 3/4" diam	1.000	Ea.	22	13	35
W-2, W	Tank, water storage, solar, elec element 2"x1/2# insul, 80 Gal	1.000	Ea.	423.50	161.50	585
X	Valve, globe, bronze, 125 lb, soldered, 3/4" diam	1.000	Ea.	20.24	12.76	33
	Copper tubing type L, 50/50 solder jnt, hngr 10' OC 3/4" diam	10.000	L.F.	12.20	58.30	70.50
	Copper tubing type M, 50/50 solder jnts hngr 10' OC 3/4" diam	60.000	L.F.	57.60	335.40	393
	Wrought copper fittings & solder, 3/4" diam	26.000	Ea.	8.58	351.52	360.10
	Sensor wire, #22-2 conductor, multistranded	1.200	C.L.F.	10.56	30.24	40.80
Q-1	Duct work, rigid fiberglass, rectangular	400.000	S.F.	188	828	1016
	Ductwork, transition pieces	8.000	Ea.	111.76	152.24	264
R-2	Shutter/damper for solar heater circulator	9.000	Ea.	227.70	348.30	576
K-1	Shutter motor for solar heater circulator blower	9.000	Ea.	514.80	259.20	774
K-1	Fan, solar energy heated air circulator, space & DHW system	1.000	Ea.	2200	925	3125
Z-1	Tank, solar energy air storage, 6'-3"H 7'x7' 306 CF/2000 Gal	1.000	Ea.	4427.50	597.50	5025
Z-2	Crushed stone 1-1/2"	11.000	C.Y.	84.70	36.30	121
C-2	Themometer, remote probe, 2" dial	1.000	Ea.	25.30	31.70	57
R-1	Solenoid valve	1.000	Ea.	49.50	28.50	78
V, R-4, R-3	Furnace, supply diffusers, return grilles, existing	1.000	Ea.			
	TOTAL			13,004.31	5279.24	18,283.55

	C8.3 -601			

8

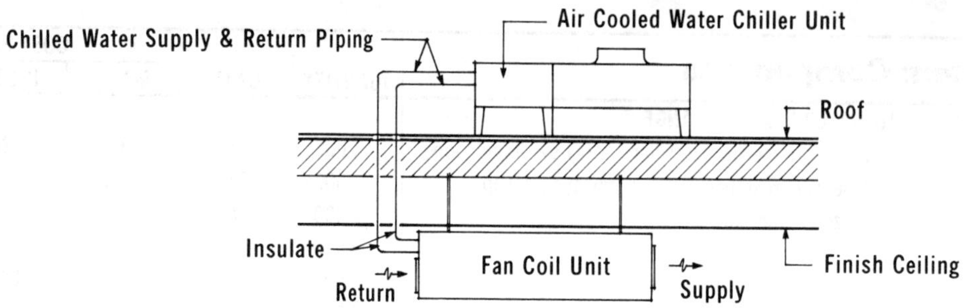

Design Assumptions: The chilled water, air cooled systems priced, utilize reciprocating hermetic compressors and propeller-type condenser fans. Piping with pumps and expansion tanks is included based on a two pipe system. No ducting is included and the fan-coil units are cooling only. (See C8.4-001 for methods of adding heat). Water treatment and balancing are not included. Chilled water piping is insulated. Area distribution is through the use of multiple fan coil units. Fewer but larger fan coil units with duct distribution would be approximately the same S.F. cost.

System Components	QUANTITY	UNIT	COST EACH MAT.	INST.	TOTAL
SYSTEMS 08.4-110-1200					
PACKAGED CHILLER, AIR COOLED, WITH FAN COIL UNIT					
APARTMENT CORRIDORS, 3,000 S.F., 5.50 TON					
Fan coil air cond unit, cabinet mntd & filters, chilled water	1.000	Ea.	2974.04	233.71	3207.75
Water chiller,air conditioning unit, reciprocating, air cooled,	1.000	Ea.	3206.50	781	3987.50
Chilled water unit coil connections	1.000	Ea.	430.78	569.22	1000
Chilled water distribution piping	440.000	L.F.	4782.80	7977.20	12,760
TOTAL			11,394.12	9561.13	20,955.25
COST PER S.F.			3.80	3.19	6.99
*Cooling requirements would lead to choosing a water cooled unit					

8.4-110	Chilled Water, Air Cooled Condenser	COST PER S.F. MAT.	INST.	TOTAL
1180	Packaged chiller, air cooled, with fan coil unit			
1200	Apartment corridors, 3,000 S.F., 5.50 ton	3.80	3.19	6.99
1240	6,000 S.F., 11.00 ton	2.80	2.55	5.35
1280	10,000 S.F., 18.33 ton	2.39	1.93	4.32
1320	20,000 S.F., 36.66 ton	2.03	1.44	3.47
1360	40,000 S.F., 73.33 ton	3.14	1.48	4.62
1440	Banks and libraries, 3,000 S.F., 12.50 ton	5.10	3.73	8.83
1480	6,000 S.F., 25.00 ton	4.66	3.03	7.69
1520	10,000 S.F., 41.66 ton	4.01	2.21	6.22
1560	20,000 S.F., 83.33 ton	5	2.11	7.11
1600	40,000 S.F., 167 ton*			
1680	Bars and taverns, 3,000 S.F., 33.25 ton	9.85	4.65	14.50
1720	6,000 S.F., 66.50 ton	8.40	4.02	12.42
1760	10,000 S.F., 110.83 ton	6.75	1.80	8.55
1800	20,000 S.F., 220 ton*			
1840	40,000 S.F., 440 ton*			
1920	Bowling alleys, 3,000 S.F., 17.00 ton	6.35	4.18	10.53
1960	6,000 S.F., 34.00 ton	5.45	3.07	8.52
2000	10,000 S.F., 56.66 ton	4.50	2.45	6.95
2040	20,000 S.F., 113.33 ton	5.60	2.21	7.81
2080	40,000 S.F., 227 ton*			
2160	Department stores, 3,000 S.F., 8.75 ton	5.15	3.35	8.50
2200	6,000 S.F., 17.50 ton	3.60	2.74	6.34
2240	10,000 S.F., 29.17 ton	3.12	1.98	5.10

8.4-110	Chilled Water, Air Cooled Condenser	COST PER S.F. MAT.	COST PER S.F. INST.	COST PER S.F. TOTAL
2280	20,000 S.F., 58.33 ton	2.56	1.62	4.18
2320	40,000 S.F., 116.66 ton	3.67	1.64	5.31
2400	Drug stores, 3,000 S.F., 20.00 ton	7.10	4.13	11.23
2440	6,000 S.F., 40.00 ton	6.35	3.26	9.61
2480	10,000 S.F., 66.66 ton	7.50	3.07	10.57
2520	20,000 S.F., 133.33 ton	6.30	2.50	8.80
2560	40,000 S.F., 267 ton*			
2640	Factories, 2,000 S.F., 10.00 ton	4.34	3.40	7.74
2680	6,000 S.F., 20.00 ton	3.98	2.79	6.77
2720	10,000 S.F., 33.33 ton	3.45	2.09	5.54
2760	20,000 S.F., 66.66 ton	4.49	1.97	6.46
2800	40,000 S.F., 133.33 ton	3.92	1.71	5.63
2880	Food supermarkets, 3,000 S.F., 8.50 ton	5.05	3.34	8.39
2920	6,000 S.F., 17.00 ton	3.53	2.72	6.25
2960	10,000 S.F., 28.33 ton	3.02	1.98	5
3000	20,000 S.F., 56.66 ton	2.47	1.59	4.06
3040	40,000 S.F., 113.33 ton	3.66	1.61	5.27
3120	Medical centers, 3,000 S.F., 7.00 ton	4.51	3.24	7.75
3160	6,000 S.F., 14.00 ton	3.17	2.63	5.80
3200	10,000 S.F., 23.33 ton	2.80	2.01	4.81
3240	20,000 S.F., 46.66 ton	2.44	1.52	3.96
3280	40,000 S.F., 93.33 ton	3.43	1.54	4.97
3360	Offices, 3,000 S.F., 9.50 ton	4.25	3.30	7.55
3400	6,000 S.F., 19.00 ton	3.93	2.86	6.79
3440	10,000 S.F., 31.66 ton	3.33	2.07	5.40
3480	20,000 S.F., 63.33 ton	4.48	2	6.48
3520	40,000 S.F., 126.66 ton	3.87	1.71	5.58
3600	Restaurants, 3,000 S.F., 15.00 ton	5.70	3.88	9.58
3640	6,000 S.F., 30.00 ton	5.10	2.96	8.06
3680	10,000 S.F., 50.00 ton	4.24	2.39	6.63
3720	20,000 S.F., 100.00 ton	5.65	2.26	7.91
3760	40,000 S.F., 200 ton*			
3840	Schools and colleges, 3,000 S.F., 11.50 ton	4.85	3.67	8.52
3880	6,000 S.F., 23.00 ton	4.42	2.97	7.39
3920	10,000 S.F., 38.33 ton	3.79	2.16	5.95
3960	20,000 S.F., 76.66 ton	4.91	2.11	7.02
4000	40,000 S.F., 153 ton*			

B8.4-120 Chilled Water, Water Cooled

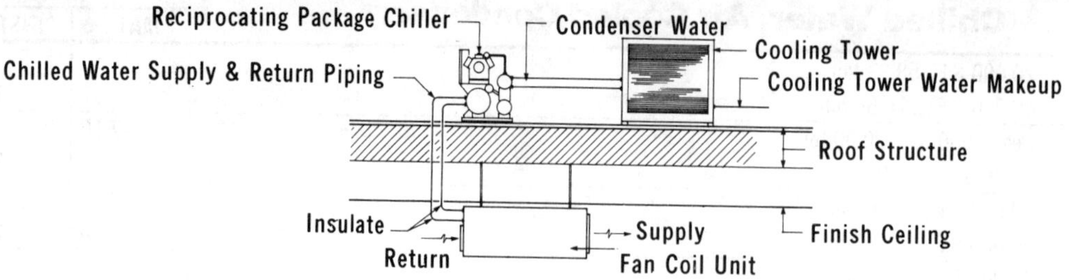

General: Water cooled chillers are available in the same sizes as air cooled units. They are also available in larger capacities.

Design Assumptions: The chilled water systems with water cooled condenser, include reciprocating hermetic compressors, water cooling tower, pumps, piping and expansion tanks and are based on a two pipe system. Chilled water piping is insulated. No ducts are included and fan-coil units are cooling only. Area distribution is through use of multiple fan coil units. Fewer but larger fan coil units with duct distribution would be approximately the same S.F. Cost (See C8.4-001 for methods of adding heat). Water treatment and balancing are not included.

System Components	QUANTITY	UNIT	COST EACH MAT.	INST.	TOTAL
SYSTEM 08.4-120-1320					
PACKAGED CHILLER, WATER COOLED, WITH FAN COIL UNIT					
APARTMENT CORRIDORS, 4,000 S.F., 7.33 TON					
Fan coil air cond unit, cabinet mntd & filters, chilled water	2.000	Ea.	3963.77	311.48	4275.25
Water chiller, reciprocating water cooled, 1 compressor semihermetic	1.000	Ea.	6146.36	1570.94	7717.30
Cooling tower, draw thru single flow, belt drive	1.000	Ea.	467.65	338.65	806.30
Cooling tower pumps & piping	1.000	System	257.21	138.61	395.82
Chilled water unit coil connections	2.000	Ea.	861.56	1138.44	2000
Chilled water distribution piping	520.000	L.F.	5652.40	9427.60	15,080
TOTAL			17,348.95	12,925.72	30,274.67
COST PER S.F.			4.34	3.23	7.57

*Cooling requirements would lead to choosing multiple chillers.

8.4-120	Chilled Water, Cooling Tower	COST PER S.F. MAT.	INST.	TOTAL
1300	Packaged chiller, water cooled, with fan coil unit			
1320	Apartment corridors, 4,000 S.F., 7.33 ton	4.34	3.23	7.57
1360	6,000 S.F., 11.00 ton	3.41	2.75	6.16
1400	10,000 S.F., 18.33 ton	2.55	2.09	4.64
1440	20,000 S.F., 26.66 ton	2.13	1.59	3.72
1480	40,000 S.F., 73.33 ton	3.40	1.63	5.03
1520	60,000 S.F., 110.00 ton	3.43	1.69	5.12
1600	Banks and libraries, 4,000 S.F., 16.66 ton	5.40	3.70	9.10
1640	6,000 S.F., 25.00 ton	4.75	3.22	7.97
1680	10,000 S.F., 41.66 ton	3.84	2.48	6.32
1720	20,000 S.F., 83.33 ton	5.55	2.44	7.99
1760	40,000 S.F., 166.66 ton	5.15	2.91	8.06
1800	60,000 S.F., 250.00 ton	5.15	3.15	8.30
1880	Bars and taverns, 4,000 S.F., 44.33 ton	9.05	4.95	14
1920	6,000 S.F., 66.50 ton	11.40	5.05	16.45
1960	10,000 S.F., 110.83 ton	11	4.42	15.42
2000	20,000 S.F., 221.66 ton	9.85	4.72	14.57
2040	40,000 S.F., 440 ton*			
2080	60,000 S.F., 660 ton*			
2160	Bowling alleys, 4,000 S.F., 22.66 ton	6.35	4.06	10.41
2200	6,000 S.F., 34.00 ton	5.85	3.62	9.47
2240	10,000 S.F., 56.66 ton	4.73	2.77	7.50
2280	20,000 S.F., 113.33 ton	6.35	2.70	9.05
2320	40,000 S.F., 226.66 ton	5.85	3.16	9.01

8

8.4-120	Chilled Water, Cooling Tower	COST PER S.F.		
		MAT.	INST.	TOTAL
2360	60,000 S.F., 340 ton*			
2440	Department stores, 4,000 S.F., 11.66 ton	4.74	3.48	8.22
2480	6,000 S.F., 17.50 ton	4	2.99	6.99
2520	10,000 S.F., 29.17 ton	3.19	2.24	5.43
2560	20,000 S.F., 58.33 ton	2.47	1.72	4.19
2600	40,000 S.F., 116.66 ton	3.98	1.82	5.80
2640	60,000 S.F., 175.00 ton	4.40	2.79	7.19
2720	Drug stores, 4,000 S.F., 26.66 ton	7	4.20	11.20
2760	6,000 S.F., 40.00 ton	6.10	3.70	9.80
2800	10,000 S.F., 66.66 ton	7.40	3.36	10.76
2840	20,000 S.F., 133.33 ton	7	2.94	9.94
2880	40,000 S.F., 266.67 ton	6.65	3.62	10.27
2920	60,000 S.F., 400 ton*			
3000	Factories, 4,000 S.F., 13.33 ton	4.66	3.48	8.14
3040	6,000 S.F., 20.00 ton	4.11	2.99	7.10
3080	10,000 S.F., 33.33 ton	3.49	2.33	5.82
3120	20,000 S.F., 66.66 ton	4.40	2.09	6.49
3160	40,000 S.F., 133.33 ton	4.29	1.93	6.22
3200	60,000 S.F., 200.00 ton	4.68	2.93	7.61
3280	Food supermarkets, 4,000 S.F., 11.33 ton	4.65	3.46	8.11
3320	6,000 S.F., 17.00 ton	3.64	2.89	6.53
3360	10,000 S.F., 28.33 ton	3.13	2.22	5.35
3400	20,000 S.F., 56.66 ton	2.54	1.73	4.27
3440	40,000 S.F., 113.33 ton	3.95	1.81	5.76
3480	60,000 S.F., 170.00 ton	4.37	2.78	7.15
3560	Medical centers, 4.000 S.F., 9.33 ton	4.03	3.17	7.20
3600	6,000 S.F., 14.00 ton	3.49	2.84	6.33
3640	10,000 S.F., 23.33 ton	2.86	2.11	4.97
3680	20,000 S.F., 46.66 ton	2.28	1.65	3.93
3720	40,000 S.F., 93.33 ton	3.65	1.73	5.38
3760	60,000 S.F., 140.00 ton	4.08	2.71	6.79
3840	Offices, 4,000 S.F., 12.66 ton	4.51	3.44	7.95
3880	6,000 S.F., 19.00 ton	4.05	3.05	7.10
3920	10,000 S.F., 31.66 ton	3.46	2.34	5.80
3960	20,000 S.F., 63.33 ton	4.35	2.09	6.44
4000	40,000 S.F., 126.66 ton	4.46	2.71	7.17
4040	60,000 S.F., 190.00 ton	4.54	2.88	7.42
4120	Restaurants, 4,000 S.F., 20.00 ton	5.75	3.76	9.51
4160	6,000 S.F., 30.00 ton	5.30	3.38	8.68
4200	10,000 S.F., 50.00 ton	4.37	2.64	7.01
4240	20,000 S.F., 100.00 ton	6.05	2.62	8.67
4280	40,000 S.F., 200.00 ton	5.30	2.97	8.27
4320	60,000 S.F., 300.00 ton	5.70	3.34	9.04
4400	Schools and colleges, 4,000 S.F., 15.33 ton	5.10	3.62	8.72
4440	6,000 S.F., 23.00 ton	4.50	3.15	7.65
4480	10,000 S.F., 38.33 ton	3.63	2.41	6.04
4520	20,000 S.F., 76.66 ton	5.35	2.38	7.73
4560	40,000 S.F., 153.33 ton	4.88	2.82	7.70
4600	60,000 S.F., 230.00 ton	4.83	3.01	7.84

Rooftop Single Zone

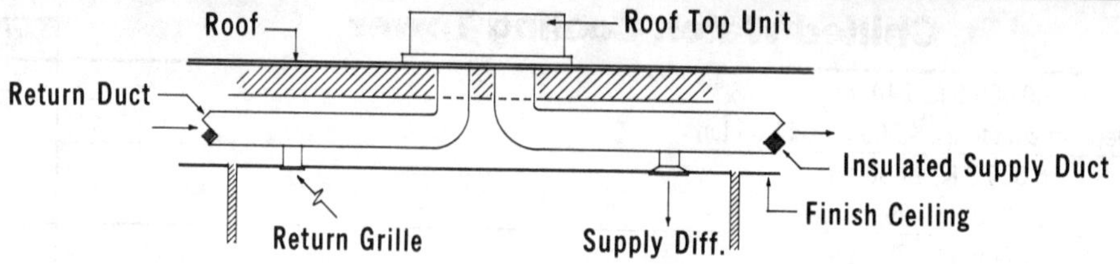

System Description: Rooftop single zone units are electric cooling and gas heat. Duct systems are low velocity, galvanized steel supply and return. Price variations between sizes are due to several factors. Jumps in the cost of the rooftop unit occur when the manufacturer shifts from the largest capacity unit on a small frame to the smallest capacity on the next larger frame, or changes from one compressor to two. As the unit capacity increases for larger areas the duct distribution grows in proportion. For most applications there is a tradeoff point where it is less expensive and more efficient to utilize smaller units with short simple distribution systems. Larger units also require larger initial supply and return ducts which can create a space problem. Supplemental heat may be desired in colder locations (See C8.4-001).

The table below is based on one unit supplying the area listed. The 10,000 S.F. unit for bars and taverns is not listed because a nominal 110 ton unit would be required and this is above the normal single zone rooftop capacity.

System Components	QUANTITY	UNIT	COST EACH MAT.	COST EACH INST.	COST EACH TOTAL
SYSTEMS 08.4-210-1280					
ROOFTOP, SINGLE ZONE, AIR CONDITIONER					
APARTMENT CORRIDORS, 500 S.F., .92 TON					
Rooftop air-conditioner, 1 zone, elec cool, standard controls, curb	1.000	Ea.	1049.95	146.05	1196
Ductwork package for rooftop single zone units	1.000	System	286.08	445.32	731.40
TOTAL			1336.03	591.37	1927.40
COST PER S.F.			2.67	1.18	3.85

8.4-210	Rooftop Single Zone Units	COST PER S.F. MAT.	COST PER S.F. INST.	COST PER S.F. TOTAL
1260	Rooftop, single zone, air conditioner			
1280	Apartment corridors, 500 S.F., .92 ton	2.67	1.18	3.85
1320	1,000 S.F., 1.83 ton	2.66	1.18	3.84
1360	1500 S.F., 2.75 ton	2.16	1.07	3.23
1400	3,000 S.F., 5.50 ton	1.92	1.20	3.12
1440	5,000 S.F., 9.17 ton	1.88	1.18	3.06
1480	10,000 S.F., 18.33 ton	1.80	1.17	2.97
1560	Banks or libraries, 500 S.F., 2.08 ton	6.05	2.67	8.72
1600	1,000 S.F., 4.17 ton	4.91	2.45	7.36
1640	1,500 S.F., 6.25 ton	4.37	2.72	7.09
1680	3,000 S.F., 12.50 ton	4.26	2.67	6.93
1720	5,000 S.F., 20.80 ton	4.08	2.65	6.73
1760	10,000 S.F., 41.67 ton	4.06	2.66	6.72
1840	Bars and taverns, 500 S.F. 5.54 ton	9.90	4.54	14.44
1880	1,000 S.F., 11.08 ton	9.60	4.43	14.03
1920	1,500 S.F., 16.62 ton	8.75	4.39	13.14
1960	3,000 S.F., 33.25 ton	8.95	4.35	13.30
2000	5,000 S.F., 55.42 ton	8.60	4.36	12.96
2040	10,000 S.F., 110.83 ton			
2080	Bowling alleys, 500 S.F., 2.83 ton	6.65	3.32	9.97
2120	1,000 S.F., 5.67 ton	5.95	3.70	9.65
2160	1,500 S.F., 8.50 ton	5.80	3.64	9.44
2200	3,000 S.F., 17.00 ton	5.35	3.62	8.97
2240	5,000 S.F., 28.33 ton	5.45	3.60	9.05
2280	10,000 S.F., 56.67 ton	5.30	3.60	8.90
2360	Department stores, 500 S.F., 1.46 ton	4.24	1.88	6.12
2400	1,000 S.F., 2.92 ton	3.43	1.71	5.14
2440	1,500 S.F., 4.37 ton	3.06	1.90	4.96

8.4-210	Rooftop Single Zone Units	COST PER S.F.		
		MAT.	INST.	TOTAL
2480	3,000 S.F., 8.75 ton	2.98	1.87	4.85
2520	5,000 S.F., 14.58 ton	2.75	1.86	4.61
2560	10,000 S.F., 29.17 ton	2.81	1.85	4.66
2640	Drug stores, 500 S.F., 3.33 ton	7.85	3.91	11.76
2680	1,000 S.F., 6.67 ton	7	4.35	11.35
2720	1,500 S.F., 10.00 ton	6.80	4.28	11.08
2760	3,000 S.F., 20.00 ton	6.55	4.25	10.80
2800	5,000 S.F., 33.33 ton	6.40	4.23	10.63
2840	10,000 S.F., 66.67 ton	6.20	4.24	10.44
2920	Factories, 500 S.F., 1.67 ton	4.83	2.14	6.97
2960	1,000 S.F., 3.33 ton	3.92	1.95	5.87
3000	1,500 S.F., 5.00 ton	3.49	2.17	5.66
3040	3,000 S.F., 10.00 ton	3.41	2.14	5.55
3080	5,000 S.F., 16.67 ton	3.14	2.13	5.27
3120	10,000 S.F., 33.33 ton	3.21	2.12	5.33
3200	Food supermarkets, 500 S.F., 1.42 ton	4.12	1.82	5.94
3240	1,000 S.F., 2.83 ton	3.31	1.66	4.97
3280	1,500 S.F., 4.25 ton	2.97	1.85	4.82
3320	3,000 S.F., 8.50 ton	2.90	1.82	4.72
3360	5,000 S.F., 14.17 ton	2.67	1.81	4.48
3400	10,000 S.F., 28.33 ton	2.73	1.80	4.53
3480	Medical centers, 500 S.F., 1.17 ton	3.39	1.50	4.89
3520	1,000 S.F., 2.33 ton	3.39	1.50	4.89
3560	1,500 S.F., 3.50 ton	2.75	1.37	4.12
3600	3,000 S.F., 7.00 ton	2.45	1.52	3.97
3640	5,000 S.F., 11.67 ton	2.39	1.50	3.89
3680	10,000 S.F., 23.33 ton	2.29	1.49	3.78
3760	Offices, 500 S.F., 1.58 ton	4.60	2.03	6.63
3800	1,000 S.F., 3.17 ton	3.73	1.86	5.59
3840	1,500 S.F., 4.75 ton	3.32	2.06	5.38
3880	3,000 S.F., 9.50 ton	3.24	2.03	5.27
3920	5,000 S.F., 15.83 ton	2.99	2.02	5.01
3960	10,000 S.F., 31.67 ton	3.05	2.01	5.06
4000	Restaurants, 500 S.F., 2.50 ton	7.25	3.21	10.46
4040	1,000 S.F., 5.00 ton	5.25	3.26	8.51
4080	1,500 S.F., 7.50 ton	5.10	3.21	8.31
4120	3,000 S.F., 15.00 ton	4.72	3.19	7.91
4160	5,000 S.F., 25.00 ton	4.91	3.19	8.10
4200	10,000 S.F., 50.00 ton	4.67	3.18	7.85
4240	Schools and colleges, 500 S.F., 1.92 ton	5.55	2.47	8.02
4280	1,000 S.F., 3.83 ton	4.51	2.25	6.76
4320	1,500 S.F., 5.75 ton	4.02	2.50	6.52
4360	3,000 S.F., 11.50 ton	3.92	2.46	6.38
4400	5,000 S.F., 19.17 ton	3.76	2.45	6.21
4440	10,000 S.F., 38.33 ton	3.73	2.45	6.18
9000	Components of ductwork packages for above systems, per ton of cooling			
9010	Ductwork; galvanized steel, 120 pounds			
9020	Insulation; fiberglass 2" thick, FRK faced, 52 SF			
9030	Diffusers; aluminum, 24" x 12", one			
9040	Registers; aluminum, return with dampers, one			

8

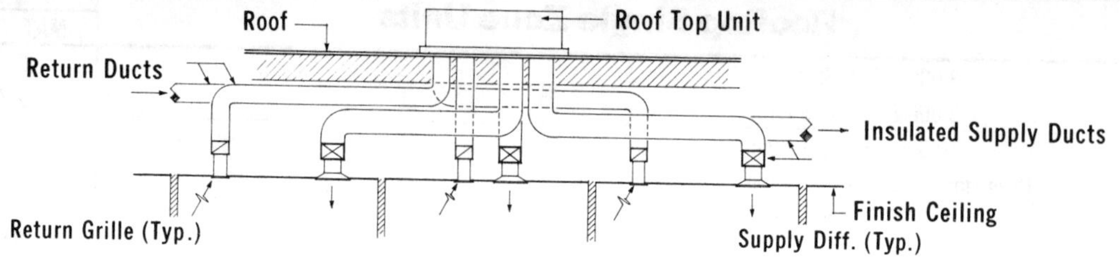

System Description: Rooftop units are multizone with up to 12 zones, and include electric cooling, gas heat, thermostats, filters, supply and return fans complete. Duct systems are low velocity, galvanized steel supply and return with insulated supplies.

Multizone units cost more per ton of cooling than single zone, however, they offer flexibility where load conditions are varied due to heat generating areas or exposure to radiational heating. For example, perimeter offices on the "sunny side" may require cooling at the same time "shady side" or central offices may require heating. It is possible to accomplish similar results using duct heaters in branches of the single zone unit. However, heater location could be a problem and total system operating energy efficiency could be lower.

System Components	QUANTITY	UNIT	COST EACH MAT.	COST EACH INST.	COST EACH TOTAL
SYSTEMS 08.4-220-1280					
ROOFTOP, MULTIZONE, AIR CONDITIONER					
APARTMENT CORRIDORS, 3,000 S.F., 5.50 TON					
Rooftop multizone unit, standard controls, curb	1.000	Ea.	10,067.20	1988.80	12,056
Ductwork package for rooftop multizone units	1.000	System	2997.28	5115.22	8112.50
TOTAL			13,064.48	7104.02	20,168.50
COST PER S.F.			4.35	2.37	6.72
Note A: Small single zone unit recommended					
Note B: A combination of multizone units recommended					

8.4-220	Rooftop Multizone Units	COST PER S.F. MAT.	COST PER S.F. INST.	COST PER S.F. TOTAL
1240	Rooftop, multizone, air conditioner			
1260	Apartment corridors, 1,500 S.F., 2.75 ton. See Note A.			
1280	3,000 S.F., 5.50 ton	4.35	2.37	6.72
1320	10,000 S.F., 18.30 ton	3.20	2.13	5.33
1360	15,000 S.F., 27.50 ton	2.97	2.11	5.08
1400	20,000 S.F., 36.70 ton	2.99	2.11	5.10
1440	25,000 S.F., 45.80 ton	2.86	2.08	4.94
1520	Banks or libraries, 1,500 S.F., 6.25 ton	9.90	5.40	15.30
1560	3,000 S.F., 12.50 ton	8.60	5.15	13.75
1600	10,000 S.F., 41.67 ton	6.50	4.73	11.23
1640	15,000 S.F., 62.50 ton	5.45	4.53	9.98
1680	20,000 S.F., 83.33 ton	5.45	4.52	9.97
1720	25,000 S.F., 104.00 ton	5.45	4.51	9.96
1800	Bars and taverns, 1,500 S.F., 16.62 ton	19.90	8.50	28.40
1840	3,000 S.F., 33.24 ton	15.05	7.55	22.60
1880	10,000 S.F., 110.83 ton	11.50	6.85	18.35
1920	15,000 S.F., 165 ton, See Note B			
1960	20,000 S.F., 220 ton, See Note B			
2000	25,000 S.F., 275 ton, See Note B			
2080	Bowling alleys, 1,500 S.F., 8.50 ton	13.45	7.30	20.75
2120	3,000 S.F., 17.00 ton	11.75	7	18.75
2160	10,000 S.F., 56.70 ton	8.85	6.45	15.30
2200	15,000 S.F., 85.00 ton	7.40	6.15	13.55
2240	20,000 S.F., 113.00 ton	7.40	6.15	13.55
2280	25,000 S.F., 140 ton, See Note B			
2360	Department stores, 1,500 S.F., 4.37 ton, See Note A			
2400	3,000 S.F., 8.75 ton	6.95	3.77	10.72
2440	10,000 S.F., 29.17 ton	4.75	3.35	8.10

8.4-220	Rooftop Multizone Units	COST PER S.F. MAT.	COST PER S.F. INST.	COST PER S.F. TOTAL
2480	15,000 S.F., 43.75 ton	4.55	3.31	7.86
2520	20,000 S.F., 58.33 ton	3.82	3.17	6.99
2560	25,000 S.F., 72.92 ton	3.81	3.17	6.98
2640	Drug stores, 1,500 S.F., 10.00 ton	15.85	8.60	24.45
2680	3,000 S.F., 20.00 ton	11.65	7.75	19.40
2720	10,000 S.F., 66.66 ton	8.70	7.25	15.95
2760	15,000 S.F., 100.00 ton	8.75	7.25	16
2800	20,000 S.F., 135 ton, See Note B			
2840	25,000 S.F., 165 ton, See Note B			
2920	Factories, 1,500 S.F., 5 ton, See Note A			
2960	3,000 S.F., 10.00 ton	7.90	4.31	12.21
3000	10,000 S.F., 33.33 ton	5.45	3.83	9.28
3040	15,000 S.F., 50.00 ton	5.20	3.78	8.98
3080	20,000 S.F., 66.66 ton	27	8.35	35.35
3120	25,000 S.F., 83.33 ton	4.36	3.62	7.98
3200	Food supermarkets, 1,500 S.F., 4.25 ton, See Note A			
3240	3,000 S.F., 8.50 ton	6.75	3.66	10.41
3280	10,000 S.F., 28.33 ton	4.59	3.26	7.85
3320	15,000 S.F., 42.50 ton	4.43	3.22	7.65
3360	20,000 S.F., 56.67 ton	3.71	3.08	6.79
3400	25,000 S.F., 70.83 ton	3.71	3.08	6.79
3480	Medical centers, 1,500 S.F., 3.5 ton, See Note A			
3520	3,000 S.F., 7.00 ton	5.55	3.01	8.56
3560	10,000 S.F., 23.33 ton	3.80	2.68	6.48
3600	15,000 S.F., 35.00 ton	3.80	2.68	6.48
3640	20,000 S.F., 46.66 ton	3.64	2.65	6.29
3680	25,000 S.F., 58.33 ton	3.05	2.54	5.59
3760	Offices, 1,500 S.F., 4.75 ton, See Note A			
3800	3,000 S.F., 9.50 ton	7.50	4.09	11.59
3840	10,000 S.F., 31.66 ton	5.15	3.64	8.79
3880	15,000 S.F., 47.50 ton	4.95	3.59	8.54
3920	20,000 S.F., 63.33 ton	4.14	3.44	7.58
3960	25,000 S.F., 79.16 ton	4.14	3.44	7.58
4000	Restaurants, 1,500 S.F., 7.50 ton	11.90	6.45	18.35
4040	3,000 S.F., 15.00 ton	10.35	6.15	16.50
4080	10,000 S.F., 50.00 ton	7.80	5.65	13.45
4120	15,000 S.F., 75.00 ton	6.55	5.45	12
4160	20,000 S.F., 100.00 ton	6.55	5.40	11.95
4200	25,000 S.F., 125 ton, See Note B			
4240	Schools and colleges, 1,500 S.F., 5.75 ton	9.10	4.95	14.05
4280	3,000 S.F., 11.50 ton	7.95	4.72	12.67
4320	10,000 S.F., 38.33 ton	6	4.35	10.35
4360	15,000 S.F., 57.50 ton	5	4.17	9.17
4400	20,000 S.F., 76.66 ton	5	4.16	9.16
4440	25,000 S.F., 95.83 ton	5	4.17	9.17
9000	Components of ductwork packages for above systems per ton of cooling:			
9010	Ductwork; galvanized steel, 240 pounds			
9020	Insulation; fiberglass, 2" thick, FRK faced, 104 SF			
9030	Diffusers; aluminum, 24" x 12", two			
9040	Registers; aluminum, return with dampers, one			

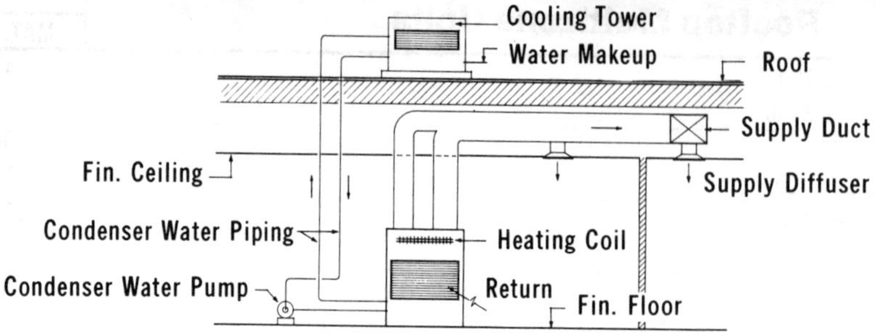

System Description: Self-contained, single package water cooled units include cooling tower, pump, piping allowance. Systems for 1000 S.F. and up include duct and diffusers to provide for even distribution of air. Smaller units distribute air through a supply air plenum, which is integral with the unit.

Returns are not ducted and supplies are not insulated.

Hot water or steam heating coils are included but piping to boiler and the boiler itself is not included.

Where local codes or conditions permit single pass cooling for the smaller units, deduct 10%.

System Components	QUANTITY	UNIT	COST EACH MAT.	COST EACH INST.	COST EACH TOTAL
SYSTEMS: 08.4-230-1300					
SELF-CONTAINED, WATER COOLED UNIT					
APARTMENT CORRIDORS, 500 S.F., .92 TON					
Self contained, water cooled, single package air conditioner unit	1.000	Ea.	785.40	239.40	1024.80
Ductwork package for water or air cooled packaged units	1.000	System	286.84	338.76	625.60
Cooling tower, draw thru single flow, belt drive	1.000	Ea.	58.70	42.50	101.20
Cooling tower pumps & piping	1.000	System	32.28	17.40	49.68
TOTAL			1163.22	638.06	1801.28
COST PER S.F.			2.33	1.28	3.61

8.4-230	Self-contained, Water Cooled Units	COST PER S.F. MAT.	COST PER S.F. INST.	COST PER S.F. TOTAL
1280	Self-contained, water cooled unit			
1300	Apartment corridors, 500 S.F., .92 ton	2.33	1.28	3.61
1320	1,000 S.F., 1.83 ton	2.32	1.27	3.59
1360	3,000 S.F., 5.50 ton	1.87	1.09	2.96
1400	5,000 S.F., 9.17 ton	1.89	1.01	2.90
1440	10,000 S.F., 18.33 ton	1.69	.92	2.61
1520	Banks or libraries, 500 S.F., 2.08 ton	3.97	1.36	5.33
1560	1,000 S.F., 4.17 ton	4.26	2.47	6.73
1600	3,000 S.F., 12.50 ton	4.29	2.30	6.59
1640	5,000 S.F., 20.80 ton	3.82	2.09	5.91
1680	10,000 S.F., 41.66 ton	3.51	2.09	5.60
1760	Bars and taverns, 500 S.F., 5.54 ton	7.80	2.47	10.27
1800	1,000 S.F., 11.08 ton	9.65	4.08	13.73
1840	3,000 S.F., 33.25 ton	7.55	3.34	10.89
1880	5,000 S.F., 55.42 ton	7.45	3.44	10.89
1920	10,000 S.F., 110.00 ton	7.30	3.30	10.60
2000	Bowling alleys, 500 S.F., 2.83 ton	5.40	1.85	7.25
2040	1,000 S.F., 5.66 ton	5.80	3.36	9.16
2080	3,000 S.F., 17.00 ton	5.20	2.85	8.05
2120	5,000 S.F., 28.33 ton	4.75	2.75	7.50
2160	10,000 S.F., 56.66 ton	4.70	2.80	7.50
2200	Department stores, 500 S.F., 1.46 ton	2.79	.95	3.74
2240	1,000 S.F., 2.92 ton	2.98	1.73	4.71
2280	3,000 S.F., 8.75 ton	3.01	1.61	4.62
2320	5,000 S.F., 14.58 ton	2.68	1.47	4.15
2360	10,000 S.F., 29.17 ton	2.44	1.42	3.86

8.4-230	Self-contained, Water Cooled Units	COST PER S.F. MAT.	INST.	TOTAL
2440	Drug stores, 500 S.F., 3.33 ton	6.35	2.18	8.53
2480	1,000 S.F., 6.66 ton	6.80	3.95	10.75
2520	3,000 S.F., 20.00 ton	6.15	3.35	9.50
2560	5,000 S.F., 33.33 ton	6.15	3.31	9.46
2600	10,000 S.F., 66.66 ton	5.55	3.29	8.84
2680	Factories, 500 S.F., 1.66 ton	3.18	1.09	4.27
2720	1,000 S.F. 3.37 ton	3.44	2	5.44
2760	3,000 S.F., 10.00 ton	3.44	1.84	5.28
2800	5,000 S.F., 16.66 ton	3.06	1.68	4.74
2840	10,000 S.F., 33.33 ton	2.79	1.62	4.41
2920	Food supermarkets, 500 S.F., 1.42 ton	2.71	.92	3.63
2960	1,000 S.F., 2.83 ton	3.59	1.97	5.56
3000	3,000 S.F., 8.50 ton	2.89	1.68	4.57
3040	5,000 S.F., 14.17 ton	2.92	1.57	4.49
3080	10,000 S.F., 28.33 ton	2.37	1.38	3.75
3160	Medical centers, 500 S.F., 1.17 ton	2.23	.76	2.99
3200	1,000 S.F., 2.33 ton	2.96	1.62	4.58
3240	3,000 S.F., 7.00 ton	2.38	1.38	3.76
3280	5,000 S.F., 11.66 ton	2.40	1.29	3.69
3320	10,000 S.F., 23.33 ton	2.14	1.17	3.31
3400	Offices, 500 S.F., 1.58 ton	3.02	1.03	4.05
3440	1,000 S.F., 3.17 ton	4.01	2.20	6.21
3480	3,000 S.F., 9.50 ton	3.26	1.75	5.01
3520	5,000 S.F., 15.83 ton	2.91	1.59	4.50
3560	10,000 S.F., 31.67 ton	2.65	1.54	4.19
3640	Restaurants, 500 S.F., 2.50 ton	4.78	1.63	6.41
3680	1,000 S.F., 5.00 ton	5.10	2.96	8.06
3720	3,000 S.F., 15.00 ton	5.15	2.76	7.91
3760	5,000 S.F., 25.00 ton	4.59	2.51	7.10
3800	10,000 S.F., 50.00 ton	3.34	2.33	5.67
3880	Schools and colleges, 500 S.F., 1.92 ton	3.66	1.25	4.91
3920	1,000 S.F., 3.83 ton	3.92	2.27	6.19
3960	3,000 S.F., 11.50 ton	3.95	2.12	6.07
4000	5,000 S.F., 19.17 ton	3.52	1.93	5.45
4040	10,000 S.F., 38.33 ton	3.23	1.92	5.15
9000	Components of ductwork packages for above systems, per ton of cooling:			
9010	Ductwork; galvanized steel, 108 pounds			
9020	Diffusers, aluminum, 24" x 12", two			

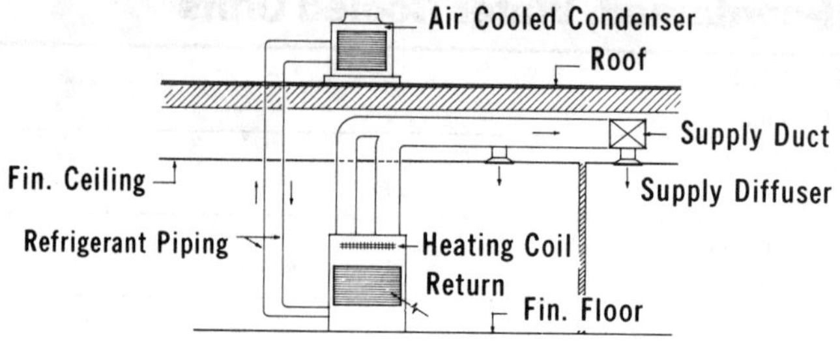

System Description: Self-contained air cooled units with remote air cooled condenser and interconnecting tubing. Systems for 1000 S.F. and up include duct and diffusers. Smaller units distribute air directly.

Returns are not ducted and supplies are not insulated.

Potential savings may be realized by using a single zone rooftop system or through-the-wall unit, especially in the smaller capacities, if the application permits.

Hot water or steam heating coils are included but piping to boiler and the boiler itself is not included.

Condenserless models are available for 15% less where remote refrigerant source is available.

System Components	QUANTITY	UNIT	COST EACH MAT.	COST EACH INST.	COST EACH TOTAL
SYSTEMS 08.4-240-1320					
SELF-CONTAINED, AIR COOLED UNIT					
APARTMENT CORRIDORS, 500 S.F., .92 TON					
Air cooled, package unit	1.000	Ea.	916.30	154.70	1071
Ductwork package for water or air cooled packaged units	1.000	System	286.84	338.76	625.60
Refrigerant piping	1.000	System	88.65	214.95	303.60
Air cooled condenser, direct drive, propeller fan	1.000	Ea.	199.49	59.37	258.86
TOTAL			1491.28	767.78	2259.06
COST PER S.F.			2.98	1.54	4.52

8.4-240	Self-contained, Air Cooled Units	COST PER S.F. MAT.	COST PER S.F. INST.	COST PER S.F. TOTAL
1300	Self-contained, air cooled unit			
1320	Apartment corridors, 500 S.F., .92 ton	2.98	1.54	4.52
1360	1,000 S.F., 1.83 ton	2.95	1.52	4.47
1400	3,000 S.F., 5.50 ton	2.54	1.42	3.96
1440	5,000 S.F., 9.17 ton	2.30	1.36	3.66
1480	10,000 S.F., 18.33 ton	1.98	1.26	3.24
1560	Banks or libraries, 500 S.F., 2.08 ton	5.35	1.92	7.27
1600	1,000 S.F., 4.17 ton	5.75	3.23	8.98
1640	3,000 S.F., 12.50 ton	5.25	3.10	8.35
1680	5,000 S.F., 20.80 ton	4.50	2.86	7.36
1720	10,000 S.F., 41.66 ton	4.34	2.83	7.17
1800	Bars and taverns, 500 S.F., 5.54 ton	10.15	4.19	14.34
1840	1,000 S.F., 11.08 ton	12.15	6.20	18.35
1880	3,000 S.F., 33.25 ton	9.50	5.40	14.90
1920	5,000 S.F., 55.42 ton	9.50	5.55	15.05
1960	10,000 S.F., 110.00 ton	9.55	5.45	15
2040	Bowling alleys, 500 S.F., 2.83 ton	7.35	2.63	9.98
2080	1,000 S.F., 5.66 ton	7.85	4.40	12.25
2120	3,000 S.F., 17.00 ton	6.15	3.89	10.04
2160	5,000 S.F., 28.33 ton	5.75	3.81	9.56
2200	10,000 S.F., 56.66 ton	5.75	3.87	9.62
2240	Department stores, 500 S.F., 1.46 ton	3.78	1.35	5.13
2280	1,000 S.F., 2.92 ton	4.04	2.26	6.30
2320	3,000 S.F., 8.75 ton	3.65	2.17	5.82
2360	5,000 S.F., 14.58 ton	3.65	2.17	5.82
2400	10,000 S.F., 29.17 ton	2.96	1.96	4.92

8.4-240	Self-contained, Air Cooled Units	COST PER S.F. MAT.	COST PER S.F. INST.	COST PER S.F. TOTAL
2480	Drug stores, 500 S.F., 3.33 ton	8.65	3.09	11.74
2520	1,000 S.F., 6.66 ton	9.20	5.15	14.35
2560	3,000 S.F., 20.00 ton	7.25	4.58	11.83
2600	5,000 S.F., 33.33 ton	6.75	4.48	11.23
2640	10,000 S.F., 66.66 ton	6.75	4.57	11.32
2720	Factories, 500 S.F., 1.66 ton	4.39	1.56	5.95
2760	1,000 S.F., 3.33 ton	4.64	2.59	7.23
2800	3,000 S.F., 10.00 ton	4.17	2.48	6.65
2840	5,000 S.F., 16.66 ton	3.58	2.28	5.86
2880	10,000 S.F., 33.33 ton	3.38	2.24	5.62
2960	Food supermarkets, 500 S.F., 1.42 ton	3.68	1.31	4.99
3000	1,000 S.F., 2.83 ton	4.54	2.35	6.89
3040	3,000 S.F., 8.50 ton	3.92	2.20	6.12
3080	5,000 S.F., 14.17 ton	3.55	2.11	5.66
3120	10,000 S.F., 28.33 ton	2.87	1.90	4.77
3200	Medical centers, 500 S.F., 1.17 ton	3.04	1.09	4.13
3240	1,000 S.F., 2.33 ton	3.77	1.94	5.71
3280	3,000 S.F., 7.00 ton	3.23	1.81	5.04
3320	5,000 S.F., 16.66 ton	2.93	1.73	4.66
3360	10,000 S.F., 23.33 ton	2.52	1.60	4.12
3440	Offices, 500 S.F., 1.58 ton	4.11	1.47	5.58
3480	1,000 S.F., 3.16 ton	5.15	2.64	7.79
3520	3,000 S.F., 9.50 ton	3.98	2.35	6.33
3560	5,000 S.F., 15.83 ton	3.43	2.17	5.60
3600	10,000 S.F., 31.66 ton	3.21	2.13	5.34
3680	Restaurants, 500 S.F., 2.50 ton	6.45	2.31	8.76
3720	1,000 S.F., 5.00 ton	6.90	3.88	10.78
3760	3,000 S.F., 15.00 ton	6.25	3.71	9.96
3800	5,000 S.F., 25.00 ton	5.05	3.36	8.41
3840	10,000 S.F., 50.00 ton	5.15	3.45	8.60
3920	Schools and colleges, 500 S.F., 1.92 ton	4.97	1.78	6.75
3960	1,000 S.F., 3.83 ton	5.35	2.98	8.33
4000	3,000 S.F., 11.50 ton	4.81	2.85	7.66
4040	5,000 S.F., 19.17 ton	4.14	2.63	6.77
4080	10,000 S.F., 38.33 ton	4	2.60	6.60
9000	Components of ductwork packages for above systems, per ton of cooling:			
9010	Ductwork; galvanized steel, 102 pounds			
9020	Diffusers; aluminum, 24" x 12", two			

8

Split System, Air Cooled

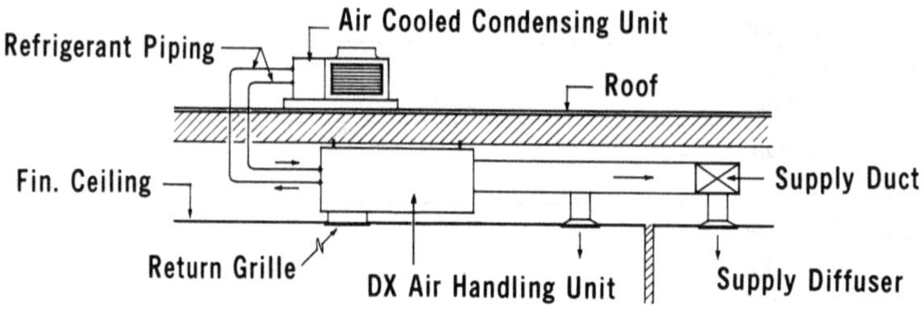

General: Split systems offer several important advantages which should be evaluated when a selection is to be made. They provide a greater degree of flexibility in component selection which permits an accurate match-up of the proper equipment size and type with the particular needs of the building. This allows for maximum use of modern energy saving concepts in heating and cooling. Outdoor installation of the air cooled condensing unit allows space savings in the building and also isolates the equipment operating sounds from building occupants.

Design Assumptions: The systems below are comprised of a direct expansion air handling unit and air cooled condensing unit with interconnecting copper tubing. Ducts and diffusers are also included for distribution of air. Systems are priced for cooling only. Heat can be added as desired either by putting hot water/steam coils into the air unit or into the duct supplying the particular area of need. Gas fired duct furnaces are also available. Refrigerant liquid line is insulated.

System Components	QUANTITY	UNIT	COST EACH		
			MAT.	INST.	TOTAL
SYSTEMS 08.4-250-1280					
SPLIT SYSTEM, AIR COOLED CONDENSING UNIT					
APARTMENT CORRIDORS, 1,000 S.F., 1.80 TON					
Fan coil air cond unit, cabinet mntd & filters direct expan air cool	1.000	Ea.	211.37	55.82	267.19
Ductwork package, for split system, remote condensing units	1.000	System	279.80	324.10	603.90
Refrigeration piping	1.000	System	83.10	356.10	439.20
Condensing unit, air cooled, incls compressor & standard controls	1.000	Ea.	691.13	223.87	915
TOTAL			1265.40	959.89	2225.29
COST PER S.F.			1.27	.96	2.23
*Cooling requirements would lead to more than one system					

8.4-250	Split Systems With Air Cooled Condensing Units	COST PER S.F.		
		MAT.	INST.	TOTAL
1260	Split system, air cooled condensing unit			
1280	Apartment corridors, 1,000 S.F., 1.83 ton	1.27	.96	2.23
1320	2,000 S.F., 3.66 ton	1.15	.97	2.12
1360	5,000 S.F., 9.17 ton	1.60	1.23	2.83
1400	10,000 S.F., 18.33 ton	1.50	1.34	2.84
1440	20,000 S.F., 36.66 ton	1.51	1.36	2.87
1520	Banks and libraries, 1,000 S.F., 4.17 ton	2.63	2.21	4.84
1560	2,000 S.F., 8.33 ton	3.64	2.79	6.43
1600	5,000 S.F., 20.80 ton	3.40	3.04	6.44
1640	10,000 S.F., 41.66 ton	3.43	3.08	6.51
1680	20,000 S.F., 83.32 ton	3.95	3.21	7.16
1760	Bars and taverns, 1,000 S.F., 11.08 ton	7.75	4.37	12.12

8.4-250	Split Systems With Air Cooled Condensing Units	COST PER S.F. MAT.	COST PER S.F. INST.	COST PER S.F. TOTAL
1800	2,000 S.F., 22.16 ton	8.25	5.20	13.45
1840	5,000 S.F., 55.42 ton	6.95	4.89	11.84
1880	10,000 S.F., 110.84 ton	8.15	5.10	13.25
1920	20,000 S.F., 220 ton*			
2000	Bowling alleys, 1,000 S.F., 5.66 ton	5.65	4.17	9.82
2040	2,000 S.F., 11.33 ton	4.95	3.79	8.74
2080	5,000 S.F., 28.33 ton	4.63	4.14	8.77
2120	10,000 S.F., 56.66 ton	4.67	4.19	8.86
2160	20,000 S.F., 113.32 ton	5.50	4.55	10.05
2320	Department stores, 1,000 S.F., 2.92 ton	1.82	1.51	3.33
2360	2,000 S.F., 5.83 ton	2.92	2.15	5.07
2400	5,000 S.F., 14.58 ton	2.55	1.95	4.50
2440	10,000 S.F., 29.17 ton	2.38	2.13	4.51
2480	20,000 S.F., 58.33 ton	2.40	2.16	4.56
2560	Drug stores, 1,000 S.F., 6.66 ton	6.65	4.90	11.55
2600	2,000 S.F., 13.32 ton	5.80	4.45	10.25
2640	5,000 S.F., 33.33 ton	5.50	4.93	10.43
2680	10,000 S.F., 66.66 ton	6.25	5.15	11.40
2720	20,000 S.F., 133.32 ton*			
2800	Factories, 1,000 S.F., 3.33 ton	2.08	1.72	3.80
2840	2,000 S.F., 6.66 ton	3.33	2.45	5.78
2880	5,000 S.F., 16.66 ton	2.73	2.43	5.16
2920	10,000 S.F., 33.33 ton	2.74	2.47	5.21
2960	20,000 S.F., 66.66 ton	3.13	2.58	5.71
3040	Food supermarkets, 1,000 S.F., 2.83 ton	1.77	1.47	3.24
3080	2,000 S.F., 5.66 ton	2.83	2.08	4.91
3120	5,000 S.F., 14.66 ton	2.48	1.89	4.37
3160	10,000 S.F., 28.33 ton	2.32	2.07	4.39
3200	20,000 S.F., 56.66 ton	2.33	2.10	4.43
3280	Medical centers, 1,000 S.F., 2.33 ton	1.42	1.18	2.60
3320	2,000 S.F., 4.66 ton	2.33	1.72	4.05
3360	5,000 S.F., 11.66 ton	2.04	1.56	3.60
3400	10,000 S.F., 23.33 ton	1.91	1.70	3.61
3440	20,000 S.F., 46.66 ton	1.92	1.73	3.65
3520	Offices, 1,000 S.F., 3.17 ton	1.97	1.64	3.61
3560	2,000 S.F., 6.33 ton	3.17	2.33	5.50
3600	5,000 S.F., 15.83 ton	2.77	2.12	4.89
3640	10,000 S.F., 31.66 ton	2.59	2.31	4.90
3680	20,000 S.F., 63.32 ton	2.97	2.45	5.42
3760	Restaurants, 1,000 S.F., 5.00 ton	5	3.68	8.68
3800	2,000 S.F., 10.00 ton	4.37	3.34	7.71
3840	5,000 S.F., 25.00 ton	4.09	3.65	7.74
3880	10,000 S.F., 50.00 ton	4.12	3.70	7.82
3920	20,000 S.F., 100.00 ton	4.84	4.01	8.85
4000	Schools and colleges, 1,000 S.F., 3.83 ton	2.41	2.03	4.44
4040	2,000 S.F., 7.66 ton	3.35	2.56	5.91
4080	5,000 S.F., 19.17 ton	3.13	2.80	5.93
4120	10,000 S.F., 38.33 ton	3.16	2.84	6
4160	20,000 S.F., 76.66 ton	3.62	2.96	6.58
9000	Components of ductwork packages for above systems, per ton of cooling:			
9010	Ductwork; galvanized steel, 102 pounds			
9020	Diffusers; aluminum, 24" x 12", two			

8

Computer rooms impose special requirements on air conditioning systems. A prime requirement is reliability, due to the potential monetary loss that could be incurred by a system failure. A second basic requirement is the tolerance of control with which temperature and humidity are regulated, and dust eliminated. As the air conditioning system reliability is so vital, the additional cost of reserve capacity and redundant components is often justified.

System Descriptions: Computer areas may be environmentally controlled by one of three methods as follows:

1. **Self-contained Units**
   These are units built to higher standards of performance and reliability. They usually contain alarms and controls to indicate component operation failure, filter change, etc. It should be remembered that these units in the room will occupy space that is relatively expensive to build and that all alterations and service of the equipment will also have to be accomplished within the computer area.
2. **Decentralized Air Handling Units**
   In operation these are similar to the self-contained units except that their cooling capability comes from remotely located refrigeration equipment as refrigerant or chilled water. As no compressors or refrigerating equipment are required in the air units, they are smaller and require less service than self-contained units. An added plus for this type of system occurs if some of the computer components themselves also require chilled water for cooling.
3. **Central System Supply**
   Cooling is obtained from a central source which, since it is not located within the computer room, may have excess capacity and permit greater flexibility without interfering with the computer components. System performance criteria must still be met.

NOTE: The costs shown below do not include an allowance for ductwork or piping

8.4-810	Computer Room Cooling Units	COST EACH MAT.	COST EACH INST.	COST EACH TOTAL
0560	Computer room unit, air cooled, includes remote condenser			
0580	3 ton	6700	940	7640
0600	5 ton	7375	1025	8400
0620	8 ton	12,700	1750	14,450
0640	10 ton	13,900	1850	15,750
0660	15 ton	16,200	2125	18,325
0680	20 ton	18,500	2525	21,025
0700	23 ton	19,600	2625	22,225
0800	Chilled water, for connection to existing chiller system			
0820	5 ton	7700	625	8325
0840	8 ton	12,100	900	13,000
0860	10 ton	13,300	990	14,290
0880	15 ton	15,000	940	15,940
0900	20 ton	17,300	1025	18,325
0920	23 ton	18,500	1125	19,625
1000	Glycol system, complete except for interconnecting tubing			
1020	3 ton	9450	1150	10,600
1040	5 ton	10,100	1175	11,275
1060	8 ton	16,200	2025	18,225
1080	10 ton	17,300	2225	19,525
1100	15 ton	20,800	2800	23,600
1120	20 ton	23,100	3000	26,100
1140	23 ton	24,200	3300	27,500
1240	Water cooled, not including condenser water supply or cooling tower			
1260	3 ton	6675	745	7420
1280	5 ton	7700	850	8550
1300	8 ton	12,500	1350	13,850
1320	15 ton	15,500	1700	17,200
1340	20 ton	16,900	1850	18,750
1360	23 ton	18,900	2075	20,975

B8.5-110 Garage Exhaust

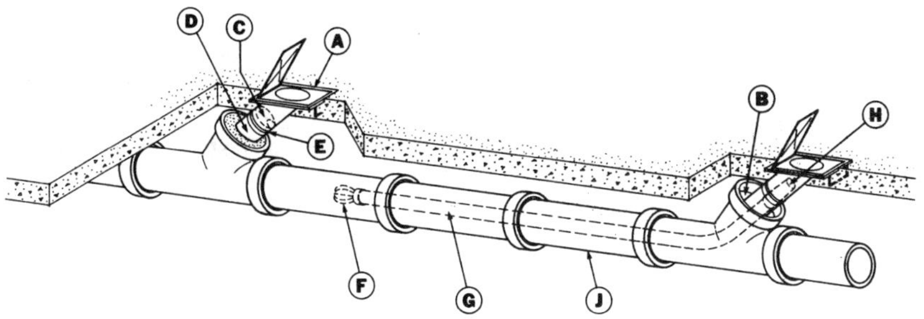

Vitrified Clay Garage Exhaust System

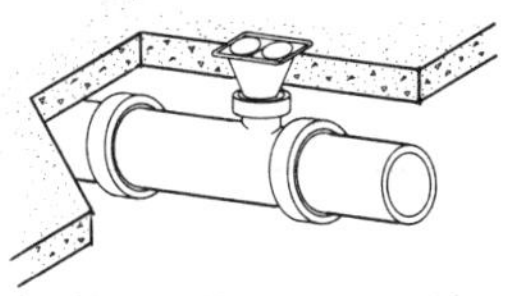

Dual Exhaust System

System Components

System Components	QUANTITY	UNIT	COST EACH MAT.	COST EACH INST.	COST EACH TOTAL
SYSTEM 085-110-1040					
GARAGE, EXHAUST, SINGLE OUTLET, 3" EXHAUST, CARS & LIGHT TRUCKS					
A Outlet top assembly for engine exhaust system, 3" diam	1.000	Ea.	100.32	7.98	108.30
B Tile adapter ring for engine exhaust system, 3" diam	1.000	Ea.	44	3.50	47.50
C Tubing adapter plate for engine exhaust system, 3" diam	1.000	Ea.	7.04	.56	7.60
D Outlet extension for engine exhaust system, 3" diam	1.000	Ea.	22.88	1.82	24.70
E Extension clamp for engine exhaust system, 3" diam	1.000	Ea.	1.76	.14	1.90
F Bullnose (guide) for engine exhaust system, 3" diam	1.000	Ea.	13.20		13.20
G Galvanized flexible tubing for engine exhaust system, 3" diam	8.000	L.F.	31.20		31.20
H Adapter for metal tubing end of engine exhaust system, 3" tail pipe	1.000	Ea.	40		40
J Sewer pipe vit. clay, premium joint, 8" diameter 4' or 5' lengths.	18.000	L.F.	69.30	83.70	153
Excavate utility trench w/chain trencher, 12 H.P., operator walking	18.000	L.F.		18.90	18.90
Backfill utility trench by hand, incl. compaction, 8" wide 24" deep	18.000	L.F.		19.62	19.62
Stand for blower, concrete over polystyrene core, 6" high	1.000	Ea.	3.08	8.07	11.15
AC&V duct, spiral reducer 10" x 8"	1.000	Ea.	20.63	14.37	35
AC&V duct, spiral reducer 12" x 10"	1.000	Ea.	27.50	19.50	47
AC&V duct, spiral preformed 45° elbow 8" diam.	1.000	Ea.	24.20	16.80	41
AC&V utility fan, belt drive, 3 phase, 2000 CFM, 1 HP	2.000	Ea.	858	262	1120
Safety switch, heavy duty fused, 240V, 3 pole, 30 amp	1.000	Ea.	61.60	78.40	140
TOTAL			1324.71	535.36	1860.07

8.5-110 Garage Exhaust Systems

8.5-110	Garage Exhaust Systems	COST PER BAY MAT.	COST PER BAY INST.	COST PER BAY TOTAL
1040	Garage, single exhaust, 3" outlet, cars & light trucks, one bay	1325	535	1860
1060	Additional bays up to seven bays	300	82	382
1500	4" outlet, trucks, one bay	1350	535	1885
1520	Additional bays up to six bays	315	82	397
1600	5" outlet, diesel trucks, one bay	1375	540	1915
1650	Additional single bays up to six	370	105	475
1700	2 adjoining bays	1400	545	1945
2000	Dual exhaust, 3" outlets, pair of adjoining bays	1550	605	2155
2100	Addtional pairs of adjoining bays	485	96	581

B12.3-110 Trenching

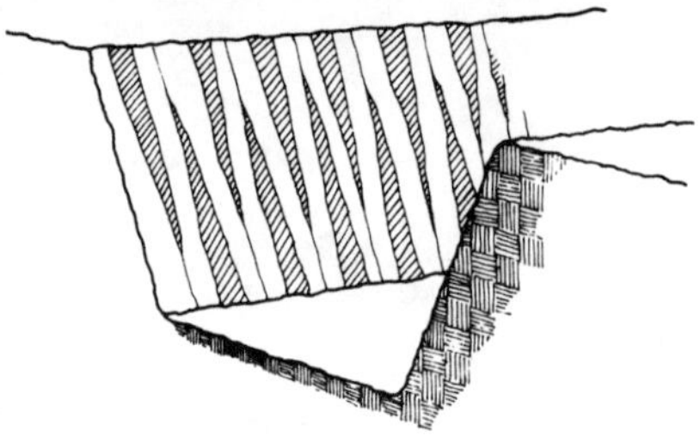

Trenching Systems are shown on a cost per linear foot basis. The systems include: excavation; backfill placed and compaction for various depths and trench bottom widths. Side slopes vary from 0:1 to 2:1.

The Expanded System Listing shows trenching systems that range from 2′ to 12′ in width. Depths range from 2′ to 24′.

System Components	QUANTITY	UNIT	COST PER L.F. EQUIP.	COST PER L.F. LABOR	COST PER L.F. TOTAL
SYSTEM 12.3-110-1310					
TRENCHING, BACKHOE, 0 TO 1 SLOPE, 2′ WIDE, 2′ DEEP, 3/8 C.Y. BUCKET					
Excavation, trench, hyd. backhoe, track mtd., 3/8 C.Y. bucket	.174	C.Y.	.19	.48	.67
Backfill and load spoil, from stockpile	.174	C.Y.	.07	.25	.32
Compaction by rammer tamper, 8″ lifts, 4 passes	.014	C.Y.	.01	.04	.05
Remove excess spoil, 6 C.Y. dump truck, 2 mile roundtrip	.160	C.Y.	.31	.24	.55
TOTAL			.58	1.01	1.59

12.3-110	Trenching	COST PER L.F. EQUIP.	COST PER L.F. LABOR	COST PER L.F. TOTAL
1310	Trenching, backhoe, 0 to 1 slope, 2′ wide, 2′ deep, 3/8 C.Y. bucket	.58	1.01	1.59
1320	3′ deep, 3/8 C.Y. bucket	.77	1.63	2.40
1330	4′ deep, 3/8 C.Y. bucket	.97	2.25	3.22
1340	6′ deep, 3/8 C.Y. bucket	1.34	3.48	4.82
1350	8′ deep, 1/2 C.Y. bucket	1.73	4.15	5.88
1360	10′ deep, 1 C.Y. bucket	2.03	4.36	6.39
1400	4′ wide, 2′ deep, 3/8 C.Y. bucket	1.22	1.98	3.20
1410	3′ deep, 3/8 C.Y. bucket	1.74	3.04	4.78
1420	4′ deep, 1/2 C.Y. bucket	1.99	3.53	5.52
1430	6′ deep, 1/2 C.Y. bucket	2.79	5.75	8.54
1440	8′ deep, 1/2 C.Y. bucket	3.59	7.95	11.54
1450	10′ deep, 1 C.Y. bucket	4.17	8.40	12.57
1460	12′ deep, 1 C.Y. bucket	4.93	10.30	15.23
1470	15′ deep, 1-1/2 C.Y. bucket	5.80	12.20	18
1480	18′ deep, 2-1/2 C.Y. bucket	7.55	13.90	21.45
1520	6′ wide, 6′ deep, 5/8 C.Y. bucket	4.58	7.55	12.13
1530	8′ deep, 3/4 C.Y. bucket	5.25	9.10	14.35
1540	10′ deep, 1 C.Y. bucket	6.15	10.45	16.60
1550	12′ deep, 1-1/4 C.Y. bucket	7.55	12.15	19.70
1560	16′ deep, 2 C.Y. bucket	9	15.15	24.15
1570	20′ deep, 3-1/2 C.Y. bucket	10.95	17.95	28.90
1580	24′ deep, 3-1/2 C.Y. bucket	17.80	31	48.80
1640	8′ wide, 12′ deep, 1-1/4 C.Y. bucket	10.75	15.35	26.10
1650	15′ deep, 1-1/2 C.Y. bucket	11.85	19.05	30.90
1660	18′ deep, 2-1/2 C.Y. bucket	14.85	21	35.85
1680	24′ deep, 3-1/2 C.Y. bucket	24	40	64
1730	10′ wide, 20′ deep, 3-1/2 C.Y. bucket	19.40	28	47.40
1740	24′ deep, 3-1/2 C.Y. bucket	31	50	81
1780	12′ wide, 20′ deep, 3-1/2 C.Y. bucket	24	34	58
1790	25′ deep, 3-1/2 C.Y. bucket	41	67	108
1800	1/2 to 1 slope, 2′ wide, 2′ deep, 3/8 C.Y. bucket	.77	1.26	2.03
1810	3′ deep, 3/8 C.Y. bucket	1.15	2.31	3.46
1820	4′ deep, 3/8 C.Y. bucket	1.56	3.62	5.18
1840	6′ deep, 3/8 C.Y. bucket	2.60	6.95	9.55

For expanded coverage of these items see *Means' Site Work Cost Data 1985*

12.3-110	Trenching	COST PER L.F.		
		EQUIP.	LABOR	TOTAL
1860	8' deep, 1/2 C.Y. bucket	4.06	9.85	13.91
1880	10' deep, 1 C.Y. bucket	5.45	11.90	17.35
2300	4' wide,2' deep, 3/8 C.Y. bucket	1.29	2.11	3.40
2310	3' deep, 3/8 C.Y. bucket	2.12	3.48	5.60
2320	4' deep, 1/2 C.Y. bucket	2.54	4.22	6.76
2340	6' deep, 1/2 C.Y. bucket	3.97	8.05	12.02
2360	8' deep, 1/2 C.Y. bucket	5.45	11.75	17.20
2380	10' deep, 1 C.Y. bucket	7.35	14.75	22.10
2400	12' deep, 1 C.Y. bucket	9.60	20	29.60
2430	15' deep, 1-1/2 C.Y. bucket	12.90	27	39.90
2460	18' deep, 2-1/2 C.Y. bucket	19.10	34	53.10
2840	6' wide, 6' deep, 5/8 C.Y. bucket	5.30	8.55	13.85
2860	8' deep, 3/4 C.Y. bucket	7.15	12.15	19.30
2880	10' deep, 1 C.Y. bucket	9.05	15.20	24.25
2900	12' deep, 1-1/4 C.Y. bucket	12	19.10	31.10
2940	16' deep, 2 C.Y. bucket	16.45	27	43.45
2980	20' deep, 3-1/2 C.Y. bucket	23	37	60
3020	24' deep, 3-1/2 C.Y. bucket	42	70	112
3100	8' wide, 12' deep, 1-1/4 C.Y. bucket	15.45	21	36.45
3120	15' deep, 1-1/2 C.Y. bucket	18.35	29	47.35
3140	18' deep, 2-1/2 C.Y. bucket	25	35	60
3180	24' deep, 3-1/2 C.Y. bucket	48	77	125
3270	10' wide, 20' deep, 3-1/2 C.Y. bucket	29	43	72
3280	24' deep, 3-1/2 C.Y. bucket	55	86	141
3370	12' wide, 20' deep, 3-1/2 C.Y. bucket	33	47	80
3380	25' deep, 3-1/2 C.Y. bucket	64	99	163
3500	1 to 1 slope, 2' wide, 2' deep, 3/8 C.Y. bucket	.90	1.47	2.37
3520	3' deep, 3/8 C.Y. bucket	1.44	2.74	4.18
3540	4' deep, 3/8 C.Y. bucket	1.98	4.36	6.34
3560	6' deep, 3/8 C.Y. bucket	3.39	8.65	12.04
3580	8' deep, 1/2 C.Y. bucket	5.50	12.40	17.90
3600	10' deep, 1 C.Y. bucket	8.50	16.90	25.40
3800	4' wide, 2' deep, 3/8 C.Y. bucket	1.53	2.52	4.05
3820	3' deep, 3/8 C.Y. bucket	2.70	4.42	7.12
3840	4' deep, 1/2 C.Y. bucket	2.98	4.43	7.41
3860	6' deep, 1/2 C.Y. bucket	4.80	9.05	13.85
3880	8' deep, 1/2 C.Y. bucket	7.10	14.65	21.75
3900	10' deep, 1 C.Y. bucket	9.35	17.15	26.50
3920	12' deep, 1 C.Y. bucket	12.35	23	35.35
3940	15' deep, 1-1/2 C.Y. bucket	17	32	49
3960	18' deep, 2-1/2 C.Y. bucket	26	40	66
4030	6' wide, 6' deep, 5/8 C.Y. bucket	6.30	9.85	16.15
4040	8' deep, 3/4 C.Y. bucket	8.75	14.60	23.35
4050	10' deep, 1 C.Y. bucket	11.45	19.05	30.50
4060	12' deep, 1-1/4 C.Y. bucket	15.55	24	39.55
4070	16' deep, 2 C.Y. bucket	22	37	59
4080	20' deep, 3-1/2 C.Y. bucket	32	51	83
4090	24' deep, 3-1/2 C.Y. bucket	60	99	159
4500	8' wide, 12' deep, 1-1/4 C.Y. bucket	19.40	26	45.40
4550	15' deep, 1-1/2 C.Y. bucket	24	37	61
4600	18' deep, 2-1/2 C.Y. bucket	33	46	79
4650	24' deep, 3-1/2 C.Y. bucket	66	105	171
4800	10' wide, 20' deep, 3-1/2 C.Y. bucket	40	56	96
4850	24' deep, 3-1/2 C.Y. bucket	72	110	182
4950	12' wide, 20' deep, 3-1/2 C.Y. bucket	44	59	103
4980	25' deep, 3-1/2 C.Y. bucket	87	135	222
5000	1-1/2 to 1 slope, 2' wide, 2' deep, 3/8 C.Y. bucket	1.04	1.71	2.75
5020	3' deep, 3/8 C.Y. bucket	1.73	3.18	4.91

For expanded coverage of these items see *Means' Site Work Cost Data 1985*

12.3-110	Trenching	COST PER L.F.		
		EQUIP.	LABOR	TOTAL
5040	4' deep, 3/8 C.Y. bucket	2.35	4.98	7.33
5060	6' deep, 3/8 C.Y. bucket	4.04	9.90	13.93
5080	8' deep, 1/2 C.Y. bucket	6.60	14	20.60
5100	10' deep, 1 C.Y. bucket	9	16.15	25.15
5300	4' wide, 2' deep, 3/8 C.Y. bucket	1.44	2.37	3.81
5320	3' deep, 3/8 C.Y. bucket	2.63	4.29	6.92
5340	4' deep, 1/2 C.Y. bucket	3.50	5.20	8.70
5360	6' deep, 1/2 C.Y. bucket	5.55	9.90	15.45
5380	8' deep, 1/2 C.Y. bucket	8.20	16.10	24.30
5400	10' deep, 1 C.Y. bucket	10.30	18.70	29
5420	12' deep, 1 C.Y. bucket	14.50	25	39.50
5450	15' deep, 1-1/2 C.Y. bucket	20	34	54
5480	18' deep, 2-1/2 C.Y. bucket	31	42	73
5660	6' wide, 6' deep, 5/8 C.Y. bucket	7.20	10.75	17.95
5680	8' deep, 3/4 C.Y. bucket	10	15.65	25.65
5700	10' deep, 1 C.Y. bucket	13.05	19.95	33
5720	12' deep, 1-1/4 C.Y. bucket	17.55	25	42.55
5760	16' deep, 2 C.Y. bucket	26	38	64
5800	20' deep, 3-1/2 C.Y. bucket	38	52	90
5840	24' deep, 3-1/2 C.Y. bucket	71	100	171
6020	8' wide, 12' deep, 1-1/4 C.Y. bucket	22	28	50
6050	15' deep, 1-1/2 C.Y. bucket	27	39	66
6080	18' deep, 2-1/2 C.Y. bucket	39	47	86
6140	24' deep, 3-1/2 C.Y. bucket	77	105	182
6300	10' wide, 20' deep, 3-1/2 C.Y. bucket	46	57	103
6350	24' deep, 3-1/2 C.Y. bucket	83	110	193
6450	12' wide, 20' deep, 3-1/2 C.Y. bucket	51	60	111
6480	25' deep, 3-1/2 C.Y. bucket	100	135	235
6600	2 to 1 slope, 2' wide, 2' deep, 3/8 C.Y. bucket	1.47	1.59	3.06
6620	3' deep, 3/8 C.Y. bucket	1.98	3.46	5.44
6640	4' deep, 3/8 C.Y. bucket	2.67	5.50	8.17
6660	6' deep, 3/8 C.Y. bucket	4.31	10.70	15.01
6680	8' deep, 1/2 C.Y. bucket	7.45	15.75	23.20
6700	10' deep, 1 C.Y. bucket	10.25	18.30	28.55
6900	4' wide, 2' deep, 3/8 C.Y. bucket	1.44	2.36	3.80
6920	3' deep, 3/8 C.Y. bucket	2.70	4.42	7.12
6940	4' deep, 1/2 C.Y. bucket	3.85	5.45	9.30
6960	6' deep, 1/2 C.Y. bucket	6.10	10.70	16.80
6980	8' deep, 1/2 C.Y. bucket	9.10	17.65	26.75
7000	10' deep, 1 C.Y. bucket	12.05	19.95	32
7020	12' deep, 1 C.Y. bucket	16.20	28	44.20
7050	15' deep, 1-1/2 C.Y. bucket	23	38	61
7080	18' deep, 2-1/2 C.Y. bucket	35	48	83
7260	6' wide, 6' deep, 5/8 C.Y. bucket	7.85	11.35	19.20
7280	8' deep, 3/4 C.Y. bucket	10.95	16.80	27.75
7300	10' deep, 3/4 C.Y. bucket	14.20	22	36.20
7320	12' deep, 1-1/4 C.Y. bucket	19.80	28	47.80
7360	16' deep, 2 C.Y. bucket	29	42	71
7400	20' deep, 3-1/2 C.Y. bucket	42	58	100
7440	24' deep, 3-1/2 C.Y. bucket	80	115	195
7620	8' wide, 12' deep, 1-1/4 C.Y. bucket	25	30	55
7650	15' deep, 1-1/2 C.Y. bucket	30	42	72
7680	18' deep, 2-1/2 C.Y. bucket	43	52	95
7740	24' deep, 3-1/2 C.Y. bucket	87	120	207
7920	10' wide, 20' deep, 3-1/2 C.Y. bucket	51	62	113
7940	24' deep, 3-1/2 C.Y. bucket	92	120	212
8060	12' wide, 20' deep, 3-1/2 C.Y. bucket	56	65	121
8080	25' deep, 3-1/2 C.Y. bucket	110	145	255

For expanded coverage of these items see *Means' Site Work Cost Data 1985*

B12.3-310 Pipe Bedding

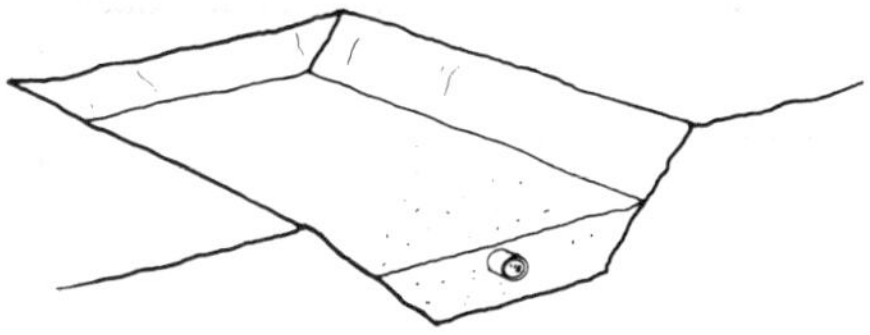

The Pipe Bedding System is shown for various pipe diameters. Compacted dead sand is used for pipe bedding and to fill 12" over the pipe. No backfill is included. Various side slopes are shown to accommodate different soil conditons.

System Components	QUANTITY	UNIT	COST PER L.F. MAT.	COST PER L.F. INST.	COST PER L.F. TOTAL
SYSTEM 12.3-310-1440					
PIPE BEDDING, SIDE SLOPE 0 TO 1, 1' WIDE, PIPE SIZE 6" DIAMETER					
Borrow, bank sand, 2 mile haul, machine spread	.067	C.Y.	.22	.22	.44
Compaction, vibrating plate	.067	C.Y.		.11	.11
TOTAL			.22	.33	.55

12.3-310	Pipe Bedding	COST PER L.F. MAT.	COST PER L.F. INST.	COST PER L.F. TOTAL
1440	Pipe bedding, side slope 0 to 1, 1' wide, pipe size 6" diameter	.22	.33	.55
1460	2' wide, pipe size 8" diameter	.49	.73	1.22
1480	Pipe size 10" diameter	.50	.74	1.24
1500	Pipe size 12" diameter	.51	.76	1.27
1520	3' wide, pipe size 14" diameter	.85	1.26	2.11
1540	Pipe size 15" diameter	.86	1.28	2.14
1560	Pipe size 16" diameter	.87	1.29	2.16
1580	Pipe size 18" diameter	.89	1.32	2.21
1600	4' wide, pipe size 20" diameter	1.28	1.90	3.18
1620	Pipe size 21" diameter	1.29	1.92	3.21
1640	Pipe size 24" diameter	1.33	1.98	3.31
1660	Pipe size 30" diameter	1.36	2.02	3.38
1680	6' wide, pipe size 32" diameter	2.37	3.53	5.90
1700	Pipe size 36" diameter	2.44	3.63	6.07
1720	7' wide, pipe size 48" diameter	3.17	4.72	7.89
1740	8' wide, pipe size 60" diameter	3.96	5.90	9.86
1760	10' wide, pipe size 72" diameter	5.70	8.50	14.20
1780	12' wide, pipe size 84" diameter	7.75	11.55	19.30
2140	Side slope 1/2 to 1, 1' wide, pipe size 6" diameter	.47	.70	1.17
2160	2' wide, pipe size 8" diameter	.77	1.15	1.92
2180	Pipe size 10" diameter	.84	1.25	2.09
2200	Pipe size 12" diameter	.89	1.33	2.22
2220	3' wide, pipe size 14" diameter	1.28	1.91	3.19
2240	Pipe size 15" diameter	1.32	1.96	3.28
2260	Pipe size 16" diameter	1.36	2.02	3.38
2280	Pipe size 18" diameter	1.44	2.14	3.58
2300	4' wide, pipe size 20" diameter	1.89	2.81	4.70
2320	Pipe size 21" diameter	1.94	2.89	4.83
2340	Pipe size 24" diameter	2.08	3.09	5.17
2360	Pipe size 30" diameter	2.33	3.47	5.80
2380	6' wide, pipe size 32" diameter	3.44	5.10	8.54
2400	Pipe size 36" diameter	3.68	5.45	9.13
2420	7' wide, pipe size 48" diameter	5.05	7.50	12.55
2440	8' wide, pipe size 60" diameter	6.55	9.75	16.30
2460	10' wide, pipe size 72" diameter	9.15	13.60	22.75
2480	12' wide, pipe size 84" diameter	12.20	18.10	30.30
2620	Side slope 1 to 1, 1' wide, pipe size 6" diameter	.71	1.06	1.77
2640	2' wide,pipe size 8" diameter	1.07	1.59	2.66

12.3-310	Pipe Bedding	COST PER L.F. MAT.	INST.	TOTAL
2660	Pipe size 10" diameter	1.17	1.74	2.91
2680	Pipe size 12" diameter	1.28	1.91	3.19
2700	3' wide, pipe size 14" diameter	1.72	2.56	4.28
2720	Pipe size 15" diameter	1.78	2.65	4.43
2740	Pipe size 16" diameter	1.84	2.74	4.58
2760	Pipe size 18" diameter	1.98	2.95	4.93
2780	4' wide, pipe size 20" diameter	2.51	3.73	6.24
2800	Pipe size 21" diameter	2.58	3.84	6.42
2820	Pipe size 24" diameter	2.82	4.20	7.02
2840	Pipe size 30" diameter	3.30	4.91	8.21
2860	6' wide, pipe size 32" diameter	4.50	6.70	11.20
2880	Pipe size 36" diameter	4.91	7.30	12.21
2900	7' wide, pipe size 48" diameter	6.85	10.20	17.05
2920	8' wide, pipe size 60" diameter	9.15	13.60	22.75
2940	10' wide, pipe size 72" diameter	12.60	18.75	31.35
2960	12' wide, pipe size 84" diameter	16.60	25	41.60
3000	Side slope 1-1/2 to 1, 1' wide, pipe size 6" diameter	.95	1.42	2.37
3020	2' wide, pipe size 8" diameter	1.35	2.01	3.36
3040	Pipe size 10" diameter	1.50	2.23	3.73
3060	Pipe size 12" diameter	1.66	2.47	4.13
3080	3' wide, pipe size 14" diameter	2.15	3.20	5.35
3100	Pipe size 15" diameter	2.24	3.33	5.57
3120	Pipe size 16" diameter	2.33	3.47	5.80
3140	Pipe size 18" diameter	2.54	3.78	6.32
3160	4' wide, pipe size 20" diameter	3.13	4.66	7.79
3180	Pipe size 21" diameter	3.23	4.80	8.03
3200	Pipe size 24" diameter	3.57	5.30	8.87
3220	Pipe size 30" diameter	4.29	6.40	10.69
3240	6' wide, pipe size 32" diameter	5.55	8.25	13.80
3260	Pipe size 36" diameter	6.15	9.15	15.30
3280	7' wide, pipe size 48" diameter	8.70	12.95	21.65
3300	8' wide, pipe size 60" diameter	11.70	17.40	29.10
3320	10' wide, pipe size 72" diameter	16	24	40
3340	12' wide, pipe size 84" diameter	21	31	52
3400	Side slope 2 to 1, 1' wide, pipe size 6" diameter	1.22	1.82	3.04
3420	2' wide, pipe size 8" diameter	1.64	2.44	4.08
3440	Pipe size 10" diameter	1.83	2.72	4.55
3460	Pipe size 12" diameter	2.04	3.04	5.08
3480	3' wide, pipe size 14" diameter	2.58	3.84	6.42
3500	Pipe size 15" diameter	2.69	4	6.69
3520	Pipe size 16" diameter	2.82	4.20	7.02
3540	Pipe size 18" diameter	3.09	4.60	7.69
3560	4' wide, pipe size 20" diameter	3.74	5.55	9.29
3580	Pipe size 21" diameter	3.87	5.75	9.62
3600	Pipe size 24" diameter	4.32	6.45	10.77
3620	Pipe size 30" diameter	5.25	7.85	13.10
3640	6' wide, pipe size 32" diameter	6.60	9.85	16.45
3660	Pipe size 36" diameter	7.40	11	18.40
3680	7' wide, pipe size 48" diameter	10.55	15.75	26.30
3700	8' wide, pipe size 60" diameter	14.30	21	35.30
3720	10' wide, pipe size 72" diameter	19.45	29	48.45
3740	12' wide, pipe size 84" diameter	25	38	63

12

For expanded coverage of these items see *Means' Site Work Cost Data 1985*

12.3-510	Drainage & Sewage Piping	COST PER L.F. MAT.	INST.	TOTAL
1000	Asbestos cement, class 2400, placed in trench, excavation and backfill not included			
1150	6" diameter	3.30	2.15	5.45
1160	8" diameter	4.29	2.46	6.75
1170	10" diameter	5.80	2.82	8.62
1180	12" diameter	7.30	3.33	10.63
1210	16" diameter	10.40	4.45	14.85
1220	Class 4000, 18" diameter	13.15	6.20	19.35
1250	24" diameter	21	8.65	29.65
1270	30" diameter	30	13.30	43.30
1280	36" diameter	40	16.40	56.40
2000	Bituminous fiber plain, placed in trench, excavation and backfill not included			
2130	4" diameter	1.38	.94	2.32
2150	6" diameter	2.75	1.05	3.80
2160	8" diameter	6.55	1.20	7.75
2900	Box culvert, precast, 8' long, placed in trench, excavation and backfill not included			
3000	6' x 3'	125	13.50	138.50
3020	6' x 7'	170	14.50	184.50
3040	8' x 3'	160	15.50	175.50
3060	8' x 8'	230	19	249
3080	10' x 3'	180	18.50	198.50
3100	10' x 8'	255	27	282
3120	12' x 3'	265	21	286
3140	12' x 8'	345	34	379
4000	Concrete, non reinforced, placed in trench, excavation and backfill not included			
4150	6" diameter	2.48	4.77	7.25
4160	8" diameter	2.70	4.65	7.35
4170	10" diameter	3.08	5.35	8.43
4180	12" diameter	3.91	5.40	9.31
4200	15" diameter	4.62	5.60	10.22
4220	18" diameter	5.95	5.55	11.50
4250	24" diameter	9.20	6.10	15.30
4400	Reinforced, no gasket, placed in trench, excavation and backfill not included			
4580	12" diameter	4.68	5.35	10.03
4600	15" diameter	5.85	5.55	11.40
4620	18" diameter	7.35	5.60	12.95
4650	24" diameter	11.55	6.10	17.65
4670	30" diameter	16.50	7.50	24
4680	36" diameter	24	8.80	32.80
4690	42" diameter	31	8.20	39.20
4700	48" diameter	40	8.40	48.40
4720	60" diameter	64	11.20	75.20
4730	72" diameter	88	12	100
4740	84" diameter	130	18	148
4800	With gasket, placed in trench, excavation and backfill not included			
4980	12" diameter	5.65	4.88	10.53
5000	15" diameter	6.80	5.05	11.85
5020	18" diameter	8.55	5	13.55
5050	24" diameter	13.20	5.65	18.85
5070	30" diameter	18.15	6.85	25
5080	36" diameter	26	26	52
5090	42" diameter	26	7.60	33.60
5100	48" diameter	42	7.20	49.20
5120	60" diameter	88	12	100
5130	72" diameter	110	10	120
5140	84" diameter	160	15.50	175.50
5700	Corrugated metal, alum, or galv. bit coated, placed in trench, excavation and backfill not included			
5760	8" diameter	4.79	2.16	6.95
5770	10" diameter	6.05	2.75	8.80

12

12.3-510	Drainage & Sewage Piping	COST PER L.F. MAT.	COST PER L.F. INST.	COST PER L.F. TOTAL
5780	12" diameter	7.10	3.40	10.50
5800	15" diameter	8.90	4.44	13.34
5820	18" diameter	10.75	4.92	15.67
5850	24" diameter	16.40	5.60	22
5870	30" diameter	21	7.45	28.45
5880	36" diameter	31	9.20	40.20
5900	48" diameter	41	17.30	58.30
5920	60" diameter	65	24	89
5930	72" diameter	91	39	130
6000	Plain			
6060	8" diameter	3.69	2.01	5.70
6070	10" diameter	4.95	2.55	7.50
6080	12" diameter	5.70	3.23	8.93
6100	15" diameter	7.10	4.25	11.35
6120	18" diameter	8.90	4.54	13.44
6140	24" diameter	13.80	5.35	19.15
6170	30" diameter	17.20	6.80	24
6180	36" diameter	26	13.60	39.60
6200	48" diameter	36	15.70	51.70
6220	60" diameter	58	23	81
6230	72" diameter	115	81	196
6300	Steel or alum. oval arch, coated & paved, placed in trench, excavation and backfill not included			
6400	15" equivalent diam	9.35	5.20	14.55
6420	18" equivalent diam	10.95	6.95	17.90
6450	24" equivalent diam	16.90	8.10	25
6470	30" equivalent diam	21	10.80	31.80
6480	36" equivalent diam	31	18.20	49.20
6490	42" equivalent diam	37	19.60	56.60
6500	48" equivalent diam	42	25	67
6600	Plain			
6700	15" equivalent diam	7.35	4.63	11.98
6720	18" equivalent diam	8.90	5.95	14.85
6750	24" equivalent diam	14.30	6.70	21
6770	30" equivalent diam	17.90	10.10	28
6780	36" equivalent diam	28	16.50	44.50
6790	42" equivalent diam	33	19	52
6800	48" equivalent diam	36	24	60
8000	Polyvinyl chloride SDR 35, placed in trench, excavation and backfill not included			
8130	4" diameter	.72	1.90	2.62
8150	6" diameter	1.60	2.04	3.64
8160	8" diameter	2.42	2.14	4.56
8170	10" diameter	3.96	2.84	6.80
8180	12" diameter	5.50	2.90	8.40
8200	15" diameter	8.65	4.91	13.56
9000	Vitrified clay, C-200, premium joint, placed in trench, excavation and backfill not included			
9030	4" diameter	1.76	2.70	4.46
9050	6" diameter	2.75	3.60	6.35
9060	8" diameter	3.85	4.65	8.50
9070	10" diameter	5.50	4.90	10.40
9080	12" diameter	7.70	6.20	13.90
9100	15" diameter	14.30	8.70	23
9120	18" diameter	21	10.10	31.10
9150	24" diameter	44	21	65
9170	30" diameter	48	34	82
9180	36" diameter	66	54	120

12

For expanded coverage of these items see *Means' Site Work Cost Data 1985*

12.3-520	Gas Service Piping	COST PER L.F. MAT.	INST.	TOTAL
2000	Polyethylene, placed in trench, excavation and backfill not included			
2070	1-1/4" diam, sdr 10	.54	1.79	2.33
2090	2" diam, sdr 11	.88	1.99	2.87
2110	3" diam, sdr 11.5	1.86	2.38	4.24
2130	4" diam, sdr 11	3.30	3.60	6.90
2150	6" diam, sdr 21	3.69	3.86	7.55
2160	8" diam, sdr 21	7	4.66	11.66
3000	Steel, schedule 40, plain end, tarred & wrapped, placed in trench, excavation and backfill not included			
3060	1" diameter	2.42	3.33	5.75
3090	2" diameter	3.85	3.60	7.45
3110	3" diameter	6.95	3.87	10.82
3130	4" diameter	9.80	7	16.80
3140	5" diameter	24	7.80	31.80
3150	6" diameter	25	9.70	34.70
3160	8" diameter	29	12.40	41.40
3170	10" diameter	42	9.20	51.20

12.3-530	Subdrainage Piping	COST PER L.F. MAT.	INST.	TOTAL
1000	Asbestos cement class 4000, perforated, placed in trench, excavation and backfill not included			
1130	4" diameter	1.98	1.83	3.81
1150	6" diameter	3.14	1.86	5
1160	8" diameter	4.95	2.50	7.45
1170	10" diameter	6.10	2.54	8.64
1180	12" diameter	8.40	2.58	10.98
2000	Bituminous fiber, perforated, placed in trench, excavation and backfill not included			
2110	3" diameter	1.16	.89	2.05
2130	4" diameter	1.32	.94	2.26
2140	5" diameter	2.31	.99	3.30
2150	6" diameter	2.86	1.05	3.91
3000	Metal, alum. or steel, perforated asphalt coated, placed in trench, excavation and backfill not included			
3150	6" diameter	2.75	1.88	4.63
3160	8" diameter	4.29	1.91	6.20
3170	10" diameter	5.40	2.61	8.01
3180	12" diameter	6.50	3.26	9.76
3220	18" diameter	9.80	4.56	14.36
4000	Porous wall concrete, placed in trench, excavation and backfill not included			
4130	4" diameter	1.43	2.14	3.57
4150	6" diameter	1.54	2.27	3.81
4160	8" diameter	2.48	3.02	5.50
4180	12" diameter	5.30	3.27	8.57
4200	15" diameter	6.05	4.05	10.10
4220	18" diameter	7.90	5.65	13.55
5000	Vitrified clay C-211, perforated, placed in trench, excavation and backfill not included			
5130	4" diameter	1.98	1.79	3.77
5150	6" diameter	3.36	2.29	5.65
5160	8" diameter	5.80	2.47	8.27
5180	12" diameter	8.10	3.36	11.46

For expanded coverage of these items see *Means' Site Work Cost Data 1985*

12.3-540	Water Distribution Piping	COST PER L.F. MAT.	COST PER L.F. INST.	COST PER L.F. TOTAL
1000	Asbestos cement, 150 psi, placed in trench, excavation and backfill not included			
1130	4" diameter	3.19	2.41	5.60
1150	6" diameter	4.57	2.48	7.05
1160	8" diameter	6.20	3.23	9.43
1170	10" diameter	9.50	3.43	12.93
1180	12" diameter	12.10	3.75	15.85
1210	16" diameter	19.30	4.69	23.99
1220	18" diameter	21	6.65	27.65
1230	20" diameter	24	7.80	31.80
1250	24" diameter	29	9.40	38.40
2000	Ductile iron class 250, mechanical joint, placed in trench, excavation and backfill not included			
2130	4" diameter	5.50	4.95	10.45
2150	6" diameter	6.05	5.70	11.75
2160	8" diameter	8.25	8.65	16.90
2170	10" diameter	11	10	21
2180	12" diameter	14.05	12.95	27
2210	16" diameter	21	19.90	40.90
2220	18" diameter	24	28	52
3000	Tyton joint			
3130	4" diameter	4.95	4.55	9.50
3150	6" diameter	5.60	5.20	10.80
3160	8" diameter	7.75	7.90	15.65
3170	10" diameter	10.45	9.30	19.75
3180	12" diameter	13.20	11.80	25
3210	16" diameter	19.55	17.45	37
3220	18" diameter	23	24	47
3230	20" diameter	25	25	50
3250	24" diameter	32	26	58
4000	Copper tubing, type K, placed in trench, excavation and backfill not included			
4050	3/4" diameter	1.28	5.20	6.48
4060	1" diameter	1.71	7.55	9.26
4080	1-1/2" diameter	2.81	9.15	11.96
4090	2" diameter	4.38	10.65	15.03
4110	3" diameter	8.40	15.60	24
4130	4" diameter	13.75	20	33.75
4150	6" diameter	47	19.70	66.70
5000	Polyvinyl chloride class 150, S.D.R. 18, placed in trench, excavation and backfill not included			
5130	4" diameter	3.22	3.58	6.80
5150	6" diameter	5.65	3.98	9.63
5160	8" diameter	8.70	5.80	14.50
5170	10" diameter	12.85	6.70	19.55
5180	12" diameter	17.05	8.95	26
6000	Polyethylene 160 psi, S.D.R. 7, placed in trench, excavation and backfill not included			
6050	3/4" diameter	.22	1.36	1.58
6060	1" diameter	.36	1.48	1.84
6080	1-1/2" diameter	.47	1.59	2.06
6090	2" diameter	.79	1.96	2.75

For expanded coverage of these items see *Means' Site Work Cost Data 1985*

B12.3-710 Manholes And Catch Basins

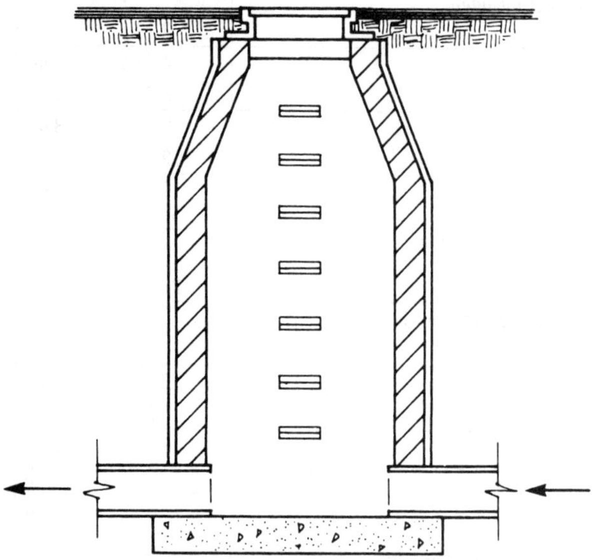

The Manhole and Catch Basin System includes: excavation with a backhoe; a formed concrete footing; brick walls; cast iron steps and compacted backfill. The system does not include a frame and cover.

The Expanded System Listing shows catch basins that have a 4-foot, 5-foot and 6-foot inside diameter riser. Depths range from 4 feet to 14 feet. Construction material shown in either concrete, concrete block, precast concrete, or brick.

System Components	QUANTITY	UNIT	COST PER EACH MAT.	INST.	TOTAL
SYSTEM 12.3-710-1920					
MANHOLE/CATCH BASIN, BRICK, 4' I.D. RISER, 4' DEEP					
Excavation, trench, hyd backhoe	14.815	C.Y.		39.11	39.11
Trim sides & bottom of trench	64.000	S.F.		23.04	23.04
Forms in place, spread footing, 4 uses	20.000	S.F.C.A.	7.20	42.60	49.80
Reinforcing in place, footings, #4 to #7 A615 grade 60	.019	Tons	10.14	9.34	19.48
Place and vibrate concrete, spread footings under 1CY, direct chute	.925	C.Y.	47.18	20.35	67.53
Catch basin or MH, brick, 4' ID, 4' dp, no frame	1.000	Ea.	330	405	735
Backfill with dozer, trench, up to 300' haul, no compaction	12.954	C.Y.		14.38	14.38
Backfill compaction, 12" lifts, air tamp	12.954	C.Y.		48.84	48.84
TOTAL			549.07	704.86	1253.93

12

12.3-710	Manholes & Catch Basins	COST PER EACH MAT.	INST.	TOTAL
1920	Manhole/catch basin, brick, 4' I.D. riser, 4' deep	550	705	1255
1940	6' deep	700	1000	1700
1960	8' deep	835	1325	2160
1980	10' deep	1025	1725	2750
3000	12' deep	1200	2125	3325
3020	14' deep	1350	2525	3875
3200	Block, 4' I.D. riser, 4' deep	435	570	1005
3220	6' deep	515	800	1315
3240	8' deep	605	1075	1680
3260	10' deep	805	1350	2155
3280	12' deep	925	1675	2600
3300	14' deep	1050	2000	3050
4620	Concrete, cast-in-place, 4' I.D. riser, 4' deep	515	520	1035
4640	6' deep	620	690	1310
4660	8' deep	760	965	1725
4680	10' deep	915	1500	2415
4700	12' deep	1050	1875	2925
4720	14' deep	1200	2450	3650

For expanded coverage of these items see *Means' Site Work Cost Data 1985*

12.3-710	Manholes & Catch Basins	COST PER EACH MAT.	COST PER EACH INST.	COST PER EACH TOTAL
5820	Concrete, precast, 4' I.D. riser, 4' deep	450	365	815
5840	6' deep	550	490	1040
5860	8' deep	645	645	1290
5880	10' deep	805	790	1595
5900	12' deep	925	1100	2025
5920	14' deep	1050	1225	2275
6000	5' I.D. riser, 4' deep	590	400	990
6020	6' deep	735	535	1270
6040	8' deep	875	705	1580
6060	10' deep	1050	900	1950
6080	12' deep	1225	1100	2325
6100	14' deep	1375	1375	2750
6200	6' I.D. riser, 4' deep	845	455	1300
6220	6' deep	1025	610	1635
6240	8' deep	1225	800	2025
6260	10' deep	1525	1025	2550
6280	12' deep	1775	1225	3000
6300	14' deep	2050	1525	3575

For expanded coverage of these items see *Means' Site Work Cost Data 1985*

TABLE OF CONTENTS

Table 8.1-031 Plumbing Approximations for Quick Estimating

Water Control
Water Meter; Backflow Preventer; Shock Absorbers; Vacuum Breakers; Mixer. .... 10 to 15% of Fixtures
Pipe And Fittings: .... 30 to 60% of Fixtures

NOTE: Lower percentage for compact buildings or larger buildings with plumbing in one area.
Larger percentage for large buildings with plumbing spread out.
In extreme cases pipe may be more than 100% of fixtures.
Percentages **do not** include special purpose or process piping.

Plumbing Labor:
1 & 2 Story Residential .... Rough-in Labor = 80% of Materials
Apartment Buildings .... Rough-in Labor = 90 to 100% of Materials
Labor for handling and placing fixtures is approximately 25 to 30% of fixtures.

Quality/Complexity Multiplier (For all installations)
Economy installation, add .... 0 to 5%
Good quality, medium complexity, add .... 5 to 15%
Above average quality and complexity, add .... 15 to 25%

Table 8.1-032 Pipe Material Consideration

1. Malleable fittings should be used for gas service.
2. Cast fittings may be broken as an aid to disassembling of heating lines frozen by long use, temperature and minerals.
3. Malleable fittings are used where there are stresses/strains due to expansion and vibration.
4. Cast iron pipe is extensively used for underground and submerged service.
5. Type M (light wall) copper tubing is available in hard temper only and is used for nonpressure and less severe applications than K and L.
6. Type L (medium wall) copper tubing, available hard or soft for interior service.
7. Type K (heavy wall) copper tubing, available in hard or soft temper for use where conditions are severe. For underground and interior service.
8. Hard drawn tubing requires fewer hangers or supports but should not be bent. Silver brazed fittings should be used.

Table 8.1-081 Valve Selection Considerations

INTRODUCTION: In any piping application, valve performance is critical. Valves should be selected to give the best performance at the lowest cost.

The following is a list of performance characteristics generally expected of valves.

1. Stopping flow or starting it.
2. Throttling flow (Modulation).
3. Flow direction changing.
4. Checking backflow (Permitting flow in only one direction).
5. Relieving or regulating pressure.

In order to properly select the right valve, some facts must be determined.

A. What liquid or gas will flow through the valve?
B. Does the fluid contain suspended particles?
C. Does the fluid remain in liquid form at all times?
D. Which metals does fluid corrode?
E. What are the pressure and temperature limits? (As temperature and pressure rise, so will the price of the valve.)
F. Is there constant line pressure?
G. Is the valve merely an on-off valve?
H. Will checking of backflow be required?
I. Will the valve operate frequently or infrequently?

Valves are classified by design type into such classifications as Gate, Globe, Angle, Check, Ball, Butterfly and Plug. They are also classified by end connection, stem, pressure restrictions and material such as bronze, cast iron,etc. Each valve has a specific use. A quality valve used correctly will provide a lifetime of trouble free service, but a high quality valve installed in the wrong service, may require frequent attention.

VALVE MATERIALS

Bronze:
Bronze is one of the oldest materials used to make valves. It is most commonly used in hot and cold water systems and other non-corrosive services. It is often used as a seating surface in larger iron body valves to ensure tight closure.

8

Carbon Steel:
Carbon steel is a high strength material. Therefore, valves made from this metal are used in higher pressure services, such as steam lines up to 600 psi at 850°F. Many steel valves are available with butt-weld ends for economy and are generally used in high pressure steam service as well as other higher pressure non-corrosive services.

Forged Steel:
Valves from tough carbon steel are used in service up to 2000 psi and temperatures up to 1000°F in Gate, Globe and Check valves.

Iron
Valves are normally used in medium to large pipe lines to control non-corrosive fluid and gases, where pressures do not exceed 250 psi at 450° or 500 psi cold water, oil or gas.

Stainless Steel:
Developed steel alloys can be used in over 90% corrosive services.

Plastic PVC:
This is used in a great variety of valves generally in high corrosive service with lower temperatures and pressures.

VALVE SERVICE PRESSURES

Pressure ratings on valves provide an indication of the safe operating pressure for a valve at some elevated temperature. This temperature is dependent upon the materials used and the fabrication of the valve. When specific data is not available, a good "rule-of-thumb" to follow is the temperature of saturated steam on the primary rating indicated on the valve body. Example, the valve has the number 150S printed on the side indicating 150 psi and hence, a maximum operating temperature of 367°F (temperature of saturated steam at 150 psi).

DEFINITIONS:

1. "WOG" - Water, oil, gas (cold working pressures).
2. "SWP" - Steam working pressure.
3. 100% area (full port) - means the area through the valve is equal to or greater than the area of standard pipe.
4. "Standard Opening" - means that the area through the valve is less than the area of standard pipe and therefore these valves should be used only where restriction of flow is unimportant.
5. "Round Port" - means the valve has a full round opening through the plug and body, of the same size and area as standard pipe.
6. "Rectangular Port" - valves have rectangular shaped ports through the plug body. The area of the port is either equal to 100% of the area of standard pipe, or restricted (standard opening). In either case it is clearly marked.
7. "ANSI" - American National Standards Institute.

STEM TYPES

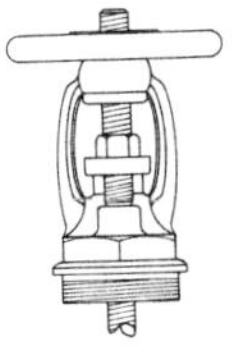

(OS & Y) - Rising Stem - Outside screw and yoke
Offers a visual indication of whether the valve is open or closed. Recommended where high temperatures, corrosives, and solids in the line might cause damage to inside-valve stem threads.
The stem threads are engaged by the yoke bushing so the stem rises through the hand wheel as it is turned.

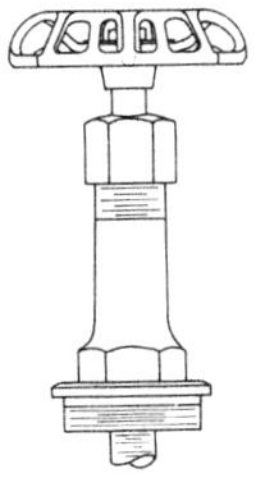

(R.S.) - Rising Stem - Inside screw
Adequate clearance for operation must be provided because both the hand wheel and the stem rise.
The valve wedge position is indicated by the position of the stem and hand wheel.

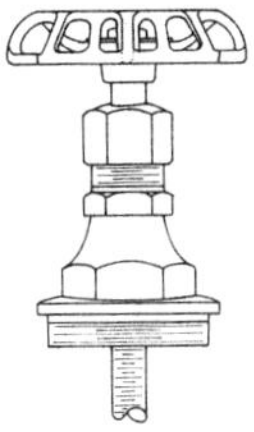

(N.R.S.) - Non-Rising stem - Inside Screw
A minimum clearance is required for operating this type of valve. Excessive wear or damage to stem threads inside the valve may be caused by heat, corrosion, and solids. Because the hand wheel and stem do no rise, wedge position cannot be visually determined.

VALVE TYPES

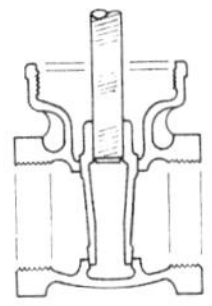

Gate Valves
Provide full flow, minute pressure drop, minimum turbulence and minimum fluid trapped in the line.
They are normally used where operation is infrequent.

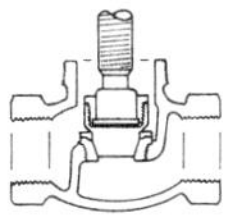

Globe Valves
Globe valves are designed for throttling and/or frequent operation with positive shut-off. Particular attention must be paid to the several types of seating materials available to avoid unnecessary wear. The seats must be compatible with the fluid in service and may be composition or metal. The configuration of the Globe valve opening causes turbulence which results in increased resistance. Most bronze Globe valves are rising stem-inside screw, but they are also available on O.S. & Y.

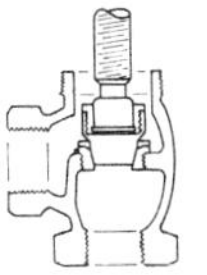

Angle Valves
The fundamental difference between the Angle valve and the Globe valve is the fluid flow through the Angle valve. It makes a 90° turn and offers less resistance to flow than the Globe valve while replacing an elbow. An Angle valve thus reduces the number of joints and installation time.

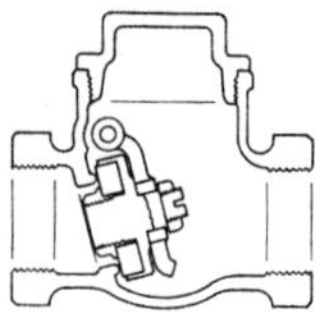

Check Valves
Check valves are designed to prevent backflow by automatically seating when the direction of fluid is reversed.
Swing Check valves are generally installed with Gate-valves, as they provide comparable full flow.
Usually recommended for lines where flow velocities are low and should not be used on lines with pulsating flow. Recommended for horizontal installation, or in vertical lines only where flow is upward.

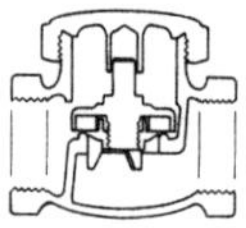
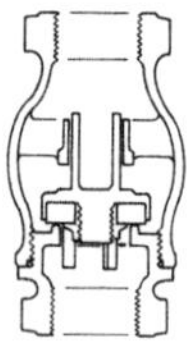

Lift Check Valves
These are commonly used with Globe and Angle valves since they have similar diaphragm seating arrangements and are recommended for preventing backflow of steam, air, gas and water, and on vapor lines with high flow velocities. For horizontal lines, horizontal lift checks should be used and vertical lift checks for vertical lines.

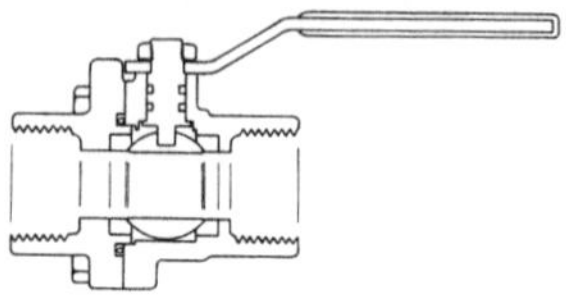

Ball Valves
Ball valves are light and easily installed, yet because of modern elastomeric seats, provide tight closure. Flow is controlled by rotating 90° a drilled ball which fits tightly against resilient seals. This ball seats with flow in either direction, and valve handle indicates the degree of opening. Recommended for frequent operation readily adaptable to automation, ideal for installation where space is limited.

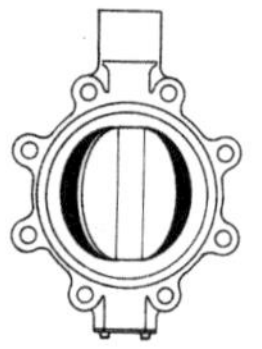

Butterfly Valves
Butterfly valves provide bubble-tight closure with excellent throttling characteristics. They can be used for full-open, closed and for throttling applications.
The Butterfly valve consists of a disc within the valve body which is controlled by a shaft.
In its closed position, the valve disc seals against a resilient seat. The disc position throughout the full 90° rotation is visually indicated by the position of the operator.
A Butterfly valve is only a fraction of the weight of a Gate valve and requires no gaskets between flanges in most cases. Recommended for frequent operation and adaptable to automation where space is limited.
Wafer and Lug type bodies, when installed between two pipe flanges, can be easily removed from the line. The pressure of the bolted flanges holds the valve in place. Locating lugs makes installation easier.

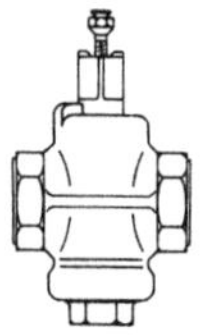

Plug Valves
Lubricated plug valves, because of the wide range of service to which they are adapted, may be classified as all purpose valves. They can be safely used on all pressure and vacuums, and on all temperatures up to the limits of available lubricants. They are the most satisfactory valves for the handling of gritty suspensions and many other destructive, erosive, corrosive and chemical solutions.

Table 8.1-101 Hot Water Consumption Rates

Type of Building	Size Factor	Maximum Hourly Demand	Average Day Demand
Apartment Dwellings	No. of Apartments:		
	Up to 20	12.0 Gal. per apt.	42.0 Gal. per apt.
	21 to 50	10.0 Gal. per apt.	40.0 Gal. per apt.
	51 to 75	8.5 Gal. per apt.	38.0 Gal. per apt.
	76 to 100	7.0 Gal. per apt.	37.0 Gal. per apt.
	101 to 200	6.0 Gal. per apt.	36.0 Gal. per apt.
	201 up	5.0 Gal. per apt.	35.0 Gal. per apt.
Dormitories	Men	3.8 Gal. per man	13.1 Gal. per man
	Women	5.0 Gal. per woman	12.3 Gal. per woman
Hospitals	Per bed	23 Gal. per patient	90 Gal. per patient
Hotels	Single room with bath	17 Gal. per unit	50 Gal. per unit
	Double room with bath	27 Gal. per unit	80 Gal. per unit
Motels	No. of units:		
	Up to 20	6.0 Gal. per unit	20.0 Gal. per unit
	21 to 100	5.0 Gal. per unit	14.0 Gal. per unit
	101 Up	4.0 Gal. per unit	10.0 Gal. per unit
Nursing Homes		4.5 Gal. per bed	18.4 Gal. per bed
Office buildings		0.4 Gal. per person	1.0 Gal. per person
Restaurants	Full meal type	1.5 Gal./max. meals/hr.	2.4 Gal. per meal
	Drive-in snack type	0.7 Gal./max. meals/hr.	0.7 Gal. per meal
Schools	Elementary	0.6 Gal. per student	0.6 Gal. per student
	Secondary & High	1.0 Gal. per student	1.8 Gal. per student

For evaluation purposes, recovery rate and storage capacity are inversely proportional. Water heaters should be sized so that the maximum hourly demand anticipated can be met in addition to allowance for the heat loss from the pipes and storage tank.

8

Table 8.1-102 Fixture Demands in Gallons Per Fixture Per Hour

FIXTURE DEMANDS:
Table below is based on 140°F final temperature except for dishwashers in public places (*) where 180°F water is mandatory.

Fixture	Apartment House	Club	Gym	Hospital	Hotel	Indust. Plant	Office	Private Home	School
Bathtubs	20	20	30	20	20			20	
Dishwashers, automatic*	15	50-150		50-150	50-200	20-100		15	20-100
Kitchen sink	10	20		20	30	20	20	10	20
Laundry, stationary tubs	20	28		28	28			20	
Laundry, automatic wash	75	75		100	150			75	
Private lavatory	2	2	2	2	2	2	2	2	2
Public lavatory	4	6	8	6	8	12	6		15
Showers	30	150	225	75	75	225	30	30	225
Service sink	20	20		20	30	20	20	15	20
Demand factor	0.30	0.30	0.40	0.25	0.25	0.40	0.30	0.30	0.40
Storage capacity factor	1.25	0.90	1.00	0.60	0.80	1.00	2.00	0.70	1.00

To obtain the probable maximum demand multiply the total demands for the fixtures (gal./fixture/hour) by the demand factor. The heater should have a heating capacity in gallons per hour equal to this maximum. The storage tank should have a capacity in gallons equal to the probable maximum demand multiplied by the storage capacity factor.

Table 8.1-201 Drainage Requirements

Drainage lines must have a slope to maintain flow for proper operation. This slope should not be less than 1/4" per foot for 3" diameter or smaller pipe and not less than 1/8" per foot for 4" diameter or larger pipe. The capacity of building drainage systems is calculated on a basis of "drainage fixture units" (d.f.u.) as per the following chart.

Type Of Fixture	d.f.u Value	Type Of Fixture	d.f.u. Value
Automatic clothes washer (2" standpipe)	3	Service sink (trap standard)	3
Bathroom group (water closet, lavatory and bathtub or shower) tank type closet	6	Service sink (P trap)	2
Bathtub (with or without overhead shower)	2	Urinal, pedestal, syphon jet blowout	6
		Urinal, wall lip	4
Clinic sink	6	Urinal, stall washout	4
Combination sink & tray with food disposal	4	Wash sink (circ. or mult.) per faucet set	2
Dental unit or cuspidor	1	Water closet, tank operated	4
Dental lavatory	1	Water closet, valve operated	6
Drinking fountain	1/2	Fixtures not listed above	
Dishwasher, domestic	2	Trap size 1-1/4" or smaller	1
Floor drains with 2" waste	3	Trap size 1-1/2"	2
Kitchen sink, domestic with one 1-1/2" trap	2	Trap size 2"	3
Kitchen sink, domestic with food disposal	2	Trap size 2-1/2"	4
Lavatory with 1-1/4" waste	1	Trap size 3"	5
Laundry tray (1 or 2 compartment)	2	Trap size 4"	6
Shower stall, domestic	2		

For continuous or nearly continuous flow into the system from a pump, air conditioning equipment or other item, allow 2 fixture units for each gallon per minute of flow.

When the "drainage fixture units" (d.f.u.) for each horizontal branch or vertical stack is computed from the table above, the appropriate pipe size for each branch or stack is determined from the table below.

8

Table 8.1-202 Allowable Fixture Units (d.f.u.) for Branches and Stacks

Pipe Diam.	Horiz. Branch (not incl. drains)	Stack Size for 3 Stories or 3 Levels	Stack Size for Over 3 Levels	Maximum for One Story Building Stack
1-1/2"	3	4	8	2
2"	6	10	24	6
2-1/2"	12	20	42	9
3"	20*	48*	72*	20*
4"	160	240	500	90
5"	360	540	1100	200
6"	620	960	1900	350
8"	1400	2200	3600	600
10"	2500	3800	5600	1000
12"	3900	6000	8400	1500
15"	7000			

*Not more than 2 water closets or bathroom groups within each branch interval nor more than 6 water closets or bathroom groups on the stack.

Stacks sized for the total may be reduced as load decreases at each story to a minimum diameter of 1/2 the maximum diameter.

Table 8.1-401 Minimum Plumbing Fixture Requirements

TYPE OF BUILDING/USE	WATER CLOSETS		URINALS		LAVATORIES		BATHTUBS OR SHOWERS		DRINKING FOUNTAIN	OTHER
	Persons	Fixtures	Persons	Fixtures	Persons	Fixtures	Persons	Fixtures	Fixtures	Fixtures
Assembly Halls Auditoriums Theater Public assembly	1-100 101-200 201-400	1 2 3	1-200 201-400 401-600	1 2 3	1-200 201-400 401-750	1 2 3			1 for each 1000 persons	1 service sink
	Over 400 add 1 fixt. for ea. 500 men: 1 fixt. for ea. 300 women		Over 600 add 1 fixture for each 300 men		Over 750 add 1 fixture for each 500 persons					
Assembly Public Worship	300 men 150 women	1 1	300 men	1	men women	1 1			1	
Dormitories	Men: 1 for each 10 persons Women: 1 for each 8 persons		1 for each 25 men, over 150 add 1 fixture for each 50 men		1 for ea. 12 persons 1 seperate dental lav. for each 50 persons recom.		1 for ea. 8 persons For women add 1 additional for each 30. Over 150 persons add 1 for each 20.		1 for each 75 persons	Laundry trays 1 for each 50 serv. sink 1 for ea. 100
Dwellings Apartments and homes	1 fixture for each unit				1 fixture for each unit		1 fixture for each unit			
Hospitals Room Ward Waiting room	Ea. 8 persons	1			Ea. 10 persons	1		1	1 for 100 patients	1 service sink per floor
Industrial Mfg. plants Warehouses	1-10 11-25 26-50 51-75 76-100	1 2 3 4 5	0-30 31-80 81-160 161-240	1 2 3 4	1-100 over 100	1 for ea. 10 1 for ea.15	1 Shower for each 15 persons subject to excessive heat or occupational hazard		1 for each 75 persons	
	1 fixture for each additional 30 persons									
Public Buildings Businesses Offices	1-15 16-35 36-55 56-80 81-110 111-150	1 2 3 4 5 6	Urinals may be provided in place of water closets but may not replace more than 1/3 required number of men's water closets		1-15 16-35 36-60 61-90 91-125	1 2 3 4 5			1 for each 75 persons	1 service sink per floor
	1 fixture for ea. additional 40 persons				1 fixture for ea. additional 45 persons					
Schools Elementary	1 for ea. 30 boys 1 for ea. 25 girls		1 for ea. 25 boys		1 for ea. 35 boys 1 for ea. 35 girls		For gym or pool shower room 1/5 of a class		1 for each 40 pupils	
Schools Secondary	1 for ea. 40 boys 1 for ea. 30 girls		1 for ea. 25 boys		1 for ea. 40 boys 1 for ea. 40 girls		For gym or pool shower room 1/5 of a class		1 for each 50 pupils	

8

Table 8.1-402 Plumbing Fixture Installation Time

Item	Rough-In	Set	Total Hours	Item	Rough-In	Set	Total Hours
Bath tub	5	5	10	Shower head only	2	1	3
Bath tub and shower, cast iron	6	6	12	Shower drain	3	1	4
Fire hose reel and cabinet	4	2	6	Shower stall, slate		15	15
Floor drain to 4 inch diameter	3	1	4	Slop sink	5	3	8
Grease trap, single, cast iron	5	3	8	Test 6 fixtures			14
Kitchen gas range		4	4	Urinal, wall	6	2	8
Kitchen sink, single	4	4	8	Urinal, pedestal or floor	6	4	10
Kitchen sink, double	6	6	12	Water closet and tank	4	3	7
Laundry tubs	4	2	6	Water closet and tank, wall hung	5	3	8
Lavatory wall hung	5	3	8	Water heater, 45 gals. gas, automatic	5	2	7
Lavatory pedestal	5	3	8	Water heaters, 65 gals. gas, automatic	5	2	7
Shower and stall	6	4	10	Water heaters, electric, plumbing only	4	2	6

Fixtures in unit pricing are based on the cost per fixture set in place. The rough-in cost, which must be added for each fixture, includes carrier, if required, some supply, waste and vent pipe, connecting fittings and stops. The lengths of rough-in pipe are nominal runs which would connect to the larger runs and stacks. The supply runs and DWV runs and stacks must be accounted for in separate entries. In the eastern half of the United States it is common for the plumber to carry these to a point 5 feet outside the building.

Table 8.1-403 Water Cooler Application

Type of Service	Requirement
Office, School or Hospital	12 persons per gallon per hour
Office, Lobby or Department Store	4 or 5 gallons per hour per fountain
Light manufacturing	7 persons per gallon per hour
Heavy manufacturing	5 persons per gallon per hour
Hot Heavy manufacturing	4 persons per gallon per hour
Hotels	.08 gallons per hour per room
Theatre	1 gallon per hour per 100 seats

Table 8.2-101 Sprinkler Systems (Automatic)
Sprinkler systems may be classified by type as follows:

l) Wet pipe System. A system employing automatic sprinklers attached to a piping system containing water and connected to a water supply so that water discharges immediately from sprinklers opened by a fire.

2) Dry pipe System. A system employing automatic sprinklers attached to a piping system containing air under pressure, the release of which as from the opening of sprinklers permits the water pressure to open a valve known as a "dry pipe valve". The water then flows into the piping system and out the opened sprinklers.

3) Pre-Action System. A system employing automatic sprinklers attached to a piping system containing air that may or may not be under pressure, with a supplemental heat responsive system of generally more sensitive characteristics than the automatic sprinklers themselves, installed in the same areas as the sprinklers; actuation of the heat responsive system, as from a fire, opens a valve which permits water to flow into the sprinkler piping system and to be discharged from any sprinklers which may be open.

4) Deluge System. A system employing open sprinklers attached to a piping system connected to a water supply through a valve which is opened by the operation of a heat responsive system installed in the same areas as the sprinklers. When this valve opens, water flows into the piping system and discharges from all sprinklers attached thereto.

5) Combined Dry pipe and Pre-Action Sprinkler System. A system employing automatic sprinklers attached to a piping system containing air under pressure with a supplemental heat responsive system of generally more sensitive characteristics than the automatic sprinklers themselves, installed in the same areas as the sprinklers; operation of the heat responsive system, as from a fire, actuates tripping devices which open dry pipe valves simultaneously and without loss of air pressure in the system. Operation of the heat responsive system also opens approved air exhaust valves at the end of the feed main which facilitates the filling of the system with water which usually precedes the opening of sprinklers. The heat responsive system also serves as an automatic fire alarm system.

6) Limited Water Supply System. A system employing automatic sprinklers and conforming to these standards but supplied by a pressure tank of limited capacity.

7) Chemical Systems. Systems using halon, carbon dioxide, dry chemical or high expansion foam as selected for special requirements. Agent may extinguish flames by chemically inhibiting flame propagation, suffocate flames by excluding oxygen, interrupting chemical action of oxygen uniting with fuel or sealing and cooling the combustion center.

8) Firecycle System. Firecycle is a fixed fire protection sprinkler system utilizing water as its extinguishing agent. It is a time delayed, recycling, preaction type which automatically shuts the water off when heat is reduced below the detector operating temperature and turns the water back on when that temperature is exceeded.
The system senses a fire condition through a closed circuit electrical detector system which controls water flow to the fire automatically. Batteries supply up to 90 hour emergency power supply for system operation. The piping system is dry (until water is required) and is monitored with pressurized air. Should any leak in the system piping occur, an alarm will sound, but water will not enter the system until heat is sensed by a Firecycle detector.

8

Area coverage sprinkler systems may be laid out and fed from the supply in any one of several patterns as shown in Figure 8.2-101. It is desirable, if possible, to utilize a central feed and achieve a shorter flow path from the riser to the furthest sprinkler. This permits use of the smallest sizes of pipe possible with resulting savings.

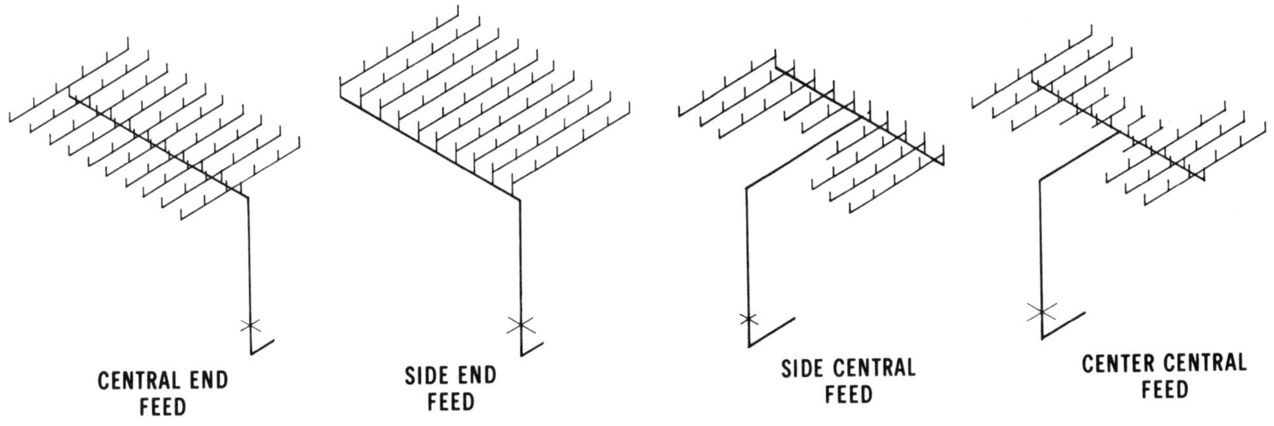

Figure 8.2-101

Table 8.2-102 System Classification

SYSTEM CLASSIFICATION

Rules for installation of sprinkler systems vary depending on the classification of occupancy falling into one of three categories as follows:

LIGHT HAZARD OCCUPANCY

The protection area allotted per sprinkler should not exceed 200 square feet with the maximum distance between lines and sprinklers on lines being 15 feet. The sprinklers do not need to be staggered. Branch lines should not exceed eight sprinklers on either side of a cross main. Each large area requiring more than 100 sprinklers and without a sub-dividing partition should be supplied by feed mains or risers sized for ordinary hazard occupancy.

Included in this group are:

Auditoriums
Churches
Clubs
Educational
Hospitals
Institutional
Libraries
(except large stack rooms)
Museums
Nursing Homes
Offices
Residential
Restaurants
Schools
Theaters

ORDINARY HAZARD OCCUPANCY

The protection area allotted per sprinkler shall not exceed 130 square feet of noncombustible ceiling and 120 square feet of combustible ceiling. The maximum allowable distance between sprinkler lines and sprinklers on line is 15 feet. Sprinklers shall be staggered if the distance between heads exceed 12 feet. Branch lines should not exceed eight sprinklers on either side of a cross main.

Included in this group are:

Automotive garages
Bakeries
Beverage manufacturing
Bleacheries
Boiler houses
Canneries
Cement plants
Clothing factories
Cold storage warehouses
Dairy products manufacturing
Distilleries
Dry cleaning
Electric generating stations
Feed mills
Grain elevators
Ice manufacturing
Laundries
Machine shops
Mercantiles
Paper mills
Printing and Publishing
Shoe factories
Warehouses
Wood product assembly

EXTRA HAZARD OCCUPANCY

The protection area allotted per sprinkler shall not exceed 90 square feet of noncombustible ceiling and 80 square feet of combustible ceiling.The maximum allowable distance between lines and between sprinklers on lines is 12 feet. Sprinklers on alternate lines shall be staggered if the distance between sprinklers on lines exceeds eight feet. Branch lines should not exceed six sprinklers on either side of a cross main.

Included in this group are:

Aircraft hangers
Chemical works
Explosives manufacturing
Linoleum manufacturing
Linseed oil mills
Oil refineries
Paint shops
Shade cloth manufacturing
Solvent extracting
Varnish works
Volatile flammable
Liquid manufacturing & use

Table 8.2-103 Sprinkler Quantities for Various Size and Types of Pipe

Sprinkler Quantities: Table below lists the usual maximum number of sprinkler heads for each size of copper and steel pipe for both wet and dry systems. These quantities may be adjusted to meet individual structural needs or local code requirements. Maximum area on any one floor for one system is: light hazard and ordinary hazard 52,000 S.F., extra hazardous 25,000 S.F.

Pipe Size	Light Hazard Occupancy		Ordinary Hazard Occupancy		Extra Hazard Occupancy	
Diameter	Steel Pipe	Copper Pipe	Steel Pipe	Copper Pipe	Steel Pipe	Copper Pipe
1 inch 1-1/4	2 sprinklers 3	2 sprinklers 3	2 sprinklers 3	2 sprinklers 3	1 sprinklers 2	1 sprinklers 2
1-1/2 2	5 10	5 12	5 10	5 12	5 8	5 8
2-1/2 3	30 60	40 65	20 40	25 45	15 27	20 30
3-1/2 4	100	115	65 100	75 115	40 55	45 65
5 6			160 275	180 300	90 150	100 170

Dry Pipe Systems: A dry pipe system should be installed where a wet pipe system is impractical, as in rooms or buildings which cannot be properly heated.
The use of an approved dry pipe system is more desirable than shutting off the water supply during cold weather.

Not more than 750 gallons of system capacity should be controlled by one dry pipe valve. Where two or more dry pipe valves are used, systems should preferably be divided horizontally.

Table 8.2-301 Standpipe Systems

The basis for standpipe system design is National Fire Protection Association NFPA 14, however, the authority having jurisdiction should be consulted for special conditions, local requirements and approval.

Standpipe systems, properly designed and maintained, are an effective and valuable time saving aid for extinquishing fires, especially in the upper stories of tall buildings, the interior of large commercial or industrial malls, or other areas where construction features or access make the laying of temporary hose lines time consuming and/or hazardous. Standpipes are frequently installed with automatic sprinkler systems for maximum protection.

There are three general classes of service for standpipe systems:

Class I for use by fire departments and personnel with special training for heavy streams (2-1/2 inch hose connections).

Class II for use by building occupants until the arrival of the fire department (1-1/2 inch hose connector with hose).

Class III for use by either fire departments and trained personnel or by the building occupants (both 2-1/2 inch and 1-1/2 inch hose connections or one 2-1/2 inch hose valve with an easily removable 2-1/2 inch by 1-1/2 inch adapter).

Standpipe systems are also classified by the way water is supplied to the system. The four basic types are:

Type 1: Wet standpipe system having supply valve open and water pressure maintained at all times.

Type 2: Standpipe system so arranged through the use of approved devices as to admit water to the system automatically by opening a hose valve.

Type 3: Standpipe system arranged to admit water to the system through manual operation of approved remote control devices located at each hose station.

Type 4: Dry standpipe having no permanent water supply.

Table 8.2-302 NFPA 14 Basic Standpipe design

Class	Design-Use	Pipe Size Minimums	Water Supply Minimums
Class I	2½ inch hose connection on each floor All areas within 30 feet of nozzle with 100 feet of hose Fire Department Trained Personnel	Height to 100 ft., 4 in. dia. Heights above 100 ft., 6 in. dia. (275 ft. max. except with pressure regulators 400 ft. max.)	For each standpipe riser 500 GPM flow For common supply pipe allow 500 GPM for first standpipe plus 250 GPM for each additional standpipe (2500) GPM max. total) 30 min. duration. 65 psi at 500 GPM
Class II	1½ inch hose connection with hose on each floor All areas within 30 feet of nozzle with 100 feet of hose. Occupant personnel	Height to 50 ft, 2 in. dia. Height above 50 ft., 2½ in. dia.	For each standpipe riser 100 GPM flow For multiple riser common supply pipe 100 GPM 30 min. duration, 65 PSI at 100 GPM
Class III	Both of above. Class I valved connections will meet Class III with addition of 2½ inch by 1½ inch adapter and 1½ inch hose.	Same as Class I	Same as Class I

Combined Systems

Combined systems are systems where the risers supply both automatic sprinklers and 2-1/2 inch hose connection outlets for fire department use. In such a system the sprinkler spacing pattern shall be in accordance with NFPA 13 while the risers and supply piping will be sized in accordance with NFPA 14. When the building is completely sprinklered the risers may be sized by hydraulic calculation. The minimum size riser for buildings not completely sprinklered is 6 inches.

The minimum water supply of a completely sprinklered, light hazard, high-rise occupancy building will be 500 GPM while the supply required for other types of completely sprinklered high-rise buildings is 1000 GPM.

General System Requirements

1. Approved valves will be provided at the riser for controlling branch lines to hose outlets.
2. A hose valve will be provided at each outlet for attachment of hose.
3. Where pressure at any standpipe outlet exceeds 100 PSI a pressure reducer must be installed to limit the pressure to 100 PSI. Note that the pressure head due to gravity in 100 feet of riser is 43.4 PSI. This must be overcome by city pressure, fire pumps, or gravity tanks to provide adequate pressure at the top of the riser.
4. Each hose valve on a wet system having linen hose shall have an automatic drip connection to prevent valve leakage from entering the hose.
5. Each riser will have a valve to isolate it from the rest of the system.
6. One or more fire department connections as an auxiliary supply shall be provided for each Class I or Class III standpipe system. In buildings having two or more zones, a connection will be provided for each zone.
7. There will be no shutoff valve in the fire department connection, but a check valve will be located in the line before it joins the system.
8. All hose connections street side will be identified on a cast plate or fitting as to purpose.

Table 8.3-001

Heating Systems

The basic function of a heating system is to bring an enclosed volume up to a desired temperature and then maintain that temperature within a reasonable range. To accomplish this, the selected system must have sufficient capacity to offset transmission losses resulting from the temperature difference on the interior and exterior of the enclosing walls in addition to losses due to cold air infiltration through cracks, crevices and around doors and windows. The amount of heat to be furnished is dependent upon the bulding size, construction, temperature difference, air leakage, use, shape, orientation and exposure. Air circulation is also an important consideration. Circulation will prevent stratification which could result in heat losses through uneven temperatures at various levels. For example, the most efficient use of unit heaters can usually be achieved by circulating the space volume through the total number of units once every 20 minutes or 3 times an hour. This general rule must, of course, be adapted for special cases such as large buildings with low ratios of heat transmitting surface to cubical volume. The type of occupancy of a building will have considerable bearing on the number of heat transmitting units and the location selected. It is axiomatic, however, that the basis of any successful heating system is to provide the maximum amount of heat at the points of maximum heat loss such as exposed walls, windows, and doors. Large roof areas, wind direction, and wide doorways create problems of excessive heat loss and require special consideration and treatment.

Heat Transmission

Heat transfer is an important parameter to be considered during selection of the exterior wall style, material and window area. A high rate of transfer will permit greater heat loss during the wintertime with the resultant increase in heating energy costs and a greater rate of heat gain in the summer with proportionally greater cooling cost. Several terms are used to describe various aspects of heat transfer. However, for general estimating purposes this book lists U values for systems of construction materials. U is the "overall heat transfer coefficient". It is defined as the heat flow per hour through one square foot when the temperature difference in the air on either side of the structure wall, roof, ceiling or floor is one degree Fahrenheit. The structural segment may be a single homogeneous material or a composite

Total heat transfer is found using the equation. $Q=AU(T_2-T_1)$

where:

- **Q** = Heat flow, BTU per hour
- **A** = Area, square feet
- **U** = Overall heat transfer coefficient
- (T_2-T_1) = Difference in temperature of air on each side of the construction component. (Also abbreviated TD)

Note that heat can flow through all surfaces of any building and this flow is in addition to heat gain or loss due to ventilation, infiltration and generation (appliances, machinery, people).

8

Table 8.3-002 Heating Approximations for Quick Estimating

Oil Tank:
- Small .... Steel: $1.00 per Gallon, Fiberglass: $2.70 per Gallon
- Large .... Steel: $.60 per Gallon, Fiberglass: $.90 per Gallon

Oil Piping & Boiler Room Piping:
- Small System .... 20 to 30% of Boiler
- Complex System with Pumps, Headers, Etc. .... 80 to 110% of Boiler

Breeching With Insulation:
- Small .... 10 to 15% of Boiler
- Large .... 15 to 25% of Boiler

Coils: .... 15 to 30% of Containing Unit

Balancing (Independent): .... ½% of H.V.A.C. Estimate

Quality/Complexity Adjustment: For all heating installations add these adjustments to the estimate to more closely allow for the equipment and conditions of the particular job under consideration.

- Economy installation, add .... 0 to 5% of System
- Good quality, medium complexity, add .... 5 to 15% of System
- Above average quality and complexity, add .... 15 to 25% of System

Table 8.3-003 The Basics of a Heating System

The function of a heating system is to achieve and maintain a desired temperature in a room or building by replacing the amount of heat being dissipated. There are four kinds of heating systems; hot-water, steam, warm-air and electric resistance. Each has certain essential and similar elements with the exception of electric resistance heating which is covered in Section 16.

The basic elements of a heating system are:

- **A.** A **combustion chamber** in which fuel is burned and heat transferred to a conveying medium.
- **B.** The **"fluid"** used for conveying the heat (water, steam or air).
- **C.** **Conductors** or pipes for transporting the fluid to specific desired locations.
- **D.** A means of disseminating the heat, sometimes called **terminal units.**

A. The combustion chamber in a furnace heats air which is then distributed. This is called a warm air system.

The combustion chamber in a boiler heats water which is either distributed as hot water or steam and this is termed a hydronic system.

The maximum allowable working pressures are limited by ASME "Code for Heating Boilers" to 15 PSI for steam and 160 PSI for hot water heating boilers, with a maximum temperature limitation of 250°F. Hot water boilers are generally rated for a working pressure of 30 PSI. High pressure boilers are governed by the ASME "Code for Power Boilers" which is used almost universally for boilers operating over 15 PSIG. High pressure boilers used for a combination of heating/process loads are usually designed for 150 PSIG.

Boiler ratings are usually indicated as either Gross or Net Output. The Gross Load is equal to the Net Load plus a piping and pickup allowance. When this allowance cannot be determined, divide the gross output rating by 1.25 for a value equal to or greater than the net heat loss requirement of the building.

B. Of the three fluids used, steam carries the greatest amount of heat per unit volume. This is due to the fact that it gives up its latent heat of vaporization at a temperature considerably above room temperature. Another advantage is that the pressure to produce a positive circulation is readily available. Piping conducts the steam to terminal units and returns condensate to the boiler.

The **steam system** is well adapted to large buildings because of its positive circulation, its comparatively economical installation and its ability to deliver large quantities of heat. Nearly all large office buildings, stores, hotels, and industrial buildings are so heated, in addition to many residences.

Hot water, when used as the heat carrying fluid, gives up a portion of its sensible heat and then returns to the boiler or heating apparatus for reheating. As the heat conveyed by each pound of water is about one-fiftieth of the heat conveyed by a pound of steam, it is necessary to circulate about fifty times as much water as steam by weight (although only one-thirtieth as much by volume). The hot water system is usually, although not necessarily, designed to operate at temperatures below that of the ordinary steam system and so the amount of heat transfer surface must be correspondingly greater. A temperature of 190°F to 200°F is normally the maximum. Circulation in small buildings may depend on the difference in density between hot water and the cool water returning to the boiler, circulating pumps are normally used to maintain a desired rate of flow. Pumps permit a greater degree of flexibility and better control.

In **warm-air** furnace systems, cool air is taken from one or more points in the building passed over the combustion chamber and flue gas passages and then distributed through a duct system. A disadvantage of this system is that the ducts take up much more building volume than steam or hot water pipes. Advantages of this system are the relative ease with which humidification can be accomplished by the evaporation of water as the air circulates through the heater, and the lack of need for expensive disseminating units as the warm air simply becomes part of the interior atmosphere of the building.

C. Conductors (pipes and ducts) have been lightly treated in the discussion of conveying fluids. For more detailed information such as sizing and distribution methods, the reader is referred to technical publications such as the American Society of Heating, Refrigerating and Air-Conditioning Engineers "Handbook of Fundamentals."

D. Terminal units come in an almost infinite variety of sizes and styles, but the basic principles of operation are very limited. As previously mentioned, warm-air systems require only a simple register or diffuser to mix heated air with that present in the room. Special application items such as radiant coils and infra-red heaters are available to meet particular conditions but are not usually considered for general heating needs. Most heating is accomplished by having air flow over coils or pipes containing the heat transporting medium (steam, hot-water, electricity). These units, while varied, may be separated into two general types, (1) radiator/convectors and (2) unit heaters.

Radiator/convectors may be cast, fin-tube or pipe assemblies. They may be direct, indirect, exposed, concealed or mounted within a cabinet enclosure, upright or baseboard style. These units are often collectively referred to as "radiators" or "radiation" although none gives off heat either entirely by radiation or by convection but rather a combination of both. The air flows over the units as a gravity "current." It is necessary to have one or more heat-emitting units in each room. The most efficient placement is low along an outside wall or under a window to counteract the cold coming into the room and achieve an even distribution.

In contrast to radiator/convectors which operate most effectively against the walls of smaller rooms, **unit heaters** utilize a fan to move air over heating coils and are very effective in locations of relatively large volume. Unit heaters, while usually suspended overhead, may be floor mounted. They also may take in fresh outside air for ventilation. The heat distributed by unit heaters may be from a remote source and conveyed by a fluid or it may be from the combustion of fuel in each individual heater. In the latter case the only piping required would be for fuel, however, a vent for the products of combustion would be necessary. The following list gives many of the advantages of unit heaters for applications other than office or residential:

a. Large capacity so smaller number of units are required, **b.** Piping system simplified, **c.** Space saved where they are located overhead out of the way, **d.** Rapid heating directed where needed with effective wide distribution, **e.** Difference between floor and ceiling temperature reduced, **f.** Circulation of air obtained, and ventilation with introduction of fresh air possible, **g.** Heat output flexible and easily controlled.

Installed costs of various individual components of heating systems are listed in SECTIONS 15.5 and 15.6.

Table 8.3-011 Factor for Determining Heat Loss for Various Types of Buildings

GENERAL: While the most accurate estimates of heating requirements would naturally be based on detailed information about the building being considered, it is possible to arrive at a reasonable approximation using the following procedure:

1. Calculate the cubic volume of the room or building.
2. Select the appropriate factor from Table 8.3-011. Note that the factors apply only to inside termperatures listed in the first column and to 0°F outside temperature.
3. If the building has bad north and west exposures, multiply the heat loss factor by 1.1
4. If the outside design temperature is other than 0°F, multiply the factor from Table 8.3-011 by the factor from Table 8.3-012
5. Multiply the cubic volume by the factor selected from Table 8.3-011. This will give the estimated BTUH heat loss which must be made up to maintain inside temperature.

Building Type	Conditions	Qualifications	Loss Factor*
Factories & Industrial Plants General Office Areas 70°F	One Story	Skylight in Roof	6.2
		No Skylight in Roof	5.7
	Multiple Story	Two Story	4.6
		Three Story	4.3
		Four Story	4.1
		Five Story	3.9
		Six Story	3.6
	All Walls Exposed	Flat Roof	6.9
		Heated Space Above	5.2
	One Long Warm Common Wall	Flat Roof	6.3
		Heated Space Above	4.7
	Warm Common Walls on Both Long Sides	Flat Roof	5.8
		Heated Space Above	4.1
Warehouses 60°F	All Walls Exposed	Skylights in Roof	5.5
		No Skylights in Roof	5.1
		Heated Space Above	4.0
	One Long Warm Common Wall	Skylight in Roof	5.0
		No Skylight in Roof	4.9
		Heated Space Above	3.4
	Warm Common Walls on Both Long Sides	Skylight in Roof	4.7
		No Skylight in Roof	4.4
		Heated Space Above	3.0

*NOTE: This table tends to be conservative particularly for new buildings designed for minimum energy consumption.

Table 8.3-012 Outside Design Temperature Correction Factor (for Degrees Fahrenheit)

Outside Design Temperature	50	40	30	20	10	0	-10	-20	-30
Correction Factor	0.29	0.43	0.57	0.72	0.86	1.00	1.14	1.28	1.43

Figures 8.3-013 and 8.3-014 provide a way to calculate heat transmission of various construction materials from their U values and the TD (Temperature Difference).

1. From the Exterior Enclosure Division or elsewhere, determine U values for the construction desired.
2. Determine the coldest design temperature. The difference between this temperature and the desired interior temperature is the TD (temperature difference).
3. Enter Figure 8.3-013 or 8.3-014 at correct U value. Cross horizontally to the intersection with appropriate TD. Read transmission per square foot from bottom of figure.
4. Multiply this value of BTU per hour transmission per square foot of area by the total surface area of that type of construction.

Figure 8.3-013 Transmission of Heat

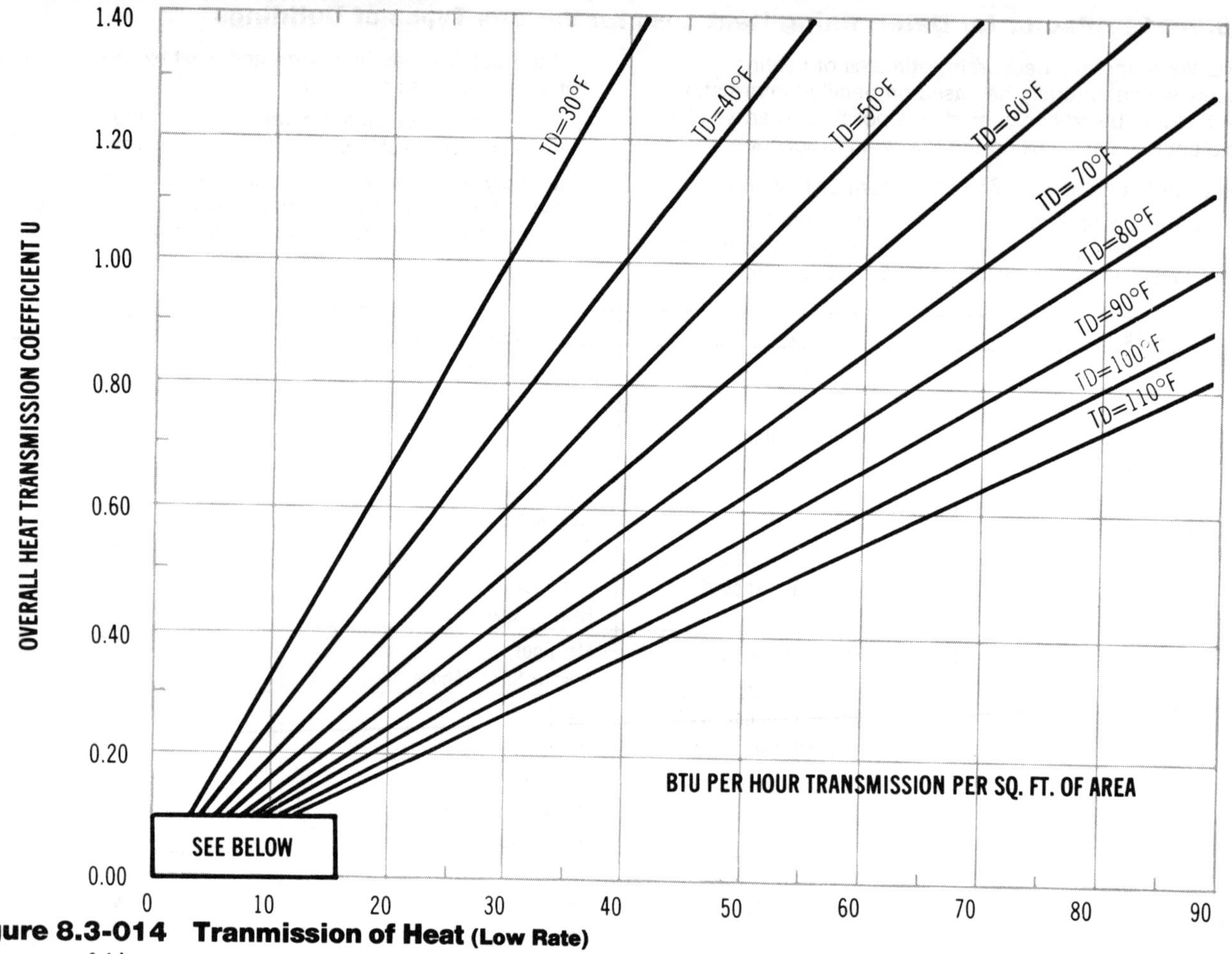

Figure 8.3-014 Tranmission of Heat (Low Rate)

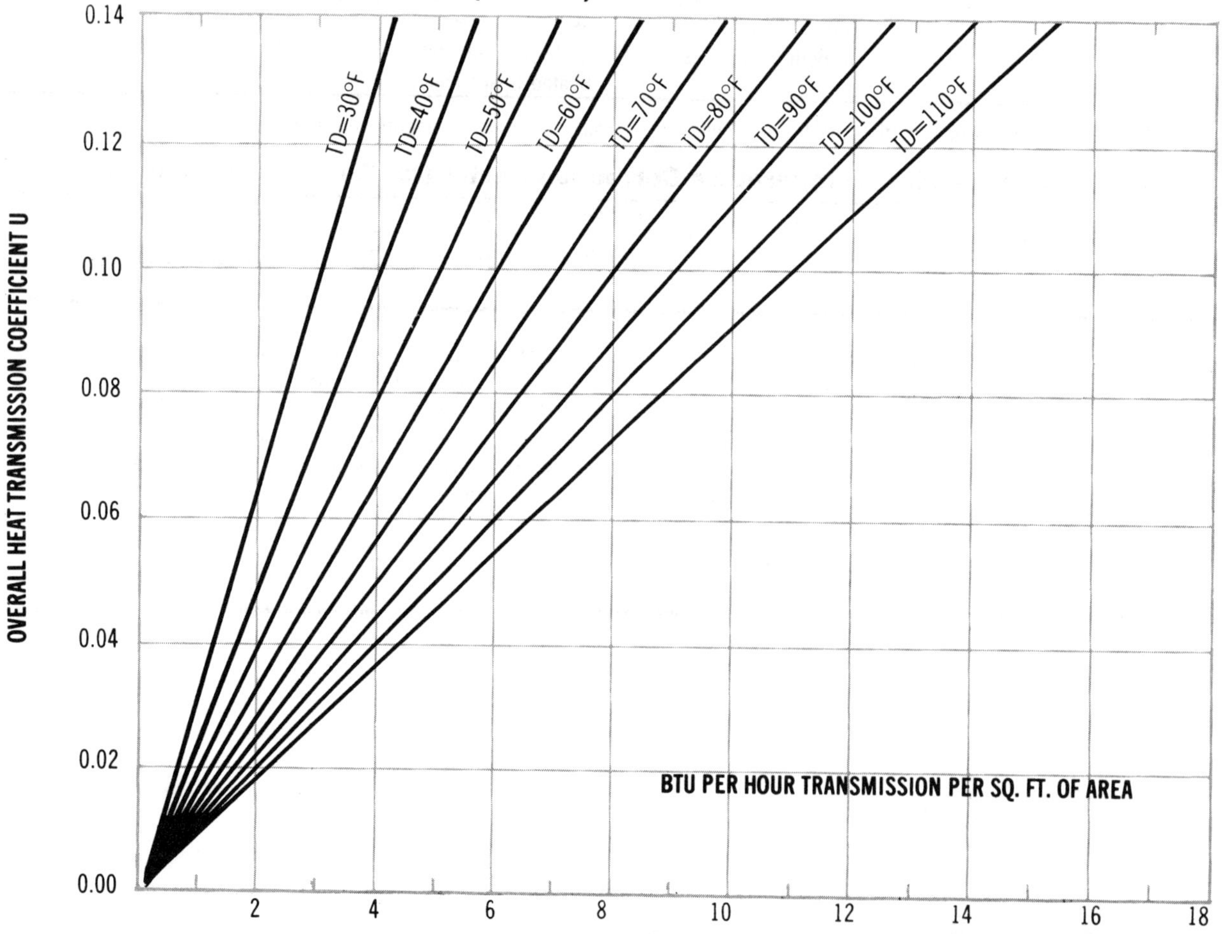

8

Table C8.3-601

Collector Tilt for Domestic Hot Water

Optimum collector tilt is usually equal to the site latitude. Variations of plus or minus 10 degrees are acceptable and orientation of 20 degrees on either side of true south is acceptable; however, local climate and collector type may influence the choice between east or west deviations.

Flat plate collectors consist of a number of components as follows: Insulation to reduce heat loss through the bottom and sides of the collector. The enclosure which contains all the components in this assembly is usually weatherproof and prevents dust, wind and water from coming in contact with the absorber plate. The cover plate usually consists of one or more layers of a variety of glass or plastic and reduces the reradiation. It creates an air space which traps the heat by reducing radiation losses between the cover plate and the absorber plate.

The absorber plate must have a good thermal bond with the fluid passages. The absorber plate is usually metallic and treated with a surface coating which improves absorptivity. Black or dark paints or selective coatings are used for this purpose, and the design of this passage and plate combination helps determine a solar system's effectiveness.

Heat transfer fluid passage tubes are attached above and below or integral with an asborber plate for the purpose of transferring thermal energy from the absorber plate to a heat transfer medium. The heat exchanger is a device for transferring thermal energy from one fluid to another. The rule of thumb of space heating sizing could be one S.F. of collector per 2.5 S.F. of floor space.

For domestic hot water the rule of thumb would be 3/4 S.F. of collector for one gallon of water used per day, on an average use 25 gallons per day per person, plus 10 gallons per dishwasher or washing machine.

Table 8.4-001

GENERAL: The purpose of air conditioning is to control the environment of a space so that comfort is provided for the occupants and/or conditions are suitable for the processes or equipment contained therein. The several items which should be evaluated to define system objectives are:

- Temperature Control
- Humidity Control
- Cleanliness
- Odor, smoke and fumes
- Ventilation

Efforts to control the above parameters must also include consideration of the degree or tolerance of variation, the noise level introduced, the velocity of air motion and the energy requirements to accomplish the desired results.

The variation in **temperature** and **humidity** is a function of the sensor and the controller. The controller reacts to a signal from the sensor and produces the appropriate suitable response in either the terminal unit, the conductor of the transporting medium (air, steam, chilled water, etc.), or the source (boiler, evaporating coils, etc.).

The **noise level** is a by-product of the energy supplied to moving components of the system. Those items which usually contribute the most noise are pumps, blowers, fans, compressors and diffusers. The level of noise can be partially controlled through use of vibration pads, isolators, proper sizing, shields, baffles and sound absorbing liners.

8

Some **air motion** is necessary to prevent stagnation and stratification. The maximum acceptable velocity varies with the degree of heating or cooling which is taking place. Most people feel air moving past them at velocities in excess of 25 FPM as an annoying draft, however, velocities up to 45 FPM may be acceptable in certain cases. Ventilation, expressed as air changes per hour and percentage of fresh air, is usually an item regulated by local codes.

Selection of the system to be used for a particular application is usually a trade-off. In some cases the building size, style, or room available for mechanical use limits the range of possibilities. Prime factors influencing the decision are first cost and total life (operating, maintenance and replacement costs). The accuracy with which each parameter is determined will be an important measure of the reliability of the decision and subsequent satisfactory operation of the installed system.

Heat delivery may be desired from an air conditioning system. Heating capability usually is added as follows: A gas fired burner or hot water/steam/electric coils may be added to the air handling unit directly and heat all air equally. For limited or localized heat requirements the water/steam/electric coils may be inserted into the duct branch supplying the cold areas. Gas fired duct furnaces are also available. **Note,** when water or steam coils are used the cost of piping and boiler must also be added. For a rough estimate use the cost per square foot of the appropriate sized hydronic system with unit heater. This will give the boiler, piping, and the unit heaters which would approximate the cost of the heating coils.The installed cost of electric and gas heaters, boilers and other heat related items on a unit basis may be located in Section 15.5 of *Means Mechanical Cost Data.*

Table 8.4-002 BTU's Per Hour Per S.F. of Floor Area and S.F. Per Ton of Air Conditioning

Air Conditioning Requirements

Type Building	BTU per S.F.	S.F. per ton	Type Building	BTU per S.F.	S.F. per ton	Type Building	BTU per S.F.	S.F. per ton
Apartments, Individual	26	450	Dormitory, Rooms	40	300	Libraries	50	240
Corridors	22	550	Corridors	30	400	Low Rise Office, Exterior	38	320
Auditoriums & Theaters	666	18*	Dress Shops	43	280	Interior	33	360
Banks	50	240	Drug Stores	80	150	Medical Centers	28	425
Barber Shops	48	250	Factories	40	300	Motels	28	425
Bars & Taverns	133	90	High Rise Office-Ext. Rms.	46	263	Office (small suite)	43	280
Beauty Parlors	66	180	Interior Rooms	37	325	Post Office, Individual Office	42	285
Bowling Alleys	68	175	Hospitals, Core	43	280	Central Area	46	260
Churches	600	20*	Perimeter	46	260	Residences	20	600
Cocktail Lounges	68	175	Hotel, Guest Rooms	44	275	Restaurants	60	200
Computer Rooms	141	85	Public Spaces	55	220	Schools & Colleges	46	260
Dental Offices	52	230	Corridors	30	400	Shoe Stores	55	220
Dept. Stores, Basement	34	350	Industrial Plants, Offices	38	320	Shop'g. Ctrs., Super Markets	34	350
Main Floor	40	300	General Offices	34	350	Retail Stores	48	250
Upper Floor	30	400	Plant Areas	40	300	Specialty Shops	60	200

*Persons per ton

12,000 BTU = 1 ton of air conditioning

Table 8.4-003 Psychrometric Table

PSYCHROMETRIC TABLE: DEWPOINT OR SATURATION TEMPERATURE (F)

Relative humidity (%)															
100	32	35	40	45	50	55	60	65	70	75	80	85	90	95	100
90	30	33	37	42	47	52	57	62	67	72	77	82	87	92	97
80	27	30	34	39	44	49	54	58	64	68	73	78	83	88	93
70	24	27	31	36	40	45	50	55	60	64	69	74	79	84	88
60	20	24	28	32	36	41	46	51	55	60	65	69	74	79	83
50	16	20	24	28	33	36	41	46	50	55	60	64	69	73	78
40	12	15	18	23	27	31	35	40	45	49	53	58	62	67	71
30	8	10	14	18	21	25	29	33	37	42	46	50	54	59	62
20	6	7	8	9	13	16	20	24	28	31	35	40	43	48	52
10	4	4	5	5	6	8	9	10	13	17	20	24	27	30	34
Dry bulb temperature (F)	**32**	**35**	**40**	**45**	**50**	**55**	**60**	**65**	**70**	**75**	**80**	**85**	**90**	**95**	**100**

This table shows the relationship between RELATIVE HUMIDITY, DRYBULB TEMPERATURE AND DEWPOINT. As an example, assume that the thermometer in a room reads 75° F, and we know that the relative humidity is 50%. The chart shows the dewpoint temperature to be 55°. That is, any surface colder than 55°F will "sweat" or collect condensing moisture. This surface could be the outside of an uninsulated chilled water pipe in the summertime, or the inside surface of a wall or deck in the wintertime. After determining the extreme ambient parameters. Table 8.4-003 is useful in determining which surfaces need insulation or vapor barrier protection.

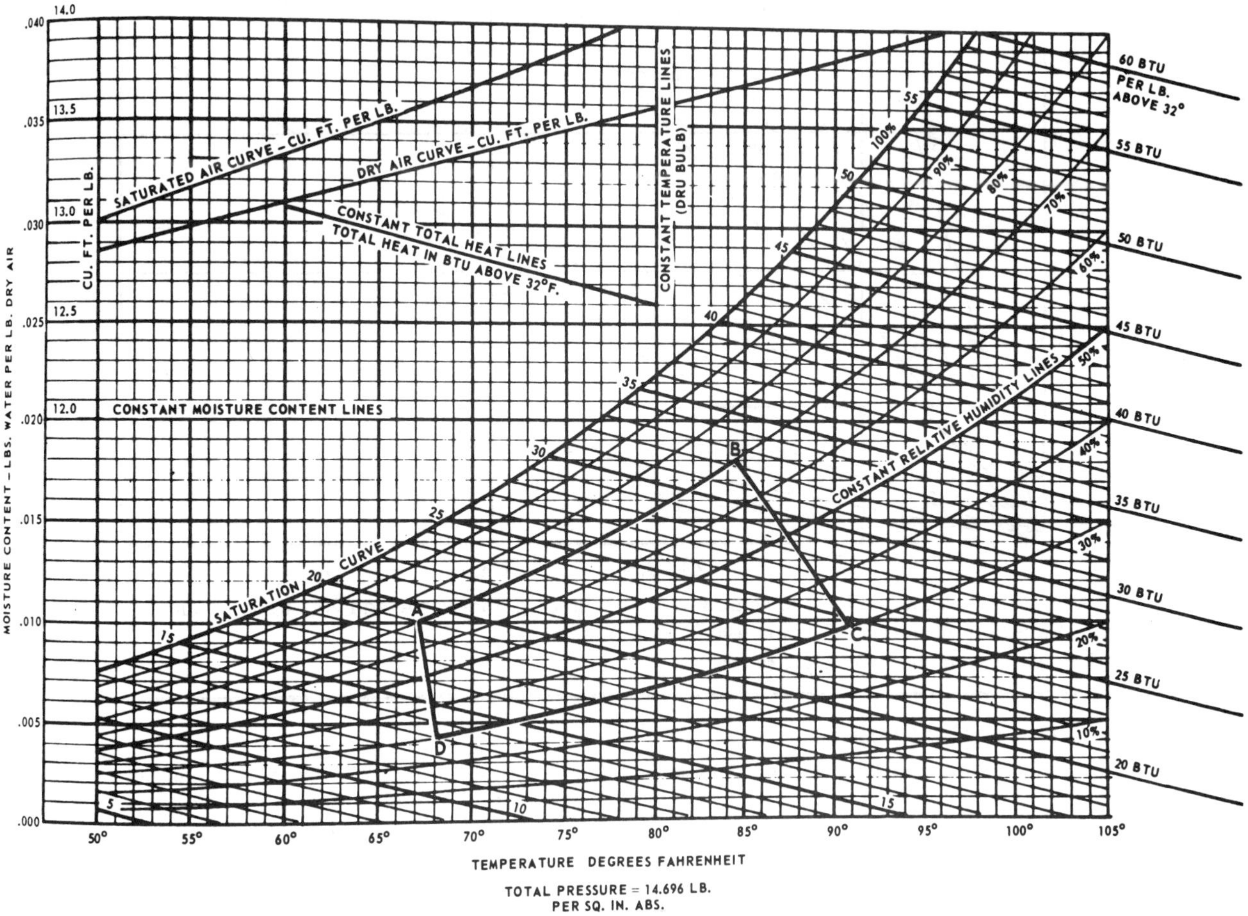

Psychrometric chart showing different variables based on one pound of dry air. Space marked A B C D is temperature-humidity range which is most comfortable for majority of people.

Table 8.4-004 Range of Time in Minutes Per Change for Various Types of Facilities

Recommended Ventilation Air Changes

Assembly Halls 2-10	Dance Halls 2-10	Laundries 1-3
Auditoriums 2-10	Dining Rooms 3-10	Markets 2-10
Bakeries 2-3	Dry Cleaners 1-5	Offices 2-10
Banks 3-10	Factories 2-5	Pool Rooms 2-5
Bars 2-5	Garages 2-10	Recreation Rooms 2-10
Beauty Parlors 2-5	Generator Rooms 2-5	Sales Rooms 2-10
Boiler Rooms 1-5	Gymnasiums 2-10	Theaters 2-8
Bowling Alleys 2-10	Kitchens-Hospitals 2-5	Toilets 2-5
Churches 5-10	Kitchens-Restaurant 1-3	Transformer Rooms 1-5

CFM air required for changes = Volume of room in cubic feet ÷ Minutes per change.

Table 8.4-005 Ductwork (Duct Weight in Pounds per L.F. Straight Runs)

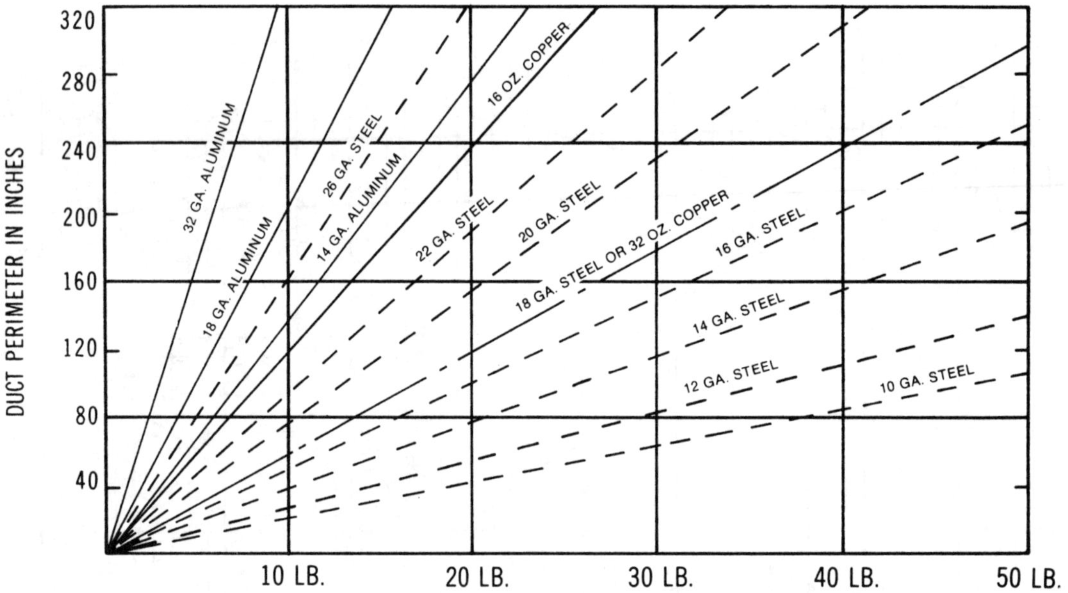

Add to the above for fittings: 90° elbow is 3 L.F.; 45° elbow is 2.5 L.F.; offset is 4 L.F.; transition offset is 6 L.F.; square-to-round transition is 4 L.F.; 90° reducing elbow is 5 L.F. For bracing and waste, add 20% to aluminum and copper, 15% to steel.

Table 8.4-006 Diffuser Evaluation

CFM = V × An × K where V = Outlet velocity in feet per minute. An = Neck area in square feet and K = Diffuser delivery factor. An undersized diffuser for a desired CFM will produce a high velocity and noise level. When air moves past people at a velocity in excess of 25 FPM, an annoying draft is felt. An oversized diffuser will result in low velocity with poor mixing. Consideration must be given to avoid vertical stratification or horizontal areas of stagnation.

Table 8.4-007 Loudness Levels for Moving Air thru Fans, Diffusers, Register, Etc. Measured in Sones

Recommended Loudness Levels (Sones)

Area	Very Quiet	Quiet	Noisy
Auditoriums			
Auditorium lobbies	3	4	6
Concert and opera halls	0.8	1	1.5
Courtrooms	2	3	4
Lecture halls	1.5	2	3
Movie theaters	1.5	2	3
Churches and Schools			
Kitchens	4	6	8
Laboratories	2	4	6
Libraries	1.5	2	3
Recreation halls	2	4	8
Sanctuaries	1	1.7	3
Schools and classrooms	1.5	2.5	4
Hospital and Clinics			
Laboratories	2	4	6
Lobbies, waiting rooms	2	4	6
Halls and corridors	2	4	6
Operating rooms	1.5	2.5	4
Private rooms	1	1.7	3
Wards	1.5	2.5	4

	Very Quiet	Quiet	Noisy
Hotels			
Banquet Rooms	1.5	3	6
Individual rooms, suites	1	2	4
Kitchens and laundries	4	7	10
Lobbies	2	4	8
Indoor Sports			
Bowling alleys	3	4	6
Gymnasiums	2	4	6
Swimming pools	4	7	10
Manufacturing Areas			
Assembly lines	5	12	30
Foreman's office	3	5	8
Foundries	10	20	40
General storage	5	10	20
Heavy machinery	10	25	60
Light machinery	5	12	30
Tool maintenance	4	7	10
Stores			
Department stores	3	6	8
Supermarkets	4	7	10

	Very Quiet	Quiet	Noisy
Offices			
Conference rooms	1	1.7	3
Drafting	2	4	8
General open offices	2	4	8
Halls and corridors	2.5	5	10
Professional offices	1.5	3	6
Reception room	1.5	3	6
Tabulation & computation	3	6	12
Public Buildings			
Banks	2	4	6
Court houses	2.5	4	6
Museums	2	3	4
Planetariums	1.5	2	3
Post offices	2.5	4	6
Public libraries	1.5	2	4
Waiting rooms	3	5	8
Restaurants			
Cafeterias	3	6	10
Night clubs	2.5	4	6
Restaurants	2	4	8

Table 8.4-008 Quality/Complexity Adjustment for Air Conditioning Systems

Economy installation, add .... 0 to 5%
Good quality, medium complexity, add .... 5 to 15%
Above average quality and complexity, add .... 15 to 25%

Add the above adjustments to the estimate to more closely allow for the equipment and conditions of the particular job under consideration.

8

TABLE 8.4-009 Sheet Metal Calculator (Weight in Lb./Ft. of Length)

Gauge	26	24	22	20	18	16	GAUGE	26	24	22	20	18	16
Wt.-Lb./S.F.	.906	1.156	1.406	1.656	2.156	2.656	WT.-Lb./S.F.	.906	1.156	1.406	1.656	2.156	2.656
SMACNA Max. Dimension - Long Side		30″	54″	84″	85″ Up		SMACNA Max. Dimension - Long Side		30″	54″	84″	85″ Up	
Sum-2 Sides							Sum-2 Sides						
2	.3	.40	.50	.60	.80	.90	56	9.3	12.0	14.0	16.2	21.3	25.2
3	.5	.65	.80	.90	1.1	1.4	57	9.5	12.3	14.3	16.5	21.7	25.7
4	.7	.85	1.0	1.2	1.5	1.8	58	9.7	12.5	14.5	16.8	22.0	26.1
5	.8	1.1	1.3	1.5	1.9	2.3	59	9.8	12.7	14.8	17.1	22.4	26.6
6	1.	1.3	1.5	1.7	2.3	2.7	60	10.0	12.9	15.0	17.4	22.8	27.0
7	1.2	1.5	1.8	2.0	2.7	3.2	61	10.2	13.1	15.3	17.7	23.2	27.5
8	1.3	1.7	2.0	2.3	3.0	3.6	62	10.3	13.3	15.5	18.0	23.6	27.9
9	1.5	1.9	2.3	2.6	3.4	4.1	63	10.5	13.5	15.8	18.3	24.0	28.4
10	1.7	2.2	2.5	2.9	3.8	4.5	64	10.7	13.7	16.0	18.6	24.3	28.8
11	1.8	2.4	2.8	3.2	4.2	5.0	65	10.8	13.9	16.3	18.9	24.7	29.3
12	2.0	2.6	3.0	3.5	4.6	5.4	66	11.0	14.1	16.5	19.1	25.1	29.7
13	2.2	2.8	3.3	3.8	4.9	5.9	67	11.2	14.3	16.8	19.4	25.5	30.2
14	2.3	3.0	3.5	4.1	5.3	6.3	68	11.3	14.6	17.0	19.7	25.8	30.6
15	2.5	3.2	3.8	4.4	5.7	6.8	69	11.5	14.8	17.3	20.0	26.2	31.1
16	2.7	3.4	4.0	4.6	6.1	7.2	70	11.7	15.0	17.5	20.3	26.6	31.5
17	2.8	3.7	4.3	4.9	6.5	7.7	71	11.8	15.2	17.8	20.6	27.0	32.0
18	3.0	3.9	4.5	5.2	6.8	8.1	72	12.0	15.4	18.0	20.9	27.4	32.4
19	3.2	4.1	4.8	5.5	7.2	8.6	73	12.2	15.6	18.3	21.2	27.7	32.9
20	3.3	4.3	5.0	5.8	7.6	9.0	74	12.3	15.8	18.5	21.5	28.1	33.3
21	3.5	4.5	5.3	6.1	8.0	9.5	75	12.5	16.1	18.8	21.8	28.5	33.8
22	3.7	4.7	5.5	6.4	8.4	9.9	76	12.7	16.3	19.0	22.0	28.9	34.2
23	3.8	5.0	5.8	6.7	8.7	10.4	77	12.8	16.5	19.3	22.3	29.3	34.7
24	4.0	5.2	6.0	7.0	9.1	10.8	78	13.0	16.7	19.5	22.6	29.6	35.1
25	4.2	5.4	6.3	7.3	9.5	11.3	79	13.2	16.9	19.8	22.9	30.0	35.6
26	4.3	5.6	6.5	7.5	9.9	11.7	80	13.3	17.1	20.0	23.2	30.4	36.0
27	4.5	5.8	6.8	7.8	10.3	12.2	81	13.5	17.3	20.3	23.5	30.8	36.5
28	4.7	6.0	7.0	8.1	10.6	12.6	82	13.7	17.5	20.5	23.8	31.2	36.9
29	4.8	6.2	7.3	8.4	11.0	13.1	83	13.8	17.8	20.8	24.1	31.5	37.4
30	5.0	6.5	7.5	8.7	11.4	13.5	84	14.0	18.0	21.0	24.4	31.9	37.8
31	5.2	6.7	7.8	9.0	11.8	14.0	85	14.2	18.2	21.3	24.7	32.3	38.3
32	5.3	6.9	8.0	9.3	12.2	14.4	86	14.3	18.4	21.5	24.9	32.7	38.7
33	5.5	7.1	8.3	9.6	12.5	14.9	87	14.5	18.6	21.8	25.2	33.1	39.2
34	5.7	7.3	8.5	9.9	12.9	15.3	88	14.7	18.8	22.0	25.5	33.4	39.6
35	5.8	7.5	8.8	10.2	13.3	15.8	89	14.8	19.0	22.3	25.8	33.8	40.1
36	6.0	7.8	9.0	10.4	13.7	16.2	90	15.0	19.3	22.5	26.1	34.2	40.5
37	6.2	8.0	9.3	10.7	14.1	16.7	91	15.2	19.5	22.8	26.4	34.6	41.0
38	6.3	8.2	9.5	11.0	14.4	17.1	92	15.3	19.7	23.0	26.7	35.0	41.4
39	6.5	8.4	9.8	11.3	14.8	17.6	93	15.5	19.9	23.3	27.0	35.3	41.9
40	6.7	8.6	10.0	11.6	15.2	18.0	94	15.7	20.1	23.5	27.3	35.7	42.3
41	6.8	8.8	10.3	11.9	15.6	18.5	95	15.8	20.3	23.8	27.6	36.1	42.8
42	7.0	9.0	10.5	12.2	16.0	18.9	96	16.0	20.5	24.0	27.8	36.5	43.2
43	7.2	9.2	10.8	12.5	16.3	19.4	97	16.2	20.8	24.3	28.1	36.9	43.7
44	7.3	9.5	11.0	12.8	16.7	19.8	98	16.3	21.0	24.5	28.4	37.2	44.1
45	7.5	9.7	11.3	13.1	17.1	20.3	99	16.5	21.2	24.8	28.7	37.6	44.6
46	7.7	9.9	11.5	13.3	17.5	20.7	100	16.7	21.4	25.0	29.0	38.0	45.0
47	7.8	10.1	11.8	13.6	17.9	21.2	101	16.8	21.6	25.3	29.3	38.4	45.5
48	8.0	10.3	12.0	13.9	18.2	21.6	102	17.0	21.8	25.5	29.6	38.8	45.9
49	8.2	10.5	12.3	14.2	18.6	22.1	103	17.2	22.0	25.8	29.9	39.1	46.4
50	8.3	10.7	12.5	14.5	19.0	22.5	104	17.3	22.3	26.0	30.2	39.5	46.8
51	8.5	11.0	12.8	14.8	19.4	23.0	105	17.5	22.5	26.3	30.5	39.9	47.3
52	8.7	11.2	13.0	15.1	19.8	23.4	106	17.7	22.7	26.5	30.7	40.3	47.7
53	8.8	11.4	13.3	15.4	20.1	23.9	107	17.8	22.9	26.8	31.0	40.7	48.2
54	9.0	11.6	13.5	15.7	20.5	24.3	108	18.0	23.1	27.0	31.3	41.0	48.6
55	9.2	11.8	13.8	16.0	20.9	24.8	109	18.2	23.3	27.3	31.6	41.4	49.1
							110	18.3	23.5	27.5	31.9	41.8	49.5

EXAMPLE: If duct is 34″ x 20″ x 15′ long, 34″ is greater than 30″ max., for 24 ga. so must be 22 ga. 34″ + 20″ = 54″ going across from 54″ find 13.5 lb. per foot. 13.5 x 15′ = 202.5 lbs. For sq. ft. of surface area 202.5 ÷ 1.406 = 144 sq. ft. **Note:** figures include an allowance for scrap.

Engineering Fees

Table 10.1-103 Mechanical and Electrical Engineering Fees

Typical **Mechanical and Electrical Engineering Fees** based on the size of the subcontract.

Type of Construction	Subcontract Size							
	$25,000	$50,000	$100,000	$225,000	$350,000	$500,000	$750,000	$1,000,000
Simple structures	6.4%	5.7%	4.8%	4.5%	4.4%	4.3%	4.2%	4.1%
Intermediate structures	8.0	7.3	6.5	5.6	5.1	5.0	4.9	4.8
Complex structures	12.0	9.0	9.0	8.0	7.5	7.5	7.0	7.0

For renovations, add 15% to 25% to applicable fee.

Builders Risk Insurance (Div. 1.1-18)

Builder's Risk Insurance is insurance on a building during construction. Premiums are paid by the owner or the contractor. Blasting, collapse & underground insurance would raise total insurance costs above those listed. Floater policy for materials delivered to the job runs $1.00 to $1.50 per $100 value. Contractor equipment insurance runs $.50 to $2.50 per $100 value.

Tabulated below are New England Builder's Risk insurance rates in dollars per $100 value for $1000 deductible. For $25,000 deductible, rates can be reduced 13% to 34%. On contracts over $500,000 rates may be lower than those tabulated. High liability limits may be difficult to obtain. Policies are written annually for the total completed value in place. For "all risk" insurance (excluding flood, earthquake and other certain perils) add $.022 to total rates below.

Table 10.1-301 New England Builder's Insurance Rates

Coverage	Frame Construction (Class 1)		Brick Construction (Class 4)		Fire Resistive (Class 6)	
	Range	Average	Range	Average	Range	Average
Fire Insurance	$.165 to .378	$.287	$.100 to .177	$.170	$.059 to .168	$.134
Extended Coverage	.077 to .111	.098	.042 to .112	.095	.029 to .075	.063
Vandalism	.008 to .011	.010	.008 to .011	.010	.008 to .011	.010
Total Annual Rate	$.250 to .500	$.395	$.150 to .300	$.275	$.096 to .254	$.207

Performance Bond (Div. 1.1-34)

The table below shows the bond charge for 24 months job; longer jobs add 1% per added month. Contractors should prequalify thru a bonding company agency before submitting a bid on a contract that requires a bond.

Table 10.1-302 Performance Bond

Contract Amount	Building Construction Projects	Highways & Bridges	
		New Construction	Highway Resurfacing
First $ 500,000 bid	$9.00 per M	$7.20 per M	$4.50 per M
Next 2,000,000 bid	$ 4,500 plus $5.85 per M	$ 3,600 plus $4.50 per M	$ 2,250 plus $4.20 per M
Next 2,500,000 bid	16,200 plus 4.90 per M	12,600 plus 3.85 per M	10,650 plus 3.60 per M
Next 2,500,000 bid	28,450 plus 4.20 per M	22,225 plus 3.40 per M	19,650 plus 3.15 per M
Over 7,500,000 bid	38,950 plus 4.10 per M	30,725 plus 3.10 per M	27,525 plus 2.80 per M

Workers' Compensation

The table below tabulates the national averages for Workers' Compensation insurance rates by trade and type of building. The average "Insurance Rate" is multiplied by the "% of Building Cost" for each trade. This produces the "Workers' Compensation Cost" by % of total labor cost, to be added for each trade by building type to determine the weighted average Workers' Compensation rate for the building types analyzed.

Table 10.2-201 Insurance Rates By Trade

Trade	Insurance Rate (% of Labor Cost)		% of Building Cost			Workers' Compensation Cost		
	Range	Average	Office Bldgs.	Schools & Apts.	Mfg.	Office Bldgs.	Schools & Apts.	Mfg.
Excavation, Grading, etc.	1.6% to 27.8%	6.9%	4.8%	4.9%	4.5%	.33%	.34%	.31%
Piles & Foundations	3.5 to 44.3	16.6	7.1	5.2	8.7	1.18	.86	1.44
Concrete	1.8 to 27.8	8.8	5.0	14.8	3.7	.44	1.30	.33
Masonry	1.4 to 18.7	7.2	6.9	7.5	1.9	.50	.54	.14
Structural Steel	3.5 to 40.5	19.5	10.7	3.9	17.6	2.09	.76	3.43
Miscellaneous & Ornamental Metals	1.2 to 15.7	6.6	2.8	4.0	3.6	.18	.26	.24
Carpentry & Millwork	2.4 to 54.1	9.6	3.7	4.0	0.5	.36	.38	.05
Metal or Composition Siding	1.7 to 16.4	7.4	2.3	0.3	4.3	.17	.02	.32
Roofing	4.1 to 44.4	17.1	2.3	2.6	3.1	.39	.44	.53
Doors & Hardware	1.4 to 15.9	5.5	0.9	1.4	0.4	.05	.08	.02
Sash & Glazing	2.5 to 33.3	7.6	3.5	4.0	1.0	.27	.30	.08
Lath & Plaster	1.6 to 22.9	7.2	3.3	6.9	0.8	.24	.50	.06
Tile, Marble & Floors	1.0 to 18.7	5.2	2.6	3.0	0.5	.14	.16	.03
Acoustical Ceilings	1.2 to 15.4	5.5	2.4	0.2	0.3	.13	.01	.02
Painting	2.0 to 16.6	7.2	1.5	1.6	1.6	.11	.12	.12
Interior Partitions	2.4 to 54.1	9.6	3.9	4.3	4.4	.37	.41	.42
Miscellaneous Items	1.1 to 54.1	8.3	5.2	3.7	9.7	.43	.31	.81
Elevators	1.3 to 15.6	5.3	2.1	1.1	2.2	.11	.06	.12
Sprinklers	1.4 to 16.8	5.6	0.5	—	2.0	.03	—	.11
Plumbing	1.2 to 12.5	4.7	4.9	7.2	5.2	.23	.34	.24
Heat., Vent., Air Conditioning	1.7 to 11.4	6.0	13.5	11.0	12.9	.81	.66	.77
Electrical	1.1 to 13.1	3.9	10.1	8.4	11.1	.39	.33	.43
Total	1.0% to 54.1%	—	100.0%	100.0%	100.0%	8.95%	8.18%	10.02%
			Overall Weighted Average				9.05%	

10

The table below lists the weighted average Workers' Compensation base rate for each state with a factor comparing this with the national average of 9.0%.

Table 10.2-202 Insurance Rates by States

State	Weighted Average	Factor	State	Weighted Average	Factor	State	Weighted Average	Factor
Alabama	6.5%	72	Kentucky	5.8%	64	North Dakota	7.1%	79
Alaska	12.0	133	Louisiana	6.5	72	Ohio	7.1	79
Arizona	8.3	92	Maine	14.5	161	Oklahoma	8.2	91
Arkansas	7.5	83	Maryland	14.2	157	Oregon	18.2	202
California	10.6	118	Massachusetts	12.7	141	Pennsylvania	10.7	118
Colorado	8.4	93	Michigan	11.2	124	Rhode Island	12.5	138
Connecticut	16.8	187	Minnesota	11.2	124	South Carolina	6.4	71
Delaware	9.9	110	Mississippi	5.5	61	South Dakota	5.4	60
District of Columbia	16.7	186	Missouri	5.0	55	Tennessee	4.9	54
Florida	6.6	73	Montana	12.2	135	Texas	5.9	65
Georgia	5.5	61	Nebraska	6.0	66	Utah	5.2	57
Hawaii	25.5	283	Nevada	10.8	12	Vermont	6.2	69
Idaho	7.7	86	New Hampshire	13.5	150	Virginia	7.2	80
Illinois	10.9	121	New Jersey	7.7	85	Washington	6.9	76
Indiana	2.2	24	New Mexico	13.2	146	West Virginia	5.0	55
Iowa	6.5	72	New York	9.1	101	Wisconsin	6.8	75
Kansas	6.4	71	North Carolina	4.8	53	Wyoming	5.5	61
Weighted Average for U.S. is 9.0% of payroll = 100								

Rates in the following table are the base or manual costs per $100 of payroll for Workers' Compensation in each state. Rates are usually applied to straight time wages only and not to premium time wages and bonuses.

The weighted average skilled worker rate for 35 trades is 9.0%.. For bidding purposes, apply the full value of Workers' Compensation directly to total labor costs, or, if labor is 32%, materials 48% and overhead and profit 20% of total cost, carry 32/80 x 9.0% = 3.6% of cost (before overhead and profit) into overhead. Rates vary not only from state to state but also with the experience rating of the contractor.

Rates are the most current available at the time of publication.

Table 10.2-203 Workers' Compensation by Trade and State

STATE	CARPENTRY — 3 stories or less 5651	CARPENTRY — interior cab. work 5437	CARPENTRY — general 5403	CONCRETE WORK—NOC 5213	CONCRETE WORK — flat (flr., sdwk.) 5221	ELECTRICAL WIRING — inside 5190	EXCAVATION — earth NOC 6217	EXCAVATION — rock 6217	GLAZIERS 5462	INSULATION WORK 5479	LATHING 5443	MASONRY 5022	PAINTING & DECORATING 5474	PILE DRIVING 6003	PLASTERING 5480	PLUMBING 5183	ROOFING 5551	SHEET METAL WORK (HVAC) 5538	STEEL ERECTION — door & sash 5102	STEEL ERECTION — inter., ornam. 5102	STEEL ERECTION — structure 5040	STEEL ERECTION — NOC 5057	TILE WORK — (interior (ceramic) 5348	WATERPROOFING 9014	WRECKING 5701
AL	6.32	3.56	6.49	6.53	3.67	3.00	6.05	6.05	5.61	4.85	3.51	4.27	6.46	16.85	5.66	2.61	8.74	5.68	3.87	3.87	8.60	11.67	4.33	3.18	11.69
AK	11.03	6.50	9.60	9.29	7.78	6.39	5.99	5.99	9.24	10.97	7.76	13.91	10.14	17.82	9.42	7.69	16.89	7.69	12.95	12.95	35.66	16.82	5.96	3.81	35.66
AZ	6.85	5.91	12.42	5.59	4.47	4.33	4.41	4.41	7.15	6.30	5.63	9.69	5.13	15.23	12.13	3.78	16.18	4.33	10.40	10.40	6.73	11.36	4.47	3.66	12.42
AR	6.84	3.77	6.90	5.77	4.03	2.72	6.41	6.41	5.41	5.21	4.14	4.00	5.23	24.28	4.35	3.62	12.25	5.33	6.14	6.14	18.84	9.55	3.30	3.20	18.84
CA	9.69	4.35	10.55	5.38	5.38	4.22	5.54	5.54	11.89	15.93	6.86	6.90	8.40	17.94	10.38	5.28	23.31	7.07	6.50	6.50	22.64	17.17	5.26	8.40	19.44
CO	6.11	4.74	7.27	7.14	5.23	3.35	7.13	7.13	5.78	7.45	5.34	9.79	5.07	23.02	7.24	5.18	16.93	5.44	4.80	4.80	12.61	14.30	4.07	3.34	14.30
CT	12.59	10.36	14.88	17.21	11.02	4.25	7.33	7.33	16.47	16.37	9.64	14.95	10.17	25.75	10.62	6.44	32.22	9.50	11.00	11.00	62.14	33.41	8.74	3.00	62.14
DE	9.39	9.39	9.39	8.44	5.76	3.28	7.51	7.51	9.36	9.39	6.96	6.04	10.20	10.90	6.96	3.35	25.31	8.00	9.19	9.19	17.99	9.19	6.43	6.04	24.81
DC	8.82	7.14	12.71	20.40	8.80	10.71	12.97	12.97	11.41	11.21	8.56	17.57	9.86	30.03	10.12	12.48	22.50	9.20	15.72	15.72	40.45	32.71	18.65	4.13	40.45
FL	8.59	4.63	10.23	10.99	7.18	4.62	8.01	8.01	7.72	7.66	6.90	7.75	7.03	14.87	7.77	5.57	19.87	6.62	6.04	6.04	16.07	15.13	4.09	4.59	16.07
GA	4.43	3.10	6.70	4.75	3.14	2.53	5.58	5.58	6.52	4.33	2.50	3.67	3.63	9.63	4.33	3.44	7.70	4.68	3.63	3.63	9.96	11.50	2.35	2.24	9.96
HI	15.42	15.87	54.13	27.08	13.77	13.13	27.75	27.75	33.28	16.18	15.39	18.68	16.60	44.30	22.92	11.57	44.35	11.43	11.46	11.46	37.11	42.12	13.90	11.62	37.11
ID	9.63	4.62	9.13	5.68	3.70	3.18	7.49	7.49	6.08	6.02	4.68	6.01	6.29	14.55	7.37	3.27	14.73	4.62	5.52	5.52	12.67	11.50	4.25	3.85	12.67
IL	7.72	5.36	9.93	15.79	5.58	4.61	5.46	5.46	7.32	7.63	4.84	7.17	8.01	23.30	7.58	5.82	18.18	6.56	8.03	8.03	33.79	25.32	4.44	2.62	33.79
IN	2.19	1.43	2.35	1.84	1.20	1.07	1.55	1.55	2.53	1.66	1.18	1.40	2.02	3.52	1.63	1.20	4.12	1.65	1.15	1.15	3.46	5.63	1.04	1.12	3.46
IA	3.82	3.28	8.04	7.19	4.07	3.81	4.23	4.23	4.32	4.60	3.66	4.20	5.92	9.32	4.17	4.07	9.62	4.35	4.38	4.38	13.16	21.70	2.77	2.41	13.16
KS	4.82	4.20	5.40	4.02	4.15	1.93	3.17	3.17	5.64	8.25	4.53	4.87	6.75	16.77	4.13	2.85	12.72	3.85	3.86	3.86	7.97	17.39	3.11	1.87	17.39
KY	6.94	3.02	6.87	3.80	3.75	2.55	4.27	4.27	4.58	5.04	4.34	4.60	5.99	9.80	4.12	2.72	11.54	3.94	4.14	4.14	10.81	7.33	3.36	4.09	10.81
LA	8.06	4.35	7.60	4.99	3.64	2.47	5.34	5.34	5.24	4.26	3.08	3.35	6.07	17.46	5.05	3.87	10.71	4.42	4.36	4.36	10.96	8.65	4.26	3.15	10.96
ME	6.59	7.64	19.90	14.58	7.71	5.73	11.59	11.59	11.18	10.36	7.54	10.04	11.33	30.32	10.32	7.39	27.20	9.41	9.07	9.07	36.50	29.85	7.24	4.96	36.50
MD	9.88	8.64	10.08	22.07	7.66	7.09	9.97	9.97	12.88	12.17	8.05	9.14	10.99	25.71	8.79	9.26	25.93	11.12	11.16	11.16	37.14	27.44	8.19	4.16	37.14
MA	8.52	4.57	18.12	12.64	5.57	3.50	5.32	5.32	8.46	6.20	6.03	10.41	11.00	14.09	7.20	5.25	52.69	7.33	7.92	7.92	30.32	28.70	5.65	5.05	30.32
MI	7.67	6.04	10.58	15.03	8.72	3.89	9.45	9.45	9.51	8.89	7.79	10.40	10.88	18.73	11.59	4.81	19.05	7.40	6.52	6.52	18.06	18.51	8.16	6.49	18.06
MN	7.87	7.87	14.47	9.78	5.80	2.96	8.82	8.82	10.05	6.29	6.54	7.14	10.14	16.79	6.54	7.87	23.44	5.75	9.92	9.92	19.77	27.02	5.75	6.09	19.77
MS	4.90	3.31	6.08	3.97	3.36	3.12	4.87	4.87	4.25	3.36	3.69	2.98	3.16	17.09	5.12	1.61	7.40	4.11	3.94	3.94	10.64	7.11	3.84	2.37	10.64
MO	3.49	3.80	3.72	5.07	2.48	1.90	3.57	3.57	4.78	5.04	2.76	3.41	4.65	11.16	3.25	2.37	8.14	3.26	3.41	3.41	8.75	12.29	2.30	2.56	12.29
MT	9.65	6.44	13.37	5.89	5.90	4.33	9.15	9.15	7.62	10.66	6.66	12.75	10.38	27.11	9.09	7.00	25.10	7.42	7.16	7.16	32.17	20.61	5.29	5.70	32.17
NE	5.22	2.76	6.08	3.99	3.89	1.91	4.59	4.59	4.17	4.79	3.73	5.01	3.73	12.27	5.44	2.92	14.80	3.18	3.25	3.25	9.16	14.45	2.76	4.21	14.45
NV	10.82	10.82	10.82	8.51	8.51	5.48	8.51	8.51	7.34	10.82	10.82	10.24	6.92	8.51	6.92	6.74	17.31	6.74	10.82	10.82	17.11	17.11	7.34	8.51	17.11
NH	11.45	5.85	11.18	15.59	7.44	5.03	15.17	15.17	8.77	10.49	9.66	8.86	10.76	26.95	12.67	6.85	36.18	6.82	8.09	8.09	24.11	19.47	6.43	4.28	24.11
NJ	6.36	5.17	6.36	6.50	5.19	2.68	6.17	6.17	5.63	6.20	6.17	7.83	9.23	10.76	6.17	3.35	14.61	4.54	8.61	8.61	18.45	7.50	3.04	3.27	23.75
NM	11.42	8.21	10.19	17.65	8.59	4.32	8.00	8.00	9.89	11.47	7.64	8.99	8.20	40.49	11.68	6.88	22.61	9.46	9.74	9.74	24.45	16.81	8.63	5.47	24.45
NY	6.63	4.03	8.89	12.97	7.04	3.92	8.40	8.40	9.92	4.73	7.20	8.65	7.28	14.54	12.46	5.39	NA	7.00	4.19	4.19	18.60	16.60	5.70	3.33	26.03
NC	3.36	2.56	5.20	3.22	2.45	2.76	3.40	3.40	3.55	3.85	2.92	2.51	4.08	13.25	3.30	2.45	8.24	3.48	2.88	2.88	16.93	5.36	2.22	1.15	16.93
ND	6.24	6.24	6.24	5.40	5.40	2.20	5.42	5.42	8.16	4.00	3.08	3.10	5.03	14.10	3.13	6.22	8.20	6.22	6.24	6.24	14.10	14.10	2.90	8.20	NA
OH	4.77	4.77	4.77	4.81	4.81	1.98	4.81	4.81	5.38	5.01	5.01	4.40	5.38	4.81	5.01	3.26	10.95	8.26	NA	NA	18.31	18.31	2.32	10.95	NA
OK	7.54	4.73	7.21	6.36	4.30	2.95	5.98	5.98	6.15	6.69	4.78	4.36	4.54	16.30	7.28	4.39	19.81	5.51	5.70	5.70	23.36	12.95	3.59	3.70	23.36
OR	20.22	10.68	18.77	16.51	10.83	5.28	11.76	11.76	11.32	17.27	12.35	16.91	16.06	29.43	16.23	7.50	36.17	10.93	12.35	12.35	35.07	30.85	15.87	11.76	35.07
PA	6.61	6.61	7.71	14.05	5.87	4.51	9.21	9.21	9.70	7.71	7.99	10.86	12.00	11.72	7.99	5.20	17.07	8.56	9.47	9.47	25.75	9.47	6.61	10.86	40.47
RI	8.14	5.37	9.28	10.02	8.89	5.14	7.64	7.64	9.93	7.82	5.97	8.37	9.18	24.41	8.30	3.74	24.12	6.21	7.33	7.33	40.04	38.15	4.77	4.11	40.04
SC	6.48	4.20	7.73	4.80	2.75	5.31	4.74	4.74	4.31	5.55	3.60	4.77	6.30	10.23	5.21	2.86	11.54	5.39	3.70	3.70	17.35	9.77	2.80	1.76	17.35
SD	4.22	3.01	6.97	4.76	3.00	2.36	5.42	5.42	4.63	4.26	3.08	3.34	3.63	10.34	3.92	3.92	11.39	3.26	3.50	3.50	9.36	9.24	2.89	1.82	9.36
TN	4.76	3.37	5.70	4.31	3.50	2.49	4.36	4.36	4.05	3.89	3.95	3.38	4.05	9.82	3.32	2.52	6.88	3.77	3.65	3.65	8.94	7.22	2.54	1.63	8.94
TX	5.33	3.83	5.33	5.02	3.49	2.89	3.59	3.59	4.60	5.86	2.33	3.64	3.69	11.16	6.85	2.91	13.85	5.55	3.28	3.28	16.57	6.45	2.82	3.47	13.77
UT	10.89	NA	4.39	6.25	2.78	2.10	2.97	2.97	4.12	5.04	3.86	4.28	4.07	10.22	4.45	2.20	11.91	3.85	3.45	3.45	NA	6.85	2.58	2.26	5.91
VT	4.15	3.14	5.32	6.92	3.25	3.38	5.24	5.24	5.02	5.07	3.81	5.30	4.24	12.40	4.75	2.69	11.71	4.41	4.64	4.64	13.07	10.62	3.13	2.51	13.07
VA	4.23	5.67	7.30	7.00	4.17	3.19	5.69	5.69	4.65	6.09	7.26	5.35	6.82	11.75	4.95	4.24	12.76	5.63	4.52	4.52	15.40	13.53	4.23	1.84	15.40
WA	6.72	6.72	6.72	5.38	4.61	2.35	6.19	6.19	7.51	5.58	6.72	7.57	6.37	12.79	6.70	2.49	6.48	2.84	6.17	6.17	8.58	8.58	5.16	6.97	7.64
WV	4.38	4.38	4.38	4.67	4.67	2.08	4.74	4.74	4.38	4.38	4.82	4.82	4.38	4.71	4.82	2.30	4.94	4.82	6.29	6.29	5.86	6.29	4.82	1.60	4.67
WI	4.20	3.54	9.07	6.59	4.75	3.39	4.21	4.21	5.07	7.23	4.66	4.37	4.58	12.39	4.77	4.06	13.11	5.09	4.66	4.66	16.01	11.72	3.67	3.37	23.35
WY	5.00	5.00	5.00	5.00	5.00	5.00	5.00	5.00	5.00	5.00	5.00	5.00	5.00	5.00	5.00	5.00	5.00	5.00	5.00	5.00	5.00	5.00	5.00	5.00	5.00
AVG.	7.39	5.49	9.56	8.77	5.46	3.87	6.87	6.87	7.60	7.35	5.78	7.20	7.20	16.57	7.24	4.71	17.13	6.01	6.60	6.60	19.53	15.89	5.16	4.43	20.66

10

Table 10.2-204 Workers' Compensation by Trade and Province

PROVINCE		Alberta	British Columbia	Manitoba	Ontario	New Brunswick	Newfndld & Labrador	Northwest Territories	Nova Scotia	Prince Edward Island	Quebec	Saskat-chewan	Yukon
CARPENTRY—3 stories or less	Rate	5.50	3.50	2.55	3.17	3.05	3.50	4.75	2.00	4.25	6.60	3.75	3.50
	Code	8-04	060412	401	062-08	403	4-3	4-41	700	401	42251	B01-11	4
CARPENTRY—interior cab. work	Rate	5.50	3.50	2.55	3.17	3.05	3.50	4.75	2.00	4.25	6.60	3.75	3.50
	Code	8-04	060412	401	062-08	403	4-3	4-41	700	401	42251	B01-11	4
CARPENTRY—general	Rate	5.50	3.50	2.55	3.17	3.05	3.50	4.75	2.00	4.25	6.60	3.75	3.50
	Code	8-04	060412	401	062-08	403	4-3	4-41	700	401	42251	B01-11	4
CONCRETE WORK—NOC	Rate	5.50	6.02	2.55	6.82	3.05	3.50	4.75	2.00	4.25	9.21	3.75	4.50
	Code	8-01	070604	401	744-09	403	4-3	4-41	700	401	42221	B01-22	3
CONCRETE WORK—flat (flr., sidewalk)	Rate	4.50	6.02	2.55	6.82	3.05	3.50	4.75	2.00	4.25	9.21	3.75	4.50
	Code	6-01	070604	401	744-09	403	4-3	4-41	700	401	42221	B01-22	3
ELECTRICAL Wiring—inside	Rate	2.60	2.69	1.40	3.72	3.30	1.90	2.75	1.00	2.00	4.58	3.75	3.50
	Code	6-06	071100	402	864-07	404	4-2	4-46	701	402	42133	B01-03	4
EXCAVATION—earth NOC	Rate	5.55	5.35	3.95	9.52	3.05	3.50	5.00	2.00	4.25	7.83	2.75	7.25
	Code	6-07	072607	407	753-13	403	4-3	4-43	700	401	42242	B01-06	2
EXCAVATION—rock	Rate	5.55	5.35	3.95	9.52	3.05	3.50	5.00	2.00	4.25	7.83	2.75	7.25
	Code	6-07	072607	407	753-13	403	4-3	4-43	700	401	42242	B01-06	2
GLAZIERS	Rate	4.75	3.33	2.55	7.58	3.30	1.90	4.75	2.00	2.00	9.21	3.75	3.50
	Code	6-03	060236	401	873-11	404	4-2	4-41	700	402	42181	B01-20	4
INSULATION WORK	Rate	4.75	5.18	2.55	7.58	3.30	1.90	4.75	2.00	4.25	6.60	3.75	4.50
	Code	6-03	070504	401	873-11	404	4-2	4-41	700	401	42272	B01-17	3
LATHING	Rate	4.75	5.18	2.55	7.02	3.30	1.90	4.75	2.00	2.00	6.60	3.75	4.50
	Code	6-03	070500	401	854-12	404	4-2	4-41	700	402	42162	B01-17	3
MASONRY	Rate	4.85	6.18	2.55	7.02	3.05	3.50	4.75	2.00	4.25	9.21	3.75	4.50
	Code	6-04	070602	401	854-12	403	4-3	4-41	700	401	42211	B01-18	3
PAINTING & DECORATING	Rate	4.75	5.18	2.55	7.58	3.30	1.90	3.25	2.00	2.00	5.52	3.75	4.50
	Code	6-03	070501	401	873-11	404	4-2	4-49	700	402	42151	B01-02	3
PILE DRIVING	Rate	4.50	8.85	3.95	9.46	3.30	3.50	5.00	2.00	4.25	16.28	3.75	7.25
	Code	6-01	072502	407	836-13	404	4-3	4-43	700	401	42298	B01-38	2
PLASTERING	Rate	4.75	5.18	2.55	7.02	3.30	1.90	4.75	2.00	2.00	6.60	3.75	4.50
	Code	6-03	070502	401	854-12	404	4-2	4-41	700	402	42162	B01-17	3
PLUMBING	Rate	3.25	4.34	1.40	3.72	3.30	1.90	2.75	1.25	2.00	5.52	3.75	3.50
	Code	6-02	070712	402	864-07	404	4-2	4-46	703	402	42114	B01-01	4
ROOFING	Rate	9.00	5.87	5.10	7.02	3.30	3.50	4.75	2.00	4.25	9.21	3.75	4.50
	Code	6-05	070600	404	854-12	404	4-3	4-41	700	401	42182	B01-28	3
SHEET METAL WORK (HVAC)	Rate	3.25	4.54	5.10	3.72	3.30	3.50	4.75	2.00	2.00	6.60	3.75	4.50
	Code	6-02	070714	404	864-07	404	4-3	4-41	700	402	42183	B01-04	3
STEEL ERECTION—door & sash	Rate	4.75	8.85	7.20	7.79	3.30	3.50	4.75	2.00	4.25	9.21	3.75	7.25
	Code	8-03	072509	405	827-09	403	4-3	4-41	700	401	42261	B01-32	2
STEEL ERECTION—inter., ornam.	Rate	4.75	8.85	7.20	7.79	3.30	3.50	4.75	2.00	4.25	9.21	3.75	7.25
	Code	8-03	072509	405	827-09	403	4-3	4-41	700	401	42261	B01-32	2
STEEL ERECTION—structure	Rate	4.75	8.85	7.20	17.81	3.30	3.50	6.00	4.35	4.25	9.21	3.75	7.25
	Code	8-03	072509	405	827-14	403	4-3	4-44	532	401	42261	B01-37	2
STEEL ERECTION—NOC	Rate	4.75	8.85	7.20	7.79	3.30	3.50	4.75	4.35	4.25	9.21	3.75	7.25
	Code	8-03	072509	405	827-09	403	4-3	4-41	532	401	42261	B01-32	2
TILE WORK—inter. (ceramic)	Rate	4.75	5.18	2.55	3.72	3.30	1.90	3.25	2.00	2.00	5.52	3.75	4.50
	Code	6-03	070506	401	864-07	404	4-2	4-49	700	402	42172	B01-16	3
WATERPROOFING	Rate	3.25	3.30	3.95	7.58	3.30	3.50	4.75	2.00	2.00	12.45	3.75	3.50
	Code	6-02	060237	407	873-11	404	4-3	4-41	700	402	42297	B01-18	4
WRECKING	Rate	10.35	5.87	2.55	19.14	3.30	3.50	5.00	2.00	4.25	20.57	3.75	7.25
	Code	6-08	070600	401	859-15	403	4-3	4-43	700	401	42288	B01-39	2

GENERAL REQUIREMENTS C10.2-300 | Unemployment & Social Security

Unemployment Taxes and Social Security Taxes (Div. 1.1-56)

Mass. State Unemployment tax ranges from 1.5% to 5.7% plus an experience rating assessment the following year, on the first $7,000 of wages. Federal Unemployment tax is 3.5% of the first $7,000 of wages. This is reduced by a credit for payment to the state. The minimum Federal Unemployment tax is .8% after all credits.

Combined rates in Mass. thus vary from 2.3% to 6.5% of the first $7,000 of wages. Combined average U.S. rate is about 5.5% of the first $7,000. Contractors with permanent workers will pay less since the average annual wages for skilled workers is $20.10 x 2,000 hours or about $40,200 per year. The average combined rate for U.S. would thus be 5.5% x 7,000 ÷ 40,200 = .96% of total wages for permanent employees.

Rates not only vary from state to state but also with the experience rating of the contractor.

Social Security (FICA) for 1985 is estimated at time of publication to be 7.05% of wages up to $40,000.

GENERAL REQUIREMENTS C10.2-400 | Overtime

Overtime (Div. 1.1-32)

The table below shows the effective hourly cost of overtime over a prolonged period of time compared to the hourly payroll cost. Short-term overtime does not have as great a reduction in efficiency and cost premiums should approach the payroll costs rather than the effective hourly costs listed below. As the total hours per week are increased, more time is lost by absenteeism and the accident rate increases.

Table 10.2-401 Overtime: Production Efficiency

Days per Week	Hours per Day	Total Hours Worked	Actual Productive Hours	Production Efficiency	Payroll Cost per Hour Overtime after 40 hrs.		Effective Cost per Hour Overtime after 40 hrs.	
					@ 1-1/2 times	@ 2 times	@ 1-1/2 times	@ 2 times
5	8	40	40.0	100.0%	100.0%	100.0%	100.0%	100.0%
	9	45	43.4	96.5	105.6	111.1	109.4	115.2
	10	50	46.5	93.0	110.0	120.0	118.3	129.0
	11	55	49.2	89.5	113.6	127.3	127.0	142.3
	12	60	51.6	86.0	116.7	133.3	135.7	155.0
6	8	48	46.1	96.0	108.3	116.7	112.8	121.5
	9	54	48.9	90.6	113.0	125.9	124.7	139.1
	10	60	51.1	85.2	116.7	133.3	137.0	156.6
	11	66	52.7	79.8	119.7	139.4	149.9	174.6
	12	72	53.6	74.4	122.2	144.4	164.2	194.0
7	8	56	48.8	87.1	114.3	128.6	131.1	147.5
	9	63	52.2	82.8	118.3	136.5	142.7	164.8
	10	70	55.0	78.5	121.4	142.9	154.5	181.8
	11	77	57.1	74.2	124.0	148.1	167.3	199.6
	12	84	58.7	69.9	126.2	152.4	180.6	218.1

10

GENERAL REQUIREMENTS C10.3-100 | General

Sales Tax (Div. 1.1-56)

State sales tax on materials is tabulated below (5 states have no sales tax). Many states allow local jurisdictions, such as a county or city, to levy additional sales tax. Some projects may be sales tax exempt, particularly those constructed with public funds.

Table 10.3-101 Sales Tax by State

State	Tax	State	Tax	State	Tax	State	Tax
Alabama	4%	Illinois	5%	Montana	0%	Rhode Island	6%
Alaska	0	Indiana	5	Nebraska	3.5	South Carolina	5
Arizona	5	Iowa	4	Nevada	5.75	South Dakota	4
Arkansas	4	Kansas	3	New Hampshire	0	Tennessee	5.5
California	4.75	Kentucky	5	New Jersey	6	Texas	4
Colorado	3	Louisiana	4	New Mexico	3.75	Utah	4.625
Connecticut	6	Maine	5	New York	4	Vermont	4
Delaware	0	Maryland	5	North Carolina	3	Virginia	3
District of Columbia	6	Massachusetts	5	North Dakota	4	Washington	6.5
Florida	5	Michigan	4	Ohio	5	West Virginia	5
Georgia	3	Minnesota	6	Oklahoma	3	Wisconsin	5
Hawaii	4	Mississippi	6	Oregon	0	Wyoming	3
Idaho	4	Missouri	4.125	Pennsylvania	6	Average	4.11%

Subcontractor's Overhead & Profit (Div. 1.1)

Listed below in the last two columns are **average** billing rates for Subcontractor's labor.

The Base Rate is for the building construction industry and includes the usual negotiated fringe benefits. These wage rates with trends are listed on the inside back cover. Workers' Compensation is a national average of states which have established rates in each trade. Average Fixed Overhead is a total of average rates for U.S. and State Unemployment, 5.5%; Social Security (FICA), 7.05%; Builders' Risk, 0.38%; and Public Liability, 0.82%. These are analyzed in C10.1-300 and C10.2-300. All the rates except Social Security vary from state to state as well as from company to company.

The Subcontractor's Overhead presumes annual billing between 0.5 and 1.5 million dollars. A Subcontractor with lower annual billing usually has a higher overhead percentage.

Subcontractor's Overhead varies greatly within each trade. Some controlling factors are Annual Volume, Type Job, Size Job, Location, Local Economic Conditions, Engineering and Logistical Support Staff and Equipment Requirements. These should be examined carefully for each job.

Abbr.	Trade	Base Rate Incl. Fringes		Workers' Comp. Ins.	Average Fixed Overhead	Subs Overhead	Subs Profit	Subs Total Overhead & Profit		Rate with Subs O & P	
		Hourly	Daily					%	Amount	Hourly	Daily
Skwk	Skilled Workers Average (35 trades)	$20.10	$160.80	9.0%	13.8%	12.8%	10%	45.6%	$ 9.15	$29.25	$234.00
	Helpers Average (5 trades)	15.20	121.60	9.3		13.0		46.1	7.00	22.20	177.60
	Foremen Average, Inside (50¢ over trade)	20.60	164.80	9.0		12.8		45.6	9.40	30.00	240.00
	Foremen Average, Outside ($2.00 over trade)	22.10	176.80	9.0		12.8		45.6	10.10	32.20	257.60
Clab	Common Building Laborers	15.50	124.00	9.6		11.0		44.4	6.90	22.40	179.20
Asbe	Asbestos Workers	22.20	177.60	7.4		16.0		47.2	10.50	32.70	261.60
Boil	Boilermakers	22.30	178.40	6.7		16.0		46.5	10.35	32.65	261.20
Bric	Bricklayers	20.05	160.40	7.2		11.0		42.0	8.40	28.45	227.60
Brhe	Bricklayer Helpers	15.65	125.20	7.2		11.0		42.0	6.55	22.20	177.60
Carp	Carpenters	19.60	156.80	9.6		11.0		44.4	8.70	28.30	226.40
Cefi	Cement Finishers	18.80	150.40	5.5		11.0		40.3	7.60	26.40	211.20
Elec	Electricians	22.10	176.80	3.9		16.0		43.7	9.65	31.75	254.00
Elev	Elevator Constructors	22.15	177.20	5.3		16.0		45.1	10.00	32.15	257.20
Eqhv	Equipment Operators, Crane or Shovel	20.65	165.20	6.9		14.0		44.7	9.25	29.90	239.20
Eqmd	Equipment Operators, Medium Equipment	20.20	161.60	6.9		14.0		44.7	9.05	29.25	234.00
Eqlt	Equipment Operators, Light Equipment	19.05	152.40	6.9		14.0		44.7	8.50	27.55	220.40
Eqol	Equipment Operators, Oilers	17.10	136.80	6.9		14.0		44.7	7.65	24.75	198.00
Eqmm	Equipment Operators, Master Mechanics	21.45	171.60	6.9		14.0		44.7	9.60	31.05	248.40
Glaz	Glaziers	19.85	158.80	7.6		11.0		42.4	8.40	28.25	226.00
Lath	Lathers	19.75	158.00	5.8		11.0		40.6	8.00	27.75	222.00
Marb	Marble Setters	19.55	156.40	7.2		11.0		42.0	8.20	27.75	222.00
Mill	Millwrights	20.35	162.80	6.4		11.0		41.2	8.40	28.75	230.00
Mstz	Mosaic and Terrazzo Workers	19.35	154.80	5.2		11.0		40.0	7.75	27.10	216.80
Pord	Painters, Ordinary	18.85	150.80	7.2		11.0		42.0	7.90	26.75	214.00
Psst	Painters, Structural Steel	19.60	156.80	26.0		11.0		60.8	11.90	31.50	252.00
Pape	Paper Hangers	19.15	153.20	7.2		11.0		42.0	8.05	27.20	217.60
Pile	Pile Drivers	19.70	157.60	16.6		16.0		56.4	11.10	30.80	246.40
Plas	Plasterers	19.60	156.80	7.2		11.0		42.0	8.25	27.85	222.80
Plah	Plasterer Helpers	16.15	129.20	7.2		11.0		42.0	6.80	22.95	183.60
Plum	Plumbers	22.20	177.60	4.7		16.0		44.5	9.90	32.10	256.80
Rodm	Rodmen (Reinforcing)	21.30	170.40	15.9		14.0		53.7	11.45	32.75	262.00
Rofc	Roofers, Composition	18.35	146.80	17.1		11.0		51.9	9.50	27.85	222.80
Rots	Roofers, Tile & Slate	18.50	148.00	17.1		11.0		51.9	9.60	28.10	224.80
Rohe	Roofer Helpers (Composition)	13.45	107.60	17.1		11.0		51.9	7.00	20.45	163.60
Shee	Sheet Metal Workers	22.20	177.60	6.0		16.0		45.8	10.15	32.35	258.80
Spri	Sprinkler Installers	22.60	180.80	5.6		16.0		45.4	10.25	32.85	262.80
Stpi	Steamfitters or Pipefitters	22.30	178.40	4.7		16.0		44.5	9.90	32.20	257.60
Ston	Stone Masons	19.85	158.80	7.2		11.0		42.0	8.35	28.20	225.60
Sswk	Structural Steel Workers	21.30	170.40	19.5		14.0		57.3	12.20	33.50	268.00
Tilf	Tile Layers (Floor)	19.25	154.00	5.1		11.0		39.9	7.70	26.95	215.60
Tilh	Tile Layer Helpers	15.25	122.00	5.1		11.0		39.9	6.10	21.35	170.80
Trlt	Truck Drivers, Light	16.00	128.00	8.4		11.0		43.2	6.90	22.90	183.20
Trhv	Truck Drivers, Heavy	16.20	129.60	8.4		11.0		43.2	7.00	23.20	185.60
Sswl	Welders, Structural Steel	21.30	170.40	19.5		14.0		57.3	12.20	33.50	268.00
Wrck	*Wrecking	15.50	124.00	20.7		11.0		55.5	8.60	24.10	192.80

*Not included in Averages.

Contractor Equipment (Div. 1.5)

Rental Rates shown in the front of the book pertain to late model high quality machines in excellent working condition, rented from equipment dealers. Rental rates from contractors may be substantially lower than the rental rates from equipment dealers depending upon economic conditions. For older, less productive machines, reduce rates by a maximum of 12%. Any overtime must be added to the base rates. For shift work, rates are lower. Usual rule of thumb is 150% of one shift rate for two shifts; 200% for three shifts.

Daily Operated rates indicated by (*) include operator (and oiler if required), fuel, maintenance and repair. For periods of less than one week, operated equipment is usually more economical to rent than renting bare equipment and hiring an operator.

Equipment moving and mobilization costs must be added to rental rates where applicable. A large crane, for instance, may take two days to erect and two days to dismantle.

Rental rates vary throughout the country with larger cities generally having lower rates. Lease plans for new equipment are available for periods in excess of six months with a percentage of payments applying toward purchase. Recently long-term rental costs for the larger cranes has not increased due to intense competition between manufacturers. During a recession, rental rates do not rise as fast as dealer costs, due to market conditions.

Monthly rental rates vary from 2% to 5% of the cost of the equipment depending on the anticipated life of the equipment and its wearing parts. Weekly rates are about 1/3 the monthly rates and daily rental rates about 1/3 the weekly rate.

The hourly operating costs for each piece of equipment includes costs to the user such as fuel, oil, lubrication, normal expendables for the equipment, and a percentage of mechanic's wages chargeable to maintenance. Hourly operating costs do not include the operator's wages and do not apply to fully operated equipment designated by an *

The daily cost for equipment used in the standard crews (foreword p. vii to xii) is figured by dividing the weekly rate by five, then adding eight times the hourly operating cost to give the total daily equipment cost, not including the operator. This figure is in the right hand column of Division 1.5 under Crew Equip. Cost.

Pile Driving rates shown for pile hammer and extractor do not include leads, crane, boiler or compressor. Vibratory pile driving requires an added field specialist at $270 per day during set-up and pile driving operation for the electric model. The hydraulic model requires a field specialist for set-up only. Up to 125 reuses of sheet piling are possible using vibratory drivers. For normal conditions, crane capacity for hammer type and size are as follows.

Crane Capacity	Hammer Type and Size		
	Air or Steam	Diesel	Vibratory
25 ton 40 ton 60 ton 100 ton	to 8,750 ft.-lb. 15,000 ft.-lb. 25,000 ft.-lb.	to 32,000 ft.-lb. 112,000 ft.-lb.	70 H.P. 170 H.P. 300 H.P.

Cranes should be specified for the job by size, building and site characteristics, availability, performance characteristics, and duration of time required.

Backhoes & Shovels rent for about the same as equivalent size cranes but maintenance and operating expense is higher. Crane operators rate must be adjusted for high boom heights as follows: for 150′ boom add 50¢ per hour; over 185′, add $1.00 per hour; over 210′, add $1.25 per hour; over 250′, add $1.90 per hour and over 295′, add $2.65 per hour.

TOWER CRANES

Capacity in Kip-Feet	Typical Jib Length in Feet	Speed at Maximum Reach and Load	Purchase Price (New)		Monthly Rental, to 6 mo.	
			Crane & 80′ Mast	Mast Sections	Crane & 80′ Mast	Mast Sections
725	100	350 FPM	$168,000	$420/L.F.	$ 4,500	$12/L.F.
900	100	500	207,000	490	4,950	13
*1100	130	1000	280,000	560	7,400	16
1450	150	1000	393,000	775	10,500	19
2150	200	1000	505,000	945	13,400	22
3000	200	1000	700,000	1,025	17,300	27

*Most widely used.

Tower Cranes of the climbing or static type have jibs from 50′ to 200′ and capacities at maximum reach range from 4,000 to 14,000 pounds. Lifting capacities increase up to maximum load as the hook radius decreases.

Typical rental rates, based on purchase price are about 2% to 3% per month.

Erection and dismantling run between $9,500 and $63,000. Climbing operation takes three men three hours per 20′ climb. Crane dead time is about five hours per 40′ climb. If crane is bolted to side of the building add cost of ties and extra mast sections. Mast sections cost $330 to $910 per vertical foot or can be rented at 2% to 3% of purchase price per month. Contractors using climbers claim savings of $1.40 per C.Y. of concrete placed, plus 8¢ per S.F. of formwork. Climbing cranes have from 80′ to 180′ of mast while static cranes have 80′ to 800′ of mast.

Truck Cranes can be converted to tower cranes by using tower attachments. Mast heights over 400′ have been used. See Division 1.5-20 for rental rates of high boom cranes.

A single 100′ high material **Hoist and Tower** can be erected and dismantled for about $11,200; a double 100′ high hoist and tower for about $16,300. Erection costs for additional heights are $75 and $95 per vertical foot respectively up to 150′ and $80 to $135 per vertical foot over 150′ high. A 40′ high portable Buck hoist costs about $4,000 to erect and dismantle. Additional heights run $65 per vertical foot to 80′ and $87 per vertical foot for the next 100′. Most material hoists do not meet local code requirements for carrying personnel.

A 150′ high **Personnel Hoist** requires about 500 to 800 man-hours to erect and dismantle with costs ranging from $10,000 to $22,400. Budget erection cost is $115 per vertical foot for all trades. Local code requirements or labor scarcity requiring overtime can add up to 50% to any of the above erection costs.

Earthmoving Equipment: The selection of earthmoving equipment depends upon the type and quantity of material, moisture content, haul distance, haul road, time available, and equipment available. Short haul cut and fill operations may require dozers only, while another operation may require excavators, a fleet of trucks, and spreading and compaction equipment. Stockpiled material and granular material is easily excavated with front end loaders. Scrapers are most economically used with hauls between 300′ and 1-1/2 miles if adequate haul roads can be maintained. Shovels are often used for blasted rock and any material where a vertical face of 8′ or more can be excavated. Special conditions may dictate the use of draglines, clamshells, or backhoes. Spreading and compaction equipment must be matched to the soil characteristics, the compaction required and the rate the fill is being supplied.

Steel Tubular Scaffolding (Div. 1.1-43)

On new construction, tubular scaffolding is efficient up to 60′ high or five stories. Above this it is usually better to use a hung scaffolding if construction permits.

In repairing or cleaning the front of an existing building the cost of tubular scaffolding per S.F. of building front increases as the height increases above the first tier. The first tier cost is relatively high due to leveling and alignment. Swing scaffolding operations may interfere with tenants. In this case the tubular is more practical at all heights.

The minimum efficient crew for erection is three men. For heights over 50′, a four-man crew is more efficient. Use two or more on top and two at the bottom for handing up or hoisting. Four men can erect and dismantle about nine frames per hour up to five stories. From five to eight stories they will average six frames per hour. With 7′ horizontal spacing this will run about 300 S.F. and 200 S.F. of wall surface, respectively. Time for placing planks must be added to the above. On heights above 50′, five planks can be placed per man-hour.

The cost per 1,000 S.F. of building front in the table below was developed by pricing the materials required for a typical tubular scaffolding system eleven frames long and two frames high. Planks were figured five wide for standing plus two wide for materials.

Frames are 2′, 4′ and 5′ wide and usually spaced 7′ O.C. horizontally. Sidewalk frames are 6′ wide. Rental rates will be lower for jobs over three months duration.

For jobs under twenty-five frames, figure rental at $6.00 per frame. For jobs over one hundred frames, rental can go as low as $2.25 per frame. These figures do not include accessories which are listed separately below. Large quantities for long periods can reduce rental rates by 20%.

Item	Unit	Purchase, Each		Monthly Rent, Each		Per 1,000 S.F. of Building Front	
		Regular	Heavy Duty	Regular	Heavy Duty	No. of Frames	Rental per Mo.
5′ Wide Frames, 3′ High	Ea.	$40	$46	$2.75	$3.00	—	—
*5′-0″ High		50	51	2.75	3.00	—	—
*6′-6″ High		63	67	2.75	3.00	24	$ 66.00
2′ & 4′ Wide, 5′ High		—	60	—	3.00	—	—
6′-0″ High		—	70	—	3.00	—	—
6′ Wide Frame, 7′-6″ High		80	88	5.50	7.00	—	—
Sidewalk Bracket, 20″		16	18	.90	1.05	12	10.80
Guardrail Post		12	17	.50	.60	12	6.00
Guardrail, 7′ section		5	6	.45	.50	11	4.95
Cross Braces		14	15	.45	.55	44	19.80
Screw Jacks & Plates		19	27	1.25	1.35	24	30.00
8″ Casters		50	—	4.00	4.20	—	—
16′ Plank, 2″ x 10″		18	—	3.50	3.75	35	122.50
8′ Plank, 2″ x 10″		9	—	1.75	2.00	7	12.25
1′ to 6′ Extension Tube		—	65	—	2.10	—	—
Shoring Stringers, steel, 10′ to 12′ long	L.F.	—	5.00	—	.30	—	—
Aluminum, 12′ to 16′ long		—	10.00	—	.50	—	—
Aluminum joists with nailers, 10′ to 22′ long		—	6.50	—	.30	—	—
Flying Truss System, Aluminum	S.F.C.A.	—	8.00	—	.45	—	—
						Total	$272.30
						2 Use/Mo.	$136.15

*Most commonly used

Scaffolding is often used as falsework over 15′ high during construction of cast-in-place concrete beams and slabs. Two ft. wide scaffolding is generally used for heavy beam construction. The span between frames depends upon the load to be carried with a maximum span of 5′.

Heavy duty scaffolding with a capacity of 10,000#/leg can be spaced up to 10′ O.C. depending upon form support design and loading.

Scaffolding used as horizontal shoring requires less than half the material required with conventional shoring.

On new construction, erection is done by carpenters.

Rolling towers supporting horizontal shores can reduce labor and speed the job. For maintenance work, catwalks with spans up to 70′ can be supported by the rolling towers.

Wood Sheet Piling (Div. 2.3-38-390)

Wood sheet piling may be used for depths to 20′ where there is no ground water. If moderate ground water is encountered T & G sheeting will help to keep it out. When considerable ground water is present, steel sheeting must be used.

For estimating purposes on trench excavation, sizes are as follows:

Depth	Sheeting	Wales	Braces	B.F. per S.F.
To 8′	3 x 12′s	6 x 8′s, 2 line	6 x 8′s, @ 10′	4.0 @ 8′
8′ x 12′	3 x 12′s	10 x 10′s, 2 line	10 x 10′s, @ 9′	5.0 average
12′ to 20′	3 x 12′s	12 x 12′s, 3 line	12 x 12′s, @ 8′	7.0 average

Sheeting to be 2′ longer than depth of wall. A four man crew with air compressor and sheeting driver can place, drive and brace 55 S.F. per hour at 8′ depth; 45 S.F. at 12′ and 40 S.F. at 16′. For normal soils, piling can be pulled in 1/3 the time to install. Pulling difficulty increases with the time in the ground.

Production can be increased by high pressure jetting. Figures below assume 50% of lumber is salvaged and includes pulling costs. Some jurisdictions require an equipment operator in addition to Crew B-31.

Daily Cost, Crew B-31		8′ Depth, 440 S.F. Installed		16′ Depth, 320 S.F. Installed	
		To Drive	To Pull	To Drive	To Pull
1 - Foreman @ $17.50 per hour	$140.00	$ 140.00	$ 46.65	$ 140.00	$ 46.65
3 - Laborers @ $15.50 per hour	372.00	372.00	124.00	372.00	124.00
1 - Carpenter @ $19.60 per hour	156.80	156.80	52.25	156.80	52.25
1 - Air Compressor & Sheeting Driver	122.65	122.65	40.90	122.65	40.90
Lumber 50% cost @ $390 per M.B.F.	—	1.76M 343.20	—	2.24M 436.80	—
Total in Place	$791.45	$1,134.65	$263.80	$1,228.25	$263.80
Total per S.F.		$ 2.58	$ .60	$ 3.84	$ .82
Grand Total per S.F.		$ 3.18		$ 4.66	

Sheeting Left in Place

Daily Cost	8′ Depth 440 S.F./Day	10′ Depth 400 S.F./Day	12′ Depth 360 S.F./Day	16′ Depth 320 S.F./Day	18′ Depth 305 S.F./Day	20′ Depth 280 S.F./Day
Crew B-31	$ 791.45	$ 791.45	$ 791.45	$ 791.45	$ 791.45	$ 791.45
Lumber @ $390 M.B.F.	1.76M 686.40	1.8M 702.00	1.8M 702.00	2.24M 873.60	2.1M 819.00	1.9M 741.00
Total in Place	$1,477.85	$1,493.45	$1,493.45	$1,665.05	$1,610.45	$1,532.45
Total per S.F.	$ 3.36	$ 3.73	$ 4.15	$ 5.20	$ 5.28	$ 5.47
Total per M.B.F. (Labor & Material)	$ 840.00	$ 830.00	$ 830.00	$ 745.00	$ 765.00	$ 805.00

When trench sheeting is left in place the cross brace and walers are salvaged. Assumed reuse 10 times.

Steel Sheet Piling (Div. 2.3-38)

Limiting weights are 22 to 38#/S.F. of wall surface with 27#/S.F. average for usual types and sizes. (Weights of piles themselves are from 30.7#/L.F. to 57#/L.F. but they are 15" to 21" wide.) Lightweight sections 12" to 28" wide from 3 ga. to 12 ga. thick are also available for shallow excavations. Piles are generally driven two at a time with an impact vibratory hammer (use vibratory to pull) hung from a crane without leads. A reasonable estimate of the life of a sheet pile is 10 uses with up to 125 uses possible if a vibratory hammer is used. Typical cost of new plain piling at the mill is $539 per ton for shallow sections; $549 per ton for Z sections. Fabricated corners, tees and crosses cost about three times as much per ton as plain material. Used piling costs from $350 to $550 per ton depending on location and market conditions. Sheet piling and H piles can be rented for about 30% of the delivered mill price or $150 per ton for the first month and $14 per month thereafter. Cleaning and trimming after driving runs about $25 per pile. Vibratory drivers are faster in wet granular soils and are excellent for pile extraction. Pulling difficulty increases with the time in the ground and may cost more than driving. It is often economical to abandon the sheet piling, especially if it can be used as the outer wall form. Allow about 1/3 additional length or more, for toeing into ground. Add bracing, waler and strut costs. These costs increase with depth and hydrostatic head. Waler costs can range up to $470 per ton of sheeting.

Cost of Sheet Piling & Production Rate by Ton & S.F.

Depth of Excavation	15′ Depth		20′ Depth		25′ Depth	
Description of Pile	22 psf = 90.9 S.F./ton		27 psf = 74 S.F./ton		38 psf = 52.6 S.F./ton	
Type of operation: Left in place or removed	Drive & Left	Drive & Extract*	Drive & Left	Drive & Extract*	Drive & Left	Drive & Extract*
Labor & Equipment to drive 1 ton	$253.00	$379.00	$211.00	$316.00	$ 144.00	$215.00
Piling, rented or bought including accessories @ $580/ton with 75% salvage value	580.00	145.00	580.00	145.00	580.00	145.00
Cost per ton, total	$833.00	$524.00	$791.00	$461.00	$ 724.00	$360.00
Production rate, tons per day	10.81	7.22	12.95	8.65	19.00	12.75
Cost per S.F. in place	$ 9.16	$ 5.74	$ 10.69	$ 6.23	$ 13.76	$ 6.84
Production Rate, S.F. per day including 33% toe-in	983.00	656.00	960.00	640.00	1,000.00	670.00

Crew B-40	Bare Cost Hr.	Bare Cost Daily	Incl. Subs O & P Hr.	Incl. Subs O & P Daily
1 Foreman	$21.70	$ 173.60	$33.95	$ 271.60
4 Pile Drivers	19.70	630.40	30.80	985.60
2 Equip. Operators (crane)	20.65	330.40	29.90	478.40
1 Oiler	17.10	136.80	24.75	198.00
One-40 ton crane & accessories		482.60		530.85
Vibratory hammer & generator		983.00		1,081.30
64 M.H. Daily Total		$2,736.80		$3,545.75

Production Rate Formula per ton
for 15′ Deep Excavation plus 33% toe-in
$2,736.80 per day ÷ 10.81 tons = $253 per ton for installation

For S.F. Production

$\frac{2000\#}{22\#/S.F.}$ = 90.9 S.F. x 10.81 tons = 983 S.F. per day

*For driving & extracting two mobilizations & demobilizations will be necessary.

12

Wellpoints (Div. 2.3-55)

A single stage wellpoint system is usually limited to dewatering to an average 15′ depth below normal ground water level. Multistage systems are employed for greater depth with the pumping equipment installed only at the lowest header level. Ejectors, with unlimited lift capacity, can be economical when two or more stages of wellpoints can be replaced or when horizontal clearance is restricted, such as in deep trenches or tunneling projects, and where low water flows are expected.

Wellpoints are usually spaced on 2-1/2′ to 10′ centers along a header pipe. Wellpoint spacing, header size, and pump size are all determined by the expected flow, as dictated by soil conditions.

In almost all points encountered in wellpoint dewatering, the wellpoints may be jetted into place. Cemented soils and stiff clays may require sand wicks about 12″ in diameter around each wellpoint to increase efficiency and eliminate weeping into the excavation. These sand wicks require 1/2 to 3 C.Y. of washed filter sand and are installed by using a 12″ diameter steel casing and hole puncher jetted into the ground 2′ deeper than the wellpoint. Rock may require predrilled holes.

Labor required for the complete installation and removal of a single stage wellpoint system is in the range of 3/4 to 2 man-hours per linear foot of header, depending upon jetting conditions, wellpoint spacing, etc.

Continuous pumping is necessary except in some free draining soil where temporary flooding is permissible (as in trenches which are backfilled after each day's work). Good practice requires provision of a stand-by pump during the continuous pumping operation.

Systems for continuous trenching below the water table should be three to four times the length of expected daily progress to insure uninterrupted digging and header pipe size should not be changed during the job.

For pervious free draining soils, deep wells may be economical because of lower installation and maintenance costs. Daily production ranges between two to three wells per day for 25′ to 40′ depths to one well per day for depths over 50′.

Detailed analysis and estimating for any dewatering problem is available at no cost from wellpoint manufacturers. Major firms will quote "sufficient equipment" quotes or their affiliates offer lump sum proposals to cover complete dewatering responsibility.

Description for 200′ System with 8″ Header		Quantities	1st Month		Thereafter	
			Unit $	Total	Unit $	Total
Equipment	Wellpoints 25′ long, 2″ diameter @ 5′ O.C.	40 Each	$ 25.00	$ 1,000	$ 17.85	$ 714
	Header pipe, 8″ diameter	200 L.F.	4.60	920	2.27	454
	Discharge pipe, 8″ diameter	100 L.F.	3.51	351	1.46	146
	8″ valves	3 Each	100.00	300	60.00	180
	Combination Jetting & Wellpoint pump (standby)	1 Each	2,125.00	2,125	1,475.00	1,475
	Operating 8″ diameter pump	1 Each	2,350.00	2,350	1,400.00	1,400
	Transportation to and from yard	Allow		500		
	Fuel 30 days x 60 G.P.D.	1800 Gallons	1.15	2,070	1.15	2,070
	Grease & oil 30 days x 2 G.P.D.	60 Gallons	5.50	330	5.50	330
	Material and Equipment Sub Total			$9,946		$ 6,769
Labor & Material	Technician to help set up	5 Man-days	$335.00	$ 1,675		
	Install & remove system	300 Man-hours	15.50	4,650		
	4 Operators regular pay @ 40 hr./wk. x 4.3 wk.	688 hr. (6 hr. shifts)	19.05	13,106	$ 19.05	$13,106
	4 Operators overtime @ 2 hr./wk. x 4.3 wk.	34 hr. (6 hr. shifts)	38.10	1,295	38.10	1,295
	Sand for points	40 C.Y.	6.50	260		
	Labor Sub Total			$20,986		$14,401
	Total Cost			$30,932		$21,170
	Monthly Cost per Header L.F.			$ 155		$ 106

Historical Cost Indexes

The table below lists both the Means City Cost Index based on Jan. 1, 1975 = 100 as well as the computed value of an index based on January 1, 1985 costs. Since the Jan. 1, 1985 figure is estimated, space is left to write in the actual index figures as they become available thru either the quarterly "Means Construction Cost Indexes" or as printed in the "Engineering News-Record". To compute the actual index based on Jan. 1, 1985 = 100, divide the Quarterly City Cost Index for a particular year by the actual Jan. 1, 1985 Quarterly City Cost Index. Space has been left to advance the index figures as the year progresses.

Year	"Quarterly City Cost Index" Jan. 1, 1975 = 100 Est.	"Quarterly City Cost Index" Jan. 1, 1975 = 100 Actual	Current Index Based on Jan. 1, 1985 = 100 Est.	Current Index Based on Jan. 1, 1985 = 100 Actual
Oct. 1985				
July 1985				
April 1985				
Jan. 1985	190.0		100.0	100.0
July 1984		187.6	98.7	
1983		183.5	96.6	
1982		174.3	91.7	
1981		160.2	84.3	
1980		144.0	75.8	
1979		132.3	69.6	
1978		122.4	64.4	
1977		113.3	59.6	
1976		107.3	56.5	
1975		102.6	54.0	
1974		94.7	49.8	
1973		86.3	45.4	

Year	"Quarterly City Cost Index" Jan. 1, 1975 = 100 Actual	Current Index Based on Jan. 1, 1985 = 100 Est.	Current Index Based on Jan. 1, 1985 = 100 Actual
July 1972	79.7	41.9	
1971	73.5	38.7	
1970	65.8	34.6	
1969	61.6	32.4	
1968	56.9	29.9	
1967	53.9	28.4	
1966	51.9	27.3	
1965	49.7	26.2	
1964	48.6	25.6	
1963	47.3	24.9	
1962	46.2	24.3	
1961	45.4	23.9	
1960	45.0	23.7	
1959	44.2	23.3	
1958	43.0	22.6	
1957	42.2	22.2	

Year	"Quarterly City Cost Index" Jan. 1, 1975 = 100 Actual	Current Index Based on Jan. 1, 1985 = 100 Est.	Current Index Based on Jan. 1, 1985 = 100 Actual
July 1956	40.4	21.3	
1955	38.1	20.1	
1954	36.7	19.3	
1953	36.2	19.1	
1952	35.3	18.6	
1951	34.4	18.1	
1950	31.4	16.5	
1949	30.4	16.0	
1948	30.4	16.0	
1947	27.6	14.5	
1946	23.2	12.2	
1945	20.2	10.6	
1944	19.3	10.2	
1943	18.6	9.8	
1942	18.0	9.5	
1941	16.8	8.8	

City Cost Indexes

Tabulated on the following pages are average construction cost indexes for 162 major U.S. and Canadian cities. Index figures for both material and installation are based on the 30 major city average of 100 and represent the cost relationship as of July 1, **1984.** The index for each division is computed from representative material and labor quantities for that division. The weighted average for each city is a weighted total of the components listed above it but does not include relative productivity between trades or cities.

The material index for the weighted average includes about 100 basic construction materials with appropriate quantities of each material to represent typical "average" building construction projects.

The installation index for the weighted average includes the contribution of about 30 construction trades with their representative man-days in proportion to the material items installed. Also included in the installation costs are the representative equipment costs for those items requiring equipment.

Since each division of the book contains many different items, any particular item multiplied by the particular city index may give incorrect results. However, when all the book costs for a particular division are summarized and then factored, the result should be very close to the actual costs for that particular division for that city.

If a project has a preponderance of materials from any particular division (say structural steel) then the weighted average index should be adjusted in proportion to the value of the factor for that division.

DIVISION	ALABAMA												ALASKA			ARIZONA		
	BIRMINGHAM			HUNTSVILLE			MOBILE			MONTGOMERY			ANCHORAGE			PHOENIX		
	MAT.	INST.	TOTAL	MAT.	INST.	TOTAL	MAT.	INST.	TOTAL	MAT.	INST.	TOTAL	MAT.	INST.	TOTAL	MAT.	INST.	TOTAL
MECHANICAL	96.4	77.1	86.5	99.6	76.8	87.9	97.5	80.9	89.0	99.1	73.8	86.1	107.7	145.6	127.1	99.9	106.8	103.4
ELECTRICAL	93.8	74.5	80.2	91.6	75.4	80.2	89.2	92.3	91.4	90.4	76.2	80.3	113.0	145.8	136.1	102.9	85.4	90.6
WEIGHTED PROJECT AVERAGE	94.1	77.4	85.0	100.0	78.3	88.2	97.1	85.8	90.9	96.1	78.8	86.6	125.4	140.9	133.9	100.0	95.9	97.7

DIVISION	ARIZONA			ARKANSAS						CALIFORNIA								
	TUCSON			FORT SMITH			LITTLE ROCK			ANAHEIM			BAKERSFIELD			FRESNO		
	MAT.	INST.	TOTAL	MAT.	INST.	TOTAL	MAT.	INST.	TOTAL	MAT.	INST.	TOTAL	MAT.	INST.	TOTAL	MAT.	INST.	TOTAL
MECHANICAL	98.9	96.1	97.5	97.5	74.8	85.9	97.1	72.7	84.6	96.9	122.4	109.9	95.9	116.3	106.4	97.3	123.5	110.7
ELECTRICAL	102.2	95.3	97.3	99.2	75.0	82.1	93.1	75.7	80.8	98.6	124.6	116.9	106.3	116.5	113.5	110.1	123.0	119.2
WEIGHTED PROJECT AVERAGE	99.2	94.1	96.4	97.1	77.4	86.3	99.0	79.3	88.2	100.9	121.9	112.3	98.8	117.4	109.0	99.8	121.2	111.5

DIVISION	CALIFORNIA																	
	LOS ANGELES			OXNARD			RIVERSIDE			SACRAMENTO			SAN DIEGO			SAN FRANCISCO		
	MAT.	INST.	TOTAL	MAT.	INST.	TOTAL	MAT.	INST.	TOTAL	MAT.	INST.	TOTAL	MAT.	INST.	TOTAL	MAT.	INST.	TOTAL
MECHANICAL	100.3	122.0	111.4	98.9	122.3	110.9	96.6	119.8	108.5	98.3	132.5	115.8	102.6	121.1	112.1	100.4	158.4	130.1
ELECTRICAL	98.3	127.5	118.9	98.6	124.3	116.7	98.1	124.1	116.4	110.1	132.8	126.1	104.1	118.1	113.9	109.4	145.9	135.2
WEIGHTED PROJECT AVERAGE	97.7	122.0	111.0	99.2	120.3	110.7	99.0	121.1	111.1	99.6	127.5	114.9	101.4	117.3	110.1	105.5	137.0	122.7

DIVISION	CALIFORNIA									COLORADO						CONNECTICUT		
	SANTA BARBARA			STOCKTON			VALLEJO			COLORADO SPRINGS			DENVER			BRIDGEPORT		
	MAT.	INST.	TOTAL	MAT.	INST.	TOTAL	MAT.	INST.	TOTAL	MAT.	INST.	TOTAL	MAT.	INST.	TOTAL	MAT.	INST.	TOTAL
MECHANICAL	98.9	122.0	110.7	97.2	124.1	111.0	95.9	134.0	115.5	98.0	91.5	94.7	97.1	94.4	95.7	104.1	92.5	98.2
ELECTRICAL	97.8	122.9	115.5	100.6	129.0	120.6	110.3	134.5	127.3	97.8	92.1	93.7	99.0	101.6	100.9	103.0	91.3	94.7
WEIGHTED PROJECT AVERAGE	104.1	121.3	113.5	98.5	121.7	111.2	98.9	129.8	115.8	98.3	92.8	95.3	101.3	99.9	100.6	103.5	94.3	98.5

DIVISION	CONNECTICUT												DELAWARE			D.C.		
	HARTFORD			NEW HAVEN			STAMFORD			WATERBURY			WILMINGTON			WASHINGTON		
	MAT.	INST.	TOTAL	MAT.	INST.	TOTAL	MAT.	INST.	TOTAL	MAT.	INST.	TOTAL	MAT.	INST.	TOTAL	MAT.	INST.	TOTAL
MECHANICAL	102.0	98.6	100.3	102.3	100.0	101.1	101.8	98.7	100.2	100.9	94.8	97.8	100.3	97.2	98.7	101.3	89.5	95.2
ELECTRICAL	98.5	93.7	95.1	92.7	95.7	94.8	92.6	92.9	92.8	91.3	89.3	89.9	105.8	102.1	103.2	102.6	91.6	94.9
WEIGHTED PROJECT AVERAGE	100.2	97.4	98.7	101.0	96.6	98.6	103.7	95.2	99.1	101.0	93.9	97.1	98.5	100.8	99.7	101.8	92.9	97.0

DIVISION	FLORIDA															GEORGIA		
	FT. LAUDERDALE			JACKSONVILLE			MIAMI			ORLANDO			TAMPA			ATLANTA		
	MAT.	INST.	TOTAL	MAT.	INST.	TOTAL	MAT.	INST.	TOTAL	MAT.	INST.	TOTAL	MAT.	INST.	TOTAL	MAT.	INST.	TOTAL
MECHANICAL	101.0	82.9	91.7	100.2	78.4	89.0	97.8	92.1	94.9	97.2	79.9	88.3	97.4	81.1	89.0	101.5	78.4	89.6
ELECTRICAL	98.1	86.1	89.6	99.3	75.4	82.4	99.9	88.8	92.0	92.7	76.2	81.0	92.7	83.0	85.8	94.9	80.5	84.8
WEIGHTED PROJECT AVERAGE	97.2	87.3	91.8	98.2	79.6	88.1	95.7	90.8	93.0	94.9	77.2	85.2	99.0	85.3	91.5	98.9	79.4	88.2

DIVISION	GEORGIA									HAWAII			IDAHO			ILLINOIS		
	COLUMBUS			MACON			SAVANNAH			HONOLULU			BOISE			CHICAGO		
	MAT.	INST.	TOTAL	MAT.	INST.	TOTAL	MAT.	INST.	TOTAL	MAT.	INST.	TOTAL	MAT.	INST.	TOTAL	MAT.	INST.	TOTAL
MECHANICAL	98.6	70.4	84.1	98.9	72.9	85.6	98.6	73.7	85.9	111.8	104.4	108.0	96.7	92.9	94.8	96.7	94.5	95.5
ELECTRICAL	101.8	63.4	74.7	107.8	68.6	80.1	100.8	76.7	83.8	105.3	101.2	102.4	95.6	93.0	93.7	99.7	96.1	97.1
WEIGHTED PROJECT AVERAGE	99.0	72.3	84.4	97.1	76.5	85.9	99.4	78.4	87.9	115.5	108.1	111.4	98.5	93.3	95.7	97.0	100.2	98.8

DIVISION	ILLINOIS									INDIANA								
	PEORIA			ROCKFORD			SPRINGFIELD			EVANSVILLE			FORT WAYNE			GARY		
	MAT.	INST.	TOTAL	MAT.	INST.	TOTAL	MAT.	INST.	TOTAL	MAT.	INST.	TOTAL	MAT.	INST.	TOTAL	MAT.	INST.	TOTAL
MECHANICAL	96.2	98.7	97.5	102.1	96.9	99.4	97.3	97.0	97.1	98.8	101.6	100.2	100.0	93.5	96.7	98.3	108.6	103.6
ELECTRICAL	94.9	98.7	97.5	91.6	99.9	97.4	94.3	96.5	95.8	93.9	99.6	97.9	96.5	93.5	94.4	100.8	91.9	94.5
WEIGHTED PROJECT AVERAGE	98.4	96.5	97.4	103.3	99.1	101.0	100.2	96.8	98.3	98.9	98.4	98.7	98.7	92.9	95.5	99.4	107.0	103.5

DIVISION	INDIANA									IOWA						KANSAS		
	INDIANAPOLIS			SOUTH BEND			TERRE HAUTE			DAVENPORT			DES MOINES			TOPEKA		
	MAT.	INST.	TOTAL	MAT.	INST.	TOTAL	MAT.	INST.	TOTAL	MAT.	INST.	TOTAL	MAT.	INST.	TOTAL	MAT.	INST.	TOTAL
MECHANICAL	101.0	96.7	98.8	98.2	97.5	97.8	98.3	94.0	96.1	100.1	97.5	98.7	96.8	89.4	93.0	99.4	88.7	93.9
ELECTRICAL	95.5	96.7	96.4	96.9	97.4	97.2	108.6	93.6	98.0	97.2	97.5	97.4	93.7	89.3	90.6	97.7	80.5	85.6
WEIGHTED PROJECT AVERAGE	99.1	97.7	98.3	98.3	98.6	98.5	98.0	94.6	96.1	99.1	97.2	98.1	98.6	90.3	94.1	98.6	89.0	93.4

DIVISION	KANSAS			KENTUCKY						LOUISIANA								
	WICHITA			LEXINGTON			LOUISVILLE			BATON ROUGE			NEW ORLEANS			SHREVEPORT		
	MAT.	INST.	TOTAL	MAT.	INST.	TOTAL	MAT.	INST.	TOTAL	MAT.	INST.	TOTAL	MAT.	INST.	TOTAL	MAT.	INST.	TOTAL
MECHANICAL	97.1	82.4	89.6	100.7	89.7	95.1	101.4	91.9	96.5	99.6	86.8	93.0	99.7	86.6	93.0	99.6	86.9	93.1
ELECTRICAL	102.4	82.4	88.3	102.9	88.0	92.4	94.4	91.0	92.0	97.3	92.1	93.6	95.6	85.5	88.4	94.3	81.7	85.4
WEIGHTED PROJECT AVERAGE	98.6	86.2	91.8	97.6	89.8	93.3	97.3	92.7	94.8	100.1	87.6	93.3	100.4	87.3	93.3	97.4	84.2	90.2

DIVISION	MAINE						MARYLAND			MASSACHUSETTS								
	LEWISTON			PORTLAND			BALTIMORE			BOSTON			LAWRENCE			LOWELL		
	MAT.	INST.	TOTAL	MAT.	INST.	TOTAL	MAT.	INST.	TOTAL	MAT.	INST.	TOTAL	MAT.	INST.	TOTAL	MAT.	INST.	TOTAL
MECHANICAL	99.8	76.5	87.8	98.3	77.7	87.7	101.2	88.7	94.8	103.6	104.3	104.0	99.8	103.6	101.8	98.0	105.1	101.6
ELECTRICAL	93.3	76.0	81.1	97.0	76.0	82.2	98.5	92.7	94.4	97.4	109.3	105.8	100.1	97.8	98.4	100.1	98.1	98.6
WEIGHTED PROJECT AVERAGE	97.7	80.1	88.1	99.0	80.5	88.9	100.7	86.0	92.7	107.6	107.0	107.3	101.5	103.5	102.6	101.4	104.0	102.8

DIVISION	MASSACHUSETTS						MICHIGAN											
	SPRINGFIELD			WORCESTER			ANN ARBOR			DETROIT			FLINT			GRAND RAPIDS		
	MAT.	INST.	TOTAL	MAT.	INST.	TOTAL	MAT.	INST.	TOTAL	MAT.	INST.	TOTAL	MAT.	INST.	TOTAL	MAT.	INST.	TOTAL
MECHANICAL	98.9	92.8	95.8	101.3	96.3	98.7	99.8	112.3	106.2	102.0	101.0	101.5	99.2	102.2	100.7	98.3	81.7	89.8
ELECTRICAL	94.5	91.3	92.3	96.3	96.5	96.4	98.4	112.4	108.2	106.7	100.0	102.0	98.4	102.3	101.1	99.3	89.4	92.3
WEIGHTED PROJECT AVERAGE	99.5	94.7	96.9	101.4	99.5	100.3	97.2	106.2	102.1	98.6	103.5	101.3	99.5	99.1	99.3	101.2	88.7	94.4

DIVISION	MICHIGAN									MINNESOTA						MISSISSIPPI		
	KALAMAZOO			LANSING			SAGINAW			DULUTH			MINNEAPOLIS			JACKSON		
	MAT.	INST.	TOTAL	MAT.	INST.	TOTAL	MAT.	INST.	TOTAL	MAT.	INST.	TOTAL	MAT.	INST.	TOTAL	MAT.	INST.	TOTAL
MECHANICAL	98.1	93.0	95.5	99.8	101.2	100.5	99.8	98.9	99.3	99.3	94.7	97.0	100.8	98.8	99.8	99.8	73.7	86.4
ELECTRICAL	93.1	93.0	93.0	99.7	101.3	100.8	98.4	98.9	98.7	96.9	90.4	92.3	98.2	103.1	101.7	94.4	73.6	79.7
WEIGHTED PROJECT AVERAGE	99.1	93.2	95.9	104.3	98.9	101.4	101.9	96.7	99.0	102.0	92.3	96.7	99.6	99.8	99.7	97.4	76.5	86.0

DIVISION	MISSOURI						MONTANA						NEBRASKA					
	KANSAS CITY			ST. LOUIS			BILLINGS			GREAT FALLS			LINCOLN			OMAHA		
	MAT.	INST.	TOTAL	MAT.	INST.	TOTAL	MAT.	INST.	TOTAL	MAT.	INST.	TOTAL	MAT.	INST.	TOTAL	MAT.	INST.	TOTAL
MECHANICAL	101.4	93.0	97.1	98.0	100.7	99.4	99.4	96.7	98.0	99.2	93.3	96.1	100.0	83.1	91.4	100.3	90.1	95.1
ELECTRICAL	103.4	100.1	101.0	99.3	106.8	104.6	97.8	96.1	96.6	97.2	90.7	92.6	95.5	83.1	86.7	91.2	90.1	90.4
WEIGHTED PROJECT AVERAGE	101.1	97.0	98.9	96.7	100.3	98.6	100.0	92.1	95.7	100.5	90.8	95.2	98.6	85.0	91.2	101.3	89.8	95.0

DIVISION	NEVADA						NEW HAMPSHIRE						NEW JERSEY					
	LAS VEGAS			RENO			MANCHESTER			NASHUA			JERSEY CITY			NEWARK		
	MAT.	INST.	TOTAL	MAT.	INST.	TOTAL	MAT.	INST.	TOTAL	MAT.	INST.	TOTAL	MAT.	INST.	TOTAL	MAT.	INST.	TOTAL
MECHANICAL	100.6	115.6	108.3	103.1	105.2	104.2	100.7	84.3	92.3	98.8	84.3	91.3	100.2	105.2	102.8	97.4	105.0	101.3
ELECTRICAL	102.3	115.7	111.7	108.1	115.5	113.3	101.2	82.8	88.2	97.2	82.8	87.0	95.6	113.6	108.3	96.4	104.8	102.3
WEIGHTED PROJECT AVERAGE	99.7	115.7	108.5	101.6	110.1	106.2	100.7	84.0	91.6	98.7	85.5	91.5	103.1	106.5	105.0	101.6	105.0	103.4

DIVISION	NEW JERSEY						NEW MEXICO			NEW YORK								
	PATERSON			TRENTON			ALBUQUERQUE			ALBANY			BINGHAMTON			BUFFALO		
	MAT.	INST.	TOTAL	MAT.	INST.	TOTAL	MAT.	INST.	TOTAL	MAT.	INST.	TOTAL	MAT.	INST.	TOTAL	MAT.	INST.	TOTAL
MECHANICAL	100.2	103.8	102.0	102.6	105.2	103.9	100.4	81.7	90.8	96.1	89.0	92.4	100.7	82.0	91.1	99.7	99.8	99.7
ELECTRICAL	95.6	103.5	101.1	93.6	96.1	95.4	93.7	81.7	85.2	91.2	89.5	90.0	91.5	82.0	84.8	103.9	106.5	105.7
WEIGHTED PROJECT AVERAGE	103.1	105.7	104.5	101.4	103.4	102.5	101.1	85.7	92.7	96.6	93.0	94.6	98.4	85.2	91.2	102.9	105.4	104.2

13

DIVISION	NEW YORK															NORTH CAROLINA		
	NEW YORK			ROCHESTER			SYRACUSE			UTICA			YONKERS			CHARLOTTE		
	MAT.	INST.	TOTAL	MAT.	INST.	TOTAL	MAT.	INST.	TOTAL	MAT.	INST.	TOTAL	MAT.	INST.	TOTAL	MAT.	INST.	TOTAL
MECHANICAL	99.7	118.3	109.2	99.6	96.7	98.1	101.1	92.3	96.6	99.5	89.4	94.3	96.6	111.3	104.1	97.4	65.6	81.1
ELECTRICAL	99.8	118.3	112.8	97.4	96.7	96.9	97.9	92.3	93.9	94.8	89.4	90.9	103.4	111.4	109.0	97.0	65.5	74.7
WEIGHTED PROJECT AVERAGE	105.0	119.0	112.7	101.0	96.6	98.6	100.3	90.3	94.8	100.6	88.6	94.0	103.4	112.1	108.1	96.9	71.1	82.8

DIVISION	NORTH CAROLINA						OHIO											
	GREENSBORO			RALEIGH			AKRON			CANTON			CINCINNATI			CLEVELAND		
	MAT.	INST.	TOTAL	MAT.	INST.	TOTAL	MAT.	INST.	TOTAL	MAT.	INST.	TOTAL	MAT.	INST.	TOTAL	MAT.	INST.	TOTAL
MECHANICAL	96.9	67.8	82.0	97.3	67.2	81.9	99.7	102.6	101.2	99.5	96.7	98.1	99.5	96.7	98.0	101.9	110.7	106.4
ELECTRICAL	96.6	66.7	75.5	98.5	66.0	75.5	93.4	101.2	98.9	92.3	96.7	95.4	97.3	105.2	102.9	101.0	112.4	109.1
WEIGHTED PROJECT AVERAGE	94.0	71.4	81.7	94.4	71.8	82.1	99.5	107.4	103.8	99.2	97.9	98.5	97.2	99.5	98.5	102.2	117.3	110.4

DIVISION	OHIO															OKLAHOMA		
	COLUMBUS			DAYTON			LORAIN			TOLEDO			YOUNGSTOWN			OKLAHOMA CITY		
	MAT.	INST.	TOTAL	MAT.	INST.	TOTAL	MAT.	INST.	TOTAL	MAT.	INST.	TOTAL	MAT.	INST.	TOTAL	MAT.	INST.	TOTAL
MECHANICAL	100.5	100.2	100.3	101.4	94.2	97.8	99.2	106.7	103.0	99.3	109.3	104.4	100.0	98.8	99.4	96.8	84.8	90.6
ELECTRICAL	101.4	86.8	91.1	103.7	94.3	97.0	96.3	105.0	102.4	98.1	108.5	105.4	96.3	98.8	98.0	100.9	84.7	89.4
WEIGHTED PROJECT AVERAGE	99.9	96.0	97.8	100.2	95.2	97.4	100.1	106.9	103.8	101.1	107.4	104.5	97.4	101.8	99.8	100.9	85.3	92.3

DIVISION	OKLAHOMA			OREGON						PENNSYLVANIA								
	TULSA			EUGENE			PORTLAND			ALLENTOWN			ERIE			HARRISBURG		
	MAT.	INST.	TOTAL	MAT.	INST.	TOTAL	MAT.	INST.	TOTAL	MAT.	INST.	TOTAL	MAT.	INST.	TOTAL	MAT.	INST.	TOTAL
MECHANICAL	96.9	87.0	91.8	98.9	110.8	105.0	98.4	113.8	106.3	100.7	96.3	98.4	100.2	94.7	97.4	101.5	85.7	93.4
ELECTRICAL	99.8	86.5	90.4	102.0	110.6	108.0	102.0	113.9	110.4	97.6	90.6	92.6	89.1	99.6	96.5	91.8	85.3	87.2
WEIGHTED PROJECT AVERAGE	99.0	88.4	93.2	101.7	112.6	107.6	103.4	111.2	107.7	100.9	95.6	98.0	98.6	96.0	97.2	97.8	89.1	93.0

DIVISION	PENNSYLVANIA												RHODE ISLAND			SOUTH CAROLINA		
	PHILADELPHIA			PITTSBURGH			READING			SCRANTON			PROVIDENCE			CHARLESTON		
	MAT.	INST.	TOTAL	MAT.	INST.	TOTAL	MAT.	INST.	TOTAL	MAT.	INST.	TOTAL	MAT.	INST.	TOTAL	MAT.	INST.	TOTAL
MECHANICAL	100.3	106.1	103.3	97.9	97.7	97.8	99.6	96.0	97.8	99.4	86.9	93.0	100.0	92.2	96.0	98.6	62.9	80.3
ELECTRICAL	98.5	106.7	104.3	96.3	90.8	92.5	93.4	83.6	86.5	104.0	86.8	91.8	98.4	91.5	93.5	97.4	68.5	77.0
WEIGHTED PROJECT AVERAGE	99.3	105.6	102.7	100.2	103.4	102.0	97.8	88.4	92.6	99.9	92.3	95.8	100.2	95.1	97.4	99.6	70.3	83.6

DIVISION	SOUTH CAROLINA			SOUTH DAKOTA			TENNESSEE											
	COLUMBIA			SIOUX FALLS			CHATTANOOGA			KNOXVILLE			MEMPHIS			NASHVILLE		
	MAT.	INST.	TOTAL	MAT.	INST.	TOTAL	MAT.	INST.	TOTAL	MAT.	INST.	TOTAL	MAT.	INST.	TOTAL	MAT.	INST.	TOTAL
MECHANICAL	99.9	67.1	83.1	99.5	65.1	81.8	100.0	79.1	89.3	100.3	73.9	86.8	99.4	86.5	92.8	98.1	76.3	86.9
ELECTRICAL	103.8	68.5	78.9	95.7	80.9	85.2	94.3	78.8	83.3	95.1	71.9	78.7	95.6	88.1	90.3	98.8	70.1	78.6
WEIGHTED PROJECT AVERAGE	97.4	72.5	83.8	100.8	78.0	88.3	98.2	81.0	88.8	97.7	76.6	86.2	96.1	84.6	89.8	97.2	73.6	84.3

DIVISION	TEXAS																	
	AMARILLO			AUSTIN			BEAUMONT			CORPUS CHRISTI			DALLAS			EL PASO		
	MAT.	INST.	TOTAL	MAT.	INST.	TOTAL	MAT.	INST.	TOTAL	MAT.	INST.	TOTAL	MAT.	INST.	TOTAL	MAT.	INST.	TOTAL
MECHANICAL	99.2	83.5	91.1	99.0	83.0	90.8	99.0	93.4	96.2	98.7	70.4	84.2	97.5	88.2	92.7	101.8	70.6	85.8
ELECTRICAL	103.2	83.4	89.2	101.6	81.7	87.5	105.1	95.7	98.5	99.7	67.6	77.0	98.1	87.0	90.3	93.2	69.8	76.7
WEIGHTED PROJECT AVERAGE	100.6	84.3	91.7	96.6	84.8	90.2	100.2	92.6	96.0	100.0	74.8	86.2	98.3	92.0	94.8	100.5	72.2	85.0

DIVISION	TEXAS												UTAH			VERMONT		
	FORT WORTH			HOUSTON			LUBBOCK			SAN ANTONIO			SALT LAKE CITY			BURLINGTON		
	MAT.	INST.	TOTAL	MAT.	INST.	TOTAL	MAT.	INST.	TOTAL	MAT.	INST.	TOTAL	MAT.	INST.	TOTAL	MAT.	INST.	TOTAL
MECHANICAL	100.3	88.7	94.4	98.9	95.1	96.9	99.8	80.8	90.1	98.9	85.1	91.9	104.5	93.7	98.9	101.0	76.4	88.4
ELECTRICAL	97.7	88.6	91.3	100.5	96.3	97.5	100.7	79.9	86.0	107.8	76.4	85.7	101.6	94.7	96.7	100.4	76.3	83.4
WEIGHTED PROJECT AVERAGE	101.5	89.9	95.1	101.0	92.0	96.1	98.9	81.9	89.6	96.1	82.7	88.8	98.5	93.7	95.9	102.0	81.0	90.5

DIVISION	VIRGINIA												WASHINGTON					
	NEWPORT NEWS			NORFOLK			RICHMOND			ROANOKE			SEATTLE			SPOKANE		
	MAT.	INST.	TOTAL	MAT.	INST.	TOTAL	MAT.	INST.	TOTAL	MAT.	INST.	TOTAL	MAT.	INST.	TOTAL	MAT.	INST.	TOTAL
MECHANICAL	103.1	70.3	86.3	102.9	69.4	85.8	103.3	72.9	87.7	103.7	69.6	86.2	100.5	111.5	106.2	103.7	115.7	109.9
ELECTRICAL	105.5	69.6	80.1	105.5	69.8	80.3	102.6	73.8	82.2	101.5	69.4	78.8	101.9	118.7	113.8	103.5	108.0	106.6
WEIGHTED PROJECT AVERAGE	99.5	72.9	85.0	99.0	73.2	84.9	99.9	76.3	87.0	99.7	70.6	83.8	99.8	109.5	105.1	103.6	107.4	105.7

DIVISION	WASHINGTON			WEST VIRGINIA						WISCONSIN						WYOMING		
	TACOMA			CHARLESTON			HUNTINGTON			MADISON			MILWAUKEE			CHEYENNE		
	MAT.	INST.	TOTAL	MAT.	INST.	TOTAL	MAT.	INST.	TOTAL	MAT.	INST.	TOTAL	MAT.	INST.	TOTAL	MAT.	INST.	TOTAL
MECHANICAL	102.1	114.5	108.5	101.9	90.1	95.8	104.6	94.2	99.3	102.2	87.8	94.8	98.7	91.4	94.9	103.4	96.7	99.9
ELECTRICAL	103.5	114.6	111.3	92.3	96.5	95.2	93.7	93.5	93.5	87.6	87.9	87.8	92.2	94.1	93.6	97.4	96.7	96.9
WEIGHTED PROJECT AVERAGE	102.6	111.5	107.4	103.4	95.3	99.0	102.9	96.2	99.2	98.2	90.4	93.9	95.8	94.7	95.2	99.7	97.1	98.3

DIVISION	CANADA																	
	EDMONTON			MONTREAL			QUEBEC			TORONTO			VANCOUVER			WINNIPEG		
	MAT.	INST.	TOTAL	MAT.	INST.	TOTAL	MAT.	INST.	TOTAL	MAT.	INST.	TOTAL	MAT.	INST.	TOTAL	MAT.	INST.	TOTAL
MECHANICAL	99.7	98.6	99.1	101.0	86.7	93.7	100.3	86.7	93.3	103.7	96.4	100.0	98.1	105.2	101.8	98.9	95.0	96.9
ELECTRICAL	107.5	97.6	100.5	104.3	86.1	91.5	105.2	86.1	91.7	104.0	100.1	101.2	99.9	105.3	103.7	111.0	94.5	99.4
WEIGHTED PROJECT AVERAGE	104.3	98.0	100.8	102.9	91.6	96.8	101.6	91.6	96.2	108.0	102.6	105.0	104.8	104.6	104.7	106.0	92.5	98.8

Square Foot and Cubic Foot Building Costs (Div. 17.1)

The cost figures in division 17.1 were derived from more than 9500 projects contained in the Means Data Bank of Construction Costs and include the contractor's overhead and profit, but do not include architectural fees or land costs. The figures have been adjusted to January 1, 1985. New projects are added to our files each year and projects over ten years old are discarded. For this reason, certain costs may not show a uniform annual progression. In no case are all subdivisions of a project listed.

These projects were located throughout the U.S. and reflect tremendous differences in S.F. and C.F. costs. This is due to both differences in labor and material costs, plus differences in the owner's requirements. For instance, a bank in a large city would have different features than one in a rural area. This is true of all different types of buildings analyzed. As a general rule, the projects on the low side did not include any site work or equipment, but the projects on the high side may include both equipment and site work. The median figures do not generally include site work.

None of the figures "go with" any others. All individual cost items were computed and tabulated separately. Thus the sum of the median figures for Plumbing, H.V.A.C. and Electrical will not normally total up to the total Mechanical and Electrical costs arrived at by separate analysis and tabulation of the projects.

Each building was analyzed as to total and component costs and percentages. The figures were arranged in ascending order with the results tabulated as shown. The 1/4 column shows that 25% of the projects had lower costs; 75% higher. The 3/4 column shows that 75% of the projects had lower costs; 25% had higher. The median column shows that 50% of the projects had lower costs, 50% had higher.

There are two times when square foot costs are useful. The first is in the conceptual stage when no details are available. Then square foot costs make a useful starting point. The second is after the bids are in and the costs can be worked back into their appropriate units for information purposes. As soon as details become available in the project design, the square foot approach should be discontinued and the project priced as to its particular components. When more precision is required or for estimating the replacement cost of specific buildings, the "Means Square Foot Costs 1985" should be used.

In using the figures in division 17.1, it is recommended that the median column be used for preliminary figures if no additional information is available. The median figures, when multiplied by the total city construction cost index figures on pages 411 to 413 and then multiplied by the project size modifier on the next page, should present a fairly accurate base figure, which would then have to be adjusted in view of the estimator's experience, local economic conditions, code requirements and the owner's particular requirements. There is no need to factor the percentage figures as these should remain constant from city to city. All tabulations mentioning air conditioning had at least partial air conditioning.

The editors of this book would greatly appreciate receiving cost figures on one or more of your recent projects which would then be included in the averages for next year. All cost figures received will be kept confidential except that they will be averaged with other similar projects to arrive at S.F. and C.F. cost figures for next year's book.

SQUARE FOOT | C14.1-100 | S.F., C.F., & % of Total Costs

The figures in the table below indicate typical ranges in square feet as a function of the "occupant" unit. This table is best used in the preliminary design stages to help determine the probable size requirement for the total project. See next page for the typical total size ranges for various types of buildings.

Table 14.1-101 Unit Gross Area Requirements

BUILDING TYPE	UNIT	GROSS AREA IN S.F. 1/4	MEDIAN	3/4
Auditorium & Play Theaters	Seat	18	25	38
Apartments	Unit	660	860	1100
Bowling Alleys	Lane		940	
Churches & Synagogues	Seat	20	28	39
Dormitories	Bed	200	230	275
Fraternity & Sorority Houses	Bed	220	315	370
Garages, Parking	Car	325	355	385
Housing for the Elderly	Unit	515	635	755
Housing, Public	Unit	700	875	1030
Hospitals	Bed	685	850	1075
Hotels	Rntl Unit	475	600	710
Ice Skating Rinks	Total	27,000	30,000	36,000
Motels	Rntl Unit	360	465	620
Nursing Homes	Bed	290	350	450
Restaurants	Seat	23	29	39
Schools, Elementary	Pupil	65	77	90
Junior High & Middle	↓	85	110	129
Senior High	↓	102	130	145
Vocational	↓	110	135	195
Shooting Ranges	Point		450	
Theaters & Movies	Seat		15	

14

Square Foot Project Size Modifier (Div. 17.1)

One factor that affects the S.F. cost of a particular building is the size. In general, for buildings built to the same specifications in the same locality, the larger building will have the lower S.F. Cost. This is due mainly to the decreasing contribution of the exterior walls plus the economy of scale usually achievable in larger buildings. The Area Conversion Scale shown below will give a factor to convert costs for the typical size building to an adjusted cost for the particular project.

The Square Foot Base Size lists the median costs, most typical project size in our accumulated data and the range in size of the projects.

The Size Factor for your project is determined by dividing your project area in S.F. by the typical project size for the particular Building Type. With this factor, enter the Area Conversion Scale at the appropriate Size Factor and determine the appropriate cost multiplier for your building size.

EXAMPLE: Determine the cost per S.F. for a 100,000 S.F. Mid-rise apartment building.

$$\frac{\text{Proposed building area} = 100{,}000 \text{ S.F.}}{\text{Typical size from below} = 50{,}000 \text{ S.F.}} = 2.00$$

Enter Area Conversion scale at 2.0, intersect curve, read horizontally the appropriate cost multiplier of 0.97. Size adjusted cost becomes 0.97 x $49.00 = $47.53 based on national average costs.

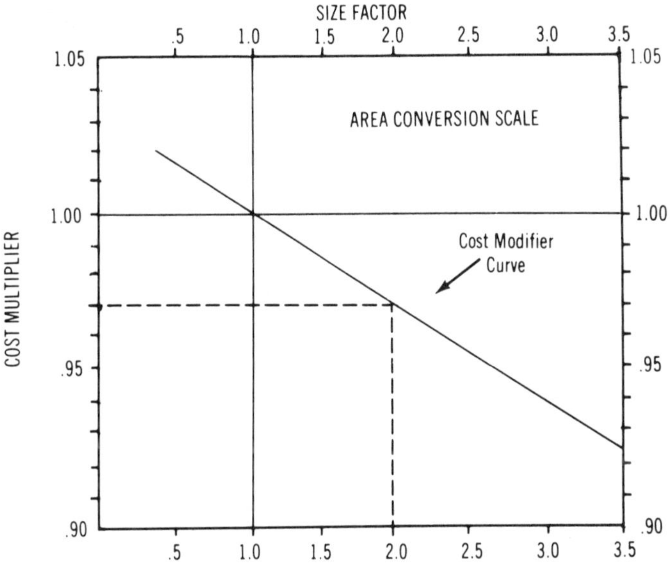

Building Type	Median Cost Per S.F.	Typical Size Gross S.F.	Typical Range Gross S.F.	Building Type	Median Cost Per S.F.	Typical Size Gross S.F.	Typical Range Gross S.F.
	Square Foot Base Size						
Apartments, Low Rise	$ 39.00	21,000	9,700 - 37,200	Jails	$113.00	13,700	7,500 - 28,000
Apartments, Mid Rise	49.00	50,000	32,000 - 100,000	Libraries	68.90	12,000	7,000 - 31,000
Apartments, High Rise	53.40	310,000	100,000 - 650,000	Medical Clinics	66.65	7,200	4,200 - 15,700
Auditoriums	64.70	25,000	7,600 - 39,000	Medical Offices	62.85	6,000	4,000 - 15,000
Auto Sales	40.90	20,000	10,800 - 28,600	Motels	48.70	27,000	15,800 - 51,000
Banks	89.45	4,200	2,500 - 7,500	Nursing Homes	66.45	23,000	15,000 - 37,000
Churches	59.35	9,000	5,300 - 13,200	Offices, Low Rise	52.75	8,600	4,700 - 19,000
Clubs, Country	57.40	6,500	4,500 - 15,000	Offices, Mid Rise	58.15	52,000	31,300 - 83,100
Clubs, Social	56.50	10,000	6,000 - 13,500	Offices, High Rise	69.85	260,000	151,000 - 468,000
Clubs, YMCA	60.70	28,300	12,800 - 39,400	Police Stations	87.75	10,500	4,000 - 19,000
Colleges (Class)	78.25	50,000	23,500 - 98,500	Post Offices	66.40	12,400	6,800 - 30,000
Colleges (Science Lab)	91.55	45,600	16,600 - 80,000	Power Plants	437.00	7,500	1,000 - 20,000
College (Student Union)	84.30	33,400	16,000 - 85,000	Religious Education	48.95	9,000	6,000 - 12,000
Community Center	61.40	9,400	5,300 - 16,700	Research	88.45	19,000	6,300 - 45,000
Court Houses	81.75	32,400	17,800 - 106,000	Restaurants	77.85	4,400	2,800 - 6,000
Dept. Stores	36.40	90,000	44,000 - 122,000	Retail Stores	37.90	7,200	4,000 - 17,600
Dormitories, Low Rise	59.80	24,500	13,400 - 40,000	Schools, Elementary	57.95	41,000	24,500 - 55,000
Dormitories, Mid Rise	75.10	55,600	36,100 - 90,000	Schools, Jr. High	57.35	92,000	52,000 - 119,000
Factories	31.90	26,400	12,900 - 50,000	Schools, Sr.High	56.20	101,000	50,500 - 175,000
Fire Stations	63.05	5,800	4,000 - 8,700	Schools, Vocational	54.45	37,000	20,500 - 82,000
Fraternity Houses	55.75	12,500	8,200 - 14,800	Sports Arenas	44.70	15,000	5,000 - 40,000
Funeral Homes	55.85	7,800	4,500 - 11,000	Supermarkets	37.40	20,000	12,000 - 30,000
Garages, Commercial	42.50	9,300	5,000 - 13,600	Swimming Pools	64.50	13,000	7,800 - 22,000
Garages, Municipal	41.55	8,300	4,500 - 12,600	Telephone Exchange	99.15	4,500	1,200 - 10,600
Garages, Parking	18.95	163,000	76,400 - 225,300	Terminals, Bus	43.45	11,400	6,300 - 16,500
Gymnasiums	54.75	19,200	11,600 - 41,000	Theaters	53.10	10,500	8,800 - 17,500
Hospitals	111.00	55,000	27,200 - 125,000	Town Halls	63.60	10,800	4,800 - 23,400
House (Elderly)	54.70	37,000	21,000 - 66,000	Warehouses	25.15	25,000	8,000 - 72,000
Housing (Public)	45.15	36,000	14,400 - 74,400	Warehouse & Office	29.05	25,000	8,000 - 72,000
Ice Rinks	43.45	29,000	27,200 - 33,600				

Repair and Remodeling (Div. 18.1)

Cost figures in BUILDING CONSTRUCTION COST DATA are based on new construction utilizing the most cost-effective combination of labor, equipment and material with the work scheduled in proper sequence to allow the various trades to accomplish their work in an efficient manner.

The costs for repair and remodeling work must be modified due to the following factors that may be present in any given repair and remodeling project:

1. Equipment usage curtailment due to the physical limitations of the project, with only hand-operated equipment being used.
2. Increased requirement for shoring and bracing to hold up the building while structural changes are being made and to allow for temporary storage of construction materials on above-grade floors.
3. Material handling becomes more costly due to having to move within the confines of an enclosed building. For multi-story construction, low capacity elevators and stairwells may be the only access to the upper floors.
4. Large amount of cutting and patching and attempting to match the existing construction is required. It is often more economical to remove entire walls rather than create many new door and window openings. This sort of trade-off has to be carefully analyzed.
5. Cost of protection of completed work is increased since the usual sequence of construction usually can not be accomplished.
6. Economies of scale usually associated with new construction may not be present. If small quantities of components must be custom fabricated due to job requirements, unit costs will naturally increase. Also, if only small work areas are available at a given time, job scheduling between trades becomes difficult and subcontractor quotations may reflect the excessive start-up and shut-down phases of the job.
7. Work may have to be done on other than normal shifts and may have to be done around an existing production facility which has to stay in production during the course of the repair and remodeling.
8. Dust and noise protection of adjoining non-construction areas can involve substantial special protection and alter usual construction methods.
9. Job may be delayed due to unexpected conditions discovered during demolition or removal. These delays ultimately increase construction costs.
10. Piping and ductwork runs may not be as simple as for new construction. Wiring may have to be snaked through walls and floors.
11. Matching "existing construction" may be impossible because materials may no longer be manufactured. Substitutions may be expensive.
12. Weather protection of existing structure requires additional temporary structures to protect building at opening.
13. On small projects, because of local conditions, it may be necessary to pay a tradesman for a minimum of four hours for a task that is completed in one hour.

All of the above areas can contribute to increased costs for a repair and remodeling project. Each of the above factors should be considered in the planning, bidding and construction stage in order to minimize the increased costs associated with repair and remodeling jobs.

Abbreviations

A	Area Square Feet; Ampere
ABS	Acrylonitrile Butadiene Styrene; Asbestos Bonded Steel
A.C.	Alternating Current ; Air Conditioning; Asbestos Cement
A.C.I.	American Concrete Institute
Addit.	Additional
Adj.	Adjustable
af	Audio-frenquency
A.G.A.	American Gas Association
Agg.	Aggregate
A.H.	Ampere Hours
A hr	Ampere-hour
A.I.A.	American Institute of Architects
AIC	Ampere Interrupting Capacity
Allow.	Allowance
alt.	Altitude
Alum.	Aluminum
a.m.	Ante Meridiem
Amp.	Ampere
Approx.	Approximate
Apt.	Apartment
Asb.	Asbestos
A.S.B.C.	American Standard Building Code
Asbe.	Asbestos Worker
A.S.H.R.A.E.	American Society of Heating, Refrig. & AC Engineers
A.S.M.E.	American Society of Mechanical Engineers
A.S.T.M.	American Society for Testing and Materials
Attchmt.	Attachment
Avg.	Average
Bbl.	Barrel
B.&B.	Grade B and Better; Balled & Burlapped
B.&S.	Bell and Spigot
B.&W.	Black and White
b.c.c.	Body-centered Cubic
B.F.	Board Feet
Bg. Cem.	Bag of Cement
BHP	Brake Horse Power
B.I.	Black Iron
Bit.; Bitum.	Bituminous
Bk.	Backed
Bkrs.	Breakers
Bldg.	Building
Blk.	Block
Bm.	Beam
Boil.	Boilermaker
B.P.M.	Blows per Minute
BR	Bedroom
Brg.	Bearing
Brhe.	Bricklayer Helper
Bric.	Bricklayer
Brk.	Brick
Brng.	Bearing
Brs.	Brass
Brz.	Bronze
Bsn.	Basin
Btr.	Better
BTU	British Thermal Unit
BTUH	BTU per Hour
BX	Interlocked Armored Cable
c	Conductivity
C	Hundred; Centigrade
C/C	Center to Center
Cab.	Cabinet
Cair.	Air Tool Laborer
Calc	Calculated
Cap.	Capacity
Carp.	Carpenter
C.B.	Circuit Breaker
C.C.F.	Hundred Cubic Feet
cd	Candela
cd/sf	Candela per Square Foot
CD	Grade of Plywood Face & Back
CDX	Plywood, grade C&D, exterior glue
Cefi.	Cement Finisher
Cem.	Cement
CF	Hundred Feet
C.F.	Cubic Feet
CFM	Cubic Feet per Minute
c.g.	Center of Gravity
CHW	Commercial Hot Water
C.I.	Cast Iron
C.I.P.	Cast in Place
Circ.	Circuit
C.L.	Carload Lot
Clab.	Common Laborer
C.L.F.	Hundred Linear Feet
CLF	Current Limiting Fuse
CLP	Cross Linked Polyethylene
cm	Centimeter
CMP	Corr. Metal Pipe
C.M.U.	Concrete Masonry Unit
Col.	Column
CO_2	Carbon Dioxide
Comb.	Combination
Compr.	Compressor
Conc.	Concrete
Cont.	Continuous; Continued
Corr.	Corrugated
Cos	Cosine
Cot	Cotangent
Cov.	Cover
CPA	Control Point Adjustment
Cplg.	Coupling
C.P.M.	Critical Path Method
CPVC	Chlorinated Polyvinyl Chloride
C. Pr.	Hundred Pair
CRC	Cold Rolled Channel
Creos.	Creosote
Crpt.	Carpet & Linoleum Layer
CRT	Cathode-ray Tube
CS	Carbon Steel
Csc	Cosecant
C.S.F.	Hundred Square Feet
C.S.I.	Construction Specification Institute
C.T.	Current Transformer
CTS	Copper Tube Size
Cu	Cubic
Cu. Ft.	Cubic Foot
cw	Continuous Wave
C.W.	Cool White
Cwt.	100 Pounds
C.W.X.	Cool White Deluxe
C.Y.	Cubic Yard (27 cubic feet)
C.Y./Hr.	Cubic Yard per Hour
Cyl.	Cylinder
d	Penny (nail size)
D	Deep; Depth; Discharge
Dis.; Disch.	Discharge
Db.	Decibel
Dbl.	Double
DC	Direct Current
Demob.	Demobilization
d.f.u.	Drainage Fixture Units
D.H.	Double Hung
DHW	Domestic Hot Water
Diag.	Diagonal
Diam.	Diameter
Distrib.	Distribution
Dk.	Deck
D.L.	Dead Load; Diesel
Do.	Ditto
Dp.	Depth
D.P.S.T.	Double Pole, Single Throw
Dr.	Driver
Drink.	Drinking
D.S.	Double Strength
D.S.A.	Double Strength A Grade
D.S.B.	Double Strength B Grade
Dty.	Duty
DWV	Drain Waste Vent
DX	Deluxe White, Direct Expansion
dyn	Dyne
e	Eccentricity
E	Equipment Only; East
Ea.	Each
Econ.	Economy
EDP	Electronic Data Processing
E.D.R.	Equiv. Direct Radiation
Eq.	Equation
Elec.	Electrician; Electrical
Elev.	Elevator; Elevating
EMT	Electrical Metallic Conduit; Thin Wall Conduit
Eng.	Engine
EPDM	Ethylene Propylene Diene Monomer
Eqhv.	Equip. Oper., heavy
Eqlt.	Equip. Oper., light
Eqmd.	Equip. Oper., medium
Eqmm.	Equip. Oper., Master Mechanic
Eqol.	Equip. Oper., oilers
Equip.	Equipment
ERW	Electric Resistant Weld
Est.	Estimated
esu	Electrostatic Units
E.W.	Each Way
EWT	Entering Water Temperature
Excav.	Excavation
Exp.	Expansion
Ext.	Exterior
Extru.	Extrusion
f.	Fiber stress
F	Fahrenheit; Female; Fill
Fab.	Fabricated
FBGS	Fiberglass
F.C.	Footcandles
f.c.c.	Face-centered Cubic
f'c.	Compressive Stress in Concrete; Extreme Compressive Stress
F.E.	Front End
FEP	Fluorinated Ethylene Propylene (Teflon)
F.G.	Flat Grain
F.H.A.	Federal Housing Administration
Fig.	Figure
Fin.	Finished
Fixt.	Fixture
Fl. Oz.	Fluid Ounces
Flr.	Floor
F.M.	Frequency Modulation; Factory Mutual
Fmg.	Framing
Fndtn.	Foundation
Fori.	Foreman, inside
Foro.	Foreman, outside

Abbreviations

Fount.	Fountain
FPM	Feet per Minute
FPT	Female Pipe Thread
Fr.	Frame
F.R.	Fire Rating
FRK	Foil Reinforced Kraft
FRP	Fiberglass Reinforced Plastic
FS	Forged Steel
FSC	Cast Body; Cast Switch Box
Ft.	Foot; Feet
Ftng.	Fitting
Ftg.	Footing
Ft. Lb.	Foot Pound
Furn.	Furniture
FVNR	Full Voltage Non Reversing
FXM	Female by Male
Fy.	Minimum Yield Stress of Steel
g	Gram
G	Gauss
Ga.	Gauge
Gal.	Gallon
Gal./Min.	Gallon Per Minute
Galv.	Galvanized
Gen.	General
Glaz.	Glazier
GPD	Gallons per Day
GPH	Gallons per Hour
GPM	Gallons per Minute
GR	Grade
Gran.	Granular
Grnd.	Ground
H	High; High Strength Bar Joist; Henry
H.C.	High Capacity
H.D.	Heavy Duty; High Density
H.D.O.	High Density Overlaid
Hdr.	Header
Hdwe.	Hardware
Help.	Helper average
HEPA	High Efficiency Particulate Air Filter
Hg	Mercury
H.O.	High Output
Horiz.	Horizontal
H.P.	Horsepower; High Pressure
H.P.F.	High Power Factor
Hr.	Hour
Hrs./Day	Hours Per Day
HSC	High Short Circuit
Ht.	Height
Htg.	Heating
Htrs.	Heaters
HVAC	Heating, Ventilating & Air Conditioning
Hvy.	Heavy
HW	Hot Water
Hyd.; Hydr.	Hydraulic
Hz.	Hertz (cycles)
I.	Moment of Inertia
I.C.	Interrupting Capacity
ID	Inside Diameter
I.D.	Inside Dimension; Identification
I.F.	Inside Frosted
I.M.C.	Intermediate Metal Conduit
In.	Inch
Incan.	Incandescent
Incl.	Included; Including
Int.	Interior
Inst.	Installation
Insul.	Insulation
I.P.	Iron Pipe
I.P.S.	Iron Pipe Size
I.P.T.	Iron Pipe Threaded
J	Joule
J.I.C.	Joint Industrial Council
K.	Thousand; Thousand Pounds
K.D.A.T.	Kiln Dried After Treatment
Kg	Kilogram
KG	Kilogauss
Kgf	Kilogram force
KHz	Kilohertz
Kip.	1000 Pounds
KJ	Kiljoule
K.L.	Effective Length Factor
Km	Kilometer
K.L.F.	Kips per Linear Foot
K.S.F.	Kips per Square Foot
K.S.I.	Kips per Square Inch
K.V.	Kilo Volt
K.V.A.	Kilo Volt Ampere
K.V.A.R.	Kilovar (Reactance)
KW	Kilo Watt
KWh	Kilowatt-hour
L	Labor Only; Length; Long
Lab.	Labor
lat	Latitude
Lath.	Lather
Lav.	Lavatory
lb.; #	Pound
L.B.	Load Bearing; L Conduit Body
L. & E.	Labor & Equipment
lb./hr.	Pounds per Hour
lb./L.F.	Pounds per Linear Foot
lbf/sq in.	Pound-force per Square Inch
L.C.L.	Less than Carload Lot
Ld.	Load
L.F.	Linear Foot
Lg.	Long; Length; Large
L. & H.	Light and Heat
L.H.	Long Span High Strength Bar Joist
L.J.	Long Span Standard Strength Bar Joist
L.L.	Live Load
L.L.D.	Lamp Lumen Depreciation
lm	Lumen
lm/sf	Lumen per Square Foot
lm/W	Lumen Per Watt
L.O.A.	Length Over All
log	Logarithm
L.P.	Liquefied Petroleum; Low Pressure
L.P.F.	Low Power Factor
Lt.	Light
Lt. Ga.	Light Gauge
L.T.L.	Less than Truckload Lot
Lt. Wt.	Lightweight
L.V.	Low Voltage
M	Thousand; Material; Male; Medium Wall Copper
m/hr	Manhour
mA	Milliampere
Mach.	Machine
Mag. Str.	Magnetic Starter
Maint.	Maintenance
Marb.	Marble Setter
Mat.	Material
Mat'l.	Material
Max.	Maximum
MBF	Thousand Board Feet
MBH	Thousand BTU's per hr.
M.C.F.	Thousand Cubic Feet
M.C.F.M.	Thousand Cubic Feet per Minute
M.C.M.	Thousand Circular Mils
M.C.P.	Motor Circuit Protector
MD	Medium Duty
M.D.O.	Medium Density Overlaid
Med.	Medium
MF	Thousand Feet
M.F.B.M.	Thousand Feet Board Measure
Mfg.	Manufacturing
Mfrs.	Manufacturers
mg	Milligram
MGD	Million Gallons per Day
MGPH	Thousand Gallons per Hour
MH	Manhole; Metal Halide; Man Hour
MHz	Megahertz
Mi.	Mile
MI	Malleable Iron; Mineral Insulated
mm	Millimeter
Mill.	Millwright
Min.	Minimum
Misc.	Miscellaneous
ml	Milliliter
M.L.F.	Thousand Linear Feet
Mo.	Month
Mobil.	Mobilization
Mog.	Mogul Base
MPH	Miles per Hour
MPT	Male Pipe Thread
MRT	Mile Round Trip
ms	millisecond
M.S.F.	Thousand Square Feet
Mstz.	Mosaic & Terrazzo Worker
M.S.Y.	Thousand Square Yards
Mtd.	Mounted
Mthe.	Mosaic & Terrazzo Helper
Mtng.	Mounting
Mult.	Multi; Multiply
MV	Megavolt
MW	Megawatt
MXM	Male by Male
MYD	Thousand yards
N	Natural; North
nA	nanoampere
NA	Not Available; Not Applicable
N.B.C.	National Building Code
NC	Normally Closed
N.E.M.A.	National Electrical Manufacturers Association
NEHB	Bolted Circuit Breaker to 600V.
N.L.B.	Non-Load-Bearing
nm	nanometer
No.	Number
NO	Normally Open
N.O.C.	Not Otherwise Classified
Nose.	Nosing
N.P.T.	National Pipe Thread
NQOB	Bolted Circuit Breaker to 240V.
N.R.C.	Noise Reduction Coefficient
N.R.S.	Non Rising Stem
ns	nanosecond
nW	nanowatt
OB	Opposing Blade
OC	On Center
OD	Outside Diameter
O.D.	Outside Dimension
ODS	Overhead Distribution System
O & P	Overhead and Profit
Oper.	Operator
Opng.	Opening
Orna.	Ornamental
O.S.&Y.	Outside Stem and Yoke
Ovhd	Overhead
Oz.	Ounce

Abbreviations

P.	Pole; Applied Load; Projection
p.	Page
Pape.	Paperhanger
PAR	Weatherproof Reflector
Pc.	Piece
P.C.	Portland Cement; Power Connector
P.C.F.	Pounds per Cubic Foot
P.E.	Professional Engineer; Porcelain Enamel; Polyethylene; Plain End
Perf.	Perforated
Ph.	Phase
P.I.	Pressure Injected
Pile.	Pile Driver
Pkg.	Package
Pl.	Plate
Plah.	Plasterer Helper
Plas.	Plasterer
Pluh.	Plumbers Helper
Plum.	Plumber
Ply.	Plywood
p.m.	Post Meridiem
Pord.	Painter, Ordinary
pp	Pages
PP; PPL	Polypropylene
P.P.M.	Parts Per Million
Pr.	Pair
Prefab.	Prefabricated
Prefin.	Prefinished
Prop.	Propelled
PSF; psf	Pounds per Square Foot
PSI; psi	Pounds per Square Inch
PSIG	Pounds per Square Inch Gauge
PSP	Plastic Sewer Pipe
Pspr.	Painter, Spray
Psst.	Painter, Structural Steel
P.T.	Potential Transformer
P. & T.	Pressure & Temperature
Ptd.	Painted
Ptns.	Partitions
Pu	Ultimate Load
PVC	Polyvinyl Chloride
Pvmt.	Pavement
Pwr.	Power
Q	Quantity Heat Flow
Quan.; Qty.	Quantity
Q.C.	Quick Coupling
r	Radius of Gyration
R	Resistance
R.C.P.	Reinforced Concrete Pipe
Rect.	Rectangle
Reg.	Regular
Reinf.	Reinforced
Req'd.	Required
Resi	Residential
Rgh.	Rough
R.H.W.	Rubber, Heat & Water Resistant; Residential Hot Water
rms	Root Mean Square
Rnd.	Round
Rodm.	Rodman
Rofc.	Roofer, Composition
Rofp.	Roofer, Precast
Rohe.	Roofer Helpers (Composition)
Rots.	Roofer, Tile & Slate
R.O.W.	Right of Way
RPM	Revolutions per Minute
R.R.	Direct Burial Feeder Conduit
R.S.	Rapid Start
RT	Round Trip
S.	Suction; Single Entrance; South
Scaf.	Scaffold
Sch.; Sched.	Schedule
S.C.R.	Modular Brick
S.D.R.	Standard Dimension Ratio
S.E.	Surfaced Edge
S.E.R.; S.E.U.	Service Entrance Cable
S.F.	Square Foot
S.F.C.A.	Square Foot Contact Area
S.F.G.	Square Foot of Ground
S.F. Hor.	Square Foot Horizontal
S.F.R.	Square Feet of Radiation
S.F.Shlf.	Square Foot of Shelf
S4S	Surface 4 Sides
Shee.	Sheet Metal Worker
Sin	Sine
Skwk.	Skilled Worker
SL	Saran Lined
S.L.	Slimline
Sldr.	Solder
S.N.	Solid Neutral
S.P.	Static Pressure; Single Pole; Self-Propelled
Spri.	Sprinkler Installer
Sq.	Square; 100 square feet
S.P.D.T.	Single Pole, Double Throw
S.P.S.T.	Single Pole, Single Throw
SPT	Standard Pipe Thread
Sq. Hd.	Square Head
S.S.	Single Strength; Stainless Steel
S.S.B.	Single Strength B Grade
Sswk.	Structural Steel Worker
Sswl.	Structural Steel Welder
St.; Stl.	Steel
S.T.C.	Sound Transmission Coefficient
Std.	Standard
STP	Standard Temperature & Pressure
Stpi.	Steamfitter; Pipefitter
Str.	Strength; Starter; Straight
Strd.	Stranded
Struct.	Structural
Sty.	Story
Subj.	Subject
Subs.	Subcontractors
Surf.	Surface
Sw.	Switch
Swbd.	Switchboard
S.Y.	Square Yard
Syn.	Synthetic
Sys.	System
t.	Thickness
T	Temperature; Ton
Tan	Tangent
T.C.	Terra Cotta
T.D.	Temperature Difference
TFE	Tetrafluoroethylene (Teflon)
T. & G.	Tongue & Groove; Tar & Gravel
Th.; Thk.	Thick
Thn.	Thin
Thrded	Threaded
Tilf.	Tile Layer, Floor
Tilh.	Tile Layer Helper
THW	Insulated Strand Wire
THWN; THHN	Nylon Jacketed Wire
T.L.	Truckload
Tot.	Total
T.S.	Trigger Start
Tr.	Trade
Transf.	Transformer
Trhv.	Truck Driver, Heavy
Trlr.	Trailer
Trlt.	Truck Driver, Light
TV	Television
T.W.	Thermoplastic Water Resistant Wire
UCI	Uniform Construction Index
UF	Underground Feeder
U.H.F.	Ultra High Frequency
U.L.	Underwriters Laboratory
Unfin.	Unfinished
URD	Underground Residential Distribution
V	Volt
VA	Volt/amp
V.A.T.	Vinyl Asbestos Tile
VAV	Variable Air Volume
Vent.	Ventilating
Vert.	Vertical
V.G.	Vertical Grain
V.H.F.	Very High Frequency
VHO	Very High Output
Vib.	Vibrating
V.L.F.	Vertical Linear Foot
Vol.	Volume
W	Wire; Watt; Wide; West
w/	With
W.C.	Water Column; Water Closet
W.F.	Wide Flange
W.G.	Water Gauge
Wldg.	Welding
Wrck.	Wrecker
W.S.P.	Water, Steam, Petroleum
WT, Wt.	Weight
WWF	Welded Wire Fabric
XFMR	Transformer
XHD	Extra Heavy Duty
Y	Wye
yd	Yard
yr	Year
△	Delta
%	Percent
~	Approximately
Ø	Phase
@	At
#	Pound; Number
<	Less Than
>	Greater Than

INDEX

INDEX

D

E

INDEX

H

I

INDEX

INDEX

T

INDEX

U

V

INDEX

NEW . . . Boating Cost Guide

1st Annual Edition

Now for the first time ever . . . authoritative cost data for boats up to 65 feet, Sail and Power.

Over 300 pages, illustrated

Over 11,000 actual cost items covering virtually every expense associated with boat maintenance, repair and improvement

The **BOATING COST GUIDE** gives you **all the cost facts** you need. This includes **both** the material or equipment cost **and** installation expenses. These costs, where appropriate, are shown three ways . . . the **minimum, average, or maximum** you might expect to pay.

- make realistic estimates for boat repairs
- discover cost-saving options
- make better planning decisions
- budget accurately for improvements
- compare price quotations
- check bills for work done

Book No. 50185 . . . . $29.95/copy

Labor Rates for the Construction Industry 1985

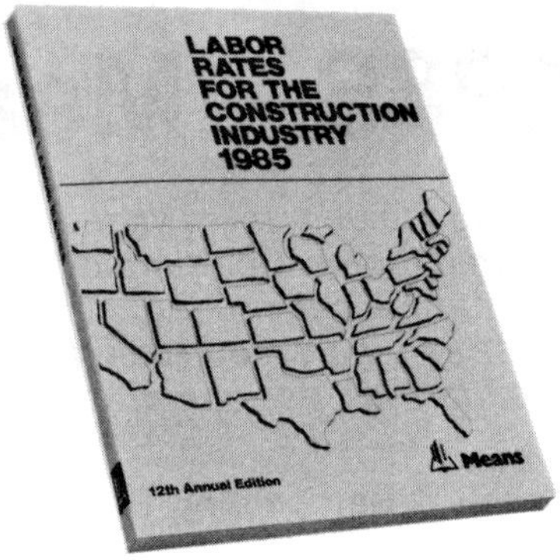

Over 300 pages

12th Annual Edition

CITY

STATE

NATIONAL

- detailed wage rates by trade for over 300 U.S. and Canadian cities
- forty-six construction trades listed by local union number in each city
- base hourly wage rates plus fringe benefit package costs gathered from reliable sources
- dependable estimates for the trade wage rates not reported at press time
- effective dates for newly negotiated union contracts for both 1985 and 1986
- factors for comparing each trade rate by city, state, and national averages
- historical 1983-1984 wage rates also included for comparison purposes
- each city chart is alphabetically arranged with handy visual flip tabs for quick reference

Book No. 60285 . . . . $34.95/copy

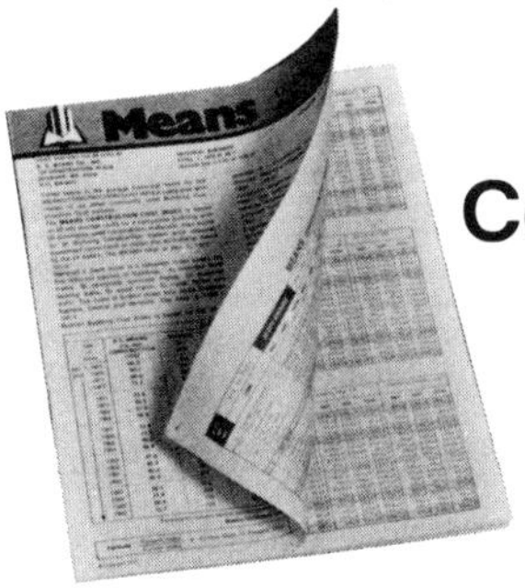

Means Construction Cost Indexes 1985

$47.95 per year

(Individual and back issues available at $15.95 per copy)

The index service providing updated cost adjustment factors.

Whether updating construction cost estimates based on information in the Means cost manuals, or those from other sources, the construction cost index service is the efficient way to assure 90-day cost accuracy.

Published quarterly (January, April, July, October), this handy report provides cost adjustment factors for the preparation of more precise estimates no matter how late in the year. It's also the ideal method for making continuous cost revisions on on-going projects as the year progresses.

The report comes in four unique sections:

- Breakdowns for 209 major cities
- National averages for 30 key cities
- Five large city averages
- Historical construction cost indexes

Book No. 61185 . . . . $47.95/year

Means Historical Cost Indexes 1985

$15.95 per copy

(20-page booklet)

Your latest estimating resource for historical cost information.

This fifth annual edition of *Means Historical Cost Indexes* is designed especially for professionals with estimating activities involving construction costs not only in various *cities* but in various *years*. The booklet indexes historical construction costs for 209 U.S. and Canadian cities from 1940 through 1985.

These historical indexes are the result of careful monitoring of specific material quantities, man-hours, building trades, equipment rental, and installation. This extensive coverage can be depended upon for functional and reliable assistance in all of your historical cost estimating activities. *Means Historical Cost Indexes* includes detailed explanations of uses, with complete examples.

Book No. 61485 . . . . $15.95/copy

1985—Means Seminar Program—1985

During the upcoming year, R.S. Means offers you a series of 2-day seminars oriented to a wide range of construction-related topics. Conducted throughout the United States, scheduled seminars thoroughly cover the subjects of cost estimating, scheduling, management, and computerized systems for construction. R.S. Means is proud to present these seminars, each specifically designed to give you the expertise you need to be successful in your profession. All seminars include comprehensive workbooks plus the latest innovations in computerized estimating and scheduling.

Repair & Remodeling Estimating
(two days) #104

Here are the specialized techniques you need to approach repair and remodeling projects with new skill and confidence. This seminar sorts out and discusses practical solutions to the difficult and varied cost estimating problems associated with the reuse and conversion of existing structures. A wealth of eye-opening procedures is presented, including information about computerized cost estimating as applied to repair and remodeling projects. Based on optimum use of the Means cost manual *Repair & Remodeling Cost Data*. A "must" for new estimators.

Unit Price Estimating
(two days) #100

For those with estimating experience, this course shows you how today's advanced estimating techniques and cost information sources (such as computerized estimating systems and cost data banks) can be used to develop unit price estimates for projects of any size. You'll get down-to-earth know-how and easy-to-apply directions on organizing information effectively, using plans efficiently, ensuring error-free results, and maximizing your use of cost data resources, including *Building Construction Cost Data*.

Square Foot Cost Estimating
(two days) #106

For estimators with experience, here's the expert guidance you've asked for. We'll help you develop the practical know-how for producing rapid, reliable square foot cost estimates using only minimal design details. You'll perfect your skills for "before plans" estimating, replacement cost estimating, and assessing buildings of all types. In square foot costing methods, participants receive clear-cut explanations and suggestions for dealing with several types of construction projects at various stages of completion. Centered around techniques used in *Means Systems Costs, Means Square Foot Costs,* and *Square Foot Estimating.*

Scheduling and Project Management
(two days) #102

This seminar helps you successfully establish project priorities, develop realistic schedules, and apply today's advanced management techniques to your construction projects. Included are the "network" approach, Precedence, and Critical Path Methods. Special emphasis is placed on cost interface with various monitoring setups such as computer-based control systems. Through this seminar you'll perfect your scheduling and management skills, ensuring completion of your projects *on time* and *within budget.* Includes hands-on application of *Means Scheduling Manual* and *Building Construction Cost Data.*

Computers in Construction
(two days) #112

Prepared for those interested in using computers for construction activities, this seminar describes and demonstrates the latest available hardware and software products for construction cost estimating, project scheduling, and accounting applications. Actual demonstrations of the Means line of construction-oriented software — as well as other construction and business software — highlight this Seminar. Optimum use is made of *Building Construction Cost Data.*

1985 SPRING SEMINARS

Location	Seminar No. & Name	Dates (1985)
TAMPA Bay Harbor Inn 7700 Courtney Campbell Causeway Tampa, FL 33607 (813) 885-2541	10453TA Repair & Remodeling Est.	Mar. 18 & 19
	10653TA Square Foot Cost Est.	Mar. 18 & 19
	10053TA Unit Price Estimating	Mar. 20 & 21
	10253TA Sched. & Proj. Mgmt.	Mar. 20 & 21
	11253TA Computers in Constr.	Mar. 20 & 21
BEVERLY HILLS Ramada Inn 1150 S. Beverly Drive Beverly Hills, CA 90035 (213) 553-6561	10453BH Repair & Remodeling Est.	Mar. 25 & 26
	10653BH Square Foot Cost Est.	Mar. 25 & 26
	10053BH Unit Price Estimating	Mar. 27 & 28
	10253BH Sched. & Proj. Mgmt.	Mar. 27 & 28
	11253BH Computers in Constr.	Mar. 27 & 28
BOSTON Natick Hilton Inn Route 9, Speen Street Natick, MA 01760 (617) 653-5000	10454BO Repair & Remodeling Est.	Apr. 1 & 2
	10654BO Square Foot Cost Est.	Apr. 1 & 2
	10054BO Unit Price Estimating	Apr. 3 & 4
	10254BO Sched. & Proj. Mgmt.	Apr. 3 & 4
	11254BO Computers in Constr.	Apr. 3 & 4
DENVER Holiday Inn 1475 S. Colorado Blvd. Denver, CO 80222 (303) 757-7731	10454DE Repair & Remodeling Est.	Apr. 15 & 16
	10654DE Square Foot Cost Est.	Apr. 15 & 16
	10054DE Unit Price Estimating	Apr. 17 & 18
	10254DE Sched. & Proj. Mgmt.	Apr. 17 & 18
	11254DE Computers in Constr.	Apr. 17 & 18
WASHINGTON, DC Howard Johnson's-Airport 2650 Jefferson Davis Arlington, VA 22202 (703) 684-7200	10454AR Repair & Remodeling Est.	Apr. 29 & 30
	10654AR Square Foot Cost Est.	Apr. 29 & 30
	10054AR Unit Price Estimating	May 1 & 2
	10254AR Sched. & Proj. Mgmt.	May 1 & 2
	11254AR Computers in Constr.	May 1 & 2
PHOENIX Holiday Inn Metro Center Holidome 2532 W. Peoria Ave. Phoenix, AZ 85029 (602) 943-2341	10455PX Repair & Remodeling Est.	May 20 & 21
	10655PX Square Foot Cost Est.	May 20 & 21
	10055PX Unit Price Estimating	May 22 & 23
	10255PX Sched. & Proj. Mgmt.	May 22 & 23
	11255PX Computers in Constr.	May 22 & 23
NEW YORK Westchester Marriott 670 White Plains Rd. Tarrytown, NY 10591 (914) 631-2200	10456TT Repair & Remodeling Est.	Jun 3 & 4
	10656TT Square Foot Cost Est.	Jun 3 & 4
	10056TT Unit Price Estimating	Jun 5 & 6
	10256TT Sched. & Proj. Mgmt	Jun 5 & 6
	11256TT Computers in Constr.	Jun 5 & 6
ATLANTA Omni International Hotel One Omni International Plaza Atlanta, GA 30303 (404) 659-0000	10456AT Repair & Remodeling Est.	Jun24&25
	10656AT Square Foot Cost Est.	Jun24&25
	10056AT Unit Price Estimating	Jun26&27
	10256AT Sched. & Proj. Mgmt.	Jun26&27
	11256AT Computers in Constr.	Jun26&27

1985 FALL SEMINARS

Location	Seminar No. & Name	Dates (1985)
HYANNIS Sheraton Regal Inn Rt. 132 & Bearse Way Hyannis, MA 02601 (617) 771-3000	10459HY Repair & Remodeling Est.	Sept. 9 & 10
	10659HY Square Foot Cost Est.	Sept. 9 & 10
	10059HY Unit Price Estimating	Sept. 11 & 12
	10259HY Sched. & Proj. Mgmt.	Sept. 11 & 12
	11259HY Computers in Constr.	Sept. 11 & 12
SAN FRANCISCO San Franciscan Hotel 1231 Market Street at Civic Center San Francisco, CA 94103 (415) 626-8000	1045ASF Repair & Remodeling Est.	Oct. 7 & 8
	1065ASF Square Foot Cost Est.	Oct. 7 & 8
	1005ASF Unit Price Estimating	Oct. 9 & 10
	1025ASF Sched. & Proj. Mgmt.	Oct. 9 & 10
	1125ASF Computers in Constr.	Oct. 9 & 10
WASHINGTON, DC Howard Johnson's-Airport 2650 Jefferson Davis Arlington, VA 22202 (703) 684-7200	1045BAR Repair & Remodeling Est.	Nov. 4 & 5
	1065BAR Square Foot Cost Est.	Nov. 4 & 5
	1005BAR Unit Price Estimating	Nov. 6 & 7
	1025BAR Sched. & Proj. Mgmt.	Nov. 6 & 7
	1125BAR Computers in Constr.	Nov. 6 & 7
DALLAS Amfac Hotel & Resort P.O. Box 61025 Dallas/Ft. Worth Airport Dallas, TX 75261 (214) 453-8400	1045CDA Repair & Remodeling Est.	Dec. 2 & 3
	1065CDA Square Foot Cost Est.	Dec. 2 & 3
	1005CDA Unit Price Estimating	Dec. 4 & 5
	1025CDA Sched. & Proj. Mgmt.	Dec. 4 & 5
	1125CDA Computers in Constr.	Dec. 4 & 5

1985 — Seminar Registration Form and Information — 1985

Registration Fees

- One 2-day seminar, $550 per person
- One 2-day seminar, two or more people registering at the same time, $500 per person
- Two 2-day seminars, $900 per person
- Two 2-day seminars, two or more people registering at the same time, $815 per person

Full payment must be received at least 15 days prior to seminar dates in order to confirm attendance (U.S. Funds).

How to Register

Complete the registration form and return with your full fee (or minimum deposit of $220/person for one 2-day seminar, or minimum deposit of $360/person for two 2-day seminars) to:

Seminar Division
R.S. Means Company, Inc.
100 Construction Plaza
P.O. Box 800
Kingston, MA 02364-0800

Class sizes are limited so please return your registration form as soon as possible.

Refunds

Advance payments are fully refundable up to 15 days prior to the seminar dates. After that time, a 2-day seminar cancellation is subject to a service charge of $160/person and two 2-day seminar cancellations $270/person.

Tax Deduction

A federal income tax deduction is allowed for expenses (including registration fees, travel, meals and lodging) for your attendance at a Means Seminar (see Treas. Reg. 1.162-5).

AACE Approved Courses:

The R.S. Means Construction Estimating and Management Seminars described and offered to you here have each been approved for 14 hours (1.4 CEU) of credit by the American Association of Cost Engineers (AACE, Inc.) Certification Board toward meeting the continuing education requirements for re-certification as a Certified Cost Engineer/-Certified Cost Consultant.

Coffee/Cola Breaks

Your registration includes the cost of a coffee break in the morning and an afternoon cola break. These informal segments will allow you to discuss topics of mutual interest with other members of the seminar.

Late Telephone Registrations

If you wish to register for any seminar with less than two weeks remaining before the seminar date, please call our Kingston office (617) 747-1270 and ask for the Seminar Registrar. Arrangements may be made if space is available.

Daily Course Schedule

The first day of each seminar session begins at 8:30 A.M. and ends at 4:30 P.M. The second day is 8:00 - 4:00. Participants are urged to bring a hand-held calculator since many actual problems will be worked out in each session.

Transportation/Hotel Accommodations

For your convenience, you can obtain assistance in making transportation and hotel arrangements by calling Waterfront Travel in Boston, MA at 1-800-343-6524, and notifying them of your planned attendance at the Means Seminar of your choice. Reservations for lodging can be made through the Waterfront Travel Service. Airline reservations, hotel accommodations and car rental arrangements will be made for you by Waterfront to fit your schedule. We suggest that you contact Waterfront Travel as soon as you return your registration form, in order to reserve the lodging of your choice. Reservations should be received by Waterfront Travel Service one month prior to the seminar date.

In-House Presentation

All R.S. Means seminars are available for in-house presentation and can be tailored to meet individual client/company needs. If you are interested in having a program conducted at your location, just contact Dan Braman, P.E., Director of Sales, R.S. Means Co., Inc. 151 Construction Plaza, P.O. Box 800, Kingston, MA 02364-0800. Telephone: (617) 747-1270.

Registration Form

Please register the following people for the Means Construction Seminars as shown here. Full payment or deposit is enclosed, and we understand we must make our own hotel reservations if overnight stays are necessary.

☐ Full payment of $ ____________ enclosed.

☐ Deposit of $ ____________ enclosed.

Balance due $ ____________ (U.S. Funds)

Firm Name ____________

Address ____________

City/State/Zip ____________

Telephone Number ____________

☐ Charge our registration(s) to:

___American Express ___Visa ___MasterCard

Account No. ____________ Exp. Date ____________

____
CARDHOLDER'S SIGNATURE

NAME OF REGISTRANT(S) (To appear on certificate of completion)	SEMINAR NAME AND NUMBER	CITY	DATES

Please mail check to: **R.S. MEANS CO., INC., 151 Construction Plaza, P.O. Box 800, Kingston, MA 02364-0800**

GALAXY

Automated Quantity Takeoff and Pricing System

Modern technology can save you money by cutting takeoff time at least 50%

Quantity takeoffs don't have to be difficult – not when you can take advantage of modern technology. GALAXY Automated Takeoff and Pricing System lets you determine the detailed cost information you need with the aid of a computer.

You can reduce the time it takes to do an estimate by as much as 50 percent, and more for many projects, thanks to GALAXY's electronic digitizer table. Working directly from a set of plans, an estimator easily feeds the information into the computer, which then makes the proper calculations to determine length, height, width, volume, area or numeric count – all in a fraction of the time it normally would take.

Using a special hand-held cursor, specific points are selected from the plans and "read" into the computer. The estimator chooses the items in convenient order and is able to quickly determine the quantities as desired. Designed for use with the STAR General Estimating Program, GALAXY provides you with access to your own files of known costs as well as any of the R.S. Means standard cost information files of up-to-date unit price information.

Best of all, GALAXY is easy to operate. Any construction professional can learn to run this system after attending the training workshop. It's all part of The Solution by R.S. Means Company, Inc., the people you depend on for your construction cost data.

GALAXY Main Benefits

- Saves valuable time during takeoffs
- Easy to operate
- Increases productivity
- Reduces need for additional manpower
- Streamlines the estimating process
- Produces detailed estimates
- Makes changes easily
- Customizes estimates to specific requirements

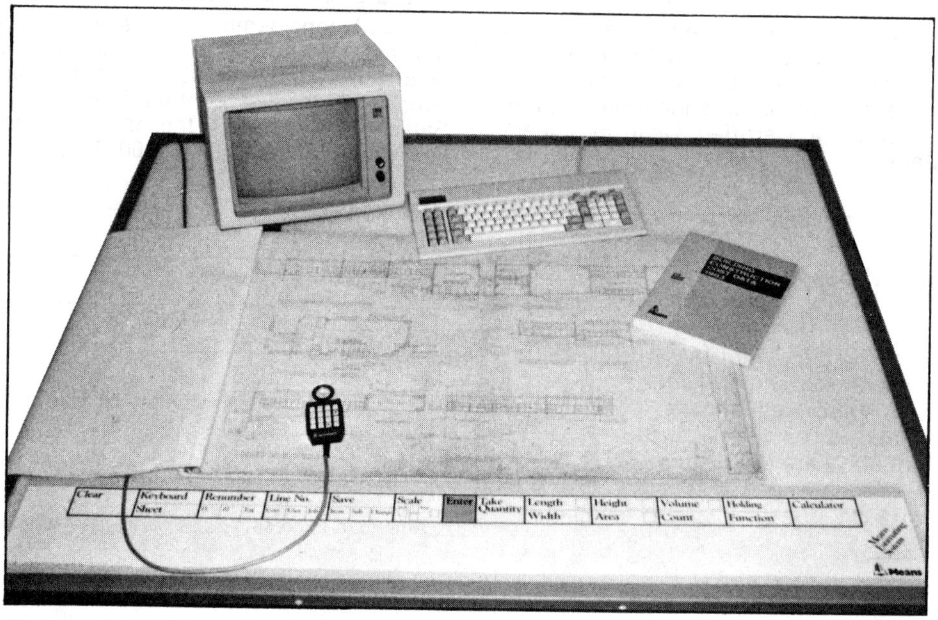

"Takeoffs" right from drawings.

Detail of Digitizer Menu

Clear	Keyboard / Sheet	Renumber: #1, #2, Tag	Line No.: Cost, User, Job	Save: Item, Sub, Change	Scale: Arch, Eng
Take Quantity	Length / Width	Height / Area	Volume / Count	Holding / Function	Calculator

STAR

General Estimating Program

Helps develop fast and accurate estimates and manage vital cost data

A computerized estimating service from the people you can depend on

Now you can use the speed and dependability of a computer to simplify your estimating process. STAR General Estimating Program is easy to use, reduces errors, provides more accurate estimates and enables you to increase productivity and improve profit margins.

Using known construction costs, which reflect your marketplace, STAR helps you to quickly develop your data file, which is organized in the UCI-16 division format. This file then gives you, the estimator, access to necessary information, such as item description, crew make-up, daily productivity, unit of measure and unit prices for material, labor, equipment and total.

In addition STAR can also access any of the Means standard cost files.

With STAR, you can store and update estimates with ease and make modifications with the touch of a button. Changes are no problem because the computer does all the work.

More importantly, STAR enables you to save time–and money! An estimator can now evaluate more jobs, which means more work, increased productivity and higher profits. It's all part of The Solution by R.S. Means Company, Inc., the people you can depend on for your construction cost data.

STAR Main Benefits

- Develop your own cost file–quickly
- Provides fast, accurate estimates
- Easy to operate
- Provides up-to-date unit prices
- Saves hours of tedious calculating
- Customizes estimates to specific requirements
- Two methods–daily output and man-hour analysis
- Helps in scheduling and planning
- Evaluated overhead and profit

Unburdened Job Report Summary
Here you see a summary report of the entire estimate by division, including your evaluation of overhead and profit.

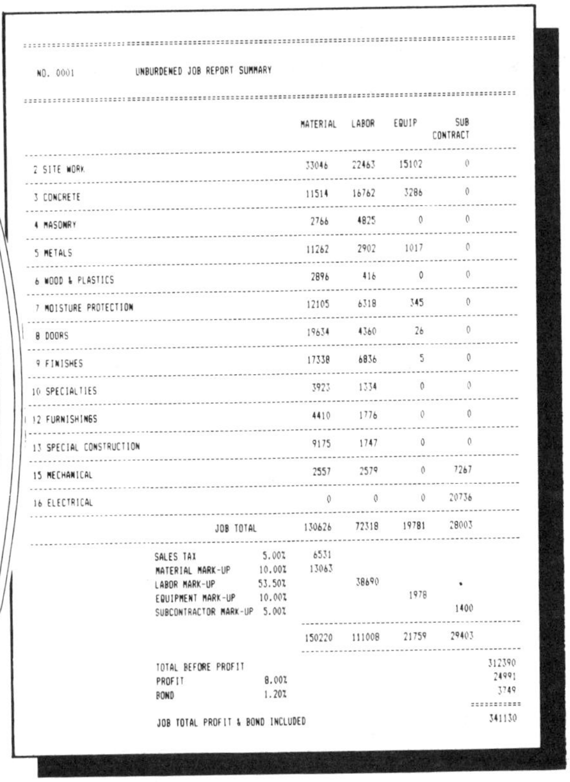

NO. 0001 UNBURDENED JOB REPORT SUMMARY

		MATERIAL	LABOR	EQUIP	SUB CONTRACT	
2 SITE WORK		33046	22463	15102	0	
3 CONCRETE		11514	16762	3286	0	
4 MASONRY		2766	4825	0	0	
5 METALS		11262	2902	1017	0	
6 WOOD & PLASTICS		2896	416	0	0	
7 MOISTURE PROTECTION		12105	6318	345	0	
8 DOORS		19634	4360	26	0	
9 FINISHES		17338	6836	5	0	
10 SPECIALTIES		3923	1334	0	0	
12 FURNISHINGS		4410	1776	0	0	
13 SPECIAL CONSTRUCTION		9175	1747	0	0	
15 MECHANICAL		2557	2579	0	7267	
16 ELECTRICAL		0	0	0	20736	
JOB TOTAL		130626	72318	19781	28003	
SALES TAX	5.00%	6531				
MATERIAL MARK-UP	10.00%	13063				
LABOR MARK-UP	53.50%		38690			
EQUIPMENT MARK-UP	10.00%			1978		
SUBCONTRACTOR MARK-UP	5.00%				1400	
		150220	111008	21759	29403	
TOTAL BEFORE PROFIT						312390
PROFIT	8.00%					24991
BOND	1.20%					3749
JOB TOTAL PROFIT & BOND INCLUDED						341130

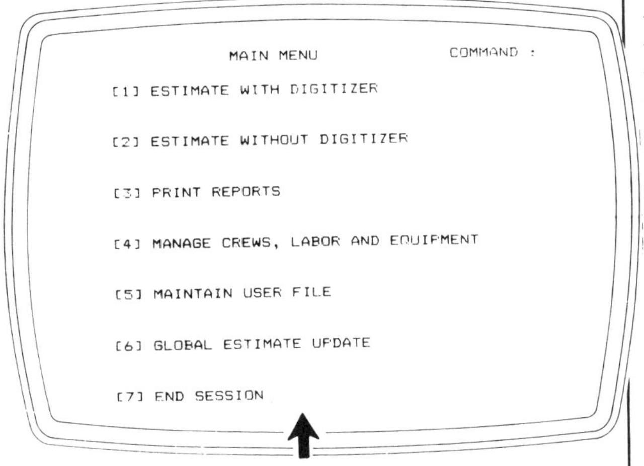

Sample Menu
This menu shows you the seven different directions you can go in the program.

UNIT PRICE COST FILES

Specific construction cost files help you create detailed estimates

Rapidly changing costs in the construction industry are no longer a problem thanks to a series of specific building information files from R.S. Means Company, Inc. The Unit Price Cost Files provide you with up-to-date cost data in several different contractor and subcontractor construction categories that help to eliminate much of the uncertainty of creating detailed estimates and bids.

These cost data files, which can be purchased together or separately, contain more than 20,000 thoroughly-researched unit prices for labor, materials and installation. They can be used in conjunction with the STAR General Estimating Program and the GALAXY Automated Quantity Takeoff and Pricing System to determine construction costs in specific building applications, including mechanical, electrical, sitework and more.

Each construction task is defined by providing productivity and crew make-up assumptions as well as material costs. These can be easily adjusted to reflect your market place needs.

The Unit Price Cost Files provide the estimator with quick access to the cost figures he needs to know in the UCI format. National average union labor rates and equipment rental rates are included. It saves time and money in the preparation of estimates and gives you the confidence of knowing that your bid is as accurate as it can be.

R.S. Means also provides you with a valuable annual update service for these important Unit Price Cost Files. Offered on a subscription basis, each priced item is checked and adjusted as needed before the files are sent to you each year.

The Unit Price Cost Files can be changed and modified to your estimate without sacrificing accuracy. It's all part of The Solution by R.S. Means Company, Inc., the people you depend on for all your construction cost data.

UNIT PRICE COST FILES

- Building Construction
- Repair and Remodeling
- Concrete and Masonry
- Residential/Light Commercial
- Mechanical
- Electrical
- Sitework
- Interiors

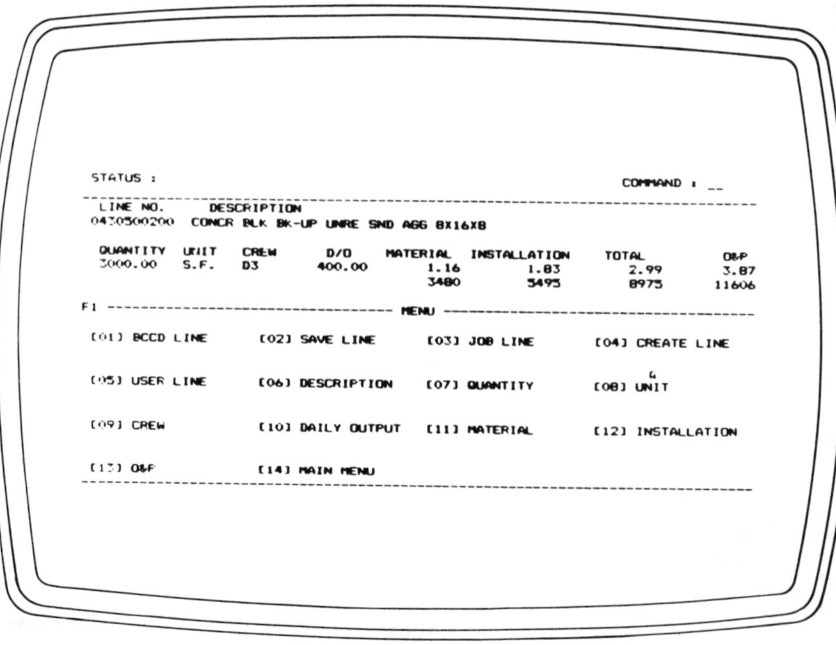

Screen showing typical line item.

You can add, delete, or change any item when used with STAR or GALAXY. Fourteen options are available.

Concise, authoritative, highly acclaimed . . .

ESTIMATING FOR THE GENERAL CONTRACTOR

by Paul J. Cook

The first book prepared specifically to help contractors and estimators in medium-sized companies estimate more efficiently and accurately

Light on theory, heavy on practical estimating methods and ideas, here's powerful help for the contractor/estimator who wants to evaluate and polish every facet of their estimating procedure.

No matter how much estimating experience you've had, no matter what your particular construction specialty, there's something of importance for you in this extraordinary volume.

Prepared by one of the industry's most successful estimating consultants, this guide is designed to give your estimating a refreshing breath of new ideas . . . insights . . . better techniques. It's a book you'll use over and over again, providing the "extra" dimension you need to make your estimates more reliable and usable than ever before.

First Edition	over 225 pages, illustrated

Hardcover Book No. 67160 . . . $35.95/copy
Softcover Book No. 67170 . . . $30.95/copy

Square Foot Estimating

A new generation of techniques for conceptual and design stage cost planning

by F. William Horsley and Billy Joe Cox

First Edition	over 300 pages, illustrated

Square Foot Estimating gives you the expertise to transform conceptual and design-stage estimating into a powerful tool for new business.

You'll see how to maximize the use of your time and effort to achieve new heights of creativity in designs, materials, and construction approaches . . . all within given budget limitations!

This fast-moving new book begins with a clear look at four estimating techniques used in design stage costing. Each is evaluated by degree of accuracy in relation to time and cost.

It spotlights the use of daring new departures from traditional thinking about square foot costs. It shows how one method — systems estimating — achieves remarkable reliability, typically in a day's work.

The editors demonstrate how building systems can be tried in an infinite number of combinations to quickly determine the direction to take for budgets, codes, loads, insulation, fire-proofing, acoustics, energy use, and special requirements.

Hardcover Book No. 67145 . . . $36.95/copy
Softcover Book No. 67146 . . . $31.95/copy

...planned, written and produced to solve — with remarkable speed — the scheduling problems of the eighties...

Means Scheduling Manual

Now with special new emphasis on using microcomputers and programs for scheduling and project controls.

$32.95 per copy | ***2nd edition over 200 pages, illustrated***

NEWLY REVISED SECOND EDITION . . .

Tightly written in one easy-to-use spiral-bound volume, here are today's most advanced methods for scheduling and managing building construction projects in the intermediate size range. This is a masterfully-prepared construction management tool equal to today's — and tomorrow's — burgeoning scheduling problems.

Means Scheduling Manual filled with the kind of sound practical advice and guidance you've always wanted in a scheduling book. This is a scheduling aid that's truly one step ahead of any other you've ever seen.

Book No. 67152 . . . . $32.95/copy

Means Systems Costs 1985

The Means illustrated systems cost manual...preferred by more construction professionals for its superb organization and price reliability

$38.95 per copy | **10th annual edition over 525 pages illustrations**

This indispensable tool provides you with thousands of detailed systems costs, accompanied by hundreds of drawings, explanations, component breakdowns and tables -- all within one easy-to-use format.

You'll find separate design aids to assist you in the conceptual phase of estimating any job. Tables of unit prices allow you to cost-out virtually any combination of building systems in logical, step-by-step sequence, customizing your project as more information becomes available.

Now uses the UNIFORMAT numbering system as a guideline for consistent estimating and reporting. It enables you to price and compare various framing and envelope systems with ease.

Use the manual's complete examples, suggestions, and standardized format to compare past and future jobs and build your own cost data file.

Book No. 60385 . . . . $38.95/copy

MEANS SQUARE FOOT COSTS 1985

$39.95 per copy | 6th Annual Edition with over 425 pages photographs illustrations

Designed for effective use at all levels of building cost budgeting, estimating and valuation . . . the planning guide and replacement cost resource for architects, design/build contractors, real estate developers, corporate planners, appraisers and assessors.

Residential • Commercial
Industrial • Institutional

Here's what you'll find packed into the 1985 edition:

- over 6,000 costs for typical buildings and attachments . . . residential, commercial, industrial, institutional
- reliable costs for use in 1985
- four classes of residential quality: economy, average, custom, luxury
- over 350 illustrations of unit-in-place components and costs, fully detailed for systems style estimates
- worked-out examples and clear instructions
- cost adjustment factors for all construction-related variables
- over 6,500 unit-in-place costs illustrated for detailed budgets
- factors for local cost differences by Zip Code
- depreciation guidelines
- samples of convenient work sheets for preparing the estimate
- nearly 150 photographs of typical building types and qualities

Useful for anyone who needs rapid budget cost estimates in the office, with a client, or in the field, this manual provides clear descriptions, photographs, and illustrations of hundreds of residential, commercial, industrial, and institutional buildings. Corresponding costs by square foot are conveniently located on the same page or adjoining pages, minimizing effort for "on-the-spot" estimates.

Convenient tables provide calculation directions and costs for attached structures and special kinds of additions or adjustments.

For those preparing more detailed, itemized estimates, costs are broken down into "systems style" component specifications and unit costs. This fully illustrated unit-in-place section contains over 6,500 costs simplifying construction component identification and pricing.

You can use this new manual to achieve any degree of cost precision you might need.

Book No. 60585 . . . . $39.95/copy

Concrete & Masonry Cost Data 1985

$36.95 per copy

3rd annual edition over 350 pages, illustrated

Full-scope concrete/masonry cost coverage . . .

- thousands of new unit prices
- the most popular systems prices
- in-depth labor cost data
- advice, assistance, ideas

The first all-inclusive estimating manual devoted to the broad range of today's advanced concrete and masonry methods and materials

a "must" resource for:

general contractors
concrete subcontractors
masonry subcontractors
architects
builders
estimators
engineers
foundation contractors
designers
carpentry subcontractors
governmental agencies

Prices for every facet of concrete/masonry activity —from complicated formwork to lavish brickwork — are provided in exhaustive detail . . . efficiently presented in the famous Means unit and systems cost format.

The massive unit cost section shows crew, daily output, and man-hour information, with bare costs for materials, labor and equipment for each component. The "most-popular" systems cost section actually pictures the concrete or masonry system you are estimating with drawings and diagrams. Unit costs are shown with each system for quick, tailor-made substitutions.

Cost adjustment factors for geographical areas by major cities enable you to adjust prices for your area of the U.S. or Canada. In most mid-sized construction projects nearly a quarter of all costs are concentrated in concrete and masonry work. You simply can't afford to use incomplete or inadequate cost information in this important area. The Means concrete/masonry cost guide gives you this specialized information quickly and reliably. So why take unnecessary risks with other methods?

Order your copy *now.*

Book No. 63085 . . . . . . $36.95/copy

Means Site Work Cost Data 1985

. . . the hard-to-find site work cost information you've always wanted in one truly comprehensive resource . . .

- ***complete unit and systems costs***
- ***illustrations and explanations***
- ***new time-saving features***

$36.95 per copy	***4th annual edition over 375 pages, illustrated***

Completely updated for 1985!

Thumb through the new 1985 site work cost guide and you'll be convinced instantly you have the tool that's equal to today's demanding estimating problems.

It's super-inclusive. Perfectly organized. And, of course, fully reliable in the Means tradition.

Its goal is to give you a cost resource that *actually does* cover every aspect of modern site work estimating. . . exploration, pollution controls, soil retention. . . hidden utilities, security, site enhancement. . . *you name it*!

You'll marvel at the 200+ pages of unit costs.

You'll appreciate the 50+ site work systems costs which illustrate the work for you.

You'll be delighted with the abundance of descriptions, notes and helpful guidance touching upon just the right information you need.

Features:

comprehensive unit prices

crew
daily output
man-hours
units of measure
material
labor
equipment
total costs

complete systems prices

cost per lineal foot or square foot
quantity
description
illustrations

Book No. 60985 . . . . $36.95/copy

Destined to become the "professional's choice" for uncompromised trade and labor productivity data.

$89.95 per copy	**First Edition over 575 pages, Hardcover**

Means Man-Hour Standards

Here is the working encyclopedia of labor productivity information construction professionals have been looking forward to... *Means Man-Hour Standards.*

This massive achievement is by far the most comprehensive, easy-to-use reference of its type available anywhere.

And because it's *Means*, you can be sure it's reliable.

You'll find every bit of labor data you may need...all in a superb layout with "quick-find" indexes, and handy visual flip tabs.

The efficient format permits very rapid comparisons of labor requirements for thousands of construction functions in the convenient U.C.I. arrangement.

Specifically, it offers you . . .

- — precise descriptions of over 12,000 construction tasks with wide ranges of related variations and specifications
- — detailed crew information with equipment required
- — productivity and crew data shown together
- — man-hours with units
- — introductory information and guidance with each section
- — U.C.I. format

With *Means Man-Hour Standards*, you simply turn to the page containing the construction activity you're estimating . . . and there in one place is the data you need. No shuffling back and forth.

Never before has obtaining and using man-hour information been so easy.

Table of Contents

Book No. 67140 . . . . $89.95/copy

Repair & Remodeling Cost Data 1985

Accuracy, efficiency, and thoroughness for estimating rebuilding jobs

$37.95 per copy | **6th annual edition over 425 pages illustrated**

Everything you need to make better rebuilding estimates

- comprehensive unit costs under UCI groupings with crew, man-hours, unit, and bare materials, labor, equipment and totals with adjustable overhead and profit columns
- illustrated systems prices for most widely-used remodeling methods with two additional sections for popular alternative systems and custom installations . . . presented in UNIFORMAT numbering system
- comprehensive labor cost information . . . union rates, crew output, productivity, every needed variation to calculate this vital cost area
- city cost modifiers for 162 U.S. and Canadian metro regions to factor in costs for your locality
- costs for the latest materials, methods, and equipment used in renovation work
- detailed diagrams, illustrations, explanations for ease of understanding and use . . . even a visual table of contents
- adjustable costs for contractor advantages such as owned equipment, low-cost materials purchases, low overhead, minimum daily charges, etc.

For contractors, subcontractors, builders, estimators, engineers, architects, governmental and insurance agencies, corporate facilities managers, firms of all sizes . . .

Reliable unit and systems cost guidance for repair and remodeling estimating problems

The Means *Repair & Remodeling Cost Data* book is the undisputed choice of anyone who must make accurate estimates for renovation projects. It's the most versatile, easy-to-use preliminary cost and bidding reference available from any source.

With over 425 pages containing 10,000 unit and building systems prices, this exciting guidebook handles almost any repair and remodeling challenge. With it you can accurately assemble rapid early-stage estimates as well as comprehensive competitive bids.

Most important of all, the Means repair and remodeling manual is designed to admirably fulfill varied user needs — whether a small-scale builder seeking to protect profits on remodeling work — right up to the most discriminating user needing prices for custom installations.

It's easy to use.Every section provides thorough, easy-to-understand explanations to achieve the desired level of precision. Its superb layout of helpful diagrams, notes, and tips allows you to put the information to work for you quickly.

Book No. 60885 .... $37.95/copy

Means Electrical Cost Data 1985

$36.95 per copy | **8th annual edition over 325 pages, illustrated**

Completely updated and expanded, with 100% focus on electrical estimating cost solutions . . .

With today's unpredictable costs for electrical construction, you need a fast, convenient way to keep up with it all.

That's what **Means Electrical Cost Data 1985** does for you. It gives you nearly 18,000 unit and systems costs covering every kind of electrical — or related — construction question you're likely to have.

For 1985 there are many additional price entries, new estimating models, and a better man-hour format which helps you in job scheduling as well as estimating.

For every phase of the electrical estimating process you'll have sharp, to-the-point answers. You'll have clear specifications, accurate materials and labor costs, and complete guidance in the dozens of notes, formulas and explanations.

For you, the busy professional, it's the fast, efficient way to estimate electrical work.

In fact, it will more than pay for itself the first time you use it for an electrical estimate!

- provides you with more electrical estimating data for better application to your work
- 10,000 fully current unit costs with additional three dimensional drawings of parts for faster identification
- man-hour format allows you to more easily interface with material deliveries and project work scheduling
- detailed estimating procedural models, many with complete diagrams, give you basic electrical frameworks to adapt to your projects
- expanded design reference tables covering every possible code, requirement, or fact you might need
- includes demolition and remodeling, square foot costs, geographical materials and labor cost factors, crew data, and electrical "systems" costs

Book No. 62085 .... **$36.95/copy**

R.S. Means® Co., Inc.
151 Construction Plaza (617) 747-1270
P.O. Box 800 Kingston, MA 02364-0800

1985 ORDER FORM

Enclosed is our check for the following publications:

QTY.	BOOK NO.	TITLE	UNIT PRICE	TOTAL
	60185	Building Construction Cost Data 1985	$ 34.95	$
	61085	Means Mechanical Cost Data 1985	36.95	
	62085	Means Electrical Cost Data 1985	36.95	
	60885	Repair & Remodeling Cost Data 1985	37.95	
	60585	Means Square Foot Costs 1985	39.95	
	60385	Means Systems Costs 1985	38.95	
	60985	Means Site Work Cost Data 1985	36.95	
	60785	Residential/Light Commercial Cost Data 1985	34.95	
	64985	Interior Cost Data 1985	35.95	
	63085	Concrete & Masonry Cost Data 1985	36.95	
	60285	Labor Rates for the Construction Industry 1985	34.95	
	61485	Means Historical Cost Indexes 1985	15.95	
	61185	Means Construction Cost Indexes *(Yearly Subscription)*	47.95	
	67145	Square Foot Estimating *(Hardcover)*	36.95	
	67146	Square Foot Estimating *(Softcover)*	31.95	
	67140	Means Man-Hour Standards *(Hardcover only)*	89.95	
	67152	Means Scheduling Manual *(Softcover only)*	32.95	
	67160	Estimating for the General Contractor *(Hardcover)*	35.95	
	67170	Estimating for the General Contractor *(Softcover)*	30.95	
	50185	Boating Cost Guide	29.95	

Note: Prices subject to change without notice. Mass. residents add 5% state sales tax:
Postage and Handling extra when billed.

TOTAL $

Canadian customers ONLY — Please order books from:
Southam Communications, Ltd.
1450 Don Mills Road
Don Mills, Ontario, Canada, M3B-2X7
(Canadian Prices are higher.)

SEND MY ORDER TO:
(Please print)

Name ______________________

Company ______________________

Address ______________________

City/State/Zip ______________________

Please Indicate If Company ☐ Or Home ☐ Address)

R.S. Means® Co., Inc.
151 Construction Plaza (617) 747-1270
P.O. Box 800 Kingston, MA 02364-0800

1986 ORDER FORM

Enclosed is our check for the following publications:

QTY.	BOOK NO.	TITLE	UNIT PRICE	TOTAL
	60016	Building Construction Cost Data 1986	$ 36.95	$
	60026	Means Mechanical Cost Data 1986	38.95	
	60036	Means Electrical Cost Data 1986	38.95	
	60046	Repair & Remodeling Cost Data 1986	38.95	
	60056	Means Square Foot Costs 1986	40.95	
	60066	Means Systems Costs 1986	41.95	
	60076	Means Site Work Cost Data 1986	38.95	
	60086	Residential/Light Commercial Cost Data 1986	36.95	
	60096	Interior Cost Data 1986	36.95	
	60116	Concrete & Masonry Cost Data 1986	38.95	
	60126	Labor Rates for the Construction Industry 1986	36.95	
	60136	Means Historical Cost Indexes 1986	16.95	
	60146	Means Construction Cost Indexes *(Yearly Subscription)*	49.95	
	60153	Means Man-Hour Standards *(Hardcover only)*	89.95	
	60164	Means Scheduling Manual *(Softcover only)*	32.95	
	60173	Square Foot Estimating *(Hardcover)*	36.95	
	60183	Square Foot Estimating *(Softcover)*	31.95	
	60192	Estimating for the General Contractor *(Hardcover)*	35.95	
	60212	Estimating for the General Contractor *(Softcover)*	30.95	
	60236	Boating Cost Guide	29.95	

Note: Prices subject to change without notice. Mass. residents add 5% state sales tax:
Postage and Handling extra when billed.

TOTAL $

Canadian customers ONLY — Please order books from:
Southam Communications, Ltd.
1450 Don Mills Road
Don Mills, Ontario, Canada, M3B-2X7
(Canadian Prices are higher)

SEND MY ORDER TO:
(Please print)

Name ______________________

Company ______________________

Address ______________________

City/State/Zip ______________________

Please Indicate If Company ☐ Or Home ☐ Address)

Building Construction Cost Data 1985

$34.95 per copy | 43rd anniversary edition over 425 pages

America's foremost construction cost information guide with over 20,000 unit prices for labor, materials and installation

Building Construction Cost Data, now in its 43rd year of publication, offers the reliability of over 20,000 thoroughly-researched unit prices, plus the most efficient, easy-to-use UCI format in the industry. Even the most complicated estimates can be accurately prepared in less time than with comparable references.

Astute construction professionals prefer *Building Construction Cost Data* as their primary estimating resource. It's the original, most sought-after book of its kind . . . known and depended upon by its users for unparalleled accuracy and versatility . . . a cost resource prepared with you — the user — in mind.

Completely updated for 1985!

- new materials, fixtures, hardware, and equipment items included in each section
- all items updated to reflect 1985 costs and construction techniques
- unit prices divided into material and installation costs with overhead and profit shown separately
- hourly and daily wage rates for installation crews with crew sizes, equipment, and average daily crew output listed
- city cost adjustment factors for material and labor costs in 19 categories for each of 162 major U.S. and Canadian metro areas
- factors to compute historical costs for all cities dating back to 1940
- square foot and cubic foot cost section showing range and median costs for 57 common building types with plumbing, HVAC, and electrical percentages tabulated separately
- easiest-to-use format in contrasting typefaces and shaded columns
- costs are the national average and can be easily adjusted for local wage scales
- special cost advantages such as owned equipment, low-cost materials purchases, and low overhead, can be identified and computed separately
- over 13 index pages for quick item location and cross-reference
- helpful examples, instructions, illustrations, and explanations of how costs were developed for each major division

Book No. 60185 . . . . $34.95/copy